sasol

The Larger Illustrated Guide to
BIRDS
of Southern Africa

Ian Sinclair • Phil Hockey • Warwick Tarboton • Niall Perrins • Dominic Rollinson • Peter Ryan

Illustrated by Norman Arlott, Peter Hayman, Alan Harris and Faansie Peacock

Published by Struik Nature
(an imprint of Penguin Random House South Africa (Pty) Ltd)
Reg. No. 1953/000441/07
The Estuaries No. 4, Oxbow Crescent, Century Avenue, Century City 7441
PO Box 1144, Cape Town, 8000, South Africa

Visit www.struiknature.co.za and join the Struik Nature Club
for updates, news, events and special offers.

First published in 1996
Second impression 1998
Second edition 2005
Third edition 2014
Fourth edition 2020

3 5 7 9 10 8 6 4 2

Publisher: Pippa Parker
Managing editor: Roelien Theron
Editor: Colette Alves
Designer: Dominic Robson
Proofreader: Emsie du Plessis

Cover: Ground Woodpeckers illustrated by Faansie Peacock

Reproduction done by Studio Reproduction Co., Cape Town and Hirt & Carter Cape (Pty) Ltd
Printed and bound in China by C&C Offset Printing Co., Ltd.

MIX
Paper from
responsible sources
FSC® C018179

ISBN 978 1 77584 7304

Also available:
Sasol Birds of Southern Africa 5th edition
ISBN 978 1 77584 668 0 (softcover – regular field guide size)
ISBN 978 1 77584 669 7 (ePub)

Sasol Voëls van Suider-Afrika 5de uitgawe
ISBN 978 1 77584 671 0 (softcover– regular field guide size)
ISBN 978 1 77584 672 7 (ePub)

The app, *Sasol eBirds Southern Africa,* is available for Apple iOS or Android

Download on the
App Store

GET IT ON
Google Play

Data for distribution maps and status bars were mostly taken from the South African Bird Atlas Project 2
(SABAP2; http://sabap2.birdmap.africa) and the 7th edition of *Roberts' Birds of Southern Africa*
(John Voelcker Bird Book Fund, 2005). Please consult these sources for more information.

CONTENTS

STRUIK NATURE CALL APP

DOWNLOAD FREE APP to access the calls in this book by scanning a QR code or visiting the relevant app store.

See page 16 for further information on call app.

Apple App Store

Google Play Store

ACKNOWLEDGEMENTS

AUTHORS' ACKNOWLEDGEMENTS

A key goal for the fifth edition was to update as much of the artwork as possible. Principal artist Norman Arlott provided many new images, with additional new artwork supplied by Faansie Peacock (seabirds, nightjars) and Alan Harris (raptors and sundry other species). We thank Peter Hayman for the use of his images from the original edition. Our sincere thanks go to the Struik Nature team – in particular Pippa Parker, Colette Alves and Dominic Robson – for their hard work and all the effort that went into this fifth edition. We also thank our generous sponsors, Sasol, for their ongoing support.

Since the first Sasol guide was published in 1993, many friends and colleagues have contributed to improving the text, maps and illustrations. Their constructive criticism and readiness to give of their time is much appreciated. In particular, we are grateful to David Allan, Rodney Cassidy, Alvin Cope, Adrian Craig, Richard Dean, Wayne Delport, Trevor Hardaker, Rob Little, Alan Kemp, Joris Komen, Rob Leslie, Etienne Marais and Morné du Plessis, and John Graham and Mike Buckham, who helped check proofs. We are also grateful to the following friends who have since passed on, whose comments greatly improved earlier editions: Richard Brooke, Tony Harris, Terry Oatley and Barrie Rose. Finally, without the understanding and patience of family and friends, who adapted to our frequent long absences in the field, this volume would still be in the making.

ARTISTS' ACKNOWLEDGEMENTS

My thanks go to the following for their help in supplying answers to queries that cropped up during the preparation of my artwork: Roger Mitchell, Dr John Fanshawe, the late Dr Richard Liversidge, and the staff of the Natural History Museum at Tring, Hertfordshire, United Kingdom – especially Mark Adams and Dr Robert Prys-Jones. I also owe a great debt to the late John G Williams, who inspired my fanatic interest in Africa and its birds, and an unpayable debt to my wife, Marie, for managing a home and family while I spent days and weeks researching in museum and field. – **Norman Arlott**

Even though I never left my studio, it was an honour to spend a few months mentally among the ocean swells while preparing the pelagic plates. It is my sincere hope that increased interest in this spectacular group of birds will help birders unravel the mysteries surrounding seabirds – many of which face severe pressures at sea and at their remote breeding colonies. I am grateful to both the team at Struik Nature and the team of authors for providing valuable references, feedback and comments. – **Faansie Peacock**

I thank Malcolm Wilson for providing invaluable references, and for linking me to his wide network of colleagues, including Dr Mark Brown and Niel Cillié, who were both most helpful in their specialist areas. I am also grateful to the authors for their feedback and guidance in developing the plates. – **Alan Harris**

I acknowledge the help of the following people and institutions for their assistance: the staff of the Natural History Museum at Tring, Hertfordshire, United Kingdom, in particular Peter Colston, Michael Walters, Jo Bailey, Mark Adams and Mrs FE Warr; Dr Alan Kemp, former curator of the Ornithology Department of the Ditsong National Museum of Natural History (formerly the Transvaal Museum) in Pretoria; Joris Komen, former curator of birds, State Museum of Namibia; Dr Clem Fisher, Liverpool Museum; the Library of BirdLife International (ICBP), Cambridge; the Iziko South African Museum, Cape Town; David Allan and Rob Hume. – **Peter Hayman**

PUBLISHER'S ACKNOWLEDGEMENTS

The publisher would like to thank Sasol Limited for its continued generous sponsorship, without which the production of this book at a competitive price would not have been possible.

The publisher gratefully acknowledges FH Chamberlain Trading (Pty) Ltd for permission to re-use some illustrations from *Chamberlain's Birds of Africa south of the Sahara*, 2nd edition.

SPONSOR'S FOREWORD

Over two decades have elapsed since Sasol first started sponsoring the production of natural history publications.

Southern Africa boasts a magnificent biodiversity, with a rich and varied birdlife that plays a vital role in maintaining the balance of nature. It has long been recognised that birds contribute to the ecosystems in which they live. As birds are a common sight in our daily lives, it is sometimes easy to forget the important role they play in maintaining and restoring the balance of natural ecosystems; they are also reliable 'feathered barometers' of the state of our biodiversity. So important is their place in our world, that we would not be able to cherish spring without their abundance of colours and sounds, or admire the flocks that undertake seasonal migrations.

Sasol has been sponsoring a number of environmental publications – with a particular focus on birds and bird-related activities – for many years as part of our commitment to the conservation of our region's natural heritage for both present and future generations. Our support for environmental publications and projects is aimed at promoting an appreciation of our heritage.

This fifth edition of *Sasol Birds of Southern Africa* helps us achieve this; it builds on the success of its previous editions and is certain to remain the region's most comprehensively illustrated and trusted field guide.

Through partnerships with publishers and birding experts, Sasol is proud of the difference our support has made in promoting environmental education and conservation, leading in turn to a new appreciation for nature, and a better understanding of the need to protect this beautiful heritage.

Our sincere thanks go to the authors Ian Sinclair, Phil Hockey, Warwick Tarboton, Peter Ryan, Niall Perrins and Dominic Rollinson, some of southern Africa's most respected and celebrated birders, who worked on the book, as well as to the team at Penguin Random House.

We are confident that this long-anticipated edition will live up to expectations and make a meaningful contribution to growing tourism in southern Africa by inspiring bird lovers to visit our beautiful shores.

Cindy Mogotsi
Senior Vice President: Corporate Affairs
Sasol

Giant Kingfisher

INTRODUCTION

SOUTHERN AFRICA'S BIRDS IN PERSPECTIVE

Southern Africa is defined as the area south of the Kunene, Okavango and Zambezi rivers, encompassing Namibia, Botswana, Zimbabwe, South Africa, Lesotho, Eswatini (Swaziland) and southern and central Mozambique, as well as the oceanic waters within 200 nautical miles of the coast. Mozambique north of the Zambezi River is excluded, as its avifauna has closer ties to that of East Africa than it does to that of the regions further south. Southern Africa covers a land area of approximately 3.5 million square kilometres and has a high bird diversity: more species breed here than in the United States of America and Canada combined. The region's bird list currently stands at 989 species, of which 96 are endemic (occurring only in the region) and 78 are near-endemic (having ranges that extend only slightly outside the region). Two bird families are endemic to southern Africa: the rockjumpers (Chaetopidae) and sugarbirds (Promeropidae).

Bokmakierie

Bokmakierie and Jackal Buzzard are two of the 96 endemic bird species confined to southern Africa.

Jackal Buzzard

Climatic and topographical diversity

One of the reasons for southern Africa's high bird diversity is its climatic and topographical diversity. The climate ranges from cool-temperate in the southwest to tropical in the north. The southwest of the region has mainly winter rains, the north and east have summer rains and some of the central parts have aseasonal rainfall. In general, the amount of rainfall increases from west to east. Winter snows are regular on the higher mountains of South Africa and Lesotho, which rise to 3 500 metres above sea level. The major centre of endemism is in the southwest arid zone, centred on the Karoo and the Namib Desert. This area is largely restricted to southern Africa, isolated from the arid zones in north and northeast Africa. However, numerous endemics are associated with the fynbos of the Western and Eastern Cape, the grasslands of the eastern plateau of South Africa and Lesotho, and the eastern highlands of Zimbabwe. Although endemics tend to be concentrated in the south and west, total diversity increases in the more moist and tropical north and east of the region.

Habitat-based distribution

Most birds tend to occur in specific habitats, and in terrestrial systems these typically follow the distributions of the main plant communities or biomes. Southern Africa supports a diversity of biomes, illustrated in the map (opposite), linked to rainfall patterns and, to a lesser extent, soil types. The hyper-arid Namib Desert along the Namibian coast gradually merges into Karoo habitats, and thence into arid Kalahari savannas in the north, and grassland further south. Further north and east, the arid savannas merge into mesic savannas and woodland. Forests are confined to areas of relatively high rainfall, with afromontane forest occurring patchily along the eastern mountain chain from the Cape Peninsula to the eastern highlands of Zimbabwe and Mount Gorongoza, Mozambique. Forests occur to near sea level in the south, but are confined to higher elevations further north. Thicket and coastal forest occur along the coastal lowlands from the Eastern Cape into Mozambique, and locally up the Zambezi River to northern Botswana and into Namibia's Zambezi Region (Caprivi Strip). In the extreme southwest, the floristically diverse fynbos is largely confined to the winter-rainfall region of the Western and Eastern Cape.

Major biomes of southern Africa

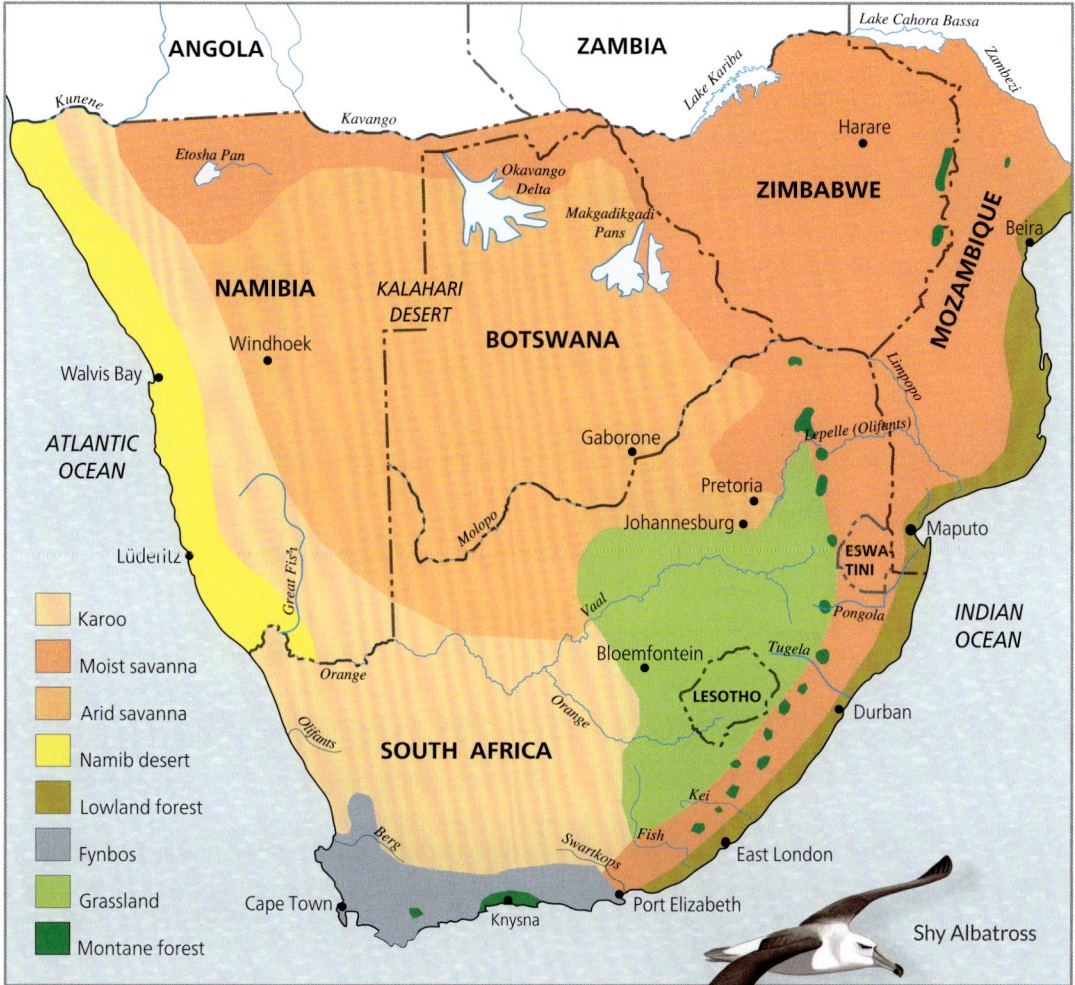

Legend:
- Karoo
- Moist savanna
- Arid savanna
- Namib desert
- Lowland forest
- Fynbos
- Grassland
- Montane forest

Shy Albatross

Marine environment

A second reason for the region's high bird diversity lies offshore, in the marine environment. Southern Africa separates two of the world's major oceans – the Atlantic and the Indian. In the northeast, the Mozambique Current flows south through the Mozambique Channel, and to the south of this lies the warm, southwestward-flowing Agulhas Current, which reverses direction south of the Agulhas Bank. These warm-water currents typically have strongly stratified surface waters that offer few nutrients for phytoplankton, and thus support low abundances of fish and their predators. They are best known for attracting tropical species, such as boobies and frigatebirds, especially in the north as one approaches the tropical seabird island of Europa, which lies just outside the region's territorial waters.

The western seaboard is characterised by the Benguela Upwelling Ecosystem, where southerly winds drive upwelling of cold, nutrient-rich bottom water along the coast, resulting in massive blooms of phytoplankton, and large abundances of small pelagic fish such as anchovies and sardines. These fish in turn support a large diversity of predators, including huge numbers of seabirds. In addition to locally breeding seabirds (including seven endemic species), the Benguela attracts hundreds of thousands of migrant seabirds, ranging from the great albatrosses of the Southern Ocean to tiny storm petrels from high northern latitudes. Pelagic birding in this corner of Africa is a must, possibly paralleled only by parts of eastern Australasia and Peru.

A growing list

Birding as a hobby has seen rapid growth in southern Africa, thanks in no small part to the role played by the four previous editions of *Sasol Birds*. In recent years there have been several additions to the region's bird list, including Short-tailed Shearwater *Ardenna tenuirostris*, Tahiti Petrel *Pseudobulweria rostrata*, Trindade Petrel *Pterodroma arminjoniana*, Red-necked Buzzard *Buteo auguralis*, Grasshopper Buzzard *Butastur rufipennis*, Sharp-tailed Sandpiper *Calidris acuminata*, White Tern *Gygis alba*, Saunders's Tern *Sternula saundersi*, White Wagtail *Motacilla alba*, Yellow-throated Leaflove *Atimastillas flavicollis*, Rufous-tailed Scrub Robin *Cercotrichas galactotes*, Upcher's Warbler *Hippolais languida*, European Pied Flycatcher *Ficedula hypoleuca*, Red-tailed Shrike *Lanius phoenicuroides* and Ortolan Bunting *Emberiza hortulana*. Other species have been added to the list after genetic work has split subspecies to full-species status (see Bird Classification and Nomenclature).

Even as we write, compelling evidence of two new subregion bird species has recently come to light: Madagascan Pratincole *Glareola ocularis* (Eastern Cape, South Africa, Feb 2005 and Jan 2007) and Forbes-Watson's Swift *Apus berliozi* (central Mozambique, Mar 2017).

It is fun to speculate about 'what might arrive next'! In terms of trans-Atlantic vagrants from the New World, there may be plenty of possibilities. Many vagrants to Tristan da Cunha and Gough Island have not yet been recorded in southern Africa, including Speckled Teal *Anas flavirostris*, Cocoi Heron *Ardea cocoi,* several waders and rails, Common Nighthawk *Chordeiles minor* and Eastern Kingbird *Tyrannus tyrannus*. Closer to home, species such as Sooty Chat *Myrmecocichla nigra* occur just north of the region's northern boundary in Angola and Zambia.

Conservation concerns

Although an ever-growing list of species is exciting for birders, we should not lose sight of the fact that, worldwide, increasing numbers of birds are becoming of conservation concern. The fates of threatened birds are tracked by BirdLife International and the IUCN (International Union for Conservation of Nature), which jointly produce the global Red Data list for birds.

Southern Africa has seven species that are Critically Endangered and that face a very high risk of extinction in the near future (all of which are scarce or vagrant seabirds), 28 Endangered (at high risk of extinction), 32 Vulnerable (at risk of becoming Endangered) and 52 Near-Threatened (likely to become Threatened in the near future). In southern Africa, vultures are particularly at risk, with four of the nine local vulture species now being regarded as Critically Endangered due to widespread poisoning, as well as electrocution, and collisions with powerlines and wind turbines; visit www.birdlife.org for more information. Birds face numerous threats, ranging from habitat loss and degradation to climate change, persecution and disease. Species introduced outside their native range by humans also constitute a key threat. In southern Africa, alien plants are a huge concern, but introduced birds also threaten some native birds through predation, competition (for nest sites and food, for example) and hybridisation (for instance, Mallards hybridise with several native ducks). Many birds have been introduced into southern Africa in the past, deliberately or accidentally, but only 10 species have free-ranging,

Trindade Petrel

Scopoli's Shearwater

Trindade Petrel was recorded at sea off Port Elizabeth in January 2014, while Scopoli's Shearwater is a recent addition to the list after being split from Cory's Shearwater.

Common Myna

Common Chaffinch

Introduced species differ in their impacts. Common Myna is now widespread in cities and towns throughout the subregion and competes with native birds for breeding resources, while Common Chaffinch is confined to the Cape Peninsula.

feral populations today: Mallard, Common Peafowl, Chukar Partridge, Feral Pigeon, Rose-ringed Parakeet, House Crow, Common Myna, Common Starling, House Sparrow and Common Chaffinch. Of these, at least the starling, myna, crow, sparrow and Mallard either do or have the potential to pose significant threats to native birds.

BIRD CLASSIFICATION AND NOMENCLATURE

The classification of the world's birds into orders, families and genera attempts to summarise their evolutionary history. Until recently, classification was based largely on morphological similarity, but in recent decades a variety of molecular tools, and particularly the ability to sequence DNA, the blueprint of life, has revolutionised our understanding of bird relationships. This remains a dynamic field of ornithology and the 'final word' is still far from being written, but the advent of genomics, which allows entire genomes to be sequenced, means we are converging on a new, stable classification for birds. The most basic split is between the Paleognaths (ratites – ostriches, rheas, Emu, cassowaries and kiwis; tinamous – gamebird-like birds confined to South and central America) and the Neognaths (all other birds). Among this latter group, gamebirds and ducks are distinctly different from the rest, which appear to have radiated rapidly, shortly after the demise of the dinosaurs some 66 million years ago. We have seen most change among the passerines, the largest order, which contains more than half of all bird species. Here, many families have been reorganised following molecular evidence showing that birds that look the same often have quite different evolutionary histories.

However, one of the principal aims of any bird field guide is to be user friendly rather than to present a scientifically accurate treatise on the evolutionary and taxonomic relationships of birds. For this reason we have largely retained the traditional ordering system with which the majority of birders are familiar. However, we have commented on the relationships among birds, where appropriate, in the text.

Southern Africa has a long history of ornithology, so it would be reasonable to expect that the avifauna is well known. However, several new species have been recognised as a result of 'splits' of well-defined geographical races or subspecies, mainly based on molecular data. For example, Scopoli's Shearwater *Calonectris diomedea* has been split from Cory's Shearwater *C. borealis*, Subantarctic Shearwater *Puffinus elegans* from Little Shearwater *P. assimilis*, Kirk's Francolin *Ortygornis rovuma* from Crested Francolin *O. sephaena*, Stuhlmann's Francolin *Campocolinus stuhlmanni* from Coqui Francolin *C. coqui*, and Kunene Francolin *Scleroptila jugularis* from Orange River Francolin *S. levalliantoides*. The local subspecies of Long-billed Pipit *Anthus similis* has been split from subspecies further north and is now a near-endemic known as Nicholson's Pipit *A. nicholsoni*. However, other species have been lumped: Black *Psalidoprocne pristoptera holomelas* and Eastern Saw-wings *P. p. orientalis* are not genetically distinct from other races from further north in Africa and so are combined under Black Saw-wing *P. pristoptera*. In some cases, genetic evidence is still needed, and decisions whether to split or lump are more a matter of opinion. An example of this is the lumping together again of Damara *Serinus alario leucolaemus* and Black-headed Canaries *S.a. alario*. Ultimately, evolution is a dynamic, ongoing process, and it is not always feasible to place all individuals or populations into neat species 'boxes'. Given the number of recent splits among, for example, northern hemisphere warblers, which were thought to be well known, it is likely that other cryptic species remain to be described.

BIRDING LITERATURE

There is a vast literature on the identification, distributions and biology of the world's birds, in media ranging from the Internet and DVDs, to field guides, handbooks and scientific papers. A comprehensive listing of these is far beyond the scope of a regional field guide. We merely list the key reference works of immediate value to birders in southern Africa, either because they contain key background information about southern African birds, or they deal with the identification of bird groups that regularly cause 'problems' for birders in the region. We have also listed a few books that detail key birding localities and how to get there.

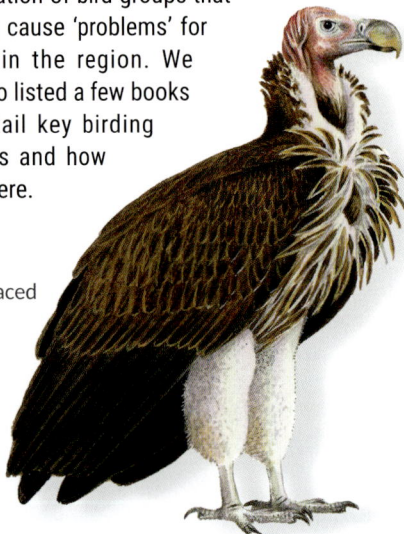

Lappet-faced Vulture

Identification

Cleere, N. 2010. *Nightjars, Potoos, Frogmouths, Oilbird, and Owlet-nightjars of the World.* Princeton University Press.

Forsman, D. 2003. *The Raptors of Europe and the Middle East.* Christopher Helm/A&C Black.

Kennerley, P. and Pearson, D. 2010. *Reed and Bush Warblers.* Christopher Helm.

Mullarney, K., Svensson, L., Zetterström, D. and Grant, P.J. 1999. *Collins Bird Guide.* Harper-Collins. Guide to the birds of Europe, including extensive information on many migratory species and potential vagrants visiting our region.

Olsen, K.M. and Larsson, H. 1997. *Skuas and Jaegers.* Pica Press.

Peacock, F. 2012. *Chamberlain's LBJs: The definitive guide to Southern Africa's Little Brown Jobs.* Pavo Publishing.

Peacock, F. 2012. *Chamberlain's Waders: The definitive guide to Southern Africa's Shorebirds.* Pavo Publishing.

Ryan, P.G. 2017. *Guide to Seabirds of Southern Africa.* Struik Nature.

Shirihai, H., Gargallo, G. and Helbig, A.J. 2001. *Sylvia Warblers.* Christopher Helm/A&C Black.

Sinclair, I. and Ryan, P. 2009. *Complete Photographic Field Guide: Birds of Southern Africa.* Struik Nature. Large-format guide featuring a comprehensive collection of photographs of southern African birds.

Sinclair, I. and Ryan, P. 2010. *Chamberlain's Birds of Africa South of the Sahara.* Struik Nature. Covers species throughout sub-Saharan Africa.

Taylor, B. and Van Perlo, B. 1998. *Rails: A Guide to the Rails, Crakes, Gallinules and Coots of the World.* Pica Press.

Distribution, biology and conservation

Barnes, K.N. (ed.). 1998. *The Important Bird Areas of Southern Africa.* BirdLife South Africa. Reference framework presenting estimated numbers of threatened species in the region's most important bird conservation localities.

Carnaby, T. 2008. *Beat About the Bush: Birds.* Jacana Media. A comprehensive reference guide that provides answers to everyday questions on a myriad southern African bird-related topics.

Hancock, P. and Weiersbye, I. 2016. *Birds of Botswana.* Princeton University Press.

Hockey, P.A.R., Dean, W.R.J. and Ryan, P.G. (eds). 2005. *Roberts Birds of Southern Africa.* 7th edition. Trustees of the John Voelcker Bird Book Fund. A fully-referenced handbook to the region's birds.

Maclean, G.L. 1990. *Ornithology for Africa.* University of Natal Press. Fascinating insights into a wide range of topics relating to the ecology, behaviour, biology and physiology of birds.

Parker, V. 1997. *The Atlas of the Birds of Central Mozambique.* Endangered Wildlife Trust and Avian Demography Unit.

Ryan, P. 2006. *Birdwatching in Southern Africa.* Struik. An introduction to birds and birding – what you need, what to look for and the basics of identification.

Tarboton, W. 2011. *Roberts Guide to the Nests and Eggs of Southern African Birds.* Jacana Media. Details the breeding habits of each locally breeding species, accompanied by photographs of nests and eggs at actual size.

Taylor, M.T., Peacock, F. and Wanless, R.M. 2015. *The 2015 Eskom Red Data Book of Birds of South Africa, Lesotho and Swaziland.* BirdLife South Africa.

Site guides

Cohen, C., Spottiswoode, C. and Rossouw, J. 2006. *Southern African Birdfinder*. Struik Nature. A guide to over 330 birding sites in southern Africa, as well as Angola, Zambia, Malawi and Madagascar. Includes pull-out map.

Marais, E. and Peacock, F. 2008. *The Chamberlain Guide to Birding Gauteng*. Mirafra. A detailed guide to 101 birding sites in and around Johannesburg and Pretoria, as well as top sites further afield.

Useful websites

FitzPatrick Institute of African Ornithology (www.fitzpatrick.uct.ac.za)
Sasol Birds website (www.sasolbirds.co.za) for downloadable checklists
Southern African Bird Atlas Project, SABAP2 (sabap2.birdmap.africa) AFRING

BIRD SOCIETIES AND CLUBS

Bird-watching and conservation organisations and clubs play an important role in communication and co-ordination between birders. Most also hold regular meetings and outings, and many produce their own newsletters. Most bird conservation organisations in the region are affiliated to BirdLife International, a global organisation that promotes bird research, and conservation of birds and their habitats (visit www.birdlife.org). BirdLife South Africa is the local partner of BirdLife International. It boasts a membership of 6 000 birders and has about 42 bird clubs nationwide. It is highly recommended that you join BirdLife South Africa and your local bird club, which will give you the opportunity to join birding outings, participate in conservation projects and attend fascinating courses on the ecology and identification of birds.

> BirdLife South Africa,
> Isdell House, 17 Hume Road,
> Dunkeld West 2196,
> Johannesburg, South Africa
> Tel: (011) 789-1122;
> info@birdlife.org.za
> www.birdlife.org.za/

Other partners in the region include BirdLife Botswana (www.birdlifebotswana.org.bw) and BirdLife Zimbabwe (www.birdlifezimbabwe.co.zw).

SABAP2

Following the success of the Southern African Bird Atlas Project, the second Southern African Bird Atlas Project (SABAP2) aims to record changes in the distribution and abundance of birds throughout South Africa, Lesotho and Eswatini in 5 x 5 minute grid cells or 'pentads'. SABAP2 welcomes the participation of all birders; to register, visit their website http://sabap2.birdmap.africa. The introduction of the app BirdLasser has had a huge impact on atlassing throughout Africa with atlas records having increased drastically since the app's release. Distributional data from SABAP2 was the primary source for updating the species distribution maps in this guide. We thank the many citizen scientists who submitted atlas data, making SABAP2 the most important resource for bird distribution data in southern Africa.

African Paradise Flycatcher

ENDEMISM AMONG SOUTHERN AFRICAN BIRDS

Endemics (96) – confined to southern Africa; **Near-endemics** (78) – largely restricted to southern Africa (most species extending their ranges into southern Angola); **Breeding endemics** (5) – breed only in southern Africa, but some leave the region outside the breeding season.

ENDEMICS

Apalis, Chirinda
Babbler, Southern Pied
Blackcap, Bush
Bokmakierie
Boubou, Southern
Bulbul, Cape
Buttonquail, Hottentot
Buzzard, Forest
Buzzard, Jackal
Canary, Black-headed
Canary, Cape
Canary, Forest
Canary, Protea
Chat, Ant-eating
Chat, Buff-streaked
Chat, Sickle-winged
Cormorant, Bank
Cormorant, Crowned
Crane, Blue
Eremomela, Karoo
Flycatcher, Fairy
Flycatcher, Fiscal
Francolin, Grey-winged
Francolin, Orange River
Francolin, Red-winged
Grassbird, Cape
Gull, Hartlaub's

Harrier, Black
Ibis, Southern Bald
Korhaan, Blue
Korhaan, Karoo
Korhaan, Northern Black
Korhaan, Southern Black
Lark, Agulhas Long-billed
Lark, Barlow's
Lark, Botha's
Lark, Cape Clapper
Lark, Cape Long-billed
Lark, Dune
Lark, Eastern Long-billed
Lark, Karoo
Lark, Karoo Long-billed
Lark, Large-billed
Lark, Melodious
Lark, Red
Lark, Rudd's
Lark, Sclater's
Lark, Short-clawed
Longclaw, Cape
Mousebird, White-backed
Parrot, Cape
Pipit, African Rock
Pipit, Yellow-breasted
Prinia, Drakensberg
Prinia, Karoo

Robin-Chat, Chorister
Robin-Chat, White-throated
Rock Thrush, Cape
Rock Thrush, Sentinel
Rockjumper, Cape
Rockjumper, Drakensberg
Scrub Robin, Brown
Scrub Robin, Karoo
Shelduck, South African
Siskin, Cape
Siskin, Drakensberg
Sparrow-Lark, Black-eared
Spurfowl, Cape
Starling, Pied
Sugarbird, Cape
Sugarbird, Gurney's
Sunbird, Greater
   Double-collared
Sunbird, Neergaard's
Sunbird, Orange-breasted
Sunbird, Southern
   Double-collared
Tchagra, Southern
Thrush, Karoo
Tit, Grey
Turaco, Knysna
Twinspot, Pink-throated
Vulture, Cape

Black Harrier

Blue Korhaan

Sclater's Lark

Warbler, Barratt's
Warbler, Cinnamon-breasted
Warbler, Knysna
Warbler, Layard's
Warbler, Namaqua
Warbler, Roberts'
Warbler, Rufous-eared
Warbler, Victorin's
Waxbill, Swee
Weaver, Cape
Weaver, Sociable
White-eye, Cape
White-eye, Orange River
Woodpecker, Ground
Woodpecker, Knysna

NEAR-ENDEMICS
Apalis, Rudd's
Babbler, Bare-cheeked
Babbler, Black-faced
Barbet, Acacia Pied
Batis, Cape
Batis, Pririt
Batis, Woodwards'
Bulbul, African Red-eyed
Bunting, Cape
Bunting, Lark-like
Bushshrike, Olive
Bustard, Ludwig's
Canary, Lemon-breasted
Canary, White-throated
Canary, Yellow
Chat, Boulder
Chat, Herero
Chat, Karoo

Chat, Tractrac
Cisticola, Cloud
Cisticola, Grey-backed
Cisticola, Rufous-winged
Cormorant, Cape
Coucal, Burchell's
Courser, Burchell's
Finch, Red-headed
Flycatcher, Chat
Flycatcher, Marico
Francolin, Kunene
Goshawk, Pale Chanting
Hornbill, Bradfield's
Hornbill, Damara Red-billed
Hornbill, Monteiro's
Hornbill, Southern
    Yellow-billed
Korhaan, Red-crested
Korhaan, Rüppell's
Lark, Benguela Long-billed
Lark, Eastern Clapper
Lark, Fawn-coloured
Lark, Gray's
Lark, Monotonous
Lark, Pink-billed
Lark, Sabota
Lark, Spike-heeled
Lark, Stark's
Lovebird, Rosy-faced
Oystercatcher, African
Parrot, Rüppell's
Penduline Tit, Cape
Pipit, Nicholson's
Rock Thrush, Short-toed
Rockrunner

Sandgrouse, Burchell's
Sandgrouse, Double-banded
Sandgrouse, Namaqua
Scrub Robin, Kalahari
Shoveler, Cape
Shrike, Crimson-breasted
Shrike, Southern White-crowned
Shrike, White-tailed
Sparrow, Cape
Sparrow, Great
Sparrow-Lark, Grey-backed
Spurfowl, Hartlaub's
Spurfowl, Natal
Spurfowl, Red-billed
Starling, Burchell's
Starling, Pale-winged
Sunbird, Dusky
Swift, Bradfield's
Tit, Ashy
Tit, Carp's
Warbler, Chestnut-vented
Waxbill, Violet-eared
Weaver, Scaly-feathered
Wheatear, Mountain
Whydah, Shaft-tailed
Wren-Warbler, Barred

BREEDING ENDEMICS
Gannet, Cape
Penguin, African
Pipit, Mountain
Swallow, South African Cliff
Tern, Damara

Burchell's Starling

Hartlaub's Spurfowl

Cape Gannet

HOW TO USE THIS BOOK

The species accounts that follow are introduced by group headers ❶ that point out common features and other potentially confusing groups. Species descriptions ❷ indicate each bird's size and mass, then ❸ briefly describe the main plumages, and highlight the most important characters for separating the species in question from similar species. Adult plumages are described first (male, then female), followed by juvenile and immature plumages, if these are distinct. For species with seasonal plumage differences, breeding plumage is usually described first, except where breeding plumage is very short lived (for example, some herons and cormorants) or where non-breeding plumage is the norm in southern Africa (for example, Palearctic migrant waders and terns).

Calls and songs are then described under 'Voice' ❹, augmented in most cases by barcode links ❺ to bird calls in the free Struik Nature Call App (downloadable for Android and Apple devices).

Each text ends with a section summarising the status (including Red List status for threatened species), habitat preferences and other behavioural

information ❻ that might prove useful in identifying a given bird. Category labels for abundance are, in descending order: 'abundant', 'common', 'fairly common', 'uncommon', 'scarce', 'rare' and 'vagrant'. 'Locally common' is used for species with restricted distributions (either in terms of specific habitat requirements or absolute range size).

The annotations ❼ on the illustrative plates highlight features that help to differentiate similar species from one another.

Abbreviations

♂	male	pre-br.	pre-breeding
♀	female	E	endemic
ad.	adult	NE	near-endemic
sub-ad.	sub-adult	BE	breeding endemic
imm.	immature	N/n	North/northern
juv.	juvenile	E/e	East/eastern
br.	breeding	S/s	South/southern
non-br.	non-breeding	W/w	West/western

❶ **LARGE FLYCATCHERS** Mainly insectivorous passerines, either hawking prey in the air or on the ground. The species on this page are medium-sized birds, rather chat-like in demeanour, that perch erect on low branches and scan the ground below for prey. Bills slender; tails rather long and square-ended or slightly notched. Sexes alike, except for Fiscal Flycatcher. Occur singly, in pairs or family groups.

❷ **FISCAL FLYCATCHER** *Melaenornis silens*
18–20 cm; 22–36 g A striking, pied flycatcher with white panels in otherwise black wings and tail. Resembles Southern Fiscal (p. ❸) in broad colour pattern, but bill more slender, legs longer, white in wings confined to secondaries (not wing coverts) and white windows in the tail; lacks white outer-tail feathers. Larger and longer tailed than Collared Flycatcher (p. 408) and nape black (not white). Female is browner above and washed darker grey on breast and belly. Juv. ❻ much browner above, with buff spots; underparts mottled grey-b... ❹ **Voice:** A string of high-pitched, y notes, sometimes extended by mimicking calls of other birds, *'tssisk'* alarm call. **Status and biology:** Endemic. Common resident in south and east of range, a non-br. winter visitor (Mar–Sep) in north of range. Found in woodland and thickets, scrub, gardens and plantations. (Fiskaalvlieëvanger) ❺

SOUTHERN BLACK FLYCATCHER *Melaenornis pammelaina*
18–20 cm; 24–32 g An entirely black, somewhat glossy flycatcher that is most easily confused with a Common Square-tailed Drongo (p. 338). Has blackish-brown (not red) eyes, a taller stance and more slender body. Easily told from Fork-tailed Drongo (p. 338) by its smaller size and square or slightly forked (not deeply forked) tail. Juv. is sooty-brown, heavily spotted above and below with buff. **Voice:** High-pitched, wheezy 2 or 3-note song, often preceded by a shrill note, *'tziiii tsooo-tsoo'*. **Status and biology:** Common resident in broadleafed woodland, savanna and forest edges. (Swartvlieëvanger)

PALE FLYCATCHER *Melaenornis pallidus*
17 cm; 20–24 g A rather uniformly buff-coloured flycatcher, darkest on the wings and palest on the throat, with a thin white eye-ring bisected by a dark eye stripe. Told from Marico Flycatcher by its pale buff (not white) underparts. Range overlaps slightly with Chat Flycatcher, but Pale Flycatcher is smaller and shorter tailed; typically shows more contrast between upper- and underparts; Chat Flycatcher shows a slight pale panel on folded secondaries; habitats differ. Juv. has fine buff spots above, with broad, buff margins to flight feathers; underparts streaked dark brown. **Voice:** Song is a melodious warbling interspersed with harsh chitters; alarm call is a soft *'churr'*. **Status and biology:** Locally common resident in moist, broadleafed woodland. (Muiskleurvlieëvanger)

CHAT FLYCATCHER *Melaenornis infuscatus*
20 cm; 39 g A large, chat-like flycatcher with long wings and tail. Rather nondescript, with paler edges to wing feathers forming slight panel on folded secondaries. Best told from chats by its long, uniformly brown tail. Larger than Marico and Pale Flycatchers, with darker brown underparts that contrast less with the tail. Spends more time on ground than other flycatchers; often raises its wings. Juv. is spotted buff above; flight feathers edged rufous, breast mottled dark brown. **Voice:** Song is a deep, warbled *'cher-cher-cherrip'*, interspersed with harsh, grating notes. **Status and biology:** Near-endemic. Common resident or local nomad in semi-arid and arid shrublands. (Grootvlieëvanger)

MARICO FLYCATCHER *Melaenornis mariquensis*
18 cm; 22–28 g A sandy brown-and-white flycatcher, easily told from Pale and Chat Flycatchers by the sharp divide between brown upperparts and white underparts; wing feathers have pale edges. Juv. is dull brown above, spotted with buff, with rufous edges to flight feathers. Underparts off-white, with dark brown streaking; probably indistinguishable from juv. Pale Flycatcher. **Voice:** Song is a series of sparrow-like *'tsii-cheruk-tukk'* chirps; soft *'tsee tsee'* alarm call. **Status and biology:** Near-endemic. Common resident in semi-arid acacia savanna and sparse woodland. Where its range overlaps with Pale Flycatcher, Marico Flycatcher is confined to thornveld, Pale Flycatcher to broadleafed woodland. (Maricovlieëvanger)

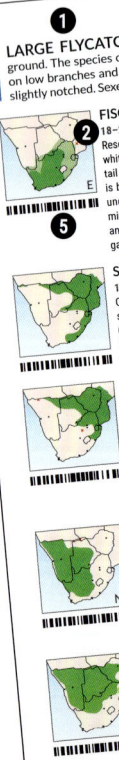

FISCAL FLYCATCHER

white-edged secondaries

♂ — juv.

white tail patches ❼ — dark eye

slim bill (cf. drongos)

SOUTHERN BLACK FLYCATCHER

PALE FLYCATCHER

ad. — juv.

square tail — ad.

ad.

juv. — ad.

pale secondary panel

pale brown below

juv.

CHAT FLYCATCHER

Distribution maps

Distribution is shown on the accompanying range maps. The maps summarise where each species occurs. Where there are marked differences in the frequency of occurrence of a species across its range, two tones are used, the darker tone indicating where the species is more commonly found. Different colours are used to distinguish resident birds from visitors: **green** for residents, **red** for summer visitors and **blue** for winter visitors. Species that generally migrate are treated as seasonal visitors even if some individuals remain year-round. Endemic species are indicated with an 'E' and vagrant records are marked with an 'X'.

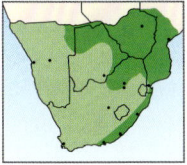

Resident species

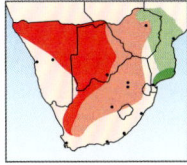

Partial migrant

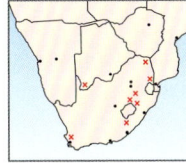

Vagrant

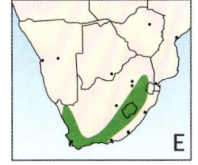

Endemic

Winter visitor

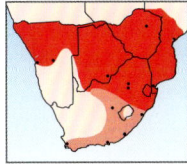

Summer visitor

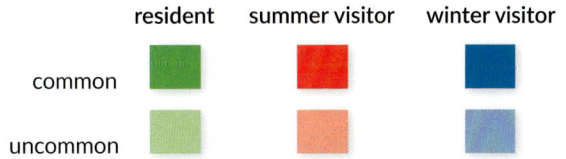

	resident	summer visitor	winter visitor
common			
uncommon			

Seasonality bars

Seasonality of occurrence and breeding are shown in the calendar bars; they indicate the months when species are present in the region, with pale shading showing presence in reduced numbers (for example, in some migratory shorebirds, many juveniles overwinter). The bars also show the timing of breeding throughout the region, 'B' indicating months of peak laying and 'b' months when laying has been recorded. Note: this cannot reflect regional differences in laying dates. In most species, breeding is aligned to rainfall, and so occurs earliest in the southwest, winter-rainfall area, and latest in the northwest, where summer rainfall tends to fall later than it does in the eastern half of the region.

J F M A M J J A S O N D

month

■ common ■ uncommon X vagrant □ absent

b egg-laying recorded B peak egg-laying

Bird calls

Bird calls can be accessed by using the free Struik Nature Call App to scan the barcodes below the distribution maps. The app can be downloaded onto a smartphone or tablet from Google Play (minimum Android v5.0*) or Apple App Store (minimum iOS 11.4*), by either searching for the app in the app store or using the QR codes on page 3.

Once downloaded, launch the app on your device and hover the camera over the barcode in the book. The barcode will scan and bring up calls for that species. Press 'play' to listen. Internet connectivity is not required to access the calls, as all are contained within the app. For most species, the app features multiple calls. Where calls are not included, there is no barcode below the species' map.

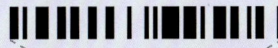

Scan barcode to play call

* Minimum operating systems correct at time of print. Operating systems are updated annually and minimum system requirements will change accordingly.

GLOSSARY

Allopatric Describes species or populations with non-overlapping ranges

Arboreal Tree dwelling

Breeding endemic A species that breeds only in a particular region, but undertakes movements or migrations during the non-breeding season, such that a measurable proportion of the population leaves the region

Brood parasite A bird that lays its eggs in the nest of another species, and leaves them to raise its offspring

Cere A fleshy covering at the base of the upper mandible in some birds (e.g. raptors, parrots, pigeons)

Colonial Associating in close proximity while roosting or nesting

Crepuscular Active at dawn and dusk

Culmen The upper ridge of a bird's beak

Cryptic Well camouflaged

Dimorphic Describes species that show two distinct morphotypes (in plumage, size or both), usually linked to sexual differences

Diurnal Active during daylight hours

Eclipse plumage Dull plumage attained by some male birds during a transitional moult after the breeding season, before they acquire brighter plumage (e.g. some ducks, sunbirds, bishops)

Endemic A species whose breeding and non-breeding ranges are confined to a particular region

Feral Describes species that have escaped from captivity and now live in the wild

Flight feathers The longest feathers on the wings and tail

Flush Put to flight

Frons Forehead

Fulvous Reddish-yellow or tawny

Gape The angle of the jaw where the upper and lower mandibles meet; characterised by fleshy, yellow flanges in chicks and recently fledged juveniles of many bird species

Gorget A band of distinctive colour on the throat

Holarctic A biogeographical region encompassing both the Nearctic and the Palearctic

Immature A bird that has undergone its first moult from juvenile plumage but has not attained adult plumage

Irruption A rapid expansion of a species' normal range

Jizz A general impression of size, shape and behaviour

Juvenile The first fully feathered plumage of a young bird

Leucistic Describes a whitish form of a particular species, with reduced feather melanin (albinism is a total lack of melanin and is also associated with pink eyes)

Melanistic Describes a dark morph of a species, the colour resulting from high levels of the pigment melanin. In the region, most frequently encountered among raptors, especially hawks

Migrant A species that undertakes (often) long-distance flights between its breeding and non-breeding areas

Yellow-breasted Pipit is a southern African **endemic**.

Feral populations of House Crow exist in a number of southern African cities.

A **melanistic** Western Cattle Egret.

Miombo Mesic broadleafed woodland dominated by *Brachystegia* trees that dominates much of south-central Africa

Montane Pertaining to mountains

Morph A colour variant within a species; the colour variation may or may not be linked to sub-specific status

Moult The replacement of old, worn feathers with new ones

Nearctic Temperate and Arctic N America, Canada and Greenland

Near-endemic A species whose range is largely restricted to a region but extends slightly outside the region's borders (in southern Africa, this category includes mostly species whose ranges extend into the arid regions of southwestern Angola)

Neotropical South and Central American

Nocturnal Active at night

Non-passerine Any bird that is not part of the order Passeriformes

Nuchal Of the back or nape of the neck (usually used in the context of 'nuchal collar')

Overwintering (of a bird) Remaining in the subregion instead of migrating to its breeding grounds

Palearctic North Africa, Europe, Asia north of the Himalayas, and southern China

Passerine A member of the largest order of birds (Passeriformes), comprising mostly small, perching songbirds, but also some larger species such as crows and ravens

Pelagic Ocean dwelling

Polymorphic Describes species that have two or more plumage types

Precocial Describes a chick able to walk and feed itself within a few hours of hatching (for example, plovers)

Race A geographical population of a species; a subspecies

Range A bird's distribution

Raptor A bird of prey

Resident A species not prone to migration, remaining in the same area year-round

Speculum A panel of distinctive colour on a bird's wing, most often applied to ducks

Sub-adult A bird intermediate in age and plumage, between immature and adult

Subterminal (of a band or other mark) Close to, but not at, the tip of a feather (most often applied to the tail)

Supercilium A stripe above the eye of a bird

Sympatric Describes species or populations with overlapping ranges

Taxon Any category used in classification, from subspecies and species to families, orders and kingdoms (plural = taxa)

Tertial Innermost flight feather on the wing, attached to the elbow and humerus

Territory The area a bird establishes and then defends against others for breeding, feeding, or both

Vagrant Rare visitor to the region that has wandered outside its normal range

European Nightjar is a **Palearctic**-breeding summer **migrant** from Eurasia.

The Olive Bushshrike is **polymorphic**, occurring in two colour morphs.

ILLUSTRATED GLOSSARY

Head and bill features

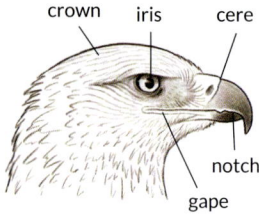

crown · iris · cere

notch
gape

Raptor (eagle)

nostril
tube

latericorn

culminicorn

ramicorn · nail

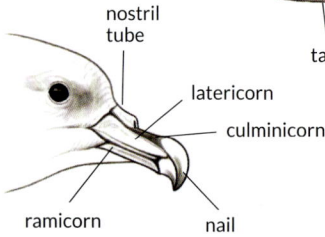

Seabird (petrel)

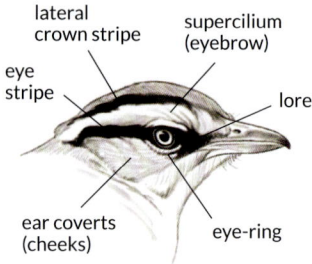

Body features

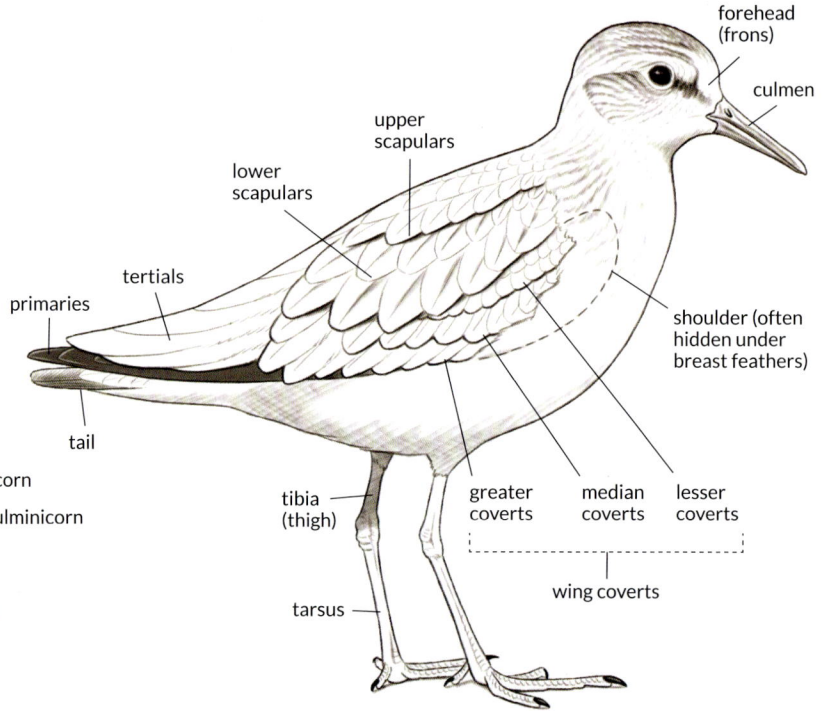

forehead
(frons)

culmen

upper
scapulars

lower
scapulars

tertials

primaries

tail

shoulder (often
hidden under
breast feathers)

tibia
(thigh)

greater
coverts

median
coverts

lesser
coverts

wing coverts

tarsus

Wader

lateral
crown stripe

supercilium
(eyebrow)

eye
stripe

lore

ear coverts
(cheeks)

eye-ring

Passerine (tchagra)

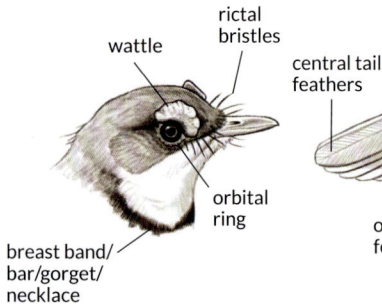

crown

upper
mandible

nape

scapulars

mantle

back

rump

uppertail
coverts

lower
mandible

gape

malar
stripe

moustachial
stripe

breast

flanks

belly

undertail
coverts

vent

tibia

tarsus

rictal
bristles

wattle

central tail
feathers

orbital
ring

outer tail
feathers

breast band/
bar/gorget/
necklace

Passerine (wattle-eye)

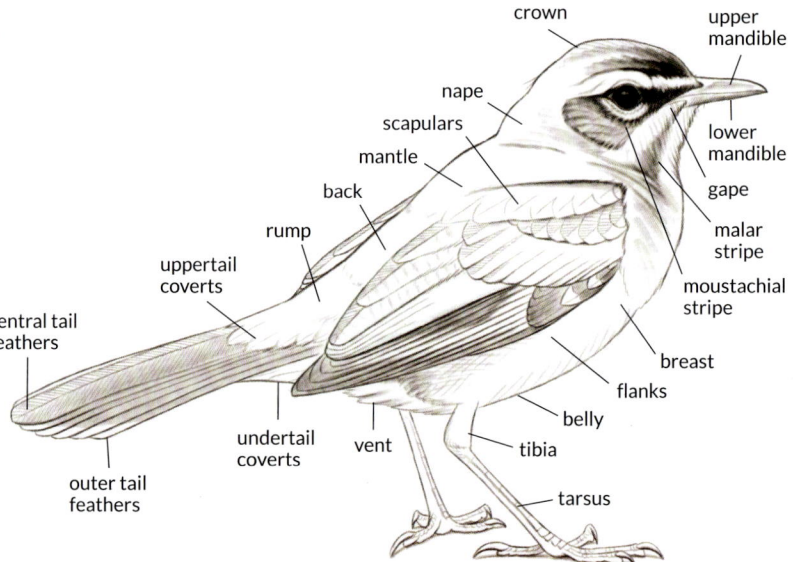

Passerine (scrub robin)

Underparts

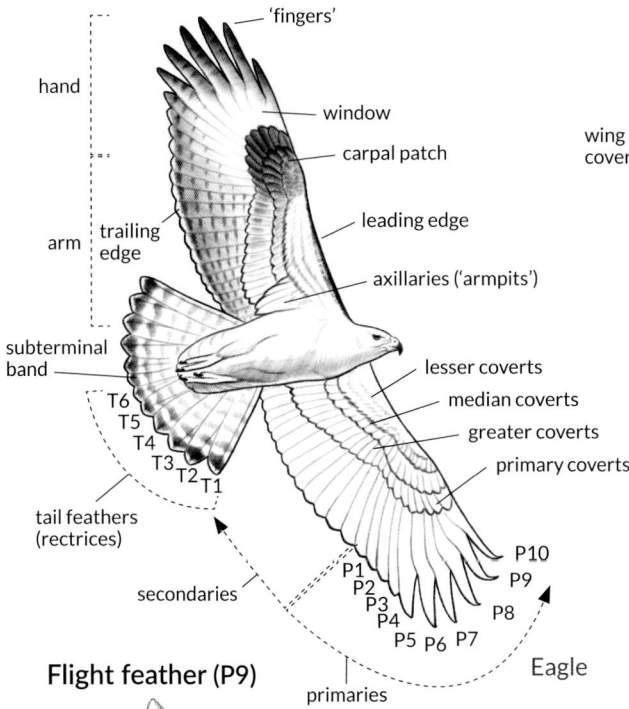

'fingers'

window

carpal patch

leading edge

axillaries ('armpits')

lesser coverts

median coverts

greater coverts

primary coverts

hand

arm

trailing edge

subterminal band

T6
T5
T4
T3
T2
T1

tail feathers (rectrices)

secondaries

P1
P2
P3
P4
P5 P6 P7

P10
P9
P8

primaries

Eagle

Upperparts

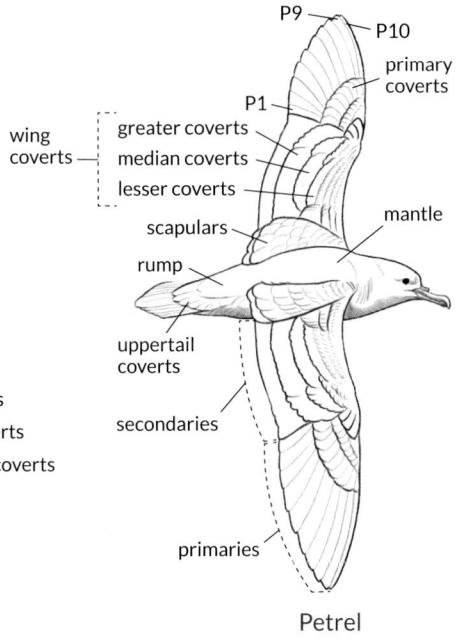

P9

P10

primary coverts

P1

greater coverts

median coverts

lesser coverts

scapulars

mantle

rump

uppertail coverts

secondaries

wing coverts

primaries

Petrel

Flight feather (P9)

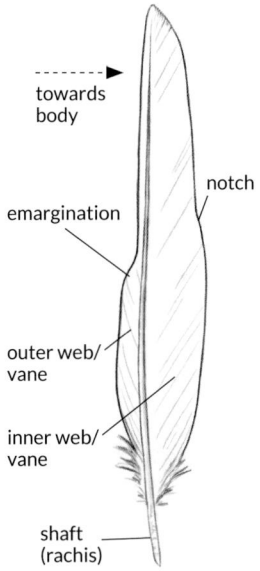

towards body

emargination

notch

outer web/ vane

inner web/ vane

shaft (rachis)

Closed wing

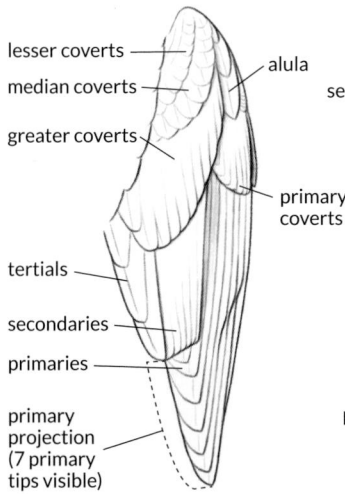

lesser coverts

median coverts

greater coverts

alula

primary coverts

tertials

secondaries

primaries

primary projection (7 primary tips visible)

Open wing

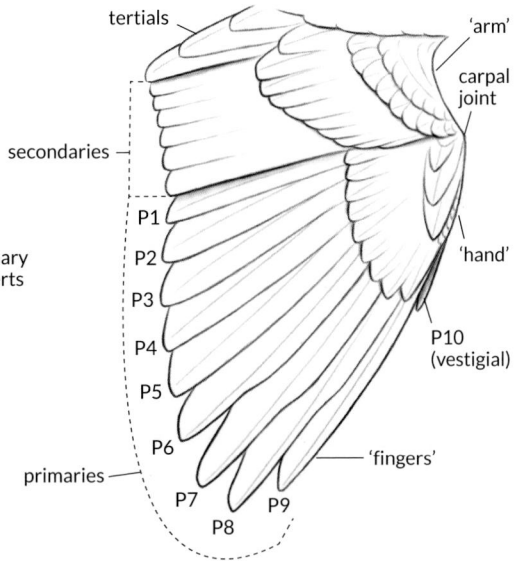

tertials

'arm'

carpal joint

secondaries

P1
P2
P3
P4
P5
P6
P7
P8
P9

'hand'

P10 (vestigial)

'fingers'

primaries

Feather markings

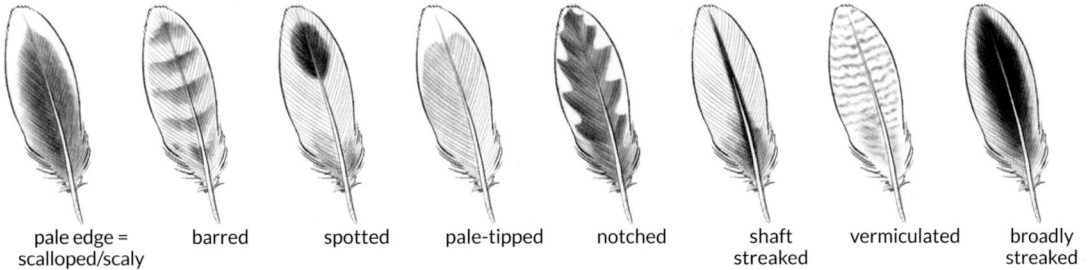

pale edge = scalloped/scaly

barred

spotted

pale-tipped

notched

shaft streaked

vermiculated

broadly streaked

PENGUINS

PENGUINS Flightless seabirds, supremely adapted for pursuit diving; closest relatives are the albatrosses and petrels. Wings reduced to small, rigid flippers for 'flying' underwater. Sexes alike; juv. plumage distinct. Only the African Penguin breeds in Africa; all others are vagrants, often coming ashore to moult. All penguins come ashore for 3–4 weeks to replace their feathers simultaneously, relying on stored resources during this moult fast.

AFRICAN PENGUIN *Spheniscus demersus*

b	b	B	B	B	B	B	B	b	b	b	b
J	F	M	A	M	J	J	A	S	O	N	D

60–70 cm; 2.2–3.5 kg The only penguin that breeds in the region. Ad. has diagnostic striped head and sides; some have a double breast band, as in Magellanic Penguin, but upper band is narrower in African Penguin and it lacks the narrow, white line around the base of the lower mandible. Juv. lacks bold patterning, with whitish underparts and upperparts varying from greyish-blue to brown; some juvs moult part or all of their heads to ad. plumage before their first complete moult. **Voice:** Loud, donkey-like braying at colonies, especially at night. **Status and biology:** Br. endemic. ENDANGERED due to competition with pelagic fisheries, pollution, predation and disturbance at br. colonies. Regular within 20 km of shore; vagrant >50 km offshore. Population has fallen by half since 2004; currently 26 000 pairs breed at islands and a few coastal sites along the west and south coasts. Breeds year-round. (Brilpikkewyn)

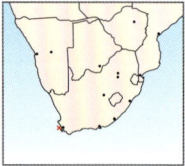

MAGELLANIC PENGUIN *Spheniscus magellanicus*

		X									
J	F	M	A	M	J	J	A	S	O	N	D

70–76 cm; 3–5.2 kg Vagrant. Slightly larger than African Penguin, with 2 breast bands; upper band broader. Narrow white line fringes the base of the lower mandible. Juv. similar to juv. African Penguin, but often shows faint stripes on breast and flanks. **Voice:** Braying call, similar to African Penguin. **Status and biology:** NEAR-THREATENED despite a breeding population of more than 1 million pairs. Very rare vagrant from br. range in S America: 1 record in Cape Town harbour (1998) was probably ship-assisted, but vagrants have reached Tristan da Cunha and Marion Island. (Magellaanpikkewyn)

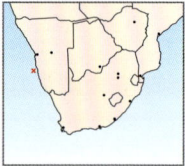

LITTLE PENGUIN *Eudyptula minor*

			X								
J	F	M	A	M	J	J	A	S	O	N	D

40–45 cm; 800–1 600 g Vagrant. A diminutive penguin resembling a tiny, newly fledged African Penguin. Differs in having pale, creamy-grey (not dark grey) feet, a plain face with no bare facial skin, and a smaller, weaker bill. Eyes usually paler; divide between blue-grey upperparts and white underparts more defined than in African Penguin. **Voice:** Loud growls and groans at colonies at night. **Status and biology:** Very rare vagrant from s Australia and New Zealand: 1 record from Ichaboe Island, Namibia, Apr 2005. With no other records of such extremely long-distance dispersal, the possibility of ship-assisted transit cannot be excluded. (Kleinpikkewyn)

KING PENGUIN *Aptenodytes patagonicus*

X	X		X			X		X			
J	F	M	A	M	J	J	A	S	O	N	D

88–95 cm; 9–15 kg Vagrant. A very large, long-flippered penguin with distinctive orange neck patches and a pinkish-orange plate on each side of the long, slender, slightly decurved bill. Juv. is duller, with creamy neck patch. **Voice:** Occasional *'dhuu'* at sea. **Status and biology:** Rare vagrant from sub-Antarctic br. islands, with <10 records from the coast between Cape Town and St Lucia. Although some birds have been confiscated from fishing vessels docking in Cape Town, records of moulting individuals probably are genuine vagrants. Usually lands on beaches or sheltered, gently shelving rocky shores. Listed as Least Concern globally, but northern br. populations, such as at Marion Island, are threatened by climate change because ads have to travel increasingly far to forage at the polar front, south of the island. (Koningpikkewyn)

AFRICAN PENGUIN

imm.

imm.
head
moulting

juv.

ad.

1 breast band
(exceptionally
2 bands)

MAGELLANIC PENGUIN

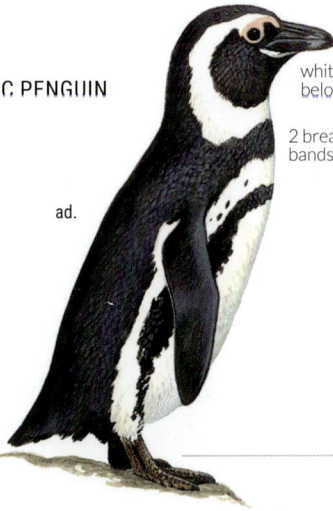

ad.

white line
below bill

2 breast
bands

LITTLE PENGUIN

small
bill

pale
feet

KING PENGUIN

imm.

ad.

23

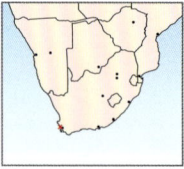

GENTOO PENGUIN *Pygoscelis papua*

J F M A M J J A S O N D

75–80 cm; 4.2–7.2 kg Vagrant. A large, elegant, long-tailed penguin with distinctive white flecking above and behind eye and a narrow, white eye-ring. Feet and bill orange. Flippers long, with pale yellowish-orange wash on underside. Juv. has paler throat and smaller white 'ear' patches. **Voice:** Loud cawing at colonies; silent at sea. **Status and biology:** Rare vagrant to S Africa from sub-Antarctic br. islands, with only 1 record (Cape Town, Jun 1992). This individual was not moulting, but showed no signs of being held in captivity, so may have been a genuine vagrant. (Witoorpikkewyn)

EASTERN ROCKHOPPER PENGUIN *Eudyptes filholi*

J F M A M J J A S O N D

45–55 cm; 2–3 kg Vagrant. Slightly smaller than Northern Rockhopper Penguin; ad. has shorter crest and less extensive dark tips to underside of flipper. Ad. has a narrow, pink line around base of bill, but much thinner than prominent pink gape of Macaroni Penguin. Juv. only separable from Northern Rockhopper by underflipper pattern. **Voice:** 'Kerr-ik kerrik kerik-kerik-kerik-kerik', distinctly higher pitched than Northern Rockhopper Penguin. **Status and biology:** VULNERABLE due to ongoing population decreases. Recently split from Southern Rockhopper Penguin *E. chrysocome* from the Falklands and s South America, which has not been recorded from the region. Rare vagrant to S Africa from sub-Antarctic br. islands. Most records are of moulting juvs. (Oostelike Geelkuifpikkewyn)

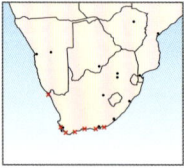

NORTHERN ROCKHOPPER PENGUIN *Eudyptes moseleyi*

J F M A M J J A S O N D

48–58 cm; 2–3.2 kg Vagrant. A fairly small penguin with a short, stubby, red bill and yellow crest that starts just in front of the eyes (does not meet on forehead, as in Macaroni Penguin). Ad. head plumes much longer and more luxuriant than Eastern Rockhopper Penguin; more extensive black on underside of flipper. Juv. browner above, with a dull red bill; yellow crest reduced or absent, but has peaked crown feathers; only separable from Eastern Rockhopper Penguin by underflipper pattern. **Voice:** Deep, raucous 'kerr-aak, kerr-aak kerrak kerrak-kerrak-kerrak' at colonies, lower pitched than call of Eastern Rockhopper Penguin. **Status and biology:** ENDANGERED due to recent population decreases, but current evidence suggests population may have stabilised. Regular vagrant to S Africa, presumably mostly from Tristan da Cunha and Gough Island. Most records are of moulting juvs. (Noordelike Geelkuifpikkewyn)

MACARONI PENGUIN *Eudyptes chrysolophus*

J F M A M J J A S O N D

68–75 cm; 3.1–5.5 kg Vagrant. Larger than rockhopper penguins, with a massive bill, broad, pink gape and golden crest meeting across forehead (but this is not obvious at sea). Juv. duller, with little or no crest; best told from juv. rockhopper penguins by its heavier bill and prominent, angled, pink gape. **Voice:** Deep, exultant braying at colonies; occasional 'harr' at sea. **Status and biology:** VULNERABLE due to ongoing population decreases. Rare vagrant to S Africa from sub-Antarctic br. islands. Most records are of moulting juvs. Recent genetic work shows that the 'Royal' Penguin *E. schlegeli*, mainly found on Macquarie Island but found in small numbers among Macaroni Penguins at colonies on Marion Island, is just a white-faced form of Macaroni Penguin. (Langkuifpikkewyn)

imm.

GENTOO PENGUIN

ad.

ımm.

short head
plumes

pink line at
base of bill

ad.

EASTERN ROCKHOPPER
PENGUIN

pale
under
flipper

ımm.

long, bushy
head
plumes

ORTHERN ROCKHOPPER
PENGUIN

ad.

imm.

heavy bill
& broad,
fleshy
gape

ad.

MACARONI
PENGUIN

darker
under
flipper

25

ALBATROSSES

ALBATROSSES Large, long-winged seabirds, adapted for dynamic soaring, that seldom come close to land. Sexes alike, but imm. and juv. plumages distinct. Primary moult complex, with most birds only replacing a subset of feathers each year. Three main groups occur in the region: great albatrosses, mollymawks and sooty albatrosses. Great albatrosses (*Diomedea*) have pink bills and mostly white underwings at all ages. Plumage whitens with age throughout life. Ads have white backs and could be confused with gannets (p. 72) and occasional Shy Albatrosses (p. 30) with white backs. All great albatrosses take so long to raise their single chick that they usually take a year off after a successful breeding attempt, freeing ads to disperse far from colonies. The three species shown on this page are known as the Wandering Albatross complex.

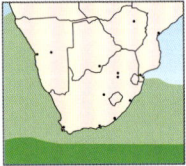

WANDERING ALBATROSS *Diomedea exulans*

J F M A M J J A S O N D

110–135 cm; 7–11 kg; wingspan 2.8–3.6 m A huge, hump-backed albatross with a pink bill. Underwing white, with black tip, trailing edge, and leading edge to carpal joint. Juv. chocolate-brown, with white face and underwings. As birds age, they become progressively whiter: body becomes mottled ('leopard' stage), then all-white, with fine vermiculations concentrated on back and breast, forming a shadow breast band. Following this stage, upperwing starts to whiten, initially from centre of wing over elbow (not from leading edge, as in Southern Royal Albatross, p. 28). Throughout these stages, birds have tail tipped black (mostly white in royal albatrosses). Males whiten faster than females; after 20 years, black only on flight feathers and a few covert tips. Old males differ from ad. Southern Royal Albatross by pinker bill with no dark cutting edge; feathering doesn't extend far onto lower mandible (unlike royal albatrosses), giving a steeper looking forehead. Often has pink mark on side of neck (absent in royal albatrosses). Probably cannot be separated reliably from Tristan Albatross at sea. **Voice:** Occasionally gives grunts and whinnies at sea. **Status and biology:** VULNERABLE, due mainly to accidental mortality on fishing gear, but population stable at most colonies. Regular, but uncommon, non-br. visitor to oceanic waters from sub-Antarctic br. islands. Occasionally visits trawlers, but seldom joins feeding mêlée. Global population around 80 000 birds. (Grootalbatros)

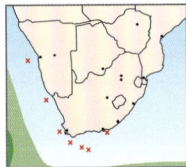

TRISTAN ALBATROSS *Diomedea dabbenena*

J F M A M J J A S O N D

100–110 cm; 6–7 kg; wingspan 2.5–3.2 m Slightly smaller than Wandering Albatross, with a shorter bill. Plumage takes longer to whiten, and does not attain fully white plumage of old male Wandering Albatrosses. Birds at sea probably not identifiable with certainty. Typical ad. males have mostly dark upperwing with pale patch on elbow, but some have more extensive white upperwings; older birds seem to have less black in tail tip relative to extent of black in upperwing than similar-plumaged Wandering Albatrosses, but more data needed to confirm this trait. Most ad. females retain some brown feathers on crown, back, breast and flanks, resembling immature Wandering Albatross, but old females resemble ad. males. Juv. like juv. Wandering Albatross. **Voice:** Occasionally gives grunts and whinnies at sea; inseparable from calls of Wandering Albatross. **Status and biology:** CRITICALLY ENDANGERED due to combination of accidental mortality on fishing gear and high chick mortality due to introduced mice at colony on Gough Island. Mainly remains in oceanic waters of S Atlantic, typically north of 35°S, but some venture across the Indian Ocean to Australia. Ringing and tracking data confirm occurrence off the west coast of s Africa, but abundance unknown. Global population <20 000 birds. (Tristangrootalbatros)

AMSTERDAM ALBATROSS *Diomedea amsterdamensis*

J F M A M J J A S O N D

100–110 cm; 6–7 kg; wingspan 2.5–3.2 m Vagrant. Slightly smaller than Wandering Albatross, with a shorter bill. Bill has a dusky tip and dark cutting edge to upper mandible (like royal albatrosses, p. 28), but these features also shown by some Tristan Albatrosses. Breeds in dark-backed plumage, similar to juv. or early imm. plumage of Wandering Albatross. **Voice:** Occasionally gives grunts and whinnies at sea; inseparable from calls of Wandering Albatross. **Status and biology:** ENDANGERED due to very small population (approx. 100 birds) on Amsterdam Island (c Indian Ocean) and accidental mortality on fishing gear, but numbers steady or increasing. Status in region poorly known: 1 record off Cape Point in Jul. 2013, and 1 satellite-tagged juv. travelled down the east coast to near Cape Agulhas. (Amsterdamgrootalbatros)

juv.

all brown except face
& underwings

ad.

WANDERING ALBATROSS

imm.

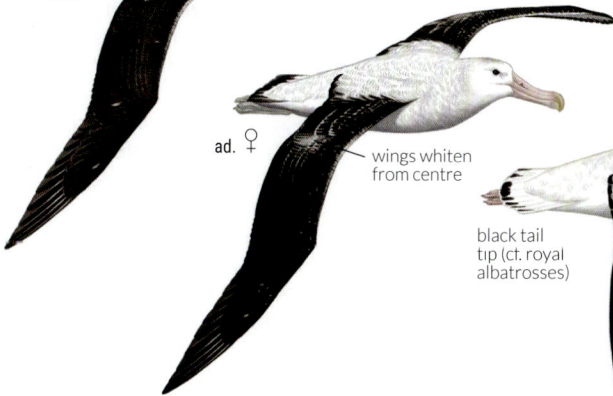

ad. ♀

wings whiten
from centre

black tail
tip (cf. royal
albatrosses)

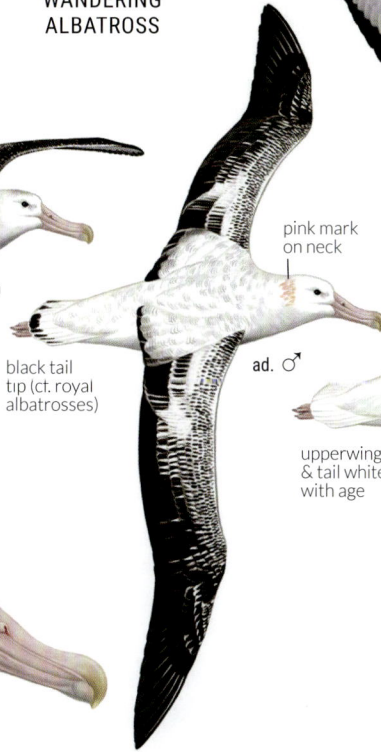

pink mark
on neck

ad. ♂

upperwings
& tail whiten
with age

old ♂

juv.

ad.

ad. ♀

ad. ♂

plumage similar
to younger
Wandering
Albatross

TRISTAN ALBATROSS

ad. ♂

ad. ♀

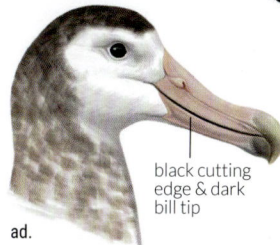

black cutting
edge & dark
bill tip

ad.

AMSTERDAM ALBATROSS

ROYAL ALBATROSSES The 2 royal albatrosses are closely related to the Wandering Albatross complex (p. 26), and are equally large, with similar underwing patterns. Their heads look sleek due to shallower forehead and feathering extending further onto lower mandible. Bills more yellow-pink, with a black cutting edge visible at close range; they lack the pink marks on the sides of the neck that are typical of the Wandering Albatross complex. Formerly considered vagrants to the region, they are now recorded at least as regularly as Wandering Albatrosses from shelf waters (but Wandering Albatrosses remain more abundant in oceanic waters). Separation of the 2 royal albatross species is not as simple as many guides suggest, and some imm. birds may be impossible to separate reliably in the field. At 1 colony the 2 species occasionally hybridise.

SOUTHERN ROYAL ALBATROSS *Diomedea epomophora*

J F M A M J J A S O N D

115–122 cm; 7–11 kg; wingspan 2.8–3.5 m Slightly larger than Northern Royal Albatross. Upperwing whitens with age from leading edge; carpal bar usually narrower. Juv. resembles juv. Northern Royal Albatross, but typically shows white coverts along leading margin of forewing; may retain mainly black upperwing and some black in outer tail and on scapulars for 10–12 years. White tail separates it from Wandering Albatross (p. 26), except for very old birds. Upperwing generally more finely marked than Wandering Albatross of equivalent age, and lacks pink mark on neck. **Voice:** Silent at sea. **Status and biology:** VULNERABLE. Rare visitor to S Africa; more common in oceanic waters than Northern Royal Albatross. Global population 8 500 pairs; all breed at islands off New Zealand. (Witvlerkkoningalbatros)

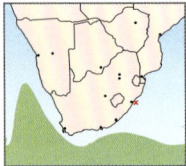

NORTHERN ROYAL ALBATROSS *Diomedea sanfordi*

J F M A M J J A S O N D

107–122 cm; 6–8 kg; wingspan 2.7–3.4 m Slightly smaller than Southern Royal Albatross. Best identified by crisp black upperwing (but beware moulting birds, which can show some white spotting due to exposed pale feather bases) contrasting with white body and tail; black upperwing coverts extend onto leading margin of forewing. From below, black carpal patch is broader than in Southern Royal Albatross and Wandering Albatross complex (p. 26). Juv. has some black in outer tail and slight scalloping on back, but has much less black in tail than any Wandering Albatross with dark upperwings. Black carpal patch narrower than ad. **Voice:** Silent at sea. **Status and biology:** ENDANGERED. Rare visitor to S Africa, mainly along the shelf edge; attends fishing vessels more than Wandering Albatross. Global population 5 200 pairs; all breed at islands off New Zealand. (Swartvlerkkoningalbatros)

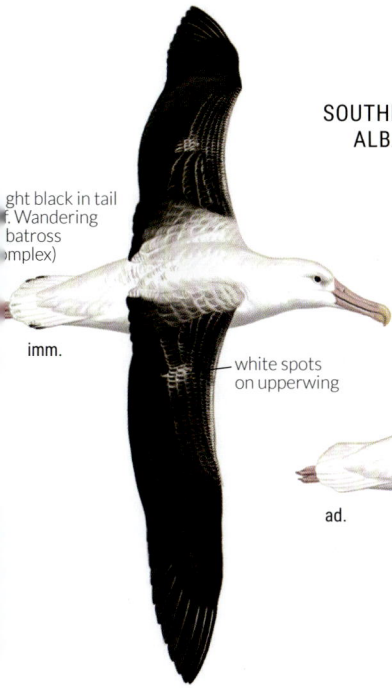

SOUTHERN ROYAL ALBATROSS

...ght black in tail
f. Wandering
batross
mplex)

imm.

white spots
on upperwing

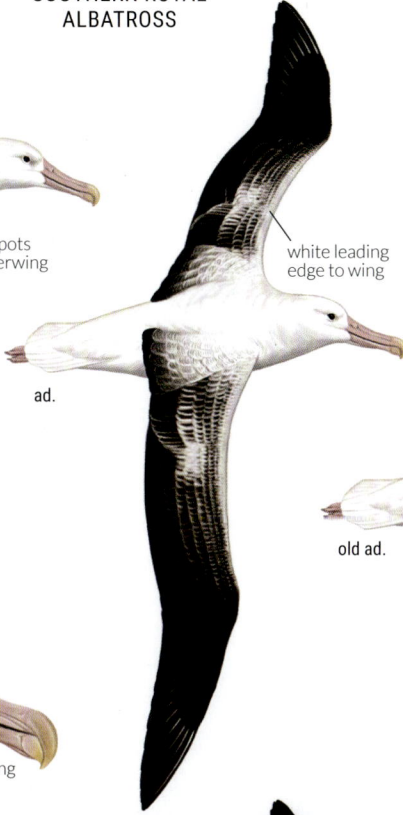

white leading
edge to wing

ad.

small black
carpal patch

wings whiten
from leading edge

old ad.

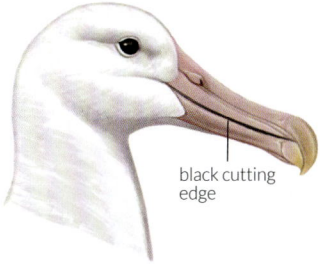

black cutting
edge

imm.

heads of two
species similar

...ght black
...tail (cf.
...andering
batross
...mplex)

imm.

entirely black
upperwing

ad.

darker tail, not
as extensive
as Wandering
Albatross
complex

fresh juv.

more extensive
black carpal patch

NORTHERN ROYAL
ALBATROSS

ad.

29

MOLLYMAWKS

MOLLYMAWKS Medium to small, dark-backed albatrosses. Generally more common than great albatrosses. Each species has diagnostic bill and underwing patterns, although bill colour changes with age and underwing whitens with age in some species. Identification of juvs and imms in the Shy Albatross complex (Shy, Chatham and Salvin's) is not well understood; probably only ads and sub-ads can be separated reliably at sea.

SHY ALBATROSS   *Thalassarche cauta*

J F M A M J J A S O N D

90–100 cm; 3–5 kg; wingspan 2.4–2.8 m The largest mollymawk, with extensive white on underwing; upperwing and mantle paler than other mollymawks; very rarely has a white back. Underwing has narrow, black border; black 'thumb-print' on leading edge near body diagnostic for the Shy Albatross complex (including vagrant Salvin's and Chatham Albatrosses). Ad. has pale grey cheeks and white crown; bill pale olive-grey with yellow tip. Imm. has grey-washed head, often with incomplete grey breast band; bill is grey with black tip. Juv. in fresh plumage has smooth grey wash on head and neck, recalling Salvin's Albatross. Sometimes treated as 2 species: most ads of Shy Albatross (*T. c. cauta*, which breeds off Tasmania) have a paler, yellowish culminicorn (top of bill) than White-capped Albatross (*T. c. steadi* from the New Zealand sub-Antarctic islands); 95% of birds in s African waters are *T. c. steadi*, with almost all records of *T. c. cauta* involving juvs or imms. **Voice:** Loud, raucous *'waak'* when squabbling over food. **Status and biology:** NEAR-THREATENED. Common non-br. visitor to fishing grounds along shelf edge, generally closer to land than other albatrosses; uncommon in oceanic waters; 120 000 pairs (90% *T. c. steadi*) breed at islands off New Zealand and Tasmania. (Bloubekalbatros)

CHATHAM ALBATROSS   *Thalassarche eremita*

X X X   X
J F M A M J J A S O N D

90–100 cm; 3–4.8 kg; wingspan 2.3–2.6 m Vagrant. Forms part of the Shy Albatross complex; underwing similar. Ad. has unmistakable yellow bill with dark tip to lower mandible, and dark, uniform grey head. Juv. similar to juv. Salvin's Albatross, but averages even darker on head and neck and has a dull yellowish bill with a dark tip. **Voice:** Silent at sea. **Status and biology:** VULNERABLE. Only 9 records, all from fishing grounds along the shelf edge of the W Cape in winter; 4 000 pairs breed on Pyramid Rock, Chatham Islands, New Zealand. (Chathamalbatros)

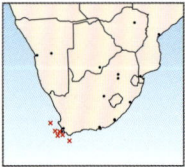

SALVIN'S ALBATROSS   *Thalassarche salvini*

X X X X X X X
J F M A M J J A S O N D

90–100 cm; 3.2–5 kg; wingspan 2.4–2.7 m Vagrant. Forms part of the Shy Albatross complex; underwing similar. Ad. has grey wash to neck and face, contrasting with pale crown. Bill grey-sided with paler, yellowish band along upper and lower mandible, and dark spot on tip of lower mandible. Juv. and imm. are similar to same ages of Shy Albatross, but black primary tips are slightly more extensive and head averages darker grey. Juv. has a broader black leading edge to underwing, potentially leading to confusion with juv. Buller's Albatross (p. 32). **Voice:** Silent at sea. **Status and biology:** VULNERABLE. Vagrant, with <40 confirmed records off W Cape, all from fishing grounds along the shelf edge; 40 000 pairs breed off New Zealand; 4 pairs on Crozets. (Salvinalbatros)

See pages 36–37 for head & bill detail

black-tipped,
grey bill

m.

SHY ALBATROSS

extensive white
in primaries
(cf. Salvin's Albatross)

ad.

yellow-tipped,
grey bill

grey of head
whitens with age

ad.

black 'thumb-
print' diagnostic
of Shy Albatross
complex

juv.

almost entirely
white underwing

head & bill can
appear paler in
bright light

SALVIN'S ALBATROSS

CHATHAM ALBATROSS

dark grey
head & crown

golden-yellow
ridge to upper
mandible

yellow bill

ad.

dark neck
& face
contrasting
with pale
crown

ad.

mostly dark
primaries

ad.

ad.

ad.

INDIAN YELLOW-NOSED ALBATROSS   *Thalassarche carteri*

J F M A M J J A S O N D

75–80 cm; 2.1–2.9 kg; wingspan 2.0–2.2 m Slightly larger than Atlantic Yellow-nosed Albatross, but head appears smaller. Ad. has only a faint grey wash on the cheek; at close range, base of yellow bill stripe is pointed (not rounded). Juv. similar to juv. Atlantic Yellow-nosed Albatross, but entire head and neck are white. In the hand, can detect difference in shape of culminicorn base. **Voice:** Throaty 'waah' and 'weeeeh' calls when squabbling over food. **Status and biology:** ENDANGERED. Fairly common year-round; ventures furthest north in winter. Most common albatross off KZN and often abundant off E Cape; 36 000 pairs breed in summer at s Indian Ocean islands. (Indiese Geelneusalbatros)

ATLANTIC YELLOW-NOSED ALBATROSS
Thalassarche chlororhynchos

J F M A M J J A S O N D

72–78 cm; 1.8–2.8 kg; wingspan 1.9–2.1 m A small, slender albatross with a relatively long bill. Underwing has crisp black border, with leading edge roughly twice as broad as trailing edge. Ad. has black bill with yellow stripe along upper mandible, becoming reddish towards tip. Differs from Indian Yellow-nosed Albatross by having grey wash to the head and nape (slightly paler on forecrown), with more extensive grey feathering around the eyes. Some individuals in worn plumage may show almost entirely white heads, making separation from Indian Yellow-nosed Albatross difficult. At close range, base of yellow stripe on upper mandible is broad and rounded (not pointed). Juv. bill all-black; head white; hard to separate from juv. Indian Yellow-nosed Albatross, but some show grey wash on mantle. In the hand, can detect difference in shape of culminicorn base. **Voice:** Throaty 'waah' and 'weeeeh' calls when squabbling over food. **Status and biology:** ENDANGERED. Fairly common year-round in small numbers; often the most abundant albatross off Namibia; 30 000 pairs breed in summer in loose colonies at Tristan da Cunha and Gough Island. (Atlantiese Geelneusalbatros)

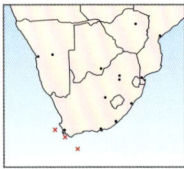

BULLER'S ALBATROSS   *Thalassarche bulleri*

X   X X X
J F M A M J J A S O N D

76–81 cm; 2.2–3.3 kg; wingspan 2.1–2.3 m Vagrant. A fairly small mollymawk with neat underwing at all ages; pattern similar to that of yellow-nosed albatrosses, but margins fractionally narrower. Head and neck washed grey, with pale forecrown. Bill of ad. has broad, yellow stripe along upper mandible and narrow stripe on lower mandible. Two subspecies often recognised: *T. b. platei* has a dark grey head with white in crown restricted to front, while *T. b. bulleri* has a paler grey head with entirely white crown. Ad. most likely confused with ad. Atlantic Yellow-nosed Albatross or Grey-headed Albatross (p. 34). Juv. has dark horn-coloured bill with black tip and a smooth grey wash on head and neck; could be confused with juv. Shy Albatross (p. 30), but has broader black leading edge to underwing (similar to juv. Salvin's Albatross, p. 30). **Voice:** Silent at sea. **Status and biology:** NEAR-THREATENED. Vagrant with <10 records from S Africa, mostly from fishing grounds along the shelf edge; 30 000 pairs breed at islands off New Zealand. (Witkroonalbatros)

ADULT 'YELLOW-NOSED' ALBATROSS BILL DETAILS

Indian Yellow-nosed
narrow yellow stripe
with pointed base;
pinkish bill tip

Atlantic Yellow-nosed
narrow yellow stripe with
rounded base; pinkish
bill tip

Buller's
broad yellow stripe,
bulging at base of upper
bill; bill tip yellow

Grey-headed
uniform yellow stripe
right to base of upper bill;
pinkish bill tip

See pages 36–37 for head & bill detail

juv.

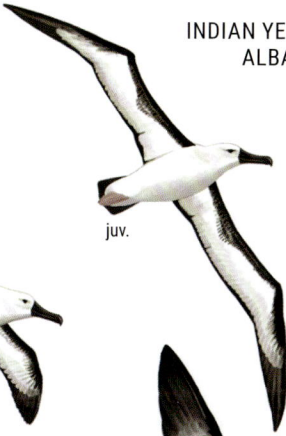

juv.

INDIAN YELLOW-NOSED ALBATROSS

ad.

white head, with a faint grey wash to cheeks

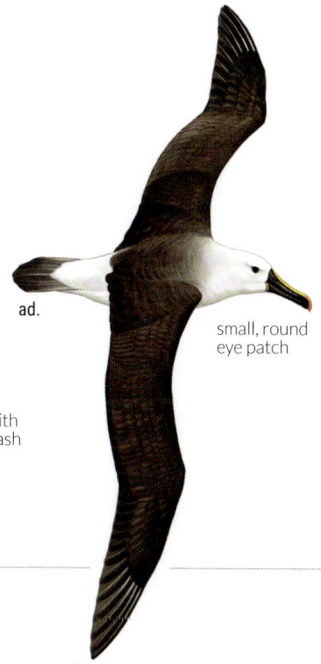

ad.

small, round eye patch

paler grey individual

ad.

ATLANTIC YELLOW-NOSED ALBATROSS

ad.

grey head & neck

dark, triangular eye patch

ad.

juv.

contrasting pale cap (all ages)

dark-tipped, pale grey or horn-coloured bill

ad.

BULLER'S ALBATROSS

broad, yellow-orange stripe on upper & lower mandibles

ad.

juv.

ad.

BLACK-BROWED ALBATROSS   *Thalassarche melanophris*

J F M A M J J A S O N D

80–95 cm; 3–4.6 kg; wingspan 2.2–2.5 m Medium-sized mollymawk. Ad. has orange bill with a reddish tip, and a small, black eyebrow. Underwing darkest of ad. mollymawks, with a broad, black leading edge and narrower trailing edge. Juv. underwing dark grey with a paler centre, lightening with age. Juv. bill dark horn-grey with black tip, gradually becoming yellow with darker tip. Amount of grey on head and neck is variable, usually forming incomplete collar. **Voice:** Grunts and squawks when squabbling over food. **Status and biology:** Fairly common year-round, but ads mostly Apr–Sep. Most abundant at shelf edge, but also in oceanic waters; usually close inshore only during storms. Formerly listed as ENDANGERED, but steady increases at the Falklands and islands in s Chile have seen it revised to Least Concern. Around 700 000 pairs breed in summer at sub-Antarctic islands; most birds in s Africa come from S Georgia, where numbers continue to decrease. (Swartrugalbatros)

GREY-HEADED ALBATROSS   *Thalassarche chrysostoma*

J F M A M J J A S O N D

75–88 cm; 3–4.2 kg; wingspan 2.1–2.5 m Similar in structure and plumage to Black-browed Albatross, showing comparable progression of underwing with age. Ad. has grey head, paler grey neck, and black bill with yellow stripe along upper and lower mandibles (see p. 32). Juv. has darker grey head, merging into mantle; cheeks almost white in some individuals. Juv. bill all dark (not black-tipped), but soon develops yellow tinge to ridge of upper mandible and tip, as imm. **Voice:** Squawks when squabbling over food. **Status and biology:** ENDANGERED. Rare visitor to s Africa, mostly juvs; joins other albatrosses feeding at trawlers; 125 000 pairs breed at sub-Antarctic islands. (Gryskopalbatros)

PHOEBASTRIA ALBATROSSES   Of the 4 *Phoebastria* albatrosses, Laysan Albatross is the only one to have been recorded in s African waters, as, apart from Waved Albatross in the Pacific Ocean, these albatrosses typically do not venture south of the equator.

LAYSAN ALBATROSS   *Phoebastria immutabilis*

J F M A M J J A S O N D

80 cm; 2–3.5 kg; wingspan 2.1–2.4 m Rare vagrant. A small, slender albatross. Superficially similar to imm. Black-browed Albatross, but with dark-washed cheeks, brown lower back extending onto rump, and distinctive underwing pattern with black streaks on underwing coverts. Pinkish feet project beyond tail in flight. Juv. as ad., but bill slightly greyer. **Voice:** Silent at sea. **Status and biology:** NEAR-THREATENED. Breeds in N Pacific; 2 regional records from sw Indian Ocean in 1980s, possibly of the same bird, are the only records in s hemisphere. (Swartwangalbatros)

See pages 36–37 for head & bill detail

BLACK-BROWED ALBATROSS

dark-tipped grey bill

broad, black outline to underwings

orange bill

underwings almost entirely dark

dark-tipped dull orange bill

juv.

imm.

ad.

ad.

dark bill, dull yellow ridge

imm.

dark grey neck & face

ad.

ad.

all-dark bill

yellow-orange stripe on upper & lower mandibles

dark underwings

GREY-HEADED ALBATROSS

broad, black outline to underwings

juv.

ad.

ad.

feet project beyond short tail

ad.

dark lower back extends onto rump

grey-washed cheeks

distinctive black-streaked underwing coverts

LAYSAN ALBATROSS

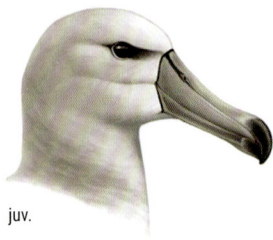  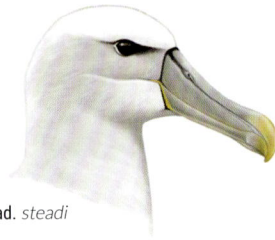 

juv.   imm.   ad. *steadi*   ad. *cauta*

SHY ALBATROSS

juv.   ad.

SALVIN'S ALBATROSS

 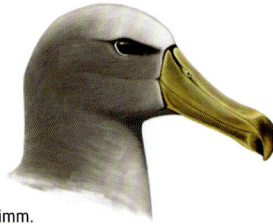 

juv.   imm.   ad.

CHATHAM ALBATROSS

juv.   imm.   ad.

BLACK-BROWED ALBATROSS

juv.   ad.

LAYSAN ALBATROSS

juv.

ad.

INDIAN YELLOW-NOSED ALBATROSS

juv.

ad.

ATLANTIC YELLOW-NOSED ALBATROSS

juv.

ad. *platei*

ad. *bulleri*

BULLER'S ALBATROSS

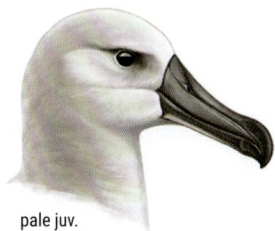

pale juv.

dark juv.

imm.

ad.

GREY-HEADED ALBATROSS

SOOTY ALBATROSSES

SOOTY ALBATROSSES Small albatrosses with sooty-brown plumage, white eye-rings and long, wedge-shaped tails. Wings exceptionally long and narrow; appear more slender than giant petrels (p. 40) or juv. gannets (p. 72). Adults easy to separate, but juvs are easily confused.

SOOTY ALBATROSS *Phoebetria fusca*

J F M A M J J A S O N D

84–90 cm; 2–3 kg; wingspan 2.0–2.2 m Ad. readily identified by its dark brown plumage, with pale shafts to primary and tail feathers. At close range, yellow stripe is visible on lower mandible (bill all-black in juvs). Juv. and imm. have conspicuous buff collar and mottling on back, but this does not extend to rump as in Light-mantled Albatross. **Voice:** Silent at sea. **Status and biology:** ENDANGERED. Rare visitor to oceanic waters off s Africa, fairly common further south; 15 000 pairs breed in summer at sub-Antarctic islands. (Bruinalbatros)

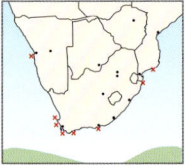

LIGHT-MANTLED ALBATROSS *Phoebetria palpebrata*

X X X X X X X X
J F M A M J J A S O N D

80–90 cm; 2.6–3.6 kg; wingspan 2.0–2.2 m Vagrant. Similar to Sooty Albatross, but with much paler neck collar shading into paler, greyish back that contrasts with upperwings. At close range, white eye-ring is shorter and broader, and bill has a lilac stripe on lower mandible. On land, head appears peaked (rounded in Sooty Albatross). Juv. only slightly paler above than juv. Sooty Albatross. Imm. has mottled body, but pale plumage extends lower on back and appears colder grey-brown than juv. Sooty Albatross. **Voice:** Silent at sea. **Status and biology:** NEAR-THREATENED. Rare vagrant to s Africa, seldom wandering north of the Subtropical Convergence; 22 000 pairs breed in summer at sub-Antarctic islands. (Swartkopalbatros)

SOOTY
ALBATROSS

pale buff
neck

imm.

juv.

uniform
mantle &
wings

ad.

ad.

ad.

yellowish
bill stripe

juv.

both species show
long, slender wings &
wedge-shaped tails

LIGHT-MANTLED
ALBATROSS

worn imm.

pale mantle
contrasts with
dark head &
upperwings

ad.

ad.

ad.

ad.

blue-grey
bill stripe

juv.

ad.

ad.

PETRELS

The largest family of tube-nosed seabirds, ranging in body size from 0.1–5.8 kg. Most are long-winged, adapted for dynamic soaring, but some have smaller wings for swimming underwater. Nostrils tube-shaped, joined on top of bill. Sexes alike; males slightly larger. Juvs resemble ads in all species except giant petrels, which are the only petrels with age-related plumage differences. Giant petrels are the largest petrels: huge, lumbering birds weighing more than small albatrosses, with massive, pale bills. All primary feathers are moulted annually, often resulting in large gaps in the outer wing.

SOUTHERN GIANT PETREL *Macronectes giganteus*

J F M A M J J A S O N D

86–100 cm; 2.5–5.8 kg; wingspan 1.8–2.0 m Relative to Northern Giant Petrel, ad. typically has paler head and breast, contrasting with dark body, but definitive identification requires seeing greenish bill tip. Juv. is dark brown, becoming lighter with age; greenish bill tip not well defined. Rare white morph has odd dark feathers; very rare leucistic birds are pure white. **Voice:** Usually silent at sea, but whinnies and neighs in conflicts. **Status and biology:** NEAR-THREATENED. Fairly common in coastal waters, scavenging at fishing boats and around seal colonies; follows ships in oceanic waters; 30 000 pairs breed at sub-Antarctic islands. Occasionally hybridises with Northern Giant Petrel. (Reusenellie)

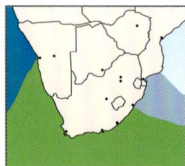

NORTHERN GIANT PETREL *Macronectes halli*

J F M A M J J A S O N D

81–98 cm; 2.8–5.8 kg; wingspan 1.8–2.0 m Very similar to Southern Giant Petrel, but has a reddish (not greenish) tip to the bill, which appears dark-tipped at a distance. Ad. has scaly grey plumage, shading from pale grey on head and breast to darker grey back and wings. Lacks a white morph. Juv. is dark brown, becoming paler with age; reddish bill tip is less marked than in ad and can be overlooked even at close range. **Voice:** Whinnies and neighs in conflicts. **Status and biology:** NEAR-THREATENED. Fairly common in coastal waters, scavenging at fishing boats; visits seal colonies more frequently than Southern Giant Petrel; 12 000 pairs breed at sub-Antarctic islands. Occasionally hybridises with Southern Giant Petrel. (Grootnellie)

SOUTHERN GIANT PETREL

head paler than rest of body

white morph (not known in Northern Giant Petrel)

ad.

juv.

greenish-tipped bill (all ages)

imm.

juv.

ad.

old ad.

juvs of both species start dark, lighten with age

NORTHERN GIANT PETREL

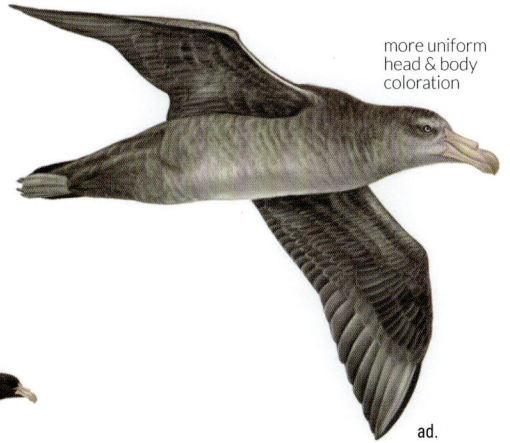

more uniform head & body coloration

ad.

moulting juv. showing large gaps in wings

reddish-tipped bill (all ages)

imm.

juv.

ad.

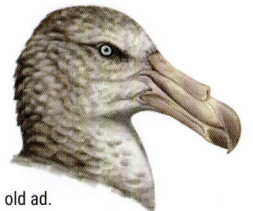

old ad.

FULMARINE PETRELS
Mid-sized petrels with boldly patterned plumage that breed at high latitudes. Ad. starts to moult while breeding.

CAPE (PINTADO) PETREL *Daption capense*

J F M A M J J A S O N D

35–40 cm; 350–550 g; wingspan 80–92 cm A small, black-and-white petrel with distinctive chequering on the back, rump and upperwing. Tail tipped black. Very rare leucistic birds may appear all-white, or have only faint greyish markings and could be confused with Snow Petrel *Pagodroma nivea*. **Voice:** High-pitched *'cheecheecheechee'* when feeding. **Status and biology:** Common non-br. visitor to coastal waters. Inveterate ship follower, scavenging galley wastes and small scraps at fishing boats. At least 200 000 pairs breed in summer on sub-Antarctic and Antarctic islands. (Seeduifstormvoël)

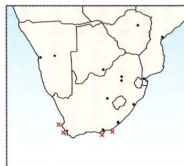

ANTARCTIC PETREL *Thalassoica antarctica*

J F M A M J J A S O N D

40–45 cm; 550–780 g; wingspan 98–105 cm Vagrant. Larger than Cape Petrel; brown above, with white subterminal band across wings and narrow, brown tip to tail. Dark feathers of upperparts often become bleached, sometimes resulting in a pale grey nuchal collar. **Voice:** Silent at sea. **Status and biology:** Rare vagrant from Antarctic waters; 250 000 pairs breed at ice-free mountains in Antarctica. (Antarktiese Stormvoël)

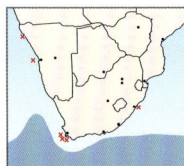

SOUTHERN FULMAR *Fulmarus glacialoides*

J F M A M J J A S O N D

45–51 cm; 670–1 000 g; wingspan 110–120 cm A pale grey petrel with white underparts and white panels in the darker grey outer wing. At close range, dark-tipped, pink bill with blue nostrils is diagnostic. **Voice:** High-pitched cackle when squabbling over food. **Status and biology:** Rare visitor to s Africa, mostly Jun–Oct. Scavenges at trawlers, often among flocks of Cape Petrels. At least 200 000 pairs breed in Antarctica and adjacent islands. (Silwerstormvoël)

CAPE PETREL

boldly patterned
upperparts highly
variable

ANTARCTIC
PETREL

worn birds show
pale collars

strikingly
patterned
upperwings

SOUTHERN
FULMAR

'gull-like'
appearance

white panels
in upperwing

PROCELLARIA PETRELS
Large petrels, superficially recalling large shearwaters, but flight is more languid, with less stiff wings held slightly bent at wrist.

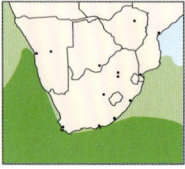

WHITE-CHINNED PETREL *Procellaria aequinoctialis*

J F M A M J J A S O N D

51–58 cm; 1–1.6 kg; wingspan 134–148 cm A large, blackish-brown petrel with a whitish bill. At close range, black 'saddle' to bill is evident. White throat is variable in extent – conspicuous in some, but reduced or absent in others. Quite often has odd white patches on head, belly or wings. **Voice:** Sometimes gives a high-pitched *'titititititi'* when sitting at sea. **Status and biology:** VULNERABLE. Common year-round, but numbers may be decreasing due to longline fishing. Most abundant in shelf waters, where it scavenges at fishing boats. Often follows ships. Two million pairs breed in summer at sub-Antarctic islands. (Bassiaan)

SPECTACLED PETREL *Procellaria conspicillata*

J F M A M J J A S O N D

50–56 cm; 1–1.3 kg; wingspan 132–146 cm Similar to White-chinned Petrel, but with a diagnostic white spectacle and dusky bill tip. Size of spectacle varies; incomplete in some birds, but most have some white on forehead. When spectacle is narrow, it is not connected to the white throat. Beware odd White-chinned Petrels with large, white throats or white head markings, often on the crown or nape, however, these birds do not show the dark bill tip of Spectacled Petrel. **Voice:** Usually silent at sea. **Status and biology:** VULNERABLE, although numbers continue to increase. Rare, year-round visitor to s African waters, but more common in summer; attends trawlers; 30 000 pairs breed in summer at Inaccessible Island in the Tristan Archipelago. (Brilbassiaan)

GREY PETREL *Procellaria cinerea*

J F M A M J J A S O N D

48–50 cm; 950–1 200 g; wingspan 115–130 cm A pale, silver-grey petrel with white underparts, dark underwings and a yellowish bill. Grey-brown of head extends far down cheeks, with only narrow, white throat. Told from pale shearwaters by dark grey (not white) underwing and heavier-chested appearance. **Voice:** Usually silent at sea. **Status and biology:** NEAR-THREATENED. Rare visitor to s African waters, seldom straying north of 40°S; 80 000 pairs breed in winter at sub-Antarctic islands. (Pediunker)

WHITE-CHINNED
PETREL

extent of white
chin variable

entirely pale bill

some birds show
white head patches

SPECTACLED
PETREL

extent of spectacle
variable, but always
some white on
forehead

dark-tipped
pale bill

GREY
PETREL

pale bill with
dark cap

old wing
coverts give
mottled
appearance

bulky build

dark grey
underwing

SHEARWATERS

SHEARWATERS Small to medium-sized petrels. Flight usually low over the water, with fairly straight, stiff wings in *Puffinus* and most *Ardenna* species, which are adapted for diving, using wings and feet underwater. *Calonectris* and some tropical *Ardenna* have larger wings, adapted more for gliding. Despite relatively small wings, several species undertake trans-equatorial migrations.

CORY'S SHEARWATER   *Calonectris borealis*

J F M A M J J A S O N D

42–50 cm; 480–750 g; wingspan 110–125 cm A large, broad-winged shearwater with slow languid flight; stays close to water, not banking and shearing as much as other shearwaters. Ash-brown above and white below with a dark-tipped, yellow bill. Upperparts paler than Great Shearwater (p. 48), but often shows pale crescent at base of tail. Told from the closely related Scopoli's Shearwater by its more robust bill, more extensive dark cap and fully dark primaries on underwing, but these features are difficult to judge on single birds; underwing pattern easily confused depending on light conditions. **Voice:** Silent at sea. **Status and biology:** Common Palearctic-br. migrant; 250 000 pairs breed at islands in ne Atlantic and w Mediterranean. (Geelbekpylstormvoël)

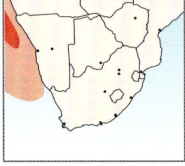

SCOPOLI'S SHEARWATER   *Calonectris diomedea*

J F M A M J J A S O N D

42–48 cm; 420–720 g; wingspan 105–120 cm Recently split from Cory's Shearwater; breeding ranges overlap narrowly in w Mediterranean. Slightly smaller with a more slender bill, smaller dark cap and pale-based primaries on underwing; features are difficult to judge on single birds; underwing pattern easily confused depending on light conditions. **Voice:** Silent at sea. **Status and biology:** Uncommon Palearctic-br. migrant mainly off Namibia; claims off S Africa require confirmation. 200 000 pairs breed at islands in the Mediterranean. (Scopolipylstormvoël)

STREAKED SHEARWATER   *Calonectris leucomelas*

J F M A M J J A S O N D

45–48 cm; 450–580 g; wingspan 108–124 cm Vagrant. Similar to Cory's and Scopoli's Shearwaters, but with white face and whiter, streaked crown, nape and cheeks. Bill is thinner and pale pink, not yellow. Underwing coverts finely streaked, appearing darker than coverts of Cory's and Scopoli's Shearwaters, especially on primary coverts. **Voice:** Silent at sea. **Status and biology:** VULNERABLE. Three records from W Cape and KZN, Aug–Oct. Breeds at islands in nw Pacific, migrating south to SE Asia, New Guinea and Australasia. (Gestreepte Pylstormvoël)

primaries may
appear pale in
bright light

heavy bill

CORY'S
SHEARWATER

all-dark
primaries

plain face
& crown

pale
primary
bases

plain face
& crown

SCOPOLI'S
SHEARWATER

more
slender bill

pale streaked
face & crown

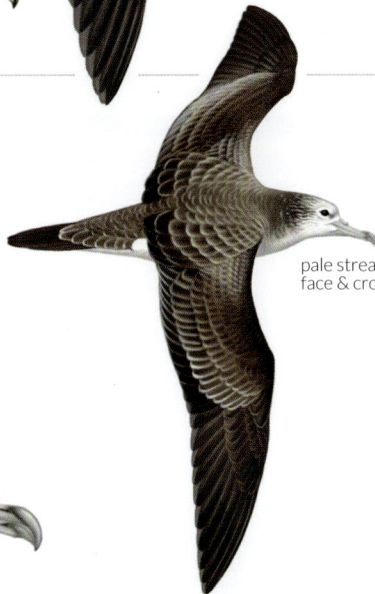

STREAKED
SHEARWATER

dark
primary
coverts

delicate bill

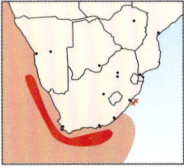

GREAT SHEARWATER *Ardenna gravis*

J F M A M J J A S O N D

45–51 cm; 800–1 100 g; wingspan 100–118 cm A dark-capped shearwater with a diagnostic dark belly patch (variable in extent), pale nuchal collar and broad, white rump. Darker above than *Calonectris* shearwaters (p. 46), with dark (not horn-coloured) bill. Underwing is mostly white, with indistinct lines across coverts. Flight action more dynamic than *Calonectris* shearwaters, with more rapid wing beats and straighter wings. **Voice:** Usually silent at sea. **Status and biology:** Common, mainly on passage (Sep–Nov and Apr–May). Five million pairs breed at Tristan da Cunha and Gough Islands, with a small colony in the Falklands. (Grootpylstormvoël)

FLESH-FOOTED SHEARWATER *Ardenna carneipes*

J F M A M J J A S O N D

45–50 cm; 550–680 g; wingspan 98–108 cm A dark brown shearwater with a dark-tipped, flesh-coloured bill and flesh-coloured legs and feet. Appears larger winged than Sooty Shearwater (p. 50); lacks silvery underwing, but has pale primary bases; flight is more petrel-like, with wings more bent at wrist. Smaller than White-chinned Petrel (p. 44), and pale bill is much more slender. Rounded tail, and pale bill, legs and feet distinguish it from Wedge-tailed Shearwater. **Voice:** Silent at sea. **Status and biology:** NEAR-THREATENED. Uncommon to locally common visitor year-round, mainly off east coast; 220 000 pairs breed in summer at south-temperate islands in Indian Ocean and off Australia and New Zealand. (Bruinpylstormvoël)

WEDGE-TAILED SHEARWATER *Ardenna pacificus*

J F M A M J J A S O N D

40–45 cm; 300–450 g; wingspan 97–105 cm A slender shearwater with a long, graduated tail that appears pointed in flight. Mostly dark-morph birds recorded from African waters: dark brown all over, with dark greyish-pink bill, legs and feet (unlike Flesh-footed Shearwater). Pale morph is brown above and white below, with clear-cut cap. Superficially resembles Great Shearwater, but lacks pale rump and nape; white underwing is mottled brown (not uniform white). Flight is light and buoyant on broad wings, held bowed forward and down. **Voice:** Silent at sea. **Status and biology:** Rare visitor to tropical east coast; occasional birds come ashore and call at night on islands in Nelson Mandela Bay; 100 000 pairs breed at tropical and subtropical islands. (Keilstertpylstormvoël)

dark bill

GREAT SHEARWATER

dusky belly patch

dark cap, white collar

FLESH-FOOTED SHEARWATER

dark-tipped pale bill

tail may appear pointed

heavy-set appearance (cf. Wedge-tailed Shearwater)

pale form

dark morph

dark morph

long, pointed tail

pale morph (extremely rare in subregion)

WEDGE-TAILED SHEARWATER

pale morph

dark morph

dark grey bill, may show a darker tip

49

SOOTY SHEARWATER *Ardenna griseus*

40–46 cm; 650–950 g; wingspan 94–106 cm A dark brown shearwater with distinctive pale, silvery underwing centres (intensity varies with light conditions). Bill long, slender and dark. Narrow, pointed wings are held straight, with little bend at wrist. Flight intersperses rapid bursts of flapping with short glides, but becomes more petrel-like in strong winds, rising higher above the sea. Told with difficulty from vagrant Short-tailed Shearwater (see below). **Voice:** Silent at sea. **Status and biology:** NEAR-THREATENED. Common in coastal waters year-round, but most abundant May–Sep. Several million pairs breed in summer at sub-Antarctic islands, mostly off New Zealand. Most of population migrates north of the equator in winter. (Malbaartjie)

SHORT-TAILED SHEARWATER *Ardenna tenuirostris*

40–45 cm; 500–800 g; wingspan 92–104 cm Vagrant. Slightly smaller and more compact than Sooty Shearwater with a distinctly shorter bill, and more rounded head with a steeper forehead; neck shorter and thicker than Sooty Shearwater; lacks dark lines across the inner underwing coverts and underwing typically appears darker than in Sooty Shearwater, but this is very dependent on light conditions. Many birds have a hooded appearance (more so than Sooty Shearwater), which is emphasised by a contrasting paler chin and collar. Often travels in small flocks, commuting at speed between feeding sites. **Voice:** Silent at sea. **Status and biology:** At least 10 million pairs breed off Tasmania and se Australia; migrates to N Pacific in winter. Only one confirmed record, s of Cape Point in Aug 2014, but large flocks moved into the S Atlantic in late summer 2017, with some migrating into the N Atlantic. (Kortstertpylstormvoël)

BALEARIC SHEARWATER *Puffinus mauretanicus*

32–38 cm; 480–550 g; wingspan 78–90 cm Vagrant. Formerly considered a race of Manx Shearwater (p. 52), but slightly larger, with brown (not black) upperparts and lacking sharp contrast on face. Typically has dark axillaries and dusky underparts and underwings, although this is variable. Some individuals are almost wholly dark, resembling a small Sooty Shearwater. Beware worn-plumage Manx Shearwaters, which can be brown above. **Voice:** Silent at sea. **Status and biology:** CRITICALLY ENDANGERED. Four records from coastal waters off west coast; 2 500 pairs breed at the Balearic Islands, Spain. (Baleariese Pylstormvoël)

sloping
forehead

long, all-dark bill

all 3 species show
silvery underwings

dark
morph

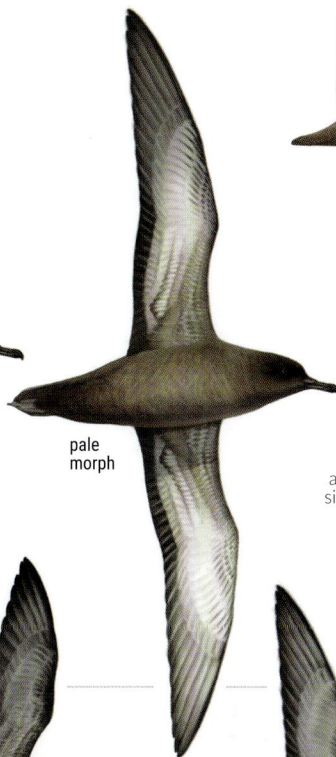

pale
morph

SOOTY
SHEARWATER

steep
forehead

shorter, all-dark bill

SHORT-TAILED
SHEARWATER

dark
morph

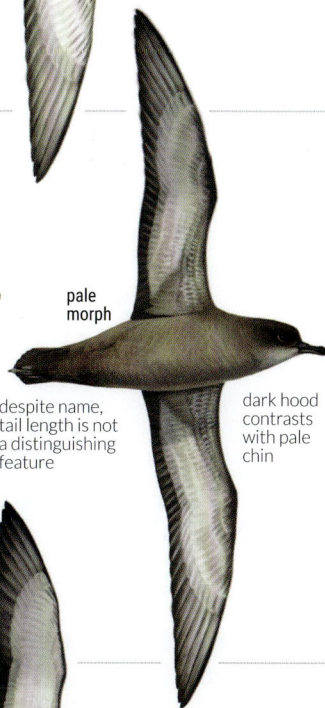

pale
morph

despite name,
tail length is not
a distinguishing
feature

dark hood
contrasts
with pale
chin

delicate
build

dark
morph

BALEARIC
SHEARWATER

very pale
underparts

pale
morph

51

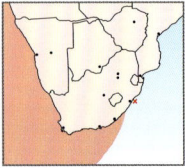

MANX SHEARWATER   *Puffinus puffinus*

J F M A M J J A S O N D

30–38 cm; 250–520 g; wingspan 76–88 cm The largest black-and-white shearwater in the region, with a relatively long, slender bill and pointed wings. Black upperparts contrast sharply with white underparts (including undertail coverts); often has white bands extending up onto sides of rump. Underwings show a much broader black trailing edge than leading edge; compare to Little/Subantarctic Shearwaters. Black cap extends below eye, with white ear coverts, creating a pale crescent behind the eye. Flight action comprises glides interspersed with rapid beats of stiff, straight wings; similar to Sooty Shearwater (p. 50), but different from rapid, fluttery flight of Little/Subantarctic Shearwaters. **Voice:** Silent at sea. **Status and biology:** Uncommon Palearctic-br. migrant to shelf waters, although in oceanic waters on migration; often found among Sooty Shearwater flocks; 300 000 pairs breed at islands in temperate Atlantic. (Swartbekpylstormvoël)

TROPICAL SHEARWATER   *Puffinus bailloni*

J F M A M J J A S O N D

28–32 cm; 140–200 g; wingspan 62–70 cm A small shearwater; intermediate in bill size, wing shape and flight action between Little/Subantarctic and Manx Shearwaters. Dark brown above, sometimes appearing blackish (blacker than Manx Shearwater). Lacks pale ear crescents of Manx Shearwater. Most records from the subregion are of the white-vented *P. b. bailloni* subspecies which breed on Europa and Reunion Islands, however birds of the dark-vented subspecies (*P. b. nicolae* and *P. b. temptator* from Seychelles and Comoros, respectively) are also known to visit our waters. Manx and Little/Subantarctic Shearwaters, can be separated from dark-vented populations as they show white vents however separation from white-vented populations requires more care. **Voice:** Silent at sea. **Status and biology:** CRITICALLY ENDANGERED Uncommon in tropical and subtropical waters off east coast; 100 000 pairs breed at tropical islands in Pacific and Indian oceans; 100 pairs at Europa Island, Mozambique Channel. (Tropiese Kleinpylstormvoël)

LITTLE SHEARWATER   *Puffinus assimilis*

J F M A M J J A S O N D

25–30 cm; 120–200 g; wingspan 56–66 cm A tiny, black-and-white shearwater with distinctive flight action, alternating rapid wing beats with short glides. Slightly smaller than Subantarctic Shearwater with more slender build and diagnostic white face extending above the eyes. Underwing pattern similar to Subantarctic Shearwater (unlike Manx Shearwater). **Voice:** Silent at sea. **Status and biology:** Scarce visitor to oceanic waters mainly off the east coast; rare in coastal waters. 40 000 pairs breed at islands off w Australia with more in the Pacific. One occupied a burrow on Bird Island, Algoa Bay, in May 1978. (Kleinpylstormvoël)

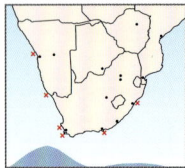

SUBANTARCTIC SHEARWATER   *Puffinus elegans*

J F M A M J J A S O N D

25–30 cm; 180–290 g; wingspan 56–66 cm A small black-and-white shearwater; slightly larger and more stocky than Little Shearwater with its dark cap extending below the eyes. Upperparts reflect silvery-grey in sunlight. Appreciably smaller than Manx Shearwater, with short, rounded wings and narrow, black trailing edge to underwing (similar in width to leading edge, not twice as wide, as in Manx Shearwater). Alternates rapid wing beats with short glides, like Little Shearwater. **Voice:** Silent at sea. **Status and biology:** Regular in oceanic waters south of Africa; scarce visitor to coastal waters mainly in winter. 200 000 pairs breed at Tristan, Gough and islands off New Zealand (Subantarktiese Pylstormvoël)

MANX SHEARWATER

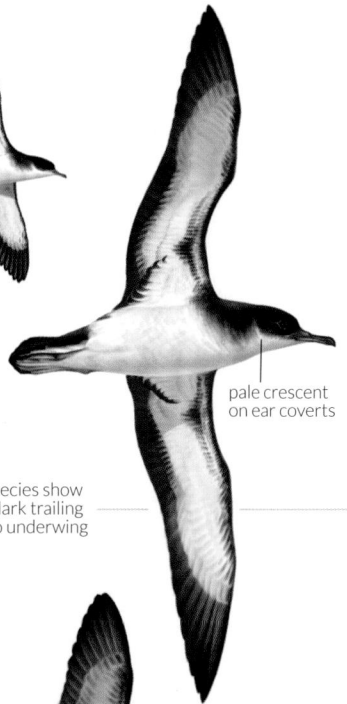

worn birds appear browner

pale crescent on ear coverts

both species show broad dark trailing edges to underwing

bailloni

TROPICAL SHEARWATER

colae

white-vented *bailloni* subsp.

dark-vented *nicolae* subsp.

whitish face

LITTLE SHEARWATER

dark face

SUBANTARCTIC SHEARWATER

both species show thin black trailing edge to underwing

PRIONS
Small, blue-grey petrels with dark 'M' marks across the upperwing, and white underparts. Flight rather fluttery and erratic. Easily separated from other petrels, but species identification is very tricky and depends on subtle differences in head and tail patterns: best achieved with photographs. Many individuals in the Broad-billed-Salvin's-Antarctic complex cannot be identified at sea.

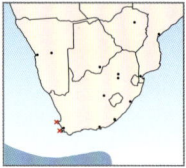

BROAD-BILLED PRION   *Pachyptila vittata*

X X X X X
J F M A M J J A S O N D

27–30 cm; 160–220 g; wingspan 56–66 cm The largest prion, with the broadest and darkest bill, but bill size of juv. overlaps narrowly with that of Salvin's Prion. Appears large-headed, with a steep forehead; whitish supercilium and dark grey face less striking than Salvin's Prion, but has a larger, darker grey carpal patch on underwing and grey breast patches are well developed. Blackish 'moustache' from gape typically curves up to join dark ear coverts. MacGillivray's Prion *P. macgillivrayi* from Gough and St Paul islands slightly smaller with bluish sides to the bill; not confirmed from the region, but a few tracked birds occur southwest of Africa mainly in winter. **Voice:** Silent at sea. **Status and biology:** Rare visitor to s Africa, but perhaps overlooked; tracked birds occur south of Africa mainly in summer. Breeds at Tristan and Gough islands (5 million pairs) and around New Zealand (1.5 million pairs). (Breëbekwalvisvoël)

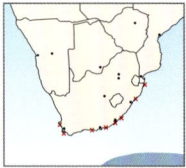

SALVIN'S PRION   *Pachyptila salvini*

X X X X
J F M A M J J A S O N D

26–29 cm; 120–200 g; wingspan 54–64 cm Intermediate in size and coloration between Broad-billed and Antarctic Prions, with narrow overlap in measurements with both species. Bill bluer than Broad-billed Prion, and facial pattern more pronounced; 'moustache' usually shorter. Appears larger headed than Antarctic Prion, usually with a distinctly broader and deeper bill. Most likely to occur in Indian Ocean. **Voice:** Silent at sea. **Status and biology:** Rare visitor to s Africa; 5 million pairs breed in summer at the Prince Edward and Crozet islands. (Marionwalvisvoël)

ANTARCTIC PRION   *Pachyptila desolata*

J F M A M J J A S O N D

25–28 cm; 120–180 g; wingspan 52–62 cm By far the most abundant prion in the region. Bill is fairly narrow and distinctly bluish. Tends to have smaller breast smudges than larger Salvin's and Broad-billed Prions, and facial pattern better defined than most Broad-billed Prions. Easily confused with Slender-billed Prion (see that species). **Voice:** Silent at sea. **Status and biology:** Common non-br. visitor, occurring in large flocks. Subject to occasional 'wrecks' when large numbers of dead and dying birds come ashore. Twenty million pairs breed in summer at sub-Antarctic islands and Antarctica, most on S Georgia. (Antarktiese Walvisvoël)

SLENDER-BILLED PRION   *Pachyptila belcheri*

J F M A M J J A S O N D

25–27 cm; 120–170 g; wingspan 52–62 cm Paler headed than Antarctic Prion, with smaller, paler grey breast patches and a more slender bill (only visible at close range). Head appears rounded, recalling Fairy Prion, with a long, white supercilium that broadens behind eye, white lores, and narrow, blue-grey (not dark grey) cheek patches. Black tail tip is reduced, with outer 2 or 3 tail feathers wholly blue-grey (only outer-tail feather all-grey in Antarctic Prion). **Voice:** Silent at sea. **Status and biology:** Rare visitor to shelf waters; abundance appears to vary between years. Three million pairs breed in summer at sub-Antarctic islands. (Dunbekwalvisvoël)

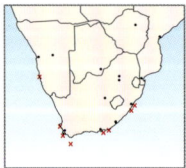

FAIRY PRION   *Pachyptila turtur*

X X X
J F M A M J J A S O N D

24–26 cm; 110–170 g; wingspan 50–60 cm Vagrant. The smallest prion; relatively easily identified by its broad, black tail tip, usually rather plain face and short, dumpy bill. Some birds show a more pronounced supercilium, recalling Slender-billed Prion, but short bill and extensive black tail tip diagnostic. **Voice:** Silent at sea. **Status and biology:** Vagrant to S Africa; most records of beached birds. Usually remains south of 40°S. Two million pairs breed in summer at sub-Antarctic islands, Tasmania and off New Zealand. (Swartstertwalvisvoël)

PRION BILL COMPARISON

| Broad-billed Prion | MacGillivray's Prion | Salvin's Prion | Antarctic Prion | Slender-billed Prion | Fairy Prion |

dark bill

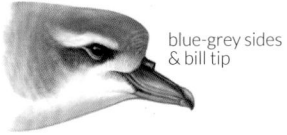

blue-grey sides
& bill tip

MacGillivray's Prion

BROAD-BILLED PRION

large-headed
appearance

steep
forehead

SALVIN'S PRION

heavier bill than
Antarctic Prion

ANTARCTIC PRION

SLENDER-BILLED PRION

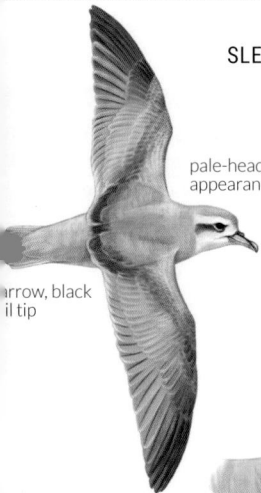

pale-headed
appearance

arrow, black
il tip

bold, white
supercilium

FAIRY PRION

broad, black
tail tip

plain face

BLUE PETREL

BLUE PETREL A prion-like petrel placed in its own genus, which usually forages in Antarctic waters. Ads breed in early summer, then undergo a very rapid moult before returning to their colonies for a few days in April–June.

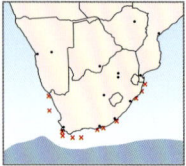

BLUE PETREL *Halobaena caerulea*

J F M A M J J A S O N D

28–30 cm; 160–240 g; wingspan 58–70 cm A small, blue-grey petrel with white underparts. Superficially similar to prions (p. 54), but slightly larger, with diagnostic, white-tipped tail; also has a white frons and black crown and nape, lacking a pale supercilium; at a distance, breast patches are darker and better defined, and body is more elongate, tapering from shoulders to relatively long tail. Flight action is petrel-like, fast and direct, rising higher above water than prions. Juv. paler on head; finely scaled white above. **Voice:** Silent at sea. **Status and biology:** Rare visitor to s Africa, occasionally irrupting in large numbers (most recently in 1984), but usually remains south of 45°S. One million pairs breed at sub-Antarctic islands in summer. (Bloustormvoël)

GADFLY PETRELS

GADFLY PETRELS Medium-sized petrels characterised by erratic, towering and very rapid flight action. Wings usually held angled. Bills dark; shorter and deeper than shearwaters and *Procellaria* petrels. Seldom occur close to shore unless there are strong onshore winds.

SOFT-PLUMAGED PETREL *Pterodroma mollis*

J F M A M J J A S O N D

32–37 cm; 220–350 g; wingspan 83–95 cm A small gadfly petrel with a variable, dark breast band and a white throat. White underparts contrast with dark underwings. Upperparts grey, with faint, darker 'M' across upperwings. Back colour and size of breast bands variable: average paler in *P. m. mollis* breeding at Tristan and Gough islands in the S Atlantic; darker in *P. m. dubia* from s Indian Ocean colonies. Some individuals with virtually no breast band claimed as Fea's Petrel *P. feae* from nw Atlantic islands, but definitive evidence of this species still needed (would be first records south of the equator). Rare dark morph lacks silvery highlights of Kerguelen Petrel (p. 60), has more slender neck, longer, more pointed wings, and is often mottled on the belly. Flight rapid and erratic, with deep wing beats. **Voice:** Silent at sea. **Status and biology:** Non-br. visitor from sub-Antarctic islands, where perhaps 700 000 pairs br. in summer. Uncommon in shelf waters, mainly in winter; common in oceanic waters year-round. (Donsveerstormvoël)

WHITE-HEADED PETREL *Pterodroma lessonii*

J F M A M J J A S O N D

40–45 cm; 500–800 g; wingspan 100–115 cm Vagrant. A large, chunky gadfly petrel with a diagnostic, whitish head and tail. Dark eyes are accentuated by blackish feathering around eyes, contrasting with pale head. Much larger and more bulky than Soft-plumaged Petrel; lacks dark cap and breast patches. **Voice:** Silent at sea. **Status and biology:** Rare visitor to s Africa from oceanic waters; 250 000 pairs breed in summer at sub-Antarctic islands. (Witkopstormvoël)

BARAU'S PETREL *Pterodroma baraui*

J F M A M J J A S O N D

38–40 cm; 400 g; wingspan 92–98 cm The only gadfly petrel in the region with mostly white underwings. Most likely to be confused with Soft-plumaged Petrel, but is larger, with a darker cap and smaller grey breast patches. From a distance could potentially be confused with *Calonectris* shearwaters (p. 46), but has a darker cap and a short, black (not pale) bill. **Voice:** Silent at sea. **Status and biology:** ENDANGERED. Scarce visitor to tropical east coast. Only recently confirmed from the region, but appears to be fairly regular off n KZN (and probably s Mozambique), with 10–20 being seen on some days. (Baraustormvoël)

BLUE
PETREL

white-tipped
tail (cf. prions)

dark grey
collar

blackish cap

SOFT-PLUMAGED
PETREL

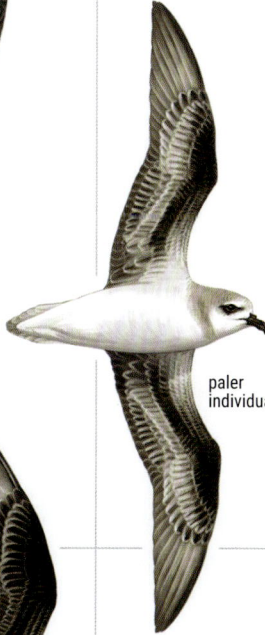

dark
underwings

dark
individual

paler
individual

rare
dark morph

WHITE-HEADED
PETREL

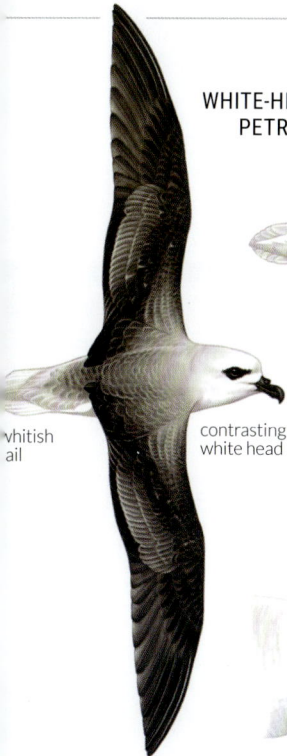

vhitish
ail

contrasting
white head

black
eye patch

BARAU'S
PETREL

black
cap

white underwings
with black carpal line

ATLANTIC PETREL *Pterodroma incerta*

J F M A M J J A S O N D

42–45 cm; 450–700 g; wingspan 95–105 cm A large, brown gadfly petrel with a conspicuous, white lower breast and belly. Larger than Soft-plumaged Petrel (p. 56), with chocolate-brown (not grey) plumage and no dark 'M' on upperwing. In worn plumage, neck and mantle can appear mottled brown. Larger than vagrant Trindade Petrel, with more uniformly dark plumage, although can appear paler due to wear. Given a poor view, could be confused with a pale morph jaeger (p. 226). **Voice:** Silent at sea. **Status and biology:** ENDANGERED due to chick predation by introduced House Mice at main br. site, Gough Island. Vagrant to coastal waters, but regular in small numbers in oceanic waters >200 km offshore. (Atlantiese Stormvoël)

TAHITI PETREL *Pseudobulweria rostrata*

J F M A M J J A S O N D

38–40 cm; 300 g; wingspan 84–90 cm Vagrant. A mid-sized brown petrel with a white lower breast and belly. Resembles a long-winged Atlantic Petrel but with white (not brown) undertail coverts, a more pronounced neck and a long, heavy bill. **Voice:** Silent at sea. **Status and biology:** NEAR-THREATENED. Five records off Maputo, Mozambique, and one off Durban, KZN, all in Nov–Dec. Only a few thousand pairs survive, breeding at islands in the S Pacific. (Tahitistormvoël)

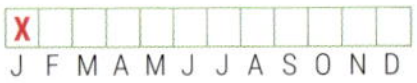

TRINDADE PETREL *Pterodroma arminjoniana*

J F M A M J J A S O N D

38–40 cm; 300–460 g; wingspan 88–102 cm Vagrant. A variable gadfly petrel with a slender body and squarish head, long wings and tail, and distinctive pale bases to the flight feathers. Pale morph is smaller than Atlantic Petrel, with pale throat and leading edge of the underwing. Dark morph most likely confused with Great-winged Petrel (p. 60), but is smaller and more slender with paler underwings and usually a pale throat. **Voice:** Silent at sea. **Status and biology:** VULNERABLE. 2 200 pairs breed at islands off Brazil; 200 pairs at Mauritius, where it hybridises with Kermadec and Herald Petrels. Only 1 record, off Port Elizabeth in Jan 2014, probably from the Mauritius population. (Trindadestormvoël)

worn birds are paler

heavy build

dark underwings

dark throat & vent

ATLANTIC PETREL

slight build

white vent

faint pale line

TAHITI PETREL

long, heavy bill

mottled whitish belly

TRINDADE PETREL

slighter build than Atlantic Petrel

palest morph

dark underwings with white flashes

plumage highly variable

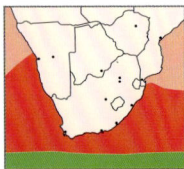

GREAT-WINGED PETREL   *Pterodroma macroptera*

J F M A M J J A S O N D

38–42 cm; 460–700 g; wingspan 95–105 cm A dark brown petrel with a short, stubby, black bill. Wings long and slender, held angled at wrist. Wing and bill shape, dark (not silvery) underwing and short neck differentiate it from Sooty Shearwater (p. 50). Smaller than White-chinned Petrel (p. 44), with black (not whitish) bill, and gadfly petrel jizz. Soars high above water in typical gadfly action, but flight tends to be more relaxed than other gadfly petrels. Appreciably larger than Jouanin's and Bulwer's Petrels (p. 62), with a shorter neck and tail. **Voice:** Silent at sea. **Status and biology:** Common non-br. visitor, mainly in summer. Often roosts on the water in small flocks, especially in calm weather. 250 000 pairs breed in winter at sub-Antarctic islands. (Langvlerkstormvoël)

KERGUELEN PETREL   *Aphrodroma brevirostris*

X X X X X
J F M A M J J A S O N D

33–36 cm; 220–400 g; wingspan 80–84 cm Vagrant. A small, compact petrel recalling a *Pterodroma*, but not closely related. Appears large-headed, with a thick 'bull neck'; eyes large. Smaller and greyer than Great-winged Petrel, with shorter, more rounded wings that show silvery highlights, especially on leading edge, and often extending onto breast. Flight fast and erratic, with rapid, stiff, shallow wing beats. Towers up to 50 m above sea, often hanging motionless or fluttering kestrel-like. **Voice:** Silent at sea. **Status and biology:** Irrupts in large numbers in some years (last major irruption in 1984), but usually remains south of 45°S; 200 000 pairs breed in summer on sub-Antarctic islands. (Kerguelense Stormvoël)

GREAT-WINGED PETREL

arcing
flight

worn birds
may show pale
covert bars

long-winged
appearance

heavy, black bill
(cf. Flesh-footed
Shearwater)

KERGUELEN PETREL

dark-hooded

bull-necked
appearance

silvery highlights
to leading edge
of wings

coloration
varies
according to
light

BULWERIA PETRELS
Fairly small, dark brown petrels with long wings and tails. No age or sex differences in plumage. Both are vagrants from colonies in tropical waters of the n hemisphere.

JOUANIN'S PETREL *Bulweria fallax*

J	F	M	A	M	J	J	A	S	O	N	D
										X	X

30–32 cm; 150–180 g; wingspan 76–83 cm Vagrant. Larger than Bulwer's Petrel, with different flight action; pale bar on upperwing coverts reduced or absent. Usually seen in calm, tropical oceans, where it flies low over the water with long glides interspersed with rapid, deep wing beats. In windy conditions, arcs and wheels high over the waves with dynamic flight like a gadfly petrel, but is smaller than Great-winged Petrel (p. 60), with longer neck and tail, and more slender wings. **Voice:** Silent at sea. **Status and biology:** NEAR-THREATENED. Regular in n Mozambique Channel, but very rare off s Africa. Thousands of pairs breed at Socotra, off the Horn of Africa. (Donkervlerkkeilstert)

BULWER'S PETREL *Bulweria bulwerii*

J	F	M	A	M	J	J	A	S	O	N	D
X	X	X								X	X

26–28 cm; 80–120 g; wingspan 68–72 cm Vagrant. A small, dark brown petrel with a diagnostic, long, wedge-shaped tail that is usually held closed and appears pointed. Superficially like a large *Oceanodroma* storm petrel (pp. 64 and 66), with deeper, stubby bill and paler grey-brown bar across upperwing coverts. Flight is buoyant and graceful, gliding low over the water. **Voice:** Silent at sea. **Status and biology:** Rare migrant to the west coast with one recent record from s Mozambique; may be more regular in oceanic waters off the west coast among large flocks of Leach's Storm Petrels; 100 000 pairs breed at islands in the N Atlantic and Pacific; small numbers in the Indian Ocean. (Bleekvlerkkeilstert)

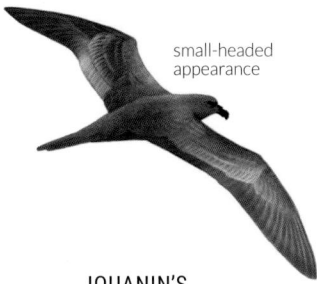

small-headed
appearance

JOUANIN'S PETREL

long
wings

long, pointed
tail

worn birds show
covert bar similar
to Bulwer's Petrel

heavy bill

wedge-
shaped tail

BULWER'S PETREL

obvious pale
covert bar

long,
slender
wings

smaller bill

63

STORM PETRELS

STORM PETRELS Small, pelagic seabirds with relatively large, broad wings and fluttering flight. Nostrils tube-shaped, joined on top of bill. Feed by surface-seizing or pattering; rarely dive. No age or sex differences in plumage. Now placed in two families; southern species (Oceanitidae) basal to all other petrels and albatrosses.

MATSUDAIRA'S STORM PETREL *Oceanodroma matsudairae*

J F M A M J J A S O N D

25 cm; 60–65 g; wingspan 52–58 cm Vagrant. A large storm petrel with long, broad wings. Flight action similar to Leach's and vagrant Swinhoe's Storm Petrels, but has a broader, yet duller bar across upperwing coverts, which reaches the leading edge of wing, and diagnostic white shafts to the outer 7 primaries which form a pale patch (evident with a reasonable view). Could be confused with Bulwer's Petrel (p. 62), but tail forked (not wedge-shaped). **Voice:** Silent at sea. **Status and biology:** VULNERABLE. Breeds at nw Pacific islands. Vagrant to KZN and s Mozambique (Jul–Sep) with 1 record off Cape Point. (Oosterse Stormswael)

SWINHOE'S STORM PETREL *Oceanodroma monorhis*

J F M A M J J A S O N D

20 cm; 38–40 g; wingspan 44–48 cm Vagrant. A fairly large storm petrel with long, broad wings and typical *Oceanodroma* flight action. Slightly smaller than vagrant Matsudaira's Storm Petrel, with a slightly thinner and paler bar across upperwing coverts, which does not reach leading edge of wings, and smaller white shafts to bases of the outer 6 primaries; flight more rapid and direct. Structure and flight action similar to Leach's Storm Petrel, and perhaps not reliably separable from dark-rumped morph of this species (although this form not known from our region). **Voice:** Silent at sea. **Status and biology:** NEAR-THREATENED. Most breed at nw Pacific islands. Vagrant to n KZN and s Mozambique (Sep–Nov). (Swinhoestormswael)

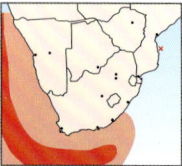

LEACH'S STORM PETREL *Oceanodroma leucorhoa*

J F M A M J J A S O N D

19–22 cm; 40–50 g; wingspan 45–50 cm Larger than Wilson's and European Storm Petrels (p. 66), with a long, forked tail and narrow, V-shaped white rump usually divided by dusky central line. Best identified by its long wings and languid flight action, gliding low over waves, with wings held forward and bent at wrist, flapping infrequently. Wing beats deep, causing erratic changes in direction. **Voice:** Rhythmical chattering and trilling at night at colonies; silent at sea. **Status and biology:** Fairly common, singly or in small groups, in oceanic waters beyond continental shelf. In calm weather, roosts on water, often with Great-winged Petrels; seldom follows ships. A few pairs breed at guano islands off W Cape, where threatened by introduced ants and vagrant Barn Owls; 8 million pairs breed in N Atlantic and N Pacific. (Swaelstertstormswael)

MATSUDAIRA'S STORM PETREL

long wings

obvious white crescent in primary bases

broad, dull covert bars reach leading edge of wing

deeply forked tail

SWINHOE'S STORM PETREL

thinner white bar in primary bases

obvious covert bars do not reach leading edge of wing

shallowly forked tail

LEACH'S STORM PETREL

rare dark-rumped morph

no pale primary bases

long, narrow wings

forked tail

white rump marginally 'folds over' to vent

often shows dark line through white rump

BAND-RUMPED STORM PETREL *Oceanodroma castro*

J F M A M J J A S O N D

20 cm; 35–48 g; wingspan 44–47 cm Vagrant. Intermediate in size and structure between Leach's and Wilson's Storm Petrels. Flight action closer to Leach's Storm Petrel (p. 64), but has shorter, broader wings, a more shallowly forked tail (appearing almost square when fanned) and a larger, white rump patch. Toes do not extend beyond tail tip in flight (unlike Wilson's Storm Petrel). **Voice:** Silent at sea. **Status and biology:** Only 1 photographed on an island near Lüderitz, Namibia; other sightings at sea off Namibia. Breeds at islands in tropical and temperate Atlantic and N Pacific; taxonomy complicated by summer and winter br. forms at the same island apparently being separate species. (Madeirastormswael)

WILSON'S STORM PETREL *Oceanites oceanicus*

J F M A M J J A S O N D

15–19 cm; 30–40 g; wingspan 38–42 cm A small, dark storm petrel with a broad, white rump that wraps around onto the flanks. Legs long; toes project beyond tail in flight, often dangling below bird when it is feeding by dancing over water, but can be retracted into belly plumage. Yellow toe webs are hard to see. Slightly larger than European Storm Petrel, with square (not rounded) tail and broader, more rounded wings; flight is more direct, with frequent glides (but varies with wind strength). Occasionally shows a paler bar on underwing, but not marked as on European Storm Petrel. **Voice:** Silent at sea. **Status and biology:** Common visitor to shelf waters, but less abundant in oceanic waters: often follows ships. Six million pairs breed at sub-Antarctic islands and Antarctica. (Gewone Stormswael)

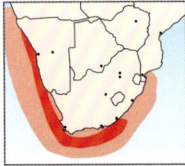

EUROPEAN STORM PETREL *Hydrobates pelagicus*

J F M A M J J A S O N D

14–18 cm; 23–31 g; wingspan 36–39 cm Slightly smaller and darker than Wilson's Storm Petrel, with a short, rounded tail and diagnostic, white underwing bar. Pale bar on upperwing coverts less pronounced than in Wilson's Storm Petrel and flight action typically more rapid and fluttery. Legs short; toes do not project beyond tail tip in flight. **Voice:** Silent at sea. **Status and biology:** Common Palearctic-br. migrant, mostly over continental shelf. Often in large flocks at trawlers; patters over water when feeding; 500 000 pairs breed in ne Atlantic. (Europese Stormswael)

BAND-RUMPED STORM PETREL

white in vent less extensive than in Wilson's Storm Petrel

toes do not extend beyond tail

long, broad wings

WILSON'S STORM PETREL

white rump 'folds over' extensively to vent

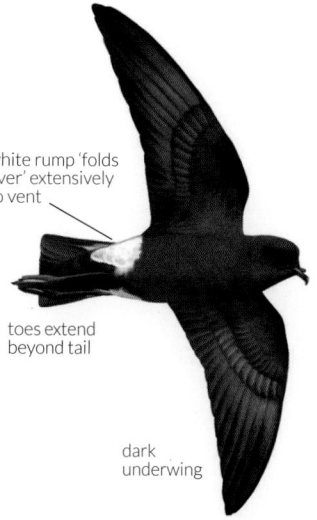

toes extend beyond tail

dark underwing

EUROPEAN STORM PETREL

plainer upperwings

toes fall short of tail

much reduced thigh patches

pale underwing bar

BLACK-BELLIED STORM PETREL *Fregetta tropica*

J F M A M J J A S O N D

19–21 cm; 50–60 g; wingspan 44–48 cm A bulky storm petrel with a broad, white rump, underwings and belly, usually with a broad, black line down the belly linking black breast and vent (but *F. t. melanoleuca* from Tristan and Gough islands has white belly – see White-bellied Storm Petrel). From the side, black belly stripe can be hard to see. Back tinged brown, matching upperwing coverts, and lacking grey scaling of nominate White-bellied Storm Petrel. Both species have a characteristic flight action, gliding over the waves, seldom flapping, regularly kicking up a line of spray with one foot or the breast. Sometimes follows ships. **Voice:** Silent at sea. **Status and biology:** Fairly common passage migrant (Apr–May and Aug–Oct); 150 000 pairs breed in summer at sub-Antarctic islands. (Swartpensstormswael)

WHITE-BELLIED STORM PETREL *Fregetta grallaria*

J F M A M J J A S O N D

19–21 cm; 45–65 g; wingspan 46–50 cm Told from nominate Black-bellied Storm Petrel by its white belly, but very similar to *F. t. melanoleuca* from Tristan and Gough islands. Australasian *F. g. grallaria* has a paler back due to broadly grey-edged mantle feathers, but *F. g. leucogaster* is almost identical to *F. t. melanoleuca*, differing in having longer wings and shorter legs; toes barely project beyond tail tip in flight; rump usually has some dark-tipped (not plain white) feathers; white belly extends further on breast, and dark vent U-shaped (not straight across). **Voice:** Silent at sea. **Status and biology:** Very rare over continental shelf and scarce in oceanic waters; seldom follows ships; 100 000 pairs breed at southern temperate islands. Breeds alongside white-bellied form of Black-bellied Storm Petrel at Tristan da Cunha. (Witpensstormswael)

GREY-BACKED STORM PETREL *Garrodia nereis*

X   X   X
J F M A M J J A S O N D

15–18 cm; 26–34 g; wingspan 36–40 cm Vagrant. A tiny storm petrel with a black head and breast merging into blue-grey back, rump and upperwing coverts; belly and underwing coverts white. Lacks white rump of *Fregetta* storm petrels. Easily overlooked as it ghosts away from ships, matching the lead-grey sea. **Voice:** Silent at sea. **Status and biology:** Usually remains south of 40°S; 50 000 pairs breed at sub-Antarctic islands. Only 3 records from s Africa, all off the Cape. (Grysrugstormswael)

WHITE-FACED STORM PETREL *Pelagodroma marina*

X   X   X X   X
J F M A M J J A S O N D

19–21 cm; 45–60 g; wingspan 42–48 cm Vagrant. A pale, long-legged storm petrel with diagnostic, white underparts and a prominent white eye stripe. Upperwing coverts pale brown; rump pale grey. Flight is erratic, with jerky wing beats; long toes extend beyond tail tip in flight. When feeding, it hovers and bounds over the water, pushing off with its feet. **Voice:** Silent at sea. **Status and biology:** One million pairs breed in summer at temperate islands in ne and S Atlantic and around Australasia, but most s African records from the Mozambique Channel. One caught ashore at night on Dyer Island, W Cape in Oct 2001. (Witwangstormswael)

toes longer than White-bellied Storm Petrel

plain white rump

black line down belly

squared-off vent

often shows white throat patch

black belly may be much reduced

BLACK-BELLIED STORM PETREL

melanoleuca

pure white belly

grallaria
paler mantle

leucogaster
darker mantle

U-shaped vent

dark throat

some rump feathers dark tipped

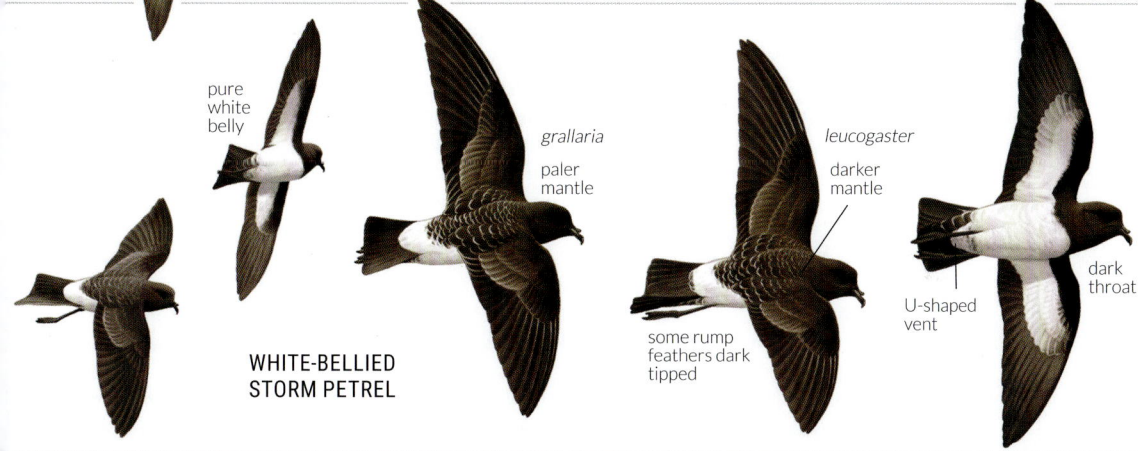

WHITE-BELLIED STORM PETREL

grey tail & rump

white belly & underwing

variable streaking on underparts

GREY-BACKED STORM PETREL

mostly white underparts

obvious white face

very long legs

WHITE-FACED STORM PETREL

69

BOOBIES AND GANNETS

Large seabirds that feed by plunge-diving. Bills dagger-like; nostrils concealed behind bill plates. Four toes webbed. Sexes alike; juv. plumage distinct.

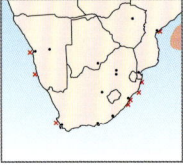

BROWN BOOBY *Sula leucogaster*

64–74 cm; 900–1 300 g; wingspan 1.3–1.5 m Vagrant. Ad. dark brown with a white lower breast and belly, and a broad, white underwing panel. Bill pale grey, washed blue-green; legs yellowish. Uniform upperparts and crisply defined white underparts distinguish it from juv. gannets. Juv. is duller brown above; has brown flecks on white belly, but lacks white speckling of juv. gannets. Differs from juv. Masked Booby by uniform brown upperparts and well-demarcated (not fuzzy) underpart pattern. **Voice:** Silent at sea. **Status and biology:** Vagrant to Namibia and east coast from br. islands in tropical S Atlantic and w Indian oceans. Most records from east coast associated with tropical cyclones. (Bruinmalgas)

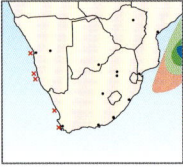

RED-FOOTED BOOBY *Sula sula*

66–77 cm; 820–950 g; wingspan 0.9–1.1 m A small, slender booby with a long, pointed tail. Ad. has bright red legs and a blue bill with a pink base; eyes dark in most birds, but some golden. Brown morph is much smaller than juv. gannets, and has plain (not speckled) plumage. White morph has yellow wash on head, but smaller than gannets, with a white tail and black carpal patches on underwing. Most common morph is white-tailed brown. Juv. is brown-streaked, with grey-brown bill and greyish-yellow feet. **Voice:** Usually silent at sea; occasional harsh *'karrk'* alarm call. **Status and biology:** Regular in Mozambique Channel, uncommon further south, often after cyclones. Rare on west coast, presumably from Atlantic colonies (1 300 pairs breed at Fernando de Noronha); 3 500 pairs breed at Europa Island, Mozambique Channel. Often accompanies ships, aerially chasing flying-fish and flying-squid scared by the ship's passage. (Rooipootmalgas)

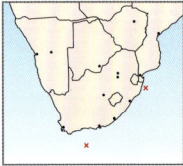

MASKED BOOBY *Sula dactylatra*

80–90 cm; 1.8–2.4 kg; wingspan 1.5–1.6 m Vagrant. A large, white booby with golden eyes and a small, black facial mask. Lacks yellow head of ad. Cape Gannet (p. 72) and has black trailing edge to wing extending to body (not confined to primaries and secondaries). Black (not white) tail and dark (not red) legs separate it from white-phase of smaller Red-footed Booby. Juv. has mottled brown head, back and rump; resembles Brown Booby, but has narrow, white nuchal collar, less extensive brown on breast, and more white on underwing which has a dark central bar (clearly defined, smaller white area in Brown Booby; juv. Red-footed Booby has all-dark underwings). **Voice:** High double honk, but generally silent at sea. **Status and biology:** Only 2 records from s Africa, off n KZN and Namibia; also 1 at sea outside territorial waters south of Cape Agulhas. Breeds at tropical islands from the Red Sea to Tanzania. (Brilmalgas)

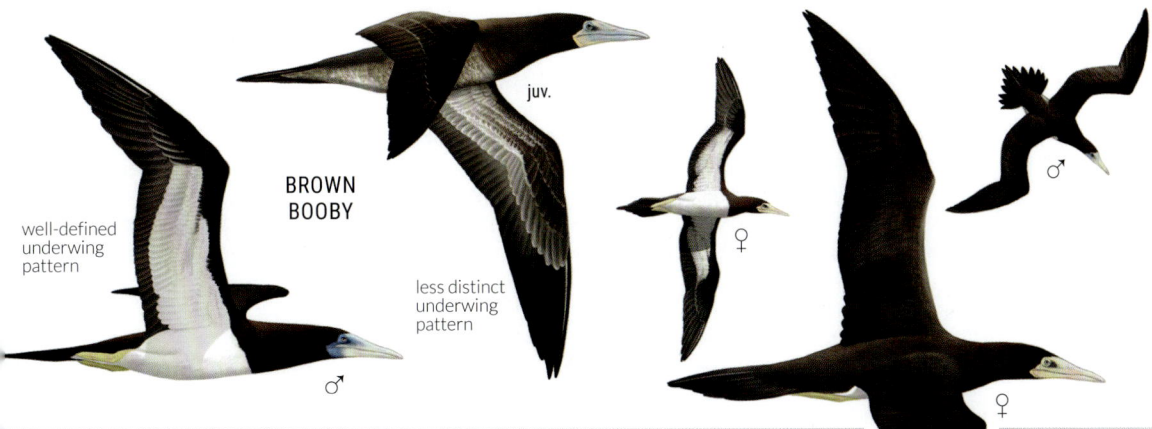

BROWN BOOBY

well-defined underwing pattern

juv.

less distinct underwing pattern

♂

♀

♂

♀

white-tailed brown morph

ad.

RED-FOOTED BOOBY

juv.

ad.

brown morph

white tertials

ad.

white tail

white 'golden' morph

ad.

imm. white morph

white-tailed brown morph

imm.

ad.

ad.

distinct mottled underwings

MASKED BOOBY

dark tail

dark tertials

dark mask

71

CAPE GANNET   *Morus capensis*

b | | | | | | | | b b B B B b
J F M A M J J A S O N D

84–95 cm; 2.3–3.1 kg; wingspan 1.6–1.7 m Ad. white with yellow wash on head, and black tail and flight feathers; some have 1–4 white outer-tail feathers. Juv. is brown with white spots, whitening gradually, typically starting with head. Juv. could be confused with Brown Booby (p. 70), but is larger and lacks clear-cut brown bib and white belly. **Voice:** Noisy *'warrra-warrra-warrra'* at colonies and when feeding at sea. **Status and biology:** ENDANGERED. Br. endemic. Largely confined to continental shelf, often feeding at trawlers; 120 000 pairs breed in dense colonies in spring and summer at 6 islands off Cape and Namibia. Large numbers follow sardine run up east coast to KZN in winter; others migrate up west coast to Gulf of Guinea. (Witmalgas)

AUSTRALASIAN GANNET   *Morus serrator*

X X X | | | | | X X X X
J F M A M J J A S O N D

83–92 cm; 2–2.8 kg; wingspan 1.6–1.7 m Vagrant. Very similar to Cape Gannet, but has a darker, greyish eye, more golden-yellow wash to head and much shorter gular stripe (but none of these features is useful for identification at sea). All confirmed records are of birds at Cape Gannet colonies, where they are best located by their higher-pitched call. Ad. has 3 outer-tail feathers white, but some Cape Gannets also show white outer tails. Juv. differs from juv. Cape Gannet only by its shorter gular stripe. **Voice:** Noisy *'warrra-warrra-warrra'* at colonies; higher pitched than Cape Gannet. **Status and biology:** Some 20 birds recorded at gannet colonies; some have bred successfully with Cape Gannets. (Australiese Malgas)

TROPICBIRDS

Medium-sized, tropical, pelagic seabirds with fairly short wings; wing beats stiff. Bills robust, with serrated edges. Four toes are webbed, but feet reduced; formerly placed with other seabirds with 4 webbed toes in Pelecaniformes, but now known to be a very distinct group, distantly related to Kagu and Sunbittern. Sexes alike, but juv. plumages distinct. Feed by plunge-diving, taking fish and squid, including flying-fish and flying-squid.

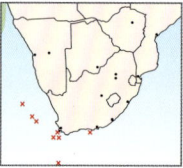

RED-BILLED TROPICBIRD   *Phaethon aethereus*

X X | | | | | | | X X
J F M A M J J A S O N D

60–65 cm (plus 40 cm streamer); 650–700 g; wingspan 100–108 cm Vagrant. Ad. has large, red bill and white tail streamers. Differs from Red-tailed Tropicbird by its finely barred back, which appears grey at a distance, and by its long, white (not dark red) tail streamers. Juv. heavily barred above, with black eye patch extending to nape to form nuchal collar; bill yellow, with a dark tip. **Voice:** Occasional piercing screams at sea. **Status and biology:** Rare vagrant to oceanic waters off west and south coasts, probably from S Atlantic population of 1 200 pairs (mainly on Ascension Island). (Rooibekpylstert)

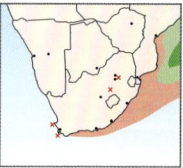

RED-TAILED TROPICBIRD   *Phaethon rubricauda*

J F M A M J J A S O N D

60–70 cm (plus 35 cm streamer); 600–800 g; wingspan 104–120 cm The palest tropicbird; ad. mostly white (tinged pink when br.), with red bill and dark red tail streamers. Compared to other ad. tropicbirds, has thin (not broad) black primary tips on outer primaries and small, black tips on scapulars. Juv. is more finely barred above than White-tailed Tropicbird, with less black in outer wing; bill is blackish. **Voice:** Deep *'kraak'*, similar to Caspian Tern (p. 234). **Status and biology:** The most common tropicbird in s African waters. Regular in Mozambique Channel, rare off KZN and vagrant further south to W Cape: occasionally recorded inland. Some birds visit cliff sites and coastal islands for days or weeks; 3 000–4 000 pairs breed at Europa Island, Mozambique Channel. (Rooipylstert)

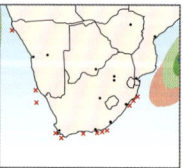

WHITE-TAILED TROPICBIRD   *Phaethon lepturus*

J F M A M J J A S O N D

45–50 cm (plus 40 cm streamer); 280–400 g; wingspan 90–98 cm The smallest tropicbird; ad. has diagnostic orange-yellow bill, neat, black face mask and extremely long, wispy, white or golden tail streamers (shafts black). In flight, has 2 black patches on each wing (outer primaries and median coverts). Subspecies from Europa Island (*P. l. europae*) shows apricot wash to upperparts. Juv. has barred upperparts and black wing tips; bill yellow, with dark tip. Barring is sparser than in juv. Red-tailed Tropicbird, and black eye patch does not form nuchal collar as in juv. Red-billed Tropicbird. **Voice:** Usually silent at sea. **Status and biology:** Regular in Mozambique Channel; vagrant further south and off west coast. Breeds at tropical islands worldwide, including 2 200 pairs at Ascension Island, Atlantic Ocean, and 500–1 000 pairs at Europa Island, Mozambique Channel. (Witpylstert)

juv.

juv.

CAPE GANNET

ad.

mainly black tail

yellow wash to head

ad.

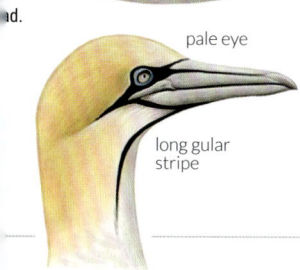

pale eye

long gular stripe

imm.

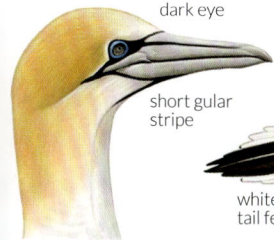

dark eye

short gular stripe

golden-yellow wash to head

white outer-tail feathers

AUSTRALASIAN GANNET

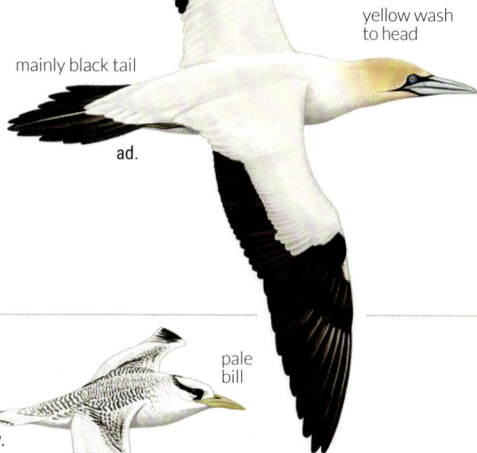

pale bill

juv.

RED-BILLED TROPICBIRD

extensive black primaries

white tail

ad.

heavily barred upperparts

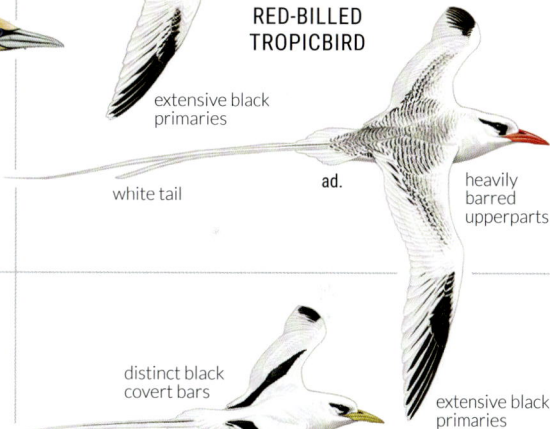

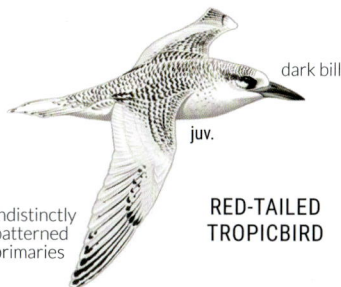

dark bill

juv.

indistinctly patterned primaries

RED-TAILED TROPICBIRD

distinct black covert bars

ad.

extensive black primaries

WHITE-TAILED TROPICBIRD

ad.

white wings & mantle

red tail

ad.

mm.

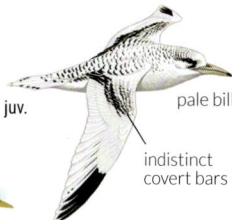

juv.

pale bill

indistinct covert bars

golden morph

FRIGATEBIRDS

Large, tropical, pelagic seabirds with extremely long, angled wings and long, deeply forked tails. Wing moult protracted, often with several waves of active moult. Males have naked, red throat pouches that are inflated in display. Females larger than males. Feet tiny, with 4 partially webbed toes. Feed by surface-seizing without landing (often taking flying-fish or flying-squid) or by stealing food from other birds. Soar to great heights. Plumage differs with age and sex.

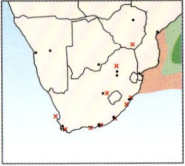

GREAT FRIGATEBIRD   *Fregata minor*

J F M A M J J A S O N D

86–104 cm; 1.1–1.5 kg; wingspan 2–2.3 m Underwing wholly dark at all ages, typically lacking white 'armpits' of Lesser Frigatebird (although some juvs have white extensions from the belly onto the underwing; see below). Ad. male black, with brown bar across upperwing coverts (often hard to see). Female has white breast and throat with a greyish chin. Imm. white from chin to belly, gradually darkening with age. Juv. has whitish or tawny head and throat, dark breast band and white oval belly patch. In some juvs the white belly extends marginally onto the underwing, but these pale patches originate from well behind the front of the white belly patch further back than the distinctive white 'armpits' of Lesser Frigatebird. **Voice:** Silent at sea. **Status and biology:** Regular in Mozambique Channel, vagrant further south to W Cape; sometimes recorded inland after cyclones. May roost in trees; 1 000 pairs breed at Europa Island, Mozambique Channel. (Grootfregatvoël)

LESSER FRIGATEBIRD   *Fregata ariel*

J F M A M J J A S O N D

70–82 cm; 700–900 g; wingspan 1.8–2.1 m Smaller and more angular than Great Frigatebird, but size hard to assess on lone birds. Ad. male black, with diagnostic, white 'armpits' extending from axillaries to sides of breast; paler brown upperwing bars often hard to see. Female has a black throat, white breast extending as collar onto neck and white 'armpits'. Differs from female Great Frigatebird by black throat and white breast extending to 'armpits'. Imm. is variable, but typically has dark brownish or black throat, more extensive black on belly than juv., and white breast extending to 'armpits'. Juv. has white or brownish head, dark breast band and triangular white belly patch with white 'armpits' originating from front of white belly patch. **Voice:** Silent at sea. **Status and biology:** Generally much less commonly seen in the region than Greater Frigatebird. Rare off east coast, seen mostly after summer cyclones but imms may remain along the coast for some time. 1 000 pairs breed at Europa Island, Mozambique Channel, but mostly forage further east than Great Frigatebird. (Kleinfregatvoël)

GREAT FRIGATEBIRD

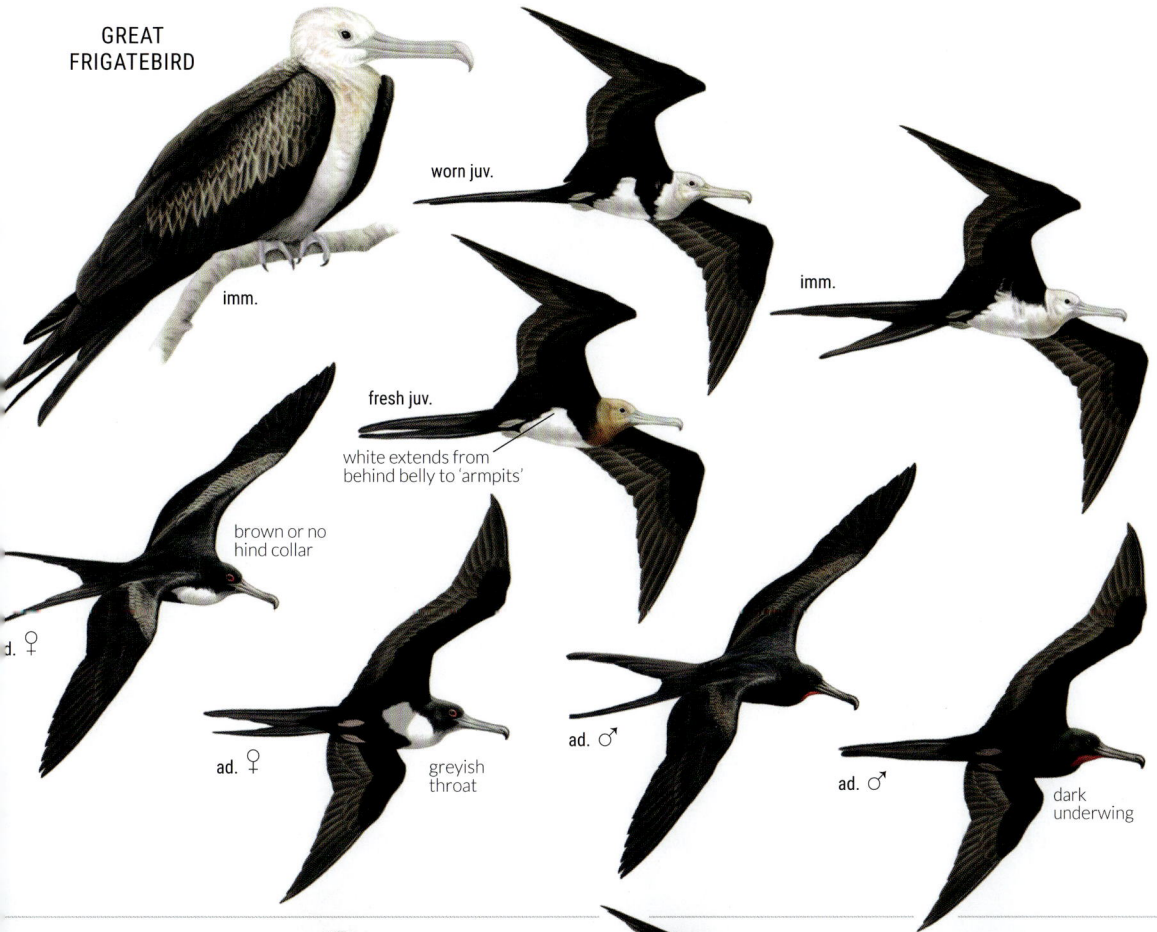

imm.

worn juv.

imm.

fresh juv.

white extends from behind belly to 'armpits'

brown or no hind collar

d. ♀

ad. ♀

greyish throat

ad. ♂

ad. ♂

dark underwing

LESSER FRIGATEBIRD

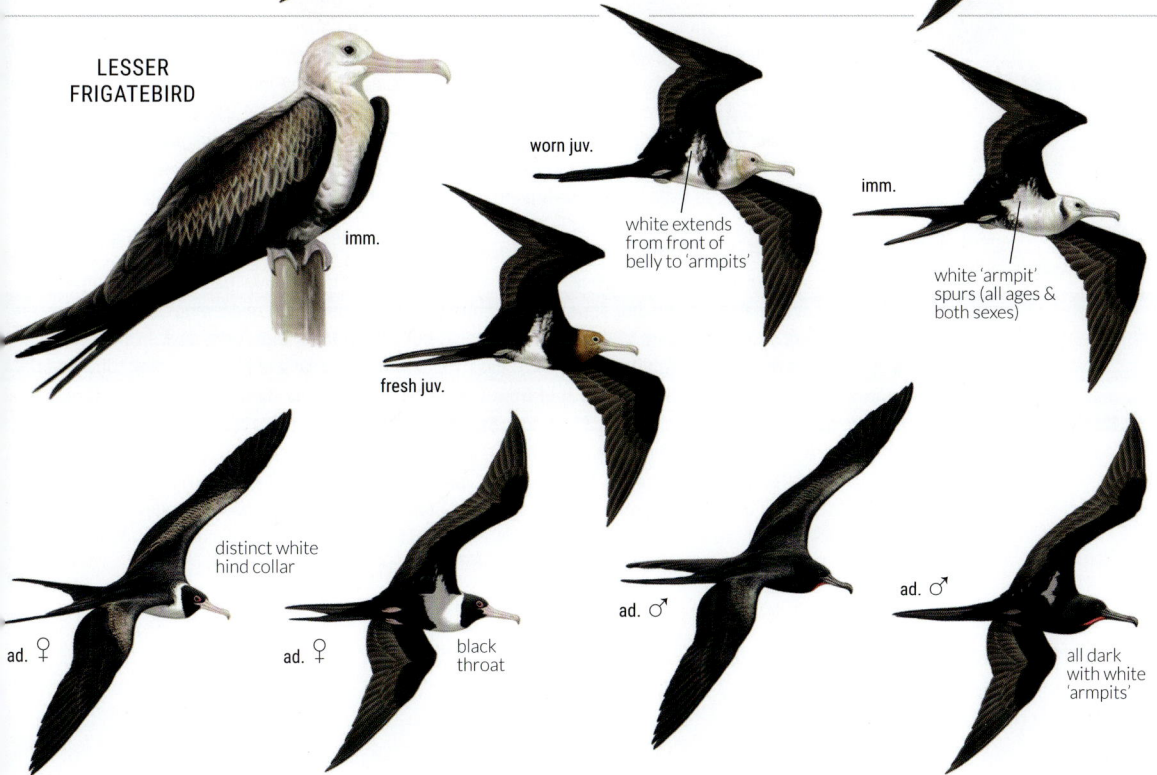

imm.

worn juv.

white extends from front of belly to 'armpits'

imm.

white 'armpit' spurs (all ages & both sexes)

fresh juv.

distinct white hind collar

ad. ♀

ad. ♀

black throat

ad. ♂

ad. ♂

all dark with white 'armpits'

PELICANS
Very large waterbirds with long bills and a distensible pouch between their lower mandibles used to trap fish and other prey. Swim with 4 webbed toes. Flight strong; often soar in thermals on broad wings with deeply slotted primaries. Moult complex, with multiple active centres in the wing. Sexes alike, but males appreciably larger.

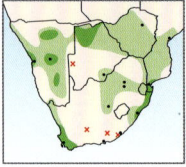

GREAT WHITE PELICAN *Pelecanus onocrotalus*

b b b b B B B b b b b b
J F M A M J J A S O N D

140–178 cm; 6–14 kg; wingspan 2.2–3.4 m A very large, white pelican with black flight feathers contrasting with white coverts in flight. Bill pouch yellow, with more of a sag in middle than Pink-backed Pelican's. Br. ad. has pink-tinged neck and underparts, bare pink-orange face, short crest, yellow breast patch and red nail at tip of bill. Non-br. ad. has yellow face. Juv. is finely mottled brown, whitening with age; lacks marked contrast between flight feathers and coverts. **Voice:** Usually silent; deep *'mooo'* given at br. colonies. **Status and biology:** Locally common (8 000 pairs) at lakes, estuaries and sheltered coastal bays, usually within 200–300 km of br. colonies. Nests colonially on ground. Often forages in flocks, working cooperatively to encircle prey. (Witpelikaan)

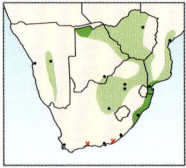

PINK-BACKED PELICAN *Pelecanus rufescens*

B | | | | b B B b | | B
J F M A M J J A S O N D

125–140 cm; 4–7 kg; wingspan 2.2–2.9 m Smaller and greyer than Great White Pelican, with a pinkish back and pinkish-yellow bill. Flight feathers are dark grey, not contrasting strongly with coverts in flight. Br. birds have grey crest. Juv. mottled brown, becoming greyer with age; best distinguished from juv. Great White Pelican by smaller size (and usually accompanied by ads). **Voice:** Usually silent; guttural calls at br. colonies. **Status and biology:** Locally common at lakes and estuaries, but typically feeds singly. Some 200 pairs nest colonially in trees at wetlands in n KZN and n Botswana. The only local pelican that perches in trees. (Kleinpelikaan)

FLAMINGOS
Peculiar waterbirds with long necks and legs, and bent bills adapted for filter feeding. Sexes alike, but males larger; juvs distinct. Breed erratically in large, dense colonies when water conditions are favourable. Closest living relatives to the grebes.

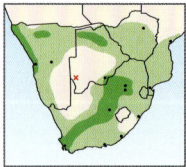

GREATER FLAMINGO *Phoenicopterus roseus*

B B b b b b b b b B B
J F M A M J J A S O N D

125–165 cm; 2.5–3.5 kg; wingspan 1.4–1.7 m The larger of the 2 African flamingos, with very long legs and neck. Ad. appears mostly white (not pink) at rest; in flight, salmon-pink wing coverts contrast with white body and black flight feathers. Face and bill pale pink, with a broad, black bill tip; eyes creamy yellow; legs pink. Juv. is dirty grey-brown, becoming paler with age; bill pale grey with a darker tip; eyes brown; legs slate-grey. **Voice:** Noisy; goose-like honking. **Status and biology:** Common resident, intra-African migrant and nomad; 50 000 birds occur at shallow lakes, saltpans, lagoons, estuaries and sandy beaches. Breeds colonially on large, temporarily flooded pans, most regularly on Sua Pan, Botswana (13 000–40 000 pairs), less frequently on Etosha Pan (up to 27 000 pairs), occasionally elsewhere. Post-breeding birds disperse widely across S Africa and Namibia. Usually in large flocks. Feeds on larger prey than Lesser Flamingo (e.g. crustaceans and molluscs), mainly filtered from bottom sediments so head and neck often submerged. (Grootflamink)

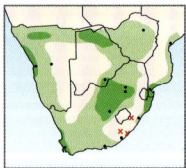

LESSER FLAMINGO *Phoenicopterus minor*

B B B b b b b b b b B B
J F M A M J J A S O N D

90–125 cm; 1.5–2 kg; wingspan 1–1.1 m Smaller and generally brighter pink than Greater Flamingo. In flight, median coverts are crimson, but appear less contrasting than upper- and underwing coverts of Greater Flamingo. Ad. has a dark red face and bill; bill tip black, but bill appears uniformly dark from a distance (not distinctly two-toned as in Greater Flamingo). Eyes golden, with purple eye-ring; legs pink. Juv. is dirty grey-brown, becoming paler with age; best told from Greater Flamingo by its dark purple-grey bill; legs grey. **Voice:** More muted honking than Greater Flamingo. **Status and biology:** NEAR-THREATENED. Locally common resident, intra-African migrant and nomad at lakes, saltpans and estuaries. Breeds colonially alongside Greater Flamingos on large, temporarily flooded pans, most regularly on Sua Pan, Botswana (up to 150 000 pairs), less frequently on Etosha Pan (1 000–10 000 pairs) and, since 2007, outside Kimberley (up to 10 000 pairs). Population fluctuates between years. Mainly filters microscopic blue-green algae and diatoms from water surface; seldom submerges head and neck, and often swims in deep water. (Kleinflamink)

GREAT WHITE PELICAN

r. ♂

imm.

imm.

ad.

sub-ad.

br.

ad.

ad.

PINK-BACKED PELICAN

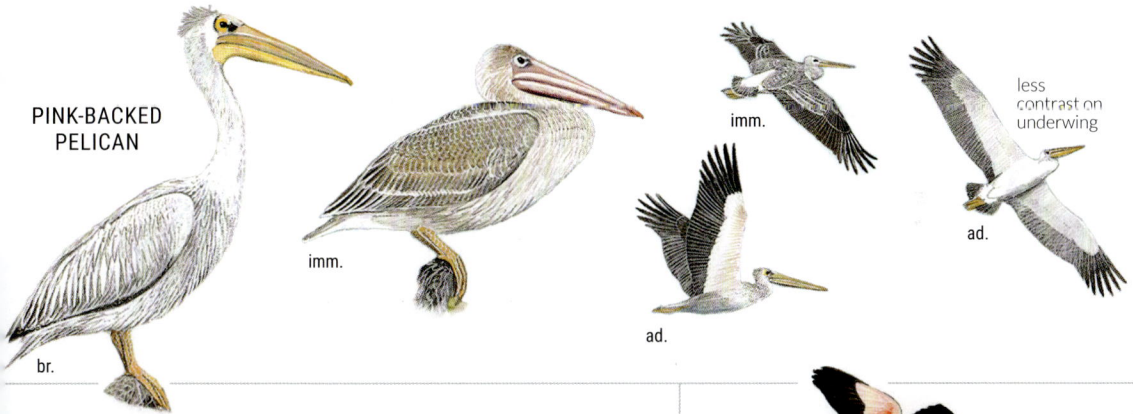

imm.

imm.

br.

ad.

less contrast on underwing

ad.

juv.

GREATER FLAMINGO

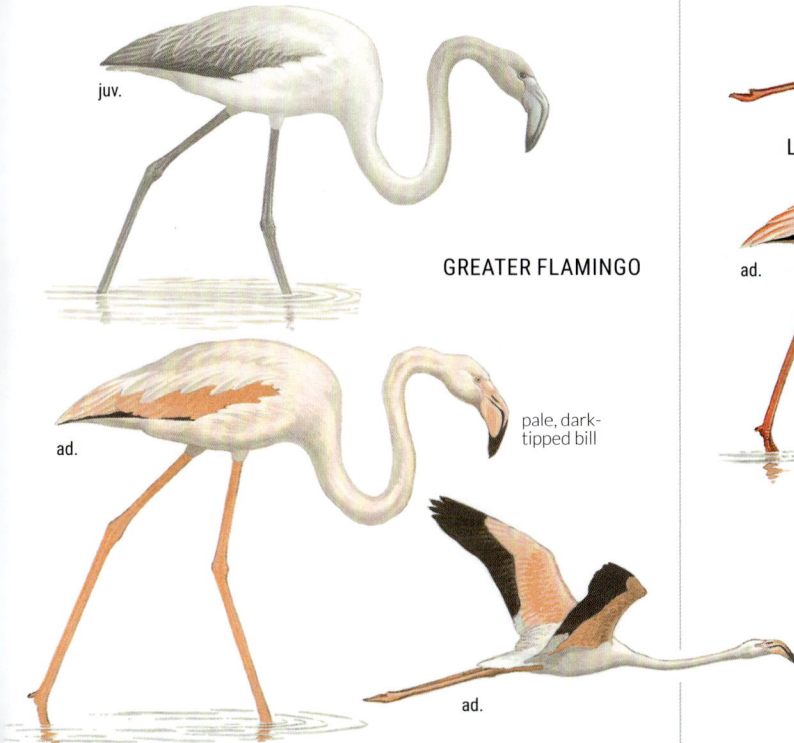

ad.

ad.

pale, dark-tipped bill

ad.

ad.

LESSER FLAMINGO

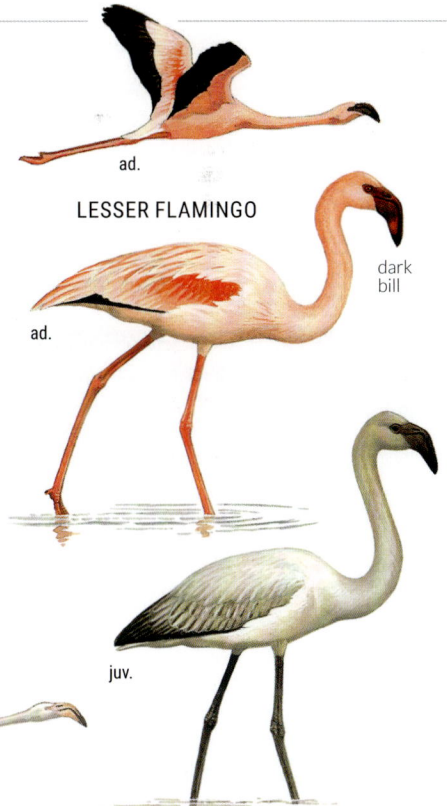

dark bill

ad.

juv.

77

CORMORANTS AND DARTERS

Medium to large waterbirds with long, slender necks; foot-propelled pursuit divers. Cormorants grasp prey with hook-tipped bills; darters spear prey. Plumage partly (cormorants) or fully (darters) wettable; have to return to land to roost. Sexes are alike; males slightly larger.

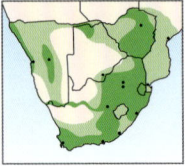

WHITE-BREASTED CORMORANT *Phalacrocorax lucidus*

b b b b b b b b b b b b
J F M A M J J A S O N D

85–95 cm; 1.8–3.2 kg; wingspan 1.3–1.6 m The largest African cormorant. Ad. has white throat and breast, washed pink in marine populations; white thigh patches in br. plumage; yellow skin at base of bill; eyes turquoise. Juv. has white underparts and browner upperparts. **Voice:** Grunts and squeaks at colony; otherwise silent. **Status and biology:** Locally common resident at dams, lakes, large rivers, estuaries and shallow coastal waters, favouring areas where the bottom sediments are sandy. Breeds colonially in trees, bushes or reeds, on cliffs or on the ground on islands; sometimes with other waterbirds. (Witborsduiker)

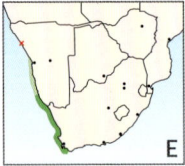

BANK CORMORANT *Phalacrocorax neglectus*

b b b b b b b b b b b b
J F M A M J J A S O N D

74–76 cm; 1.6–2.3 kg; wingspan 1.2–1.3 m Robust, all-dark, marine cormorant confined to the Benguela coast (central Namibia to Cape Agulhas). Best told from Cape Cormorant by its matt black plumage, angled head profile (steep forehead, flat crown) and heavier build. Br. ad. has white rump patch. Eyes turquoise, gradually turning brick-red from the top with age. Juv. is browner. Partial leucism fairly regular. **Voice:** Raucous *'wheeerrr'* at colony. **Status and biology:** Endemic. ENDANGERED; population has fallen by 70% over last 3 decades possibly due to shortage of food. Occurs in coastal waters; seldom more than 20 km from colonies. Breeds colonially on offshore rocks and islands. (Bankduiker)

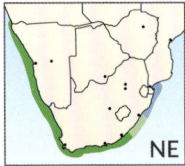

CAPE CORMORANT *Phalacrocorax capensis*

B B b b b b b b B B B B
J F M A M J J A S O N D

60–65 cm; 900–1 600 g; wingspan 1–1.1 m An all-dark, marine cormorant; slightly smaller and more slender than scarce Bank Cormorant, with paler gular skin. Often in large flocks; flies over the sea in long lines. Br. ad. glossy black, with chrome-yellow gular patch, turquoise eye and banded turquoise-and-black eye-ring; non-br. ad. duller. Juv. brown, with paler underparts, almost whitish in some birds; gular skin grey-brown. **Voice:** Nasal grunts and croaks, mainly at colony. **Status and biology:** ENDANGERED. Near-endemic. Occurs in coastal waters and adjacent wetlands. Breeds colonially on cliffs and offshore islands. Forages up to 50 km offshore on pelagic fish; also on seabed up to 30 m deep. Numbers have decreased over last 20 years; currently 100 000 pairs. (Trekduiker)

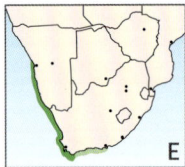

CROWNED CORMORANT *Microcarbo coronatus*

B B b b b b b b B B B B
J F M A M J J A S O N D

54–58 cm; 680–850 g; wingspan 88–98 cm A small, short-billed and long-tailed marine cormorant confined to the west and south coasts (n Namibia to Tsitsikamma, W Cape). Slightly larger than Reed Cormorant, with a shorter, less graduated tail; ranges overlap narrowly in coastal wetlands; told apart with difficulty. Ad. glossy black with prominent forecrown crest; back feathers appear more uniform than ad. Reed Cormorant, with narrower black tips. Eyes red; facial skin orange. Non-br. ad. is duller. Juv. is brown, with paler breast; lacks extensive white underparts of juv. Reed Cormorant. **Voice:** Cackles and hisses at colony. **Status and biology:** Endemic. NEAR-THREATENED; global population 3 000 pairs. Occurs in coastal waters, sometimes roosting in nearby wetlands. Typically forages close inshore. Breeds colonially on cliffs, offshore islands and disused boats. (Kuifkopduiker)

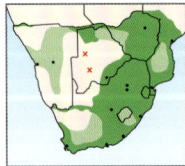

REED CORMORANT *Microcarbo africanus*

B b b b b b b B B B B B
J F M A M J J A S O N D

50–56 cm; 450–650 g; wingspan 80–90 cm A small, short-billed and long-tailed cormorant found mainly in freshwater habitats. Range overlaps narrowly with Crowned Cormorant at coastal wetlands, but it is slightly smaller, with a longer, more graduated tail. Ad. glossy black, with prominent forecrown crest in br. plumage; back feathers more contrasting than ad. Crowned Cormorant, with silver bases and broad, black tips. Eyes red; facial skin orange. Non-br. ad. is duller. Juv. is brown above, whitish below. **Voice:** Cackles and hisses at colony. **Status and biology:** Locally common resident at dams, lakes, rivers and estuaries (also shallow coastal waters in Mozambique and n Namibia). Breeds colonially in trees, bushes or reeds, often with other waterbirds. (Rietduiker)

AFRICAN DARTER *Anhinga rufa*

b b b b b b b B B B B B
J F M A M J J A S O N D

80–92 cm; 1–1.7 kg; wingspan 1.2–1.3 m Differs from cormorants by its long, slender neck and head, and long, pointed bill. When swimming, often only long neck and head are visible (hence alternative name 'snakebird'). Often glides in flight, showing broad wings, deeply slotted outer primaries, and long, broad tail. Ad. has elongate, white-striped scapulars and wing coverts. Br. male has rufous foreneck, with white stripe down side of neck. Female and non-br. male have pale brown throat. Juv. has buffy neck and lacks streaking on back. **Voice:** Croaks when breeding; otherwise silent. **Status and biology:** Common resident of lakes, dams and slow-moving rivers; rarely coastal lagoons and estuaries. Breeds in colonies in trees or reeds, often with other waterbirds. (Slanghalsvoël)

imm.

br.

WHITE-BREASTED CORMORANT

ad.

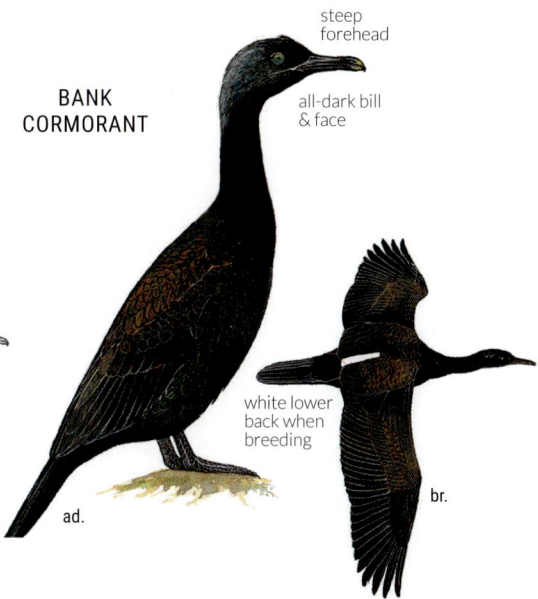

BANK CORMORANT

steep forehead

all-dark bill & face

white lower back when breeding

ad.

br.

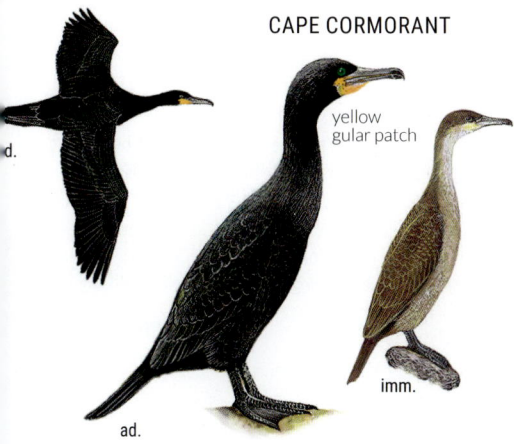

CAPE CORMORANT

ad.

yellow gular patch

imm.

ad.

CROWNED CORMORANT

less contrasting back

ad.

imm./non-br. ad.

REED CORMORANT

strongly contrasting back

ad.

imm./non-br. ad.

very long tail

AFRICAN DARTER

imm.

long neck & tail

juv.

br. ♂

79

SPOONBILLS
Large wading birds with straight, spoon-tipped bills, related to ibises. Sexes alike; juvs duller.

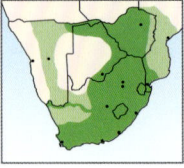

AFRICAN SPOONBILL *Platalea alba*

b b B B B B B B B b b
J F M A M J J A S O N D

86–92 cm; 1.4–1.8 kg A tall, white, slender wading bird with diagnostic, spoon-shaped bill. Flies with neck outstretched (not tucked in, as in herons and egrets). Ad. has a bare pink-red face, pale eyes, a grey bill with a pinkish border and base, and greyish-pink legs; sexes alike, but female has a shorter bill. Juv. bill yellow-horn and shorter than ad.; legs dark grey; head streaked grey, and flight feathers tipped blackish. **Voice:** Low *'kaark'*; at br. colonies emits various grunts and claps bill. **Status and biology:** Fairly common resident and local nomad at lakes, flood plains and estuaries. Feeds with characteristic side-to-side sweeping motion of bill. Breeds colonially in trees or reeds, often with ibises, egrets, herons and cormorants. (Lepelaar)

IBISES
Fairly large birds with long legs and necks and decurved bills. Most species at least partly aquatic. Sexes alike; juvs distinct. Breed singly or colonially.

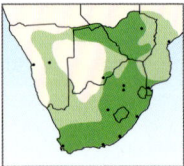

AFRICAN SACRED IBIS *Threskiornis aethiopicus*

b b b b b b b b B B B b
J F M A M J J A S O N D

66–84 cm; 1.2–1.9 kg A large, mostly white ibis with a heavy, decurved, black bill. Ad. has naked, black head and neck. Elongate, plume-like, black scapulars are fluffed out in display, making it appear black-tailed at rest. Flight feathers are tipped black, giving narrow, black edge to wing. Br. ads have dirty-yellow flanks and naked, pink skin on underwing (grey in non-br.). White plumage often appears dirty brown due to habits of scavenging at landfill sites. Juv. has white-feathered neck and greyish cast to plumage. **Voice:** Loud croaking at br. colonies. **Status and biology:** Common resident in parts of its range, a summer visitor in much of the interior, these birds over-wintering in Zambia; in open habitats, from wetlands and fields to grassland and offshore islands; often at refuse tips. Often in flocks; frequently flies in 'V' formation. Breeds colonially in trees or reeds, usually with egrets, herons and cormorants. (Skoorsteenveër)

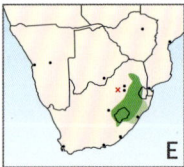

SOUTHERN BALD IBIS *Geronticus calvus*

b B B b b
J F M A M J J A S O N D

74–80 cm; 1–1.1 kg A dark, glossy ibis. Ad. easily recognised by its bald red head, long red bill, short red legs and glossy green neck ruff. Juv. duller, lacking coppery shoulder patch; head covered in short, pale brown feathers. Bill red only at base; legs brown. **Voice:** High-pitched, wheezing call. **Status and biology:** VULNERABLE. Endemic. Range has contracted in recent decades. Locally common resident, some 8 000 birds (2 000 pairs at about 200 colonies). Occurs in flocks in short grassland (often in burnt areas) and cultivated fields. Breeds colonially on cliff ledges or occasionally in trees. (Kalkoenibis)

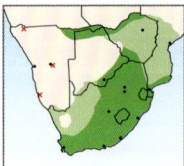

HADEDA (HADADA) IBIS *Bostrychia hagedash*

b b b   b b b b b b
J F M A M J J A S O N D

76–85 cm; 1–1.5 kg A large, stout, grey-brown ibis with glossy bronze-green wing coverts. Grey face has whitish stripe running from bill to below and behind eye. Shortish, deep bill is dark grey with a red ridge on the upper mandible. Northern populations of the subspecies *B. h. brevirostris* have white eyes and more obvious white cheek stripes. In flight, wings broad; neck and legs short compared to Glossy Ibis. Juv. has duller, shorter bill. **Voice:** Noisy; gives a raucous *'ha-da'* or *'ha-ha-da-da'* call in flight and when flushed. **Status and biology:** Common resident and nomad in forest clearings, woodland, savanna, grassland, farmland and suburbia. Range has expanded westward in recent years, apparently in response to human habitat modification, especially irrigation. Usually in small parties; roosts and breeds in trees. (Hadeda)

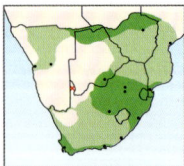

GLOSSY IBIS *Plegadis falcinellus*

b b b b   b b b B B B
J F M A M J J A S O N D

50–65 cm; 550–750 g A rather small, slender ibis with long legs; appears blackish from a distance. In flight, wings, neck and legs are much longer than Hadeda Ibis. Br. ad. has a dark chestnut head, neck and body. Wings, back and tail are dark, glossy green with bronze and purple highlights. A narrow, bluish-white line rings the base of the bill. Non-br. ad. has a pale-flecked head and neck. Juv. resembles non-br. ad., but body is dull, sooty-brown. **Voice:** Normally silent; low, guttural *'kok-kok-kok'* when br. **Status and biology:** Locally common resident and nomad at lakes, dams, pans, estuaries and flooded grassland; rarely on open, sandy shores. Breeds colonially (up to 200 pairs), often with other ibises and egrets. (Glansibis)

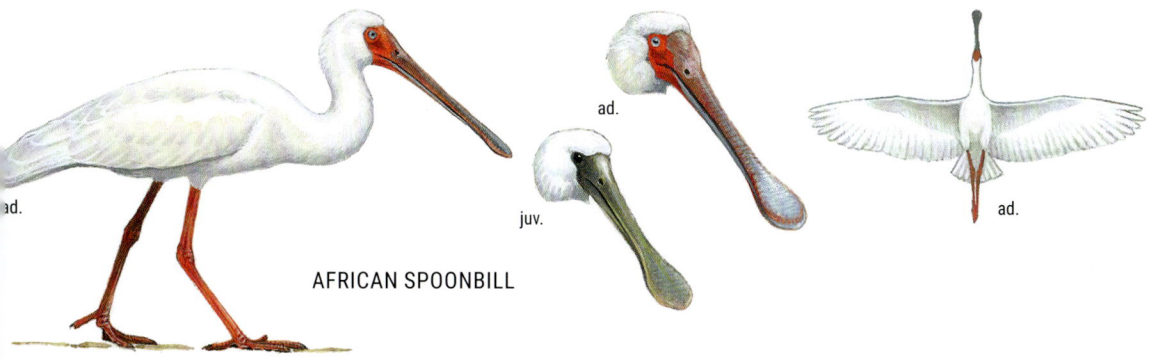

ad.

ad.

juv.

ad.

AFRICAN SPOONBILL

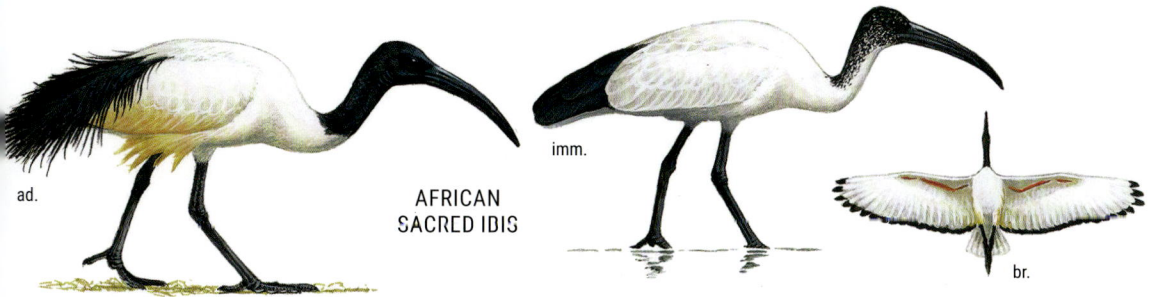

ad.

imm.

**AFRICAN
SACRED IBIS**

br.

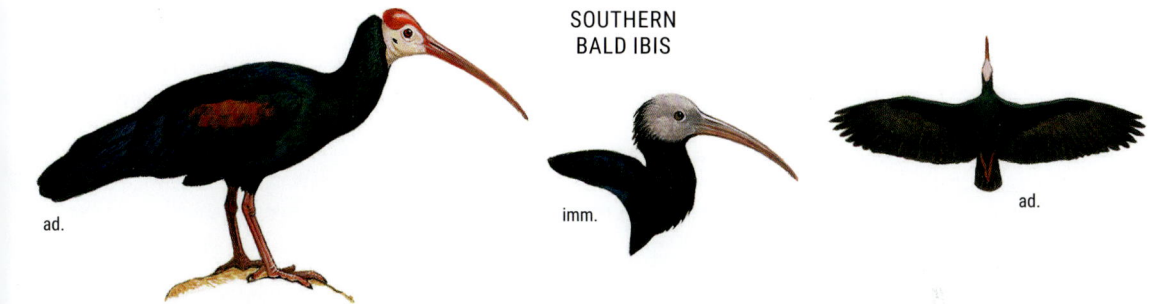

**SOUTHERN
BALD IBIS**

ad.

imm.

ad.

HADEDA IBIS

pale
eye

brevirostris

broad
wings

ad.

ad.

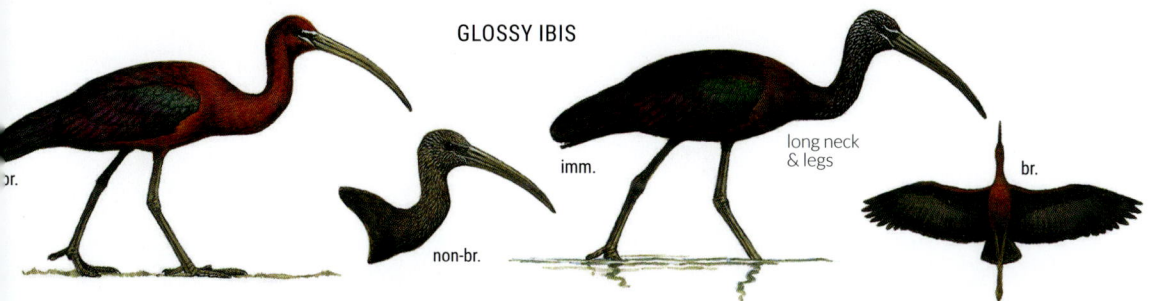

GLOSSY IBIS

br.

non-br.

imm.

long neck
& legs

br.

81

HERONS, EGRETS AND BITTERNS

Tall, slender waterbirds with long legs and, usually, long necks. Bills dagger-like. Hunt mainly aquatic animals, typically by stealth, occasionally by active pursuit. Sexes alike, but many have distinct juv. plumages. Ads often have distinctive changes to bare part coloration at the start of the br. season. Most breed in colonies; some bitterns solitary nesters.

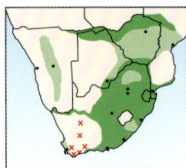

GOLIATH HERON *Ardea goliath*

b	b	b	b	b	b	b	b	B	B	B	b
J	F	M	A	M	J	J	A	S	O	N	D

135–150 cm; 3–4.3 kg The largest heron, with a massive, heavy bill. Rich chestnut head, neck and underparts recall much smaller Purple Heron, but has an unstriped head and blackish (not yellow) legs and bill. In flight, the huge, broad wings beat slowly and deeply. Juv. duller and less rufous, with foreneck, chest, belly and underwing coverts white, streaked with black. Recently fledged birds have buff margins to back feathers. **Voice:** Loud, low-pitched *'kwaaark'*. **Status and biology:** Locally common resident and nomad at lakes, dams, large rivers and estuaries, usually where there are extensive reed or papyrus beds. Range has expanded in W Cape in recent years. Breeds singly or in small colonies in trees, reed beds or on bare ground on islands, sometimes with other herons and egrets. (Reusereier)

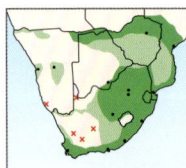

PURPLE HERON *Ardea purpurea*

B	B	b	b	b	b	b	B	B	b	b	b
J	F	M	A	M	J	J	A	S	O	N	D

78–86 cm; 550–1 200 g A medium-sized heron with a long, slender neck and bill. The black-striped, rufous head and neck and dark grey wings are distinctive. Much smaller than Goliath Heron, with black cap and paler bill and legs. Juv. duller and paler, with less well-marked head stripes and browner back. **Voice:** Hoarse *'kraaark'*. **Status and biology:** Common resident and local nomad at wetlands, typically among sedges and reeds; rarely along the coast. Seldom forages in the open. Breeds singly or in small colonies, typically in dense reeds. (Rooireier)

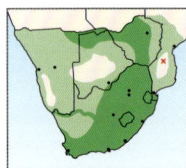

GREY HERON *Ardea cinerea*

b	b	b	b	b	b	B	B	B	B	B	b
J	F	M	A	M	J	J	A	S	O	N	D

90–100 cm; 1.1–2 kg A large, greyish heron of wetland habitats. Told in flight from Black-headed Heron by its uniform grey (not contrasting dark and pale) underwing. Ad. has mostly white head and neck, with black eye stripe ending in a wispy plume. Foreneck streaked black, with broad, black bands on either side of breast and belly, often showing as black shoulder patches. Bill yellow (orange-pink in pre-br. birds); legs yellow-brown (flushed red in pre-br. birds). Juv. duller, with rather plain head; lacks black flanks; bill yellow-brown. Differs from juv. Black-headed Heron by its white (not black) ear coverts, pale flanks and yellow (not dark) upper legs. **Voice:** Harsh *'kraaunk'* in flight. **Status and biology:** Common resident and local nomad at pans, dams, slow-flowing rivers, lagoons, estuaries and sheltered coastlines with rock pools. Breeds colonially in reeds or trees, often with other herons, egrets, cormorants and ibises. (Bloureier)

BLACK-HEADED HERON *Ardea melanocephala*

B	B	b	b	b	b	b	b	b	B	B	B
J	F	M	A	M	J	J	A	S	O	N	D

86–94 cm; 1.2–1.9 kg A large, greyish heron of terrestrial habitats. Slightly smaller than Grey Heron, with contrasting dark flight feathers and pale underwing coverts in flight. Black crown and hindneck contrast with white throat; underparts grey; bill and legs black. Juv. has slate-grey crown and hindneck and buff wash on foreneck; underparts paler. Differs from juv. Grey Heron by dark ear coverts, black legs (lacking yellow bases) and contrasting underwing pattern. **Voice:** Loud *'aaaaark'*; various hoarse cackles and bill-clapping at nest. **Status and biology:** Common resident and local nomad in grassland, fields and scrubland; also marsh fringes, but seldom forages in water. Breeds colonially (up to 50 pairs) in reeds or trees, often with other herons, egrets, cormorants and ibises. (Swartkopreier)

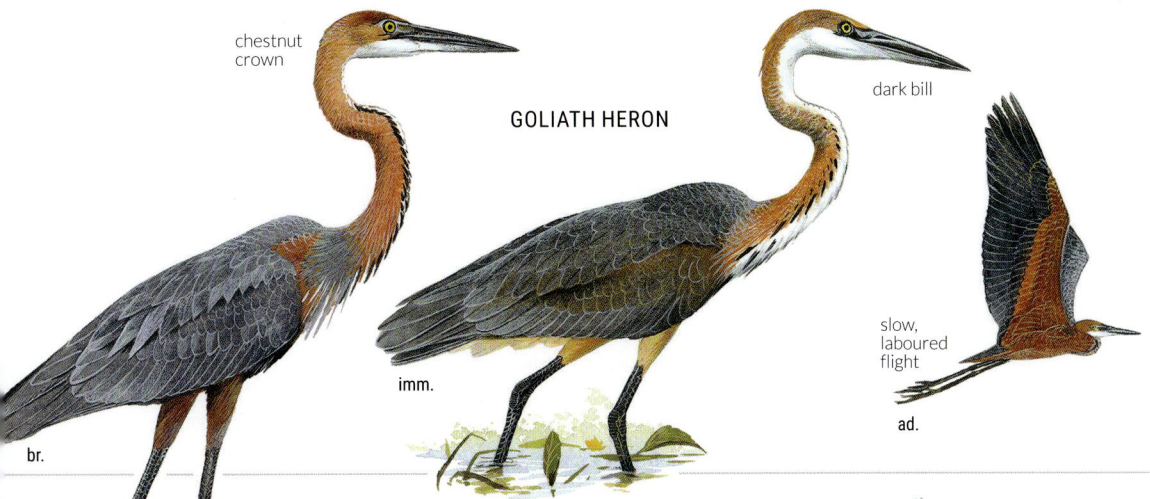

chestnut crown

GOLIATH HERON

dark bill

slow, laboured flight

imm.

ad.

br.

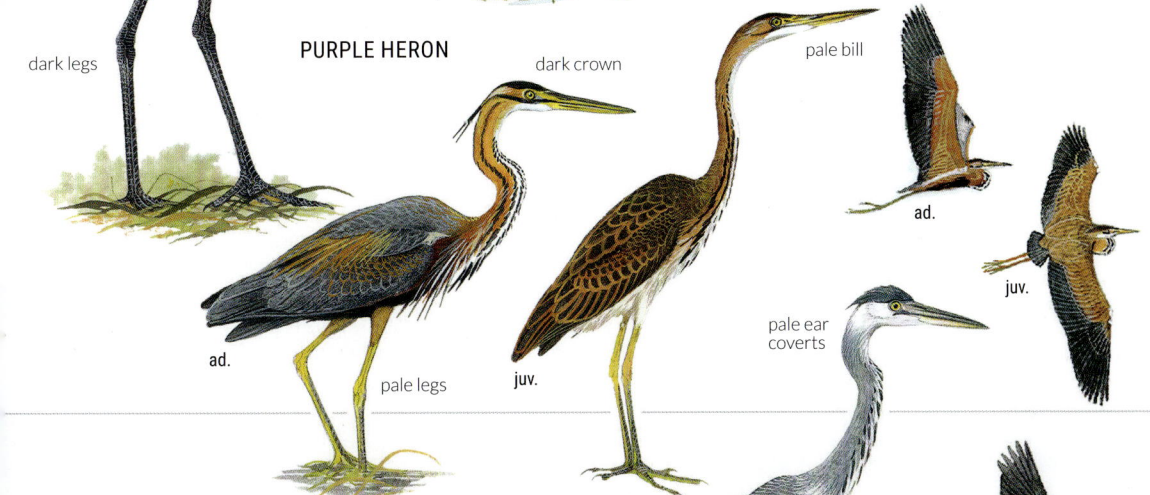

dark legs

PURPLE HERON

dark crown

pale bill

ad.

juv.

ad.

pale legs

juv.

pale ear coverts

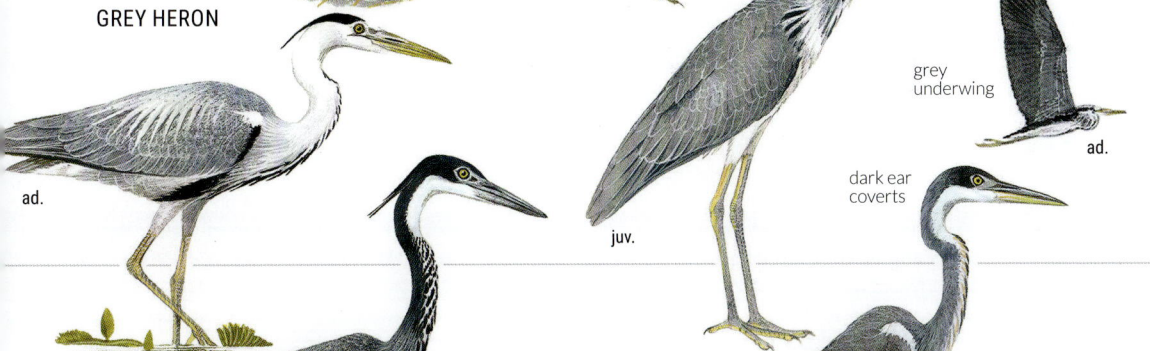

GREY HERON

grey underwing

ad.

dark ear coverts

ad.

juv.

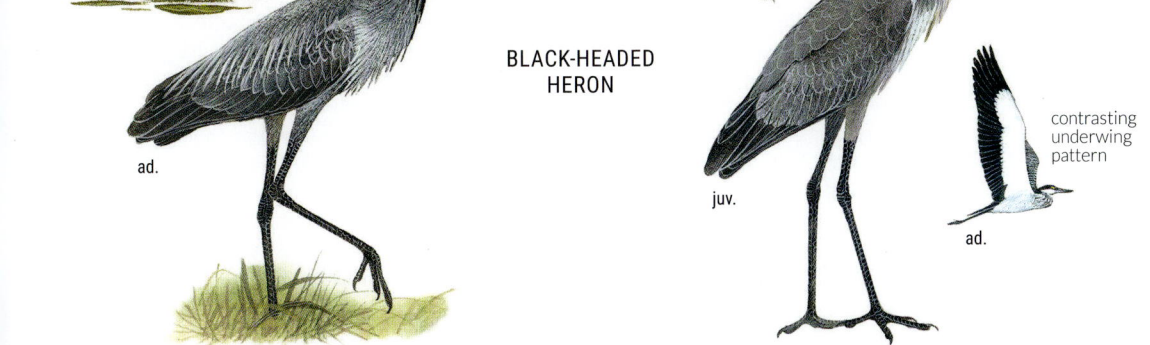

ad.

BLACK-HEADED HERON

contrasting underwing pattern

juv.

ad.

83

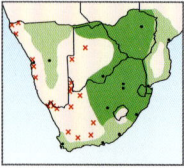

GREAT EGRET *Ardea alba*

B B B B B B B b b b b B
J F M A M J J A S O N D

85–95 cm; 800–1 400 g The largest egret, structurally similar to the large herons, with a heavy bill. Larger and heavier billed than Yellow-billed Egret. Legs and feet black; lacks yellow toes of much smaller Little Egret. Br. ad. has elaborate plumes, black bill, lime-green lores and pale cream eyes. Non-br. ad. has yellow bill and eyes; gape extends behind the eye. **Voice:** Deep, heron-like *'waaaark'*. **Status and biology:** Common resident and nomad at lakes, dams and estuaries. Breeds colonially (up to 200 pairs) in reeds or trees, usually with other egrets, herons, cormorants and ibises. (Grootwitreier)

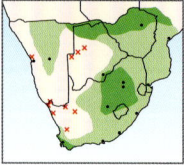

YELLOW-BILLED (INTERMEDIATE) EGRET *Ardea intermedia*

B B b b b b b b B B B B
J F M A M J J A S O N D

65–72 cm; 300–500 g A medium-sized, white egret with a yellow bill and blackish legs and toes, with paler tibia varying from bright yellow to dull, brownish yellow (often brighter on inner leg). Neck shorter than Great Egret's and not held in such a pronounced 'S' shape; bill shorter and gape ends just below eye (behind eye in Great Egret). Larger and longer necked than Western Cattle Egret, with a longer bill. In br. plumage, has long plumes on back and chest; bill and upper legs are red, and lores are lime-green. **Voice:** Typical, heron-like *'waaaark'*. **Status and biology:** Common resident and nomad at marshes and flooded grassland; rarely at estuaries. Breeds colonially (up to 70 pairs) in reeds or trees, usually with other egrets, herons, cormorants and ibises. (Geelbekwitreier)

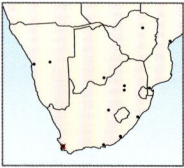

SNOWY EGRET *Egretta thula*

X X
J F M A M J J A S O N D

56–60 cm; 350–450 g Vagrant. Slightly smaller than Little Egret, with a shorter bill and more extensive yellow lores extending onto the base of the bill. Br. ad. has brown eyes and orange-yellow lores and base of upper mandible with short, bushy head plumes (fewer, elongate plumes than breeding Little Egret). Often shows yellow to backs of black legs. Non-br. ad. has duller bare parts. Juv. has yellowish-green lores and base of bill; easily confused with juv. Little Egret. **Voice:** Harsh *'aarr'*. **Status and biology:** Vagrant from the Americas; 2 records from s Africa: Lakeside, Cape Town, Apr 2002, and Black River, Cape Town, Jun 2015. (Sneeuwitreier)

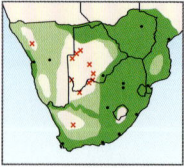

LITTLE EGRET *Egretta garzetta*

B B B b b b b b b B B B
J F M A M J J A S O N D

55–65 cm; 450–600 g A fairly small, slender egret with a long, black bill, black legs and contrasting, yellow toes. Bill is more slender and slightly shorter than that of vagrant Western Reef Heron. Lacks yellow lores extending onto the base of bill of vagrant Snowy Egret, but beware age-related variation. Br. ad. has cream eyes, mauve face and elongate plumes on nape, mantle and foreneck; toes orange-red at start of br. period. Non-br. ad. lacks plumes and has yellow eyes and a grey-green face. Juv. has greenish base to bill, duller yellow toes, and grey-green lower legs. **Voice:** Harsh *'waaark'*. **Status and biology:** Common resident and local nomad at wetlands, estuaries and along the coast. Often actively pursues fish in shallow water and frequently uses other birds or animals to scare or herd fish. Breeds colonially (up to 120 pairs) in reeds or trees, usually with other egrets, herons, cormorants and ibises. (Kleinwitreier)

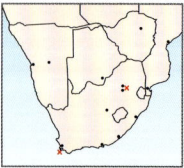

WESTERN REEF HERON *Egretta gularis*

X
J F M A M J J A S O N D

56–66 cm; 350–500 g Vagrant. White morph similar to Little Egret, but with a longer, heavier, yellow-brown bill and greenish-black legs with dull yellow lower tarsi merging into dull yellow toes. Dark morph (not recorded from s Africa) is dark grey, with a white throat and variable white patch on primary coverts, visible in flight. Imm. is greyish-brown or white, variably mottled with grey or grey-brown. **Voice:** Harsh *'gaaar'*. **Status and biology:** Rare vagrant from tropical coasts of W and E Africa; 1 record from Cape Point, Apr 2002; 1 record from Gauteng, Apr 2006. (Westelike Kusreier)

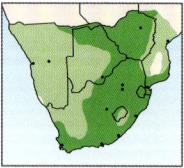

WESTERN CATTLE EGRET *Bubulcus ibis*

b b b b b b b b B B B B
J F M A M J J A S O N D

48–54 cm; 280–450 g A small, compact egret with a short bill, neck and legs. Legs olive-yellow; bill yellow. Ad. has buff plumes on crown, mantle and breast, forming a shaggy bib. Buff areas increase in br. season, but are never as extensive as in Squacco Heron (p. 86). Pre-br. birds have red bill and legs and violet lores. Melanistic birds can appear similar to Rufous-bellied Heron (p. 86), but bill is entirely orange-yellow, and generally show paler areas in the wings, tail and rump. Juv. has black legs at fledging, but these soon pale. **Voice:** Heron-like *'aaaark'* or *'pok-pok'*. **Status and biology:** Common; resident in parts of its range, migratory in others; birds ringed in Gauteng have been recovered north of its range; in a wide range of open habitats, including grassland, fields and coastline. Often uses large animals to flush prey, sometimes standing on animals and gleaning ectoparasites. Roosts in flocks, commuting up to 20 km to feeding areas. Breeds in colonies of up to 10 000 pairs in reeds or trees, usually with other egrets, herons, cormorants and ibises. (Veereier)

GREAT EGRET

gape extends beyond eye

br.

non-br.

YELLOW-BILLED EGRET

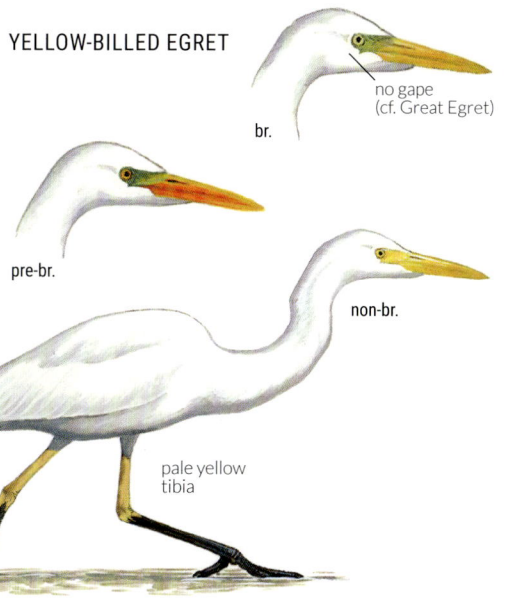

no gape (cf. Great Egret)

br.

pre-br.

non-br.

pale yellow tibia

SNOWY EGRET

shy mes

r.

chrome-yellow lores

non-br.

some yellow on tarsi

LITTLE EGRET

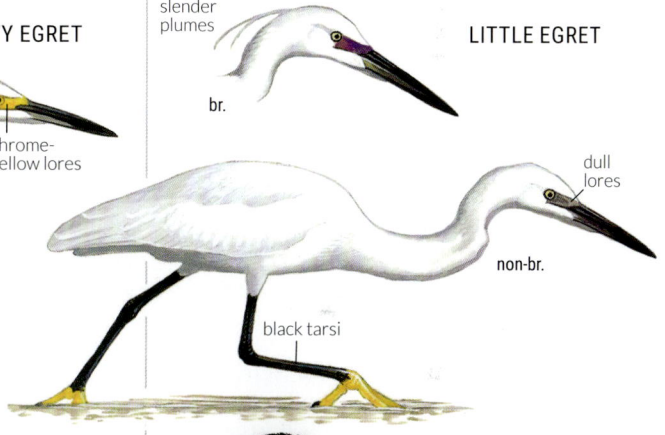

long, slender plumes

br.

dull lores

non-br.

black tarsi

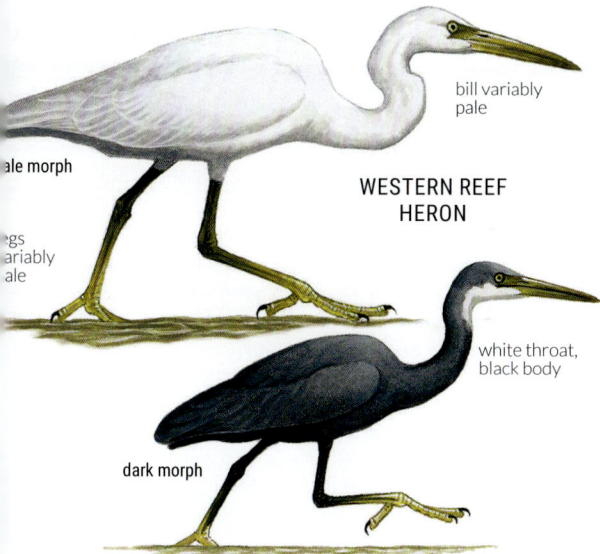

bill variably pale

WESTERN REEF HERON

ale morph

gs ariably ale

dark morph

white throat, black body

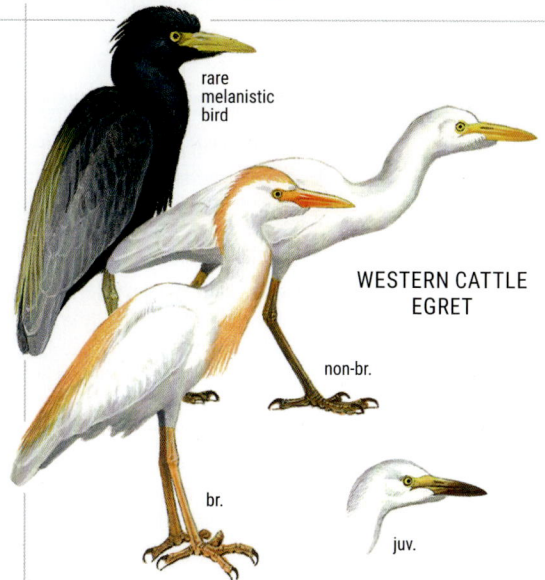

rare melanistic bird

WESTERN CATTLE EGRET

non-br.

br.

juv.

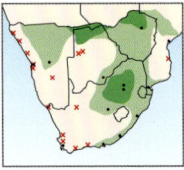

BLACK HERON *Egretta ardesiaca*

B B B b b b b b b b b B
J F M A M J J A S O N D

50–60 cm; 280–380 g A small, slate-black egret, darker than Slaty Egret and ad. of vagrant Little Blue Heron, with black legs and orange-yellow toes (not dull yellow or greenish legs). Often uses its wings to form an 'umbrella' over its head, possibly shading water to attract small fish. Ad. has slate-coloured nape, back and breast plumes, which are pronounced in br. season; toes red at start of br. season. Juv. is slightly paler, lacking plumes. **Voice:** Deep *'kraak'*. **Status and biology:** Locally common resident and nomad at lakes and marshes; occasionally estuaries. Breeds colonially (up to 20 pairs) in reeds or trees, usually with other herons, egrets, cormorants and ibises. (Swartreier)

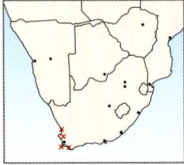

LITTLE BLUE HERON *Egretta caerulea*

X X X X X X X X X X X X
J F M A M J J A S O N D

64–70 cm; 300–400 g Vagrant. A medium-sized heron with a black-tipped, pale grey bill; legs and feet greenish-grey. Ad. blue-grey, with darker, purplish head and neck. Juv. is white; easily confused with Little Egret (p. 84), but has yellowish-green legs and a distinctive two-tone bill. Imm. has odd grey feathers, appearing mottled. **Voice:** Harsh *'thrrrr'* when disturbed. **Status and biology:** Nearctic-br. vagrant that winters in central and S America; at least 4 records from wetlands on the west and south coasts. Some individuals have remained for several years. (Kleinbloureier)

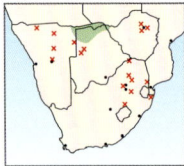

SLATY EGRET *Egretta vinaceigula*

b B B b
J F M A M J J A S O N D

48–58 cm; 250–340 g A dark grey egret with a diagnostic, rufous throat. Slightly smaller than Black Heron, with greenish-yellow (not black) legs and feet; does not use its wings to shade water when feeding. Eyes yellow (black in pre-br. birds). Juv. paler, lacking head and breast plumes; legs greenish-grey. **Voice:** Heron-like squawks and alarm *'krrr krrr krrr'*. **Status and biology:** VULNERABLE. Localised resident and nomad at marshes and along vegetated lake shores. Mostly confined to n Botswana (Okavango) and the Zambezi Region (Caprivi), but disperses widely, with vagrants appearing annually in ne S Africa. Breeds sporadically in colonies of up to 200 pairs, often with Rufous-bellied Herons and Dwarf Bitterns (p. 88), either in reeds or small trees. (Rooikeelreier)

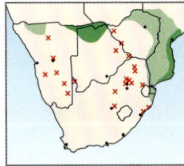

RUFOUS-BELLIED HERON *Ardeola rufiventris*

b b B B B B B B b b b b
J F M A M J J A S O N D

38–40 cm; 300 g A small heron with a sooty head and breast, and rufous belly, wings and tail. Skulking; normally seen only when flushed. In flight, bright yellow legs and feet contrast strongly with dark underparts. Bill pink-yellow with dark tip; pre-br. birds have red lores and legs. Female duller, with pale throat. Juv. dull brown with buff streaking on breast and throat. **Voice:** Heron-like *'waaaaak'*. **Status and biology:** Fairly common resident in n Botswana; uncommon migrant and nomad elsewhere. Occurs in dense marshes and flooded grassland. Breeds singly or in small colonies in reeds or low trees, often with other small herons. (Rooipensreier)

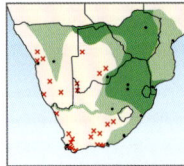

SQUACCO HERON *Ardeola ralloides*

B B B B B B b b b B b b
J F M A M J J A S O N D

42–46 cm; 250–350 g A small, buff-and-white heron, with a heavy, dark-tipped bill. At rest, appears mostly buff and brown, with white underparts. In flight, white wings and tail are prominent; could be confused with Western Cattle Egret (p. 84), but is smaller and more compact, with broad, rounded wings. Br. ad. has blue lores and base to bill, elongate crown feathers and yellow legs (red in pre-br. period). Non-br. ad. has dark brown streaking on body and greenish-yellow bill; legs duller. Juv. has grey-brown body with heavier streaking, but usually not as boldly streaked as non-br. Malagasy Pond Heron. Wings mottled brown, belly pale grey. **Voice:** Low-pitched *'kruuk'*, and rattling *'kek-kek-kek'*. **Status and biology:** Common resident and local nomad along vegetated margins of lakes, pans and slow-moving rivers; skulks in long grass, sitting motionless for long periods. Breeds in small colonies in reeds or low trees, often with other herons and egrets. (Ralreier)

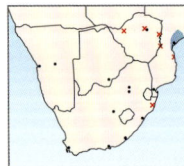

MALAGASY POND HERON *Ardeola idae*

J F M A M J J A S O N D

45–48 cm; 280–380 g Slightly larger than Squacco Heron, with a heavier bill. Br. plumage, seldom seen in Africa, is completely white. Non-br. ad. is dark brown (not buffy) above, with broader and darker streaking on throat and breast than Squacco Heron; also tends to show a sharper contrast between streaked breast and white belly; back streaked brown (uniform pale buff-brown in Squacco Heron). Juv. duller, with brown mottling in wings. **Voice:** Loud *'kruuk'*, similar to Squacco Heron. **Status and biology:** ENDANGERED. Very rare, non-br. winter visitor to Mozambique and e Zimbabwe from Madagascar. Appears to be increasingly rare in the subregion, presumably due to declining Madagascar population as a result of habitat loss and exploitation of eggs and young. Occurs along vegetated lake shores, often in more open areas than Squacco Heron. (Madagaskar-ralreier)

BLACK HERON

foraging 'umbrella'

yellow toes only

LITTLE BLUE HERON

black-tipped grey bill

imm.

juv.

ad.

chestnut throat

yellow legs & feet

SLATY EGRET

juv.

imm.

ad.

ad.

ad.

RUFOUS-BELLIED HERON

ad.

br.

dark-tipped pink-yellow bill

imm.

SQUACCO HERON

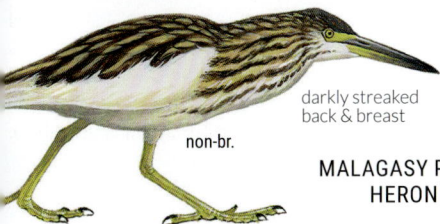

uniform brown back

non-br.

br.

non-br.

juv.

MALAGASY POND HERON

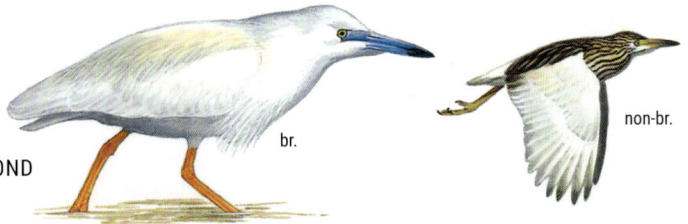

darkly streaked back & breast

non-br.

br.

non-br.

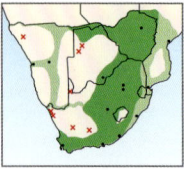

BLACK-CROWNED NIGHT HERON   *Nycticorax nycticorax*

`b b b b b b b b B B b b`
J F M A M J J A S O N D

54–60 cm; 500–700 g A stocky, compact heron with a heavy, blackish bill, reddish eyes and yellow legs. Ad. distinctive: black crown, nape and back contrast with grey wings and tail, and white underparts. Juv. is grey-brown, with white spotting above; underparts whitish, with dark brown streaks. Bill yellowish, with dark tip. Larger than juv. Green-backed Heron, with shorter, deeper bill, and heavier spotting on underparts. Juv. smaller than Eurasian Bittern; plumage colder grey-brown, lacking black streaking and barring. Yellow bill, spotted (not streaked) back, small cap and paler eyes distinguish it from juv. White-backed Night Heron. **Voice:** Harsh *'kwok'* in flight. **Status and biology:** Common resident, nomad and migrant at lakes, rivers and rocky shores. A bird ringed in Europe was recovered in Mozambique. Roosts communally in reeds or trees during the day, flying out at dusk to feed. Breeds colonially (up to 1 000 pairs) in trees or reeds, usually with other herons, egrets, cormorants and ibises. (Gewone Nagreier)

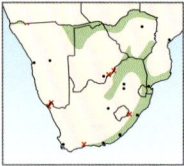

WHITE-BACKED NIGHT HERON   *Gorsachius leuconotus*

`b b b b b   b B B B B b`
J F M A M J J A S O N D

50–56 cm; 400–500 g A night heron with a large, dark head, conspicuous pale eye-ring, rufous neck and white throat. Generally appears much darker than Black-crowned Night Heron. In flight and during display, dark back and wings contrast with a small, white back patch. Bill black, with yellow base to lower mandible; eyes dark red-brown; lores yellow (blue in pre-br. birds); legs yellow. Juv. is paler brown above with a streaked neck and white-spotted wing coverts. Has black (not yellow) bill and larger dark cap than juv. Black-crowned Night Heron; mantle plain. White back develops with age. **Voice:** Sharp *'kaaark'* when disturbed. **Status and biology:** Uncommon resident at slow-moving rivers overhung with dense vegetation. Easily overlooked; roosts by day in dense cover; more nocturnal than Black-crowned Night Heron. Breeds singly on a low branch overhanging water. (Witrugnagreier)

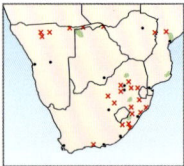

EURASIAN BITTERN   *Botaurus stellaris*

`b b b           b b b b`
J F M A M J J A S O N D

65–75 cm; 1–1.3 kg Larger than juv. Black-crowned Night Heron, with tawny, heavily streaked plumage, a black crown and broad, conspicuous moustachial stripes. Typically remains in dense cover; more often heard than seen. Flight owl-like, with bowed, rounded wings. Juv. is less heavily marked above, with reduced black cap. **Voice:** Deep, resonant, 3–5-note boom, similar to grunting of a distant lion. **Status and biology:** Rare resident and local nomad in extensive reed beds, sedges and flooded grassland. Has undergone serious population declines in s Africa, largely as a result of wetland degradation. Breeds singly. (Grootrietreier)

LITTLE BITTERN   *Ixobrychus minutus*

`b B b   b b b B B B b b`
J F M A M J J A S O N D

25–30 cm; 120–150 g A tiny, rather pale heron. Pale upperwing coverts contrast with dark flight feathers in flight (lacks uniformly dark upperwing of smaller Dwarf Bittern). Male has greenish-black back and crown; female browner above, with striped foreneck. Migrant, Palearctic-br. *I. m. minutus* is slightly larger and paler than resident *I. m. payesii*, which has more rufous face and neck. Juv. is more heavily streaked below; smaller and more buffy than Green-backed Heron, with greenish (not yellowish) legs. **Voice:** Short, barked *'rao'*, every few seconds when displaying. **Status and biology:** Uncommon resident and local nomad and Palearctic-br. migrant, mainly in reed beds. *I. m. payesii* breeds singly or in loose colonies. (Kleinrietreier)

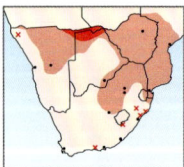

DWARF BITTERN   *Ixobrychus sturmii*

`B B B           b B`
J F M A M J J A S O N D

26–30 cm; 130–170 g A tiny, dark-backed heron with a rather broad neck; appears almost rail-like in flight. Ad. is dark slaty-blue above and buff below, with broad, dark stripes running down throat onto breast and belly. Bill blackish, with pale green base to lower mandible. Legs yellowish-green, but orange in pre-br. period. In flight, easily mistaken for Green-backed Heron unless underparts are visible. Juv. is paler, with upperparts scalloped buff; breast more rufous with finer streaking. **Voice:** Barking *'ra-ra-ra-ra-ra...'* in display, otherwise silent. **Status and biology:** Uncommon intra-African br. migrant, following seasonal rains. Occurs at lakes and ponds surrounded by grass and trees, also in mangroves. Breeds singly or in loose colonies. (Dwergrietreier)

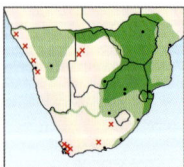

GREEN-BACKED (STRIATED) HERON   *Butorides striata*

`B B B b b b b B B B B B`
J F M A M J J A S O N D

40–44 cm; 200–230 g A small, dark heron with an erectile black crown. Stands motionless in characteristic, hunched posture; bobs and flicks tail when alarmed. Ad. has dark, grey-green back and upperwings and paler grey underparts; black, wispy nape plume is not usually visible. From behind, back is greener and more scaled than Dwarf Bittern, and underparts are plain grey (not heavily streaked). Bill and legs yellowish, but in pre-br. period bill is black and legs orange-red. Juv. is brown above, finely spotted whitish; buff below with dark brown streaks. Fledgling has pink-orange legs. **Voice:** Sharp *'baaek'* when flushed. **Status and biology:** Common resident and local nomad at sluggish rivers overhung with trees, lake shores (often in rocky areas), mangroves and rocky shores. Breeds singly or in loose colonies. (Groenrugreier)

typical resting
posture

white
spots

juv.

ad.

ad.

ad.

juv.

ad.

juv.

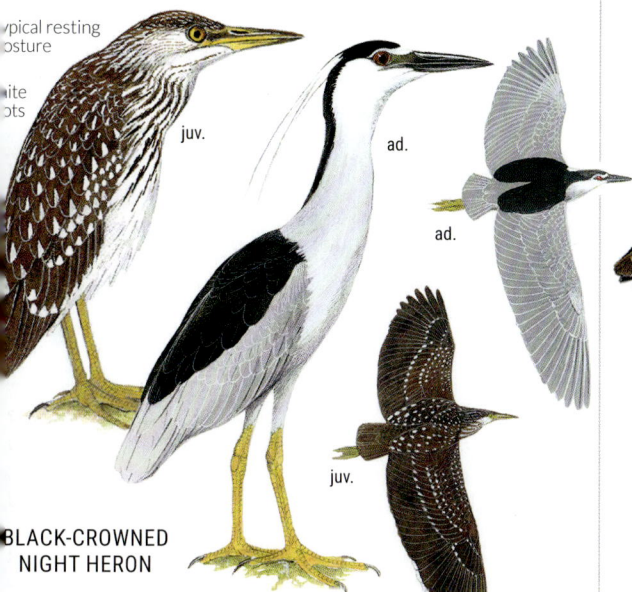

BLACK-CROWNED
NIGHT HERON

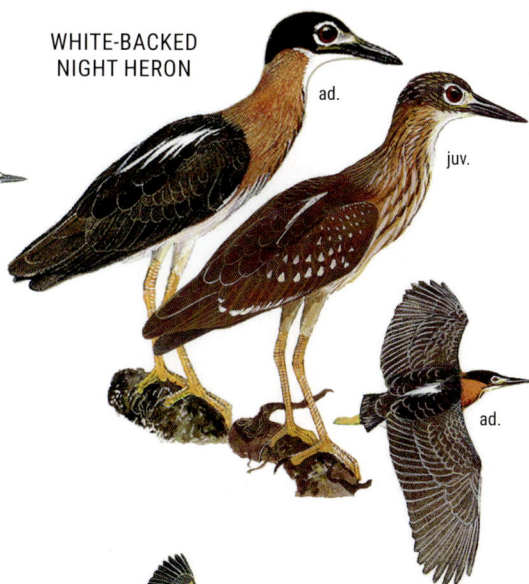

EURASIAN
BITTERN

dark
mask

black
spots

ad. ♀

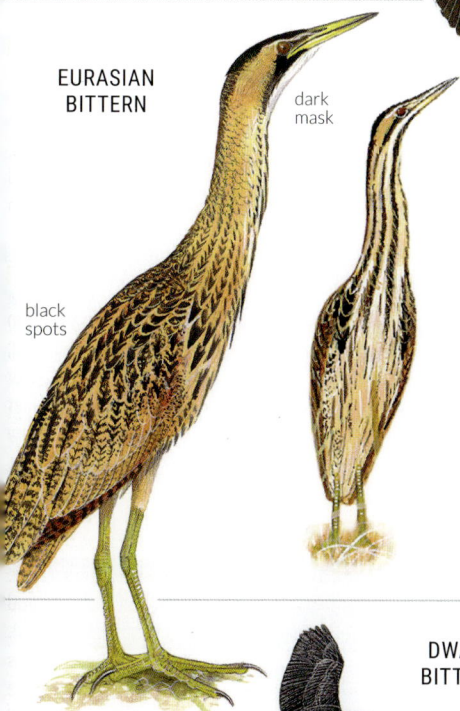

LITTLE
BITTERN

ad. ♂

ad. ♂

ad. ♀

juv.

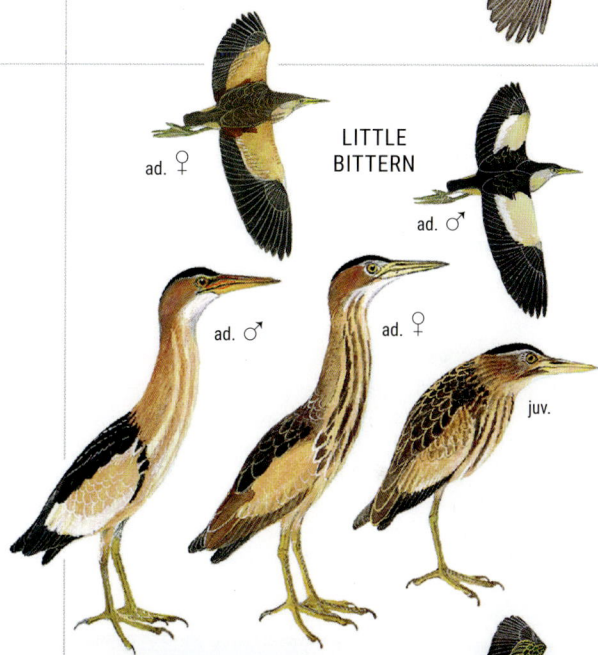

DWARF
BITTERN

dark cap,
paler back

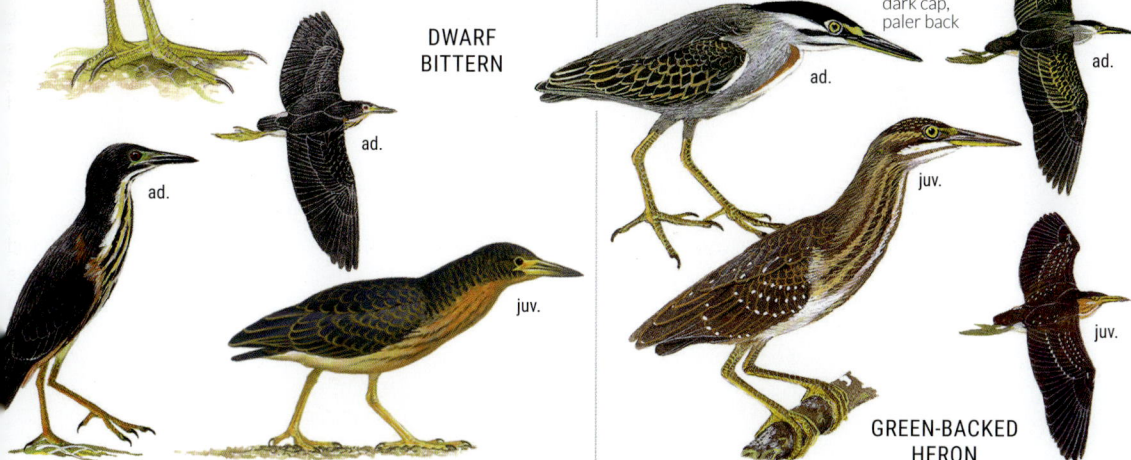

ad.

ad.

ad.

ad.

juv.

juv.

ad.

juv.

GREEN-BACKED
HERON

STORKS
Large wading birds with long legs and necks. Bills usually dagger-like. Most species at least partly aquatic. Sexes alike; juvs distinct. Unlike herons, they are regularly seen soaring, often in large flocks. Breed singly or colonially.

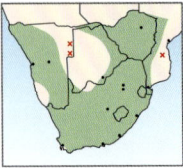

BLACK STORK *Ciconia nigra*

b B B B b b b
J F M A M J J A S O N D

95–110 cm; 2.5–3.2 kg A large, glossy black stork with a white belly and undertail. Larger than Abdim's Stork, with black (not white) rump and lower back, red (not greenish-grey) bill, and red face and legs. Juv. is browner, with olive-yellow bill and legs. **Voice:** Silent except on nest, when loud whining and bill-claps are given. **Status and biology:** Uncommon and decreasing resident and nomad at lakes, rivers, estuaries and lagoons; population around 1 000 pairs. Breeds singly on a cliff ledge; often reuses the same nest sites for many years. (Grootswartooievaar)

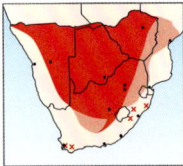

ABDIM'S STORK *Ciconia abdimii*

J F M A M J J A S O N D

76–81 cm; 1.1–1.6 kg A rather small, black stork with a white belly and undertail. Smaller than Black Stork, with white (not black) lower back and rump, and greenish-grey (not red) bill. Face is blue; legs are grey-green, with pink ankles and feet. In flight, shows less leg projection than does Black Stork. Juv. is duller and browner. **Voice:** Usually silent; weak, 2-note whistle at nests and roosts. **Status and biology:** Common intra-African migrant that breeds in the Sahel. Occurs in grassland and fields, often in large flocks and frequently with White Storks. Gathers in large numbers at termite emergences. (Kleinswartooievaar)

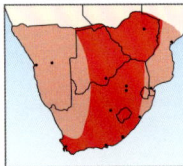

WHITE STORK *Ciconia ciconia*

b B b
J F M A M J J A S O N D

100–120 cm; 2.4–4 kg A large, mostly white stork with black primaries and secondaries. In flight, white tail differentiates it from Yellow-billed Stork. Bill and legs red, but red legs often appear white because birds excrete on them to cool down. Juv. has darker bill and legs; back and wing coverts tinged brown. **Voice:** Silent except on nest, when loud whining and bill-claps are given. **Status and biology:** Common, Palearctic-br. migrant in grassland and fields; occasionally at shallow wetlands. Numbers fluctuate between years depending on rainfall elsewhere in Africa. A few pairs breed in W Cape, building large stick platforms in a tall tree or on a building. Nests often reused, but added to each year, and may reach up to 2 m high. (Witooievaar)

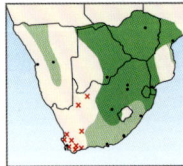

YELLOW-BILLED STORK *Mycteria ibis*

b b B B B b
J F M A M J J A S O N D

95–105 cm; 1.2–2.4 kg A white stork with black flight feathers and a long, slightly decurved, yellow bill. Differs from White Stork in flight by its black (not white) tail. Br. ad. has naked, red facial skin and pink-tinged wing coverts and back. Juv. is brownish above, washed grey-brown below, becoming whiter with age. Head is mostly feathered, and facial skin, bill and legs are duller. **Voice:** Normally silent except during br. season when it gives loud squeaks and hisses. **Status and biology:** Common resident and partial intra-African migrant at lakes, rivers and estuaries. Feeds with partly open bill, stirring up bottom sediments with its feet characteristically spreading one wing while foraging. Breeds colonially (10–50 pairs), often with other storks, herons, egrets, ibises and cormorants, building a stick platform in a tree. (Nimmersat)

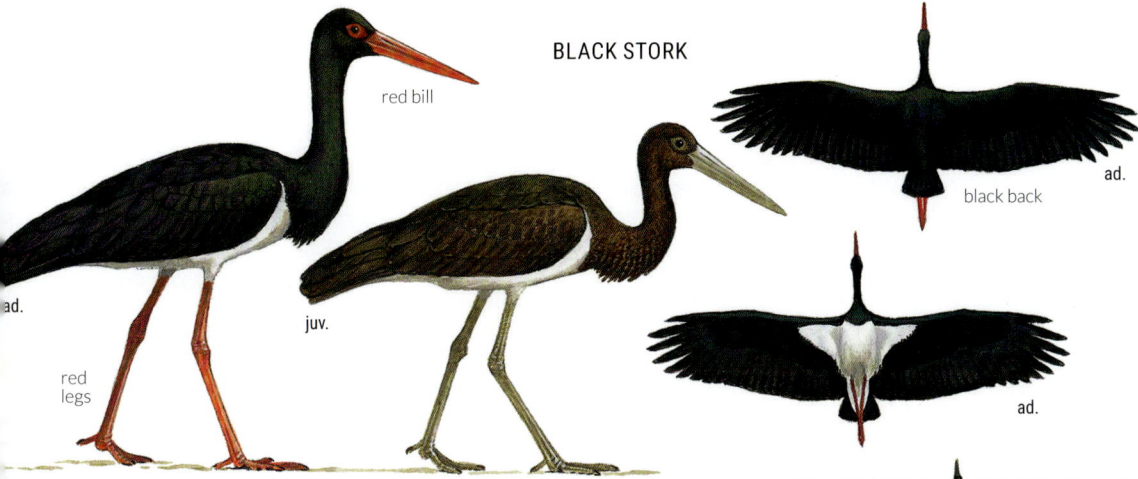

BLACK STORK

ad.

red bill

red legs

juv.

black back

ad.

ad.

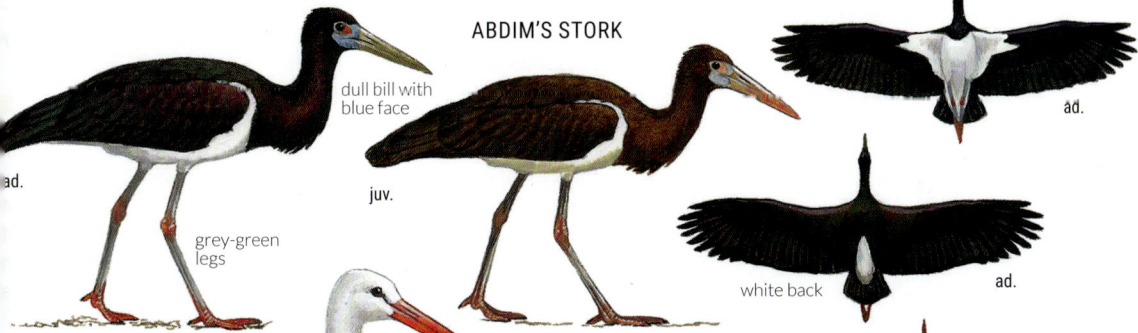

ABDIM'S STORK

ad.

dull bill with blue face

juv.

grey-green legs

ad.

white back

ad.

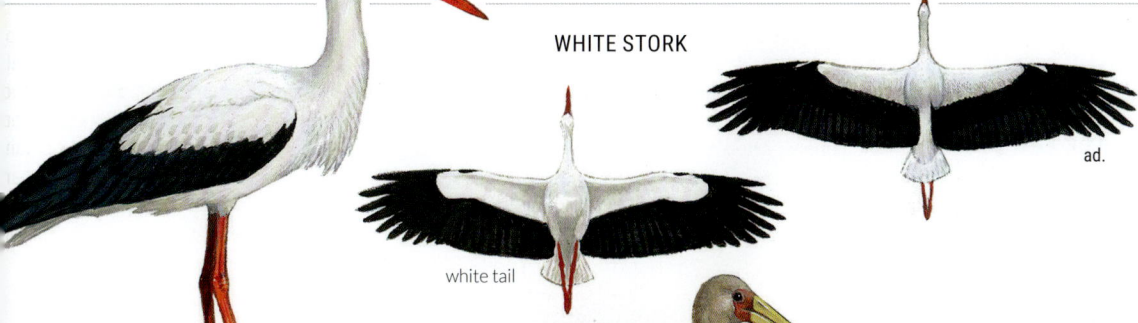

WHITE STORK

white tail

ad.

imm.

ad.

black tail

ad.

wing shading

YELLOW-BILLED STORK

ad.

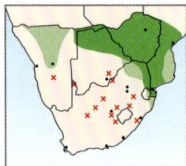

SADDLE-BILLED STORK   *Ephippiorhynchus senegalensis*

b B B B b b b b
J F M A M J J A S O N D

145–150 cm; 5–7 kg A very tall, black-and-white stork with a diagnostic, red-and-black-banded bill with a yellow 'saddle' at the base. In flight, wings are strikingly black and white. Male has brown eyes (female yellow), a small, yellow wattle (absent in female) and more black on wing in flight. Juv. is grey, with a brown head and neck; lacks yellow 'saddle' on bill. Flight feathers greyish, and leading edge of wing dark. **Voice:** Normally silent except for bill-clapping during display. **Status and biology:** Uncommon resident and nomad at freshwater dams, lakes and rivers. Usually solitary or in pairs. Breeds singly in a large platform nest on top of a tree; sometimes on top of an old raptor's nest. (Saalbekooievaar)

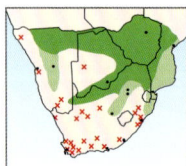

MARABOU STORK   *Leptoptilos crumenifer*

b B B B B b   b
J F M A M J J A S O N D

130–150 cm; 4.5–8.5 kg A huge stork with a grey, blade-like bill, a naked head and a pendulous throat pouch. In flight, head is tucked into shoulders, and dark grey wings contrast with white body. Juv. has head and neck covered with sparse, woolly down. **Voice:** Low, hoarse croak when alarmed; claps bill when displaying. **Status and biology:** Fairly common resident or local nomad in savanna, grassland and at wetlands; often around towns, scavenging at refuse dumps and abattoirs. Competes with vultures at carcasses. One recent irruption (2009) saw birds as far out of range as the sw W Cape, and stragglers now fairly regular outside the core range. Up to 30 pairs breed colonially, building stick platforms in tall trees. (Maraboe)

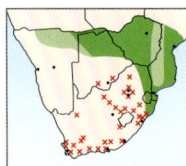

AFRICAN OPENBILL   *Anastomus lamelligerus*

B B B b b   b b b b
J F M A M J J A S O N D

74–90 cm; 700–1 200 g A rather small, all-dark stork with an ivory-horn bill that has a diagnostic wide, nutcracker-like gap between the mandibles. Ad. has glossy sheen to plumage. Juv. is duller and browner, with pale feather tips; fledges without noticeable bill gap, but this soon develops. **Voice:** Croaking 'honk', seldom uttered. **Status and biology:** Locally common resident and nomad at freshwater lakes and dams; occasionally irrupts well outside its normal range, with exceptional numbers recorded in 2010. Feeds mainly on freshwater snails, using its peculiar bill to extract the snails from their shells. Breeds colonially, occasionally up to 100 pairs. (Oopbekooievaar)

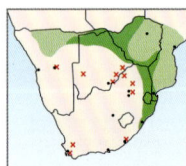

WOOLLY-NECKED STORK   *Ciconia episcopus*

b B B B b
J F M A M J J A S O N D

80–90 cm; 1.6–1.9 kg A glossy black stork with a diagnostic, woolly, white neck, white belly and undertail coverts (which extend to tip of black tail, tail appearing all-white from below). In flight, has white rump and lower back. Juv. is dull brown; black forehead extends further back on crown. **Voice:** Harsh croak; seldom uttered. **Status and biology:** VULNERABLE. Fairly common resident and intra-African migrant found at wetlands, often along rivers and streams; also mangroves, coastal mud flats, reefs, and regularly visits suburban gardens. Attends grass fires. Solitary or in pairs, seldom in flocks. Breeds singly or in loose, small colonies of up to 5 pairs; nesting with increasing frequency in suburban Durban and Pietermaritzburg; nests in a large tree, usually overhanging water. (Wolnekooievaar)

SADDLE-BILLED
STORK

ad. ♂

juv.

ad. ♀

ad. ♂

ad. ♀

MARABOU STORK

AFRICAN OPENBILL

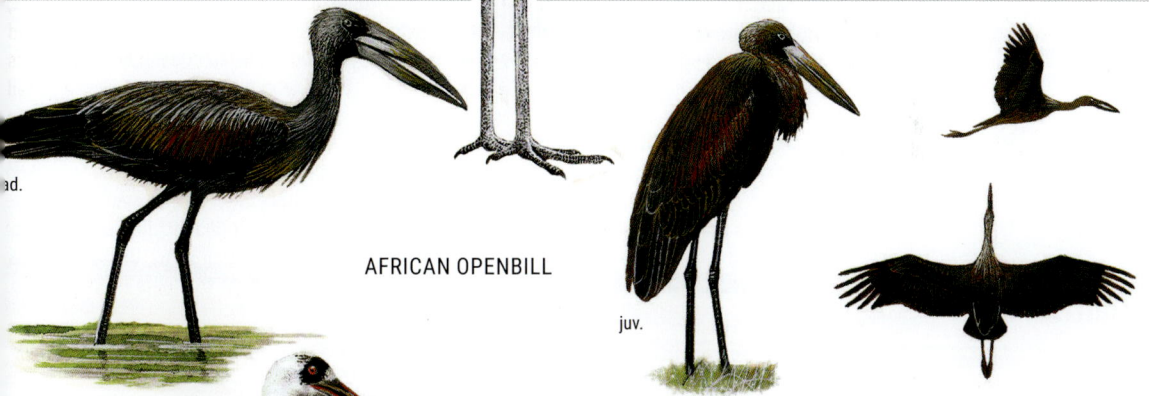

juv.

ad.

juv.

ad.

WOOLLY-NECKED
STORK

HAMERKOP
A monotypic family confined to Africa and Madagascar, most closely related to the pelicans and Shoebill *Balaeniceps rex*. Sexes alike; juv. resembles ad.

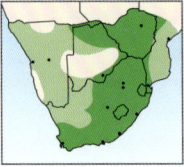

HAMERKOP *Scopus umbretta*

b b b b b b b B B b b b
J F M A M J J A S O N D

50–58 cm; 400–600 g A dark brown, heron-like bird with long, black legs, a heavy crest and a flattened, boat-shaped, black bill. Hammer-shaped head profile is diagnostic, but often obscured when neck retracted and crest rests on its back. Flight buoyant, often gliding and soaring; could be confused with a raptor or Hadeda (p. 80). **Voice:** Sharp *'kiep'* in flight; jumbled mixture of querulous squawks and frog-like croaks during courtship. **Status and biology:** Fairly common resident at lakes, dams and rivers. Nest is a huge, domed structure of sticks, with a small side entrance, usually built in a sturdy tree or on a cliff ledge. (Hamerkop)

GREBES
Small to medium-sized diving birds, with rounded bodies and very short tails; closest living relatives are flamingos. Feed by pursuit diving, propelled by lobed toes. They routinely raise their back feathers after surfacing and orient their backs towards the sun for passive heating; sleek down feathers just before diving to expel air and reduce buoyancy. Legs well back on body; clumsy on land. Build floating nests from aquatic vegetation. Sexes alike, but most species have br. and non-br. plumages. Juvs have striped heads. Ads often carry small chicks on their backs.

GREAT CRESTED GREBE *Podiceps cristatus*

b b b b b b b b b b b b
J F M A M J J A S O N D

45–56 cm; 500–750 g A large, long-necked grebe. Ad. has distinctive, dark double crest and rufous-edged ruff ringing sides of head. Ruff of non-br. birds is smaller and paler. In flight, appears long and slender, with neck extended and legs trailing; wings long and thin, with conspicuous white secondaries and lesser coverts. Juv. has black-and-white-striped head; lacks crests and head ruff. **Voice:** Barking *'rah-rah-rah'*, also various growls and snarls. **Status and biology:** Locally common resident on large lakes and pans; rarely in estuaries and sheltered bays. (Kuifkopdobbertjie)

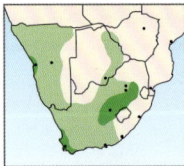

BLACK-NECKED GREBE *Podiceps nigricollis*

B B B B b b b b b b B B
J F M A M J J A S O N D

28–33 cm; 300 g Slightly larger than Little Grebe and appears more elegant thanks to its longer, more slender neck, more angular head and longer bill with an upturned tip. Eyes bright red. Br. ads have golden ear tufts, black head and throat, and chestnut flanks. Non-br. ads and imms have white cheeks, throat and flanks. On water, sits with back higher than, or level with, rump; its steep forehead is a useful identification feature. Preening birds often roll over, flashing white belly. **Voice:** Seldom calls, but gives mellow trill during display. **Status and biology:** Locally common resident and nomad at lakes, pans and sheltered coastal bays (mostly in Namibia). Often in flocks. Breeds erratically, usually colonially. (Swartnekdobbertjie)

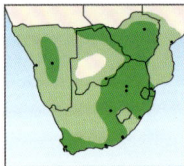

LITTLE GREBE *Tachybaptus ruficollis*

B B B b b b b b B B B B
J F M A M J J A S O N D

23–29 cm; 120–190 g A small, compact grebe with a pale, fleshy gape flange (absent in Black-necked Grebe). Bill short and relatively deep; head rounded. Br. ad. has rich chestnut sides to face and neck; non-br. ads duller. Juv. has black-and-white-striped cheeks. On water, sits with rump higher than back. **Voice:** Distinctive, whinnying trill, frequently given. **Status and biology:** Common resident on lakes, dams and other freshwater bodies; rarely in estuaries and sheltered bays. Apparently flies at night to locate new wetlands. (Kleindobbertjie)

HAMERKOP

juv.

rufous
ruff

ad.

**GREAT CRESTED
GREBE**

ad.

**BLACK-NECKED
GREBE**

orange
ear tufts

pale
flanks

pale
neck

non-br.

br.

non-br.

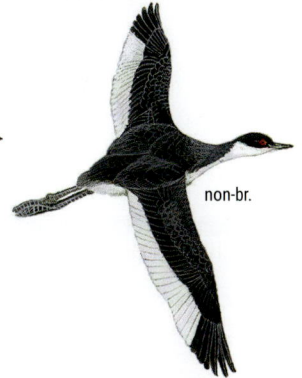

LITTLE GREBE

may carry
chicks on back

chestnut
neck

pale spot

non-br.

br.

non-br.

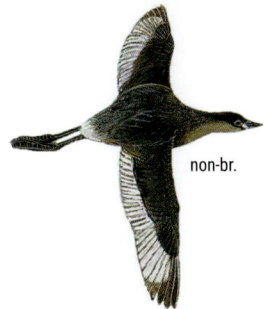

DUCKS AND GEESE
Waterbirds with flattened bills that feed by diving, dabbling or grazing ashore. Three toes are webbed for swimming. Plumage fully waterproof; flight feathers moulted simultaneously, resulting in 3–4-week flightless period. Males appreciably larger than females (except whistling ducks), and differ in plumage in some species. Chicks precocial; usually with ads, but some feathered chicks and young juvs can pose identification problems. Together with gamebirds, they form a distinct group, the Galloanserae, separate from most other birds (Neoaves).

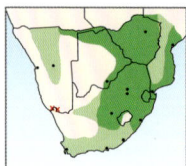

WHITE-FACED WHISTLING DUCK *Dendrocygna viduata*

BBbbb bbbbB
J F M A M J J A S O N D

44–48 cm; 550–900 g A dark brown duck with a diagnostic, white face and throat (sometimes stained brown), chestnut foreneck and blackish belly. Much darker on neck and breast than Fulvous Whistling Duck, with finely barred flanks and blackish vent. Appears all-dark in flight apart from the white face. Juv. has a dirty-brown face. **Voice:** Distinctive 3-note, wispy whistle, '*whit we-weeer*', which often betrays its presence. **Status and biology:** Common resident and nomad at freshwater lakes, lagoons and adjacent grassland, often in large flocks. Feeds on land and in water; occasionally dives for up to 10 seconds. (Nonnetjie-eend)

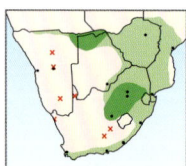

FULVOUS WHISTLING DUCK *Dendrocygna bicolor*

BBBbbbbbbbbb
J F M A M J J A S O N D

43–50 cm; 600–900 g Structure similar to White-faced Whistling Duck, but with a golden-brown head, breast and belly, a white vent, a dark line down the back of the neck, and conspicuous white flank stripes. In flight, the creamy uppertail coverts contrast with the dark brown wings, rump and tail. Juv. is paler than ad., and lacks buff scaling on mantle. **Voice:** Soft, 2-note whistle. **Status and biology:** Locally common resident and nomad at freshwater lakes and dams. Feeds mainly in water; dives for up to 20 seconds. (Fluiteend)

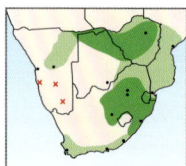

WHITE-BACKED DUCK *Thalassornis leuconotus*

BBBBBbbbbbbB
J F M A M J J A S O N D

38–42 cm; 560–700 g A fairly small, compact duck with a large head and characteristic humped back; sits low in the water, appearing somewhat grebe-like. Plumage is barred buff and dark brown, with a white spot at the base of the bill; the white back is only visible in flight. **Voice:** A low-pitched whistle, rising on the second syllable. **Status and biology:** Locally common resident and nomad found at wetlands with water lilies or other floating vegetation. It is an excellent diver; spends much of the day roosting, with its head tucked into its scapulars. Usually in pairs or small groups, but pairs remain together even when in larger flocks. (Witrugeend)

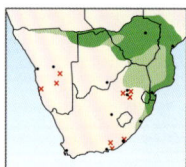

AFRICAN PYGMY GOOSE *Nettapus auritus*

BBBbb bBBB
J F M A M J J A S O N D

30–33 cm; 230–320 g A tiny duck with distinctive orange breast and flanks, a white face and dark greenish upperparts. Male has bright yellow bill and lime-green neck patch, neatly edged black. Female and juv. duller, with indistinct head markings. In flight, has a large, white wing patch formed by the inner secondaries and greater coverts; underwing coverts blackish. Often sits motionless among floating vegetation. **Voice:** A soft, repeated '*tsui-tsui*'. **Status and biology:** Locally common resident at freshwater lakes with floating vegetation, especially *Nymphaea* lilies. Breeds in a hole in a tree or cliff, usually near water. (Dwerggans)

WHITE-FACED WHISTLING DUCK

barred flanks

juv.

ad.

ad.

FULVOUS WHISTLING DUCK

white crescent

white-striped flanks

WHITE-BACKED DUCK

white spot at base of bill

AFRICAN PYGMY GOOSE

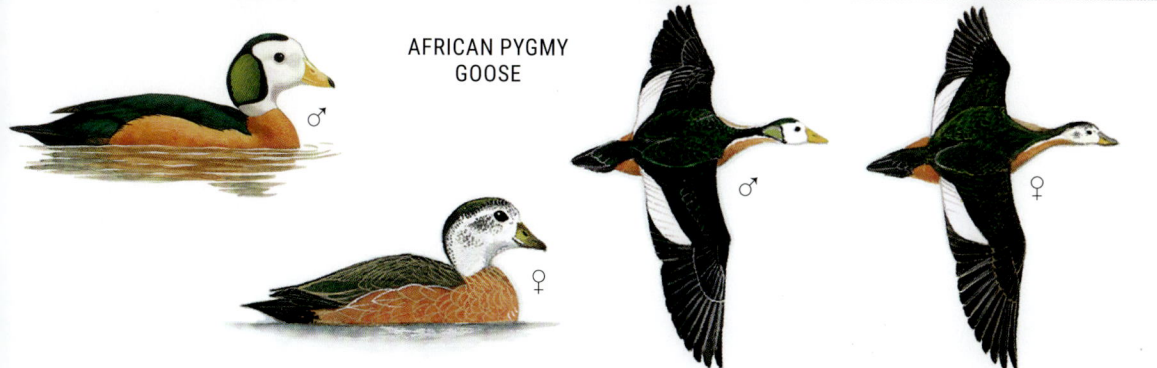

♂

♀

♂

♀

SPUR-WINGED GOOSE *Plectropterus gambensis*

B B B b b b b b b b B B
J F M A M J J A S O N D

75–100 cm; 2.5–7 kg A large, black goose with variable amounts of white on the face, throat, belly and forewings; white on head and body is most extensive in northern populations (*gambensis* subspecies), much reduced in southern birds. Bill, face and legs pink-red. In flight, its large size and white forewing separate it from Knob-billed Duck. Sexes alike, but males are much larger, with more extensive facial skin and wattles; facial skin brighter and head knob swollen in br. male. Juv. duller and browner; lacks bare facial skin, and white on forewing reduced or absent. **Voice:** A feeble, wheezy whistle in flight. **Status and biology:** Common resident and local nomad at wetlands and nearby grassland and fields. Feeds mainly on land, but flies to water to roost and moult. Often in large flocks, frequently with Egyptian Geese. (Wildemakou)

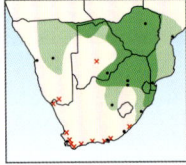

KNOB-BILLED DUCK *Sarkidiornis melanotos*

B B B b   b b b b
J F M A M J J A S O N D

56–76 cm; 1–2.5 kg A large duck with a speckled head and contrasting blue-black and white plumage. Wings black. Male has a rounded knob on the bill, which enlarges during the br. season. Female is smaller, with no bill knob and a paler patch on the lower back. Juv. is duller, with dark speckling on the breast, belly and flanks and a pale supercilium and dark eye-line. Could be confused with Garganey (p. 104), but is considerably larger and lacks scaled breast and flanks. **Voice:** Whistles, but usually silent. **Status and biology:** Locally common nomad and intra-African summer migrant at pans and lakes in woodland and along larger rivers. Feeds in water and on land, including in old croplands. Breeds in a tree hole or old Hamerkop nest. (Knobbeleend)

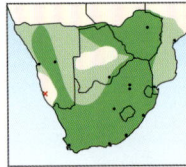

EGYPTIAN GOOSE *Alopochen aegyptiaca*

b b b b b b B B B B b b
J F M A M J J A S O N D

60–75 cm; 1.5–3.5 kg A large, goose-like duck; mostly buffy-brown with a dark brown face, neck ring (variable in extent) and breast patch. Larger and longer necked than South African Shelduck; also stands more erect, and, in flight, the white secondary coverts have a thin, black subterminal bar. Juv. is duller, only developing brown mask and breast patch after 3–5 months; forewing greyish. **Voice:** Male gives hoarse hisses, female a grunting honk; both honk repeatedly when alarmed or taking flight. **Status and biology:** Common resident and local nomad at wetlands and adjacent grassland and fields; also suburban areas, golf courses and sheltered bays on the coast. Often gathers in large flocks when not br. Nests in a shallow scrape on the ground, in a tree hole, an old bird's nest or even on a flat rooftop. (Kolgans)

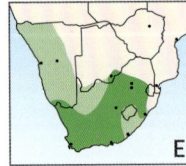

SOUTH AFRICAN SHELDUCK *Tadorna cana*

        b b B B b
J F M A M J J A S O N D

60–65 cm; 900–1 800 g Smaller than Egyptian Goose, with a shorter neck and legs, and a more horizontal stance. Plumage warm russet and brown. Male head grey; female has a white face (extent of white varies considerably). In flight, both sexes have white forewings lacking a black line across the secondary coverts (as present in Egyptian Goose). Juv. is like male but duller, with head suffused brown and dusky-brown upperwing coverts. **Voice:** Nasal honk; male voice deeper. **Status and biology:** Endemic. Common resident and local nomad, feeding mainly on terrestrial vegetation, but gathering to roost and moult at freshwater lakes and dams. Often with Egyptian Geese. Nests in holes in the ground, especially old aardvark burrows. (Kopereend)

SPUR-WINGED GOOSE

gambensis shows paler head

variable white on wings & head

br. ♂

juv.

KNOB-BILLED DUCK

♀

ad.

EGYPTIAN GOOSE

juv.

ad.

SOUTH AFRICAN SHELDUCK

♂

♀

♂

♀

99

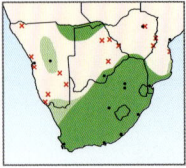

YELLOW-BILLED DUCK *Anas undulata*

B B B b b b B B B B B B
J F M A M J J A S O N D

52–58 cm; 700–1 150 g A dark brown duck with a distinctive, bright yellow bill with a black saddle. Sexes virtually alike; male has brighter speculum and all-black lower mandible. Pale feather edges give it a scaled appearance at close range; scaling less distinct in juvs. Structure similar to African Black Duck; told apart in flight by blue-green (not purple-blue) speculum narrowly edged with white, and grey (not white) underwings. **Voice:** Male gives a rasping hiss; female quacks. **Status and biology:** Common resident and nomad at freshwater lakes, ponds and flooded fields; also lagoons and estuaries. Often in flocks. Feeds mainly by dabbling or up-ending in shallow water. Hybridises with introduced Mallards (p. 104); some hybrids resemble Yellow-billed Duck apart from their orange (not blackish) legs. Also occasionally hybridises with Red-billed Teal (p. 102). (Geelbekeend)

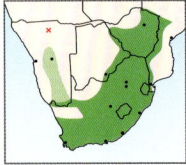

AFRICAN BLACK DUCK *Anas sparsa*

b b B B B B B B B b b b
J F M A M J J A S O N D

48–58 cm; 780–1 200 g A dark, sooty-coloured duck with white speckles on the back and orange legs. Darker than Yellow-billed Duck, with a blue-grey bill with a blackish saddle and pinkish base. Appears long-bodied when sitting on the water. In flight, the purple-blue speculum, bordered white, is bluer than that of Yellow-billed Duck, and the underwing is whitish (not grey). Juv. is paler, with a whitish belly and less brightly coloured legs. **Voice:** Female quacks in flight; male gives high-pitched, peeping whistle. **Status and biology:** Fairly common resident along streams and rivers; less frequent on ponds and dams. Feeds by dabbling or up-ending, mainly at dawn and dusk. (Swarteend)

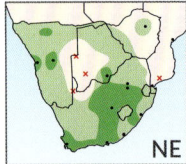

CAPE SHOVELER *Spatula smithii*

b b b b b b B B B B B b
J F M A M J J A S O N D

48–54 cm; 450–750 g An elongate duck with a long, black, spatulate bill. Both sexes have finely speckled grey-brown plumage. Male has a paler, greyer head than female, with a darker bill, paler yellow eyes, brighter orange legs, and more prominent blue-grey upperwing coverts. Female and juv. are darker and less rufous than vagrant Northern Shoveler (p. 104), especially on the head and neck, and have a smaller and darker, slate-grey bill lacking any orange tinge along the margins. Wings make distinctive whirring noise on takeoff. **Voice:** Female quacks; male makes a soft, rasping call. **Status and biology:** Near-endemic. Common resident and local nomad at freshwater lakes and ponds, often in flocks. Feeds mainly in shallow water, filtering water and mud with bill lamellae. (Kaapse Slopeend)

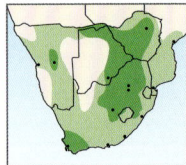

SOUTHERN POCHARD *Netta erythrophthalma*

B B B b b b b b B B B B
J F M A M J J A S O N D

48–51 cm; 550–950 g A fairly large, dark brown duck. Male glossy, with paler, chestnut-brown flanks, a pale blue bill and bright red eyes. Shape very different from Maccoa Duck (p. 102), with a more slender bill. In flight, both sexes have a distinct, white wingbar extending onto the primaries. Female is dark brown with pale facial patches, one at the base of the bill, the other a crescent extending behind the eye. Juv. resembles female, but lacks the white facial crescent. **Voice:** Male makes a whining sound; female quacks. **Status and biology:** Common resident and local nomad at lakes and dams, preferring areas with deeper water. Mainly feeds by diving (for up to 18 seconds), but also dabbles and grazes. (Bruineend)

YELLOW-BILLED DUCK

green
speculum

AFRICAN BLACK DUCK

bluish
speculum

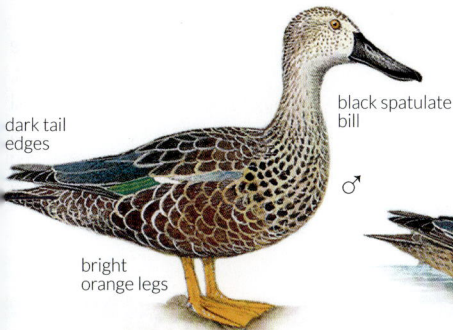

CAPE SHOVELER

dark tail
edges

black spatulate
bill
♂

bright
orange legs

♀

blue
forewing

♂

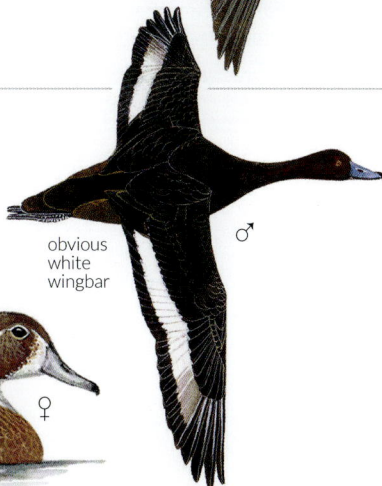

♂

obvious
white
wingbar

♂

white
crescent

♀

SOUTHERN
POCHARD

juv.

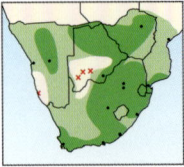

RED-BILLED TEAL *Anas erythrorhyncha*

B B B B b b b B B B B B
J F M A M J J A S O N D

43–48 cm; 400–850 g Readily identified by its dark cap, pale cheeks and red bill with a dark saddle. Larger than Hottentot Teal, with red (not blue) bill and plain, whitish cheeks lacking dark smudges. Told from Cape Teal by its dark cap and darker brown (not grey) plumage. In flight, has warm buff secondaries and lacks an iridescent speculum. Female has a broader black bill saddle than male. **Voice:** Male gives a soft, nasal whistle; female quacks. **Status and biology:** Common resident and nomad at freshwater wetlands. Feeds mainly by dabbling in shallow water. (Rooibekeend)

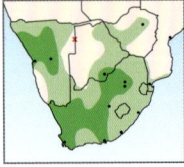

CAPE TEAL *Anas capensis*

b b b b b b B B B B B b
J F M A M J J A S O N D

44–48 cm; 350–520 g A pale grey duck with a rather plain, bulbous-looking head, bright red eye and pink bill with a black base and bluish tip. In flight, it has a dark, greenish speculum bordered by white. Easily distinguished from Red-billed Teal by its pale head that lacks a dark cap; wing patterns quite different. Sexes alike; juv. is duller, with narrower, pale feather margins. **Voice:** A thin whistle, usually given in flight. **Status and biology:** Locally common resident and nomad at fresh or saline wetlands, especially saltpans, lagoons and sewage works. Mainly feeds in shallow water, dabbling or up-ending, but occasionally dives. (Teeleend)

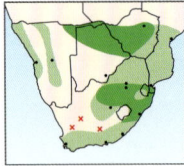

HOTTENTOT TEAL *Spatula hottentota*

B B B B b b b b b b b b
J F M A M J J A S O N D

32–36 cm; 200–280 g A tiny duck with a dark crown, cream cheeks and distinctive, dark smudges extending from the ear coverts down the neck. Much smaller than Red-billed Teal, with a blue (not red) bill. In flight has a green speculum (brighter in male) with a white trailing edge, and a black-and-white underwing. Juv. is duller. **Voice:** Gives high-pitched quacks on taking flight, but generally silent. **Status and biology:** Locally common resident and nomad at freshwater wetlands, favouring areas with emergent or floating vegetation. (Gevlekte Eend)

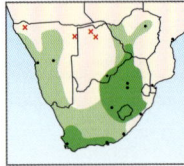

MACCOA DUCK *Oxyura maccoa*

B B B b b b b b B B B B
J F M A M J J A S O N D

48–51 cm; 550–800 g A stiff-tailed duck with a characteristic, dumpy body, short neck and broad, deep-based bill. Br. male has a chestnut body, black head and blue bill. Female and eclipse male are dark brown, with a pale stripe under the eye and a paler throat, giving the head a striped appearance (but lacks pale crescent behind eye of female Southern Pochard, p. 100). Sits low in the water, often with its stiff tail cocked at a 45° angle. In flight, the upperwing is uniform dark brown. **Voice:** A peculiar, nasal trill. **Status and biology:** VULNERABLE. Uncommon resident and nomad at freshwater lakes, dams and lagoons. Sometimes dumps its eggs in nests of other ducks and coots. Feeds mainly by diving (for up to 22 seconds). (Bloubekeend)

dark cap

RED-BILLED TEAL

tan-coloured secondaries

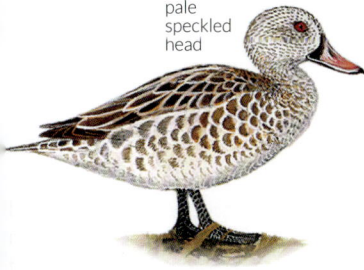

pale speckled head

CAPE TEAL

green speculum bordered white

dark cap & cheek patch

HOTTENTOT TEAL

white-tipped secondaries

♂

MACCOA DUCK

♂

tail often held erect

♀

INTRODUCED AND VAGRANT DUCKS

At least 3 Palearctic-br. ducks that migrate into Africa during the n hemisphere winter occasionally reach s Africa. However, ducks are popular ornamental birds, and frequent escapes of captive birds occur, making it hard to distinguish genuine vagrants. One escapee, the Mallard, has established feral populations in several areas, where they pose a threat to native species through hybridisation.

GARGANEY *Spatula querquedula*

37–41 cm; 300–550 g Vagrant. Br. male has a large, white supercilium and black-and-white, lanceolate back feathers. The brown breast is sharply demarcated by the white belly and pale grey flanks. Female and eclipse male have a pale supercilium and dark eye stripe. Differ in shape from female Maccoa Duck (p. 102) and juv. Knob-billed Duck (p. 98). In flight, both sexes have pale blue forewings and green speculums superficially similar to Cape (p. 100) and Northern Shovelers. **Voice:** A nasal *'quack'*, and some harsh rattles. **Status and biology:** Rare, Palearctic-br. migrant to lakes and marshes, mainly in the north of the region. Feeds by dabbling and up-ending. (Somereend)

NORTHERN PINTAIL *Anas acuta*

50–65 cm; 600–1 200 g Vagrant. Br. male is striking, with a dark, chocolate-brown head, a white stripe running down the side of the neck to the white breast, lanceolate back feathers and long central tail feathers. Female and eclipse male are speckled tan and brown, with a blue-grey bill, a long, slender neck, plain buffy face, and a brown speculum with a white trailing edge. Tail pointed in both sexes. **Voice:** Male gives a soft, nasal honk; female quacks. **Status and biology:** Very rare, Palearctic-br. summer visitor to freshwater wetlands, mostly in Zimbabwe; most winter in n Africa. Also some escapees. (Pylsterteend)

MALLARD *Anas platyrhynchos*

50–64 cm; 850–1 400 g Introduced from Europe. Br. male has a glossy, bottle-green head, white ring around the neck and chestnut breast. Females and eclipse males superficially resemble Yellow-billed Ducks (p. 100), but are paler brown with a horn-coloured bill and a dark line through the eye. The domesticated form of Mallard is larger, with a heavy 'bottom' and has a khaki (not grey) back. **Voice:** Male gives a rasping hiss; female quacks. **Status and biology:** Feral populations derived from escapees occur at some wetlands and are subject to management controls. Regularly hybridises with Yellow-billed Duck, rarely with African Black Duck and Cape Shoveler (p. 100). (Groenkopeend)

NORTHERN SHOVELER *Spatula clypeata*

44–52 cm; 350–550 g Vagrant. Male in br. plumage is unmistakable, with a green head, white breast and chestnut belly and flanks. In flight, has a powder-blue forewing. Female and eclipse male are dull rufous-brown, paler and warmer than Cape Shoveler (p. 100), especially on the head and neck; bill heavier, longer and more spatulate. Female has white edges to tail and orange bill margins. **Voice:** Male gives a nasal *'crook, crook'*; female quacks. **Status and biology:** Palearctic-br. vagrant to inland water bodies; most records probably are escapees. (Europese Slopeend)

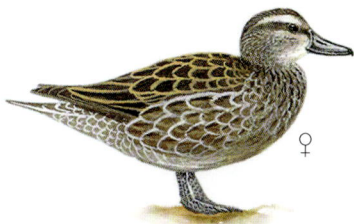

GARGANEY

NORTHERN
PINTAIL

long, slender
neck

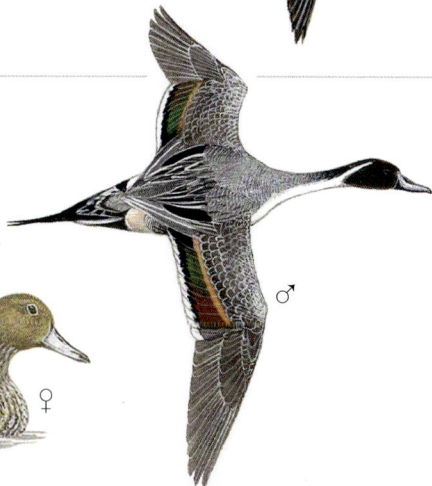

pointed
tail

MALLARD

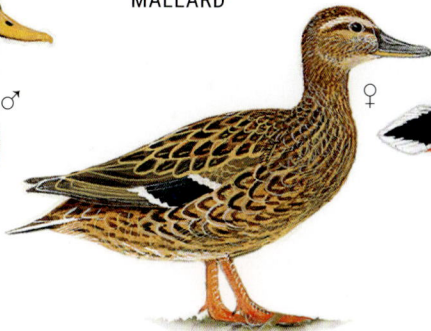

NORTHERN
SHOVELER

white tail
edges

105

EXOTIC DUCKS Escapees from captive wildfowl collections are sufficiently regular to warrant inclusion in this guide. In addition to the species illustrated here, several others have been recorded. A seemingly wild Chiloe Wigeon *Anas sibilatrix* at Cape Town has been mooted as a possible genuine trans-Atlantic vagrant. Other regularly recorded species include Black *Cygnus atratus* and Mute *C. olor* Swans and Brazilian Teal *Amazonetta brasiliensis*. A feral population of Mute Swans in the s Cape died out in the 1980s. No distribution maps are presented for these species because their occurrence is dictated by the distribution of waterfowl collections rather than by biology and geography.

TUFTED DUCK   *Aythya fuligula*

40–47 cm; 1–1.4 kg Male black, with white lower breast, belly and flanks, striking yellow eyes and blue-grey bill with a black nail; diagnostic drooping crest on hind crown. Male in eclipse plumage has reduced crest and greyish flanks. Female is browner overall (with brown flanks); crest reduced; some show patch of pale feathers at base of greyish bill. In flight, both sexes have a broad, white, upperwing bar; told from Southern Pochard (p. 100) by its mostly pale (not blackish) underwing coverts. **Voice:** Usually silent. **Status and biology:** Palearctic-br. migrant that reaches Nigeria and Kenya. Vagrancy to s Africa is possible, but records are probably all escapees. (Kuifeend)

COMMON SHELDUCK   *Tadorna tadorna*

62 cm; 800–1 450 g Both sexes have red bills (brighter in male, with fleshy knob at base when br.) and glossy, dark green heads and upper necks, contrasting with white lower necks and upper breasts. Mantle and broad breast band are chestnut; underparts mostly white, with black band running from middle of breast band to belly. In flight, white coverts contrast strongly with dark flight feathers. **Voice:** Male gives low-pitched whistling, female gives a rapid *'ga-ga-ga-ga-ga'*; mostly silent when not br. **Status and biology:** Palearctic-br. species that barely reaches sub-Saharan Africa. Popular as an ornamental species; s African records certainly are of escapees. (Bonteend)

MANDARIN DUCK   *Aix galericulata*

40–50 cm; 450–500 g Br. male's elaborate plumage is diagnostic; elongated, orange tertials form 'sails' on its back. Female much duller; told from female Wood Duck by its narrow, white eye-ring and stripe behind eye (not a large, white eye patch and broader white stripe) and whitish (not black) nail on its greyish bill. Speculum greenish-blue (more purplish in Wood Duck). Eclipse male resembles female, but retains red bill. In flight, dark upper- and underwings contrast with white trailing edge to secondaries; wings pointed and flight pigeon-like. **Voice:** Male gives whistled, wheezy *'sheeooo'*; female gives various quacking calls. **Status and biology:** Native to E Asia (where decreasing), but is a popular ornamental species that has been introduced in many parts of the world. Has bred in s Africa, but there is no evidence of a self-sustaining feral population. (Mandaryneend)

WOOD DUCK   *Aix sponsa*

43–50 cm; 480–880 g Br. male has a distinctive greenish, mane-like crest, edged white, and white cheek stripe and throat. In flight, upperwing dark, with bluish-purple speculum (more greenish in Mandarin Duck). Female much duller; told from female Mandarin Duck by its broader white eye-ring and white cheek patch, and black (not white) nail on greyish bill. Eclipse male resembles female, but retains pinkish-red bill and has more extensive white throat markings. **Voice:** Male gives a weak, ascending *'tjeeeeee'*; female gives a squealing *'ooo-eeek'*. **Status and biology:** Nearctic-br. resident and short-distance migrant. Vagrants have reached Iceland, but highly unlikely to reach s Africa. (Carolina-eend)

RINGED TEAL   *Callonetta leucophrys*

37 cm; 190–360 g Male has a buffy-grey face with a narrow, black line extending over forehead and crown and curving around to form a narrow collar; bill blue-grey. Upperparts mostly chestnut and buff. Upper breast pinkish-buff, speckled blackish; lower breast and belly white, with a black vent bar. In flight, green speculum contrasts with white secondary coverts. Female is a drab version of male, with browner upperparts and brown-barred underparts; face is brown and white, recalling female Southern Pochard (p. 100). **Voice:** Usually silent. **Status and biology:** Resident and short-distance migrant in Neotropical forest and woodland wetlands. Highly unlikely to occur as a vagrant, but occasional escapees recorded. (Ringnekeend)

RED-CRESTED POCHARD   *Netta rufina*

56 cm; 830–1 300 g Rounded chestnut head and foreneck of male diagnostic; bill pinkish-red. Hindneck, mantle, breast and vent black; flanks white. Female mostly pale brown, with darker brown cap extending below eye, and white cheeks, foreneck and vent; bill dark grey, tipped pink. Eclipse-plumage male resembles female, but bill is reddish-pink. In flight, both sexes show pale underwings and a broad, white bar on the upperwing, superficially similar to Southern Pochard (p. 100). **Voice:** Usually silent. **Status and biology:** Palearctic-br. species that barely reaches sub-Saharan Africa. A popular, ornamental species, with scattered records of escapees in Zimbabwe and S Africa. (Krooneend)

TUFTED DUCK

♀ ♂

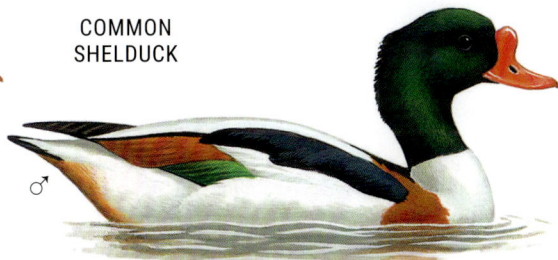

COMMON SHELDUCK

♀ ♂

MANDARIN DUCK

♀ ♂

WOOD DUCK

♀ ♂

RINGED TEAL

♀ ♂

RED-CRESTED POCHARD

♀ ♂

VULTURES

VULTURES Among the largest birds of prey, most vultures live by scavenging from animal carcasses. With wingspans between 1.4 m and 2.8 m, and deeply slotted primaries adapted for soaring in thermals, vultures spend much of the day flying at considerable heights, some species covering vast distances in search of food. Sexes alike in most species; juvs typically duller and darker, taking up to 6 years to acquire ad. plumage. The 3 *Gyps* vultures on this page, collectively known as 'griffons', are gregarious, nesting and roosting together and gathering to feed, sometimes in hundreds, at carcasses. Ads are readily identified, but juvs can be mistaken for each other.

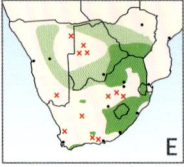

CAPE VULTURE   *Gyps coprotheres*

b B b b
J F M A M J J A S O N D

100–118 cm; 7.4–10.8 kg; wingspan 2.5–2.6 m Appreciably larger than White-backed Vulture; ad. has much paler, cream-coloured body and wing coverts, contrasting with darker tail and flight feathers. Eyes are yellow, face and throat bluish and there are patches of blue skin (not always visible) on either side of the base of the neck. Perched birds show a diagnostic line of dark spots on the wing, formed by dark centres to each greater upperwing covert. In flight, ad. appears pale from below with less contrast on the underwing than ad. White-backed Vulture, and the secondaries have a distinct dark terminal bar. Sexes alike. Juv. body and wing coverts are darker than ad., streaked with buff, becoming paler and less streaked with age; upperwing coverts lack dark spots, but are edged with white, producing a diagnostic pale line across upperwing. Underwing shows more contrast between leading and trailing edges than ad., but leading edge is never as white as in White-backed Vulture. Eyes dark brown; neck skin flushed pink. If flying with White-backed Vultures, its larger wingspan is evident. **Voice:** Silent, except at colonies and carcasses when it cackles and hisses. **Status and biology:** Endemic. ENDANGERED due to poisoning, shortage of food as well as electrocution and collision with powerlines. Range has contracted, but remains locally common in core of range. Roosts and nests on cliffs, ranging out over adjacent grassland and arid savanna in search of food. Scarce in well-wooded savanna; bush encroachment may have contributed to its extinction as a br. species in Namibia. (Kransaasvoël)

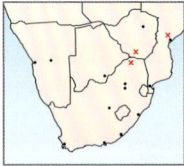

RÜPPELL'S VULTURE   *Gyps rueppelli*

X X X X X X X X X X X X
J F M A M J J A S O N D

95–107 cm; 6.8–8 kg; wingspan 2.3–2.5 m Vagrant. Intermediate in size between Cape and White-backed Vultures. Ad. has silvery-edged contour feathers that give it a diagnostic scaled or spotted appearance. Underwing appears uniformly dark (not two-tone like Cape Vulture and ad. White-backed Vulture); differs from juv. White-backed Vulture by scaled body and underwing coverts. Bill and eyes yellow. Sexes alike. Juv. has blackish bill and eyes; pale feather edges buffy and much narrower, appearing darker and streaked (not scaled); bill is black; neck reddish-pink. **Voice:** Noisy hissing and cackling. **Status and biology:** CRITICALLY ENDANGERED. Vagrant from further north in Africa; solitary birds occasionally visit Cape Vulture colonies. (Rüppellaasvoël)

WHITE-BACKED VULTURE   *Gyps africanus*

b B b b b b
J F M A M J J A S O N D

90–100 cm; 4.6–6.6 kg; wingspan 2.1–2.3 m The most common vulture in savannas. Smaller than Cape Vulture. Ad. body and wing coverts range from buff to brown, variably streaked, but always darker than ad. Cape Vulture; greater upperwing coverts lack dark spots. If visible, its white lower back is diagnostic. In flight, whitish underwing coverts contrast strongly with darker flight feathers. Face and neck usually blackish; bill and eye blackish (bill not as heavy as Cape Vulture's). Also shows patches of bare skin to either side of the base of the neck, as in Cape Vulture. Sexes alike. Juv. body and coverts darker and more heavily streaked with buff than ad. Underwing initially uniformly dark brown with narrow, white line along leading edge, this becoming progressively more extensive with age; lacks the whitish line across the upperwing coverts of juv. Cape Vulture. **Voice:** Silent, except at colonies and carcasses when it cackles and hisses. **Status and biology:** CRITICALLY ENDANGERED. Fairly common in savanna and open woodland, nesting and roosting on trees. (Witrugaasvoël)

ad.

pale eye,
dark bill

ad.

ad.

imm.

juv.

little
contrast
between
pale coverts
& secondaries

juv. & imm. have
dark eyes

CAPE VULTURE

ad.

imm.

agnostic
ark 'dotted'
e on
overts

juv.

**RÜPPELL'S
VULTURE**

ad.

bill & eyes ivory
coloured

ad.

juv.

dark bill
& eyes

narrow pale
lines on
underwing
coverts

juv.

overall
scalloped
appearance

scalloped
appearance

white
ruff

ad.

ad.

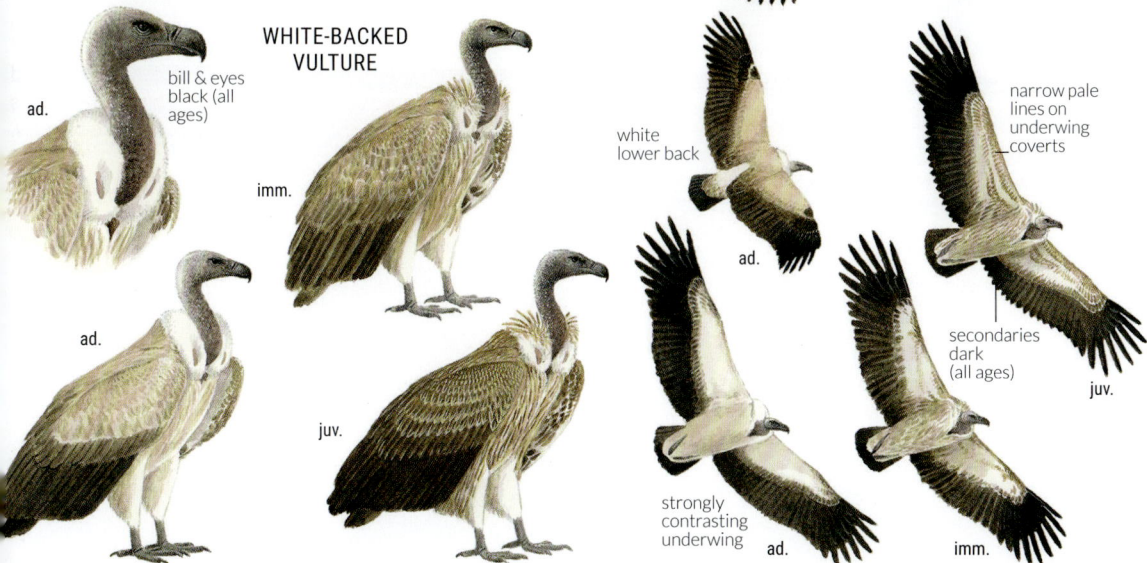

**WHITE-BACKED
VULTURE**

ad.

bill & eyes
black (all
ages)

imm.

white
lower back

narrow pale
lines on
underwing
coverts

ad.

ad.

juv.

secondaries
dark
(all ages)

juv.

strongly
contrasting
underwing

ad.

imm.

BEARDED VULTURE   *Gypaetus barbatus*

J F M A M J J A S O N D

110 cm; 5.2–6.2 kg; wingspan 2.6–2.8 m A very large, impressive vulture that has long, pointed wings and a long, wedge-shaped tail; restricted to mountainous areas. Ad. is mainly blackish-brown above, with white feather shafts giving it a faintly streaked appearance; head, breast, underparts and leggings gingery-orange; shaggy about the neck. The orange colour results from iron oxide pigmentation from soil and varies in intensity. Black facial mask ends in a short 'beard', visible only at close range. Sexes alike. Juv. is mainly dark brown with paler underparts; assumes ad. plumage through successive moults over 6 years. At all ages it is easily recognised in flight by its distinctive wing and tail shape. **Voice:** Mostly silent; a shrill whistling and chittering when displaying. **Status and biology:** NEAR-THREATENED. Uncommon resident, living in widely spaced pairs in the Lesotho highlands and adjacent Drakensberg; range has contracted in last century. Typically seen on the wing. Usually solitary, but small numbers of juvs sometimes gather at carcasses. Scavenges bones from carcasses; swallows small bones whole and breaks open large bones by dropping them onto rocks. (Baardaasvoël)

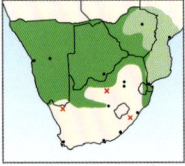

LAPPET-FACED VULTURE   *Torgos tracheliotos*

J F M A M J J A S O N D

98–115 cm; 4.5–8.5 kg; wingspan 2.6–2.8 m A very large, mainly black-plumaged vulture. Ad. has white leggings and a conspicuous white bar across the inner underwing coverts, which distinguishes it in flight from other dark-plumaged vultures. At close range, its deeply wrinkled, bare-skinned, dark pink head is diagnostic; bill is horn coloured. Sexes alike. Juv. is uniformly dark brown apart from white dappling on mantle; head pale pinkish. It is much larger than Hooded Vulture (p. 112), but in flight, without the benefit of a size comparison, they can be confused; best distinguished by its much broader head. **Voice:** Mostly silent; high-pitched whistle when displaying. **Status and biology:** ENDANGERED. Uncommon, in semi-arid savanna and desert. Arrives to scavenge at carcasses in ones and twos, occasionally in larger numbers; dominates other vultures at carcasses. Very susceptible to poisoning. Subject to vagrancy, with individuals known to move >1 000 km. (Swartaasvoël)

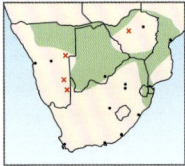

WHITE-HEADED VULTURE   *Trigonoceps occipitalis*

J F M A M J J A S O N D

75–85 cm; 3.3–5.3 kg; wingspan 2–2.3 m Strikingly coloured and patterned; mainly black plumage of ad. contrasts with the white head, belly and leggings. In flight from below, combination of white head, leggings and white line along the trailing edge of the underwing coverts is diagnostic. Female differs from male by having white inner secondaries and tertials. Angular head is white, with naked pink face; bill orange-red, with a pale blue cere. Juv. is uniformly dark brown with a dull white head; in flight from below shows a diagnostic (but rather indistinct) narrow white line along the trailing edge of the underwing coverts. **Voice:** Mostly silent; shrill whistles and chattering during interactions with other vultures. **Status and biology:** CRITICALLY ENDANGERED. Uncommon resident, living in open savanna in widely spaced pairs; single birds or pairs join other vultures at large carcasses. (Witkopaasvoël)

BEARDED VULTURE

ad.

imm.

ginger underparts

ad.

juv.

dropping bones

brown underparts

long wedge-shaped tail

juv.

pink face & massive bill

LAPPET-FACED VULTURE

ad.

juv.

juv.

white bar on underwing coverts

white leggings

ad.

angular, white head

ad. ♂

white body & leggings

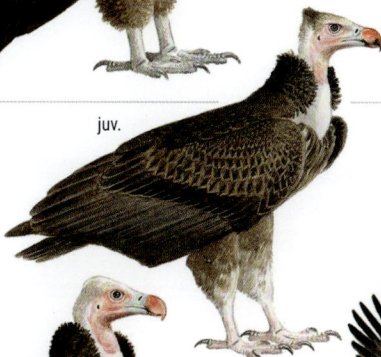

juv.

ad. ♀

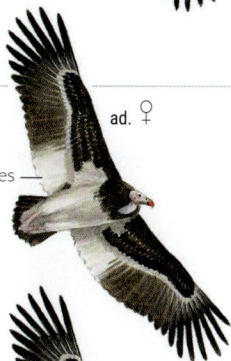

ad. ♀

white secondaries

juv.

ad. ♂

WHITE-HEADED VULTURE

narrow white line on underwing coverts

111

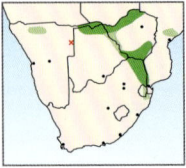

HOODED VULTURE   *Necrosyrtes monachus*

B B b b
J F M A M J J A S O N D

65–75 cm; 1.8–2.6 kg; wingspan 1.7–1.8 m A small, all-brown vulture with a small, 'beaky' head and slender bill. Ad. has crown and face bare-skinned, usually white, but flushing scarlet when agitated. At carcasses, its small size distinguishes it from other savanna vultures. In flight, the head shape is diagnostic and the flight feathers have a silvery edging, giving them a translucent appearance from below. Sexes alike. Juv. darker brown than ad., with greyish, down-covered head. Distinguished in flight from juv. Palm-nut Vulture by its slightly larger size and different head shape, and from juv. Egyptian Vulture by its rounded (not wedge-shaped) tail and broader wings. **Voice:** Mostly silent; gives a soft, whistling call at nest. **Status and biology:** CRITICALLY ENDANGERED. Uncommon to locally common resident in well-wooded savanna. Scavenges at large and small carcasses, often joining other vultures to feed; dozens and occasionally hundreds gather to feed on large carcasses. (Monnikaasvoël)

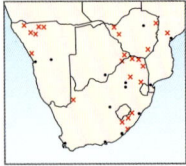

EGYPTIAN VULTURE   *Neophron percnopterus*

X X X X X X X X X X X X
J F M A M J J A S O N D

58–71 cm; 1.6–2.1 kg; wingspan 1.6–1.7 m A small, slender-billed vulture with a bare face and throat, and a long tail. Ad. is white, with black flight feathers; bare facial skin bright yellow, flushing orange when agitated; has a short crest on back of head. In flight, all-white, wedge-shaped tail and rather long, narrow wings are diagnostic. Sexes alike. Juv. is uniformly dark brown with a bare, grey-coloured face and throat. Wedge-shaped tail and long, very slender bill separate it from juv. Palm-nut and Hooded Vultures; much smaller than juv. Bearded Vulture (p. 110), with dark (not pale brown) underparts. **Voice:** Mostly silent; soft whistles, grunts and hisses when excited. **Status and biology:** Formerly a br. resident in E and W Cape; probably now extinct as a br. bird in the region, although occasional records of juvs in the company of ads suggest breeding may still occur. Scavenges on scraps as well as at mammal carcasses. (Egiptiese Aasvoël)

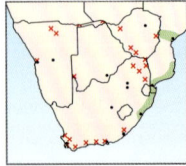

PALM-NUT VULTURE   *Gypohierax angolensis*

b B
J F M A M J J A S O N D

60 cm; 1.4–1.8 kg; wingspan 1.3–1.5 m A small, stocky, short-tailed vulture. Ad. head, body, shoulders and primaries white (but body feathers are often soiled), with contrasting black secondaries and a short, black, white-tipped tail. Eye is surrounded by pink-coloured bare skin, brightening when agitated. In flight, the strikingly patterned black-and-white wings are distinctive. Sexes alike. Juv. entirely brown in first plumage; short, rounded tail and heavy bill distinguish it from juv. Egyptian Vulture; large head and heavy bill distinguish it from juv. Hooded Vulture. Ad. plumage attained in 4 years; imms often appear mottled and scruffy, but told from juv. and imm. African Fish Eagle (p. 114) by white-tipped black tail (not white with black tip). **Voice:** Usually silent; 'kok-kok-kok' in flight. **Status and biology:** Very localised in the region; a small resident population restricted to east coast where the fruit of introduced *Raphia* palms provide its main food, although it also forages on carrion, scraps and small animals. Occasional birds move about widely, even reaching the S African west coast. Scattered records in the interior probably are vagrants from Angola and Zambia. (Witaasvoël)

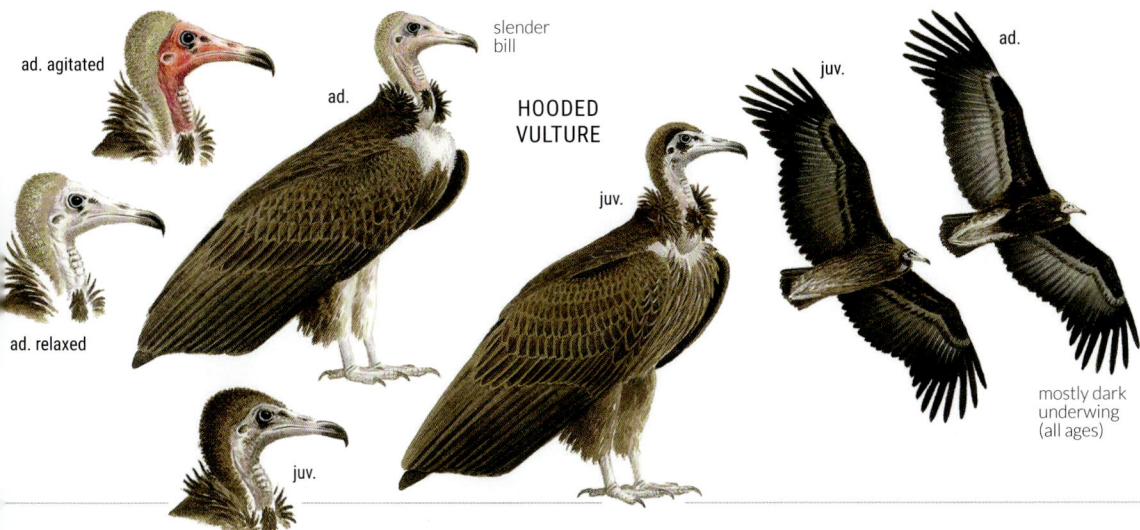

ad. agitated

ad.

slender bill

ad.

juv.

ad. relaxed

HOODED VULTURE

juv.

juv.

mostly dark underwing (all ages)

shaggy feathers

juv.

black flight feathers contrasting with white coverts

wedge-shaped tail (all ages)

juv.

imm.

ad.

ad.

EGYPTIAN VULTURE

juv.

juv.

imm.

ad.

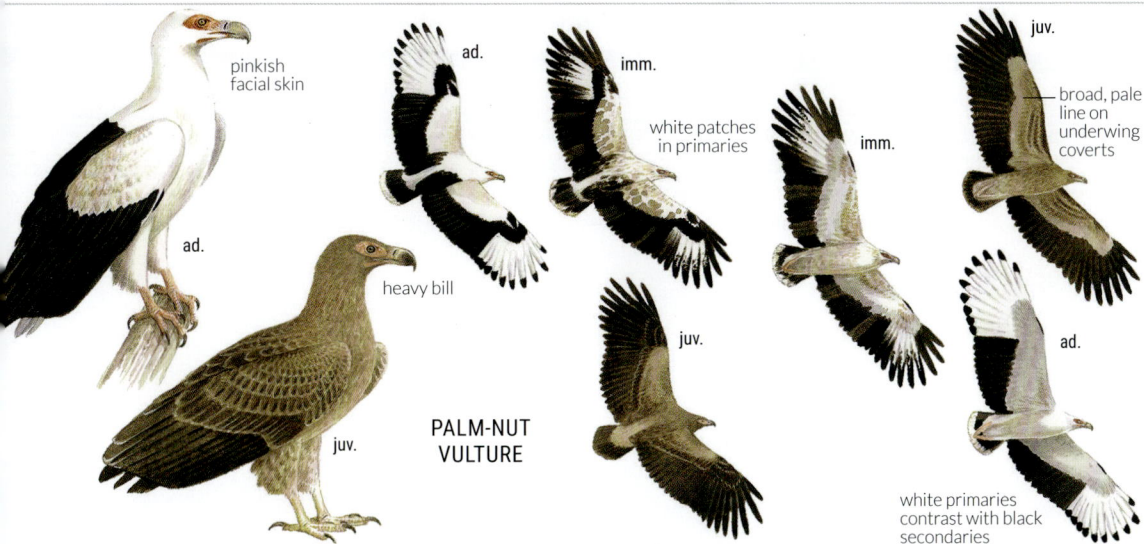

pinkish facial skin

ad.

imm.

white patches in primaries

imm.

juv.

broad, pale line on underwing coverts

ad.

heavy bill

juv.

PALM-NUT VULTURE

juv.

ad.

white primaries contrast with black secondaries

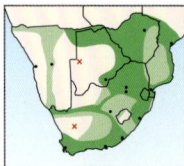

AFRICAN FISH EAGLE   *Haliaeetus vocifer*

63–73 cm; 2–3.8 kg; wingspan 1.9 m Ad. unmistakable; white head, breast and tail contrast with black and chestnut wings, back and underparts. Female larger and more extensively white on breast than male. Juv. is a motley looking brown-and-white streaked bird that is easily misidentified. The head is mostly white, streaked with brown, with a dark eye stripe; tail is distinctive – longer than ad's, white with a broad, dark brown terminal bar. Gradually neatens with age, taking 4–5 years to acquire ad. plumage. Facial mask may lead to confusion with Western Osprey, but is much larger, with broader wings, and white tail with dark terminal bar. **Voice:** Ringing *'kyow-kow-kow'* with head thrown back, given from perches or in flight; male's call is higher pitched. **Status and biology:** Common resident; pairs live year-round in territories on large rivers, lakes, estuaries and lagoons. Preys mainly on fish, but also scavenges at carcasses, robs nestlings from heronries, catches ducks and other birds, and steals fish and other prey from storks, pelicans and other waterbirds. (Visarend)

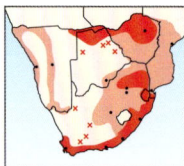

WESTERN OSPREY   *Pandion haliaetus*

52–68 cm; 1.25–1.85 kg; wingspan 1.5–1.7 m A slender, long- and narrow-winged raptor, usually associated with water, that is placed in its own subfamily. Its white underparts, dark brown upperparts and dark facial mask from the bill through the eye to the nape are distinctive. In flight, the pale underwing with black carpal patches is diagnostic at all ages. Female has a partial breast band (absent in male). Juv. similar to ad., but has pale-fringed upperpart feathers and dark-spotted crown. Facial mask may cause confusion with juv. African Fish Eagle, but Western Osprey has much narrower wings, white (not dark brown) thighs, and a brownish tail (not white with a broad, black terminal bar). **Voice:** Shrill whistles, but usually silent in s Africa. **Status and biology:** Uncommon Palearctic-br. migrant, mainly at coastal lagoons and estuaries, less frequently on inland waters (mainly in the north); regularly overwinters. One breeding record, KZN, 1963. Occurs solitarily, usually seen perched on a dead snag overlooking water. Feeds almost exclusively on fish, caught by plunging feet-first from up to 50 m in the air; sometimes hovers while hunting. Regional population <500 birds. Occurs widely in the Old and New worlds, but the Australasian population recently has been split as *P. cristatus*. (Visvalk)

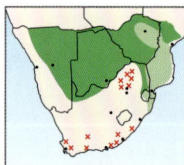

BATELEUR   *Terathopius ecaudatus*

55–70 cm; 1.9–2.9 kg; wingspan 1.7–1.8 m Ad. very distinctive and easily identified, both when perched and in flight; juv. could be confused with other brown eagles. Ad. has contrasting black, white and chestnut plumage and bare red face and legs. Back, rump and tail usually russet, but 5–10% have back and rump partially or wholly cream. At rest, rear crown feathers are elongated, giving a slightly crested appearance. Sexes differ in wing pattern: males have a broad, black terminal band along the trailing edge, females have mostly white flight feathers (visible at rest as a paler panel in the folded wing) with only a narrow, black terminal band. In flight, ads have similar underwing pattern to ad. Jackal Buzzard (p. 128); distinguished by more elongated primaries, white (not blackish) underwing coverts and breast, and shorter tail. Juv. is dark brown, slightly scaled and darker on the back, with milky-green feet and legs and pearl-grey face. Tail longer than ad., but still short compared to juv. snake eagles (p. 124) and other brown eagles (pp. 118–120); further distinguished from these species by its dark underwings, brown (not yellow) eye and bare, grey facial skin. Ad. plumage acquired after 6–7 years. By 4 years, imm. is still all-brown; feet and legs pink, face yellowish; female wing pattern (narrow, black trailing edge to white flight feathers) starts becoming visible. **Voice:** Usually silent; in aerial display gives a loud bark, *'kow-wah'*, and makes noisy wing beats. **Status and biology:** NEAR-THREATENED. Common resident in savanna and open woodland, although increasingly confined to large, protected areas due to poisoning in agricultural lands. Pairs occupy territories year-round; juvs wander widely. Often seen on ground at waterholes. Spends hours each day on the wing in search of prey, gliding rapidly on stiffly-held wings, 50–150 m above the ground. Flight largely dependent on air-lift, so they are essentially grounded on rainy days. Preys on small mammals, birds, reptiles and occasional frogs, fish and insects; also scavenges, eating roadkills and drowned creatures; juvs commonly join vultures to feed at large mammal carcasses. (Berghaan)

ad.

imm.

imm.
3–4 years

mottled,
blotchy
appearance
until attaining
adult plumage

ad.

broad wings

juv.

juv.

juv.

ad.

tail white
with dark
tip

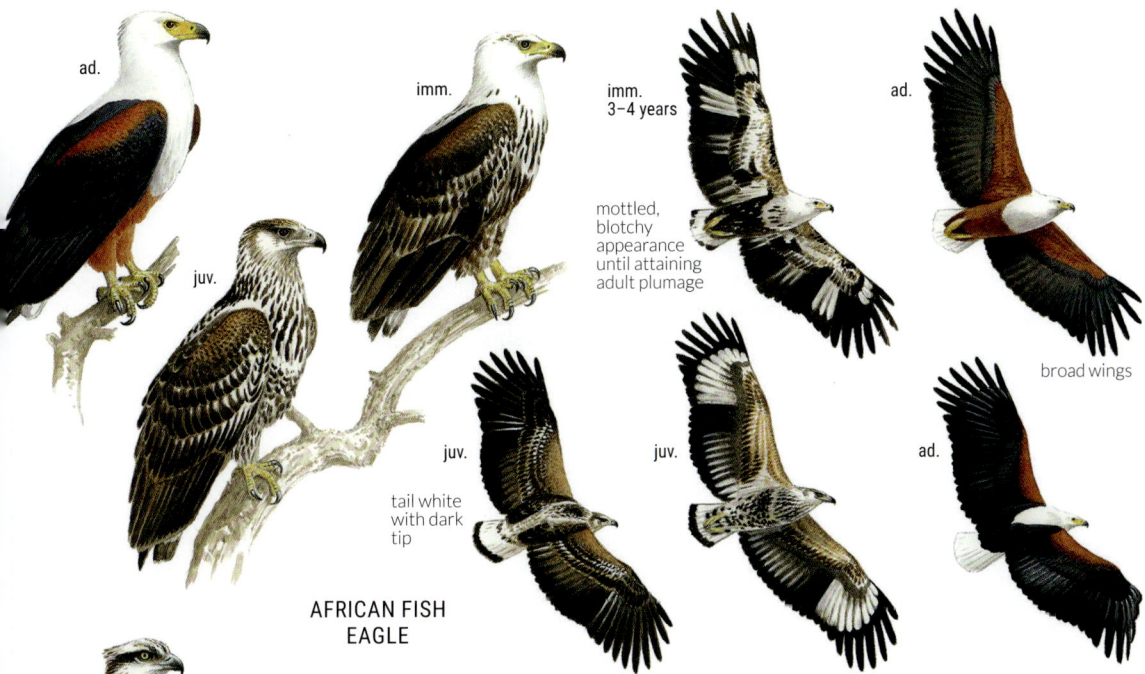

AFRICAN FISH
EAGLE

ad. ♀

juv.

ad. ♂

ad. ♂

ad. ♀

ad.

WESTERN
OSPREY

ad. ♀

dark
carpal
patch

long, angled
wings

BATELEUR

ad. ♂
cream variant

ad. ♂

ad. ♂

ad. ♀

ad. ♀

mostly
white
flight
feathers

broad black
trailing edges
to flight feathers

bare
facial
skin

juv.

juv.

imm. ♂
3–4 years

ad. ♂

juv.

ad. ♀

very short
tail often
with feet
projecting
(all ages)

LARGE EAGLES These 3 species are the largest eagles in the region; impressive raptors with long, broad wings, deeply slotted primaries and feathered legs. Sexes alike or nearly so; females larger than males. Juvs differ from ads in plumage, gradually acquiring ad. plumage through successive moults over 4–7 years. Once adulthood is reached, these species live in pairs in large territories, defended year-round. Nest sites are reoccupied annually, often for decades.

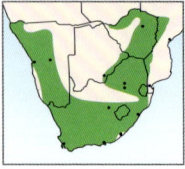

VERREAUX'S EAGLE   *Aquila verreauxii*

J F M A M J J A S O N D

80–96 cm; 3–5.6 kg; wingspan 2–2.8 m A large eagle, characteristic of mountainous country; in flight, has a very distinctive wing shape, broadest in the secondaries and tapering towards the body. At rest, ad. is coal-black with yellow talons and cere; the white rump, white 'V' on the back, and pale panels in the outer wing are usually only visible in flight. Sexes alike; female slightly larger with tail more wedge-shaped than in male. Juv. is mottled brown and buff with a diagnostic rufous crown and nape and a dark face and throat; attains ad. plumage in 4 years. **Voice:** Rarely calls, but melodious *'keee-uup'* when br. **Status and biology:** Locally common resident, especially in drier areas. Confined to hills and mountains where there are Rock Hyrax (its main prey) and nesting cliffs; occasionally nests on trees or pylons. Lives in pairs. Imms range widely before adulthood. (Witkruisarend)

MARTIAL EAGLE   *Polemaetus bellicosus*

J F M A M J J A S O N D

78–86 cm; 2.4–5.2 kg; wingspan 1.9–2.4 m A large, thickset eagle with a short crest. Ad. has dark brown head, upperparts and upper breast, with white underparts finely spotted with black. Much larger than ad. Black-chested Snake Eagle (p. 124), with broader wings and a shorter tail; underwings mainly dark brown (not white) and undertail finely barred (not white with a few dark bands). Female larger than male, and belly usually more spotted. Juv. much paler than ad., with mottled grey back and uniformly white head and underparts; told from juv. Crowned Eagle by its greyish-white (not creamy white) head and underparts, and finer dark barring on the flight feathers and tail. Imm. gradually attains ad. plumage over 5–7 years; dark brown eyes turn yellow after 3–4 years. **Voice:** Mostly silent; rapid *'klooee-klooee-klooee'* in display. **Status and biology:** VULNERABLE. Uncommon resident in savanna, grassland, semi-desert and Karoo. Population decreasing, even in large protected areas, but has benefited from powerlines providing breeding sites in treeless habitats such as the Karoo. Mostly encountered singly; spends much of the day soaring, sometimes at great heights; occasionally hovers. Preys on a variety of mammals, including hares, mongooses and small antelope, and large birds and reptiles, especially monitor lizards. (Breëkoparend)

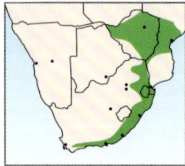

CROWNED EAGLE   *Stephanoaetus coronatus*

J F M A M J J A S O N D

80–98 cm; 2.6–4.2 kg; wingspan 1.5–1.8 m A large eagle with broad, rounded wings, a long, broad tail and a short crest. Ad. is grey-black above, with a rufous-brown head and neck and heavily barred and mottled underparts. In flight, shows rufous underwing coverts and pale flight feathers with 1–3 contrasting black bars; tail white with broad, black bars. Female larger than male and usually has darker underparts. Juv. has cream-coloured head and underparts and flanks have dark speckling; colour warmer than greyish-white juv. Martial Eagle, and has fewer, broader bars on flight feathers and tail. Imm. attains ad. plumage in 4 years. **Voice:** The most vocal eagle in the region, pairs calling year-round, often in unison. Calls in flight, a ringing *'kewee-kewee-kewee'*; male higher pitched than female. Nestlings and fledged chicks often call persistently, uttering a far-carrying *'queee-queee …'*. **Status and biology:** Locally fairly common resident in forest and mature woodland; sometimes quite tolerant of disturbance, with a large number of pairs breeding in suburban areas around Durban. Lives in pairs that maintain lifelong territories. Preys on mammals, especially hyraxes, monkeys and small antelope. (Kroonarend)

VERREAUX'S EAGLE

ad.

black face, buff crown

juv.

white back

ad.

long tail, tapering secondaries (all ages)

juv.

ad.

juv.

white panels

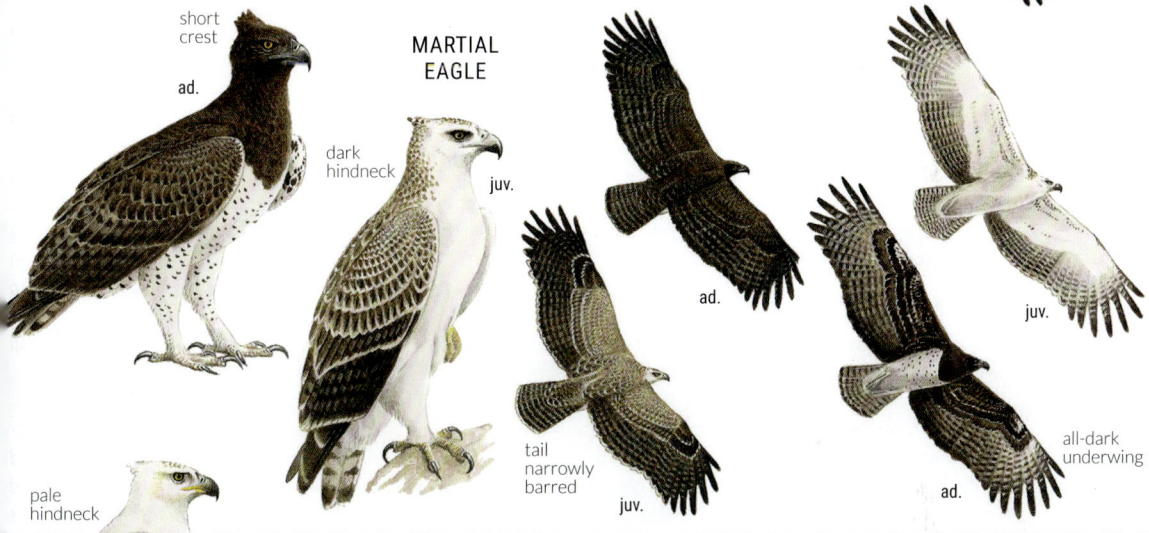

short crest

ad.

MARTIAL EAGLE

dark hindneck

juv.

ad.

juv.

tail narrowly barred

juv.

ad.

all-dark underwing

pale hindneck

juv.

juv.

ad.

ad. underwing barring is variable

ad.

CROWNED EAGLE

imm.

buff-coloured underparts

long, broadly barred tail

imm.

juv.

117

BROWN EAGLES

BROWN EAGLES Several species of mostly brown-plumaged eagles occur in the region, and identification is complicated by plumage variation and sex-linked size dimorphism (females larger than males). The larger species take several years to attain adult plumage, resulting in a series of transitional plumages, and some species exhibit plumage polymorphism. Smaller brown eagles can be confused with brown buzzards, especially larger species such as juvenile Jackal Buzzard (p. 128). At rest, check tail barring, whether or not the tarsi are feathered, eye colour, gape length and presence of a crest. In flight, observe the extent of white areas in the wing, rump and tail, as well as the proportions of the head, wing and tail. But note that wing and tail shape vary depending on flight mode (e.g. gliding versus soaring). Steppe and Lesser Spotted Eagles (p. 120) arrive in the austral summer. They are highly nomadic and often occur in flocks (10s to 100s), although their abundances vary greatly between years. Any gathering of brown eagles is likely to be one or more of these species. Most birds reaching s Africa are juvs and the distinctive wing patterns these birds display in flight provide the best means of identification.

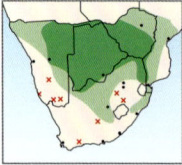

TAWNY EAGLE *Aquila rapax*

65–76 cm; 1.6–2.5 kg; wingspan 1.7–2 m A large, variably coloured brown eagle with a shaggy appearance, a heavy bill and long, broad wings. Most individuals are uniformly tawny coloured, but individuals range from uniformly pale buff to heavily streaked dark and light brown; the buff form is most prevalent in the arid west. Matches Steppe Eagle in size but has more massive bill and a shorter gape, extending only to below middle of eye (not to back, as in Steppe Eagle). Tail is unbarred or faintly barred (ad. Steppe Eagle's tail is more heavily barred). Eye pale yellow. Juv. initially ginger-brown, fading to buff; paler than juv. Steppe Eagle; eye brown. Has less prominent, pale wingbars and rump than juv. Steppe Eagle. Takes 4 years to acquire ad. plumage; transitional plumages intermediate between juv. and ad. **Voice:** Seldom calls; sharp bark, *'kyow'*. **Status and biology:** VULNERABLE. Fairly common resident, especially in semi-arid savannas; scarce in arid grasslands and eastern Karoo, numbers and range decreasing outside conserved areas due to persecution and poisoning. Pairs defend territories year-round; imms range widely before adulthood. Preys on a wide range of small animals, but also scavenges, commonly joining vultures at large mammal carcasses; often pirates food from other raptors. (Roofarend)

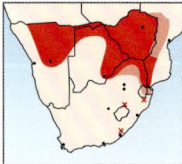

STEPPE EAGLE *Aquila nipalensis*

70–84 cm; 2–3.8 kg; wingspan 1.7–2.6 m A large, all-brown eagle; larger than Tawny Eagle but with a less massive bill that has a longer gape that extends to the back (not middle) of its eye. Sexes alike. Ad. dark brown with a finely barred tail (visible only at close range); eyes brown (yellow in Tawny Eagle). Most birds in s Africa are juvs or imms; these are paler than ad., with contrasting dark flight feathers. Juv. has pale windows in base of outer primaries, prominent pale bars along edge of upper- and underwing coverts, pale trailing edge to wing and tail, and pale uppertail coverts forming a distinct U-shaped band. Larger than juv. Lesser Spotted Eagle (p. 120) with broader wings and a longer tail; also has broader white lines on wing coverts. At rest, has shaggy leg feathers (not narrow, stovepipe legs of Lesser Spotted Eagle). **Voice:** Silent in Africa. **Status and biology:** ENDANGERED. Locally common Palearctic-br. migrant to savanna areas. Often in flocks, especially at termite emergences. (Steppe-arend)

ad. buff morph

light brown underparts

ad. streaked morph

pale eye, short gape

ad.

ad.

TAWNY EAGLE

ad.

ad.

juv. 'blond'

imm.

ad. dark

juv. 'blond'

juv. 'blond'

ad. dark

relatively small head

ad.

STEPPE EAGLE

ad.

oval nostril

long, fleshy gape

ad.

ad.

juv.

white lines on wings

heavy leggings

broad, white rump

juv.

broad, white covert line

juv.

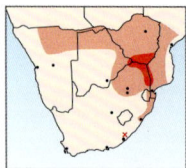

LESSER SPOTTED EAGLE *Clanga pomarina*

J F M A M J J A S O N D

58–65 cm; 1.1–2.1 kg; wingspan 1.3–1.7 m A rather small, brown eagle, only slightly larger than Wahlberg's Eagle. Sexes alike; most birds in the region are juvs or imms. Ad. uniformly brown, lacking the individual colour variation found in Wahlberg's Eagle; best distinguished by its thin, tightly feathered tarsi ('stovepipe' leggings) and lack of a crest. In flight it has broader, more rounded wings, a shorter, broader tail, and the underwing shows less contrast between coverts and flight feathers. When gliding, wings typically held bent down at the wrist. Juv. has pale tips to median and greater wing coverts, forming pale spots or bars on folded wing. In flight, juv. told from Wahlberg's Eagle by the narrow, white line along the edge of the upper- and underwing coverts, a white crescent-shaped rump and paler inner primaries. Juv. plumage is similar to juv. of larger Steppe Eagle (p. 118); best distinguished by its shorter tail, double comma mark in underwing covers, narrower white line across the underwing and less prominent white rump. **Greater Spotted Eagle** *C. clanga* claimed from Nylsvley, Limpopo, but record not confirmed. One satellite-tagged bird reached southern Zambia, but probably a hybrid with Lesser Spotted Eagle; Greater Spotted Eagle seldom reaches the equator. Greater Spotted Eagle is slightly larger and broader winged than Lesser Spotted Eagle, with dark brown plumage apart from a pale crescent along the edge of the primary underwing coverts (otherwise underwing coverts darker than flight feathers), and a pale 'C' on the lower rump at the base of the tail, similar to Steppe Eagle. Juv. has prominent white spots on the rump and wing coverts, forming a pale line along the edge of the greater coverts, and pale tips to the wing and tail feathers apart from the outer primaries. Imm. loses the white spots, but retains a pale line along the greater coverts and pale 'C' on lower rump. **Voice:** Silent in Africa. **Status and biology:** Erratic, but sometimes abundant, migrant to savanna areas between mid-Oct and Mar; presence seemingly linked to rainfall. Often flocks with Steppe Eagles, frequently seen eating termites on the ground; gathers in numbers to prey on nestling queleas. (Gevlekte Arend)

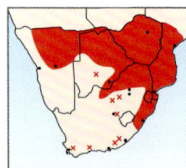

WAHLBERG'S EAGLE *Hieraaetus wahlbergi*

b B B b b
J F M A M J J A S O N D

55–60 cm; 900–1 500 g; wingspan 1.3–1.5 m A small, rather variably coloured eagle with a short crest on the hind crown (not always visible). Most individuals are uniformly warm brown; some are dark brown (appearing black at a distance), others are buff-brown or rufous-brown; 5–10% are pale morphs ('blondes'), having a creamy-white head and underparts, and buff wings and back. Rare individuals are two-toned, with dark heads and paler bodies, or vice versa, or a cream-coloured crown and dark brown body, or contrasting pale flecking on darker plumage. In perched birds, the small size, upright stance and flat-topped head with a short projecting crest are good distinguishing features. In soaring flight, the relatively long, rectangular wings and long tail (mostly held closed) are diagnostic, giving the appearance of 2 planks crossed at right-angles. In darker individuals the flight feathers are paler than the coverts, giving a two-tone effect. Eye brown; cere, gape and talons bright yellow. Juv. like ad., but with pale fringes to upperpart feathers and smaller crest. **Voice:** Drawn-out whistle while soaring; yelping *'kop-yop-yip-yip-yip'* when perched. **Status and biology:** Common intra-African migrant, breeding in s Africa in summer (Aug–Apr) in well-wooded savanna. Found in widely dispersed pairs, except during migration. Preys on birds, small mammals, reptiles, frogs and invertebrates. (Bruinarend)

BOOTED EAGLE *Hieraaetus pennatus*

b b B b b
J F M A M J J A S O N D

45–55 cm; 520–1 200 g; wingspan 1.1–1.3 m A small, unobtrusive, variably coloured eagle; all colour morphs have diagnostic white 'landing lights' on the leading edge of the wings where they meet the body. Ad. pale morph most common (80%): light brown head and upperparts contrast with creamy-white underparts; in flight, shows strong contrast between whitish underwing coverts and dark flight feathers. Dark morph is rather uniformly dark brown with slightly paler inner primaries visible in flight. Rare rufous morph has rufous underparts. In silhouette, told from Wahlberg's Eagle by its broader wings and shorter, broader tail. Juv. like respective ad. morph but upperwing coverts, secondaries and tail feathers have paler edges. Easily confused with juv. Ayres's Hawk-Eagle (p. 122), but flight feathers are much less heavily barred when viewed from below. **Voice:** Vocal when nesting, a high-pitched *'kee-keeee'* or *'pee-pee-pee-pee'*; otherwise silent. **Status and biology:** Occurs widely, but sparsely outside of br. range (W Cape to s Namibia). Locally common (Jul–Mar) in nesting areas, rare elsewhere; local population augmented in summer by Palearctic breeders. Unobtrusive and easily overlooked, favouring perches on cliffs or trees; usually encountered solitarily. Hunts on the wing, sometimes in pairs, preying mainly on birds, but also reptiles, small mammals and termite alates. (Dwergarend)

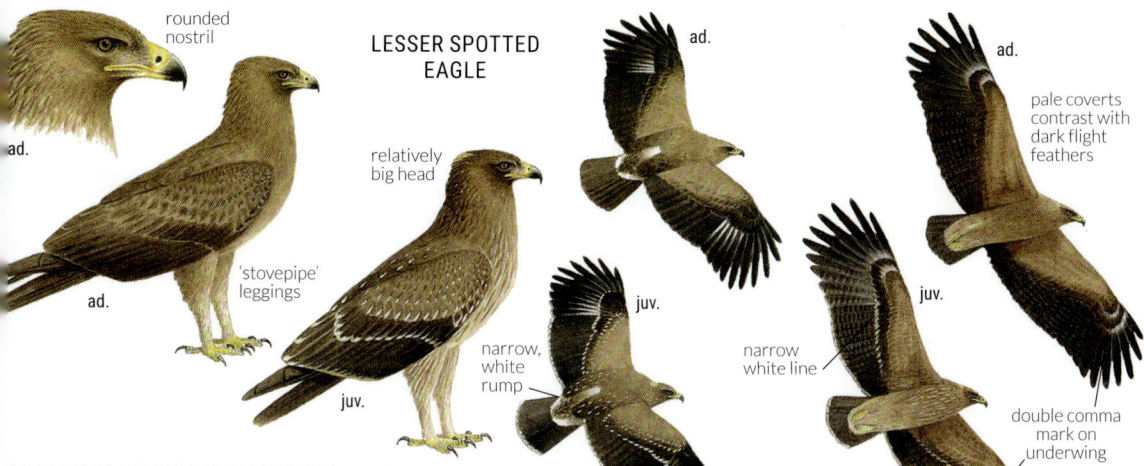

rounded
nostril

**LESSER SPOTTED
EAGLE**

ad.

ad.

pale coverts
contrast with
dark flight
feathers

ad.

relatively
big head

ad.

'stovepipe'
leggings

narrow,
white
rump

juv.

juv.

narrow
white line

juv.

double comma
mark on
underwing

**GREATER
SPOTTED EAGLE**

heavily
spotted
above

juv.

juv.

ad.

dark coverts; no
contrast with flight
feathers

upperparts
with distinct
line & heavy
spots

very dark
brown
plumage

ad. dark
morph

**WAHLBERG'S
EAGLE**

short crest

ad. medium
morph

ad.
dark-headed
morph

ad. cream-
crowned
morph

long,
rectangular
wings

ad.
dark
morph

ad. pale
morph

ad.
medium
morph

large colour
variation

ad. pale
morph

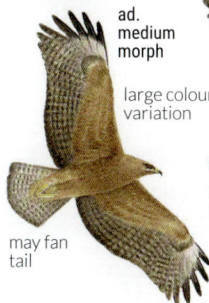

long,
narrow tail

may fan
tail

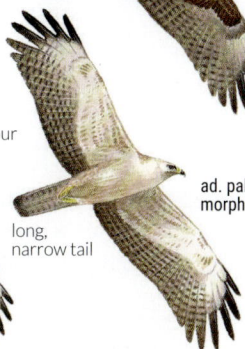

ad. pale
morph

ad. dark
morph

ad. showing
'landing lights'

distinctly pale
upperwing
coverts

ad.

ad. pale
morph

ad. dark
morph

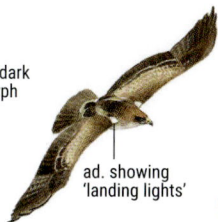

**BOOTED
EAGLE**

SMALLER EAGLES

SMALLER EAGLES Three small to medium-sized eagles with feathered tarsi. Long-crested Eagle is a sluggish species that spends much of its time perched. The other 2 are dashing hunters that catch much of their prey while on the wing. In addition to age-related plumage differences, identification of Ayres's Hawk-Eagle is complicated by several colour morphs.

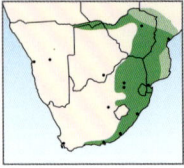

LONG-CRESTED EAGLE *Lophaetus occipitalis*

b b b b b B B B B B B b
J F M A M J J A S O N D

52–58 cm; 810–1 300 g; wingspan 1.1–1.3 m A small, brownish-black eagle with a long, floppy crest. In flight, has conspicuous white panels in the outer wings, visible from above and below, recalling Verreaux's Eagle (p. 116), but much smaller than that species and back is black (not white). Flight action is fast and direct on stiffly held wings, with shallow wing beats. Male has white leggings, female and juv. have brownish leggings. Ad. and juv. similar, differing in eye colour (yellow in ad., grey in juv.) and length of crest (shorter in juv.). **Voice:** High-pitched, screamed 'kee-ah' during display or when perched. **Status and biology:** Common resident in mature woodland, plantations and along forest edges, especially near water. Lone vagrants are regularly found outside the normal range. Lives in pairs, but usually encountered solitarily, often perched on a tall, dead tree, pole or other vantage point overlooking open ground. Preys mainly on rodents, hunting them from such perches; also takes reptiles, frogs, invertebrates and occasional birds. (Langkuifarend)

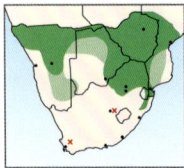

AFRICAN HAWK-EAGLE *Aquila spilogaster*

b B B b
J F M A M J J A S O N D

60–68 cm; 1.2–1.7 kg; wingspan 1.3–1.6 m A medium-sized, black-and-white eagle. Ad. is brownish-black above and white below, breast streaked with black. Sexes similar, but female typically is more heavily marked than male. In flight, primaries and secondaries are plain white with a broad, black trailing edge; from above has diagnostic whitish primary bases that form contrasting white panels in the outer wings. Larger than Ayres's Hawk-Eagle, with more boldly patterned wings and tail. Juv. brown above and rufous below, becoming black-streaked with age; easily confused with juv. Black Sparrowhawk (p. 142) but is larger, with longer wings and a shorter tail; also has pale panels in outer wing (smaller than in ad.); if seen close-up, the feathered tarsi are diagnostic. **Voice:** Seldom calls; whistled, musical 'klee-klee-klee'. **Status and biology:** Fairly common resident in well-wooded savanna, extending marginally into arid areas. Lives in widely spaced pairs which defend permanent territories; pairs frequently perch together and hunt cooperatively. A rapacious hunter, it preys mainly on gamebirds, as well as small mammals and other birds. (Grootjagarend)

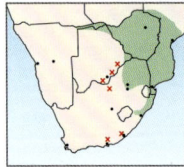

AYRES'S HAWK-EAGLE *Hieraaetus ayresii*

B B b b b
J F M A M J J A S O N D

45–58 cm; 620–1 100 g; wingspan 1.2 m A small, dashing eagle. Ad. shows much individual variation, from an all-dark morph (rare) to individuals with lightly spotted underparts; commonest morph is intermediate; some individuals have a white head or a pale supercilium. Sexes alike. Resembles a diminutive African Hawk-Eagle; when perched, best distinguished by its well-marked underparts and leggings; in flight, by absence of white outer panel in wings and more uniformly barred tail; heavily barred underwing distinguishes the dark morph from the dark morph of the Booted Eagle (p. 120). Juv. has pale rufous head and underparts; in flight the heavily barred underwing distinguishes it from similarly coloured juv. Booted Eagle and African Hawk-Eagle. **Voice:** Normally silent; shrill 'pueep-pip-pip-pueep' when displaying. **Status and biology:** Rare br. resident in the north, with non-br. ranging further south, mainly in late summer (Jan–May), although recently discovered breeding in n KZN. Usually solitary; br. pairs occur in mature woodland and forest edge. A rapacious predator, often seen in steep heart-shaped stoop as it hunts prey, which is mostly comprised of birds, especially doves and pigeons, and occasional small mammals. (Kleinjagarend)

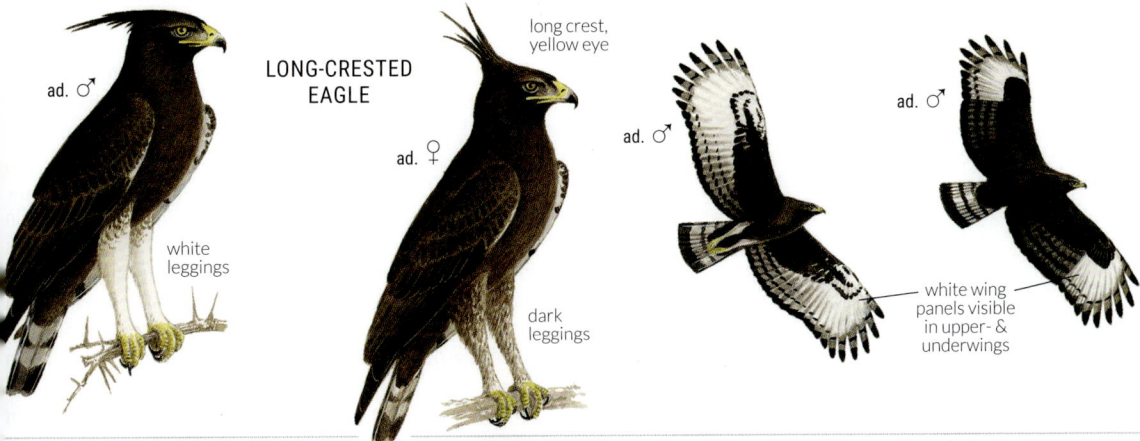

ad. ♂

LONG-CRESTED EAGLE

white leggings

long crest, yellow eye

ad. ♀

dark leggings

ad. ♂

ad. ♂

white wing panels visible in upper- & underwings

ad.

AFRICAN HAWK-EAGLE

white leggings

white wing panels

ad.

ad.

imm.

broad subterminal band

juv.

rufous underparts

juv.

juv.

ad.

AYRES'S HAWK-EAGLE

heavily spotted below

ad. dark morph

ad. pale morph

spotted leggings

juv.

no white wing panels

white 'landing lights' often evident in flight

ad.

ad.

juv.

imm.

flight feathers heavily barred (all ages)

SNAKE EAGLES
Large raptors with unusually large, broad heads and large, yellow eyes. Sit upright when perched. Labelled as eagles due to their size, but more closely related to kites, and previously were called harrier-eagles. They eat primarily snakes, which are swallowed whole; other reptiles and frogs are also eaten. Their unfeathered legs are armour-plated with hard, round scales that protect them from snake bites. Sexes alike, with very little size dimorphism; juvs and imms differ in plumage from ads.

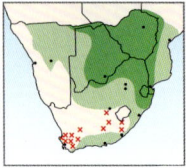

BROWN SNAKE EAGLE *Circaetus cinereus*

70–75 cm; 1.6–2.4 kg; wingspan 1.7 m A large, uniformly dark brown eagle with striking yellow eyes, well-defined barring on the tail, and pale grey, unfeathered legs. In flight, from below, the dark brown underwing coverts contrast with the unbarred white flight feathers, distinguishing it from other brown-plumaged eagles. Sexes alike. Juv. dark brown like ad. but has some pale edging to back feathers and wing coverts giving it a slightly scaled appearance; also has scattered white flecks on underparts from feather bases showing through. Could be confused with juv. Black-chested Snake Eagle, but darker brown, with a more strongly barred tail; similar to juv. Bateleur (p. 114), but has a longer, barred tail and pale eyes. **Voice:** Generally silent; croaking 'hok-hok-hok-hok' flight call. **Status and biology:** Largely sedentary, but marked individuals known to move >2 000 km. Frequents well-wooded savanna, extending marginally into semi-arid areas. A solitary species when not br., lone birds typically encountered perched on high vantage points, especially dead trees and powerline pylons. (Bruinslangarend)

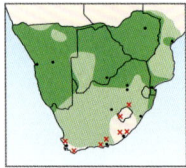

BLACK-CHESTED SNAKE EAGLE *Circaetus pectoralis*

63–68 cm; 1.2–2.2 kg; wingspan 1.7 m A medium-sized, dark brown-and-white eagle with typical snake eagle features of large head, large yellow eyes and bare, pale grey legs at all ages. From a distance, ad. may be mistaken for ad. Martial Eagle (p. 116), but white underparts are plain (not spotted); in flight, has mostly white (not dark brown) underwings. Similar to Brown Snake Eagle when seen from behind, but white underparts are diagnostic. Sexes alike. Juv. in first plumage is uniformly buff-brown, paler than juv. Brown Snake Eagle, with pale edging to body feathers giving scaled appearance. Acquires partial ad. plumage with progressive moults, showing some barring on flight feathers, mottled dark feathers on the head and back, and white belly, blotched with brown. **Voice:** Rarely calls; melodious, whistled 'kwo-kwo-kwo-kweeu'. **Status and biology:** Widespread from mesic and semi-arid savanna to desert; resident in some areas, nomadic or a seasonal visitor in others. Usually solitary when not br.; sometimes in loose aggregations of dozens of birds, roosting communally or hunting aerially within sight of each other. Commonly hunts by hovering 50–100 m above the ground for minutes, then parachuting down onto prey; snakes are usually carried into the air and swallowed whole in flight. (Swartborsslangarend)

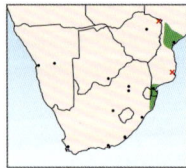

SOUTHERN BANDED SNAKE EAGLE *Circaetus fasciolatus*

54–58 cm; 920–1 100 g; wingspan 1.2 m A smallish, buzzard-like raptor. Main features are its large, rounded head, cinnamon-brown chest and strongly barred underparts, underwing flight feathers and tail. Ad. and juv. distinguished from similar Western Banded Snake Eagle by having 4–6 (not 1) pale tail bars, with narrower dark terminal band. Legs and eyes yellowish. Sexes alike. Juv. is dark brown above and pale below, with dark streaks on the head, neck and upper breast; tail is more finely barred than ad. **Voice:** Harsh 'crok-crok-crok' and high-pitched 'ko-ko-ko-ko-keear'. Often calls, especially in early morning. **Status and biology:** NEAR-THREATENED. Uncommon resident in eastern lowlands, frequenting forest, woodland and plantations, often near water. Unobtrusive and easily overlooked; usually solitary, perches quietly for long periods, often concealed within the foliage of a tree. (Dubbelbandslangarend)

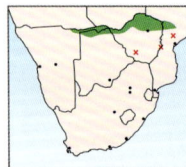

WESTERN BANDED SNAKE EAGLE *Circaetus cinerascens*

56–60 cm; 1–1.2 kg; wingspan 1.2 m A smallish, buzzard-like raptor. Very like Southern Banded Snake Eagle, but with a darker belly, darker underwing coverts and a diagnostic single, broad, white central band across the dark tail, resulting in a broader terminal tail band, visible both in flight and at rest. Legs and eyes yellowish. Sexes alike. Juv. is pale below, with a dark-streaked head, neck and breast; best told from juv. Southern Banded Snake Eagle by broad, grey tail tip. Imm. is dark brown, with a broad, grey tail band. **Voice:** Loud, descending cawing notes; also high-pitched 'kok-kok-kok-kok-kok'; a fairly vocal species, calling from perches and in flight. **Status and biology:** Uncommon, sedentary, and largely confined to riparian woodland associated with large rivers and wetlands. Occurs singly or in pairs; typically perches for long periods below the canopy of a tall riverine tree. Swoops down on prey, mainly snakes, which are brought back to a perch to eat. (Enkelbandslangarend)

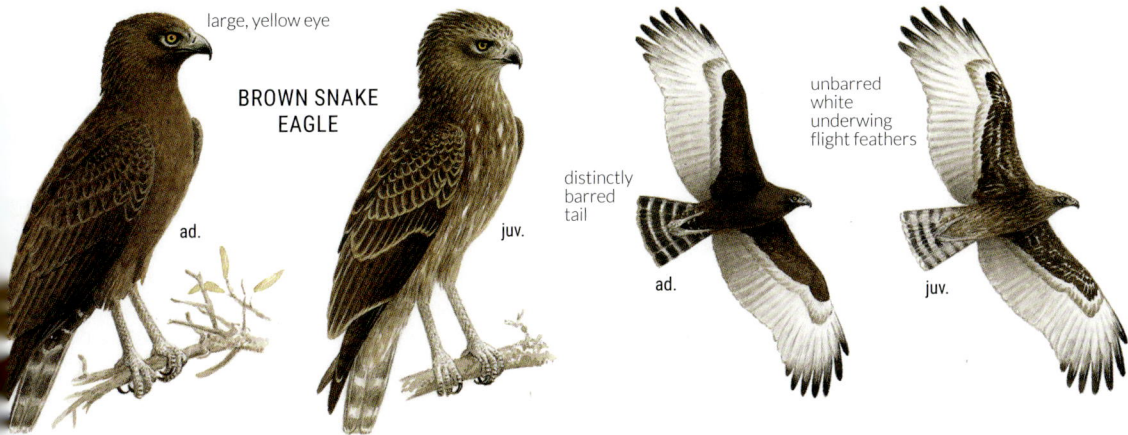

large, yellow eye

BROWN SNAKE EAGLE

ad.

juv.

distinctly barred tail

ad.

unbarred white underwing flight feathers

juv.

ad.

juv.

ad.

barred, white underwing flight feathers

imm.

juv.

BLACK-CHESTED SNAKE EAGLE

ad.

juv.

barred chest & belly

ad.

3 dark tail bars

ad.

juv.

heavily barred underwing

SOUTHERN BANDED SNAKE EAGLE

tail pattern differences between two species retained (all ages)

WESTERN BANDED SNAKE EAGLE

ad.

juv.

mostly dark breast & belly

broad white tail bar

ad.

ad.

juv.

125

MILVUS KITES

MILVUS KITES Long-winged raptors with long, forked tails and distinctive loose flight action, regularly twisting their long, shallowly forked tails. Wings held level, not canted up as in harriers, and are narrower than buzzards', with bent wrists. Widespread generalists, allied to fish eagles. The 2 species in s Africa are treated as races by some authorities, but are genetically distinct and at least ads are readily separated. Sexes alike, females slightly larger than males.

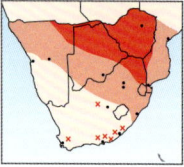

BLACK KITE *Milvus migrans*

J F M A M J J A S O N D

51–60 cm; 650–920 g; wingspan 1.35–1.7 m Marginally larger than Yellow-billed Kite, with a paler, grey head, black (not yellow) bill (but cere yellow), pale yellow (not dark reddish) eyes and a less deeply forked tail. Juv. has darker brown face than ad., with a pale-streaked neck and breast, buffy feather margins to back, brownish eyes, and slightly less forked tail. **Voice:** Usually silent in winter quarters. **Status and biology:** A locally common, Palearctic-br. summer visitor, Oct–Mar. Occurs in a wide range of habitats from forest edge to semi-desert. Commonly in flocks, often mingling with Yellow-billed Kites; regular at termite emergences. (Swartwou)

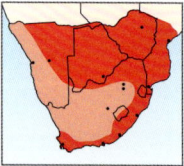

YELLOW-BILLED KITE *Milvus aegyptius*

b B B b b
J F M A M J J A S O N D

50–58 cm; 570–760 g; wingspan 1.3–1.7 m Slightly smaller than Black Kite; ad. has a dark brown (not greyish) head and bright yellow (not black) bill; eyes dark reddish (not pale yellow); tail more deeply forked. Juv. has buffy feather margins, black bill (cere yellowish), and less deeply forked tail than ad.; separated with difficulty from juv. Black Kite. **Voice:** High-pitched, shrill whinnying; seldom calls when not br. **Status and biology:** Common, intra-African br. summer visitor. Occurs singly, in pairs or in flocks; widely distributed in habitats ranging from forest edge to grassland; common in densely populated rural settlements; absent only from very arid areas. Feeds mainly by scavenging from the wing, swooping to take food items from the ground, or catching it in the air. Often fearless, snatching food from unsuspecting fishermen and picnickers. Scavenges roadkills and is attracted to fires and termite emergences. (Geelbekwou)

HONEY BUZZARDS

HONEY BUZZARDS Despite their common name, the honey buzzards (*Pernis*) are not closely related to the *Buteo* buzzards. Sexes alike, with little size dimorphism, but there is considerable individual variation in plumage. Separation from buzzards relies largely on structure.

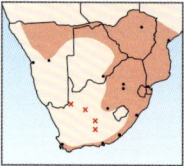

EUROPEAN HONEY BUZZARD *Pernis apivorus*

J F M A M J J A S O N D

52–60 cm; 680–810 g; wingspan 1.25–1.45 m A small-headed, long-necked, buzzard-like raptor; its variable plumage often results in it being mistaken for other raptors, especially Common Buzzard (p. 130), juv. African Harrier-Hawk (p. 146) or a small eagle. At rest, its large, bright yellow eyes, thin, 'beaky' face, weak bill, grey cere and long neck all assist identification. Tail pattern is diagnostic: pale greyish-brown with a broad, black subterminal bar edged with white and 2 narrower black bars close to the body (inner one mostly obscured by tail coverts). In flight, appears rather long-tailed and has loose flight action; glides with wings bowed down at wrist. Ad. plumage ranges from all blackish or brown to having white underparts barred dark brown or rare rufous morph. Sexes virtually alike. Juv. equally variable but has less distinct tail barring; eyes dark brown; cere yellow; best told from Common Buzzard by shape. Could be confused with juv. African Harrier-Hawk (p. 146), but wings are not as broad and outer primaries are less deeply slotted. One bird north of Maputo, Mozambique in Mar 2018 suspected to be Crested Honey Buzzard *P. ptilorhynchus* or possibly a hybrid with this species; Crested Honey Buzzard is more bulky and broader winged, with 6 (not 5) primary 'fingers' shown in flight. Underwing lacks dark carpal patches. Flight feathers show one additional bar in flight; male has only one dark band at the base of the tail. Ad. has dark eyes and well-defined throat pattern. **Voice:** Generally silent; occasionally gives a high-pitched *'meeuu'*. **Status and biology:** Uncommon, Palearctic-br. summer visitor, Nov–Apr; occurs solitarily except on passage. Appears to have increased in numbers in recent years in the subregion, possibly due to a decrease in suitable overwintering habitat further north in Africa, thus forcing birds further south. An unobtrusive bird that frequents tall woodland, forest edge and plantations. Perches quietly for long periods in well-foliaged trees; often confiding. Attracted to wasps' nests, preying on the larvae and pupae. (Wespedief)

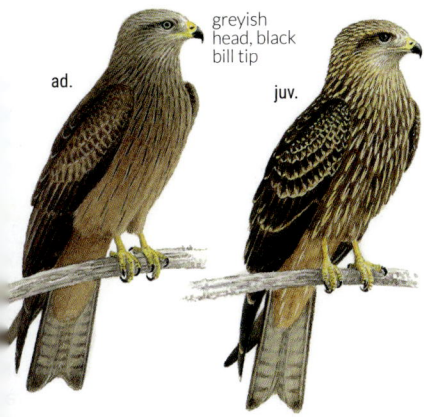

ad.

greyish head, black bill tip

juv.

BLACK KITE

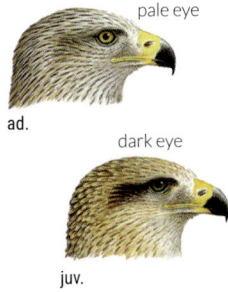

pale eye

ad.

dark eye

juv.

ad.

juv.

shallowly forked tail

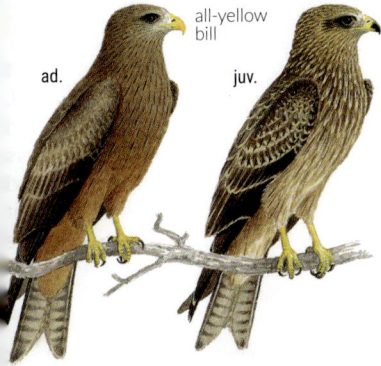

ad.

all-yellow bill

juv.

YELLOW-BILLED KITE

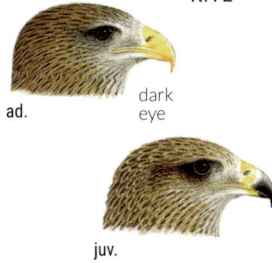

ad.

dark eye

juv.

ad.

juv.

tail deeply forked

ad. ♀ dark

ad. ♂ rufous-barred

EUROPEAN HONEY BUZZARD

ad. ♂ pale

dark, oval carpal patches

ad. ♂

ad. ♀

ad. ♂ pale

juv. pale

heavily barred underwings

ad. ♂ barred

ad. ♀ dark

uv. dark

juv. mid

small head; weak bill

juv. pale

juv. dark

juv. mid

juv. dark

BUZZARDS

Collectively referred to by their generic name as 'Buteos', buzzards are thick-set, medium-sized hawks with broad wings and rounded tails. Three other species named buzzards (European Honey Buzzard p. 126, Grasshopper Buzzard p. 132, and Lizard Buzzard p. 138) are unrelated to the genus *Buteo*. All have bare tarsi, unlike most eagles. Jackal and Augur Buzzards are sister-species that have almost non-overlapping ranges; they may hybridise occasionally at the contact zone and are treated by some authorities as subspecies. Both are largely sedentary; br. residents in s Africa. Usually single or in pairs. Ads of both species occur in several colour morphs; juvs are mostly brown, and are best told from smaller brown buzzards (p. 130) by their broader wings and larger size; they acquire ad. plumage in 2 years.

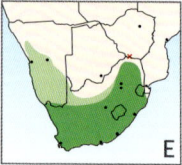

JACKAL BUZZARD *Buteo rufofuscus*

J F M A M J J A S O N D

55–62 cm; 900–1 700 g; wingspan 1.2–1.4 m A large buzzard with a short, rounded tail and broad wings, widest in the secondaries. Ad. has black head, back and upperwing coverts. Underparts variably marked: most have a chestnut breast and blackish underparts and underwing coverts mottled and barred with white; tail uniformly rufous. Sexes similar in plumage: female has mottled black throat and upper breast; male has rufous or white on breast extending to throat. Individuals occur with mainly black breast and belly, others more extensively chestnut below. Occasional birds with all-white underbody, these distinguished (in flight) from ad. Augur Buzzard by having blackish (not white) underwing coverts. In flight, ad. can be mistaken for ad. female Bateleur (p. 114) but dark underwing coverts distinguish the buzzard, together with different shape and longer tail. Juv. is brown above; in fresh plumage, pale feather edges result in a slightly scaled appearance. Underparts uniformly warm rufous, sometimes lightly streaked; tail rufous-brown, finely barred dark brown. Juv. easily mistaken for brown morph Common Buzzard (p. 130) or rufous morph Long-legged Buzzard (p. 132); best told by broad wings and large, eagle-like head. During transition to ad. plumage, underparts, back and wings become mottled with ad. feathers. **Voice:** A yelping '*weeaah-ka-ka-ka*', similar to call of Black-backed Jackal; male higher pitched than female; also a mewing call. **Status and biology:** Endemic, locally common resident in Karoo scrub, grassland and agricultural land; most numerous in hilly and mountainous terrain. Mainly sedentary, with some local movement of ads and juvs over a few hundred km. Solitary except when br., lone birds commonly encountered perched on utility poles alongside roads. When hunting, frequently hovers or hangs motionless on updrafts along ridges of hills; eats mostly small mammals and reptiles. (Rooiborsjakkalsvoël)

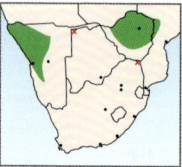

AUGUR BUZZARD *Buteo augur*

J F M A M J J A S O N D

55–60 cm; 900–1 300 g; wingspan 1.2–1.5 m A striking, black-and-white buzzard with a chestnut-coloured tail. Ad. told from ad. Jackal Buzzard by white (not black-and-rufous) underparts. In flight, the largely white underwing (including underwing coverts) distinguishes it from the white-breasted form of Jackal Buzzard. Smaller than Black-chested Snake Eagle (p. 124), with plain (not barred) flight feathers and rufous (not black-and-white-barred) tail. Sexes similar, but female has a black throat, male a white throat. A black-fronted (melanistic) morph occurs in E Africa, but has not been recorded in s Africa. Juv. has a dark brown head, back and wing coverts; underparts whitish to buff, variably streaked with scattered dark brown and black feathers. Appears much whiter below than juv. Jackal Buzzard and flight feathers from below are more extensively barred. Larger and broader winged than Common Buzzard (p. 130), with paler underwing coverts. **Voice:** A loud, barking '*kyaah-ka-ka-ka-ka*' or '*kow-kow-kow-kow*', quite different from that of Jackal Buzzard. **Status and biology:** Locally common resident of mountainous and hilly country in wooded savanna and semi-desert. Solitary or in pairs; usually perches on high vantage points from where it hunts for reptiles, small mammals and birds. (Witborsjakkalsvoël)

ad. ♂

ad. ♀

ad.

ad.

flight feathers
contrast with
black coverts

underpart
coloration
highly variable

ad. ♂

JACKAL BUZZARD

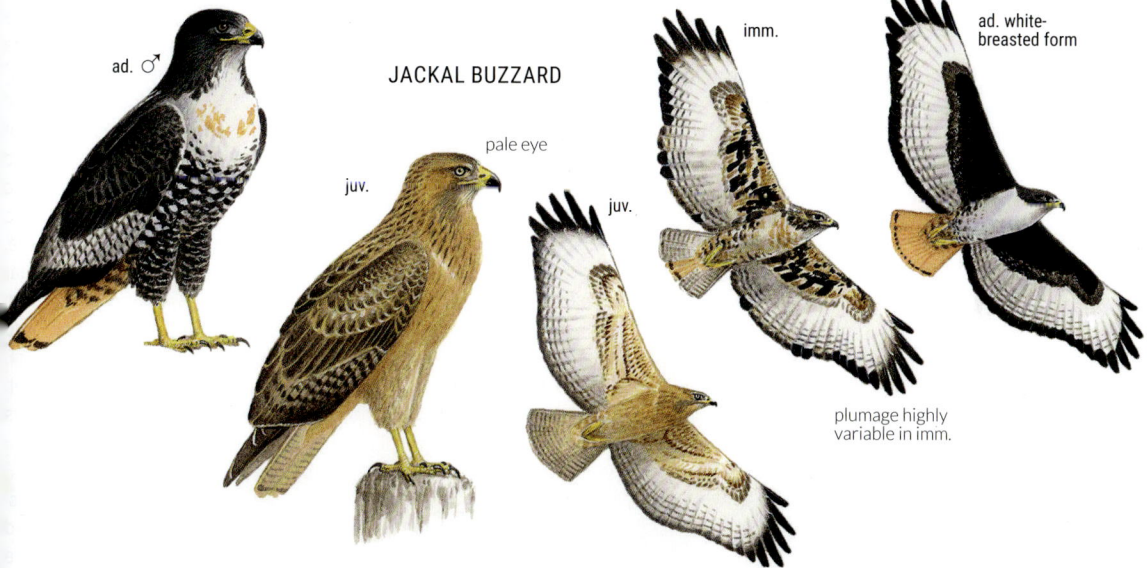

pale eye

imm.

ad. white-
breasted form

juv.

juv.

plumage highly
variable in imm.

AUGUR BUZZARD

ad.

ad.
typical coloration

entirely white
underparts

ad. ♀

juv.

white
underwing
coverts

ad. ♂

juv.

129

BROWN BUZZARDS
These mostly brown-coloured buzzards are among the most difficult of raptors to identify due to considerable variation in plumage within species. Beware also juv. Jackal and Augur Buzzards (p. 128) and European Honey Buzzard (p. 126). The widespread Common Buzzard is the most abundant species in summer; the slightly smaller Forest Buzzard is seldom encountered outside its restricted range, but confusion now caused by Common Buzzards breeding in the W Cape.

COMMON (STEPPE) BUZZARD *Buteo buteo*

b b b | | | | | | **b b b**
J F M A M J J A S O N D

46–52 cm; 540–920 g; wingspan 1.1–1.35 m The most common brown buzzard in the region, mainly in summer; told from small eagles by its bare yellow legs. Sexes alike, but plumage is highly variable, from pale brown or greyish-brown to rufous or almost black. In flight, brown underwing coverts contrast with paler flight feathers. Most, but not all birds, have a diffuse white breast band, resembling ad. Forest Buzzard, but typically have barred (not spotted or blotched) thighs, flanks and sides of breast. Juv. like ad. but more streaked below with yellowish (not brown) eyes and narrower terminal tail bar. Smaller than juv. Jackal and Augur Buzzards (p. 128); in flight, wings narrower and less contrast between lesser and other underwing coverts. **Voice:** Gull-like '*pee-ooo*'; seldom calls in Africa. **Status and biology:** Common Palearctic-br. summer migrant, mainly Oct–Mar, with occasional birds overwintering. Mostly found in open country, avoiding very arid and forested areas. Solitary, except during migration when hundreds may move in loose groups using updrafts along mountain slopes. Frequently perches on utility or fence poles. Hunts mainly insects, rodents and reptiles from perches, but also soars and hovers. Some pairs of all-brown buzzards br. in W Cape apparently are this species. (Bruinjakkalsvoël)

FOREST BUZZARD *Buteo trizonatus*

| | | | | | | **b B B b**
J F M A M J J A S O N D

45–50 cm; 510–700 g; wingspan 1.05–1.25 m Slightly smaller and more slender than Common Buzzard; most are whiter below, with irregular brown blotches (not bars) on the thighs, flanks and sides of breast. Greater and median underwing coverts typically are paler than those of Common Buzzard, making underwing coverts appear less distinct from flight feathers. Sexes alike. Juv. like ad., but underparts typically are more heavily marked, often with tear-shaped flank streaks. Smaller and whiter than juv. Common Buzzard; eye brown (not yellowish). **Voice:** Most vocal at the start of the br. season, a shrill '*peeoo*'; recently fledged juvs are vocal when begging for food from their parents. **Status and biology:** NEAR-THREATENED. Endemic, locally common resident in forests and plantations, with some seasonal movement from s Cape to mountains in ne S Africa. Occurs singly or in br. pairs. Typically hunts for prey (rodents, birds, reptiles, etc.) from perches along forest edges. (Bosjakkalsvoël)

ad. grey-brown morph

ad. pale morph

ad.

dark underparts with barred flanks

COMMON BUZZARD

ad. grey-brown morph

large variation in plumage coloration

ad. dark morph

ad. rufous morph

ad. pale morph

dark underwing coverts

juv.

ad. dark brown morph

juv.

ad. dark morph

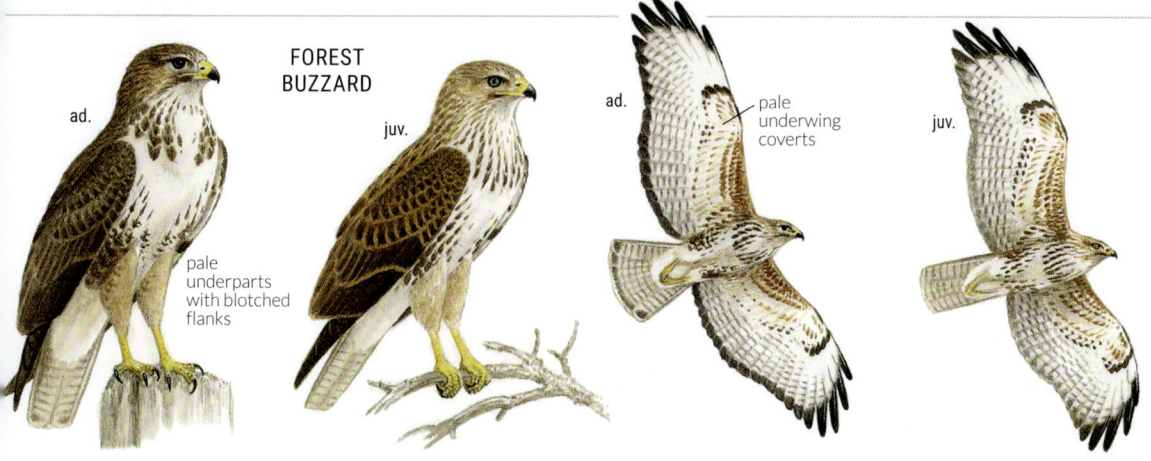

FOREST BUZZARD

ad.

juv.

ad.

pale underwing coverts

juv.

pale underparts with blotched flanks

131

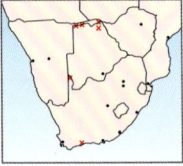

RED-NECKED BUZZARD *Buteo auguralis*

52–50 cm; 525–890 g; wingspan 1.05–1.27 m Vagrant. Smaller and more slender than Common Buzzard (p. 130); sides of head are red-brown, chin paler, with brown breast band and brown spotted pale underparts. Tail rufous with a black subterminal band. Underwing coverts paler than in Common or Forest Buzzards (p. 130). Juv. head is more rufous than adult, underparts are similar to adult but with a much whiter throat and breast, and a brown-grey (rather than rufous) tail. **Voice:** Believed to be silent in the region. **Status and biology:** Known from 7 records, in Kgalagadi Park, N Cape (Aug 2001), Ngepi Camp, Namibia (Mar 2009), Mahango GR, Namibia (Aug 2012), Chobe NP, Botswana (Jul 2014), Bwabwata NP, Namibia (Jul 2014), Kasane, Botswana (Jan 2015) and Stilbaai, W Cape (Dec 2016). (Rooinekjakkalsvoël)

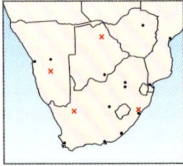

LONG-LEGGED BUZZARD *Buteo rufinus*

52–65 cm; 600–1 700 g; wingspan 1.3–1.6 m Vagrant. Larger than Common Buzzard (p. 130); identification is confounded by considerable variability in plumage among both species. Best told by longer wings and tail and larger bill, but differences are subtle and hard to confirm without direct comparison of size and shape. The most common morph has warm brown upperparts with a whitish head, neck and breast, faintly streaked with brown, a darker brown rump and belly (never barred), and a diagnostic, unbarred, pale rufous tail that appears almost translucent in flight. The pale brown underwing coverts contrast more with blackish carpal patches and darker brown belly than is typical of Common Buzzard. The upperwings show more prominent pale bases to outer primaries than in Common Buzzard. Rare dark morph has broad terminal tail bar. Sexes alike. Juv. like ad., but darker, with tail finely barred; eye yellowish-brown (brown in ad.). Wing beats are slow and deep. **Voice:** Largely silent in Africa. **Stats and biology:** A Palearctic-br. summer migrant to n Africa; vagrants occasionally wander south of the equator, rarely into s Africa (<10 records), however occurrence in the subregion remains controversial, with no photographically confirmed records. Solitary, in open grassland, arid savanna and semi-desert. (Langbeenjakkalsvoël)

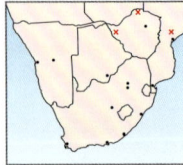

GRASSHOPPER BUZZARD *Butastur rufipennis*

39–44 cm; 310–410 g; wingspan 92–106 cm Vagrant. A mid-sized hawk, not closely related to *Buteo* buzzards. Greyish above, with rufous underparts lightly streaked with black; a white throat with a black central stripe, and striking pale yellow eyes. Cere and legs are pale yellow. Outer flight feathers and adjcent coverts rufous with black tips, forming distinctive upperwing pattern in flight; underwing paler with dark trailing edge and streaked underwing coverts. Underparts rufous at all ages. Juv. has a brown eye, paler streaked head and neck, and less distinct breast streaks. **Voice:** Loud, chattering calls, usually given only when breeding. **Status and biology:** Rare vagrant from central Africa; only 3 records, all in Nov–Dec: Hwange, Zimbabwe in 2014, north of Beira, Mozambique in 2016 and one in Mana Pools, Zimbabwe in Nov 2017. (Rooivlerkjakkalsvoël)

RED-NECKED BUZZARD

ad.

sides of neck rufous (all ages)

juv.

ad.

juv.

ad.

rufous tail, black subterminal band

pale underwings

LONG-LEGGED BUZZARD

ad. rufous

ad.

often with pale head & neck

ad. pale

pronounced black carpal patch

ad. rufous

pale rufous, unbarred tail

ad. dark

juv.

juv.

ad. dark

pale eye

ad.

GRASSHOPPER BUZZARD

juv.

ad.

juv.

ad.

dark trailing edge to underwing

obvious rufous wing patches

HARRIERS

Slender, long-winged and long-tailed raptors that have a characteristic loose, buoyant flight, mixing bouts of flapping with gliding, usually within a few metres of the ground with wings held in a shallow 'V' above the body. At close quarters the head appears rather owl-like due to facial ruff. All are species of open country, either marshland, grassland, fynbos or Karoo scrub. Prey is hunted by steadily quartering an area in slow flight, twisting and diving onto prey (rodents, small birds, amphibians). Of the 5 species occurring in the region, 3 are Palearctic-br. summer visitors and 2 are resident. Sexes similar in the resident species, dimorphic in the migrants. Juvs distinct in all species. The endemic Black Harrier is readily identified, but females and immatures of the two marsh harriers on this plate, and the two 'ringtail' harriers on the next are difficult to identify unless a good view is obtained.

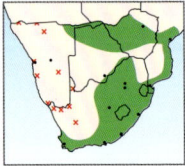

AFRICAN MARSH HARRIER *Circus ranivorus*

J F M A M J J A S O N D

44–50 cm; 360–680 g; wingspan 1–1.2 m A variably streaked, brown harrier told from Western Marsh Harrier in all plumages by the barred tail and barred underwing flight feathers. Ad. is streaked on breast, belly and underwing coverts, with considerable variation in the extent of the white breast band and the amount of white on the upper leading edge of the wing. Sexes similar; female typically is darker and more rufous below than male. Juv. chocolate-brown with a pale head and leading edge to the upperwing, similar to female or juv. Western Marsh Harrier, but has a broad, pale breast band and lightly barred flight feathers and tail (only visible at close range). **Voice:** Vocal during br. cycle; display call a high-pitched 'fee-ooo', male higher pitched than female. **Status and biology:** A locally common resident, frequenting large, permanent wetlands and adjacent open country; sedentary, with local dispersal by juvs. Solitary or in pairs. (Afrikaanse Vleivalk)

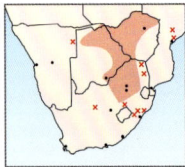

WESTERN MARSH HARRIER *Circus aeruginosus*

J F M A M J J A S O N D

45–52 cm; 410–780 g; wingspan 1.1–1.3 m The largest harrier in the region, most similar to African Marsh Harrier in appearance but distinguishable in all plumages by its unbarred tail. Ad. male (rare in s Africa) has a brown body and wing coverts, variably streaked white on face and breast, pale grey flight feathers with black tips to the outer primaries. Female is dark brown, with a creamy-white cap and throat, and usually has white-edged forewings. Unbarred flight feathers and tail separate it from juv. African Marsh Harrier. Juv. resembles female, but white crown and forewing may be absent. **Voice:** Silent in Africa. **Status and biology:** A rare Palearctic-br. summer visitor (Oct–Apr); mostly associated with large, inland sedge and reed marshes; regularly occurs alongside African Marsh Harrier. (Europese Vleivalk)

AFRICAN MARSH
HARRIER

ad.

ad.

rump can
be pale

ad.

barred tail &
underwings

juv.

juv.

juv.

very faint
barring

WESTERN MARSH
HARRIER

ad. ♂

ad. ♀

grey
unbarred
tail & flight
feathers

ad. ♂

ad. ♂
dark morph

ad. ♀

white
unbarred
underwing

ad. ♂

d. ♂
l-dark variant

juv. ♂

juv.

plain tail

juv.

lacks any barring
in tail or underwing
(all ages)

ad. ♀

BLACK HARRIER *Circus maurus*

b b b **B** b b
J F M A M J J A S O N D

42–50 cm; 350–600 g; wingspan 1–1.1 m Ad. is mostly black, with a conspicuous white rump, grey-barred tail and silvery wing panels. Sexes similar, but male has greyer primaries and a more coal-black plumage than female. Differs from very rare dark morph Montagu's Harrier by white rump and barred tail. Juv. has dark brown upperparts with buff feather edges; underparts buffy, streaked blackish on the breast and flanks. Differs from female and juv. Pallid and Montagu's Harriers by more heavily streaked underparts and by paler ground colour to the barred flight feathers. Imm. similar to ad., but with variable brown mottling on body. **Voice:** Generally silent; *'pee-pee-pee-pee'* display call; harsh *'chak-chak-chak'* when alarmed. **Status and biology:** ENDANGERED. Uncommon endemic resident, with north and eastward movements after br. in W and N Cape. Breeds mainly in fynbos, but occurs in any open habitat (Karoo scrub, grassland, agricultural lands) at other times. Solitary or in pairs, usually encountered hunting over low vegetation in characteristic flap-glide flight. (Witkruisvleivalk)

'RINGTAIL' HARRIERS

These 2 rather small, slender harriers are scarce non-br. visitors from the Palearctic, arriving in Nov and departing in Mar–Apr. They occur singly in their winter quarters, but may form small, loosely aggregated flocks on passage. They occasionally roost communally, sometimes with African Marsh Harriers (p. 134). Both favour open, treeless landscapes in which to hunt, especially natural grasslands, wetland margins and agricultural areas. They hunt prey on the wing in characteristic harrier fashion. Ad. males are easily distinguished by differences in their wing and underpart markings, but females and juvs (collectively referred to as 'ringtails') are commonly confused. They are smaller and more slender than marsh harriers, and both species have diagnostic whitish rumps. Separating ringtail Pallid and Montagu's Harriers requires a good view of the bird and an understanding of the key features to check; distinguishing them from a distant or fleeting view is impossible. Identification is confounded by moult-induced variability, especially in second- and third-year birds which exhibit intermediate plumages. Sub-ad. males are usually recognisable, provided sufficient ad. feathers have emerged, but intermediate-aged females may defy identification. Unfortunately, ringtails greatly outnumber ad. males in the region.

PALLID HARRIER *Circus macrourus*

J F M A M J J A S O N D

40–48 cm; 310–450 g; wingspan 95–115 cm Ad. male is pearl-grey, paler than ad. male Montagu's Harrier, especially on the head and breast; underparts plain. In flight, lacks dark bars on wings; black primary patch is smaller; underwings white apart from black wedge at tip. Ad. female is brown above, with white uppertail coverts, and buffy below with dark brown streaking on the breast and underwing coverts. Differs from female Montagu's Harrier in head and underwing features. Both females have a white supercilium, a white cheek patch just below the eye, a dark line through the eye, and a dark crescent below the pale cheek. In Pallid Harrier, the dark eye-line thins out behind the eye, but in Montagu's Harrier projects beyond the eye as a broad line. In Pallid Harrier the dark cheek crescent (below the eye) curves forward to the base of the bill forming a moustachial stripe; in Montagu's Harrier this crescent does not extend forward beyond the centre of the eye. The supercilium of the female Pallid Harrier is less developed than in Montagu's Harrier, being thinner and not extending forward much beyond the eye; in Montagu's Harrier it extends to the lores. In the underwing of female (and juv.) Pallid Harrier the secondaries are distinctly darker than the primaries: this contrast is less marked in Montagu's Harrier. In Pallid Harrier females the pale bars that run the length of the secondaries become less distinct and taper towards the body, but remain consistently broad in female Montagu's Harrier. Juvs of both species differ from females in being less streaked and having warm rufous-coloured heads and underparts. Their facial markings differ more between the 2 species than they do in ad. females. In Pallid Harrier, the broad, dark crescent below the eye joins the broad eye stripe, and curves forward to the base of the bill, resulting in a more striking facial pattern than that of juv. Montagu's Harrier, where the dark cheek crescent is much reduced and does not extend forward below the eye. Pallid Harrier has a narrower white supercilium and a smaller white area below the eye, making it appear darker faced than juv. Montagu's Harrier. **Voice:** Silent in the region. **Status and biology:** NEAR-THREATENED. Uncommon, Palearctic-br. summer visitor. (Witborsvleivalk)

MONTAGU'S HARRIER *Circus pygargus*

J F M A M J J A S O N D

40–46 cm; 230–440 g; wingspan 95–115 cm Ad. male is darker grey than ad. male Pallid Harrier; black wing tips are more extensive and there is a diagnostic black line across the upperwing; 2 lines across the underwing are visible at close range. Flanks are streaked chestnut; undersides of secondaries lightly barred. Female and juv. are plain brown above, apart from paler grey-brown upperwing coverts, white uppertail coverts and a banded tail. See Pallid Harrier for separating females and juvs of the 2 species. **Voice:** Silent in the region. **Status and biology:** Uncommon, Palearctic-br. summer visitor. (Blouvleivalk)

ad.

ad.

ad.

imm.

white rump
(all ages)

juv.

**BLACK
HARRIER**

imm.

juv.

ad. ♂

**PALLID
HARRIER**

ad. ♀

ad. ♂

no black line
in upperwing

ad. ♀

ad. ♀

contrast
between
secondaries
& primaries

♀

juv.

pale bases to
primaries with
inconspicuous
trailing edge

juv.

broad dark
cheeks, distinct
pale collar

ad. ♂

white
underwings
& belly

ad. ♂

ad. ♀

streaked
underwing
& belly

ad. ♂

ad. melanistic
(very rare)

**MONTAGU'S
HARRIER**

ad. ♂

black line in
upperwing

rrower
rk cheeks,
distinct
le collar

juv.

juv.

juv.

ad. ♀

ad. ♀

ad. ♀

dark bases to
primaries with
dark trailing edge

less contrast
between secondaries
& primaries

SMALL KITES AND HAWKS

Three unrelated hawks grouped together here for convenience. *Elanus* kites are rodent specialists that are among the most ancient of raptors. *Kaupifalco* is a monotypic genus of uncertain affinities, whereas the Cuckoo-Hawk is the African representative of a genus centred in SE Asia. Sexes alike in all 3 species, with little size dimorphism (females marginally larger).

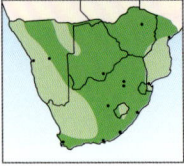

BLACK-WINGED KITE   *Elanus caeruleus*

b B B B b b b b b b b b
J F M A M J J A S O N D

30–33 cm; 210–290 g; wingspan 75–85 cm A small, grey-and-white, open-country raptor with distinctive black shoulder patches and striking red eyes. White tail often wagged up and down while perched, especially during interactions with conspecifics. Juv. is brownish above, with a brown crown, grey tail and buff-edged upperpart feathers; eyes yellowish-brown. **Voice:** High-pitched, whistled *'peeeu'*, soft *'weep'* and rasping *'wee-ah'*. **Status and biology:** Common resident and local nomad in open savanna, grassland and agricultural areas; moves widely across region to exploit rodent outbreaks. Found solitarily or in pairs during the day, gathering at dusk to roost communally (10–300 birds), especially in reed beds. Often perches on telephone poles and lines, hunting from such positions mainly for small diurnal rodents. Also hunts by hovering, parachuting onto prey with wings in deep 'V'. (Blouvalk)

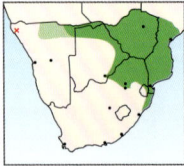

LIZARD BUZZARD   *Kaupifalco monogrammicus*

b   b b b b B B B b
J F M A M J J A S O N D

35–37 cm; 220–360 g; wingspan 80 cm A small, short-legged hawk, intermediate in size and shape between a buzzard and an accipiter. Superficially resembles Gabar Goshawk (p. 140) and the grey-coloured accipiters, but white throat with black central stripe is diagnostic at all ages. In flight, white rump and white tail bar (occasionally 2 bars) are conspicuous. Ad. has red cere and legs, and dark red eyes. Juv. similar to adult but has orange cere and legs and pale fringes to upperpart feathers. **Voice:** Noisy in br. season, both sexes calling from a perch or while soaring: a melodious whistled *'wheeo, wot-wot-wot-wot-wot'* or *'klioo, klu-klu-klu-klu'*; also a plaintive whistled *'peeeoo'*. **Status and biology:** Locally common resident in moist savanna and forest clearings; erratic or nomadic in more arid areas. Usually encountered singly, perched unobtrusively on a branch below the canopy of a tall tree. (Akkedisvalk)

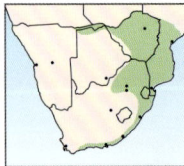

AFRICAN CUCKOO-HAWK   *Aviceda cuculoides*

b b b   b B B b
J F M A M J J A S O N D

40 cm; 220–350 g; wingspan 90 cm Grey plumage and barred underparts give it an accipiter-like appearance, but it has a more compact shape, a small crest, rufous nape patch, short legs, and long, rather broad wings. At rest, wing tips reach to tail tip (tail extends well beyond wings in accipiters). Flight slow and relaxed. Coloration recalls African Goshawk (p. 144), but distinguished by crest, short legs, longer wings and much broader barring on belly and underwing coverts. In flight, ginger 'armpits' are diagnostic. Male has red-brown eyes; female has yellow eyes. Juv. is brown above and white below, with brown-streaked and boldly spotted underparts; best told from juv. African Goshawk by its larger size and shape. Imm. is browner above than ad., with mottled brown (not grey) breast. **Voice:** Loud, far-carrying *'teee-oooo'* whistle; a shorter *'tittit-eoo'* while soaring. **Status and biology:** Uncommon resident of dense woodland and forest fringes; usually solitary, perching unobtrusively in canopy of a tree from which it makes short sorties into surrounding foliage in pursuit of diverse prey. (Koekoekvalk)

distinct
red eye

BLACK-WINGED KITE

ad.

pale underwings
with dark tips to
primaries

ad.

ad.

ad.

juv.

ad. hunting by
hovering

distinct
throat
stripe

ad.

juv.

LIZARD BUZZARD

ad.

ad.

prominent
white tail bar

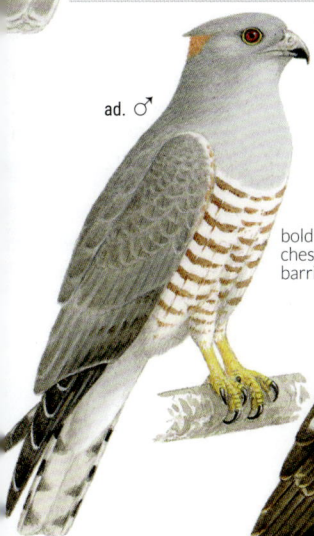

eye red
(yellow in ♀)

ad. ♂

bold
chestnut
barring

small crest

juv.

bold
spots on
underparts

juv.

AFRICAN
CUCKOO-HAWK

ad. ♂

ad. ♂

rufous
'armpits'

CHANTING GOSHAWKS
Large, long-legged hawks with rather short, rounded wings. Adults grey with red legs and ceres; juvs nondescript streaky brown. Perch conspicuously on tree tops; in the case of the arid-country Pale Chanting Goshawk, telephone poles are commonly used as perches. Their common name derives from their loud, ringing calls. Sexes alike, females slightly larger. The smaller Gabar Goshawk is sometimes placed in the same genus, but is a smaller, less conspicuous bird, perching unobtrusively and hunting by stealth, ambushing its prey from concealed positions.

PALE CHANTING GOSHAWK   *Melierax canorus*

J F M A M J J A S O N D

48–62 cm; 620–1 400 g; wingspan 1.1 m Larger than Dark Chanting Goshawk. Ad. pale grey with a finely barred belly; paler than ad. Dark Chanting Goshawk with a white (not grey) rump and very pale grey secondaries contrasting strongly with the dark primaries. Juv. is brown above with buff feather edges; barred and streaked below, with boldly barred flight feathers and tail; superficially harrier- or buzzard-like, but with long, orange legs, pale yellow eyes and pink-orange cere. Paler than juv. Dark Chanting Goshawk with a white (not barred) rump. **Voice:** Piping '*kleeu-kleeu-klu-klu-klu*', usually given at dawn. **Status and biology:** Near-endemic. Common resident of arid savanna, semi-desert and Karoo scrub. Solitary or in pairs, usually perches conspicuously on a tree top or telephone pole, scanning the ground for prey. A versatile hunter that will run after prey on foot or pursue it relentlessly in flight; sometimes uses Honey Badgers and other mammals to flush prey. (Bleeksingvalk)

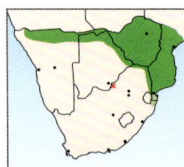

DARK CHANTING GOSHAWK   *Melierax metabates*

J F M A M J J A S O N D

43–50 cm; 480–880 g; wingspan 1 m Smaller than Pale Chanting Goshawk; ad. is darker grey with a grey (not white) rump, pale grey secondaries and darker grey upperwing coverts. In flight does not show white on the wings and rump as in Pale Chanting Goshawk. Most other accipiters are smaller, and have yellow legs. Only 2 other accipiters in the region have red legs: Gabar Goshawk is much smaller and has a white rump; Ovambo Sparrowhawk (p. 142) has a barred breast. Lacks black throat stripe of smaller, shorter-legged Lizard Buzzard (p. 138). Juv. best told from juv. Pale Chanting Goshawk by its brown-barred (not white) rump. **Voice:** Piping '*kleeu-kleeu-klu-klu-klu*'. **Status and biology:** Scarce to locally common resident of well-wooded savanna, found solitarily or in pairs, usually perched in upper branches of a tree scanning the ground for prey. Takes live prey, but also feeds on roadkills and other carrion; sometimes follows Honey Badgers, Southern Ground Hornbills and other larger animals to catch prey they flush. (Donkersingvalk)

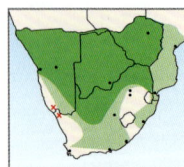

GABAR GOSHAWK   *Micronisus gabar*

J F M A M J J A S O N D

28–36 cm; 110–220 g; wingspan 60 cm A fairly small, accipiter-like hawk with a red cere and legs, prominent white rump, plain grey throat and upper breast, and finely barred belly. Easily confused with Shikra (p. 144) and Ovambo Sparrowhawk (p. 142), which both lack the white rump and differ in cere coloration, and with Lizard Buzzard (p. 138), which has a white rump but also a broad, white tail bar. At rest white-tipped secondaries form a diagnostic white line on the folded wing. Much smaller than Dark Chanting Goshawk, with shorter legs. Melanistic form (5–25% of population) is most frequent in arid areas; told from dark morph Ovambo Sparrowhawk by more boldly barred wings and tail and red bare parts. Sexes alike, female one-third larger than male. Juv. is the only juv. accipiter with a white rump. It is brown above; breast is rufous-streaked; belly and underwing coverts are barred rufous; eyes yellow; legs orange. **Voice:** Various whistled calls, especially a piping '*kew-he, kew-he, kew-heee*' and a rapidly repeated '*kji-kji-kji-kji ...*'. **Status and biology:** Locally common resident of savanna and semi-desert, where it is restricted to tree-lined watercourses. Mostly solitary; in pairs while br. A versatile predator, hunting from a covered perch from which it launches in pursuit of passing birds; commonly steals chicks from nests of weavers and other birds. (Witkruissperwer (Kleinsingvalk))

PALE CHANTING GOSHAWK

ad.

pale wing panel

juv.

ad.

white rump & secondaries

ad.

dark tips to wings

juv.

white rump

DARK CHANTING GOSHAWK

ad.

juv.

ad.

grey rump & secondaries

dark tips to wings

ad.

mottled rump

juv.

ad.

ad.

red cere

ad. melanistic

ad.

ad.

juv.

white rump

white-tipped secondaries

juv.

streaked chest, barred belly

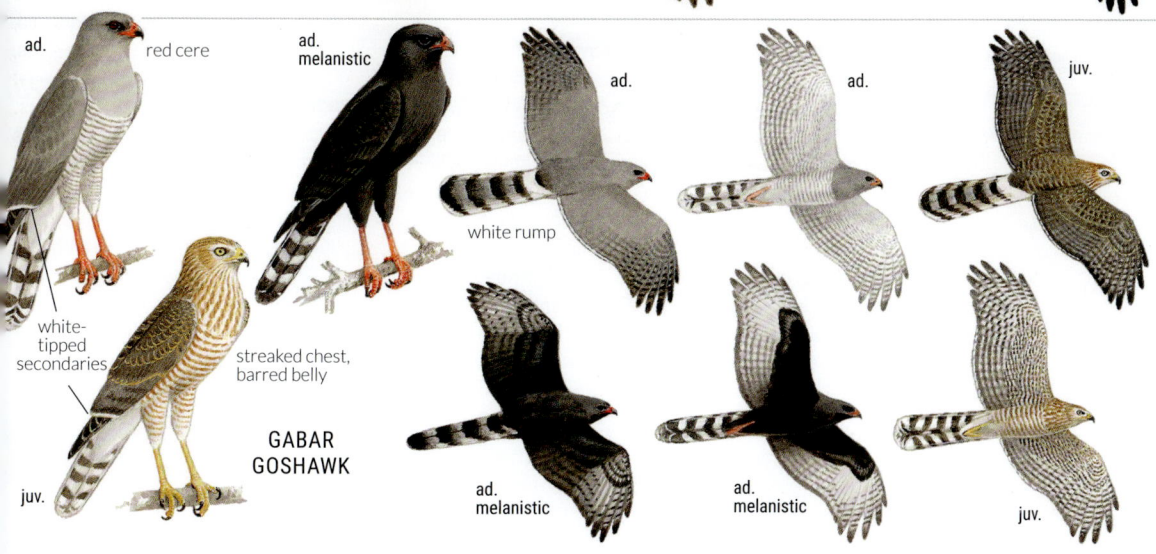

GABAR GOSHAWK

ad. melanistic

ad. melanistic

juv.

141

ACCIPITERS

ACCIPITERS Small to medium-sized raptors with short, broad wings and long tails; at rest, tail extends well beyond folded wings. Structure is adapted for rapid manoeuvring in flight. Hunts prey by stealth and ambush, often among dense vegetation, and consequently rather unobtrusive. Identification relies on size differences, colour of eyes, bill and legs, pattern on upper tail, whether or not the rump is white, and on general plumage differences. Females are appreciably larger than males, especially in species that target birds as prey. Males often have brighter plumage than females, and juvs have very different plumages than ads. The distinction between sparrowhawks and goshawks originated in Britain, where only 2 species occur: the larger was called a goshawk and the smaller, with proportionately longer legs and toes, a sparrowhawk. Extension of the use of these names to Africa (and elsewhere) has not adhered to these differences: the largest African accipiter is the Black Sparrowhawk, whereas goshawk is used arbitrarily for smaller species (the alternative name for Shikra is Little Banded Goshawk).

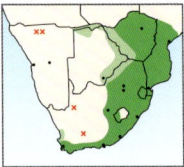

BLACK SPARROWHAWK *Accipiter melanoleucus*

b B B B B B B B b
J F M A M J J A S O N D

46–58 cm; 510–1 000 g; wingspan 90–95 cm The largest accipiter in the region, although size dimorphism between the sexes is pronounced. Ad. black above; underparts vary from white (most common form) through intermediates to virtually black (white on throat only). No consistent difference in plumage between sexes. Birds with white underparts are unmistakable; all-dark birds only likely to be confused with dark morphs of Ovambo Sparrowhawk or Gabar Goshawk (p. 140); distinguished by larger size and yellow legs and cere. Juv. dark grey-brown above, with either rufous underparts finely streaked with dark brown (rufous morph) or white below, variably streaked with rufous and dark brown (pale morph). Rufous morph told from juv. African Hawk-Eagle (p. 122) by its unfeathered tarsi; pale morph is larger than juv. African Goshawk with streaked (not spotted) underparts and no white supercilium. **Voice:** Noisy while br., otherwise silent. Male gives *'kee-yip'*; female loud *'kek-kek-kek-kek'*. **Status and biology:** Locally common resident, restricted to tall-tree woodlands, forests and plantations; often forages far from cover. Range expanded into W Cape since 1980s, and is now common, ranging widely from wooded areas. (Swartsperwer)

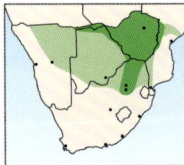

OVAMBO SPARROWHAWK *Accipiter ovampensis*

B B B b
J F M A M J J A S O N D

32–40 cm; 120–300 g; wingspan 65–70 cm A medium-sized accipiter, but dimorphism between large female and smaller male very marked. Ad. has fine, dark grey barring from throat to vent, absence of white on rump, dark red-brown eye and uppertail pattern (barred, with vertical white shaft streaks in central feathers) diagnostic; legs and cere usually yellow, but can be orange or red; breast markings, eye colour and uppertail pattern distinguish it from similar Shikra (p. 144). Very rare black morph is similar to black morph Gabar Goshawk (p. 140), but legs and cere yellow, not red, and diagnostic uppertail pattern (white shaft streaks) distinguish it from Gabar Goshawk. Juv. has 2 colour forms; both have brown upperparts and a brown, barred tail with the distinctive white shaft streaks present; underparts mottled rufous in the rufous form, and dull white, streaked and mottled with brown, in the pale form; both have a broad, white supercilium. Rufous form very like juv. Rufous-breasted Sparrowhawk but uppertail pattern and brown (not yellow) eyes distinguish it. **Voice:** A rapidly repeated *'keeep-keeep-keeep'* or *'kwee-kwee-kwee …'* when br. **Status and biology:** Uncommon resident, restricted to tall-tree woodland and savanna, including exotic plantations. Occurs singly except when br.; nesting pairs regularly include one of pair in imm. plumage. Preys almost entirely on birds up to dove size, hunted by ambush from perches and by pursuit from soaring. (Ovambosperwer)

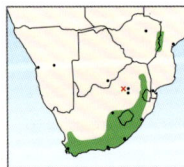

RUFOUS-BREASTED SPARROWHAWK *Accipiter rufiventris*

b B B B b
J F M A M J J A S O N D

30–38 cm; 105–210 g; wingspan 65–70 cm A smallish accipiter with uniformly coloured upperparts and no white on the rump or tail. Ad. male is dark slate-grey above and plain rufous below (including underwing coverts), with a paler throat; rufous lower cheek contrasts with dark crown, giving a capped appearance, unlike the uniform head of African Goshawk (p. 144). Tail barred grey-brown with a narrow pale tip. Eyes, cere and legs are yellow. Female appreciably larger and browner above; underparts duller rufous in some birds, but not in others. Juv. is brown above, with slight pale supercilium and underparts mottled rufous, variably streaked white. Best told from juv. Ovambo Sparrowhawk by its darker head, more uniform upperparts and yellow (not brown) eyes. **Voice:** Sharp, staccato *'kee-kee-kee'* or *'kew-kew-kew'* during display. **Status and biology:** Fairly common resident of montane forest and plantations; often forages far from cover. An agile aerial hunter preying mainly on small birds which it pursues in flight. (Rooiborssperwer)

BLACK SPARROWHAWK

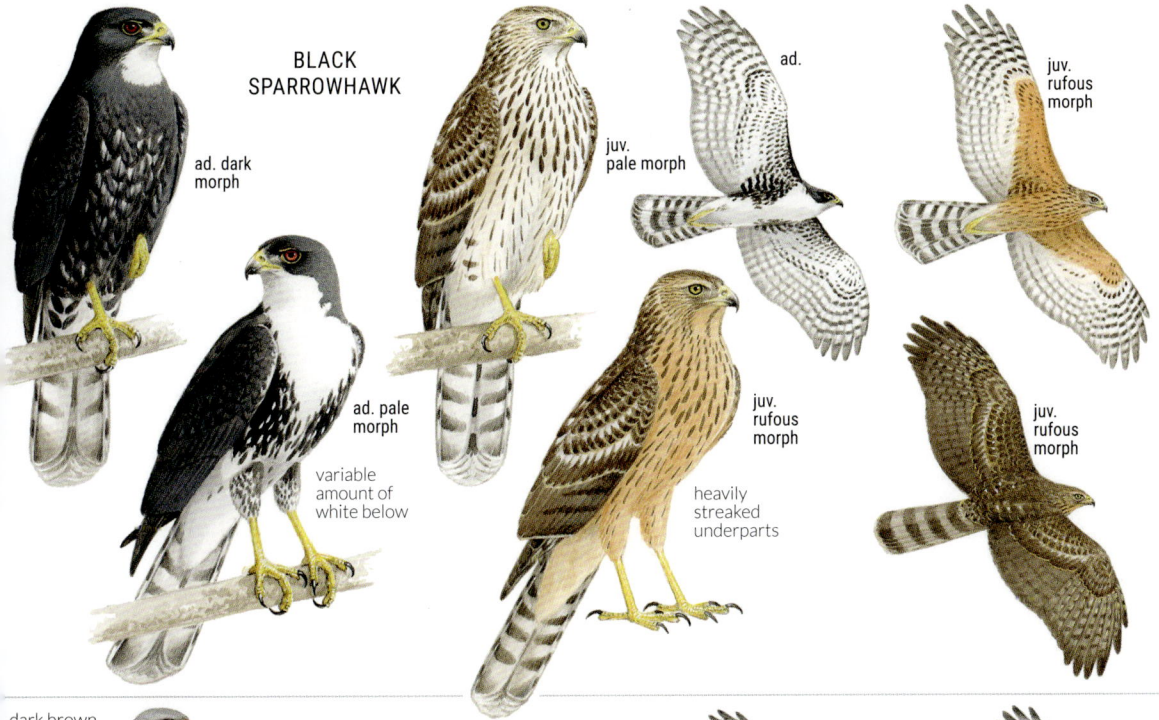

ad. dark morph

ad. pale morph

variable amount of white below

juv. pale morph

ad.

juv. rufous morph

juv. rufous morph

heavily streaked underparts

juv. rufous morph

OVAMBO SPARROWHAWK

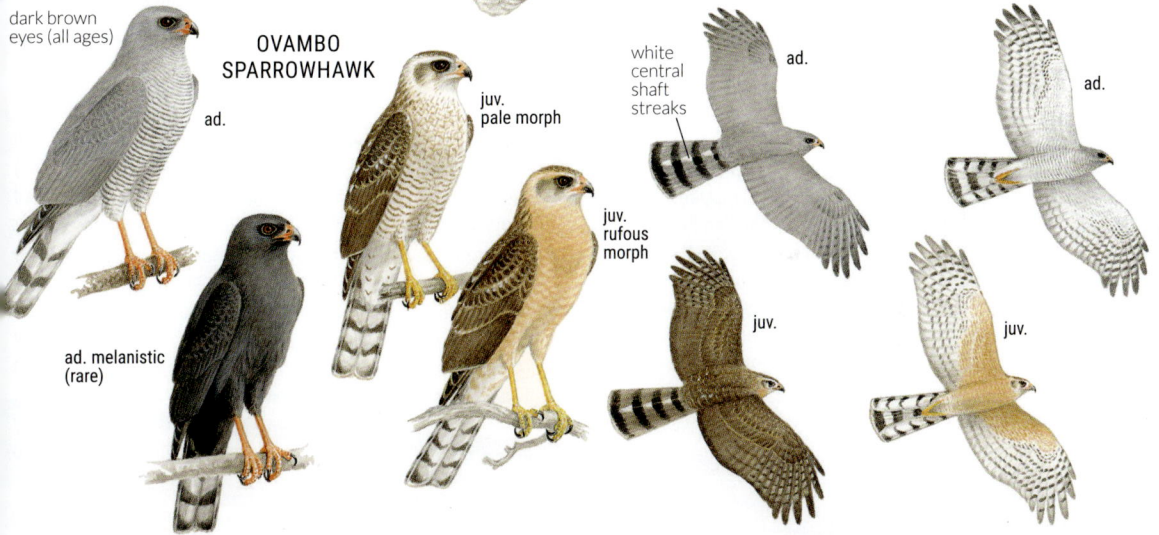

dark brown eyes (all ages)

ad.

ad. melanistic (rare)

juv. pale morph

juv. rufous morph

white central shaft streaks

ad.

ad.

juv.

juv.

RUFOUS-BREASTED SPARROWHAWK

ale yellow yes (all ages)

uniformly rufous underparts

ad.

long, thin tail

heavily mottled underparts

juv.

ad.

ad.

uniformly rufous underparts & coverts

juv.

juv.

AFRICAN GOSHAWK *Accipiter tachiro*

b b B B B b
J F M A M J J A S O N D

36–44 cm; 180–500 g; wingspan 70–75 cm A medium-sized accipiter, but size dimorphism between the sexes is pronounced. Sexes differ in plumage, but yellow legs and eyes and a pale grey (sometimes green-yellow) cere diagnostic in both. Ad. male has a plain grey head and upperparts, and white underparts finely barred rufous with rufous wash to flanks; underparts can appear uniformly pale rufous from a distance. Tail dark grey-brown, indistinctly barred, with 2 conspicuous white spots on central feathers. Larger than Little Sparrowhawk, although a small male African Goshawk can appear similar in size to a large female Little Sparrowhawk, with a dark (not white) rump and grey (not yellow) cere. Female is brownish-grey above and white below, barred with brown; tail brown with 4 darker bars and no white spots. Adults of both sexes show a blue-grey base to the mandible, which is all black in Little Sparrowhawk. Juv. dark brown above with an indistinct white supercilium; underparts white with a dark median throat stripe and bold, tear-shaped spots on the breast and belly; flight feathers and tail boldly barred dark brown. **Voice:** Calls year-round, mainly around sunrise, in flight or from a perch, a sharp '*quick*' repeated every 2–5 seconds. **Status and biology:** Common resident of forest, dense woodland and wooded suburbia; often forages far from cover. (Afrikaanse Sperwer)

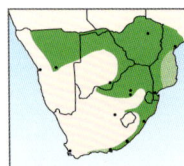

LITTLE SPARROWHAWK *Accipiter minullus*

b B B B
J F M A M J J A S O N D

23–27 cm; 68–120 g; wingspan 40–45 cm The smallest accipiter in the region. Ad. best identified by its small size, white rump, 2 conspicuous spots on its upper tail and bright yellow eyes, legs, cere and eye-ring. Slightly smaller than Shikra, but easily told in flight by its white rump and tail spots (not plain grey rump and tail) and yellow (not cherry-red) eyes. Sexes alike. All-dark bill of Little Sparrowhawk distinguishes it from blue-grey-based bill of African Goshawk. Juv. is brown above and white below, with large, dark brown spots on breast and belly; resembles a diminutive juv. African Goshawk, but lacks this species' pale supercilium and dark median throat stripe. **Voice:** Male high-pitched '*tu-tu-tu-tu-tu*' during br. season; female softer '*kew-kew-kew*'. **Status and biology:** Locally common but secretive resident of forest, woodland and plantations. Preys mainly on small birds caught by ambush on the wing; occasionally takes other small vertebrates and insects. (Kleinsperwer)

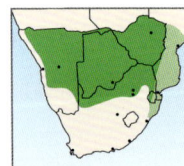

SHIKRA *Accipiter badius*

b b b B B B b
J F M A M J J A S O N D

28–30 cm; 80–170 g; wingspan 55–60 cm A small accipiter. Ad. has diagnostic red eye (deep red in male; orange-red in female), pale rufous (not grey) chest barring, grey (not white) rump, and plain grey central tail feathers (not barred or spotted). Absence of white rump distinguishes it from Gabar Goshawk (p. 140) and Little Sparrowhawk; smaller than Ovambo Sparrowhawk (p. 142), which has grey-barred underparts, a dark red-brown eye, and conspicuous white streaks on the central tail feathers. Sexes alike in plumage. Juv. has yellow eye, is brown above, with streaked breast and barred belly like juv. Gabar Goshawk, but its rump is brown (not white). **Voice:** Male high-pitched '*keewik-keewik-keewik*'; female softer '*kee-uuu*'. **Status and biology:** Common resident and local nomad in savanna and tall woodland. Occurs singly or in pairs, typically encountered perched, watching the ground from a concealed position in a tree; preys mainly on lizards and large insects, but also small birds and other small vertebrates. (Gebande Sperwer)

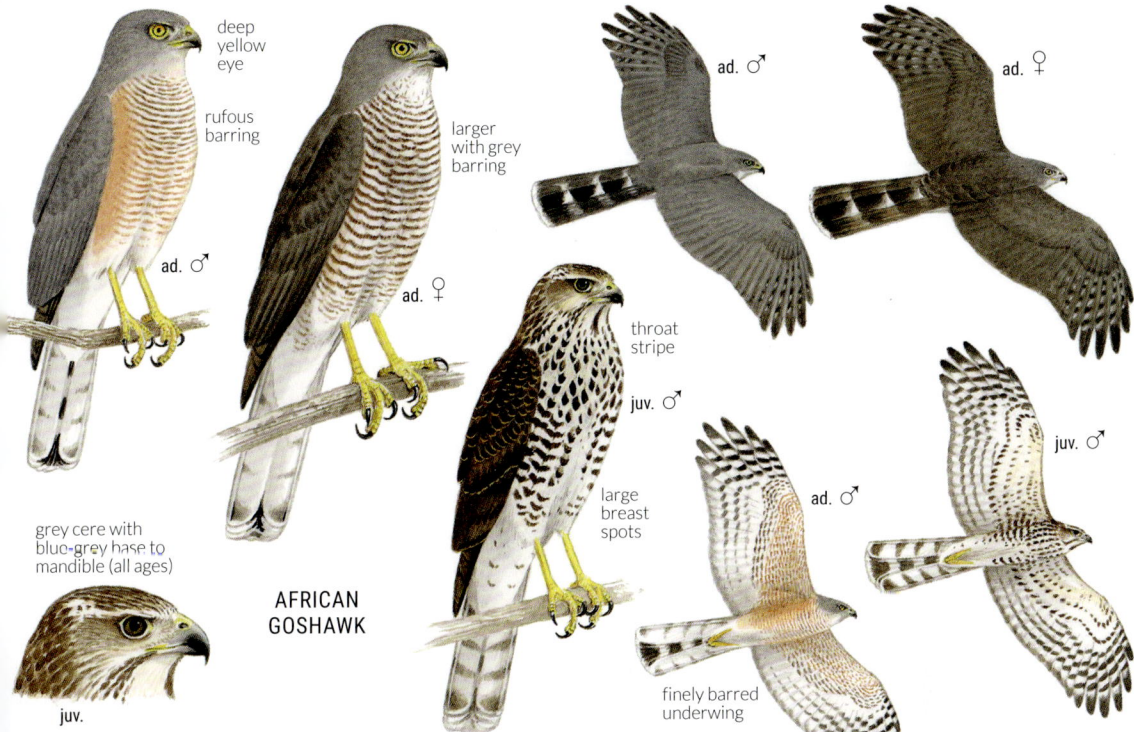

deep yellow eye

rufous barring

ad. ♂

larger with grey barring

ad. ♀

ad. ♂

ad. ♀

throat stripe

juv. ♂

large breast spots

grey cere with blue-grey base to mandible (all ages)

juv.

AFRICAN GOSHAWK

ad. ♂

juv. ♂

finely barred underwing

yellow cere with all-black mandible

juv.

yellow eye & eye-ring

no throat stripe

LITTLE SPARROWHAWK

spotted underparts

juv.

ad.

white rump & 2 tail spots

juv.

ad.

ad.

cherry-red eye, yellow cere

ad.

juv.

throat stripe

ad.

plain grey central tail feathers

heart-shaped blotching

SHIKRA

ad.

juv.

HARRIER-HAWKS

The 2 species of *Polyboroides* (one in Africa and one in Madagascar) are large hawks with very broad, deeply slotted wings, long tails and long, slender legs which have the unique ability of being able to bend backwards and sideways at the tarsal joint. This enables them to insert their feet into holes and cracks in rocks and tree stems to pull out bats, lizards, nestling birds and other prey. Their unusually narrow, small head also facilitates probing into holes. They have bare facial skin which changes colour from red to yellow according to the bird's state of agitation. Sexes alike.

AFRICAN HARRIER-HAWK (GYMNOGENE) *Polyboroides typus*

b B B b
J F M A M J J A S O N D

60–66 cm; 620–950 g; wingspan 1.2–1.5 m A large, broad-winged hawk with a small head and long legs. Ad. is grey above and finely barred below, with a bare facial skin that can rapidly change from yellow to dark pink depending on its reaction to external stimuli. Has a slow, buoyant flight with leisurely flapping; broad black tips to flight feathers and black tail with a central white band are distinctive. Juv. and imm. are variably coloured from dark brown to buff, and variably streaked and spotted; easily confused with other brown-plumaged raptors, but gangly appearance, long, bare, yellow tarsi and slender head with grey facial skin distinguish it in flight and when perched. Juv. most likely to be mistaken for European Honey Buzzard (p. 126), however, shows broader wings, often with noticeably bulging secondaries and more deeply slotted outer primaries (visible in flight) and longer tarsi (visible at rest). Sub-ad. birds show variable amounts of grey (ad.) plumage between juv. feathers. **Voice:** Whistled '*peeer*' or '*suuu-eeee-ooo*', mainly in the br. season. **Status and biology:** Fairly common resident in woodland, forests and more open, scrubby habitats. Occurs singly or in pairs. Frequently encountered scrabbling about on rock faces or large tree stems, or hanging from a branch with spread wings and 1 leg inserted into a crack. Also clings to weaver nests to extract nestlings. (Kaalwangvalk)

BAT HAWK

A peculiar, crepuscular hawk allied to the kites that ranges from tropical Africa to Indo-Malaysia. It specialises in hunting bats, catching and eating them on the wing at dawn and dusk. Its long, pointed wings and rapid flight give it a falcon-like appearance. Sexes alike.

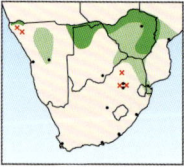

BAT HAWK *Macheiramphus alcinus*

b b b B B B b
J F M A M J J A S O N D

45 cm; 600–650 g; wingspan 95–120 cm A dark brown raptor, appearing black in the field, with large, yellow eyes, pale grey legs and bill, and small but variable amounts of white on the throat and belly. At close quarters, white eyelids (conspicuous when eyes are closed) and pair of white nape patches are distinctive. Juv. has more white on underparts, and pale-spotted underwing and tail. Juv. could be confused with sub-ad. Black Sparrowhawk (p. 142), however, shows sharper wings and a shorter tail. **Voice:** High-pitched whistling, similar to that of thick-knees. **Status and biology:** Uncommon resident and nomad; occurs mostly in tropical lowland savanna, especially where baobabs are present, and along forested escarpments where pairs may nest in tall eucalypts; vagrants may turn up almost anywhere. Solitary or in pairs, perching quietly in the upper branches of tall, well-foliaged trees during daylight. Active at dusk, hunting through the evening and again at dawn. Preys almost exclusively on bats, catching them on the wing; small bats swallowed in the air. (Vlermuisvalk)

ad.

narrow, 'beaky' head

ad.

colour variation in facial skin

ad.

shaggy feathers

juv. pale breasted

juv. dark

ad.

ad.

white bar on black tail

juv.

variably coloured light buff to rufous

AFRICAN HARRIER-HAWK

juv. pale

juv. dark

imm.

bulging secondaries

bare facial skin

juv. dark

BAT HAWK

ad.

bright yellow eye

ad.

ad.

juv.

juv.

weak bill

spots n nape

white eyelids

sleeping

long-winged, falcon-like flight

147

FALCONS Small to medium-sized raptors, most with long, pointed wings. Genetic evidence indicates that they are more closely related to parrots than other raptors. The falcons on this page are fast-flying, streamlined raptors with consummate flying ability, hunting aerial prey by diving ('stooping') on it with wings almost closed, at speeds of up to 300 km/h. As a result, Peregrine and Lanner Falcons are favoured for falconry. Sexes alike in plumage, but males one-third smaller than females in these 2 species. Red-necked Falcon lives in open, park-like savanna; also primarily a bird-hunter, it waits in ambush on a perch, often at a drinking site, dashing out to catch its prey in flight.

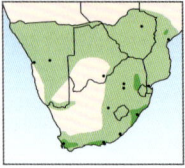

PEREGRINE FALCON   *Falco peregrinus*

J F M A M J J A S O N D

34–44 cm; F. p. minor 500–850 g; F. p. calidus 680–1 300 g; wingspan 80–112 cm A large, chunky falcon with broad-based, pointed wings and a relatively short tail. Smaller and more compact than Lanner Falcon, with a dark cap and broad, black moustachial stripes. Flight is dashing and direct; soars less than Lanner Falcon. Ad. of resident *F. p. minor* has dark slate-grey upperparts, mostly white breast and finely barred belly. Palearctic-br. *F. p. calidus* is larger and paler above, with a spotted breast and narrower moustachial stripes. Juv. dark brown above with variable paler streaking on crown and pale nape patches; underparts heavily streaked brown (but not as strongly marked as juv. Lanner). **Voice:** A loud '*krrChuck krrChuck*' in display and strident '*kak-kak-kak-kak-kak*' around nesting cliff; also whining and chopping notes. **Status and biology:** Locally common to uncommon resident (*F. p. minor*) and uncommon, Palearctic-br. summer migrant (*F. p. calidus*), Oct–Mar. Br. birds require high cliffs and gorges, or tall buildings, but migrant birds occur in wide range of open habitats; often forage at wetlands. Solitary or in pairs. Hunts from high vantage points on cliffs, buildings or while soaring, stooping on passing birds. (Swerfvalk)

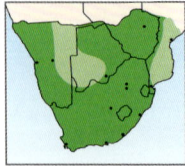

LANNER FALCON   *Falco biarmicus*

J F M A M J J A S O N D

36–48 cm; 420–800 g; wingspan 88–113 cm A large falcon, less compact than Peregrine Falcon, with broader and longer wings. Soars more than Peregrine, often fanning tail. Ad. easily distinguished from Peregrine Falcon by its plain, pinkish-cream underparts and rufous (not black) cap. Sexes alike in plumage. Juv. is dark grey-brown above, with creamy-brown crown and nape, and heavy brown streaking on breast and belly (juv. Peregrine Falcon has darker cap and finer breast streaking). Plumage recalls Taita Falcon (p. 150), however, lacks rufous nape patches and is considerably larger. **Voice:** Harsh '*kak-kak-kak-kak-kak*', similar to call of Peregrine Falcon; also whining and chopping notes. **Status and biology:** Fairly common resident and local nomad in a wide range of habitats from mountains to deserts and open grassland; avoids forests. Solitary or in pairs; often perches on roadside poles; more often seen on the ground than Peregrine. Preys on a wide range of birds caught in dashing flight; some prey scooped up from the ground. (Edelvalk)

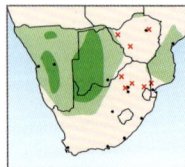

RED-NECKED FALCON   *Falco chicquera*

J F M A M J J A S O N D

30–36 cm; 150–265 g; wingspan 55–70 cm A small, dashing, long-tailed, savanna falcon. Ad. has chestnut crown and nape, uniformly barred, slate-grey upperparts, and a broad, subterminal black tail band. Throat and upper breast are creamy-buff, sometimes with indistinct rufous breast band; lower breast and belly finely barred. Head usually appears clean-cut, with narrow, dark brown moustachial stripes on white cheeks. Juv. is duller, with dark brown head, 2 buff patches on nape, and pale rufous underparts, finely barred brown. Could be confused with Taita Falcon (p. 150), but has much paler head and underparts, and tail and wings are longer. **Voice:** Shrill '*ki-ki-ki-ki-ki*' during br. season. **Status and biology:** NEAR-THREATENED. Locally fairly common resident in arid savanna and woodland, often near *Borassus* palms. Solitary or in pairs, usually encountered perched within tree canopy. Preys mainly on small birds caught on the wing after rapid aerial pursuit. (Rooinekvalk)

dark
crown

ad. *calidus*
migratory race

PEREGRINE
FALCON

juv. *calidus*

ad. *minor*

ad. *minor*

more compact
build than
Lanner Falcon,
with shorter tail

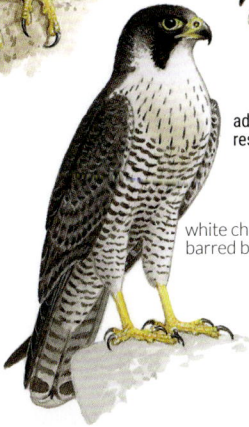

ad. *minor*
resident race

white chest,
barred belly

juv. *minor*

streaked
underparts

juv.

evenly
streaked
underparts

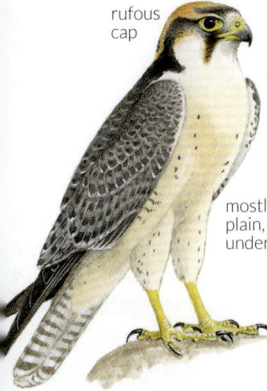

rufous
cap

ad.

mostly
plain, buff
underparts

juv.

heavy
streaking

LANNER
FALCON

uniformly
buff
underparts

ad.

broad, long
wings & tail

ad.

thighs, lower
belly & vent
noticeably pale

juv.

coverts
darker
than flight
feathers

rufous
head

ad.

barred
back

dull
brown
head

juv.

RED-NECKED
FALCON

ad.

long tail,
broad black
subterminal
band

ad.

densely
barred
underwings
& tail

juv.

149

SMALL FALCONS AND HOBBIES

These 4 falcons are all fast-flying, rapacious aerial hunters. Taita Falcon is a diminutive version of the Peregrine Falcon (p. 148), restricted to high cliffs; it preys on passing birds by stooping on them at great speed. Eleonora's Falcon is a rare, Palearctic-br. summer migrant to the region from the Mediterranean. The name 'hobby' is used for slender, long-winged falcons that hunt aerial insects, bats and small birds, mainly at dawn and dusk. The Eurasian Hobby is a fairly common non-br. migrant from the Palearctic, whereas the African Hobby is a more rare resident species.

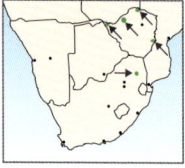

TAITA FALCON *Falco fasciinucha*

25–30 cm; 195–340 g; wingspan 61–72 cm A small, compact, cliff-dwelling falcon with a short tail, broad shoulders and large head. In flight, chunky shape, short tail, and rapid wing beat recall a small parrot. Ad. dark slate above, with broad moustachial stripes and rufous-washed breast and belly; rufous underwing coverts contrast with grey-and-black-barred flight feathers. Told from African Hobby by its rufous nape patches, mainly white cheeks and throat, stocky build and shorter tail. Juv. has buff fringes to back feathers, rufous face and throat, and is lightly streaked below; best told from juv. Red-necked Falcon by its stocky build, short tail and preference for cliff habitats. **Voice:** High-pitched '*kree-kree-kree*' and '*kek-kek-kek*'. **Status and biology:** VULNERABLE. Rare and localised resident closely associated with cliffs and gorges in woodland. Hunts prey, mostly small birds, on the wing from cliff perches. (Taitavalk)

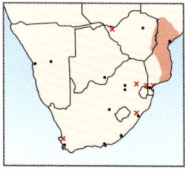

ELEONORA'S FALCON *Falco eleonorae*

36–42 cm; 340–450 g; wingspan 84–103 cm Vagrant. A large, slender falcon with very long wings and tail and darker underwings than other falcons. Ad. of more common pale morph is dark sooty-brown above with white cheeks and throat contrasting with dark-streaked rufous breast and belly. Legs are greenish-yellow; male has a yellow cere; female blue. Larger than hobbies, with longer, darker wings and a longer tail; in good light, rufous extends further onto belly than Eurasian Hobby. Dark morph is almost black; much darker and larger than Sooty Falcon (p. 152). Juv. has buffy fringes to upperpart feathers, appearing scaled at close range; underparts creamy buff to rufous with dark grey-brown chevrons. Larger and richer coloured below than juv. Sooty Falcon; rufous underwing coverts contrast with paler flight feathers (coverts uniform or paler in juv. Sooty Falcon). **Voice:** Silent in Africa. **Status and biology:** Very rare, Palearctic-br. summer migrant; most winter in Madagascar, but recent satellite tracking studies show that juvs occur widely throughout Africa. Most likely to be found in Mozambique, foraging over savanna, woodland and forest. (Eleonoravalk)

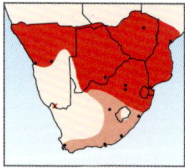

EURASIAN HOBBY *Falco subbuteo*

28–36 cm; 135–320 g; wingspan 68–84 cm A mid-sized, slender falcon with long, pointed wings and dashing flight. Ad. diagnostic features are its white, heavily streaked breast, rufous leggings and vent, and black hood and 'moustache'. African Hobby is similar but has all-rufous underparts; juv. Amur and Red-footed Falcons (p. 156) have red (not yellow) legs and they lack rufous 'trousers'. Juv. has creamy, heavily streaked underparts; lacks rufous vent and leggings; most similar to juv. Amur and Red-footed Falcons, distinguished by its yellow (not orange-red) legs, longer wings and more dashing flight. **Voice:** Silent in Africa. **Status and biology:** Fairly common, Palearctic-br. summer visitor, mainly in woodland and savanna. Usually solitary, but may hunt in small groups at dusk, flying fast and low over open ground. (Europese Boomvalk)

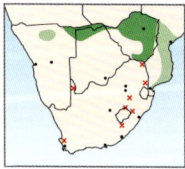

AFRICAN HOBBY *Falco cuvierii*

28–30 cm; 150–220 g; wingspan 60–73 cm A rather small, compact falcon; smaller than Eurasian Hobby, with unstreaked, rufous underparts (including throat and face). Can appear all dark in poor light. More slender than Taita Falcon, with longer wings and tail and a dark head lacking rufous nape patches. Juv. has dark-streaked, rufous breast and belly; head and throat colour varies from rufous to white. White-throated juv. resembles pale morph Eleonora's Falcon, but is much smaller and shorter tailed. **Voice:** High-pitched '*kik-kik-kik-kik*' display call. **Status and biology:** Uncommon, intra-African br. summer migrant in woodland, forest and adjoining open country. Hunts mainly at dawn and dusk. (Afrikaanse Boomvalk)

TAITA FALCON

fous patterning
n hindneck

ad.

ad.

juv.

lightly
streaked
below

short
tail

ad.

ad.

white throat
contrasts with rufous
underparts

ELEONORA'S FALCON

ad. dark
morph

ad. pale
morph

single moustachial
stripe (all ages)

juv.

light rufous,
extending
to belly
& breast

wing tips
reach tip of tail at rest

long tail &
wings

ad. pale
morph

ad. pale
morph

white
throat &
cheeks

dark coverts
contrast with pale
flight feathers
(all ages)

ad. dark
morph

dark trailing
edge to
wings

juv.

EURASIAN HOBBY

double
moustachial
stripes
(all ages)

ad.

pale nape spots

juv.

ad.

ad.

juv.

buff-yellow
leggings, vent
& undertail

rufous restricted
to leggings, vent &
undertail

uniform
patterning to
underwings
with thin pale
trailing edges

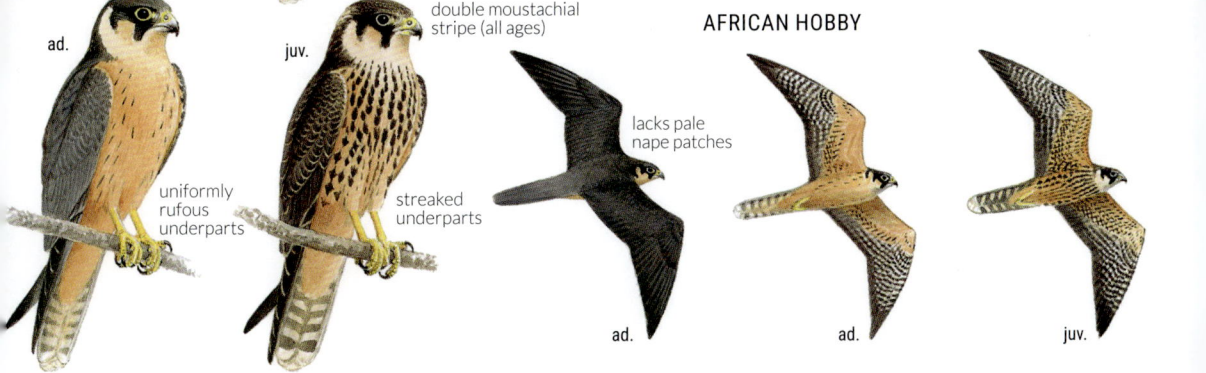

AFRICAN HOBBY

double moustachial
stripe (all ages)

ad.

juv.

lacks pale
nape patches

uniformly
rufous
underparts

streaked
underparts

ad.

ad.

juv.

151

GREY-PLUMAGED KESTRELS AND FALCONS

Sooty Falcon is a long-winged, fast-flying aerial hunter allied to the hobbies and Eleanora's Falcon (p. 150). It is placed with the 2 grey-plumaged kestrels because of their similarity in plumage. Kestrels are shorter-winged than true falcons and lack their supreme flying skills; they eat largely ground-dwelling prey, hunting from perches or by hovering. Sexes alike in these 2 kestrels, but differ in many other species.

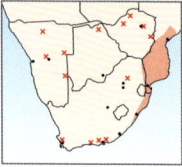

SOOTY FALCON *Falco concolor*

J F M A M J J A S O N D

32–36 cm; 200–250 g; wingspan 75–88 cm A sleek, long-winged falcon, with folded wings extending, usually crossed over, beyond tail tip (unlike shorter, broader wings of Grey Kestrel). Ad. is slate-grey with blackish primaries; lacks chestnut vent of Red-footed and Amur Falcons (p. 156), and has yellow (not orange-red) legs and cere. In flight, resembles Eurasian Hobby (p. 150) or Eleanora's Falcon (p. 150): in poor light the 3 are easily mistaken for each other. Sexes alike. Juv. is darker grey above than ad., with narrow, pale feather margins in fresh plumage; throat is creamy; rest of underparts buffy, streaked soft grey. Told from juv. Eurasian Hobby by its narrower moustachial streak and paler breast streaking. Smaller and paler below than juv. Eleanora's Falcon, with pale underwing coverts paler than flight feathers (not rufous underwing coverts darker than flight feathers). **Voice:** Mostly silent in the region. **Status and biology:** NEAR-THREATENED. Uncommon, non-br. summer migrant from ne Africa and Arabia, mainly to coastal Mozambique and n KZN. Frequents tall-tree savanna, often along margins of wetlands; solitary or in small groups, hunting aerially at dusk. (Roetvalk)

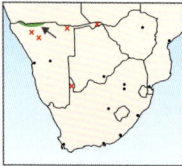

GREY KESTREL *Falco ardosiaceus*

J F M A M J J A S O N D

30–35 cm; 190–300 g; wingspan 58–72 cm A small, compact, uniformly grey kestrel with barred flight feathers. Wings shorter and broader than those of Sooty Falcon, not reaching the tip of the square tail at rest. Flat crown gives rather square-headed look (head of Sooty Falcon is rounded), and has more extensive yellow skin around eye. Lacks contrasting pale head and rump and strongly barred tail of Dickinson's Kestrel. Sexes alike. Juv. is lightly washed brown; facial skin and cere are greenish. **Voice:** High-pitched, rasping trill. **Status and biology:** Uncommon resident; solitary or in pairs, usually perched on a high vantage point in open woodland, often around palms. Most easily seen along the Kunene River west of Ruacana, Namibia. (Donkergrysvalk)

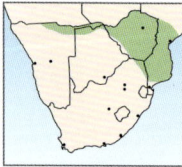

DICKINSON'S KESTREL *Falco dickinsoni*

J F M A M J J A S O N D

28–30 cm; 170–240 g; wingspan 61–68 cm A small, compact, grey kestrel with contrasting pale grey head and rump (not uniform grey as in Grey Kestrel) and strongly grey-and-black-barred tail, with broader black subterminal bar. In flight from below, flight feathers conspicuously barred. Head is square, with flat crown; appears large-headed. Ad. has extensive bare yellow skin around eye, joining with cere. Sexes alike. Juv. is lightly washed brown, with small, greenish eye-ring and cere; head and rump darker; best separated from Grey Kestrel by its strongly barred tail and underwings (ranges overlap marginally in n Namibia). **Voice:** A tremulous, high-pitched '*keee-keee-keee*'. **Status and biology:** Uncommon resident in low-lying tropical savanna, especially where baobabs or tall palms are abundant. Subject to erratic winter influxes in n Botswana and Zimbabwe. Solitary or in pairs; typically perches high in a dead tree. Most prey taken from the ground. (Dickinsongrysvalk)

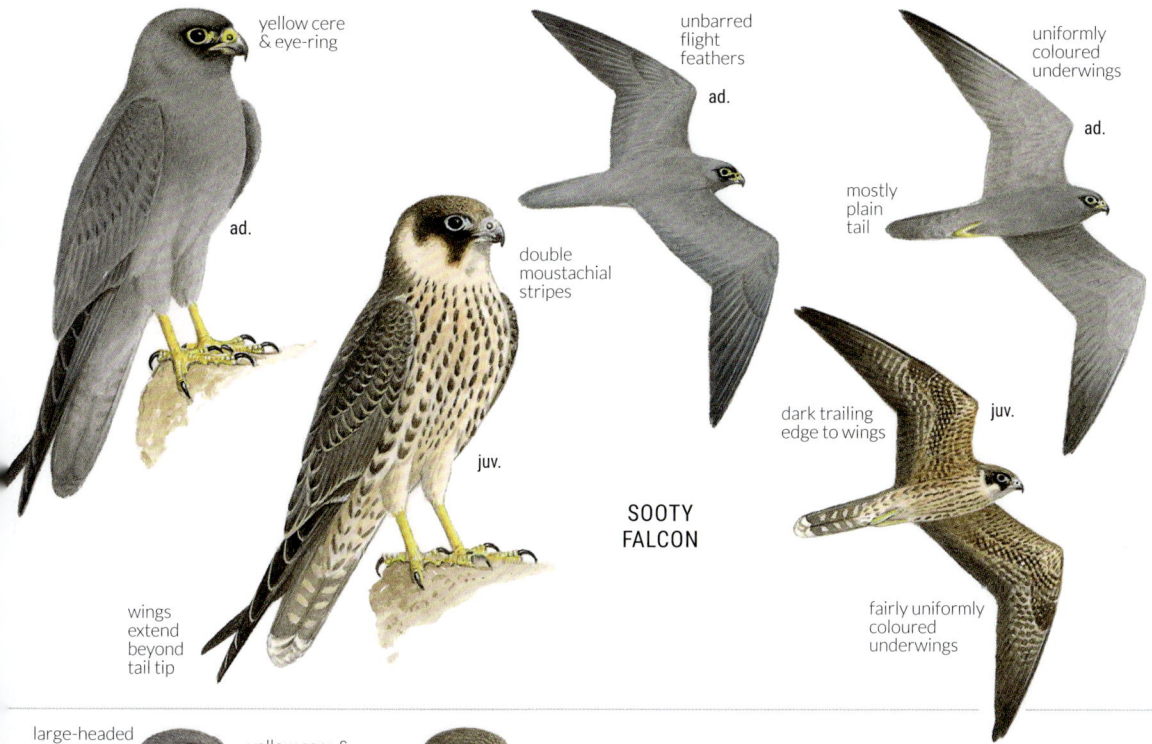

yellow cere & eye-ring

ad.

unbarred flight feathers

ad.

uniformly coloured underwings

ad.

mostly plain tail

double moustachial stripes

dark trailing edge to wings

juv.

fairly uniformly coloured underwings

juv.

SOOTY FALCON

wings extend beyond tail tip

juv.

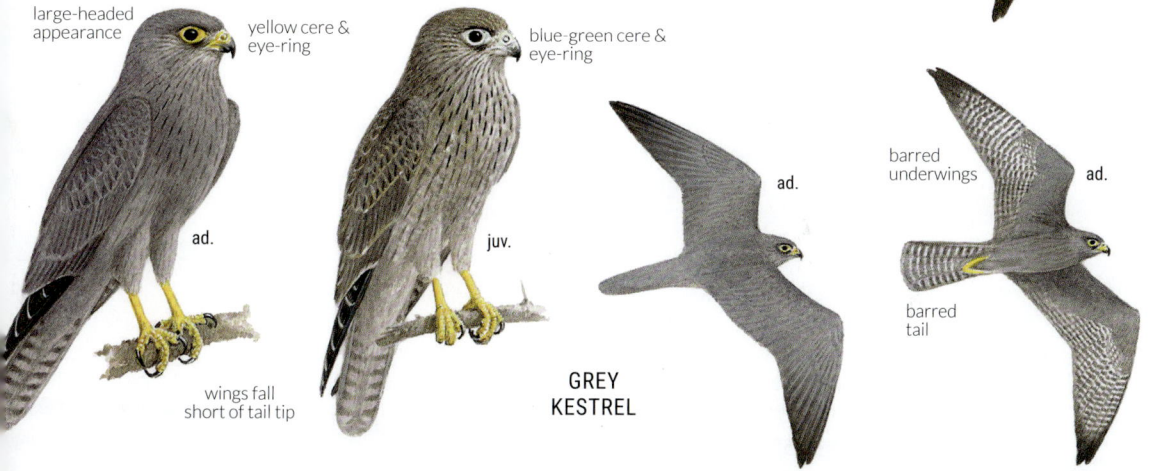

large-headed appearance

yellow cere & eye-ring

ad.

blue-green cere & eye-ring

juv.

barred underwings

ad.

ad.

barred tail

wings fall short of tail tip

GREY KESTREL

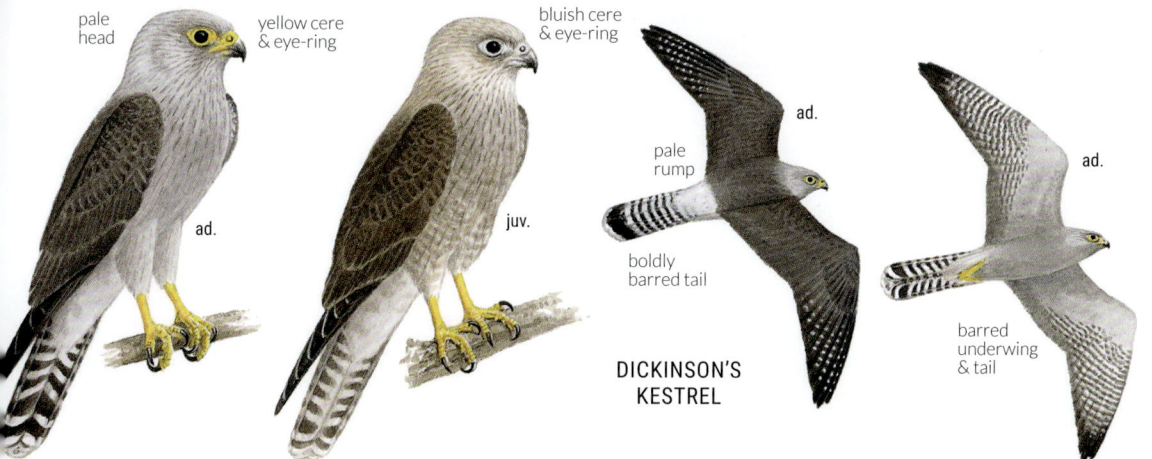

pale head

yellow cere & eye-ring

ad.

bluish cere & eye-ring

juv.

ad.

pale rump

boldly barred tail

ad.

barred underwing & tail

DICKINSON'S KESTREL

153

KESTRELS

KESTRELS These 3 kestrels are primarily insectivorous, preying especially on grasshoppers, although they also take rodents and other vertebrates. All are accomplished at hunting on the wing by hovering. Lesser Kestrel is a Palearctic-br. summer visitor, always in flocks, and often mixing with Amur and/or Red-footed Falcons (p. 156). Rock and Greater Kestrels are resident species, living in pairs that defend territories through the br. season and dispersing singly at other times of the year, although may gather in small groups at favoured hunting sites (e.g. along ridges where updrafts aid hovering).

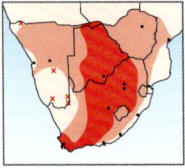

LESSER KESTREL *Falco naumanni*

J F M A M J J A S O N D

26–32 cm; 110–180 g; wingspan 62–73 cm A small, slender kestrel with well-marked sexual dimorphism. Male easily identified by its unmarked chestnut back, grey greater coverts and rather plain, warm buff underparts. Female has cream underparts with dark streaks; face is rather pale, with dark malar stripes. Resembles juv. Rock Kestrel but is barred (not spotted) above, has malar stripes and a rufous-and-black (not grey-and-black) barred tail. If perched birds are seen well, the pale claws of Lesser Kestrel are a good distinguishing feature (black/dark claws in Rock Kestrel). Also similar to Greater Kestrel, especially juv., but is smaller, with greater contrast between upper- and underparts and, in flight, blackish-grey outer wing contrasts with brown inner wing (upper wing of Greater Kestrel more uniformly brown). Juv. is more rufous than female, with less well-defined malar stripe and rufous (not grey) rump. **Voice:** Silent by day, but noisy at roosts, giving high-pitched '*kiri-ri-ri-ri*'. **Status and biology:** Locally abundant, Palearctic-br. summer visitor. Restricted to open country, including grassland, agricultural lands and semi-arid scrub. Invariably in flocks, often mingling with Amur and Red-footed Falcons (p. 156). Commonly encountered perched on telephone and fence wires. Roosts communally in traditional sites in tall trees, gathering noisily at dusk in large numbers. (Kleinrooivalk)

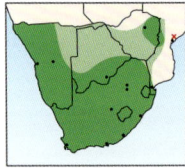

ROCK KESTREL *Falco rupicolus*

J F M A M J J A S O N D

30–34 cm; 185–275 g; wingspan 58–78 cm A fairly small, slender kestrel; grey head and chestnut back, spotted with black, is diagnostic in ads of both sexes. Female differs from male by having more prominent black bars on tail and rump and some dark streaking on head. Juv. similar to female, but has a brownish head and more heavily streaked upper- and underparts. Smaller than Greater Kestrel, with greater contrast between upper- and underparts and, in flight, blackish (unbarred) outer upperwing contrasts with browner inner wing. Colour and barring in tail distinguish it from juv. Lesser Kestrel. **Voice:** High-pitched '*kik-ki-ki*'. **Status and biology:** Common resident in parts of its range and a winter visitor or nomad in other areas. Frequents open country, especially grassland and semi-arid scrub; when br., usually in hilly country with cliffs available for nest sites. Usually solitary in non-br. season. (Kransvalk)

GREATER KESTREL *Falco rupicoloides*

J F M A M J J A S O N D

34–38 cm; 200–300 g; wingspan 68–84 cm A rather large, round-headed, pale kestrel; ad. has a diagnostic cream-coloured eye; upper- and underparts are uniformly pale rufous, densely barred on the back and flanks, and streaked on the breast. Lacks malar stripe; grey-and-black-barred tail is tipped white. Larger size, pale coloration and cream-coloured eye distinguish it from other kestrels. Juv. has rufous tail barred dark brown, dark eyes and streaked (not barred) flanks. **Voice:** Shrill, repeated '*kee-ker-rik*'. **Status and biology:** Locally common resident, especially in open, semi-arid and arid country. Solitary or in pairs, usually perched on a high vantage point, often a roadside telephone pole. Hunts from perches or by hovering. (Grootrooivalk)

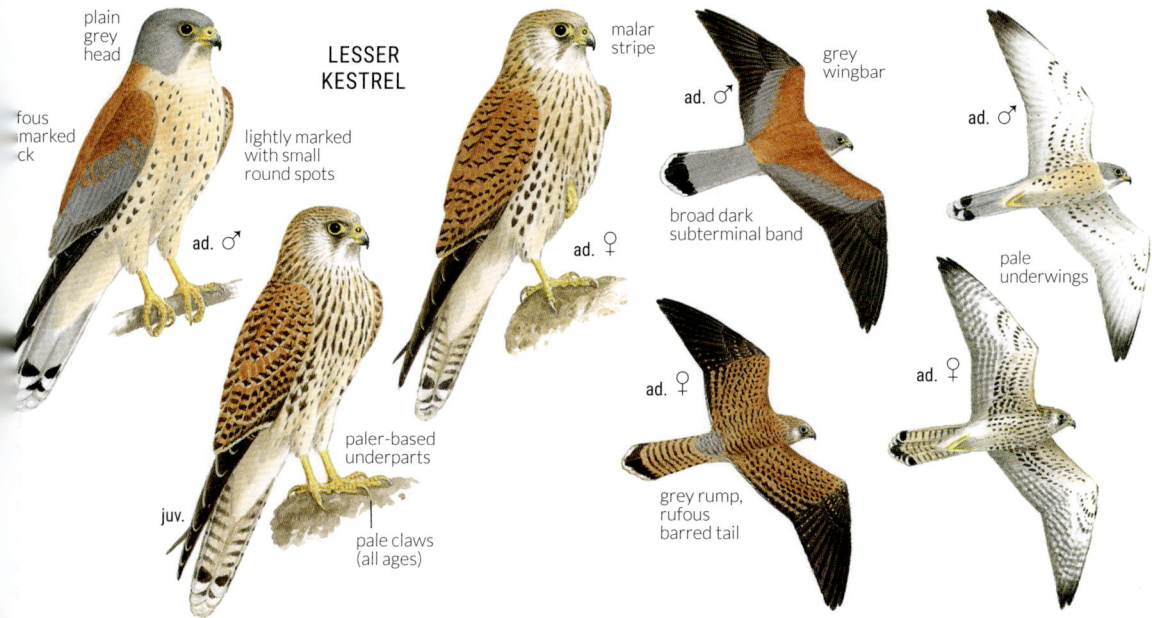

plain grey head

fous marked ck

LESSER KESTREL

lightly marked with small round spots

ad. ♂

malar stripe

grey wingbar

ad. ♂

ad. ♂

ad. ♀

broad dark subterminal band

pale underwings

paler-based underparts

juv.

pale claws (all ages)

ad. ♀

grey rump, rufous barred tail

ad. ♀

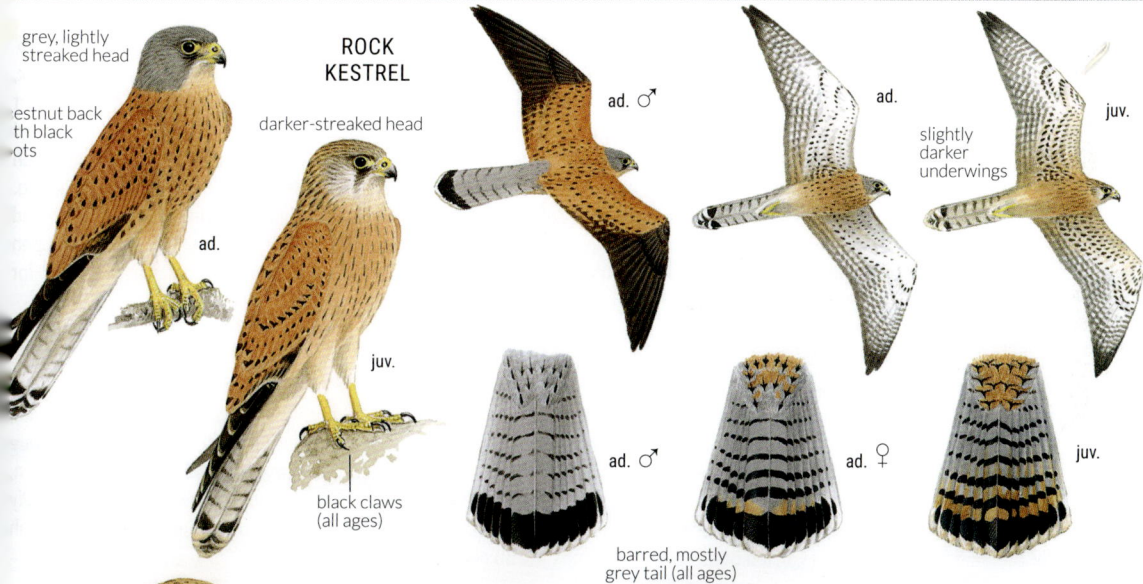

grey, lightly streaked head

ROCK KESTREL

estnut back th black ots

darker-streaked head

ad. ♂

ad.

juv.

slightly darker underwings

ad.

juv.

ad. ♂

ad. ♀

juv.

black claws (all ages)

barred, mostly grey tail (all ages)

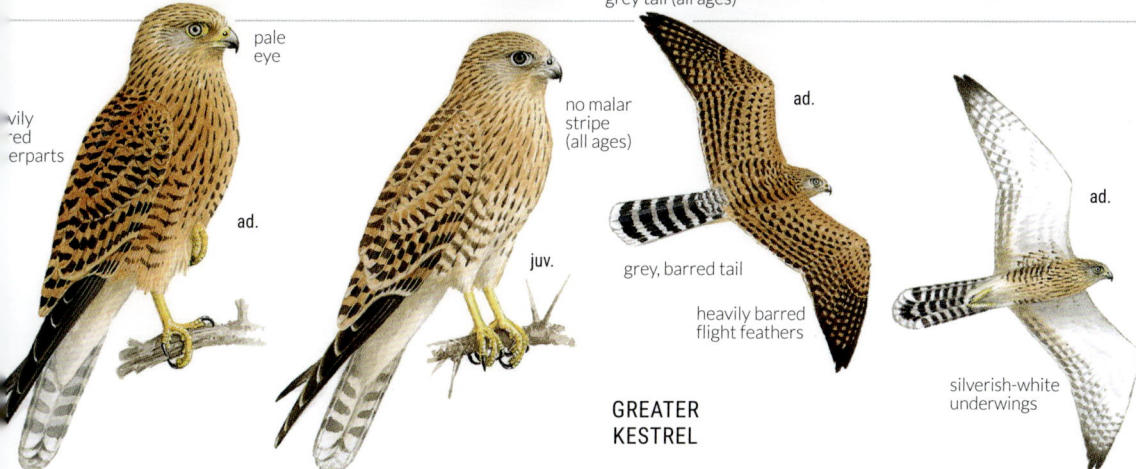

pale eye

vily ed erparts

ad.

no malar stripe (all ages)

ad.

ad.

juv.

grey, barred tail

heavily barred flight feathers

silverish-white underwings

GREATER KESTREL

155

SMALL FALCONS Amur and Red-footed Falcons are essentially kestrels, lacking the dashing flight and aerial hunting skills of true falcons; instead they hunt insects and other ground-dwelling prey from perches and by hovering. Sexes differ markedly in both species. They are non-br. summer visitors, moving into s Africa in nomadic flocks, often mingling with migratory Lesser Kestrels (p. 154). The Pygmy Falcon is the African representative of the falconets, diminutive, shrike-like falcons. It lives in sedentary pairs in semi-arid areas, closely associated in s Africa with the distribution of Sociable Weavers, which provide its nest and roost sites.

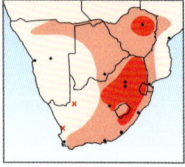

AMUR FALCON *Falco amurensis*

J F M A M J J A S O N D

28–30 cm; 100–185 g; wingspan 63–71 cm A small, kestrel-like falcon. Male is dark slate-grey, paler below than above, with chestnut vent and bright red legs, cere and eye-ring; very similar to male Red-footed Falcon at rest, but in flight its white (not dark grey) underwing coverts are diagnostic. Female is whitish below, variably streaked, spotted and barred with dark grey; vent is buffy and unmarked. Superficially resembles Eurasian Hobby (p. 150), but has a white (not dark) forehead, paler underwing and orange-red (not yellow) legs, cere and eye-ring. Juv. resembles female, but is streaked below and has buffy edges to upperpart feathers. **Voice:** Generally silent, except at roosts where it makes a shrill chattering. **Status and biology:** Common, Asian-br. summer visitor, thought to migrate across the Indian Ocean from India to Africa, but returns north via the Horn of Africa. Frequents open country, especially grasslands and agricultural areas, invariably in flocks and often associated with Lesser Kestrels (p. 154). Commonly perches on telephone and fence wires. Roosts communally in tall trees. (Oostelike Rooipootvalk)

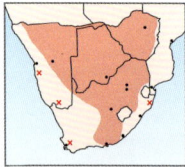

RED-FOOTED FALCON *Falco vespertinus*

J F M A M J J A S O N D

29–31 cm; 120–200 g; wingspan 66–75 cm Closely related to Amur Falcon; male distinguishable only in flight by dark grey (not white) underwing coverts. Rufous vent and red legs, cere and eye-ring distinguish it from other grey-plumaged falcons. Female has distinctive, ginger-buff head, underparts and underwing coverts, a brownish cap, black facial mask and moustachial stripe, and red or orange legs, cere and eye-ring. Juv. like female but streaked with dark grey below, back feathers edged with buff and has a whitish head, dark face mask and indistinct cap; paler than female or juv. Amur Falcon. **Voice:** Generally silent, except at roosts where it makes a shrill chattering. **Status and biology:** NEAR-THREATENED. An uncommon, Palearctic-br. summer migrant; main non-br. range west of that of Amur Falcon. Occurs mainly in arid savanna; invariably in flocks, often associated with other migrant kestrels. Roosts communally in tall trees. (Westelike Rooipootvalk)

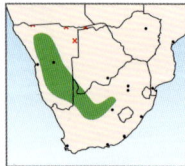

PYGMY FALCON *Polihierax semitorquatus*

J F M A M J J A S O N D

18–20 cm; 55–66 g; wingspan 34–40 cm A tiny, shrike-like falcon; unmistakable. Male is grey above and white below; female has chestnut back. Juv. has dull brown back and buff-washed underparts. Flight undulating and rapid, reminiscent of a woodpecker; white-spotted black flight feathers and tail contrast with white rump. **Voice:** Noisy; high-pitched '*chip-chip*' and '*kik-kik-kik-kik*'. **Status and biology:** Locally common resident in arid savanna; in pairs or family groups, typically near nests of Sociable Weavers (or occasionally nests of Red-billed Buffalo Weaver, Wattled Starling or White-browed Sparrow-Weaver). Sits upright on exposed perches; often bobs head and pumps tail up and down. (Dwergvalk)

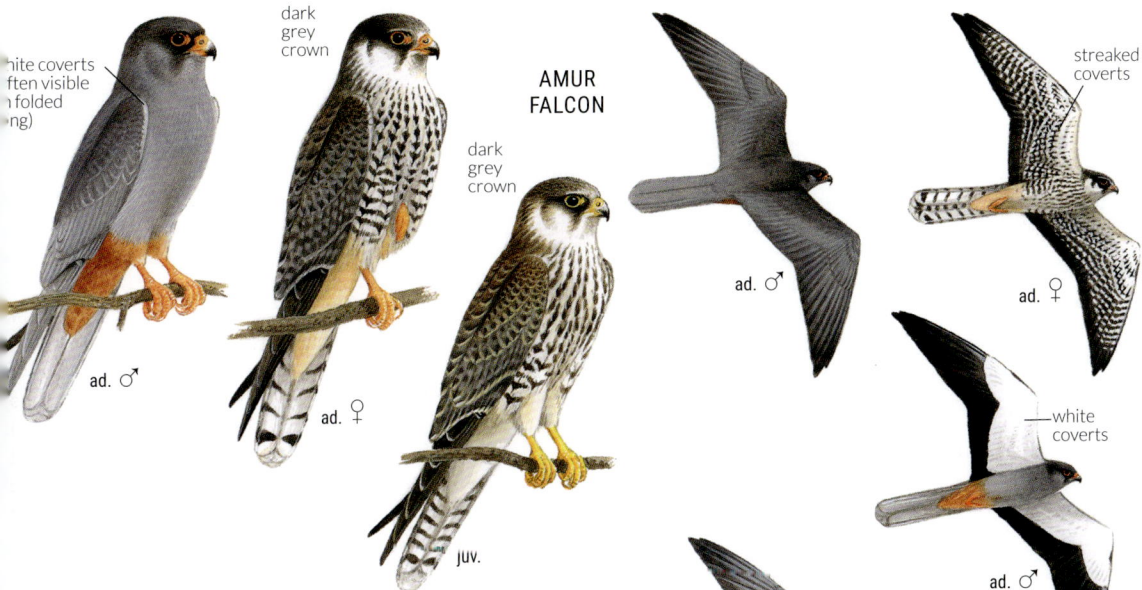

white coverts
(often visible
on folded
wing)

dark grey crown

AMUR FALCON

dark grey crown

streaked coverts

ad. ♂

ad. ♀

white coverts

ad. ♂

ad. ♂

juv.

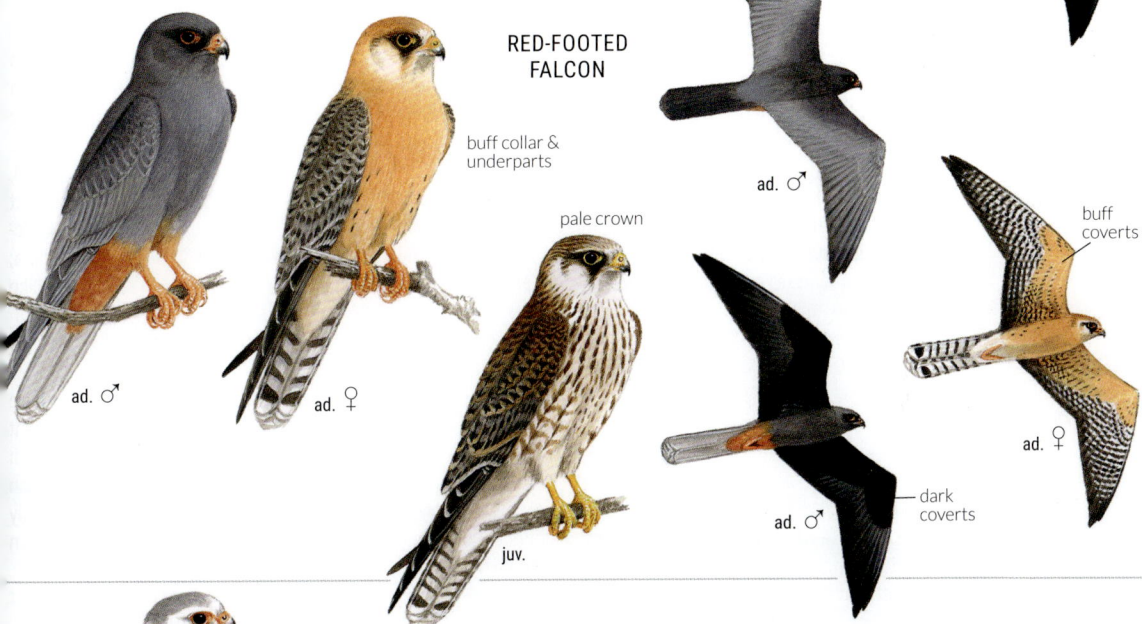

RED-FOOTED FALCON

buff collar & underparts

pale crown

buff coverts

ad. ♂

ad. ♀

ad. ♂

dark coverts

ad. ♂

ad. ♀

juv.

grey mantle

PYGMY FALCON

white rump

chestnut back

shrike-like posture

ad. ♀

ad. ♂

ad. ♂

ad. ♀

juv. ♂

ad.

OSTRICHES
The largest birds; flightless; adapted for running, with long legs and only 2 toes. Sexes differ; male larger and darker.

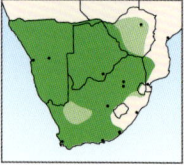

COMMON OSTRICH *Struthio camelus*

b	b	b	b	b	b	B	B	B	B	B	B
J	F	M	A	M	J	J	A	S	O	N	D

1.2–2 m; 60–80 kg Huge, flightless bird with long, bare legs and neck. Male has blackish plumage with white wings and tail (often stained chestnut). Female and imms grey-brown. Juv. banded tan and brown, resembling small, downy bustard or long-legged gamebird. While genuine Common Ostriches occur widely across the semi-arid and arid regions of Namibia and Botswana, with few exceptions, ostriches found across S Africa originate from domesticated hybrids originally bred for the feather industry and derived from Somali Ostrich stock. These birds, which are smaller and have differently coloured bare parts to wild Common Ostriches, have been very widely introduced into game-farming areas. **Voice:** Booming, lion-like roar, mostly at night. **Status and biology:** Wild populations scarce; restricted to large reserves and wilderness areas. Feral birds derived from farming operations are widespread. Occurs in savanna and semi-desert plains. (Volstruis)

GUINEAFOWL
An endemic African family of large gamebirds with mostly naked heads and necks. Often in flocks; roost communally in trees. Sexes alike. Chicks precocial, able to fly within 2–3 weeks of hatching, while still appreciably smaller than ads; at this stage could be confused with other gamebirds.

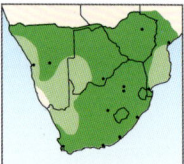

HELMETED GUINEAFOWL *Numida meleagris*

B	b	b						b	b	B	B
J	F	M	A	M	J	J	A	S	O	N	D

55–60 cm; 1.1–1.8 kg A large, well-known gamebird with blue-grey plumage, uniformly spotted with white. Head pattern varies geographically, but generally naked blue and red with cheek wattles and a pale casque on the crown; male has a larger casque than female. Birds with white faces and yellowish toes are usually hybrids with domesticated strains. Juv. plumage browner; head partly feathered, with dull blue skin and greatly reduced casque and wattles. **Voice:** Soft 'kek' contact calls given during the day, but noisy at dawn and dusk, making raucous 'kek-ek-ek kaaaaa' and 'eerrrrk' calls. **Status and biology:** Common resident in grassland, woodland, savanna and fields; may flock in hundreds. (Gewone Tarentaal)

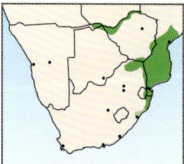

CRESTED GUINEAFOWL *Guttera pucherani*

b	b									b	B
J	F	M	A	M	J	J	A	S	O	N	D

46–52 cm; 800–1 500 g Similar to Helmeted Guineafowl, but with a tuft or topknot of curly black feathers on the head, naked blue-grey face, red eyes, black neck and pale blue (not white) spots on the body feathers. The outer secondaries have white outer webs, producing a white wing panel in flight. Juv. duller, with short topknot and fine black-and-white barring on its body feathers. **Voice:** A 'chik-chik-chil-urrrrr' and a soft 'keet-keet-keet' contact call. **Status and biology:** Locally common resident of forest edge, thicket and dense woodland. Often follows troops of monkeys, picking at scraps of fruit and faeces. (Kuifkoptarentaal)

PEAFOWL
Large, strongly dimorphic gamebirds with ornate head plumes; ad. males have showy, iridescent plumage.

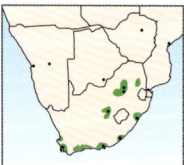

INDIAN PEAFOWL *Pavo cristatus*

					b	b	b				
J	F	M	A	M	J	J	A	S	O	N	D

90–120 cm (220 cm incl. train); 2.8–6 kg Introduced. A large, distinctive gamebird. Male readily identified by its crest, long, ocellated train and rufous flight feathers, and iridescent blue head, neck and breast. Female is smaller and duller with greenish neck, black-spotted breast and white belly; lacks long train. **Voice:** Very loud, trumpeting 'kee-ow' (he-elp), mostly given at dawn and dusk. **Status and biology:** Feral populations on Robben Island and in some suburban areas; free-ranging domestic birds often found around farm buildings. (Makpou)

COMMON OSTRICH

ad. ♂

ad. ♀

chick

juv.
ca. 80 days

HELMETED GUINEAFOWL

juv.

bony casque

whitish face

ad.

ad.
domesticated form

yellow toes

CRESTED GUINEAFOWL

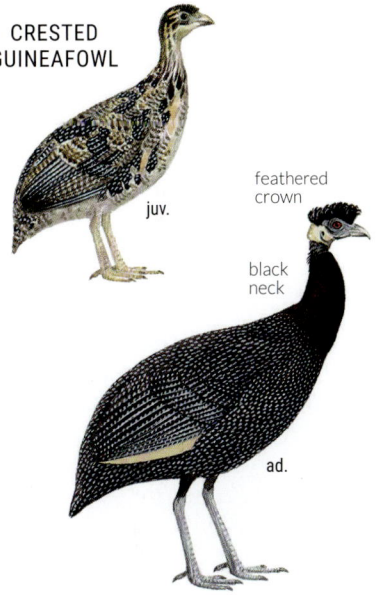

juv.

feathered crown

black neck

ad.

INDIAN PEAFOWL

ad. ♂
display

ad. ♂

ad. ♀

SPURFOWL AND FRANCOLINS

The galliform gamebirds form a lineage of birds related to waterfowl (Galloanserae) that are distinct from most other birds (Neoaves). The francolins and spurfowl are small to medium-sized chicken-like gamebirds. Sexes alike in most species; male has a spur on each tarsus. Precocial chicks can flutter within 2 weeks of hatching, when still much smaller than ads.

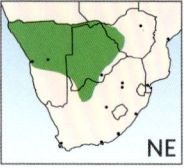

RED-BILLED SPURFOWL *Pternistis adspersus*

b b **B B B B** b b b b b b
J F M A M J J A S O N D

35–38 cm; 350–620 g A fairly large, dark brown spurfowl with a red bill, dull red legs and a diagnostic yellow eye-ring. Head, neck and underparts are finely barred dark brown and white; lores blackish; back very finely vermiculated, appearing plain brown from a distance. Juv. browner above than ad., with more prominent dark brown barring; bill and legs brown; yellow around eye reduced. First-plumage chick (half size of ad.) has a broad, buff supercilium, distinctive whitish stripes on each body feather, a dull horn-coloured bill and pale yellow legs. **Voice:** Loud, harsh *'chaa-chaa-chek-chek'* given at dawn and dusk. **Status and biology:** Near-endemic. Locally common resident in arid savanna and open, broadleafed woodland, often in thickets along watercourses. Less skulking than most other spurfowl, becoming tame around camp sites. (Rooibekfisant)

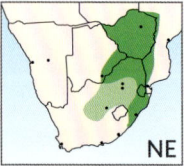

NATAL SPURFOWL *Pternistis natalensis*

B B B B B b b b b b b **B**
J F M A M J J A S O N D

30–38 cm; 330–650 g A medium-sized, brown spurfowl with orange-red bill and legs and strikingly vermiculated underparts. Smaller than Red-necked Spurfowl, lacking bare red face and throat. Best told from Cape Spurfowl by its much paler lower flanks and belly, although ranges do not overlap. Juv. richly barred above and buff below with white feather-shaft streaks; bill greenish-brown; legs dull red-brown. **Voice:** Raucous, screeching *'krr kik-ik-ik'*. **Status and biology:** Near-endemic. Common resident in woodland, especially riparian thicket. Occasionally hybridises with Swainson's Spurfowl, and possibly Red-necked and Red-billed Spurfowl. (Natalse Fisant)

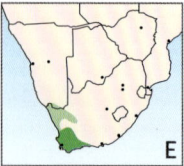

CAPE SPURFOWL *Pternistis capensis*

b b b   b **B B B B**
J F M A M J J A S O N D

40–43 cm; 650–1 000 g A large, dark brown spurfowl with a slightly paler supercilium, sub-loral spot and throat. Bill is dull red with a dark top to upper mandible and yellowish base; legs orange-red. Lacks bare red face and throat of Red-necked Spurfowl. At close range, body feathers are finely vermiculated, with bolder buffy-white stripes on lower breast and flanks, but lower flanks lack the broad, creamy feather margins of Natal Spurfowl; ranges do not overlap. Juv. duller, with mainly brown bill and dull red-brown legs. **Voice:** Male advertising call, given mostly in the morning, is a series of loud *'ka-ke-KAA'* notes, ending in a guttural laugh. **Status and biology:** Endemic. Common resident in lowland fynbos, pastures, fields, large gardens and riparian thickets in the Karoo. Often confiding. (Kaapse Fisant)

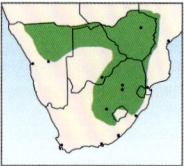

SWAINSON'S SPURFOWL *Pternistis swainsonii*

B B B B B b b b b b **B**
J F M A M J J A S O N D

33–38 cm; 350–820 g A large, dark brown spurfowl with black legs and bill and a bare red face and throat. Neck and mantle feathers are fringed pale grey, and breast and flank feathers have a prominent blackish central stripe. Juv. is duller and paler, with less extensive red facial skin and throat patch covered in small, white feathers. Legs yellowish; belly white, finely barred blackish. In flight, shows barring in flight feathers. **Voice:** Raucous *'krraae-krraae-krraae'* given by males at dawn and dusk. **Status and biology:** Common resident in dry savanna and fields, usually in groups of 3–6 birds. Its range has expanded southwards across the highveld in the past half-century. Occasionally hybridises with Natal, Red-billed and Red-necked Spurfowl. (Bosveldfisant)

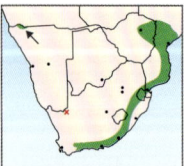

RED-NECKED SPURFOWL *Pternistis afer*

B B B b **B B B**   b **B**
J F M A M J J A S O N D

35–40 cm; 430–850 g A large, brown-backed spurfowl with red bill and legs, and a bare red face and throat. Plumage varies regionally: in the east of S Africa, *P. a. castaneiventer* has a dark brown head and blackish underparts, with 2 white stripes on each feather (width of white stripes variable). In Mozambique and e Zimbabwe, *P. a. humboldtii* has a mostly white head and black belly patch; upperparts paler grey-brown. In nw Namibia, *P. a. afer* has a white supercilium and moustachial stripe and broad, white margins to feathers of lower breast, flanks and belly. Juv. duller, with whitish feathers on face and throat; breast and flanks barred buff, white and dark brown; bill and legs dull yellow. **Voice:** Loud *'kwoor-kwoor-kwoor-kwaaa'* given at dusk and dawn. **Status and biology:** Fairly common resident of forest edges, thicket, riparian scrub and adjoining grassland. Occasionally hybridises with Swainson's Spurfowl. (Rooikeelfisant)

RED-BILLED
SPURFOWL

yellow
eye-ring

finely
barred
belly

ad.

NATAL
SPURFOWL

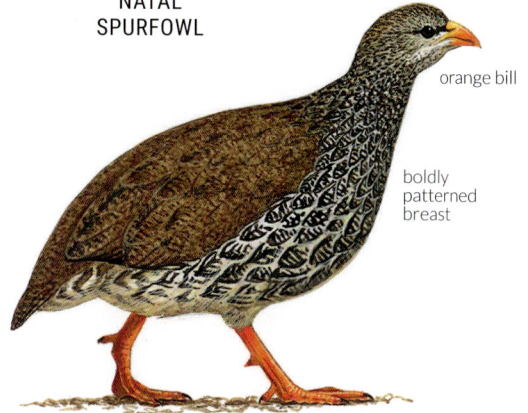

orange bill

boldly
patterned
breast

CAPE SPURFOWL

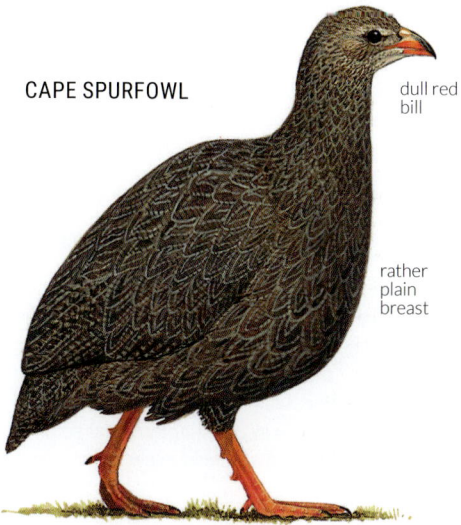

dull red
bill

rather
plain
breast

SWAINSON'S
SPURFOWL

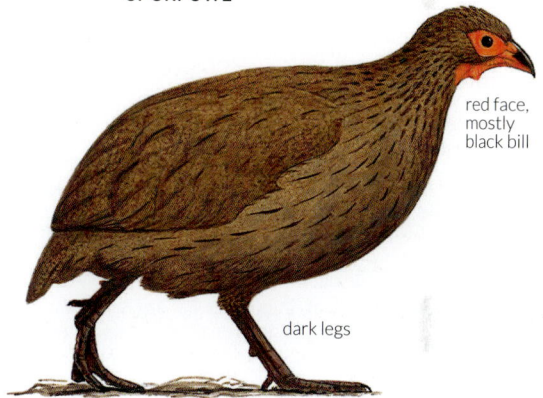

red face,
mostly
black bill

dark legs

castaneiventer

red face
& bill

red legs

afer

mostly white,
dark-streaked
underparts

RED-NECKED
SPURFOWL

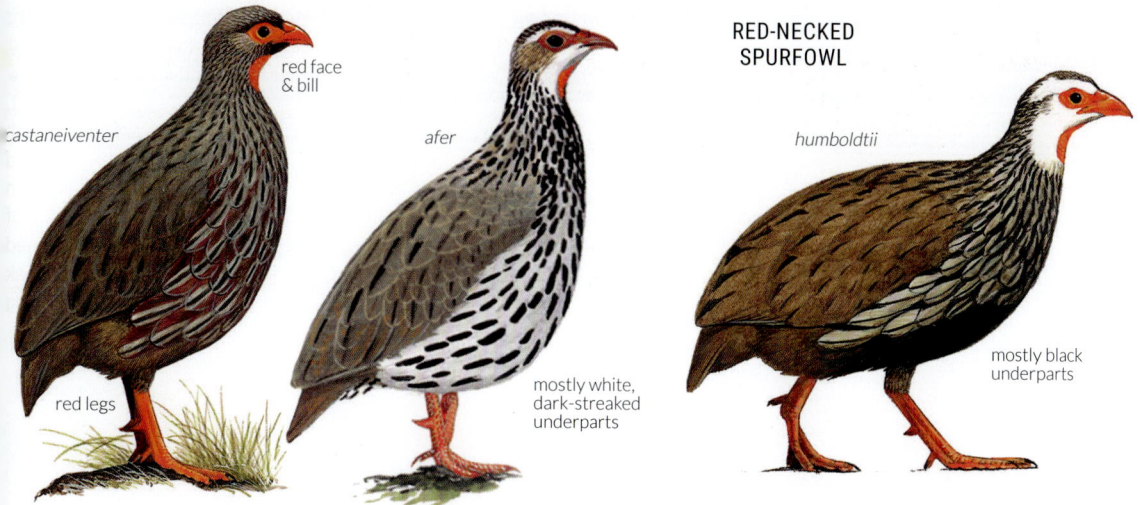

humboldtii

mostly black
underparts

161

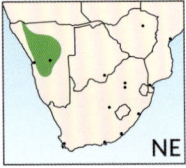

HARTLAUB'S SPURFOWL *Pternistis hartlaubi*

b b b b b B B B b b b
J F M A M J J A S O N D

25–28 cm; 210–290 g A small, strongly dimorphic spurfowl with a heavy, decurved, yellow-horn bill and dull yellow legs. Occurs alongside Orange River Francolin (p. 164), but unlikely to be confused given an adequate view. Upperparts mottled blackish-brown and rufous. Male has a striking black-and-white supercilium and chestnut cheeks; underparts whitish, heavily streaked brown. Female has face and underparts dull orange-brown. Juv. resembles dull, scruffy female; juv. male develops a white supercilium after about 3 months. **Voice:** Distinctive duet, *'ke-rak, keer-a keer-a kew'*, led by female; calls mostly at dawn. **Status and biology:** Near-endemic. Locally common resident of boulder-strewn slopes and rocky outcrops. Pairs remain together year-round. (Klipfisant)

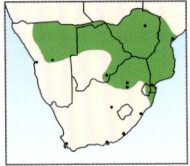

CRESTED FRANCOLIN *Ortygornis sephaena*

B B B b b   b B B
J F M A M J J A S O N D

30–35 cm; 240–460 g A distinctive francolin with a rather long, black tail that is often cocked at a 45° angle, imparting a bantam-like appearance. Dark cap contrasts with broad, white supercilium; neck and breast with small white chevrons and heavy, dark streaks, while lower breast and belly lack dark streaks and have thin pale streaks and very light barring; back and wing coverts have distinctive white stripes. Female is duller than male; back finely barred with buff (not white) stripes. Juv. paler above than ad. Kirk's Francolin of central Mozambique is elevated as a separate species and is separated by having dark brown streaks (not fine barring and light streaks) on belly and flanks. **Voice:** Rattling duet, *'chee-chakla, chee-chakla'*. **Status and biology:** Common resident in woodland and well-wooded savanna. (Bospatrys)

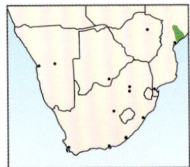

KIRK'S FRANCOLIN *Ortygornis rovuma*

B B B b b   b B B
J F M A M J J A S O N D

28–33 cm; 200–450 g Slightly smaller than Crested Francolin, with similar plumage, however, this species has dark teardrop markings (not white streaks) on the lower breast and lacks any barring below. **Voice:** Similar rattling duet, *'chee-chakla, chee-chakla'*, to Crested Francolin but more compact. **Status and biology:** Common shy resident in lowland woodland and well-wooded savanna of central Mozambique. (Mosambiekpatrys)

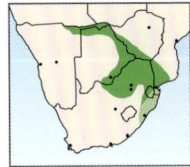

COQUI FRANCOLIN *Campocolinus coqui*

B B b b B b b b b B B
J F M A M J J A S O N D

20–26 cm; 200–300 g A small francolin with marked sexual dimorphism; bill black with yellow base; legs yellow. Male has a plain, tawny head with darker crown; breast and belly heavily barred. Female has neatly defined pale throat and supercilium, and plain, buffy breast; superficially resembles a female quail, but is appreciably larger. Difficult to flush; shows chestnut wings and outer tail in flight. Juv. resembles ad. female, but paler above and buffy below with reduced black barring. **Voice:** Distinctive, disyllabic *'ko-ki, ko-ki'*; also a fast, trumpeted *'ker-aak, aak, kara-kara-kara'* with last notes fading away, given in territorial defence. **Status and biology:** Locally common to common resident of woodland and savanna, especially on sandy soils. (Swempie)

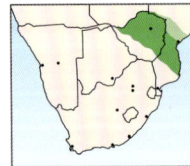

STUHLMANN'S FRANCOLIN *Campocolinus stuhlmanni*

B B b b B b b b b B B
J F M A M J J A S O N D

20–26 cm; 200–300 g Recently split from Coqui Francolin; differs in having much reduced barring on the belly in both sexes; ranges not known to overlap. **Voice:** Similar to that of Coqui Francolin. **Status and biology:** Locally common resident in tall grasslands and well-grassed miombo (*Brachystegia*) woodland of central Mozambique and Zimbabwe. Generally occurs in moister habitats than Coqui Francolin. (Miomboswempie)

white
supercilium

heavy
decurved
bill

HARTLAUB'S SPURFOWL

plain face &
underparts

boldly
streaked
underparts

♂

♀

yellow legs

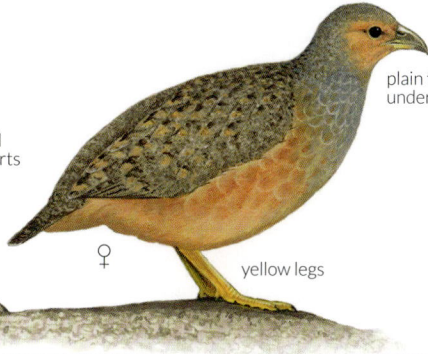

both species
cock their
tails

bold white
supercilium

CRESTED FRANCOLIN

KIRK'S FRANCOLIN

bold white
supercilium

dark-streaked
belly, no
barring

pale-streaked
belly, very faint
barring

COQUI FRANCOLIN

STUHLMANN'S FRANCOLIN

tawny
head

heavily
barred
lower
belly

white
supercilium
& throat

♂

♀

yellow legs

♂

♀

mostly unbarred lower
belly (both sexes)

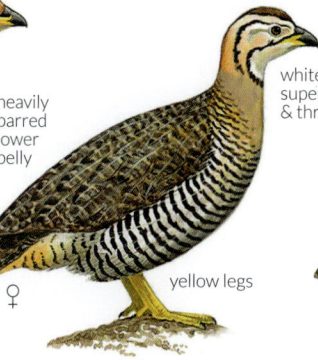

GREY-WINGED FRANCOLIN *Scleroptila afra*

b b b | | | | | B B B B b
J F M A M J J A S O N D

30–33 cm; 370–520 g The most distinctive of the red-winged francolins, identified by its grey-speckled (not white or buff) throat and finely mottled black-and-white belly. Rufous wing patch smaller and duller than other species; largely confined to primary bases. Juv. duller, with whitish throat; acquires grey throat within 2 months of hatching. **Voice:** Whistled, rather variable *'wip wip wip kipeeo, wip kipeeoo'*. **Status and biology:** Endemic. Fairly common resident of strandveld, fynbos, southern Karoo and montane grassland. Usually in coveys of 3–8 birds. (Bergpatrys)

RED-WINGED FRANCOLIN *Scleroptila levaillantii*

b b b b b b b b b b B B
J F M A M J J A S O N D

33–38 cm; 360–550 g The largest red-winged francolin; separated from Grey-winged Francolin by its white (not grey-speckled) throat, buffy (not black-and-white) belly and much more extensive rufous in wing (primaries, outer seondaries and primary coverts). Lacks the moustachial stripe of Shelley's and Orange River Francolins, and white throat is bordered by a rufous (not black) necklace. Further told from Shelley's Francolin by buffy (not black-and-white) belly, and broad, blackish mottling across upper breast. Juv. duller, with less distinct blackish breast mottling. **Voice:** Piping *'wip-tilleee'*, sometimes preceded by several short *'wip'* notes. **Status and biology:** Endemic (after split from *S. crawshayii*, which replaces it north of the Zambezi River). Uncommon to locally common resident in grassland and fields, usually on lower slopes, and in valleys in mountainous terrain; numbers have decreased in many areas due to unfavourable agricultural practices. (Rooivlerkpatrys)

ORANGE RIVER FRANCOLIN *Scleroptila levalliantoides*

b b b b b b B B b b b
J F M A M J J A S O N D

32–35 cm; 360–530 g Slightly smaller than Red-winged Francolin, with a black moustachial stripe extending into a thin, black necklace around its white throat; lacks a broad, dark breast band. Belly buffy, with variable dark streaks; lacks black-and-white-barred belly patch of Shelley's Francolin. Told from Grey-winged Francolin by its white (not grey-freckled) throat. Plumage varies regionally, with paler birds of subspecies *S. l. pallidior* in Botswana differing from Kunene Francolin by lacking the well-developed black-and-white necklace that extends onto the breast. **Voice:** Rather strident *'kibitele'*, faster and higher pitched than that of Shelley's Francolin; mostly given at dawn. **Status and biology:** Endemic. Locally common resident in grassland and semi-arid savanna. (Kalaharipatrys)

KUNENE FRANCOLIN *Scleroptila jugularis*

b b b b b b B B B b b
J F M A M J J A S O N D

32–35 cm; 360–530 g Recently split from Orange River Francolin; paler overall with a well-developed black-and-white necklace, which extends onto the breast. Differs further by showing whiter underparts with fewer chestnut markings on the upper flanks. **Voice:** Similar to Orange River Francolin but followed by a sigh-like whistle at the end. **Status and biology:** Near-endemic. Found on sandy Kalahari sweet grassland and boulder-strewn grassy hills in nw Namibia, extending into se Angola. (Kaokopatrys)

SHELLEY'S FRANCOLIN *Scleroptila shelleyi*

b b B B b b b b B B B B
J F M A M J J A S O N D

30–36 cm; 380–590 g Superficially resembles Red-winged Francolin, but has a boldly barred black-and-white (not buffy) belly and a moustachial stripe bordering its white throat, extending into a narrow, black necklace. In flight, primaries and outer secondaries are rufous. Juv. duller, with scruffy plumage; chequered belly patch less prominent. **Voice:** Rhythmic, repeated *'til-it, til-leoo'* ('I'll drink your beer'); slower and less varied than other red-winged francolins. **Status and biology:** Locally common resident in savanna and open woodland, often associated with rocky ground. (Laeveldpatrys)

CHUKAR PARTRIDGE *Alectoris chukar*

| | | | | | | | | b b b
J F M A M J J A S O N D

32–34 cm; 380–580 g Introduced. A handsome partridge, superficially similar to spurfowl, but not closely related. Black necklace and black-and-rufous flank stripes are diagnostic. In flight, grey rump and central tail contrast with rufous outer tail. Juv. is rather drab and plain, lacking bold necklace and flank stripes. **Voice:** A dry *'chuk-chuk-chuk-chukar'*, hence the common name. **Status and biology:** Small, feral population of some 300 birds on Robben Island, W Cape, following release there in 1964. (Asiatiese Patrys)

GREY-WINGED FRANCOLIN

rufous patch confined to primaries

all-dark bill

finely speckled, grey throat

RED-WINGED FRANCOLIN

bill has yellow base

rufous hindneck

white throat with heavy mottling below

rufous flight feathers

ORANGE RIVER FRANCOLIN

thin black necklace

mostly plain buff belly & underparts

KUNENE FRANCOLIN

broad black necklace

pale belly

SHELLEY'S FRANCOLIN

pure white throat

barred/ mottled grey-brown belly & underparts

CHUKAR PARTRIDGE

plain back

boldly barred flanks

BUTTONQUAILS
Small, quail-like shorebirds that freeze when alarmed; typically only seen when flushed. Flight weaker than quails. Polyandrous; females larger and more brightly coloured; males incubate and care for precocial chicks.

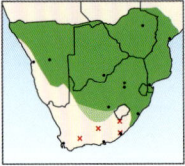

COMMON BUTTONQUAIL *Turnix sylvaticus*

B B B b b b b b b B B B
J F M A M J J A S O N D

14–16 cm; 30–70 g A rather pale buttonquail, appreciably smaller than Common or Harlequin Quails. In flight, has a longer neck and more rapid wing beats; pale upperwing coverts contrast with darker flight feathers. Paler than Black-rumped Buttonquail; best told in flight by the lack of a dark rump and back. On the ground, told by black spots on the sides of the breast and paler, buffy (not rufous) face. Female is larger and more boldly marked than male. **Voice:** Female advertises with a repeated, low-pitched hoot, *'hmmmm'*. **Status and biology:** Locally common resident and nomad in tall grassland, old fields and open savanna. (Bosveldkwarteltjie)

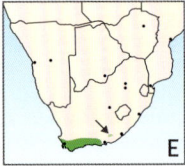

HOTTENTOT BUTTONQUAIL *Turnix hottentottus*

b B b b
J F M A M J J A S O N D

14–15 cm; 40–60 g A richly coloured buttonquail; paler above than Black-rumped Buttonquail, lacking a clearly defined dark back; belly has more extensive black spotting. Legs yellow (bright chrome-yellow in female), not flesh-white. Male is less richly marked. **Voice:** A low, flufftail-like booming. **Status and biology:** ENDANGERED. Endemic. Locally common resident and nomad in fynbos and strandveld. Usually found in open areas dominated by restios; prefers areas 2–4 years after fires. (Kaapse Kwarteltjie)

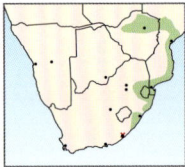

BLACK-RUMPED BUTTONQUAIL *Turnix nanus*

B b b b b b B
J F M A M J J A S O N D

14–15 cm; 40–60 g More richly coloured than Common Buttonquail, with a plain ginger face and throat. In flight, shows a diagnostic dark back and rump that contrast with the paler wing coverts. Best separated from female Blue Quail by flight action and contrast between rump and upperwing coverts. Range not known to overlap with Hottentot Buttonquail; differs by flesh-white (not yellow) legs and unmarked belly. Female is less barred on breast and flanks. Juv. has a spotted breast. **Voice:** A flufftail-like *'ooooop-ooooop'*. **Status and biology:** Scarce to locally common nomad in moist grassland, often around temporary wetlands, but also damp areas in hilly and mountainous terrain. (Swartrugkwarteltjie)

QUAILS
Small, sexually dimorphic gamebirds. Run swiftly in a hunched position. More likely to be heard than seen when they call from concealed positions in grasslands and croplands. Usually only seen when flushed; flight action rapid and stronger than buttonquails. Usually monogamous.

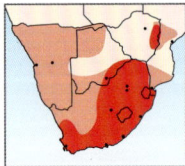

COMMON QUAIL *Coturnix coturnix*

B b b b B B B
J F M A M J J A S O N D

16–20 cm; 80–115 g A small, pale buff gamebird, streaked black and white above with a prominent white supercilium; underparts buffy, breast streaked brown. Male has a black or russet throat. Slightly larger than Harlequin Quail, with paler plumage. Juv. less streaked than ad. **Voice:** A repeated, high-pitched *'whit wit-wit'* ('wet my lips'), slower and deeper than Harlequin Quail; also a sharp *'crwee-crwee'* given in flight. **Status and biology:** Locally abundant intra-African migrant in grassland, fields and croplands. (Afrikaanse Kwartel)

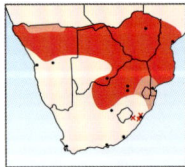

HARLEQUIN QUAIL *Coturnix delegorguei*

B B b b b B
J F M A M J J A S O N D

14–18 cm; 68–90 g Male is darker than Common Quail, with chestnut-and-black underparts. Female has buffy underparts, darker than Common Quail, with contrasting white throat; supercilium buff. Juv. paler than female. **Voice:** A high-pitched *'whit, wit-wit'*, more metallic than Common Quail; squeaky *'kree-kree'* given in flight. **Status and biology:** Locally common nomad and intra-African summer migrant in grassland, damp fields and savanna, often in moister areas than Common Quail. (Bontkwartel)

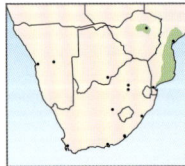

BLUE QUAIL *Excalfactoria adansonii*

b b b b
J F M A M J J A S O N D

12–14 cm; 40–50 g A small quail, similar in size to buttonquails; told from them in flight by its more compact shape and uniform upperwing (lacking paler upperwing coverts). Male told from male Harlequin Quail in flight by its small size and chestnut wing coverts, and blue underparts that typically appear black. On the ground, female and juv. can be distinguished by their barred underparts. In flight they appear small and dark. **Voice:** A high-pitched, whistled *'teee-ti-ti'*; much less vocal than other quails. **Status and biology:** Rare resident and nomad in damp and flooded grassland. Range contracted northwards in the last century. (Bloukwartel)

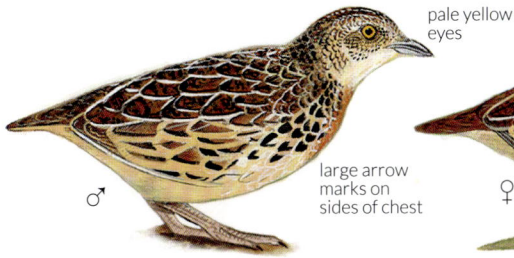

pale yellow
eyes

rufous
breast

COMMON
BUTTONQUAIL

♂

large arrow
marks on
sides of chest

♀

HOTTENTOT
BUTTONQUAIL

♂

scaled flanks

♀

yellow legs

pale blue
eyes

BLACK-RUMPED
BUTTONQUAIL

♂

♀

rufous
face &
breast

dark rump &
lower back

COMMON QUAIL

HARLEQUIN QUAIL

brown
throat

♂

♀

whitish
belly

♂

black &
chestnut
underparts

♀

buffy belly

whitish
throat

BLUE QUAIL

chestnut wing
coverts

♂

♂

blue
underparts

♀

barred
flanks

♀

167

FLUFFTAILS

FLUFFTAILS Tiny, rail-like birds, now placed in their own family, together with the 2 Madagascan wood rails. Renowned for their secretive behaviour, they are best located by their hooting or rattling calls. Hooting calls could be confused with those of the buttonquails (p. 166). Many species are crepuscular and call at night. Sexes differ; females difficult to identify.

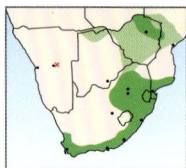

RED-CHESTED FLUFFTAIL *Sarothrura rufa*

b b b b | | | | b B B B b
J F M A M J J A S O N D

15–17 cm; 30–46 g Male is distinguished by red of head extending to lower breast, and by dark tail. Belly and back are uniformly dark with small, white speckles. Darker than Streaky-breasted Flufftail, with more extensive red on underparts. Female and juv. are darker than other female flufftails, especially on throat and upper breast. **Voice:** Low hoot, *'woop'*, repeated 1x per second; more rapid *'gu-duk, gu-duk, gu-duk'*; and metallic *'tuwi-tuwi-tuwi'*. **Status and biology:** Common resident of dense reeds and sedges around marshes and streams. (Rooiborsvleikuiken)

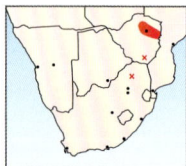

STREAKY-BREASTED FLUFFTAIL *Sarothrura boehmi*

B b b | | | | | | | b b
J F M A M J J A S O N D

14–16 cm; 25–40 g Male has less red on breast than male Red-chested Flufftail, very short tail, with a paler throat and streaked lower breast and belly; when flushed, appears much paler. Female and juv. are paler below than female Red-chested Flufftail. **Voice:** Low hoot, *'gawooo'*, repeated 20–30 times at 0.5-second intervals. **Status and biology:** Uncommon intra-African br. migrant, Nov–Apr, in rank, seasonally flooded grassland. Largely confined to Zimbabwe's Mashonaland Plateau. (Streepborsvleikuiken)

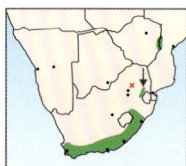

STRIPED FLUFFTAIL *Sarothrura affinis*

b b b | | | | | b b b b
J F M A M J J A S O N D

14–15 cm; 25–30 g Male has a plain red head and tail; body is black, boldly striped white. Female and juv. are finely barred buff above and have chestnut-washed tail. **Voice:** Low *'oooooop'* hoot lasting 1 second and repeated every 2 seconds; also high-pitched, chattering call, reminiscent of sugarbird. **Status and biology:** Locally common resident and nomad in montane grassland and fynbos; sometimes associated with wetlands. (Gestreepte Vleikuiken)

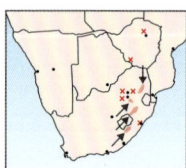

WHITE-WINGED FLUFFTAIL *Sarothrura ayresi*

| | | | | | | | | | b b
J F M A M J J A S O N D

14 cm; 26–34 g Both sexes show diagnostic white secondaries in flight and have black-and-chestnut-barred tails and white bellies. Outer web of outer primary white, forming a thin, white line on the folded wing. Flight fast and direct, with whirring wing beats. At rest, female differs from female Striped Flufftail by chestnut wash on neck and barred (not plain) tail. **Voice:** Low, deep hoot, repeated every second, often in duet, with differences in pitch between birds. **Status and biology:** CRITICALLY ENDANGERED. Rare and localised summer visitor, recently confirmed breeding locally, to upland marshes and vleis, where sedges and aquatic grasses grow in shallow water. (Witvlerkvleikuiken)

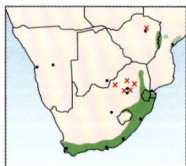

BUFF-SPOTTED FLUFFTAIL *Sarothrura elegans*

B b B B B | | | b b B B
J F M A M J J A S O N D

15–17 cm; 40–60 g Back of male has large, buff spots; tail barred black and rufous. Red on head extends only onto upper breast, not onto lower breast and back as in Red-chested Flufftail. Female and juv. rich brown above, with buff breast and paler, barred belly. **Voice:** Low, foghorn-like *'dooooooooooo'* lasting 3–4 seconds, mainly at night and on overcast and rainy days. Calls from elevated perches at night, on or near the ground during the day. **Status and biology:** Locally common resident and nomad in dense forest understorey, adjacent scrub and well-wooded gardens. (Gevlekte Vleikuiken)

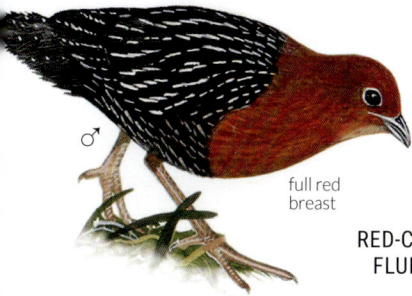
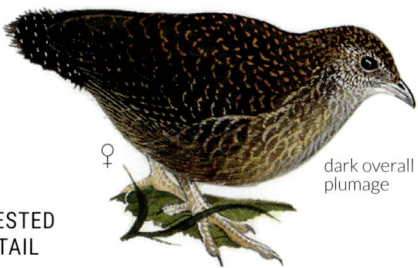

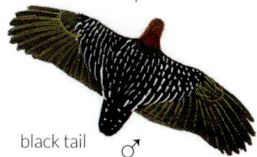

full red breast

dark overall plumage

dark tail ♀

black tail ♂

RED-CHESTED FLUFFTAIL

striped lower breast ♂

♀ whitish belly

dark tail ♀

black tail ♂

STREAKY-BREASTED FLUFFTAIL

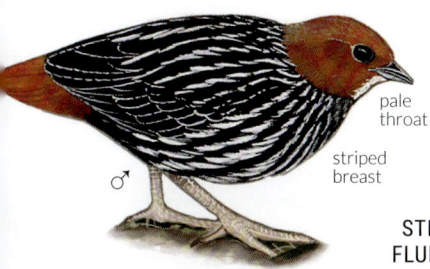

pale throat

striped breast ♂

♀

chestnut tail ♀

chestnut tail ♂

STRIPED FLUFFTAIL

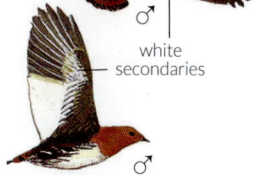

♂ white belly

♀

barred tail ♂

white secondaries ♂

WHITE-WINGED FLUFFTAIL

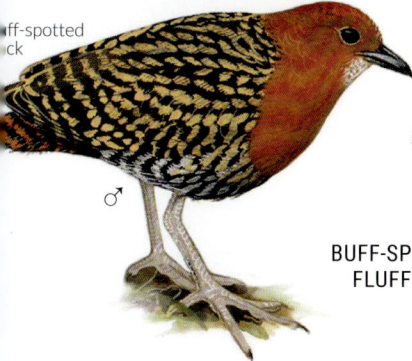

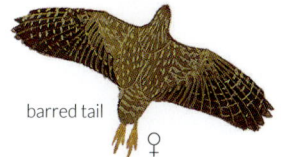

ff-spotted
ck ♂

warm brown upperparts

well-barred belly ♀

barred tail ♂

barred tail ♀

BUFF-SPOTTED FLUFFTAIL

169

FINFOOTS Peculiar, grebe-like birds with lobed toes and stout, dagger-like bills that are sister to the flufftails. Three species occur worldwide: 1 in Africa, 1 in SE Asia and the Sungrebe *Heliornis fulica* in C and S America. Sexes differ in plumage; males appreciably larger than females.

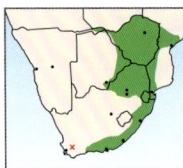

AFRICAN FINFOOT *Podica senegalensis*

`B b b b | | | b B B B B`
`J F M A M J J A S O N D`

52–65 cm; 350–650 g A reclusive waterbird; swims low in the water, snaking head back and forth. Superficially recalls African Darter (p. 78), but with a shorter, thicker neck and a heavy, red bill; seldom dives. Out of the water, the bright red-orange legs and feet are conspicuous. Male is greyer, with a plain face; female browner with boldly patterned brown-and-white face. Juv. has a dark bill and pale belly. **Voice:** Normally silent; occasionally gives a short, frog-like *'krork'*. **Status and biology:** Uncommon resident of rivers and streams with well-vegetated banks; usually remains among, or close to, vegetation overhanging water; easily overlooked. (Watertrapper)

COOTS, MOORHENS, SWAMPHENS, GALLINULES, RAILS AND CRAKES The
Rallidae is a large, diverse family of mostly secretive birds of dense vegetation, often associated with wetlands. Coots and, to a lesser extent, moorhens are unusual in spending more time in open habitats than other members of the family, and are thus generally easier to locate and observe than smaller crakes and rails, which are often secretive and best located by calls. Sexes alike in most species (except Allen's Gallinule and Little and Striped Crakes); males generally larger than females. Breed singly; usually territorial; mating systems varied, although usually monogamous. Chicks semi-precocial, cared for by one or both parents.

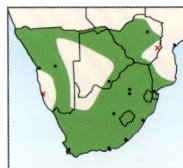

RED-KNOBBED COOT *Fulica cristata*

`b b b b b b b b b b b b`
`J F M A M J J A S O N D`

36–44 cm; 500–900 g The most aquatic rail; sometimes mistaken for a duck; swims well, aided by long, lobed toes. Ad. black with a white bill and frontal shield. The dull red knobs on top of the shield are larger in br. birds. Juv. is grey, with small frontal shield; lacks whitish flank line and undertail coverts of smaller juv. Common Moorhen. **Voice:** Harsh, metallic *'claak'*. **Status and biology:** Common to abundant resident at dams, lakes, slow-flowing rivers and estuaries. Often feeds by upending or diving. Nest is a floating platform of vegetation, anchored to waterweed or emergent vegetation. (Bleshoender)

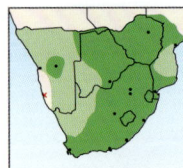

COMMON MOORHEN *Gallinula chloropus*

`b b b b b b b b b b b b`
`J F M A M J J A S O N D`

30–36 cm; 180–350 g A dull, sooty-black moorhen with greenish-yellow legs and red frontal shield. Larger and darker than Lesser Moorhen, with mostly red (not yellow) bill. Juv. is greyer above than juv. Lesser Moorhen, with darker belly and dull brown bill. **Voice:** Sharp *'krrik'*. **Status and biology:** Common resident and nomad at wetlands with fringing vegetation. Bolder than other gallinules, often swimming in open water. (Grootwaterhoender)

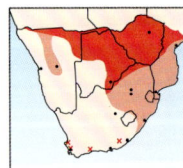

LESSER MOORHEN *Paragallinula angulata*

`B B B b | | | | | | b b`
`J F M A M J J A S O N D`

22–26 cm; 100–160 g Smaller and more secretive than Common Moorhen, with less conspicuous white flank feathers and mainly yellow (not red) bill. Juv. is sandy buff (juv. Common Moorhen is grey) with a pale belly and dull, yellowish-green bill and legs. **Voice:** A series of hollow notes, *'do do do do do do do'*. **Status and biology:** Locally common intra-African br. migrant, Nov–May (mostly Dec–Apr), in flooded grassland and small, secluded ponds. (Kleinwaterhoender)

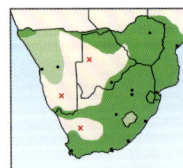

BLACK CRAKE *Amaurornis flavirostra*

`B B B b b b b b b b B B`
`J F M A M J J A S O N D`

18–22 cm; 70–115 g A fairly small, plain black crake with a bright yellow bill and red eyes and legs. Juv. greyer, with a black bill and dark legs. **Voice:** A throaty *'chrrooo'* and a hysterical, bubbling, wheezy duet. Alarm call is a repeated, soft *'chuk'*. **Status and biology:** Common resident in marshes with dense cover. Often bold, foraging out in the open. Sometimes responds strongly to pishing, drooping its wings and stamping its feet. (Swartriethaan)

AFRICAN FINFOOT

juv.

not to scale

♂

♀

RED-KNOBBED COOT

ad.

ad.

ad.

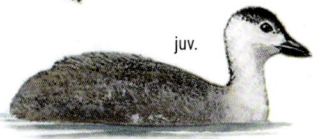

juv.

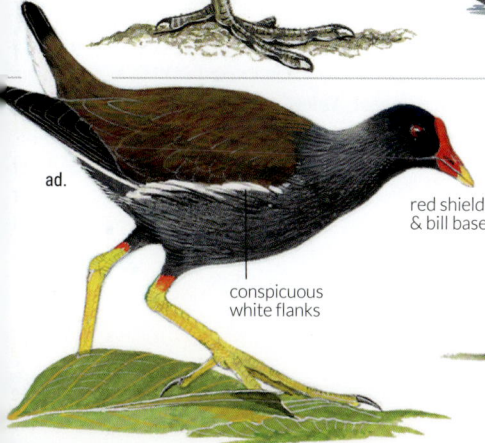

ad.

red shield & bill base

conspicuous white flanks

COMMON MOORHEN

juv.

dark bill

ad.

ad.

mostly yellow bill

LESSER MOORHEN

lighter grey underparts

pale bill

juv.

jet-black plumage

bright yellow bill

ad.

BLACK CRAKE

imm.

171

AFRICAN SWAMPHEN   *Porphyrio madagascariensis*

b b b b b b b b b b b b
J F M A M J J A S O N D

38–46 cm; 500–700 g By far the largest gallinule in the region; easily identified by its massive red bill and frontal shield, long, red legs, and purplish coloration, with turquoise face and foreneck and greenish back. Juv. is dull brown above and grey below, with a large, reddish-brown bill. **Voice:** A variety of harsh shrieks, wails and booming notes. **Status and biology:** Common resident in reed beds, marshes and flooded grassland. Holds plant stems with foot while eating soft stem pulp. (Grootkoningriethaan)

ALLEN'S GALLINULE   *Porphyrio alleni*

B B B b b   b
J F M A M J J A S O N D

25–28 cm; 110–165 g Smaller and darker than African Swamphen, with blue (in br. male) or green (in br. female) frontal shield. Non-br. birds have dull brown shield. Ad. differs from ad. Purple Gallinule by red (not yellow) legs and lack of yellow tip to bill. Juv. is pale buff brown, often scaly on upperparts, and lacks white flank stripes of juv. Common Moorhen (p. 170). Has pale, fleshy legs (not greenish-brown as in juv. Common Moorhen or olive as in juv. American Gallinule). **Voice:** Six or more rapidly uttered, sharp clicks, *'duk duk duk duk duk duk'*. **Status and biology:** Locally common resident and intra-African br. migrant, Dec–Apr, in flooded grassland. Holds plant stems and fruit with foot while eating. (Kleinkoningriethaan)

PURPLE GALLINULE   *Porphyrio martinica*

        X X X X X X
J F M A M J J A S O N D

27–35 cm; 120–300 g Vagrant. Smaller than African Swamphen, with bright yellowish-green (not red) legs and yellow tip to bill; frontal shield of ad. is pale blue. Most records are juvs, which are similar to juv. Allen's Gallinule but have olive (not flesh-coloured) legs; differ from juv. Common Moorhen (p. 170) by lack of white flank stripes. **Voice:** Harsh, barking calls; usually silent. **Status and biology:** Vagrant from Americas, mostly Apr–Jun, but some remain for many months. Usually in reed beds, but recent arrivals may occur virtually anywhere; regular vagrant at Tristan da Cunha, and several records from ships at sea. (Amerikaanse Koningriethaan)

AFRICAN RAIL   *Rallus caerulescens*

B B b b b b b B B B B B
J F M A M J J A S O N D

27–30 cm; 125–200 g A fairly large rail with a long, decurved, red bill and red legs, grey breast, and black-and-white-barred flanks. With a very poor view could be confused with African Crake (p. 174), but has plain brown (not streaked) back. Juv. has a brown bill and a buff breast. **Voice:** An explosive, high-pitched trill, *'trrreee-tee-tee-tee-tee-tee'*, descending in pitch. **Status and biology:** Common resident of marshes, reed beds and flooded grassland, usually in areas with standing water. Often ventures into the open, especially at dawn and dusk. (Grootriethaan)

AFRICAN SWAMPHEN

ad.

red shield & robust bill

imm.

juv.

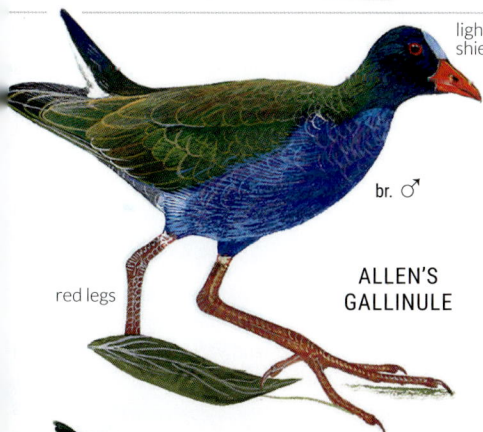

light blue shield

red legs

br. ♂

ALLEN'S GALLINULE

br. ♀

green-blue shield

non-br.

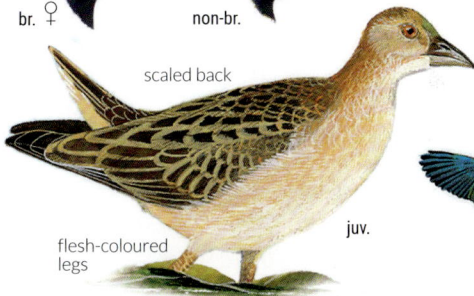

scaled back

flesh-coloured legs

juv.

ad.

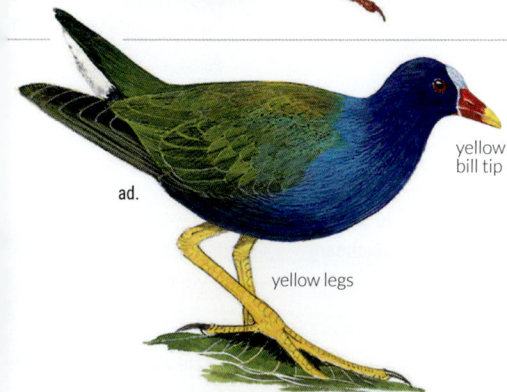

ad.

yellow bill tip

yellow legs

PURPLE GALLINULE

imm.

olive legs

black-&-white-barred undertail

ad.

long, decurved bill

AFRICAN RAIL

ad.

juv.

sometimes cocks tail

173

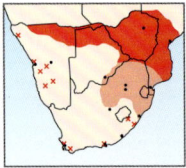

AFRICAN CRAKE *Crex egregia*

BBbb   bBB
J F M A M J J A S O N D

19–23 cm; 120–160 g A medium-sized crake; smaller and darker than Corn Crake. Smaller than African Rail (p. 172), with a short, stubby bill and streaked (not plain brown) back. Flanks and belly boldly barred black and white. When flushed, it flies a short distance with legs dangling, showing the brown, mottled upperparts and barred flanks. Juv. is browner, with less boldly barred flanks. **Voice:** A monotonous, hollow-sounding series of notes, *'krrr-krrr-krrr'*. **Status and biology:** Uncommon to locally common intra-African migrant, Oct–Apr, br. in damp grassland and at seasonal wetlands. (Afrikaanse Riethaan)

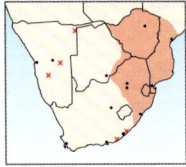

CORN CRAKE *Crex crex*

J F M A M J J A S O N D

25–30 cm; 130–200 g A large, pale sandy crake showing conspicuous chestnut-orange wing coverts in flight. Paler and larger than African Crake, with a buffy (not grey) breast. Rarely seen except when flushed, when it flies with whirring wings and dangling legs. Juv. more buffy above, with indistinct barring on vent. **Voice:** A harsh *'krek krek'* (hence its scientific name), but seldom calls in Africa; occasionally gives a *'tsuck'* call when flushed. **Status and biology:** Uncommon Palearctic-br. migrant in rank grassland and open savanna. Occurs around the edges of marshes, but seldom in areas with standing water. (Kwartelkoning)

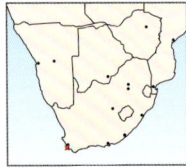

LITTLE CRAKE *Porzana parva*

X
J F M A M J J A S O N D

18–20 cm; 40–60 g Vagrant. A small crake; male resembles Baillon's Crake but is slightly larger with a reddish spot at the base of the bill and a more boldly patterned back with pale brown stripes; appears more slender due to longer, more pointed wings. Female distinctive with a pale face and buff neck and underparts (not blue-grey as Baillon's Crake). **Voice:** Mostly silent in wintering area; occasional sharp *'tjuck'* call. **Status and biology:** Breeds in Europe and w Asia; winters from Senegal to Ethiopia and nw India. Rarely reaches the equator, although previous vagrant records from Zambia. Only one s African record: a well-watched female on the Cape Peninsula in March 2012. (Europese Kleinriethaan)

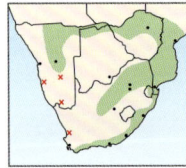

BAILLON'S CRAKE *Porzana pusilla*

BBBbb   bbbb
J F M A M J J A S O N D

16–18 cm; 30–45 g Much smaller than African Crake, with white-spotted upperparts and less contrasting black-and-white barring on the flanks and undertail. Barred undertail separates it from larger Spotted and Striped Crakes. Juv. paler, mottled and barred below. **Voice:** A soft *'qurrr-qurrr'* and various frog-like croaks. **Status and biology:** Uncommon to locally common resident and nomad in wetlands, reed beds and flooded grassland. Secretive; rarely seen in the open except at dawn or dusk. (Kleinriethaan)

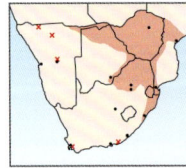

SPOTTED CRAKE *Porzana porzana*

J F M A M J J A S O N D

21–24 cm; 70–130 g A medium-sized crake with distinctive, white-spotted upperparts and whitish leading edge to wing in flight; bill mostly yellow. Flanks barred, but finer and less boldly marked than African Crake; undertail buff. Told from slightly smaller Striped Crake by spotted (not striped) upperparts, yellowish (not greenish) legs, barred (not plain) flanks and paler undertail coverts. Often flicks its tail, showing buffy undertail coverts. **Voice:** A short *'kreck'* when alarmed; generally silent in Africa. **Status and biology:** Uncommon Palearctic-br. migrant in flooded grassland and wetland. (Gevlekte Riethaan)

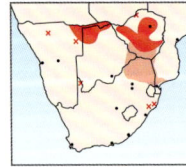

STRIPED CRAKE *Aenigmatolimnas marginalis*

Bbb   b
J F M A M J J A S O N D

18–22 cm; 40–60 g A rather plain, medium-sized crake. Combination of white stripes on its buffy upperparts, plain flanks and russet vent is diagnostic. Female has a blue-grey breast and belly. Male and juv. have buff breast; juv. warmer buff, but with reduced stripes. **Voice:** A rapid *'tik-tik-tik-tik-tik'*, which may continue for a minute or more. **Status and biology:** Uncommon, intra-African br. migrant, Dec–Mar, in seasonally flooded grassland and marshes. Abundance and range varies in relation to rainfall. (Gestreepte Riethaan)

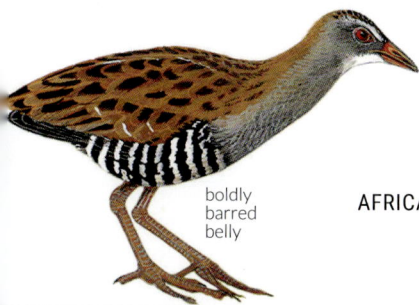

boldly
barred
belly

AFRICAN CRAKE

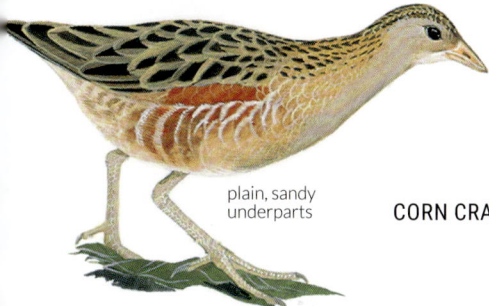

plain, sandy
underparts

CORN CRAKE

conspicuous
chestnut wing
coverts

arred
ndertail

red spot
at base of
bill

ad. ♂

LITTLE CRAKE

ad. ♀

ad. ♂

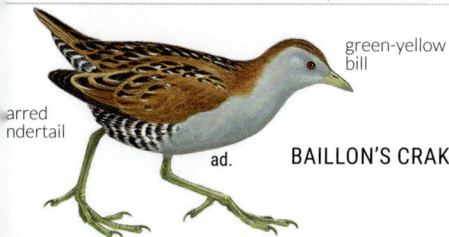

arred
ndertail

green-yellow
bill

ad.

BAILLON'S CRAKE

juv.

ad.

uffy
ndertail

yellow &
red bill

spotted
breast

STRIPED CRAKE

ad. ♂

plain
flanks

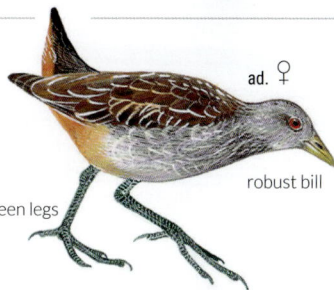

ad. ♀

green legs

robust bill

SPOTTED CRAKE

pale leading
edge to wing

white stripes on
upperparts

russet
undertail

juv.

ad. ♂

ad. ♂

CRANES

Tall, stately birds of open grassland and wetland, placed in the same order as the rails, finfoots and flufftails. Could possibly be confused with herons (e.g. Black-headed Heron, p. 82). Sexes alike, but males slightly larger. Gather in flocks when not breeding, and roost communally, often at wetlands. Displays include 'dances' with raised wings.

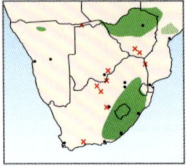

GREY CROWNED CRANE   *Balearica regulorum*

B B b b b b | | | b B B
J F M A M J J A S O N D

100–110 cm; 3–4 kg A distinctive crane with long, golden crown feathers, white wing coverts, black primaries and chestnut secondaries. Bare cheek patch is white with red top. Imm. has feathered cheek patch and shorter crown feathers. **Voice:** Trumpeting flight call, *'may hem'*; deep *'huum huum'* when br. **Status and biology:** ENDANGERED. Locally common resident and nomad, moving in response to rainfall. Some 6 000 birds in s Africa. Occurs at shallow wetlands, and in grassland and agricultural lands. Commutes to wetlands to roost. (Mahem)

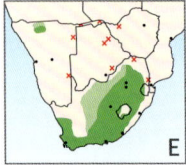

BLUE CRANE   *Grus paradisea*

b b b b | | | b b b B B
J F M A M J J A S O N D

100–120 cm; 4–5.5 kg A blue-grey crane with a bulbous head and long, trailing 'tail' (formed by elongated inner secondaries and tertials). In flight, shows less contrast between flight feathers and coverts than other cranes. Juv. lacks long 'tail' and is paler grey, especially on head. **Voice:** Loud, nasal *'kraaaank'*. **Status and biology:** VULNERABLE. Endemic. Resident and local nomad; population some 20 000 birds, most in agricultural lands in W Cape and adjacent Karoo; scarce and decreasing in eastern grassland; small isolated population in n Namibia. In pairs or family groups when br., but non-br. flocks contain hundreds of birds. Occasionally hybridises with Wattled Crane. (Bloukraanvoël)

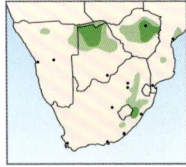

WATTLED CRANE   *Grus carunculata*

b b b b B B B B B B b b
J F M A M J J A S O N D

120–170 cm; 7–8.5 kg A very large crane with a white neck and long facial wattles. Can be identified even at long range by white neck and upper breast contrasting with black underparts and grey back. Juv. has pale crown. **Voice:** Loud *'kwaarnk'*; seldom calls. **Status and biology:** VULNERABLE. 8 000 birds globally, 2 000 in s Africa. Uncommon and localised at shallow wetlands and adjoining grassland. Usually in pairs or small groups, but non-br. birds gather in flocks of up to 50. Occasionally hybridises with Blue Crane. (Lelkraanvoël)

SECRETARYBIRD

A large, distinctive raptor with long legs and a long neck. Endemic to Africa; placed in its own family. Has long, broad wings and soars strongly, but usually searches for prey by walking through open vegetation; often raises wings when running. Many prey killed by stamping. Although renowned for killing snakes and other reptiles, it also eats a wide range of other small animals, including insects, especially grasshoppers. Sexes alike; juvs duller.

SECRETARYBIRD   *Sagittarius serpentarius*

b b B B B B B B B B b b
J F M A M J J A S O N D

1.25–1.50 m; 2.8–5 kg Could possibly be mistaken for a crane, but its long legs, head plumes, long central tail feathers, characteristic long-striding gait and horizontal body posture are diagnostic. In flight, the 2 central tail feathers project well beyond legs, and black flight feathers and thighs contrast with pale grey coverts and body. Juv. has shorter tail and yellow (not red) bare facial skin. **Voice:** Normally silent, but gives deep croak during aerial display. **Status and biology:** VULNERABLE. Uncommon to locally common resident of savanna and open grassland, usually singly or in pairs. (Sekretarisvoël)

ort, bushy
il

GREY CROWNED
CRANE

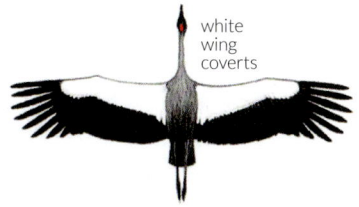

white
wing
coverts

grey
neck

ad.

juv.

long
plumes

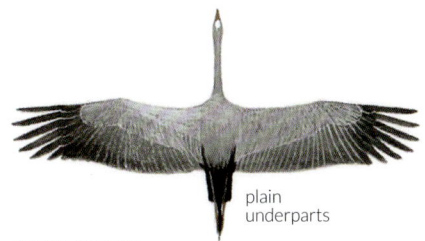

plain
underparts

BLUE CRANE

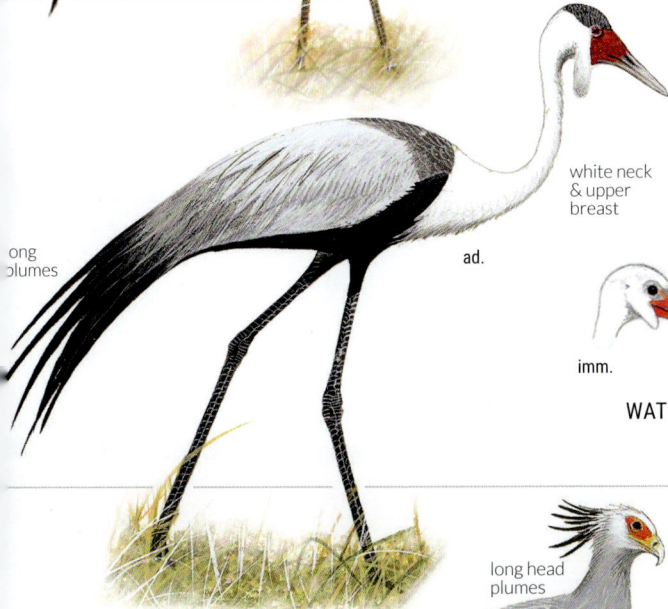

ong
plumes

ad.

white neck
& upper
breast

imm.

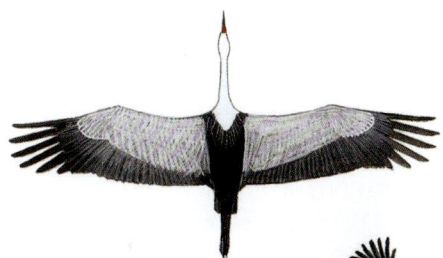

WATTLED CRANE

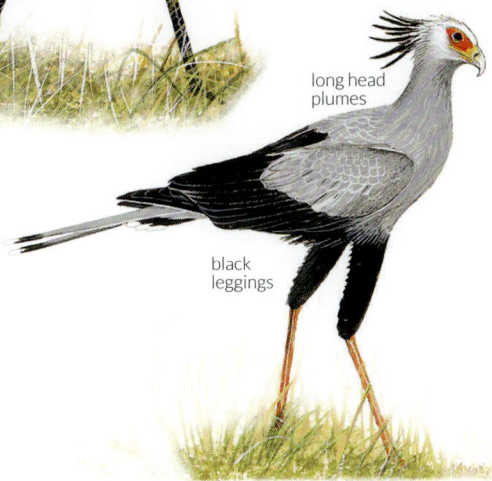

long head
plumes

black
leggings

SECRETARYBIRD

barred
undertail

177

LARGE BUSTARDS
Bustards are large cursorial birds of open country, formerly linked to cranes and their relatives, but now placed in their own order, distantly related to cuckoos and turacos. Southern Africa has a large proportion of the world's 26 species. The 3 species on this page are the largest in the region. Sexes differ; males larger, with throat pouches that are inflated in spectacular white 'balloon' displays. Females select males at dispersed leks. Chicks precocial; cared for by female alone.

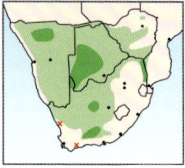

KORI BUSTARD *Ardeotis kori*

B	B	b				b	b	b	B	B	B
J	F	M	A	M	J	J	A	S	O	N	D

110–140 cm; 4.5–18 kg The largest bustard; flight heavy and laboured. Neck finely barred grey (lacking orange hindneck); crest on rear of crown. Displaying male raises crest, inflates white throat and throws tail forward onto its back, fluffing out its white undertail coverts. Female much smaller. Juv. has paler head, shorter crest and browner back. **Voice:** Deep, resonant *'oom-oom-oom'* given by displaying male. **Status and biology:** NEAR-THREATENED. Generally scarce resident and nomad in semi-arid savanna and grassland, usually near cover of trees. Numbers probably decreasing largely due to collisions with power lines and fences, but still fairly common in some protected areas. (Gompou)

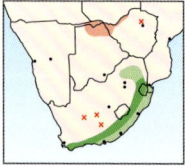

DENHAM'S BUSTARD *Neotis denhami*

									b	B	b	b
J	F	M	A	M	J	J	A	S	O	N	D

90–120 cm; 4–14 kg Larger than Ludwig's Bustard, with a pale grey foreneck, chestnut (not orange) hindneck and black-and-white-striped crown. Usually shows more white on folded wing, but this varies considerably. Displaying male inflates throat to form conspicuous white 'balloon'. Female is smaller, with a paler, more mottled back and less white in wings. Juv. has crown mottled brown. **Voice:** Displaying male utters a subdued, sharp *'pop'* at intervals. **Status and biology:** NEAR-THREATENED. Occurs in open grassland and agricultural land. *N. d. stanleyi* is a locally common resident and nomad in S Africa; *N. d. jacksoni* is a scarce non-br. visitor to n Botswana and Zimbabwe. (Veldpou)

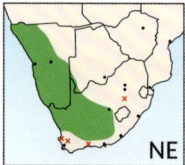

LUDWIG'S BUSTARD *Neotis ludwigii*

b	b	b	b	b		b	B	B	b	b	b
J	F	M	A	M	J	J	A	S	O	N	D

80–100 cm; 2.2–6 kg Smaller than Denham's Bustard with a dark grey-brown (not pale grey) foreneck, more orange (not chestnut) hindneck and largely uniform, dark head. Normally shows less white in folded wing, but amount of white on wings of both species is variable. Male display 'balloon' is grey (not white). Female is markedly smaller. Juv. is paler on head and neck. **Voice:** Displaying male gives explosive, deep *'woodoomp'* every 15–30 seconds. **Status and biology:** ENDANGERED due to high rates of collisions with power lines. Near-endemic. Locally common nomad and partial migrant in Karoo scrub and arid savanna; frequents drier areas than Denham's Bustard. (Ludwigpou)

NE

KORI BUSTARD

barred grey neck

♂

displaying ♂

upperwing

DENHAM'S BUSTARD

♀

pale grey foreneck

♂

displaying ♂

♂ upperwing

♀ upperwing

LUDWIG'S BUSTARD

♀

dark grey-brown foreneck

♂

displaying ♂

upperwing

SMALL BUSTARDS AND KORHAANS

Most smaller bustards are known in s Africa as korhaans. The species on this page are polygynous, with marked sexual differences in size and plumage; males have loud, distinctive calls and display flights.

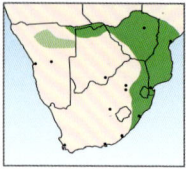

BLACK-BELLIED BUSTARD   *Lissotis melanogaster*

B	b	b							b	B	B	B
J	F	M	A	M	J	J	A	S	O	N	D	

58–65 cm; 1.1–2.7 kg A fairly large bustard, with a long, thin neck and a boldly mottled back. Male has a black belly, extending as a thin line up the front of the neck. Male distinctive in flight: underwing black with white panel in outer primaries; upperwing has mainly white primaries, greater and lesser coverts, black secondaries, and brown median coverts. Female is nondescript, off-white below with duller wings (flight feathers mainly dark brown with white spots in outer primaries; coverts brown with white central bar; underwing blackish); best separated from other small bustards by long, thin neck and white belly. **Voice:** In display, male slowly raises head, giving a sharp *'chikk'*, then pulls head down, giving a loud *'pop'*. **Status and biology:** Fairly common resident in woodland and tall, open grassland. (Langbeenkorhaan)

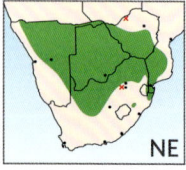

RED-CRESTED KORHAAN   *Lophotis ruficrista*

B	b	b							b	B	B	B
J	F	M	A	M	J	J	A	S	O	N	D	

48–50 cm; 400–900 g A rather nondescript, small bustard; both sexes have black bellies. Male has grey-washed neck; bushy red crest is visible only during display. Female has mottled brown crown and neck; differs from female Northern Black Korhaan by black belly extending onto lower breast, chevrons (not barring) on back, and yellow-brown (not reddish) bill. In aerial display, male flies straight up, then suddenly tumbles to ground as though shot, before gliding to land. **Voice:** Song protracted, starting with a long series of accelerating clicks, *'tic-tic-tic...'*, switching to an extended series of loud, piping whistles, *'pi-pi-pi-pipity-pipity...'*. **Status and biology:** Near-endemic. Common resident in dry woodland and semi-desert grassland. Polygynous; males display in traditional areas. (Boskorhaan)

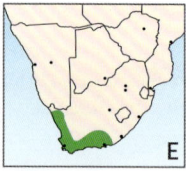

SOUTHERN BLACK KORHAAN   *Afrotis afra*

b									b	B	B	b
J	F	M	A	M	J	J	A	S	O	N	D	

48–52 cm; 600–900 g Slightly bulkier than Northern Black Korhaan; best identified in flight when distinguished by all-dark (not black and white) flight feathers; lacks white panels in primaries. Female and imm. lack bold black-and-white head and neck patterning. **Voice:** Male gives raucous *'kerrrak-kerrrak-kerrrak'* in flight and on ground. **Status and biology:** Endemic. Common resident of coastal fynbos and Karoo scrub. Polygynous; males have traditional calling sites and display by calling in flight or from the ground. (Swartvlerkkorhaan)

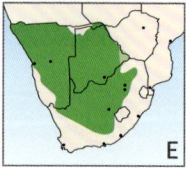

NORTHERN BLACK KORHAAN   *Afrotis afraoides*

B	b	b	b	b	b	b	b	b	b	B	B
J	F	M	A	M	J	J	A	S	O	N	D

48–52 cm; 520–950 g Male striking, with black head and neck and white cheek patch. Female and imm. duller, with buff-brown head and neck; differ from female Red-crested Korhaan by black barring (not chevrons) on upperparts and reddish (not yellow-brown) bill. Both sexes differ from Southern Black Korhaan by white inner webs to primaries, forming conspicuous white window on upper- and underwing in flight. **Voice:** Male gives raucous *'kerrrak-kerrrak-kerrrak'* in flight and on ground. **Status and biology:** Endemic. Common resident in Karoo grassland and arid savanna. Polygynous; males display by calling in flight or from the ground. (Witvlerkkorhaan)

BLACK-BELLIED BUSTARD

black foreneck

♀

♂

♀

extensive white in wings

RED-CRESTED KORHAAN

chevrons on back

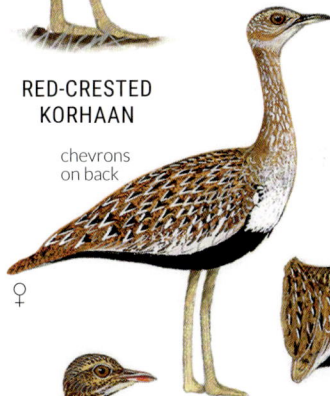

displaying ♂

chevrons on back

♂

♀

buff-barred inner primaries

SOUTHERN BLACK KORHAAN

barred back

♂

♀

black primaries

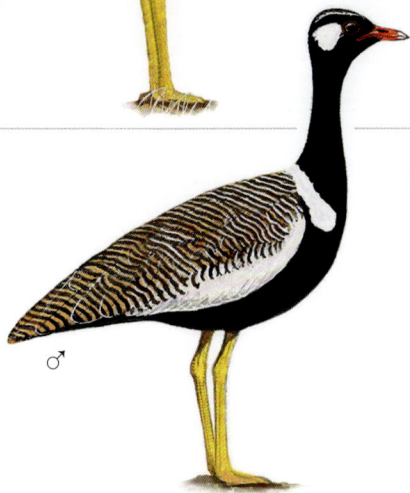

NORTHERN BLACK KORHAAN

barred back

♂

♀

extensive white in primaries

181

EUPODOTIS KORHAANS

Eupodotis korhaans are monogamous, occurring in pairs or family groups year-round; plumage only slightly dimorphic; both sexes call together; lack the aerial displays of other korhaan genera.

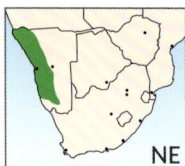

RÜPPELL'S KORHAAN   *Eupodotis rueppellii*

b	B	B	B	B	b	b	b	b	b	b	b
J	F	M	A	M	J	J	A	S	O	N	D

50–58 cm; 1–1.35 kg Slightly smaller than Karoo Korhaan, with warmer brown upperparts, conspicuous black line down foreneck, blue-tinged neck and contrasting facial pattern, including a black moustachial stripe and line behind the eye. Female is less boldly marked than male; lacks black moustachial stripe. Juv. duller; lacks black facial markings and has white spots on neck and wing coverts. **Voice:** Similar *'wrok-rak'* to Karoo Korhaan's, but slightly higher pitched. **Status and biology:** Near-endemic. Common resident on gravel plains and in arid scrub. Usually in pairs or family groups. (Woestynkorhaan)

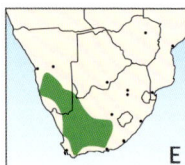

KAROO KORHAAN   *Eupodotis vigorsii*

B	b			b	b	b	b	B	B	B
J	F	M	A	M	J	J	A	S	O	N	D

56–62 cm; 1.1–2 kg A drab, grey-brown korhaan with a black throat and nape patch, and contrasting buff and black wings in flight. Female and imm. have smaller, less-defined throat and nape patches. Overlaps with Rüppell's Korhaan in s Namibia; has a more subdued face pattern, and lacks bluish neck and black line down its throat. Is also darker brown, but northern race *E. v. namaqua* is paler than birds illustrated here. Lacks blue body and wing patch of Blue Korhaan. **Voice:** Deep, frog-like duet, *'wrok-rak'* or *'wrok-rak-rak'*, mostly at dawn and dusk; male utters deeper first syllable, female responds. **Status and biology:** Endemic. Common resident in Karoo scrub; locally in open fields in s Cape. Pairs remain together year-round, often accompanied by young from the most recent br. attempt. (Vaalkorhaan)

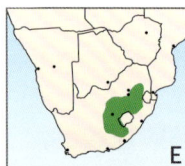

BLUE KORHAAN   *Eupodotis caerulescens*

b	b	b					b	B	B	b
J	F	M	A	M	J	J	A	S	O	N	D

52–60 cm; 1.2–1.6 kg Blue-grey underparts, neck and wing (in flight) contrasting with chestnut back are diagnostic. Male White-bellied Bustard also has blue-grey foreneck, but has rufous hindneck and white belly. Male has white face with pale grey ear coverts. Female is less brightly coloured, with brown ear coverts. Juv. similar to female, with black-streaked ear coverts. Overlaps with closely related Karoo Korhaan in W, but easily distinguished by its blue plumage. **Voice:** A deep, throaty, two-syllabled *'krok kau'* initiated by one bird and often taken up by others in the group; calls mainly at dawn. **Status and biology:** NEAR-THREATENED. Endemic. Locally common resident in grassland and e Karoo. Usually in groups of 3–5 birds. (Bloukorhaan)

WHITE-BELLIED BUSTARD   *Eupodotis senegalensis*

b								b	B	B
J	F	M	A	M	J	J	A	S	O	N	D

52–60 cm; 1.2–1.6 kg Male's combination of blue foreneck, rufous hindneck and white belly is diagnostic. Female and juv. are duller, with rufous-brown necks; superficially resemble female Black-bellied Bustard (p. 180), but necks and legs are shorter, and underwing coverts are pale. S African endemic *E. s. barrowi* is sometimes treated as a separate species. Central African *E. s. mackenziei* occurs occasionally in n Namibia. **Voice:** Rhythmic, crowing *'takwarat'*, higher pitched and less hoarse than Blue Korhaan; calls mainly at dawn and dusk. **Status and biology:** Uncommon resident in open grassland and lightly wooded savanna; prefers taller grass than most other korhaans. (Witpenskorhaan)

large creamy
wing patch

black
foreneck
♂

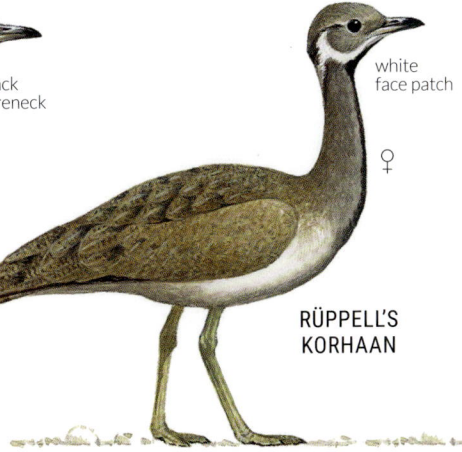
white
face patch
♀

RÜPPELL'S
KORHAAN

buffy
wing patch

plain face
& neck

KAROO KORHAAN

orange
cheeks
♀

blue wing patch

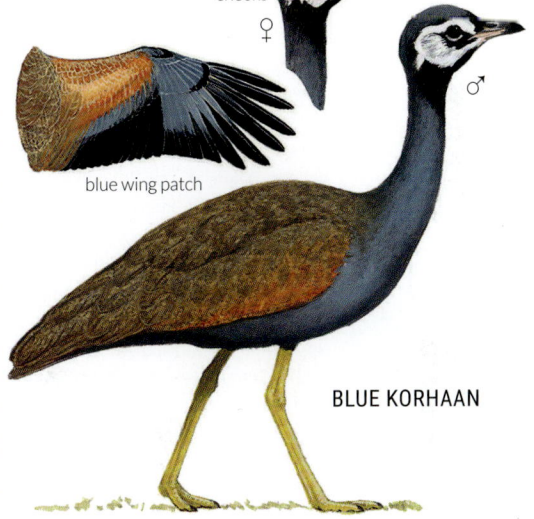
♂

BLUE KORHAAN

WHITE-BELLIED
BUSTARD

buffy
wing patch

♂
more rufous
than blue

♀

♀
white
throat

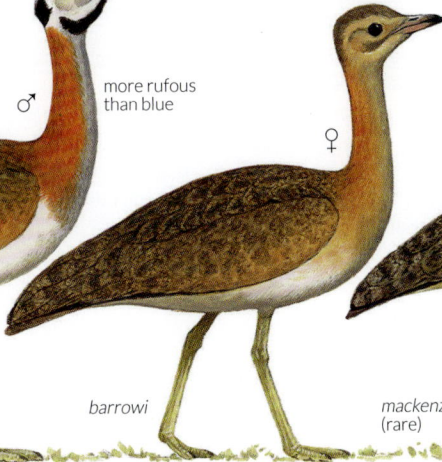

♂
more
blue than
rufous

white belly

barrowi

barrowi

mackenziei
(rare)

CRAB-PLOVER

A peculiar wader in its own family, which is allied to oystercatchers. Breeds colonially; lays a single white egg in a nest burrow. Sexes alike.

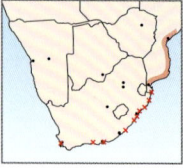

CRAB-PLOVER *Dromas ardeola*

J F M A M J J A S O N D

33–38 cm; 250–350 g; wingspan 76 cm A large, predominantly white wader with a black back and flight feathers. Easily told from Pied Avocet by its heavy, black, dagger-shaped bill (not long, slender and recurved). Long, greyish legs extend well beyond the tail in flight. Juv. and imm. are greyer than ad. on upperwings and tail, with grey backs and streaking on crown and hindneck. Juvs have upperparts suffused with brown. **Voice:** A variety of metallic calls while foraging; *'kwa-daaa-dak'*, *'kwa-da-dak'* or *'grr-kwo-kwo-kwo-kwo'*. Flight call a 2-note *'kwa-da'*. **Status and biology:** Uncommon summer migrant from br. grounds in ne Africa and Arabia; some 200+ in Mozambique (especially Ponta da Barra, Ponta São Sebastião and the Bazaruto Archipelago); uncommon and erratic in KZN (mainly Richards Bay); vagrant to E Cape. In coastal areas and estuaries, especially mangrove stands and sandy areas rich in crabs. Usually in small flocks; joins other waders and terns in high-tide roosts. (Krapvreter)

STILTS AND AVOCETS

A small family of large, long-legged waders with long straight (stilts) or recurved (avocets) bills, related to plovers and lapwings. Sexes alike.

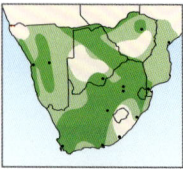

BLACK-WINGED STILT *Himantopus himantopus*

b b b b b b b B B B B b
J F M A M J J A S O N D

35–40 cm; 150–195 g; wingspan 75 cm A fairly large, black-and-white wader with very long, red legs and a thin, pointed, black bill. In flight, black underwings contrast with white underparts, and long legs trail conspicuously. Head and neck vary from pure white to predominantly dusky; occasionally has black line down nape. Sexes alike, but female has duller wings. Juv. and imm. have greyish nape, greyish-pink legs and brownish wings with pale trailing edges. **Voice:** A harsh *'kik-kik-kik-kik'*, especially when alarmed; very vocal in defence of nest and young. **Status and biology:** Common resident and local nomad at estuaries, marshes, vleis, saltpans and on flooded ground. Often common at man-made impoundments, especially sewage works. Regional population some 15 000 birds. (Rooipootelsie)

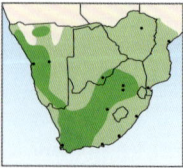

PIED AVOCET *Recurvirostra avosetta*

b b b b b B B B B B b b
J F M A M J J A S O N D

42–45 cm; 270–380 g; wingspan 78 cm An unmistakable white-and-black wader with a long, very thin, upturned bill and a black cap and hindneck. In flight, pied pattern is striking, with 3 black patches on each upperwing; underwing white except for black tip. Long, bluish-grey legs and toes extend beyond the tail in flight. Crab-Plover has similar coloration, but has a much shorter, heavier bill. Juv. has the black replaced by mottled brown. **Voice:** A clear *'kooit...kooit'*; also a *'kik-kik'* alarm call. **Status and biology:** Common and widespread resident and local nomad at lakes, estuaries, vleis, saltpans and temporary pools, both coastal and inland; occasionally on sandy beaches. Usually in small flocks (but sometimes numbering hundreds). Feeds by wading or swimming, using a sweeping, side-to-side bill movement, or by upending, duck-style, in deeper water. (Bontelsie)

SHEATHBILLS

A family of 2 plump, white, pigeon-like Antarctic waders with rounded wings, rather short, stout legs and heavy bills that have large nasal sheaths in ads. Sexes alike.

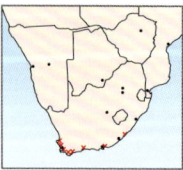

SNOWY SHEATHBILL *Chionis albus*

X X X X X X
J F M A M J J A S O N D

38–40 cm; 500–780 g; wingspan 78 cm Ship-assisted vagrant. A plump, white wader with bare pink skin around the eyes and a yellowish sheath and base to bill. Juv. has smaller facial patch; bill lacks well-developed sheath. **Voice:** Silent in Africa. **Status and biology:** Rare visitor to coastal areas, especially seabird roosts and harbours; most records Apr–Jun. Breeds in summer on Antarctic Peninsula and migrates to s S America. South African records assumed to be ship-assisted; birds land on ships during their northward migration and fly ashore when the ship rounds Africa. Most records from 1986–early 2000s, when oriental fishing vessels frequently left the Patagonian Shelf in autumn to fish off S Africa. Two recent records from E Cape in Jun 2013 and W Cape in Jun 2017. (Grootskedebek)

CRAB-PLOVER

stout,
black bill

imm.

d.

ad.

ad.

ad.

thin,
pointed bill

ad.

imm.

long, red
legs

juv.

**BLACK-WINGED
STILT**

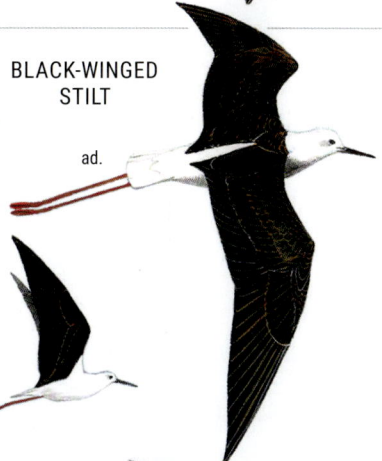

ad.

ad.

juv.

PIED AVOCET

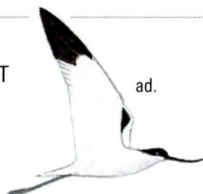

ad.

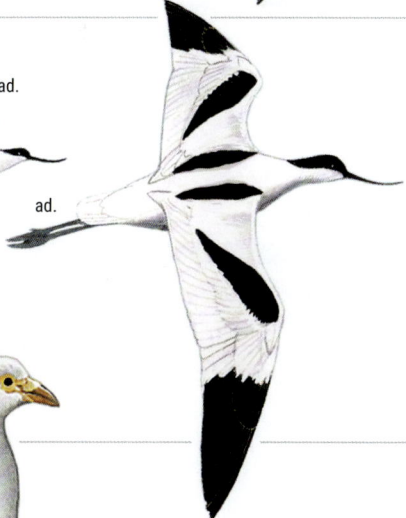

ad.

ad.

long, upturned
bill

**SNOWY
SHEATHBILL**

185

JACANAS
Distinctive, rail-like waders with extremely long toes for walking on floating vegetation. Sexes alike in plumage, but females larger than males in most species, linked to polyandry.

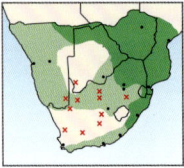

AFRICAN JACANA *Actophilornis africanus*

B	B	B	b	b	b	b	b	b	b	B	B
J	F	M	A	M	J	J	A	S	O	N	D

25–32 cm; 120–250 g Ad. has a distinctive rich chestnut body, white neck, yellow upper breast, black-and-white head and a blue frontal shield. Female appreciably larger than male. Juv. paler, with white belly and no frontal shield; considerably larger than Lesser Jacana, with dark brown (not grey) upperwing coverts, brown (not black) underwing coverts. Lacks a white trailing edge to the wing at all ages. **Voice:** Noisy; sharp '*krrrek*', rasping '*krrrrrrk*' and barking '*yowk-yowk*'. **Status and biology:** Common resident and nomad at wetlands with floating vegetation, especially water lilies. Polyandrous; males perform all parental duties after female lays. (Grootlangtoon)

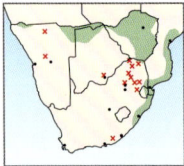

LESSER JACANA *Microparra capensis*

								b	b	b	b
J	F	M	A	M	J	J	A	S	O	N	D

15–17 cm; 40–42 g Much smaller than African Jacana, with white underparts and eyebrow and pale buffy-grey-brown upperparts. Much smaller than juv. African Jacana; in flight has white trailing edge to secondaries and pale upperwing coverts contrast with dark flight feathers; underwing coverts black (not brown). Female only slightly larger than male. Juv. less rufous above than adult. **Voice:** Soft, flufftail-like '*poop-oop-oop-oop*'; scolding '*ksh-ksh-ksh*'; high-pitched '*titititititi*'. **Status and biology:** Uncommon resident and local nomad at flood plains and wetlands with emergent grasses and sedges. The only jacana to share parental duties. (Dwerglangtoon)

OYSTERCATCHERS
Large waders with black or black-and-white plumage. Sexes alike. Despite their common name, oystercatchers feed mostly on mussels, obtained from rocky coastlines.

AFRICAN (BLACK) OYSTERCATCHER *Haematopus moquini*

B	b	b	b					b	b	B	B
J	F	M	A	M	J	J	A	S	O	N	D

42–45 cm; 580–820 g; wingspan 84 cm A large, dumpy all-black wader with short, pink legs and an orange-red bill. Birds occasionally have white patches on belly or entire plumage dove-grey. Male's bill averages shorter and less pointed than female's. Juv. is duller, faintly buff-scaled; bill duller with brown tip; legs greyish-pink. **Voice:** A '*klee-kleeep*' call and a fast '*peeka-peeka-peeka*' alarm call. Display calls include rapid trilling. **Status and biology:** NEAR-THREATENED. Near-endemic, resident at islands, coast and adjacent wetlands from n Namibia to n KZN; vagrant to s Mozambique. Usually in pairs or small flocks; global population 6 600. (Swarttobie)

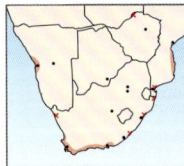

EURASIAN OYSTERCATCHER *Haematopus ostralegus*

J	F	M	A	M	J	J	A	S	O	N	D

40–45 cm; 400–820 g; wingspan 83 cm Regular vagrant. Slightly smaller and more slender than African Oystercatcher, with white belly and bold, white wingbars, back and rump visible in flight. Juv. and most non-br. ads have a white throat crescent and brown-tinged black feathering. Juv. has duller bill and legs than ad. **Voice:** A sharp, high-pitched '*klee-kleep*'. **Status and biology:** NEAR-THREATENED. Rare, Palearctic-br. migrant chiefly in summer, to coast, estuaries and lagoons; vagrant inland; some remain for extended periods. Most common in central Namibia, E Cape and KZN. (Bonttobie)

THICK-KNEES (DIKKOPS)
Large, mainly terrestrial waders with short, stout bills and long yellow legs. Plumage cryptic at rest, but with bold wing patterns in flight. Active at night; roost during the day. Sexes alike.

SPOTTED THICK-KNEE *Burhinus capensis*

B	b	b	b				b	B	B	B	B
J	F	M	A	M	J	J	A	S	O	N	D

43 cm; 380–600 g; wingspan 80 cm Larger than Water Thick-knee, with dark brown and buff-spotted upperparts; lacks grey wing coverts. In flight, shows small, white patches at base of primaries; remainder of flight feathers and greater coverts blackish. Underwing white with black leading and trailing edges and a median stripe. Juv. more streaked above than ad. **Voice:** Rising then falling '*whi-whi-whi-WHI-WHI-WHI-whi-whi*'; usually calls at night. **Status and biology:** Common resident in open areas, including fields and parks. Often in pairs; predominantly nocturnal. (Gewone Dikkop)

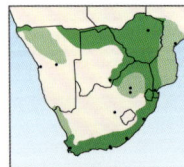

WATER THICK-KNEE *Burhinus vermiculatus*

b							b	b	B	B	B	b
J	F	M	A	M	J	J	A	S	O	N	D

38–41 cm; 280–430 g; wingspan 76 cm Smaller than Spotted Thick-knee, with plainer, grey-brown plumage finely streaked dark brown, and a grey wing panel visible at rest and in flight. Juv. more streaked above. **Voice:** Mournful '*ti-ti-ti-tee-teee-tooo*', slowing and dropping in pitch; usually calls at night. **Status and biology:** Common resident and local nomad along rivers, lake shores and lagoons. Usually in pairs; mainly nocturnal but more active by day than Spotted Thick-knee. (Waterdikkop)

ad.

large size

dark crown & eye-stripe

juv.

juv.

juv.

plainer upperwing (cf. Lesser Jacana)

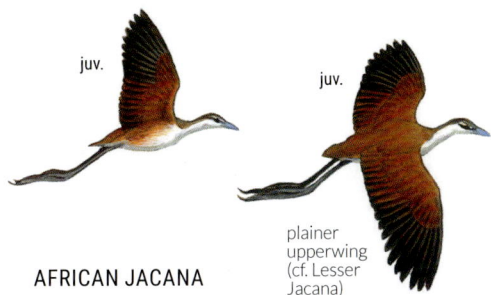

AFRICAN JACANA

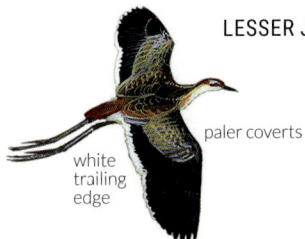

tiny size

buff-brown crown & eye-stripe

buffy-orange sides of neck

LESSER JACANA

paler coverts

white trailing edge

AFRICAN OYSTERCATCHER

ad.

ad.

juv.

non-br.

non-br.

EURASIAN OYSTERCATCHER

SPOTTED THICK-KNEE

spotted above

pale wingbar

WATER THICK-KNEE

PRATINCOLES
Peculiar waders, closely related to coursers (p. 190) but with long, pointed wings, short legs and short, broad bills adapted for catching aerial insect prey. Sexes alike, with distinct br. plumages in some species. Breed singly or in loose colonies.

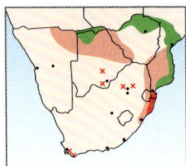

COLLARED PRATINCOLE *Glareola pratincola*

J F M A M J J A S O N D

24–25 cm; 60–100 g; wingspan 62 cm A large, long-winged pratincole with a pale buffy-yellow throat, thinly edged black in br. plumage. Tail extends beyond wing tips at rest. In non-br. plumage, collar absent or reduced to a few speckles. Paler above than Black-winged Pratincole, with more distinct white eye-ring and gradual merging of grey breast into whitish belly. Best separated in flight by pale trailing edge to secondaries and dark rufous (not black) underwing coverts (but in poor light, underwing can appear all dark). Flight buoyant and graceful, showing conspicuous white rump and deeply forked tail. Juv. lacks clearly defined throat markings and has buff edges to mantle feathers; underwing coverts rufous. **Voice:** 'Kik-kik', especially in flight. **Status and biology:** Locally common resident and intra-African migrant, usually at wetland margins and open areas near water. (Rooivlerksprinkaanvoël)

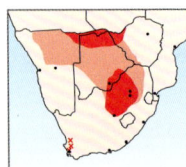

BLACK-WINGED PRATINCOLE *Glareola nordmanni*

J F M A M J J A S O N D

23–25 cm; 90–110 g; wingspan 64 cm Slightly darker and longer legged than Collared Pratincole, with less distinct eye-ring and sharper divide between grey breast and whitish belly. Buff throat patch often poorly defined. Best distinguished in flight, when it lacks white trailing edge to secondaries and has black (not rufous) underwing coverts. Juv. is drabber and more scalloped on the upperparts than non-br. ad; sometimes with patchy rufous flecking on the underwing coverts (but never has wholly rufous underwing coverts of juv. Collared Pratincole). **Voice:** An often-repeated, single- or double-noted 'pik'. **Status and biology:** NEAR-THREATENED. Locally common Palearctic-br. summer migrant; usually in large, nomadic flocks in grassland, fallow lands and at the edges of wetlands. Southern Africa supports most of the world population during the southern summer. Numbers and flock sizes decreased dramatically during the 20th century, but occasional large flocks still occur: one flock in the Free State in Dec 1991 estimated at more than 250 000 birds. (Swartvlerksprinkaanvoël)

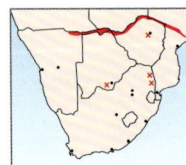

ROCK PRATINCOLE *Glareola nuchalis*

J F M A M J J A S O N D

17–20 cm; 42–55 g; wingspan 50 cm A small, dark pratincole with relatively short wings and tail; much smaller than Collared and Black-winged Pratincoles. A diagnostic white line extends from the eye down across the lower part of the nape to form a nuchal collar. Legs and base of bill red. Appears dark in flight, with a conspicuous white rump; from below, lower belly, vent and base of tail white; small, white wingbar at base of secondaries (underwing only). Juv. is duller; lacks white eye stripe, has darker legs and is lightly speckled buff above and on breast. **Voice:** A loud, repeated, plover-like 'kik-kik'. **Status and biology:** Uncommon intra-African migrant br. at rocky areas (occasionally on sand bars) mostly on Zambezi and Kavango rivers; vagrant elsewhere. Regional population <2 000 birds. Numbers decreased in the second half of the 20th century due mainly to impoundments and water abstraction along large rivers; has disappeared from some sites. (Withalssprinkaanvoël)

COLLARED PRATINCOLE

non-br.

br.

tail longer
than wings

juv.

non-br.

red 'armpits'

br.

white trailing
edge

juv.

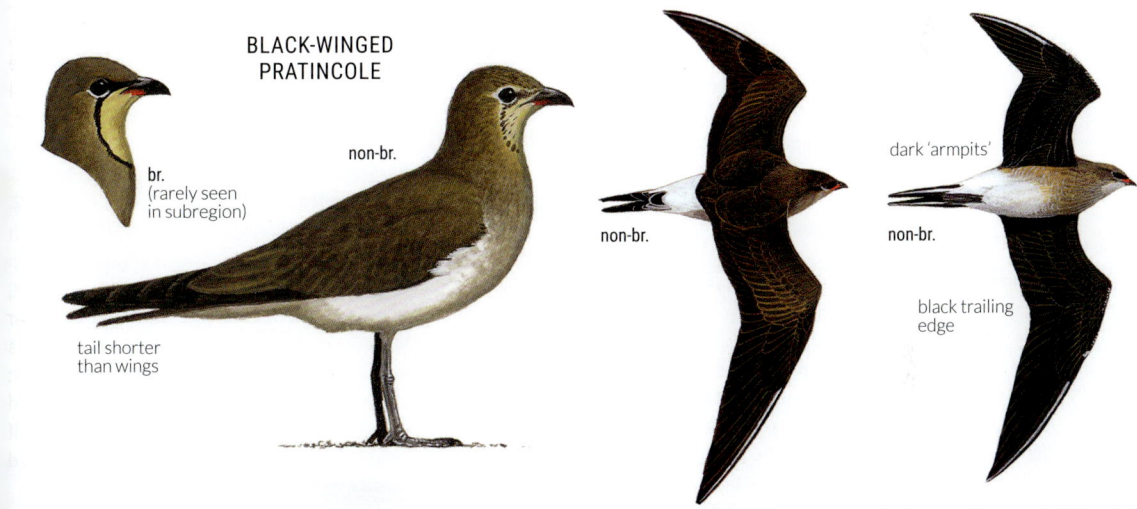

BLACK-WINGED
PRATINCOLE

br.
(rarely seen
in subregion)

non-br.

non-br.

tail shorter
than wings

dark 'armpits'

non-br.

black trailing
edge

white collar

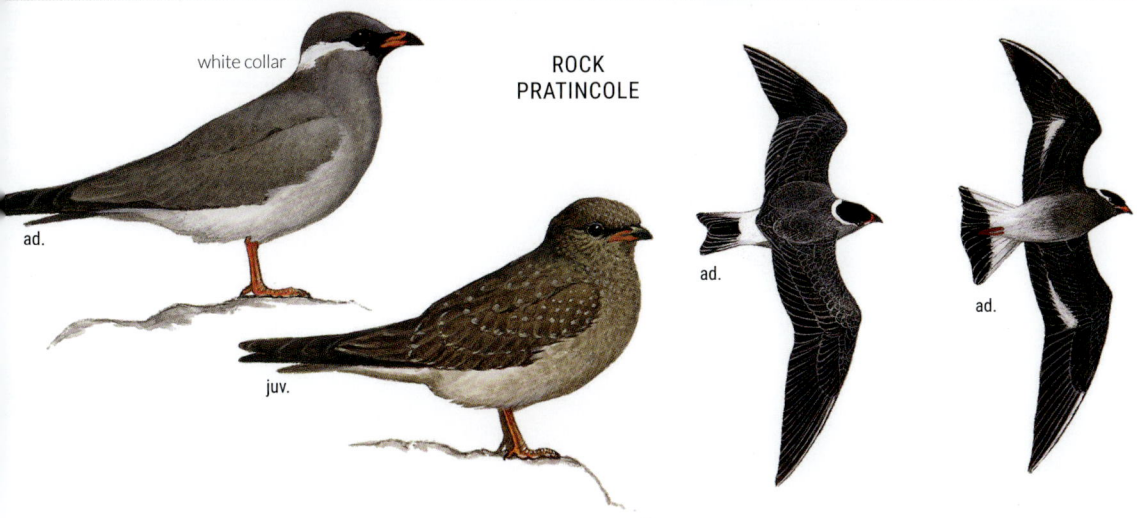

ROCK
PRATINCOLE

ad.

ad.

ad.

juv.

189

COURSERS

Fairly small, long-legged, terrestrial waders. Sexes alike. Some care is needed to separate the 2 *Cursorius* species, which are active mainly by day. The other 3 species are readily identified, but are easily overlooked as they are largely nocturnal.

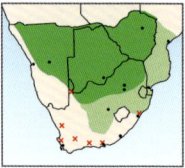

TEMMINCK'S COURSER   *Cursorius temminckii*

b				b	b	b	B	B	B	B	b
J	F	M	A	M	J	J	A	S	O	N	D

19–21 cm; 65–80 g; wingspan 42 cm A plain, rufous courser with a broad, black patch behind the eye. Differs from Burchell's Courser by having a rufous (not grey) hind crown and a blackish patch (not bar) on the belly; upperparts darker grey-brown, contrasting more with paler underparts. In flight, underwing black, with a narrow white trailing edge to secondaries, but secondaries appear blackish from above. Juv. duller, with lightly speckled underparts and scalloped upperparts. **Voice:** Grating 'keerkeer'. **Status and biology:** Locally common resident, nomad and intra-African migrant found in dry, sparsely grassed and recently burnt areas. (Trekdrawwertjie)

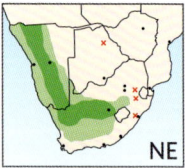

BURCHELL'S COURSER   *Cursorius rufus*

b	b	b	b			b	B	B	B	B	B
J	F	M	A	M	J	J	A	S	O	N	D

21–23 cm; 70–110 g; wingspan 46 cm A plain rufous courser with a narrow, black line behind the eye and a blue-grey hind crown and nape. Paler than Temminck's Courser, with less contrast between upperparts and breast. Blackish bar (not patch) separates brown upper belly from white lower belly and vent. In flight, has broad white trailing edge to secondaries. Underwing coverts rufous; secondaries mostly white; primaries black. Juv. mottled above, with barred tail tip and poorly defined black belly bar. **Voice:** Harsh, repeated 'wark'. **Status and biology:** Uncommon, near-endemic nomad in dry, sparsely grassed plains and open fields. Population possibly decreasing. Small numbers move into W Cape in late winter. (Bloukopdrawwertjie)

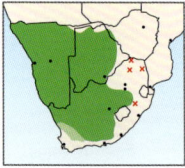

DOUBLE-BANDED COURSER   *Rhinoptilus africanus*

B	B	b	b	b	b	b	b	b	B	B	B
J	F	M	A	M	J	J	A	S	O	N	D

20–24 cm; 70–100 g; wingspan 45 cm A whitish courser with 2 narrow, black breast bands (diagnostic); head plain, with prominent creamy supercilium. Upperparts are scaled; dark back and wing coverts have broad, creamy-buff edges. In flight, shows white uppertail coverts and conspicuous chestnut secondaries and inner primaries contrasting with dark outer primaries. Juv. has chestnut breast bands. **Voice:** Thin, falling and rising 'teeu-wee' whistle; repeated 'kee-kee', mostly at night. **Status and biology:** Fairly common resident and local nomad of semi-arid and desert plains, usually in stony sites. Favours areas of pale calcrete. (Dubbelbanddrawwertjie)

THREE-BANDED COURSER   *Rhinoptilus cinctus*

	b	b	b	b	b	B	B	B	b	b
J	F	M	A	M	J	J	A	S	O	N	D

25–28 cm; 120–150 g; wingspan 52 cm Larger than Double-banded Courser, with diagnostic rufous, black, and white bands on breast and neck. White supercilium forks behind eye and extends into hind collar. Upper breast is conspicuously streaked. In flight shows white uppertail coverts, but upperwings are relatively uniform, with darker brown flight feathers than Double-banded Courser. Juv. like ad., but with barred secondaries and central tail feathers. **Voice:** Repeated 'kika-kika-kika' at night. **Status and biology:** Uncommon to locally common in arid and semi-arid savanna, especially mopane woodland; mostly resident. Largely nocturnal; roosts under bushes during the day. (Driebanddrawwertjie)

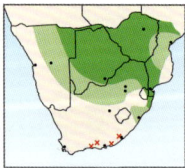

BRONZE-WINGED COURSER   *Rhinoptilus chalcopterus*

						b	b	B	B	b	b
J	F	M	A	M	J	J	A	S	O	N	D

25–28 cm; 120–165 g; wingspan 52 cm A distinctive courser with a broad, dusky band across the breast and lower neck separated from the pale lower breast and belly by a black bar. In flight, white uppertail coverts and narrow, white wingbars contrast with dark upperparts. Also known as Violet-tipped Courser due to the iridescent violet tips to its flight feathers, but these are seldom visible in the field. Juv. has rufous-tipped feathers on upperparts. **Voice:** Ringing 'ki-kooi' at night. **Status and biology:** Fairly common resident and intra-African summer migrant in woodland and savanna, but also in more open habitats on migration. Largely nocturnal; roosts under bushes during the day, often in quite exposed sites. (Bronsvlerkdrawwertjie)

rufous hind crown

black trailing edge

dark belly patch

TEMMINCK'S COURSER

grey hind crown

white trailing edge

dark bar on belly

BURCHELL'S COURSER

rufous wing patches

DOUBLE-BANDED COURSER

streaked lower neck & upper breast

THREE-BANDED COURSER

plain lower neck

BRONZE-WINGED COURSER

black & white bands in tail

violet tips to flight feathers

LAPWINGS
Large waders with short bills and necks; body plumage ranges from cryptic among terrestrial species to strongly patterned among some wetland species, but all have striking wing and tail patterns. Like other plovers, they feed mainly visually, picking prey; they often stand and scan, then dash forward to grab prey. Sexes alike.

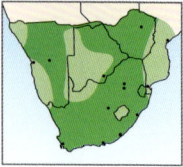

BLACKSMITH LAPWING   *Vanellus armatus*

b b b b b b B B B B b b
J F M A M J J A S O N D

28–31 cm; 130–200 g; wingspan 78 cm A boldly marked, black, white and grey lapwing with blackish legs and dark red eyes. In flight, pale grey coverts contrast with black flight feathers. Juv. is duller, with greyish brown upperparts, and mottled brown and white head and neck. Sometimes confused with Spur-winged Lapwing, but back is spangled (not plain brown) and head and neck plumage is rather scruffy, with whitish (not black) throat; lacks white wingbar; usually accompanied by ads. **Voice:** Very vocal, with a loud, ringing '*tink, tink, tink*' alarm call. **Status and biology:** Common resident and nomad at wetland margins and adjoining grassland and fields. Often in flocks when not br. (Bontkiewiet)

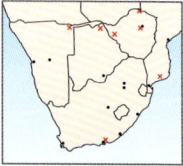

SPUR-WINGED LAPWING   *Vanellus spinosus*

X X X X X X X X X X X X
J F M A M J J A S O N D

25–28 cm; 135–175 g; wingspan 75 cm Vagrant. A small lapwing with a pale brown mantle, white neck, and black breast, crown and throat stripe; legs blackish. In flight, wing pattern similar to Crowned Lapwing (p. 194) but readily identified by its black belly and white sides to neck. Could be confused with juv. Blacksmith Lapwing but has a plain brown (not spangled or blackish) back and crisp black-and-white head and neck pattern; white wingbar distinctive. Juv. has browner crown with fine white spotting, and back feathers edged buff. **Voice:** A loud, screeching 3- or 4-note '*ti-ti-tirri-ti*'. **Status and biology:** Vagrant to large rivers and flood plains from further north in Africa; with records from n Namibia, Botswana, Zimbabwe and Mozambique. Increasing numbers in Zambia and Malawi suggest ongoing southward range expansion, highlighted by a breeding record from Lake Chivero, Zimbabwe, in Nov 2019. (Spoorvlerkkiewiet)

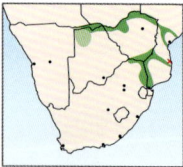

WHITE-CROWNED LAPWING   *Vanellus albiceps*

b b B B B b
J F M A M J J A S O N D

28–32 cm; 150–210 g; wingspan 80 cm A striking lapwing, with large, pendulous, yellow wattles and a diagnostic white median crown stripe from the forehead to the nape; legs greenish-yellow. White (not brown) breast and plain grey (not streaked) sides of the neck separate it from African Wattled Lapwing. In flight upperwings mostly white, with only the outermost primaries and inner coverts black. Underwings white, except for black outer primaries. Tail is predominantly black. Distinguished from Long-toed Lapwing in flight from below by lack of broad, black breast band. Juv. resembles ad. but has barred upperparts, less white on crown and throat, and smaller wattles. **Voice:** A repeated, ringing '*peek-peek*'. **Status and biology:** Uncommon resident and nomad of sandbanks and sand bars along major rivers; numbers may be decreasing. (Witkopkiewiet)

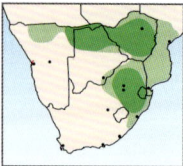

AFRICAN WATTLED LAPWING   *Vanellus senegallus*

b b B B b
J F M A M J J A S O N D

34–35 cm; 180–280 g; wingspan 86 cm The largest African lapwing with a fairly large, yellow face wattle. At rest, appears mostly brown, with red, black and white forehead, streaked head and neck, blackish band across belly and white vent. In flight, has white wingbar between brown inner coverts and blackish flight feathers; tail has broad, black subterminal band contrasting with white tail tip and rump. Differs from White-crowned Lapwing in having a dark brown (not white) breast bordered by a black belly line, and streaked (not plain) grey sides to head and upper neck; yellow wattles are smaller. Juv. has much smaller wattles and less distinct head markings than ad. **Voice:** A high-pitched, ringing '*keep-keep*'; regularly calls at night. **Status and biology:** Fairly common resident, nomad and local migrant at wetland margins and adjacent grassland. Regional population around 15 000 birds. (Lelkiewiet)

ad.

juv.

juv.

dark cheek;
white belly

BLACKSMITH LAPWING

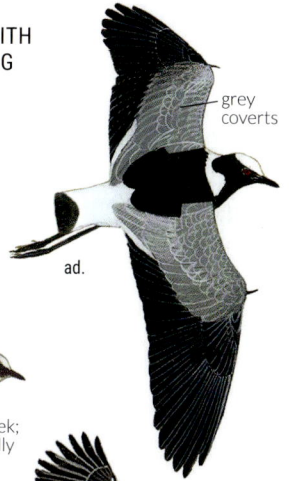

grey coverts

ad.

SPUR-WINGED LAPWING

white cheek;
black belly

white coverts

plain grey neck

white breast & belly

WHITE-CROWNED LAPWING

AFRICAN WATTLED LAPWING

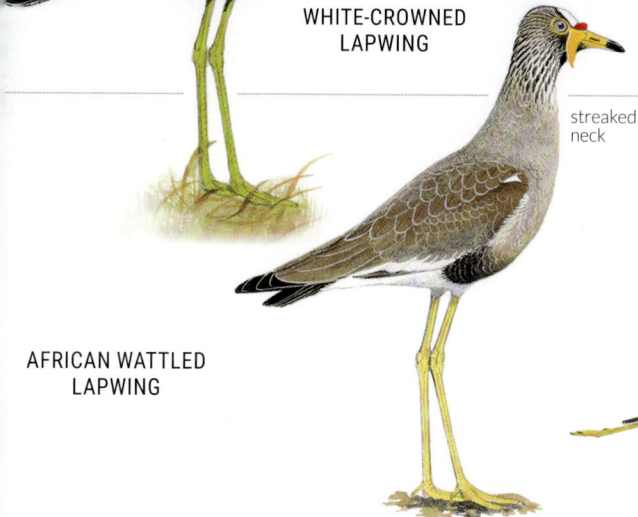

streaked neck

brown breast,
black belly bar

193

CROWNED LAPWING   *Vanellus coronatus*

29–31 cm; 150–220 g; wingspan 78 cm A large, mostly brown lapwing, with a black cap interrupted by a white 'halo'. Legs and base of bill bright pinkish-red. Sandy-brown breast separated from white belly by a black band. In flight, brown back and lesser/median coverts contrast with white greater primary coverts, and with blackish flight feathers. Juv. much like ad., but has scalloped upperparts, buff-barred crown, and duller red legs and base to bill. **Voice:** Noisy; loud, grating *'kreeeep'* given day and night. **Status and biology:** Common resident and local nomad of open country, including short grassland (either grazed or burnt); also on golf courses, playing fields and fallow land; seldom associated with water. Aggregates in small flocks, especially when not br.; regularly associates with Black-winged Lapwing. Regional population perhaps as large as 100 000 birds, but numbers have decreased locally in W Cape. (Kroonkiewiet)

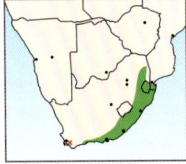

BLACK-WINGED LAPWING   *Vanellus melanopterus*

26–28 cm; 150–200 g; wingspan 78 cm Larger and bulkier than Senegal Lapwing, with more extensive white on the forehead (almost reaching the eye), redder legs and a broader black border separating the breast from the belly. Female has narrower breast band, less white on forehead and a duller grey crown. Flight pattern similar to Crowned Lapwing, but with black (not white) primary coverts. In flight from above, distinguished from Senegal Lapwing by the broad, white wingbar and black (not white) secondaries. From below distinguished by black (not white) secondaries, and wholly white underwing coverts (mid and outer primary coverts black in Senegal Lapwing). Non-br. birds have eye-ring and legs blackish, not red (Senegal Lapwing always has blackish legs). Juv. is browner, with back and wing feathers edged buff. **Voice:** A shrill, piping *'ti-tirree'*, higher pitched than that of Senegal Lapwing; often calls at night. **Status and biology:** Uncommon resident and local migrant, with movements between uplands and the coast in north of range. Favours short grassland; frequently associates with Crowned Lapwing. Regional population around 2 500 birds. (Grootswartvlerkkiewiet)

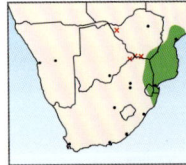

SENEGAL LAPWING   *Vanellus lugubris*

22–26 cm; 100–140 g; wingspan 72 cm Smaller and more slender than Black-winged Lapwing, with a narrow, black border to the grey breast, less white on the forehead, a darker grey crown and a slight greenish tinge to the upperparts. Legs blackish. In flight from above, distinguished from Black-winged Lapwing by the lack of a broad, white wingbar and the white (not black) secondaries. From below distinguished by white (not black) secondaries, and black (not white) mid and outer primary coverts. Juv. is less clearly marked than ad. and has upperparts spotted buff. **Voice:** A clear *'tee-yoo, tee-yoo'*; often calls at night. **Status and biology:** Uncommon resident and local migrant in dry, open, short-grass areas in savanna; favours recently burnt patches. Most easily seen in s Kruger National Park and adjacent regions of s Mozambique and e Eswatini. (Kleinswartvlerkkiewiet)

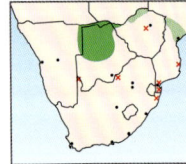

LONG-TOED LAPWING   *Vanellus crassirostris*

29–31 cm; 160–220 g; wingspan 76 cm A localised species and the only lapwing in the region with a white face, throat and foreneck. The black of the nape extends down the sides of the neck to form a broad breast band. Legs and base of bill pinkish-red. Striking in flight, with a grey-brown back and mostly white wings with black outer primaries and a variable brown bar across the median coverts. Tail black; rump white. Juv. resembles ad., but upperparts and breast mottled with buff. **Voice:** A repeated, high-pitched *'pink-pink'*. **Status and biology:** Locally common resident and local nomad in marshes and flood plains. (Witvlerkkiewiet)

CROWNED LAPWING

ad.

juv.

ad.

black & white cap

ad.

white belly

broad, white forehead

ad.

juv.

ad.

white covert bars

BLACK-WINGED LAPWING

pale purplish legs

ad.

narrower white forehead

SENEGAL LAPWING

ad.

ad.

dark grey legs

juv.

ad.

white trailing edge

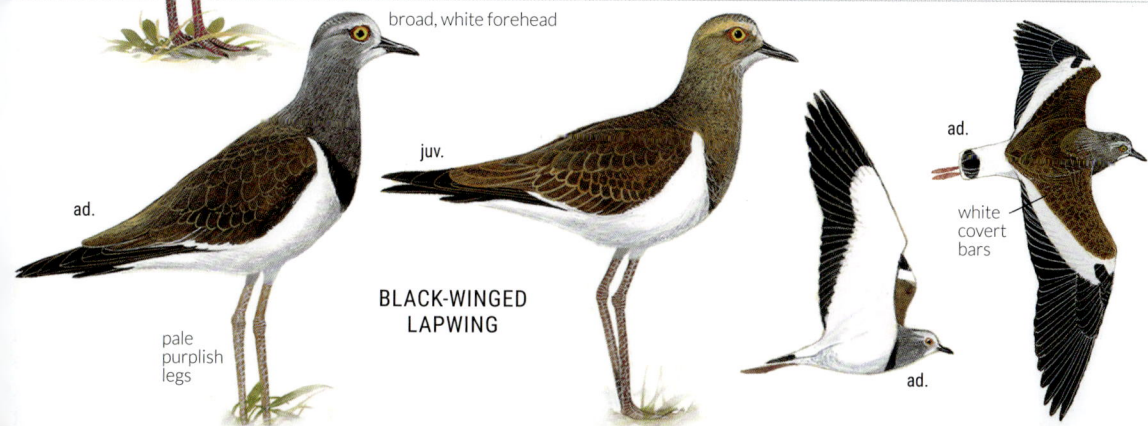

LONG-TOED LAPWING

mostly white wings

white face; black breast band

PLUVIALIS PLOVERS

PLUVIALIS **PLOVERS** Three fairly large plovers that appear to be basal to both lapwings and *Charadrius* plovers. The larger Grey Plover is easily identified in flight by its white rump and black axillaries, but the 2 vagrant species are hard to distinguish; superficially resemble Red Knot (p. 216) in flight. Usually forage singly, but roost communally. Feed visually, standing and scanning, then dashing forward to grab prey. Sexes alike, with distinct br. plumages.

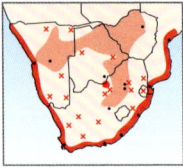

GREY PLOVER *Pluvialis squatarola*

J F M A M J J A S O N D

27–31 cm; 180–300 g; wingspan 76 cm A large, stubby-billed plover. Best told from smaller, vagrant American and Pacific Golden Plovers by black (not grey) axillaries, contrasting with the whitish underwings and white (not dark brown) rump in flight. Has a heavier bill than American or Pacific Golden Plovers. At rest (non-br. plumage) has grey (not buff or golden) speckling on the back and wing coverts. In br. plumage, underparts are black and upperparts are speckled white (not golden). A wide, white forecrown and supercilium extends to sides of upper breast. Juv. has upperparts with buffy-yellow markings. **Voice:** A clear, slightly mournful *'tluui'*, lowest in pitch in the middle. **Status and biology:** Common Palearctic-br. summer visitor to the open shore and coastal wetlands; inland only during migration; many overwinter. Regional population around 25 000 birds. (Grysstrandkiewiet)

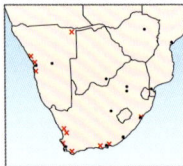

AMERICAN GOLDEN PLOVER *Pluvialis dominica*

J F M A M J J A S O N D

24–26 cm; 120–180 g; wingspan 68 cm Vagrant. Slightly smaller and more slender than Grey Plover, with a more upright stance; easily differentiated in flight by plain grey underwing (lacking black axillaries) and brown (not white) rump. Main challenge is to separate it from slightly smaller Pacific Golden Plover. American Golden Plover best identified by its heavier build (appearing shorter-legged), less yellowish speckling on mantle (appears colder grey-brown from a distance), bolder supercilium, lack of buff on the breast and, importantly, tertials not extending to tail tip. In br. plumage, distinguished from Pacific Golden Plover by its black (not white) vent and lack of a white flank stripe. **Voice:** A whistled *'kuue-eep'*, or *'oodle-oo'* similar to Pacific Golden Plover. **Status and biology:** Nearctic-br. summer vagrant to coastal wetlands and short grasslands; rarely inland. (Amerikaanse Goue Strandkiewiet)

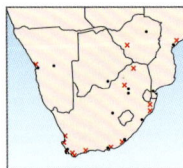

PACIFIC GOLDEN PLOVER *Pluvialis fulva*

J F M A M J J A S O N D

23–25 cm; 110–170 g; wingspan 65 cm Vagrant. Easily told from larger Grey Plover by its plain grey underwing (lacking black axillaries) and brown (not white) rump. Smaller, more slightly built and more 'leggy' than American Golden Plover with a less prominent supercilium; with distinct rufous ear coverts, structure recalls Red Knot (p. 216) in flight. From a distance, appears warmer brown due to its more buffy breast and more extensive yellow spangling on the mantle; tertials extend almost to the tail tip (longer than in American Golden Plover). In br. plumage, has bold, white flank stripe and white (not black) vent. **Voice:** A whistled *'kuue-eep'*, or *'oodle-oo'* similar to American Golden Plover. **Status and biology:** Palearctic-br. summer vagrant to coastal estuaries and lagoons; rarely inland. (Asiatiese Goue Strandkiewiet)

GREY PLOVER

non-br.

weak supercillium

non-br.

black 'armpits'

white rump

non-br.

white undertail coverts

br.

br.

br.

prominent supercillium

dark rump

AMERICAN GOLDEN PLOVER

grey-brown upperparts

no toe projection

tertials fall short of tail tip

black undertail coverts

non-br.

br.

non-br.

non-br.

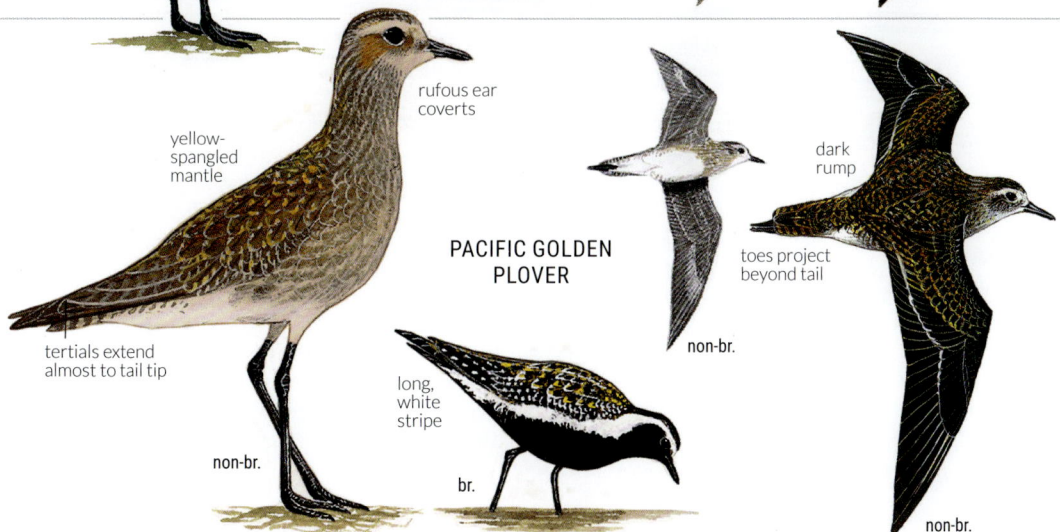

rufous ear coverts

yellow-spangled mantle

dark rump

PACIFIC GOLDEN PLOVER

toes project beyond tail

tertials extend almost to tail tip

long, white stripe

non-br.

non-br.

non-br.

br.

197

CHARADRIUS PLOVERS

CHARADRIUS PLOVERS Small to medium-sized waders with short bills and necks. Feed mainly visually, picking prey from surface; often stand and scan, then dash forward to grab prey; some species use their feet to disturb prey, vibrating one foot on the mud surface. Most forage singly, not in flocks like stints and sandpipers, which forage by touch. Sexes alike, with distinct br. plumages in migratory species and some resident species.

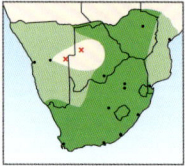

THREE-BANDED PLOVER *Charadrius tricollaris*

b b b b b b B B B B B B
J F M A M J J A S O N D

18 cm; 28–45 g; wingspan 46 cm A small, brown-backed plover with 2 black breast bands, a white ring around the crown, grey cheeks and conspicuous red eye-ring and base to the bill. Tail fairly long, extending beyond folded wings; bobs tail when alarmed like Common Sandpiper (p. 212). In flight, shows a narrow, white wingbar and white sides to rump, outer tail and tail tip: rump and tail appear elongated, as in Common Ringed Plover (p. 200). Juv. has duller red eye-rings and legs than ad.; breast bands diffusely mottled with white; eyes dark. **Voice:** A penetrating, high-pitched *'weee-weet'* whistle, usually in flight. **Status and biology:** Widespread, common resident on most types of freshwater wetlands, including streams and farm dams. Favours muddy areas close to vegetation; rare on the open coast, usually only where there are freshwater seeps. Regional population at least 25 000 birds. (Driebandstrandkiewiet)

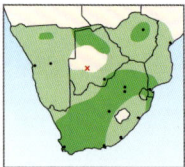

KITTLITZ'S PLOVER *Charadrius pecuarius*

b b b b b b B B B B B B
J F M A M J J A S O N D

14–16 cm; 25–45 g; wingspan 44 cm Ad. has distinctively patterned head; banded black and white in br. plumage; dark brown and buff in non-br. plumage. Has a dark eye patch at all ages and seasons. Breast creamy-buff to chestnut; dark shoulder patch. Juv. duller than non-br. ad., with buff frons and broad buff supercilium joining across nape. Told from juv. White-fronted Plover by the broad, buffy supercilium extending onto the buffy (not white) nape. Sometimes confused with non-br. Caspian Plover (p. 200), but is smaller and has pale band across nape; white wingbar more prominent in flight. Pale band across nape also helps distinguish juv. from similar Greater and Lesser Sand Plovers. **Voice:** A short, clipped trill, *'kittip'* and a thin *'tee-peep'*. **Status and biology:** Common resident, local nomad and intra-African migrant found on dry mud or short grass, usually near water; also on mud flats in estuaries and coastal lagoons, kelp-strewn areas of open coast, ploughed fields and bare areas in fallow lands. Often in flocks outside br. season. Regional population at least 55 000 birds. (Geelborsstrandkiewiet)

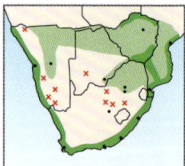

WHITE-FRONTED PLOVER *Charadrius marginatus*

B b b b b b B B B B B B
J F M A M J J A S O N D

16–17 cm; 35–60 g; wingspan 44 cm A pale, grey-brown plover with broad, white frons, blackish line from base of bill through eye (reduced or almost absent in juv.) and white nuchal collar. Ad. has black band above white frons (broader in male). Breast varies from white to rich buff. Superficially resembles Chestnut-banded Plover but is slightly larger, with a white collar and only a partial breast band; appears longer-tailed. Head pattern and paler upperparts distinguish it from Kittlitz's Plover. White collar separates it from larger Lesser Sand Plover (p. 200). Juv. has upperpart feathers edged pale buff; breast usually white with variable dusky, lateral patches. Some imms have small, dusky pectoral patches and could be confused with Kentish Plover *C. alexandrinus*, but no confirmed records of this species from the region. Inland race *C. m. mechowi* is darker brown above and rich buffy below. **Voice:** A clear *'wiiit'*; *'tukut'* alarm call. **Status and biology:** Common resident and local migrant on sandy and rocky shores, muddy coastal areas and larger inland rivers and pans. Regional population at least 20 000 birds. (Vaalstrandkiewiet)

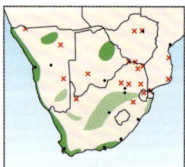

CHESTNUT-BANDED PLOVER *Charadrius pallidus*

b b b b b b B B B B B B
J F M A M J J A S O N D

15 cm; 32–45 g; wingspan 42 cm The smallest, palest plover of the region. Ad. has a narrow, chestnut breast band that extends across the hind cheeks and onto the crown. Male has neat black markings on forehead and lores. Female has chestnut lores and a paler breast band. Distinguished from larger White-fronted Plover by the breast band and lack of a white, hindneck collar; appears more compact and shorter-tailed. Imm. resembles female, but with narrower, more buffy breast band; no chestnut on sides of neck. Juv. lacks any black or chestnut coloration; breast band pale grey, often incomplete. Juv. White-fronted Plover is similar but lacks the pale grey breast band. **Voice:** A single *'prrp'* or *'tooit'*. **Status and biology:** NEAR-THREATENED. Uncommon, localised resident and partial migrant found predominantly at saltpans in summer; some move to estuaries and coastal wetlands in winter, when several thousand present at Walvis Bay and Sandwich Harbour, Namibia. In the south of the range, most birds are confined to commercial saltpans. Regional population around 11 000 birds. (Rooibandstrandkiewiet)

white crown ring

red eye-ring

2 black
breast bands

**THREE-BANDED
PLOVER**

**KITTLITZ'S
PLOVER**

br.

br.

buffy collar

buffy collar

non-br.

juv.

br.

♀

♂

**WHITE-FRONTED
PLOVER**

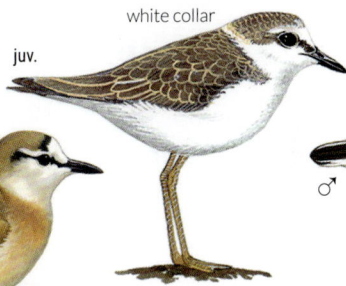

white collar

juv.

♂

♂

inland race
mechowi

rich buff
below

♂

**CHESTNUT-BANDED
PLOVER**

chestnut
band

♂

faint,
incomplete
band

juv.

♀

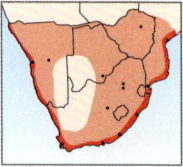

COMMON RINGED PLOVER   *Charadrius hiaticula*

J F M A M J J A S O N D

18–20 cm; 45–70 g; wingspan 50 cm A small, short-legged, brown-backed plover with a complete white collar above a blackish-brown breast band; breast band often incomplete in juvs and non-br. plumage. In flight, narrow, white wingbar and collar are conspicuous. Br. ad. has crisply patterned head with a black forehead with a narrow, white band across the frons, black face and white patch above and behind the eyes; bill orange with black tip. Non-br. ad. has duller head and breast markings; bill blackish. Juv. is even duller, usually with an incomplete breast band. Slightly larger than Little Ringed Plover, with an orange (not yellow) eye-ring, heavier bill, orange (not pinkish-grey) legs and a conspicuous wingbar. **Voice:** A fluty *'too-li'*. **Status and biology:** Common Palearctic-br. summer visitor; some overwinter. Coastal and inland wetlands, preferring patches of soft, fine mud; also on rocky shores. Regional population at least 10 000 birds. (Ringnekstrandkiewiet)

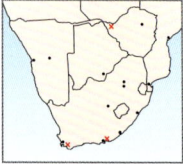

LITTLE RINGED PLOVER   *Charadrius dubius*

X X X J F M A M J J A S O N D

15–17 cm; 30–48 g; wingspan 44 cm Vagrant. Superficially similar to Common Ringed Plover, but slightly smaller, with a narrower breast band, more slender bill, conspicuous yellow (not orange) eye-ring and pinkish-grey or greenish-yellow (not orange) legs; lacks obvious white wingbar in flight. Juv. has narrow, yellow eye-ring and brown breast band (often incomplete); told from all other small plovers of the region by its uniform upperwings. **Voice:** Descending *'pee-oo'*, or *'cloo'*. **Status and biology:** Palearctic-br. vagrant, with most remaining north of the equator; 3 subregion records: Hwange, w Zimbabwe, Jan 2002, 1 at Port Elizabeth, E Cape, August 2017 and at Onrus, W Cape, shortly thereafter. Favours dry mud along the shores of lakes and rivers. (Kleinringnekstrandkiewiet)

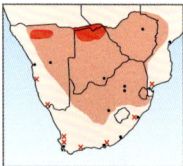

CASPIAN PLOVER   *Charadrius asiaticus*

J F M A M J J A S O N D

18–22 cm; 60–78 g; wingspan 58 cm A medium-sized plover of mainly arid habitats with a slender bill. In br. plumage, differs from Greater and Lesser Sand Plovers in having a black lower border to the chestnut breast band and a broad, white supercilium; lacks an extensive, dark eye patch. Male has brighter breast band than female. In all other plumages, has complete (or almost complete) grey-brown wash across the breast (rather than breast side patches of non-br. Greater and Lesser Sand Plovers), a broad, buffy supercilium, and pale lores (generally darker in Greater and Lesser Sand Plovers). In flight, upperparts are uniform, apart from pale bases to inner primaries. Appreciably larger than Kittlitz's Plover (p. 198); supercilium does not extend onto nape as a nuchal collar and wingbar less prominent. Smaller than American and Pacific Golden Plovers (p. 196) with much smaller pale wingbars. Juv. is similar to non-br. ad. but buffy tips to mantle feathers impart scaled appearance. **Voice:** A loud, sharp *'tyut'*. **Status and biology:** Locally common Palearctic-br. summer visitor, mainly in north, favouring sparsely grassed areas and wetland fringes; usually in flocks. Regional distribution centred on the Kalahari sandveld of Botswana and e Namibia. Historically was common further south. (Asiatiese Strandkiewiet)

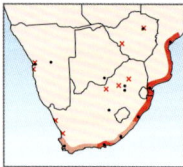

GREATER SAND PLOVER   *Charadrius leschenaultii*

J F M A M J J A S O N D

22–25 cm; 75–130 g; wingspan 55 cm A medium-sized plover, separated with difficulty from Lesser Sand Plover. Greater Sand Plover is larger and taller, with a longer, more robust bill and more angular head. In br. plumage, the brown shoulder patches become rufous and extend across the breast, but breast band narrower than Lesser Sand Plover. Leg colour variable; usually greyish-green, very rarely black (legs of Lesser Sand Plover are almost invariably black). In flight, the more extensive white on the sides of the rump and the dark subterminal tail band aid distinction from Lesser Sand Plover. Much larger than White-fronted Plover (p. 198), with a heavier bill; lacks a white nuchal collar. Told from Caspian Plover by its much heavier bill, extensive white on the sides of the rump and less prominent pale supercilium. Juv. has upperpart feathers edged buff. **Voice:** A short, musical *'trrri'*. **Status and biology:** Uncommon to locally common Palearctic-br. summer visitor to coastal wetlands, most regular in the east, especially Mozambique; vagrant inland. When foraging, averages about 10 paces between searching pauses; Lesser Sand Plover typically averages 2 or 3 paces. Regional population around 2 000 birds. (Grootstrandkiewiet)

LESSER SAND PLOVER   *Charadrius mongolus*

J F M A M J J A S O N D

19–21 cm; 45–90 g; wingspan 50 cm Compared to Greater Sand Plover, this species is smaller, has shorter legs, a smaller body and a shorter, less robust bill; head appears softer and more rounded. Legs almost always black or very dark grey (usually greyish-green in Greater Sand Plover). In flight, little colour contrast between tail, rump and back (Greater Sand Plover has a dark, subterminal tail band). In br. plumage, the rufous breast band is broader than that of Greater Sand Plover. Slightly larger than White-fronted and Kittlitz's Plovers (p. 198) and lacks these species' pale hindneck collars. Juv. has upperpart feathers edged buff. **Voice:** Harsh *'chittick'*, usually given in flight. **Status and biology:** Uncommon Palearctic-br. summer visitor to coastal areas of Mozambique; rare coastal visitor elsewhere, with some inland records. When foraging, averages 2 or 3 paces between searching pauses; Greater Sand Plover averages about 10 paces. Regional population probably <1 000 birds. (Mongoolse Strandkiewiet)

broad forehead band

dark eye-ring

juv.

br.

white wingbar

br.

bright yellow-orange legs

non-br.

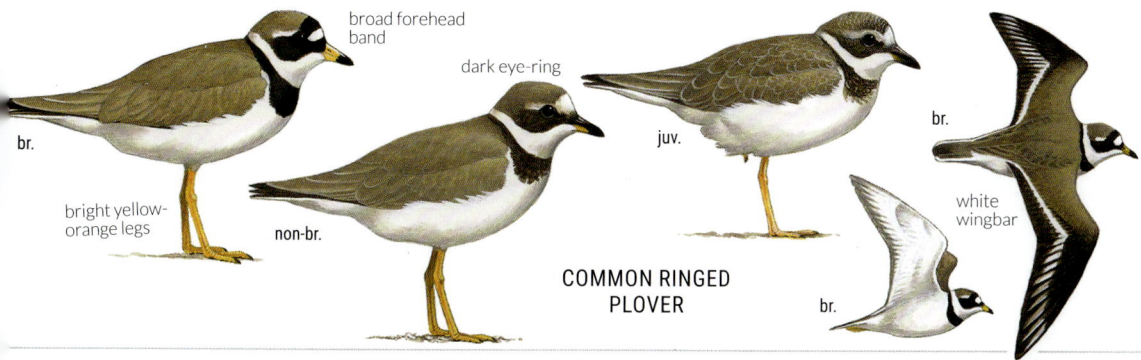

COMMON RINGED PLOVER

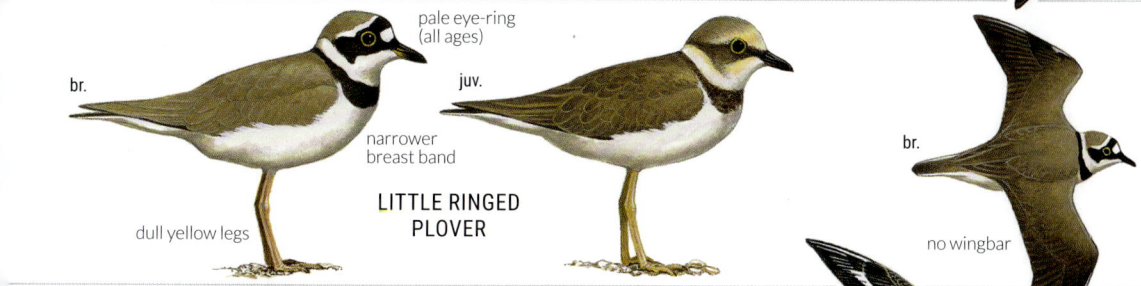

pale eye-ring (all ages)

br.

juv.

narrower breast band

br.

LITTLE RINGED PLOVER

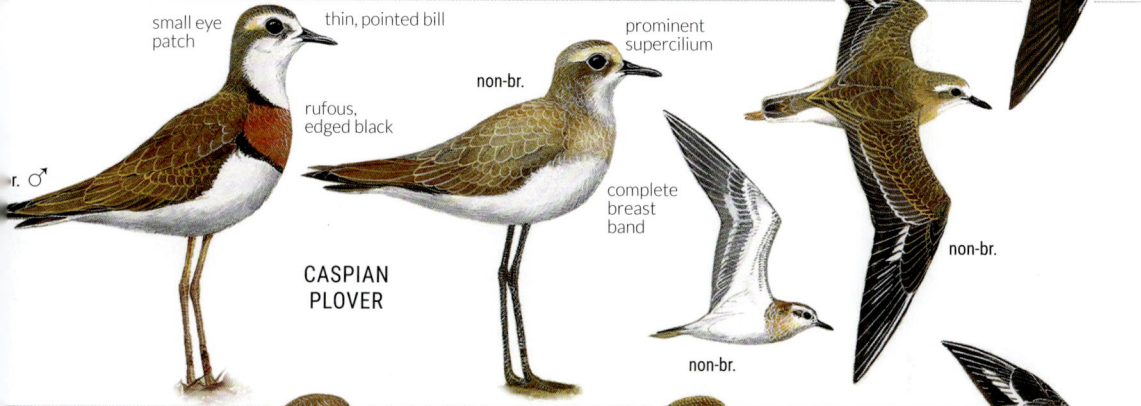

dull yellow legs

no wingbar

small eye patch

thin, pointed bill

prominent supercilium

non-br.

rufous, edged black

r. ♂

complete breast band

non-br.

CASPIAN PLOVER

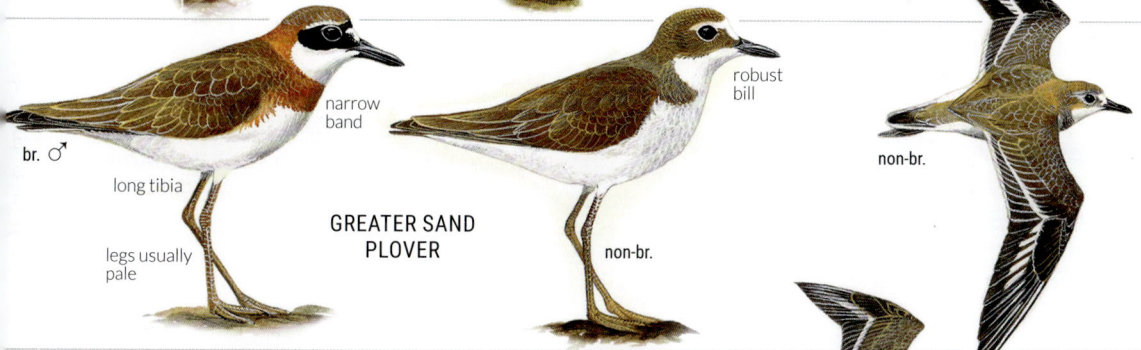

non-br.

narrow band

robust bill

br. ♂

non-br.

long tibia

non-br.

legs usually pale

GREATER SAND PLOVER

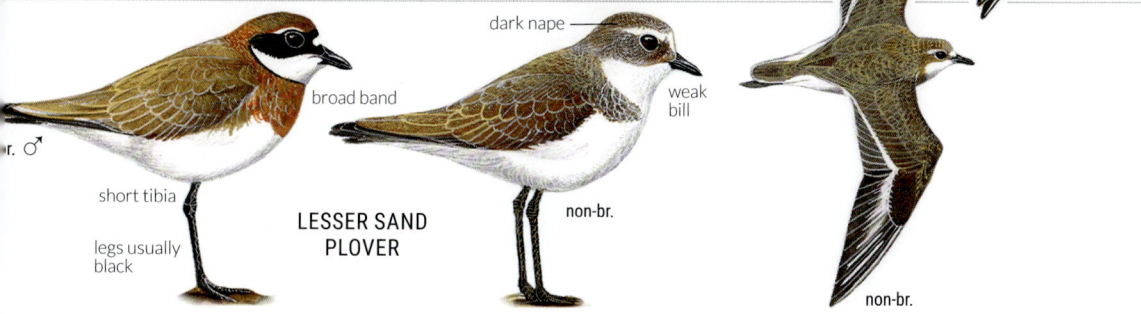

dark nape

broad band

weak bill

r. ♂

non-br.

short tibia

LESSER SAND PLOVER

legs usually black

non-br.

non-br.

PAINTED-SNIPES
Peculiar, long-billed waders of marshy, freshwater habitats and only distantly related to true snipes. Sexes differ; females larger and more brightly coloured than males. Polyandrous, like phalaropes and most jacanas, with males responsible for incubation and care of the chicks.

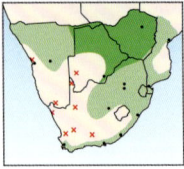

GREATER PAINTED-SNIPE *Rostratula benghalensis*

b b B b | b b B B b b
J F M A M J J A S O N D

24–26 cm; 90–140 g; wingspan 52 cm Differs from true snipes by having a white eye-ring extending onto ear coverts, dark breast band and obvious white breast extending up onto shoulder. Horn-coloured bill shorter than true snipes and slightly decurved; legs longer. Female is the dominant sex and is more strikingly marked, with rich chestnut neck and breast and uniform upperparts. Male and juv. are more cryptic, with buff spotting on upperparts and conspicuous golden 'V' on back. Flight is slow, on large, broad wings, often with legs trailing. **Voice:** Silent when flushed; male gives trilling call, female gives soft 'wuk-oooooo', repeated monotonously, often at night. **Status and biology:** Uncommon resident and local nomad at marshes and flooded grassland. Skulks among reeds in marshes and on the edges of wetlands. Numbers and range decreasing in sw of the region. (Goudsnip)

SNIPES
Long-billed waders in the same family as sandpipers (Scolopacidae). Confined to marshy habitats where they usually remain among vegetation, relying on their cryptic plumage to escape detection. Sexes alike. Feed by probing; detect prey with sensitive bill tip.

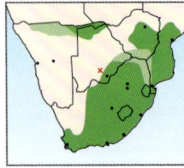

AFRICAN SNIPE *Gallinago nigripennis*

b b b b B B B b
J F M A M J J A S O N D

28–30 cm; 110–140 g; wingspan 45 cm A long-billed snipe, only likely to be confused with increasingly rare Great Snipe. Differs in having a mainly white belly, lacking strongly marked flanks and vent, white (not barred) central underwing, and more uniform upperwing lacking prominent white tips to coverts. When flushed, typically calls, jinks from side to side and flies a considerable distance compared to Great Snipe's usually silent take-off and shorter, more direct flight. Outer tail partly barred (white in ad. Great Snipe, but barred in juv.). Juv. plainer and duller above. **Voice:** Sucking 'scaap' when flushed; male produces whirring, drumming sound with stiffened outer-tail feathers during aerial display flights. **Status and biology:** Common resident and local nomad at marshes and flooded grassland, usually in muddy areas with short vegetation. (Afrikaanse Snip)

GREAT SNIPE *Gallinago media*

J F M A M J J A S O N D

28–30 cm; 130–220 g; wingspan 48 cm A large, chunky snipe with a relatively short, heavy bill that is held more level in flight than that of African Snipe. Flight rather heavy; tends to fly directly away when flushed, not jinking like African Snipe; usually silent on take-off. Best distinguished by bold, blackish chevrons on breast, flanks and vent, uniformly grey-barred underwing (lacking white central bar of African Snipe) and 2 well-defined, white wingbars formed by large, white tips to upperwing coverts. Pale 'braces' on scapulars less marked than African Snipe. Adult has white (not partly barred) outer-tail feathers. Juv. more rufous above with brown crescents below; white spots on wing coverts reduced; outer tail barred. **Voice:** Generally silent; occasionally utters a soft croak (lower pitched than African Snipe) when flushed. **Status and biology:** NEAR-THREATENED. Formerly quite common and widespread, but now only a scarce Palearctic-br. summer migrant to marshes and flooded grassland, mainly in north-central parts of the region. (Dubbelsnip)

GREATER
PAINTED-SNIPE

♂

♀

♂

white
covert
bars

flexible, slightly
decurved bill

♀

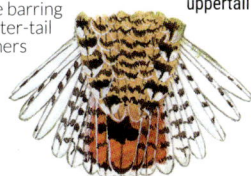

uppertail

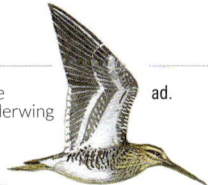

very long bill

more barring
in outer-tail
feathers

pale
underwing
bar

ad.

ad.

juv.

AFRICAN
SNIPE

ad.

narrow
white
covert tips

white
belly

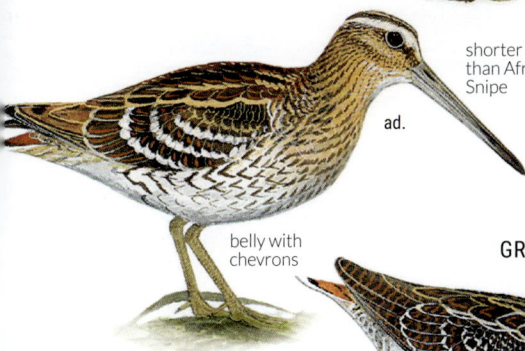

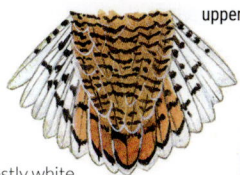

uppertail

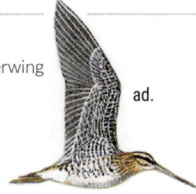

dark
underwing

ad.

shorter bill
than African
Snipe

ad.

mostly white
outer-tail feathers

GREAT SNIPE

belly with
chevrons

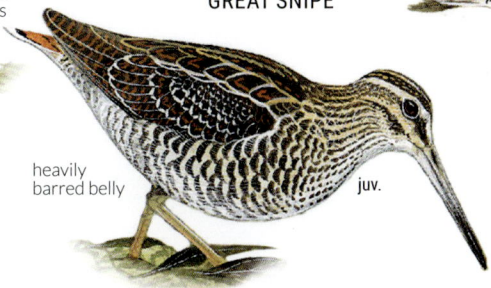

ad.

prominent
white
covert tips

heavily
barred belly

juv.

CURLEWS, GODWITS AND SANDPIPERS

Waders that mostly breed at high latitudes in the n hemisphere, migrating south for the boreal winter. Undertake the most impressive long-distance flights of any birds, with some species migrating non-stop across the Pacific Ocean, but usually break their migration at several stopover sites where they rapidly put on weight to fuel their next flight. Their dependence on secure breeding, wintering areas, plus stopover sites renders them particularly susceptible to habitat loss and climate change, and numbers of many species reaching s Africa have decreased dramatically in the last few decades. Juvs of larger species remain in s Africa year-round; numbers vary in relation to Arctic lemming cycles (greatest in years following boreal summers when lemmings are abundant so few wader chicks are eaten by predators), but these cycles greatly dampened down by rapid climate change in tundra regions. Feed by probing or picking. Sexes alike, although females are longer billed in some species. Most species have distinct br. plumages; juvs have pale-fringed upperpart feathers. The two *Numenius* species are the largest members of the family in s Africa; the dowitcher is a rare vagrant that is most likely to be confused with a godwit (p. 206).

WHIMBREL *Numenius phaeopus*

J F M A M J J A S O N D

40–45 cm; 340–500 g; wingspan 82 cm A large, grey-brown wader with a strongly decurved bill. Smaller and darker than Eurasian Curlew, with a shorter bill (2–3, not 3–4 times head length), curved along its full length (Eurasian Curlew bill decurved mostly towards tip); head striped (not uniform). In flight, shows white rump and back, and more uniform upperwing than Eurasian Curlew. Female larger with longer bill. Juv. has scalloped upperparts. 'Steppe Whimbrel' *N. p. alboaxillaris*, until recently thought to be an extinct subspecies, was rediscovered in Mozambique in 2016, with individuals since also found at Richards Bay, KZN. It is slightly larger and bigger-bodied, with wider and longer wings than the nominate. The most important difference between the two subspecies is the patterning of the axillary feathers ('armpit' area); axillaries of 'Steppe Whimbrel' are almost pure white, compared to the barred and scalloped axillaries of the nominate. **Voice:** Bubbling, whistled *'whiri-iri-iri-iri-iri'*, often given in flight; highly vocal in non-br. season. **Status and biology:** Common Palearctic-br. summer migrant at coastal wetlands and, to a lesser extent, open shores; vagrant inland, often in pastures. Roosts communally at high tide. Regional population some 15 000, with 10 000 in Mozambique; many overwinter. Was less common than Eurasian Curlew until the 1950s, but that trend now dramatically reversed. (Kleinwulp)

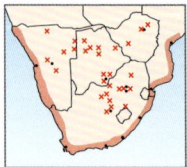

EURASIAN CURLEW *Numenius arquata*

J F M A M J J A S O N D

53–59 cm; 600–800 g; wingspan 90 cm The largest wader, with a very long, decurved bill (curvature primarily towards the tip), proportionally much longer (3–4, not 2–3 times head length) than that of Common Whimbrel. Eurasian Curlew is paler, with a plain (not striped) head, although a faint supercilium is sometimes visible. In flight, shows a conspicuous white rump and back; inner upperwings markedly paler than outer wings. Female larger with longer bill. Juv. has a relatively short bill, but lacks head stripes of Common Whimbrel and is paler; back feathers have buff fringes, but these lost by Dec. **Voice:** A loud *'cur-lew'*. **Status and biology:** NEAR-THREATENED. Uncommon, Palearctic-br. migrant at large estuaries and lagoons; vagrant inland. Roosts communally at high tide. Regional population 1 000, mostly at a few key sites; regularly overwinters. Numbers decreased greatly during 20th century: Langebaan Lagoon, W Cape, is now probably the only site that regularly supports more than 100 birds. (Grootwulp)

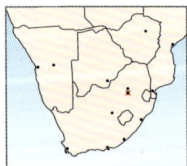

ASIAN DOWITCHER *Limnodromas semipalmatus*

J F M A M J J A S O N D

34–36 cm; 160–200 g; wingspan 70 cm Vagrant. In both br. and non-br. plumage superficially resembles Bar-tailed Godwit (p. 206), but is slightly smaller and shorter legged with a heavy, straight, all-black bill that appears bulbous-tipped and lacks the pinkish base of Bar-tailed Godwit (although a few birds show a pale pinkish base to the lower mandible). In flight, rump is duller with fine barring. Juv. has buff-fringed back feathers; underparts washed buff. **Voice:** Soft *'chewp'* in flight. **Status and biology:** NEAR-THREATENED. Breeds in central Asia and winters in SE Asia and n Australia. 1 record; Gauteng, Nov. 2004 (only the second record for Africa). (Asiatiese Snipgriet)

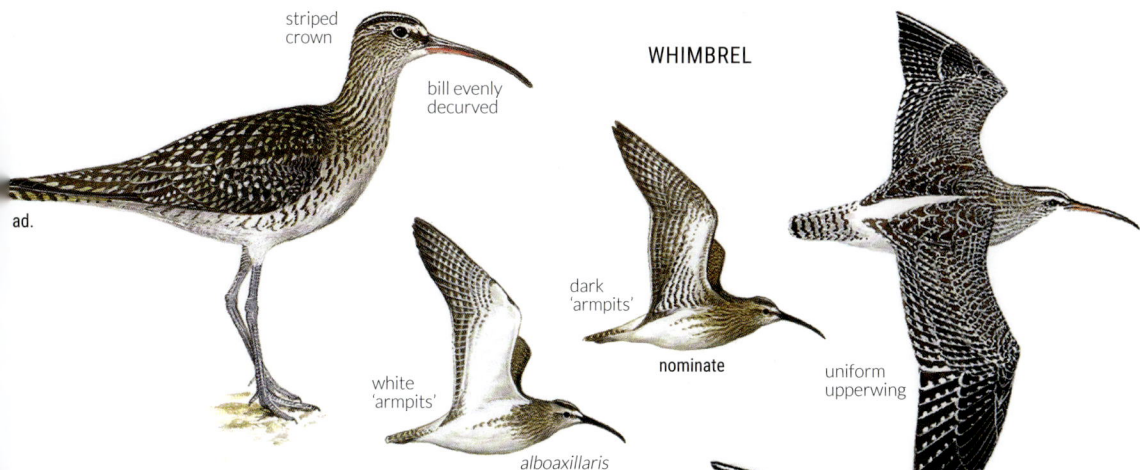

striped
crown

bill evenly
decurved

WHIMBREL

dark
'armpits'

white
'armpits'

nominate

uniform
upperwing

ad.

alboaxillaris

plain crown

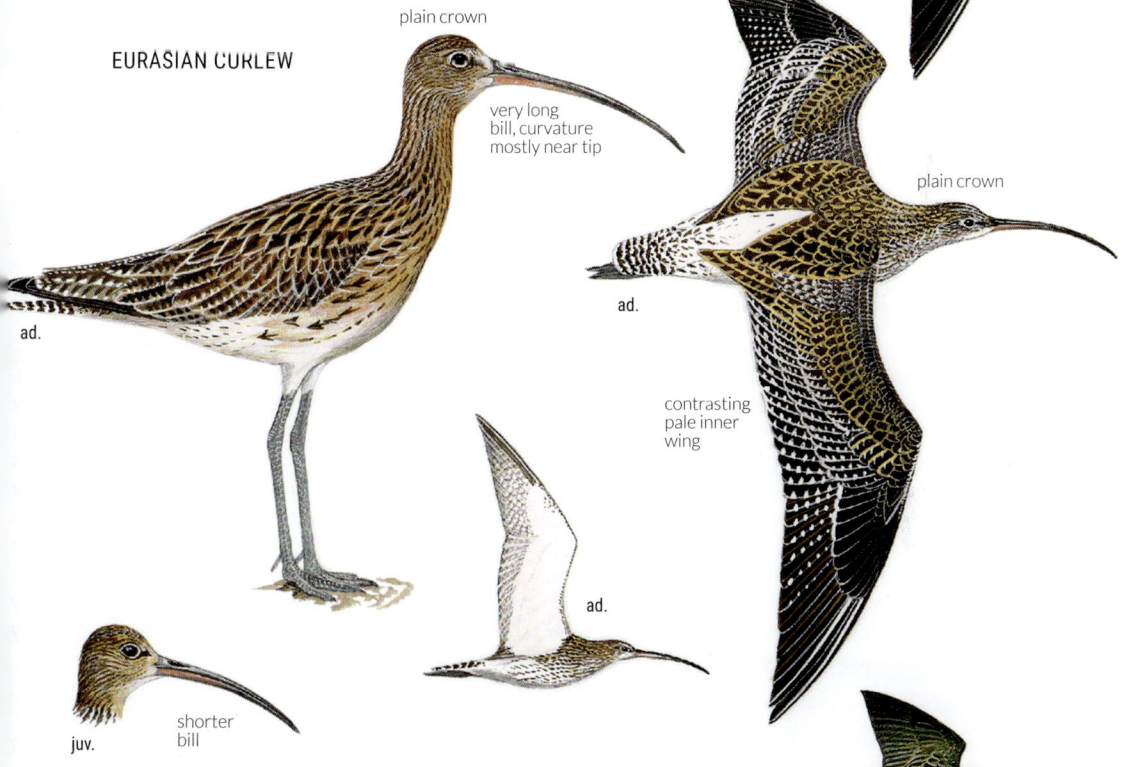

EURASIAN CURLEW

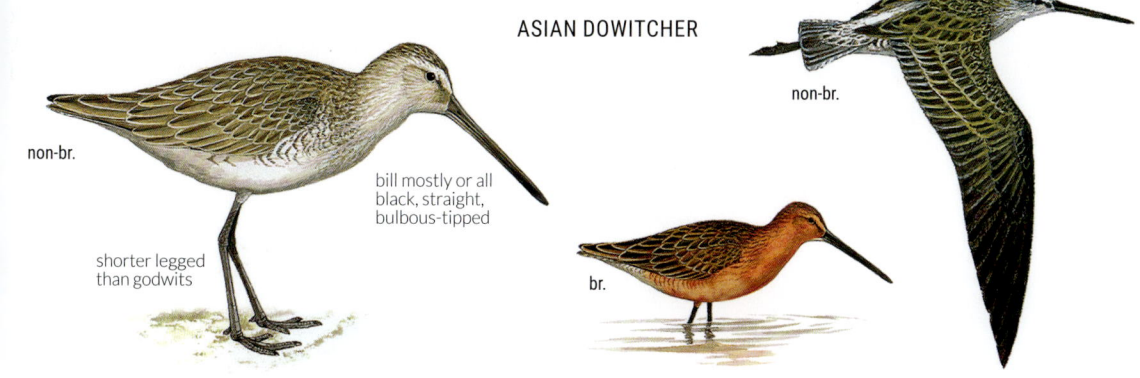

very long
bill, curvature
mostly near tip

plain crown

ad.

ad.

contrasting
pale inner
wing

ad.

ad.

juv.

shorter
bill

ASIAN DOWITCHER

non-br.

non-br.

bill mostly or all
black, straight,
bulbous-tipped

shorter legged
than godwits

br.

BAR-TAILED GODWIT *Limosa lapponica*

J F M A M J J A S O N D

37–40 cm; 200–300 g; wingspan 72 cm A large, pale, long-legged wader with a long, slightly upturned bill. Slightly smaller than Black-tailed Godwit, with appreciably shorter legs and a shorter bill with a less extensive pinkish base. Back and wing coverts have distinct pale fringes, making upperparts appear streaked (more uniform in Black-tailed Godwit). At rest, structure similar to vagrant Hudsonian Godwit; best separated in flight, when Bar-tailed Godwit has thin, brown tail bars, and white rump and back; lacks bold, white wingbar and black-tipped tail of Black-tailed and Hudsonian Godwits; underwing coverts whitish (not black as Hudsonian Godwit). In br. plumage, head, neck and underparts rich chestnut, finely barred on sides of breast; back black with chestnut feather margins. Female larger with longer bill; br. plumage duller. In both br. and non-br. plumages superficially resembles vagrant Asian Dowitcher (p. 204), but is slightly larger and longer legged; bill quite distinct (not straight, all-black with a bulbous tip). Juv. darker above and more buffy below. **Voice:** A *'wik-wik'* or *'kirrik'* call, often given in flight. **Status and biology:** NEAR-THREATENED. Locally common Palearctic-br. migrant to large, coastal wetlands; vagrant inland. Regional population some 10 000 birds, most on the west coast and in Mozambique; regularly overwinters. (Bandstertgriet)

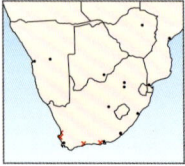

HUDSONIAN GODWIT *Limosa haemastica*

X X X X X   X   X
J F M A M J J A S O N D

40 cm; 200–320 g; wingspan 74 cm Vagrant. Structure similar to Bar-tailed Godwit; hard to separate at rest, but in flight has an obvious white wingbar, black tail with a narrow, creamy tip and diagnostic black underwing coverts. Smaller than Black-tailed Godwit, with shorter legs, slightly upturned bill and dark underwing coverts. Further differs in br. plumage from Black-tailed Godwit by having dark chestnut underpart coloration extending to belly, and from both Bar-tailed and Black-tailed Godwits by having a whitish face, streaked grey. Juv. resembles non-br. ad., but is more buffy above and below, with upperparts appearing scaled. **Voice:** High-pitched *'ta-wit'* flight call. **Status and biology:** Nearctic-br. summer vagrant to large coastal wetlands in W and E Cape; unconfirmed record from Walvis Bay, Namibia. (Hudsonbaaigriet)

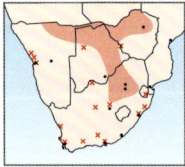

BLACK-TAILED GODWIT *Limosa limosa*

J F M A M J J A S O N D

40–44 cm; 250–400 g; wingspan 76 cm Superficially resembles Bar-tailed and Hudsonian Godwits, but is markedly larger with longer legs, appearing distinctly taller than other godwits in mixed flocks. Upperparts plainer grey-brown than other godwits; bill longer, almost straight, with a more extensive pink base. In flight the broad, white wingbar and black tail are conspicuous. All age classes distinguished from Hudsonian Godwit in flight by larger wingbar extending across all secondaries (outer secondaries only in Hudsonian Godwit) and white (not black) underwing coverts. In br. plumage, neck and upper breast chestnut; belly white, barred black; differs from br. plumage of Hudsonian Godwit by having dark chestnut underpart coloration extending only to breast (not belly), and in having a reddish (not whitish) face. Female larger with longer bill. Juv. has upperpart feathers blackish, edged cinnamon; neck and upper breast buffy-orange. **Voice:** Repeated *'weeka-weeka'*, given especially in flight. **Status and biology:** NEAR-THREATENED. Rare, but annual Palearctic-br. summer migrant to inland and coastal wetlands; mostly juvs that overwinter. Forages in deep water more regularly than does Bar-tailed Godwit; sometimes in small groups. (Swartstertgriet)

BAR-TAILED GODWIT

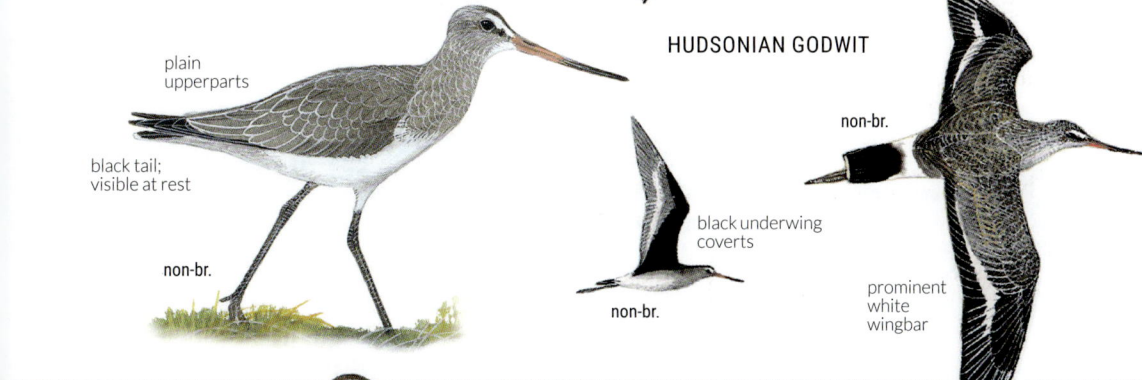

patterned
upperparts

slightly
upturned bill

on-br.

barred tail;
visible at rest

juv.

white back

non-br.

br.

no
wingbar

non-br.

br.

non-br.

HUDSONIAN GODWIT

plain
upperparts

non-br.

black tail;
visible at rest

black underwing
coverts

non-br.

non-br.

prominent
white
wingbar

straight bill

BLACK-TAILED GODWIT

plain
upperparts

straight bill

juv.

ck tail;
ible at rest

non-br.

non-br.

white underwing
coverts

non-br.

prominent
white
wingbar

br.

br.

TRINGA 'SHANKS'
Large to medium-sized waders with fairly long legs that mostly feed by visually picking prey. Most species feed mainly in the water, but Wood and Green Sandpipers spend more time on the shoreline.

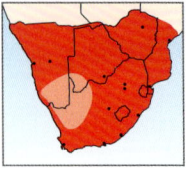

COMMON GREENSHANK   *Tringa nebularia*

J F M A M J J A S O N D

30–34 cm; 140–250 g; wingspan 68 cm A fairly large wader with long, greenish (rarely yellowish) legs and toes. Larger than Marsh Sandpiper (p. 210), with a heavier, slightly upturned bill with grey-green base (not all dark, needle-like bill of Marsh Sandpiper) and relatively shorter legs; in flight, toes don't extend as far beyond the tail tip. In flight, both species have rather plain upperwings contrasting with white back and rump; tail mainly white with fine brown bars along outer edge. Br. ad. is more heavily streaked on head and neck. Juv. is darker and browner above with buff feather fringes in fresh plumage. Bar-tailed Godwit (p. 206) is larger, browner and heavier, and has relatively short legs and an upturned bill with pinkish base. Differs from vagrant Greater and Lesser Yellowlegs in having green or yellowish-green (not orange-yellow) legs and by its white, not brown, back. **Voice:** A loud, ringing *'chew-chew-chew'*, usually given when flushed or in flight. **Status and biology:** Common Palearctic-br. summer migrant at coastal and freshwater wetlands; many overwinter. Roosts communally at high tide in coastal localities. Regional population perhaps 30 000 birds. (Groenpootruiter)

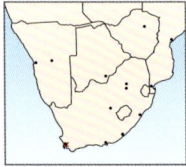

GREATER YELLOWLEGS   *Tringa melanoleuca*

J F M A M J J A S O N D

30–32 cm; 130–180 g; wingspan 70 cm Vagrant. Structure recalls Common Greenshank, but bill straighter and has yellow (not grey-green) legs; upperparts more obviously spotted. In flight, grey-brown (not white) back is diagnostic. Larger than vagrant Lesser Yellowlegs with longer bill (1.2–1.3, not 1 times head length) that has a paler grey-green base (not all dark). Br. ad. and juv. more strongly spangled on upperparts; br. Ad. has blackish chevrons on flanks (reduced in non-br. ads and juvs). **Voice:** A loud, ringing *'tew-tew-tew-tew'*, similar to that of Common Greenshank, but slightly softer. **Status and biology:** Nearctic-br. vagrant. Regularly wades in deep water. Only 1 record; Cape Peninsula, Dec 1971. (Grootgeelpootruiter)

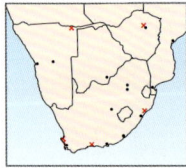

LESSER YELLOWLEGS   *Tringa flavipes*

J F M A M J J A S O N D

22–26 cm; 65–95 g; wingspan 60 cm Vagrant. Overall shape and behaviour similar to Common Redshank (p. 210), but has yellow (not red) legs, mostly dark (not red-based) bill and dark (not white) back and secondaries in flight. Structure also recalls Marsh Sandpiper (p. 210), but is slightly more robust with a heavier bill, yellow (not grey-green) legs and white-notched greater coverts and tertials. Distinguished from Marsh Sandpiper in transitional plumage (when the legs can become bright yellow) by its relatively short, heavier bill and, in flight, by its grey-brown (not white) back. Smaller Wood Sandpiper (p. 212) has shorter legs, less heavily streaked breast and a more prominent supercilium. Appreciably smaller than vagrant Greater Yellowlegs with shorter (1, not 1.2–1.3 times head length), straighter, all-dark bill. Juv. has brown upperparts, spotted with buff; breast greyish. **Voice:** High-pitched *'teu'*, or *'teu-teu'*, faster and less strident than Greater Yellowlegs. **Status and biology:** Nearctic-br. summer vagrant. Rare, with records from Harare, Zimbabwe, Rundu, Namibia, Stanger, KZN, as well as a handful of records from W Cape. (Kleingeelpootruiter)

non-br.

dull yellow-
green legs

juv.

white
back

non-br.

pale-based,
slightly
upturned bill

COMMON GREENSHANK

long bill

**GREATER
YELLOWLEGS**

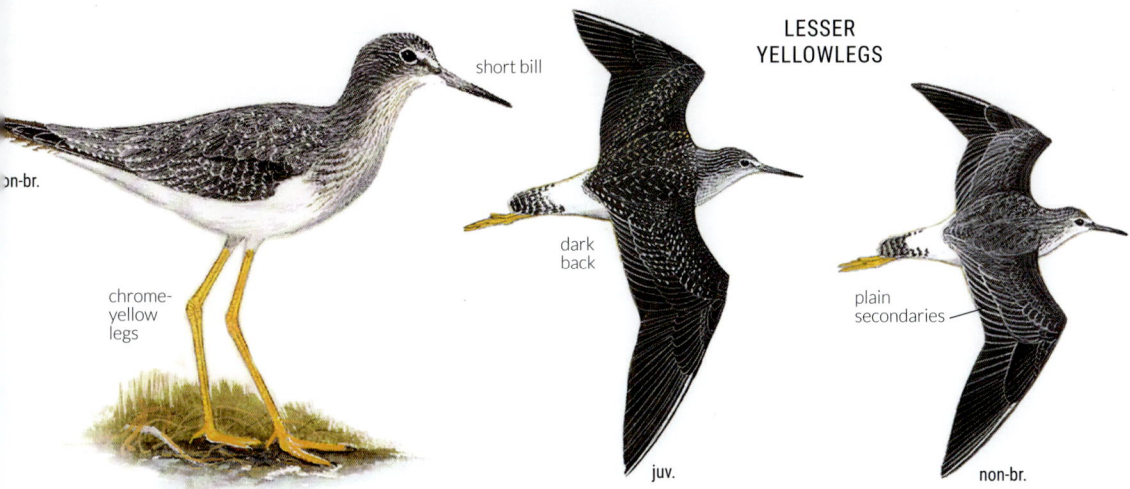

on-br.

chrome-
yellow legs

dark
back

white-
notched
secondaries

juv.

non-br.

**LESSER
YELLOWLEGS**

on-br.

short bill

chrome-
yellow
legs

dark
back

plain
secondaries

juv.

non-br.

209

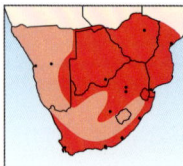

MARSH SANDPIPER   *Tringa stagnatilis*

J F M A M J J A S O N D

22–26 cm; 55–95 g; wingspan 56 cm A medium-sized, pale grey wader. Smaller and more slender than Common Greenshank (p. 208), with a much thinner, straight, black bill, and proportionally longer legs; toes extend further beyond tail tip in flight. Rather plain upperwing contrasts with white rump and back in flight (back grey-brown in vagrant Lesser Yellowlegs, p. 208, and Wilson's Phalarope, p. 224). The legs of some Marsh Sandpipers become brighter yellow prior to northward migration, increasing the risk of confusion with Lesser Yellowlegs, but Marsh Sandpiper has a longer, thinner bill. Non-br. Wilson's Phalarope has shorter, yellow (not grey-green) legs. Br. ad. has blackish mottling on upperparts. Juv. darker and browner above; back feathers edged buff when fresh. **Voice:** High-pitched *'yeup'*, often repeated; higher pitched and less strident than Common Greenshank; also a single *'tchuk'*. **Status and biology:** Fairly common Palearctic-br. summer migrant at freshwater and coastal wetlands. Roosts communally at high tide in coastal localities. Regional population perhaps 10 000 birds. (Moerasruiter)

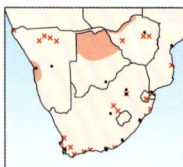

COMMON REDSHANK   *Tringa totanus*

J F M A M J J A S O N D

27–29 cm; 90–150 g; wingspan 62 cm Told from most other waders by its bright red or orange-red legs and red base of the bill; Spotted Redshank and some Ruffs (p. 214) share these features, but both lack the striking white secondaries and inner primaries. Ruff further lacks the white back and finely barred white tail. At rest, Common Redshank appears plain brownish-grey above, whereas a Ruff is scaled and mottled. It is smaller and darker than Spotted Redshank, with brighter legs, a rather plain face and a shorter bill (1.1–1.2, not 1.4–1.5 times head length) that has red at the base of both mandibles. Much longer legged and darker brown above than Terek Sandpiper (p. 212), which has a grey back and rump and a longer, upturned bill. Juv. is generally browner than ad., with the legs and base of bill duller red. **Voice:** *'Tiw-hu-hu'*, first syllable higher pitched. **Status and biology:** Rare Palearctic-br. summer migrant at coastal and freshwater wetlands, occasionally in small flocks; very small numbers overwinter. In recent years, most reliably encountered at Walvis Bay and other wetlands on the central Namibian coast, as well as at Langebaan Lagoon, W Cape. (Rooipootruiter)

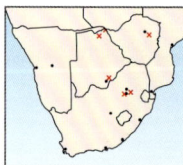

SPOTTED REDSHANK   *Tringa erythropus*

X   X X
J F M A M J J A S O N D

29–32 cm; 125–200 g; wingspan 64 cm Vagrant. A large, long-billed wader with long, dark red legs. Larger and paler (in non-br. plumage) than Common Redshank, with a longer bill (1.4–1.5 times head length); only the base of the lower mandible is red. The tip of the bill is downturned. Head is more strikingly patterned, with a pale supercilium and black lores contrasting with white supra-loral patch. In flight, white back contrasts with finely barred, white rump and more heavily barred tail; upperwings grey, darker on primaries; lacks white secondaries and inner primaries of Common Redshank. In br. plumage, body is black, finely spotted and scalloped white; legs black; in flight, dark body contrasts strongly with white underwings. Juv. is more extensively barred grey below than non-br. ad. **Voice:** Clear, double-noted *'tu-wik'*. **Status and biology:** Palearctic-br. summer vagrant. Occurs at freshwater and coastal wetlands; only 6 regional records. (Gevlekte Rooipootruiter)

MARSH SANDPIPER

long, needle-shaped, all-black bill

n-br.

very long legs

br.

ad. worn plumage

white back

non-br.

COMMON REDSHANK

mottled breast

br. ♂

non-br.

white trailing edge

non-br.

non-br.

SPOTTED REDSHANK

long bill; slightly decurved tip

br.

white breast

non-br.

non-br.

br.

dark trailing edge

non-br.

211

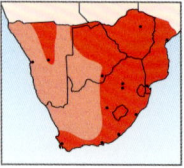

WOOD SANDPIPER   *Tringa glareola*

J F M A M J J A S O N D

19–21 cm; 45–80 g; wingspan 56 cm A medium-sized sandpiper with a grey-brown back and white spotting along margins of tertials, upperwing coverts and back feathers. Intermediate in size between Common and Green Sandpipers, which it superficially resembles. Distinguished from Green Sandpiper by the paler upperparts with much more prominent pale flecking, yellower legs, less streaked breast and white supercilium extending behind eye. In flight, has dark back, white rump and finely (not broadly) barred tail; underwings pale grey (not blackish). Lacks the white shoulder patch and finely barred upperparts, dark rump, tail and striking white wingbar of Common Sandpiper. Darker and browner than Common Greenshank (p. 208) or Marsh Sandpiper (p. 210), and has a square, white rump and dark (not white) back. Slightly smaller than vagrant Lesser Yellowlegs (p. 208) with shorter, yellow-green legs, yellowish base to bill, and longer, bolder supercilium. Juv. resembles ad., but upperparts warmer brown. **Voice:** Very vocal, with a high-pitched, slightly descending *'chiff-iff-iff'*. **Status and biology:** Common Palearctic-br. summer migrant at freshwater wetlands; scarce at estuaries and coastal habitats. Occurs singly or in small flocks. Regional population perhaps 50 000 birds. (Bosruiter)

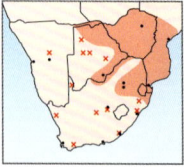

GREEN SANDPIPER   *Tringa ochropus*

J F M A M J J A S O N D

19–21 cm; 35–75 g; wingspan 58 cm Slightly larger than Wood Sandpiper, with darker greenish upperparts, only finely spotted whitish, and broadly (not narrowly) barred tail; appears a little dumpier due to shorter legs. Sides of breast and flanks boldly streaked. Has a prominent white eye-ring, but lacks white supercilium extending behind the eye. In flight, blackish underwings diagnostic (pale in Wood Sandpiper). Larger and longer legged than Common Sandpiper, which lacks white spots on upperparts; easily differentiated in flight by its white rump and lack of a white wingbar. Darker above and much shorter legged than either Common Greenshank (p. 208) or Marsh Sandpiper (p. 210); easily differentiated in flight by its dark (not white) back and black underwings. Juv. is browner above with buff spotting; breast streaking paler. **Voice:** A rather mellow, whistled *'tew-a-tew'*. **Status and biology:** Rare Palearctic-br. summer migrant favouring freshwater wetlands, principally along tropical rivers and at small waterbodies such as sewage ponds. Largely confined to the north and east of the region; vagrant elsewhere. Nearctic-br. Solitary Sandpiper *T. solitaria* is a potential vagrant (1 record from Zambia) that is similar to Green Sandpiper, but has a dark (not white) rump. (Witgatruiter)

▍OTHER SMALL SANDPIPERS   These 2 species superficially resemble *Tringa* 'shanks', but are placed in different genera.

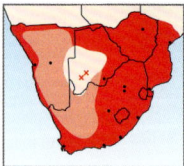

COMMON SANDPIPER   *Actitis hypoleucos*

J F M A M J J A S O N D

19–21 cm; 35–75 g; wingspan 40 cm A fairly small sandpiper with olive-brown upperparts, finely barred dark brown, and an obvious white shoulder in front of the closed wing. Tail projects well beyond wing tips at rest; bobs tail regularly. Legs dull green. In flight, has prominent white wingbar, dark rump and barred sides to dark tail. Flight comprises rapid bursts of shallow wing beats interspersed with short glides on slightly bowed wings. Smaller and shorter legged than Wood and Green Sandpipers; lacks white spots on upperparts; easily told in flight by its prominent white wingbar and dark (not white) rump. **Voice:** A shrill *'ti-ti-ti'*, usually given in flight. **Status and biology:** Common Palearctic-br. summer migrant at wetlands and along the coast. Often rather tame, allowing close approach. Regional population at least 10 000 birds. (Gewone Ruiter)

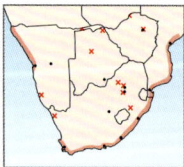

TEREK SANDPIPER   *Xenus cinereus*

J F M A M J J A S O N D

22–25 cm; 65–100 g; wingspan 50 cm The only wader with yellow-orange legs and a long, upturned, dark brown bill with an orange base. Appears rather dumpy; legs much shorter than Common Redshank (p. 210) or Ruff (p. 214). From a distance, appears pale with a dark shoulder. In flight, has whitish trailing edges to secondaries, similar to Common Redshank, but is much paler grey overall and has a grey (not white) back and rump. Br. ad. is slightly darker and more streaked above. Juv. is more buffy above than ad. **Voice:** A series of fluty, uniformly pitched *'weet-weet-weet'* notes. **Status and biology:** Uncommon Palearctic-br. summer migrant at muddy estuaries and lagoons, favouring mangroves and areas with eel grass (*Zostera*); a few overwinter. Regional population some 4 000 birds, mainly in Mozambique. In S Africa, most easily seen at Richards Bay, the Sundays, Swartkops, Gamtoos and Gouritz estuaries, and Langebaan Lagoon, W Cape. Rare in Namibia. (Terekruiter)

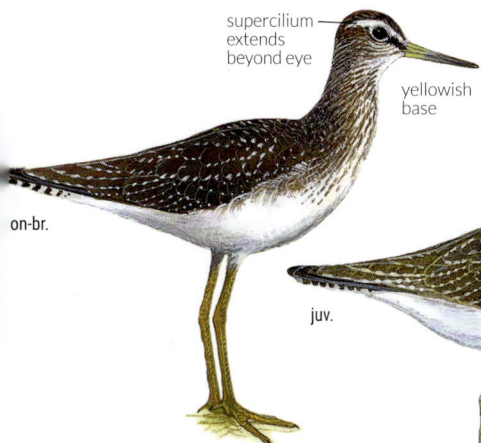

supercilium extends beyond eye

yellowish base

WOOD SANDPIPER

on-br.

juv.

narrow bars

dusky white belly

non-br.

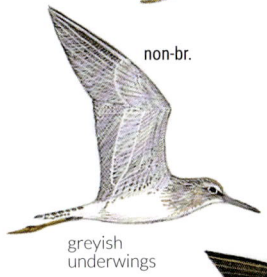

greyish underwings

non-br.

very dark back

supercilium does not extend beyond eye

GREEN SANDPIPER

non-br.

broad bars

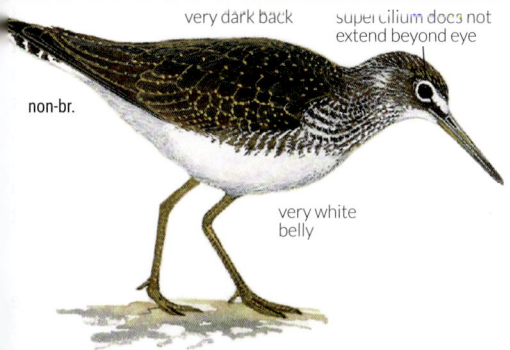

very white belly

blackish underwings

non-br.

non-br.

...eds with ...ominent tail-...obbing action

long tail

white shoulder patch

prominent wingbar

COMMON SANDPIPER

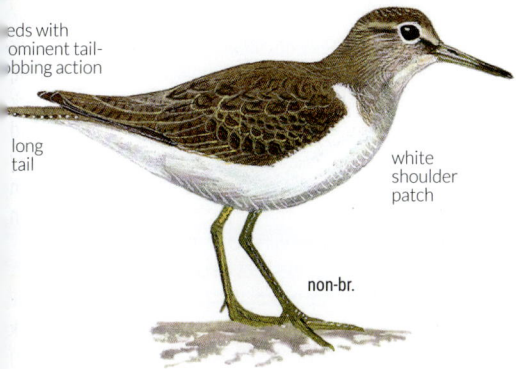

non-br.

non-br.

white trailing edge

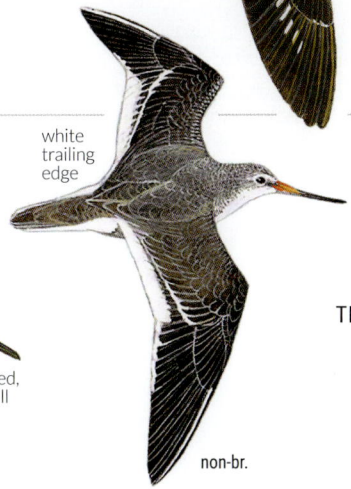

non-br.

short, orange legs

orange-based, upturned bill

TEREK SANDPIPER

non-br.

TURNSTONES
Mid-sized, stocky shorebirds with strikingly patterned plumage that are largely confined to coastal habitats away from their breeding grounds.

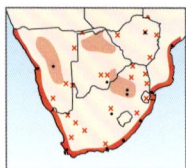

RUDDY TURNSTONE *Arenaria interpres*

J F M A M J J A S O N D

21–25 cm; 90–150 g; wingspan 54 cm A stocky wader with a short, black bill and orange legs. White belly and lower breast contrast with blackish patches on upper breast. In flight, appears boldly black and white above, with a strong white wingbar, white back stripes and rump, and black tail with narrow white tip. In br. plumage (often seen before northward migration) has rich chestnut back and striking black-and-white-patterned head and neck. Juv. has upperparts browner and more conspicuously scaled than non-br. ad. **Voice:** A hard '*kttuck*', especially in flight. **Status and biology:** Common Palearctic-br. migrant at rocky shores and estuaries, occasionally on sandy shores and at coastal lagoons; rare inland, mainly on migration. Regional population some 40 000 birds; many overwinter. Usually in small flocks. Flicks over small stones and weeds with its bill in search of food. (Steenloper)

CALIDRIS SANDPIPERS
The largest genus of sandpipers, which poses the greatest identification challenges among shorebirds because most species are rather drab brown above and paler below in non-br. plumage. Feed mostly by probing in soft wet sediments, sometimes in the water. The 2 species on this page used to be placed in different genera, thanks in part to their peculiar breeding systems, but genetic evidence indicates that they belong in *Calidris*.

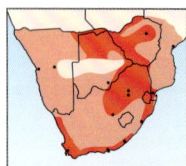

RUFF *Calidris pugnax*

J F M A M J J A S O N D

20–30 cm; male 120–200 g, female 80–140 g; wingspan 50–55 cm A medium-sized to large wader with conspicuously scaled upperparts. The black, slightly decurved bill may have an orange or reddish base: often has patch of paler feathers at base of bill. Males are less common than females in the region, and are appreciably larger than females; some have a white head and neck during the non-br. season. Leg colour varies from greenish-yellow to orange-red, occasionally black. In flight, oval white patch on either side of dark rump is diagnostic. Easily told from Common Redshank (p. 210) by its slightly shorter bill that droops at the tip, and its scaly upperparts; in flight, lacks that species' striking white secondaries and white back. Juv. resembles non-br. ad. but is much buffier. Juv. female could be confused with vagrant Buff-breasted Sandpiper, but is larger with a longer bill (roughly head length) and legs, distinct rump pattern, stronger white wingbar and lacks that species' dark underwing crescent at base of primaries. Sometimes feed by swimming like a phalarope (p. 224). **Voice:** Generally silent; occasional '*tooi*' flight call. **Status and biology:** Common Palearctic-br. summer visitor, mainly at vleis, lakes, estuaries and adjacent grassy areas; occasionally in farmyards, fallow fields or on open coast. Some overwinter, at which time a few gain partial br. plumage of elaborate head and neck plumes. Regional population at least 50 000 birds. (Kemphaan)

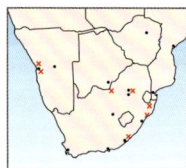

BUFF-BREASTED SANDPIPER *Calidris subruficollis*

X X X X   X   X
J F M A M J J A S O N D

18–20 cm; 45–65 g; wingspan 45 cm Vagrant. Coloration most closely resembles that of juv. female Ruff, but it is smaller and proportionally shorter legged (legs typically more yellow), and has a shorter and straighter bill (bill roughly 0.6, not 1 times head length). Crown spotted (not streaked), head appears bulbous on a thin neck. Face is rather plain with narrow white eye-ring. In flight, lacks oval white patches at sides of rump and wingbar virtually absent. Underwing predominantly white, with a small, dark crescent at base of primaries. Juv. has a paler lower belly; upperparts more prominently scalloped. **Voice:** Low '*preeet*', but usually silent outside br. season. **Status and biology:** NEAR-THREATENED. Nearctic-br. summer vagrant to open habitats, including short grassy areas and wetlands; probably frequently overlooked inland. (Taanborsstrandloper)

br.

non-br.

prominent
black & white
plumage in
flight

non-br.

bright orange
legs

**RUDDY
TURNSTONE**

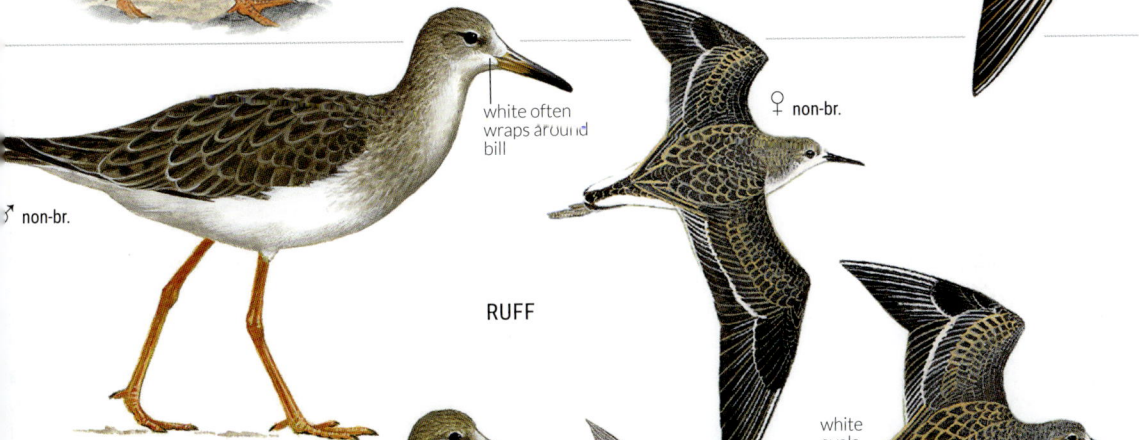

white often
wraps around
bill

♀ non-br.

♂ non-br.

RUFF

bill ± head
length

white
coverts

white
ovals

♂ non-br.

ad. ♀ non-br.

ad. ♂ non-br.
white morph

lollipop-like
head

bill <
head length

scalloped
back

thin neck

dark
crescent

short
legs

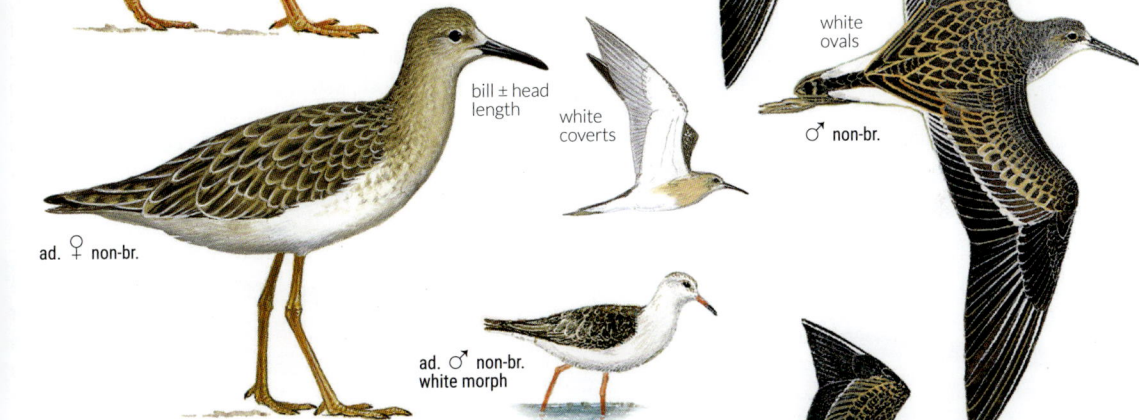

**BUFF-BREASTED
SANDPIPER**

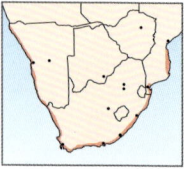

RED KNOT *Calidris canutus*

J F M A M J J A S O N D

23–25 cm; 110–200 g; wingspan 58 cm A short-legged, dumpy and rather plain wader. In flight, shows pale wingbar and white rump flecked with pale grey. Larger and more heavily built than Curlew Sandpiper (p. 218), with a shorter (1, not 1.2–1.5 times head length), almost straight (not decurved) bill and greenish (not blackish) legs. Smaller than Grey Plover (p. 196), with a plain grey or slightly scaled (not speckled) back and a longer, less bulbous bill; in flight, lacks striking white rump and black 'armpits' (axillaries). Only slightly smaller than American and especially Pacific Golden Plovers (p. 196), but is colder grey above, with a paler rump (not uniform with back and tail) and appreciably longer bill. Slightly smaller than vagrant Great Knot, with a shorter, less deep bill (see that species for other differences). In br. plumage, often seen prior to northward migration in autumn, underparts deep chestnut-orange; upperparts spangled chestnut and black; male brighter than female. Juv. has neatly scalloped upperparts. **Voice:** A low-pitched '*knut*'. **Status and biology:** Locally common Palearctic-br. summer visitor to estuaries and coastal lagoons; vagrant inland; regularly overwinters. Usually in flocks; roosts communally at high tide. Regional population around 8 000 birds, mostly at Walvis Bay and Sandwich Harbour in Namibia, and Langebaan Lagoon, W Cape. (Knoet)

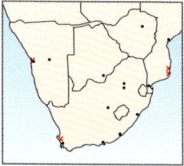

GREAT KNOT *Calidris tenuirostris*

X X X | | | | | | | X X X
J F M A M J J A S O N D

26–28 cm; 125–220 g; wingspan 64 cm Vagrant. Slightly larger than Red Knot, with a longer bill (1.1–1.3, not 1 times head length) that has a deeper base and more drooped tip; legs darker grey. Wings slightly longer, extending well beyond tail tip at rest (level with tail tip in Red Knot); wingbar slightly less marked in flight. Non-br. Great Knot is overall slightly paler, with marked speckling on breast and streaking on flanks. In br. plumage has spangled rufous on mantle; underparts whitish (not chestnut), with heavy black mottling and streaking. **Voice:** Soft '*nyut*'. **Status and biology:** ENDANGERED. Rare vagrant from br. grounds in far east Palearctic to coastal estuaries in Mozambique; on west coast 3 records from Langebaan Lagoon, W Cape (possibly the same individual), and 2 from Walvis Bay, Namibia. (Grootknoet)

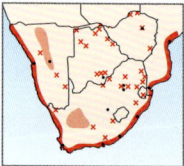

SANDERLING *Calidris alba*

J F M A M J J A S O N D

19–21 cm; 45–90 g; wingspan 42 cm A fairly small sandpiper with a short, stubby, black bill and a dark carpal patch. In non-br. plumage it is the palest sandpiper of the region: pale grey-brown above, finely streaked darker brown; white below. In flight, bold white wingbar contrasts strongly with blackish flight feathers; dark line through white rump. Slightly larger than Broad-billed Sandpiper (p. 220), with shorter bill; lacks head stripes. In br. plumage, upperpart feathers are blackish, barred and edged with rufous-chestnut and tipped grey; upper breast has a broad, diffuse chestnut band, streaked black. Juv. has upperpart feathers darker than non-br. ad; feathers spangled with buffy-yellow and white. **Voice:** A single, decisive '*wick*' given in flight. **Status and biology:** Common Palearctic-br. summer visitor, mainly to sandy beaches, but also rocky coasts, estuaries and lagoons; occasionally inland during migration. Often in flocks; runs up and down the beach following receding waves, searching sand for prey. Regional population around 85 000 birds, but decreasing; the W Cape population decreased by 90% from 1981–2011. (Drietoonstrandloper)

robust bill = head length

non-br.

short legs

RED KNOT

non-br.

rump extensively flecked grey

br.

r.

plump body

br.

bill > head length

non-br.

GREAT KNOT

rump lightly marked grey

spots & chevrons on flanks

non-br.

br.

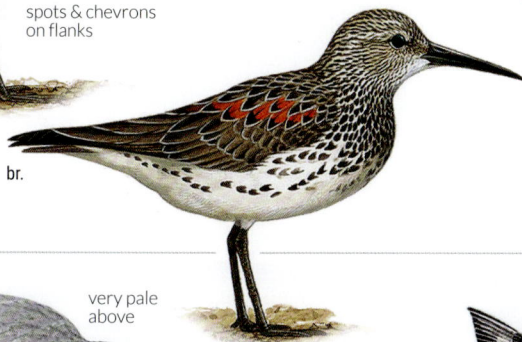

very pale above

non-br.

SANDERLING

dark shoulder patch

blunt-tipped bill

broad, white wingbar

br.

br.

non-br.

br.

217

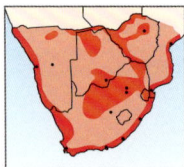

CURLEW SANDPIPER   *Calidris ferruginea*

J F M A M J J A S O N D

18–23 cm; 45–85 g; wingspan 44 cm The most common small to medium-sized sandpiper with a long, decurved bill (longer in female) and fairly long, blackish legs. In flight, square white rump separates it from all other small sandpipers except vagrant White-rumped Sandpiper (p. 220), which is smaller with much shorter legs and a shorter bill. Told with difficulty from vagrant Dunlin unless white rump is seen; averages slightly larger, longer legged and longer billed, but individual variation in structure can cause confusion. In br. plumage, underparts and face are chestnut, fringed white and finely barred black; upperparts spangled chestnut and black. Juv. has buffy edges to upperpart feathers, appearing browner and more obviously scaled above in fresh plumage. **Voice:** Short trill, '*chirrup*'. **Status and biology:** NEAR-THREATENED. Once an abundant Palearctic-br. summer visitor to coastal and freshwater wetlands (estimated 300 000 birds in 1980s), but numbers have decreased dramatically at least at coastal sites in the W Cape (90% decrease from 1981–2011). Many birds overwinter, moving to favoured wintering wetlands. Usually in flocks; roosts communally at high tide in coastal locations. (Krombekstrandloper)

DUNLIN   *Calidris alpina*

J F M A M J J A S O N D

15–22 cm; 40–70 g; wingspan 42 cm Vagrant. Resembles a compact Curlew Sandpiper, with shorter legs and a shorter, less decurved bill (but beware sex-linked differences in bill size among both species); pale supercilium is less marked. Bill less evenly curved than that of Curlew Sandpiper, with curvature mostly at the tip. Key distinguishing feature is the dark stripe down the centre of the rump. Broad-billed Sandpiper (p. 220) has markedly shorter legs, a striped head and a flattened bill tip. In br. and transitional plumages, Dunlin has a black patch on the belly and chestnut feathering on the back. Juv. has blackish upperpart feathers edged buff and chestnut; white 'V' conspicuous on mantle. **Voice:** A weak '*treep*'. **Status and biology:** Rare Palearctic-br. vagrant that seldom crosses equator; only 2 well-substantiated records – Kruger National Park and Langebaan Lagoon. (Bontstrandloper)

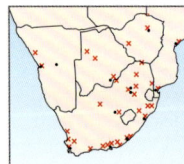

PECTORAL SANDPIPER   *Calidris melanotos*

J F M A M J J A S O N D

19–23 cm; 45–75 g; wingspan 45 cm Regular vagrant. A dark-backed wader, similar in size to Curlew Sandpiper, but typically darker brown above with shorter, greenish or yellowish (not black) legs and a yellowish base to its slightly decurved bill. The abrupt division between the streaked breast and white underparts is distinctive. In flight, shows dark-centred rump with white sides. Main identification challenge is vagrant Sharp-tailed Sandpiper. Larger and longer-necked than the stints, with a dark-capped appearance and darker upperparts. Smaller than Ruff (p. 214), with shorter legs and more heavily marked breast. Larger than vagrant Baird's Sandpiper (p. 220), which has an all-black bill and legs. Much larger than vagrant Long-toed Stint (p. 222). Juv. resembles ad., but is brighter. **Voice:** A harsh '*trrup*'. **Status and biology:** Rare but regular summer visitor that breeds in the tundra of N America and locally in e Siberia; found in margins of freshwater wetlands and estuaries, favouring muddy areas. Records widely scattered across the region, suggesting multiple arrival routes. (Geelpootstrandloper)

SHARP-TAILED SANDPIPER   *Calidris acuminata*

J F M A M J J A S O N D

17–21 cm; 40–115 g; wingspan 44 cm Vagrant. A mid-sized sandpiper, similar in size and behaviour to Pectoral Sandpiper, but appears slightly longer legged and flatter crowned. It has a more rufous cap, more prominent white eye-ring and a broad white supercilium that widens behind the eye and meets on the forehead. Bill is slightly shorter and less decurved; mainly black, sometimes with a paler pinkish base to the lower mandible; legs more greenish-yellow. The lower margin of the breast streaking is less well defined from the pale belly, and the flanks typically show distinctive dark chevrons. In flight, the tail feathers have sharp points, unlike the rounded tips on Pectoral Sandpiper. **Voice:** Musical twitter '*trrt-trrt-twhitwhit*'. **Status and biology:** First recorded in Mar 2018, when 2 birds were seen north of Maputo, Mozambique; 1 bird returned in Feb 2019. (Spitsstertstrandloper)

CURLEW SANDPIPER

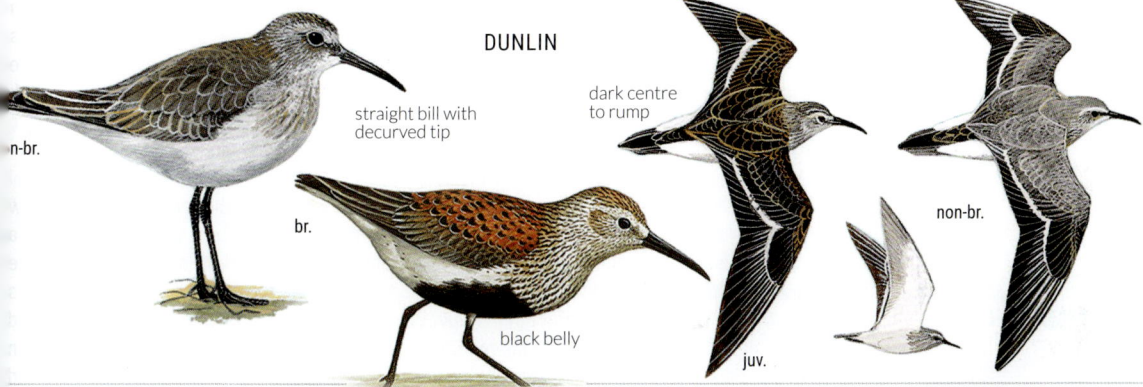

n-br.

mostly
white
rump

br.

juv.

evenly
decurved bill

br.

juv.

non-br.

DUNLIN

n-br.

straight bill with
decurved tip

dark centre
to rump

br.

non-br.

black belly

juv.

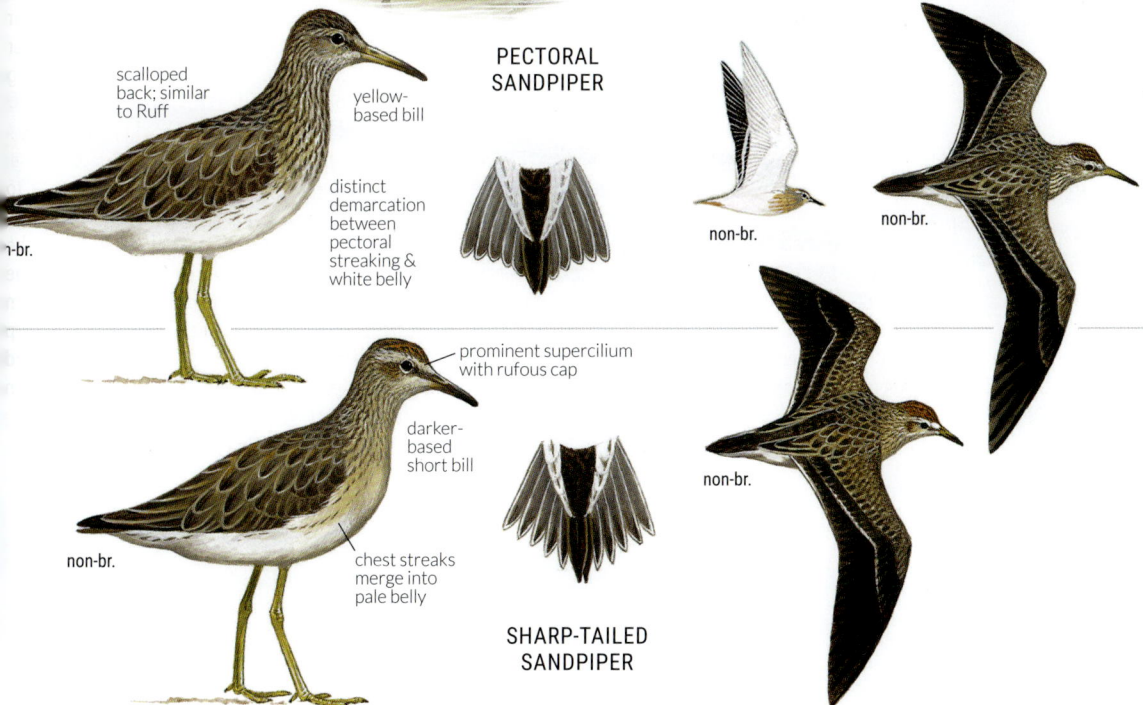

PECTORAL
SANDPIPER

scalloped
back; similar
to Ruff

yellow-
based bill

distinct
demarcation
between
pectoral
streaking &
white belly

non-br.

non-br.

n-br.

prominent supercilium
with rufous cap

darker-
based
short bill

non-br.

non-br.

chest streaks
merge into
pale belly

SHARP-TAILED
SANDPIPER

219

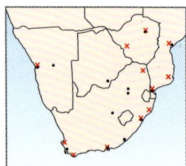

BROAD-BILLED SANDPIPER *Calidris falcinellus*

17 cm; 30–45 g; wingspan 38 cm Vagrant. A small, short-legged wader, intermediate in size between Little Stint (p. 222) and Curlew Sandpiper (p. 218). Bill is relatively long and broad, with drooped, flattened tip, distinguishing it from Little Stint. Legs dark grey, sometimes greenish. Non-br. plumage pale grey above, with dark shoulder patch (less pronounced than in Sanderling, p. 216). The whitish forked supercilium is diagnostic (but vagrant Red-necked Stint, p. 222, occasionally shows partial second supercilium). In br. plumage, upperparts blackish, feathers with narrow, buffy margins. Juv. is similar to br. ad., but upperpart feathers edged paler. **Voice:** A low-pitched, short trill, 'drrrt'. **Status and biology:** Rare, Palearctic-br. summer migrant to estuaries, less often vleis and lakes; favours areas of soft mud where it forages by rapidly driving its bill vertically into the ground. (Breëbekstrandloper)

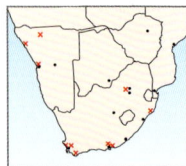

WHITE-RUMPED SANDPIPER *Calidris fuscicollis*

15–17 cm; 30–45 g; wingspan 42 cm Vagrant. Intermediate in size between Curlew Sandpiper (p. 218) and Little Stint (p. 222); it is the only sandpiper, other than Curlew Sandpiper, with a white rump. Smaller than Curlew Sandpiper, with shorter legs and a much shorter, straighter, black bill, usually with brownish base to lower mandible. Distinctly larger than Little Stint, with a more elongate shape due to its wings extending beyond the tail tip at rest; lacks a dark central rump stripe. Flanks usually show some light streaking (flanks generally lack streaking in Curlew Sandpiper and Little Stint). Its elongated shape and posture are similar to Baird's Sandpiper, but it is paler grey above with dark streaking extending further onto breast and flanks; bill broader tipped, usually with a paler base; best told by its white rump, lacking a dark central stripe (usually only visible in flight). Juv. has a chestnut crown and darker brown back feathers broadly fringed with chestnut. **Voice:** A thin, mouse-like 'jeet', given in flight. **Status and biology:** Nearctic-br. summer vagrant to coastal wetlands, rarely inland. (Witrugstrandloper)

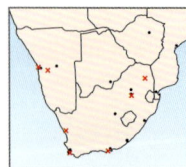

BAIRD'S SANDPIPER *Calidris bairdii*

15–17 cm; 30–45 g; wingspan 42 cm Vagrant. Intermediate in size between Curlew Sandpiper (p. 218) and Little Stint (p. 222). Smaller than Curlew Sandpiper with shorter legs and a much shorter, straight, black bill; easily told in flight by its black line down the centre of its white rump (most easily seen in flight). Larger than Little Stint, with a more elongate shape due to its wings extending beyond the tail tip at rest. Its elongated shape and posture are similar to White-rumped Sandpiper, but has a finer, straighter, all-dark bill (lacking pale base to lower mandible) and dark central rump stripe. Upperparts scaled, essentially brown, rarely grey; breast with buffy wash and light vertical streaking, but little or no streaking on lower breast and flanks. Juv. more buffy-brown, strongly scaled on its upperparts. **Voice:** A short trill, 'kreep'. **Status and biology:** Nearctic-br. summer vagrant to open shores and wetland margins. Recorded at least 15 times; most records coastal, but one each from Gauteng and Kruger National Park. (Bairdstrandloper)

STINTS The smallest of the *Calidris* waders; notoriously difficult to identify. Little Stint is the only common species in the region and is quite variable according to age, sex and season; jizz varies with posture and behaviour. Identification of vagrant stints requires a sound knowledge of the Little Stint variations.

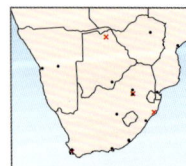

TEMMINCK'S STINT *Calidris temminckii*

13–15 cm; 18–30 g; wingspan 35 cm Vagrant. A plain-backed stint with yellow-green legs and a diagnostic white outer tail. Bill has slight droop at tip and dull yellow base. Appears more elongate than Little Stint (p. 222), with tail projecting beyond wing tips. Distinguished by combination of greenish-grey or olive legs, white outer-tail feathers and uniform grey-brown upperparts. Long-toed Stint (p. 222) is longer necked, much darker and browner above, and generally has (longer) yellow legs. Juv. resembles ad. but upperparts are scaled with buff. **Voice:** A shrill 'prrrrtt'. **Status and biology:** Palearctic-br. summer vagrant, usually wintering north of equator. Favours muddy edges of freshwater wetlands, often in small openings among vegetation. At least 7 subregion records from Gauteng, KZN, W Cape, Namibia and Botswana. (Temminckstrandloper)

scalloped back

striped crown

forked supercilium

-br.

long, drooped bill

broad tip

non-br.

juv.

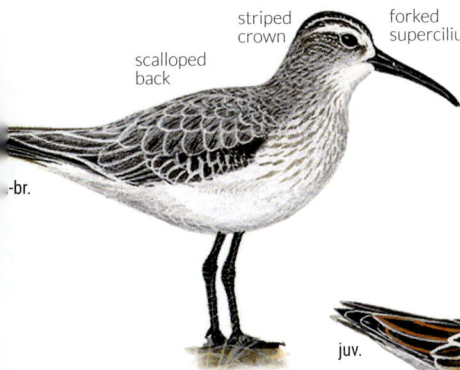

BROAD-BILLED SANDPIPER

WHITE-RUMPED SANDPIPER

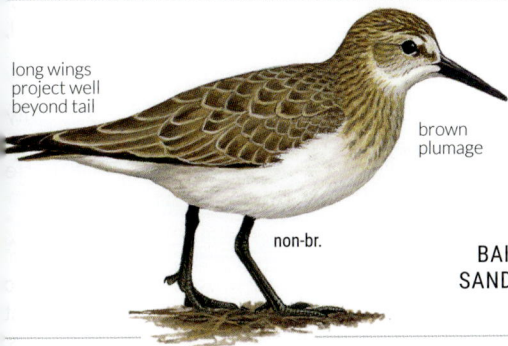

long wings project well beyond tail

non-br.

bill longer than in stints

lightly streaked flanks

white rump

non-br.

long wings project well beyond tail

brown plumage

non-br.

dark rump

non-br.

BAIRD'S SANDPIPER

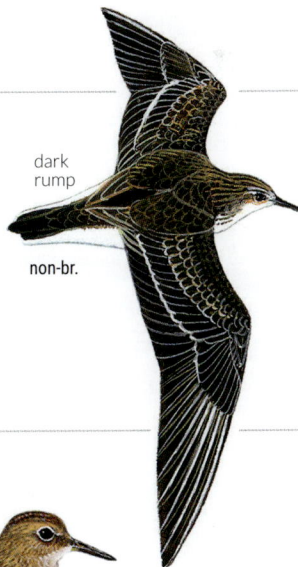

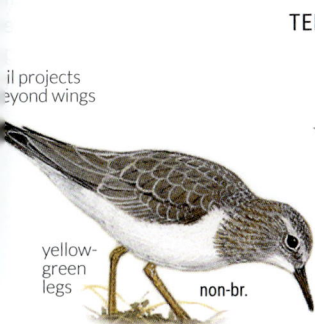

white outer tail

non-br.

TEMMINCK'S STINT

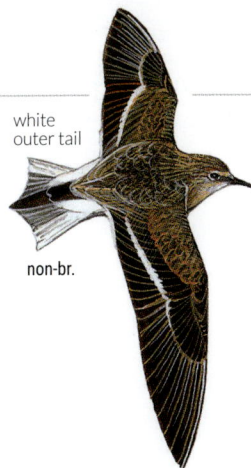

il projects eyond wings

juv.

dark, clear-cut breast band

yellow-green legs

non-br.

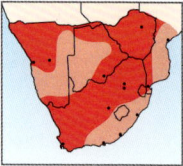

LITTLE STINT *Calidris minuta*

12–15 cm; 18–35 g; wingspan 34 cm The most abundant stint in the region; plumage and structure rather variable. Non-br. birds have diffuse grey-brown streaks on sides of breast and pale grey-brown upperparts, feathers with dark shafts; most show dark area around feather shaft. In flight, has narrow white wingbar and white sides to rump. Br. ad. and juv. rich brown above and on breast; juv. has fine buff margins to feathers. Very similar to much rarer Red-necked Stint, but averages slightly longer legged and with a longer, finer bill. In non-br. plumage, Red-necked Stint lacks black patches around upperpart feather shafts. In br. plumage Little Stint has a white (not rufous) chin. Legs usually black, but occasionally greenish-yellow, causing possible confusion with Temminck's (p. 220) and Long-toed Stints. Temminck's Stint has plainer upperparts and white (not grey) outer-tail feathers. Long-toed Stint has darker upperparts, a thinner bill and greenish-yellow legs. Smaller than Broad-billed Sandpiper (p. 220) and vagrant Dunlin (p. 218), which have longer bills that droop slightly at the tip, and also smaller than Baird's and White-rumped Sandpipers (p. 220), which appear more attenuated because their wings extend beyond their tail tips at rest. **Voice:** A short, sharp '*schit*'. **Status and biology:** Common Palearctic-br. summer migrant at freshwater wetlands, estuaries and lagoons; rarely on rocky shoreline; few overwinter. Often forages in flocks; feeding action typically very rapid probing. Regional population some 100 000 birds, but numbers decreasing. (Kleinstrandloper)

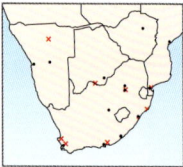

RED-NECKED STINT *Calidris ruficollis*

13–16 cm; 22–35 g; wingspan 35 cm Rare visitor or vagrant. Extremely difficult to distinguish from Little Stint, except in br. plumage when the throat, neck and cheeks are plain rufous (lacking vertical dark streaking and whitish throat of Little Stint); breast whitish with black chevrons. On average, bill is shorter and stubbier than Little Stint; also appears heavier bellied and shorter legged, with little leg showing above the ankle joint. Wings slightly longer than Little Stint. In non-br. plumage, upperparts are uniform pale grey, with narrow, dark feather shafts (lacking surrounding dark suffusion of Little Stint). Juv. has contrasting upperpart coloration: mantle and upper scapulars blackish, edged rufous; lower scapulars and tertials pale grey, edged white (juv. Little Stint has darker grey-brown tertials, often edged rufous). **Voice:** A short '*chit*' or '*prrp*'. **Status and biology:** Rare Palearctic-br. summer vagrant found at muddy fringes of wetlands and saltpans. Usually with Little Stints or Curlew Sandpipers. Small numbers recorded regularly in KZN and on the west coast in the 1980s, but few recent records: status uncertain. (Rooinekstrandloper)

LONG-TOED STINT *Calidris subminuta*

14 cm; 20–35 g; wingspan 32 cm Vagrant. A small, slender stint; legs and neck appear relatively long; bill slender. Yellow legs distinctive (but see Temminck's Stint (p. 220), and beware aberrant pale-legged juv. Little Stints). Only at very close range can the long toes be seen. Generally darker and browner above than respective plumages of Little Stint; upperpart feathers have broad, dark centres in non-br. plumage and are broadly edged rufous in br. plumage. Back more heavily mottled than Temminck's Stint, with more strongly defined supercilium and dark (not white) outer tail. Juv. is more brightly patterned than ad. **Voice:** Flight call a shrill '*kreeep*' or '*prrrt*'. **Status and biology:** Palearctic-br. summer vagrant favouring muddy fringes of freshwater wetlands. Four records: E Cape, Gauteng, Botswana and Mozambique. (Langtoonstrandloper)

br.

white 'V' on mantle

white chin

br.

br.

ad. transitional plumage

juv. worn plumage

non-br.

sharper-tipped bill

LITTLE STINT

non-br.

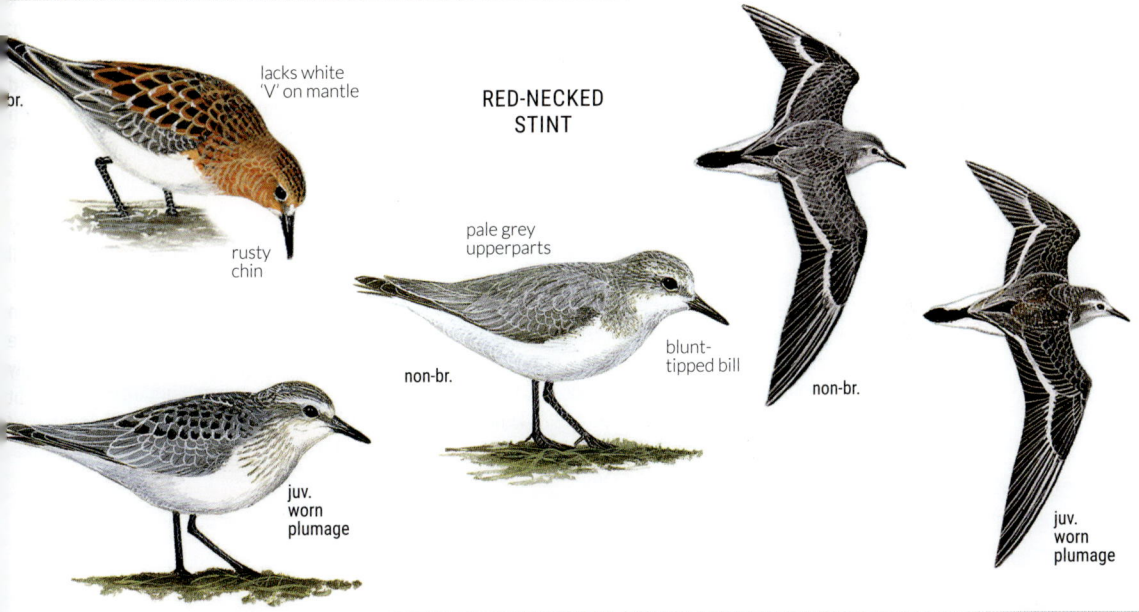

br.

lacks white 'V' on mantle

RED-NECKED STINT

rusty chin

pale grey upperparts

non-br.

blunt-tipped bill

non-br.

juv. worn plumage

juv. worn plumage

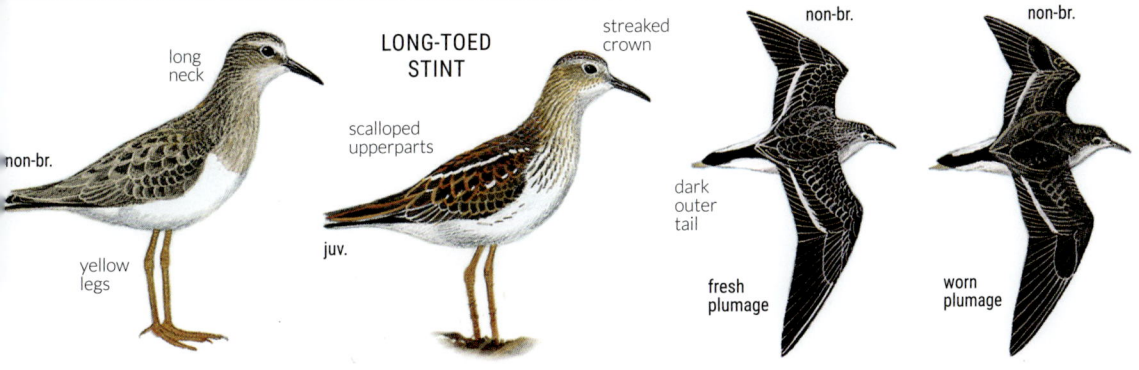

long neck

LONG-TOED STINT

streaked crown

non-br.

non-br.

non-br.

scalloped upperparts

juv.

dark outer tail

fresh plumage

worn plumage

yellow legs

PHALAROPES
Largely aquatic waders with long, dumpy bodies and lobed toes that are adapted for swimming. Sexes differ; females larger and more brightly coloured in br. plumage, linked to polyandry.

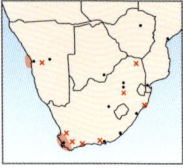

RED-NECKED PHALAROPE *Phalaropus lobatus*

J F M A M J J A S O N D

17–20 cm; 28–45 g; wingspan 30–35 cm Slightly smaller than Red Phalarope with a darker grey, more distinctly streaked back, obvious in flight in darker upperparts (especially upperwing coverts), black rump fringed with white (Red Phalarope has a grey rump). At close range, the longer, thinner, all-black bill (lacking a yellow base) is distinctive; bill finely tapered (not wide along its entire length). Vagrant Wilson's Phalarope has a longer bill, yellow (not black) legs, and lacks a distinct black eye patch. In flight, easily told from Wilson's Phalarope by its dark (not white) rump and prominent wingbar. In br. plumage, upperpart feathers blackish with extensive buffy-yellow fringes. Body dark grey; both sexes have a chestnut gorget, white chin and buff-fringed back feathers, but female much brighter than male. Juv. darker and browner above than non-br. ad., with rufous-fringed feathers; breast washed buff. **Voice:** Low 'tchick', given in flight. **Status and biology:** Uncommon Palearctic-br. summer migrant at lakes, saltpans and sewage works; rarely at sea. In recent years most reliably seen at Walvis Bay saltworks, Namibia, and Velddrif saltworks on the S African west coast, where small numbers also overwinter. Sometimes spins in tight circles in the water when feeding. (Rooihalsfraiingpoot)

RED PHALAROPE *Phalaropus fulicarius*

J F M A M J J A S O N D

20–22 cm; 40–65 g; wingspan 40 cm The common phalarope at sea, usually in small flocks. Larger than Red-necked Phalarope, with a shorter, thicker bill that is broad to the tip and may show a yellow base. Non-br. ads are more plain and paler grey above (but appear mottled in moulting birds); head mainly white with a black eye patch. Distinguished from Red-necked Phalarope in flight by its grey (not black, fringed with white) rump. Smaller than Wilson's Phalarope, with a shorter, broader bill and indistinct black eye patch; in flight told by its grey (not white) rump and prominent wingbar. In br. plumage (rarely seen in s Africa) the chestnut underparts, white face mask and black crown are diagnostic. Female brighter than male. Juv. has dark brown upperparts with feathers edged buff. Face, neck and upper breast patchily suffused with pinkish-buff. **Voice:** A soft, low 'wiit'. **Status and biology:** Fairly common Palearctic-br. summer migrant at sea off west coast; rare on south and east coasts; vagrants occur along the coast and on inland lakes. (Grysfraiingpoot)

WILSON'S PHALAROPE *Steganopus tricolor*

X X X X X X X X X
J F M A M J J A S O N D

22–24 cm; 40–80 g; wingspan 42 cm Vagrant. Larger than other phalaropes; told in non-br. plumage by its longer bill, yellow (not dark) legs and lack of a distinct black eye patch. In flight, differs in having a square white rump contrasting with grey back and tail; no wingbar. Superficially resembles Marsh Sandpiper (p. 210) but much shorter legged with grey (not white) back. In br. plumage, both sexes have a distinct, dark maroon stripe running through their eye and down the sides of their neck; sides of breast and mantle washed rufous; legs blackish. Female brighter than male, with pale grey crown; male's crown dark brown. Juv. has darker and more scalloped upperparts than non-br. ad. **Voice:** Short, grunting 'grrg'. **Status and biology:** Nearctic-br. summer vagrant with scattered records from coastal wetlands and saltpans, mainly in the west. Lunges from side to side with the bill to retrieve food close to the surface; swims less than other phalaropes and rarely spins. Usually less approachable than other phalaropes. (Bontfraiingpoot)

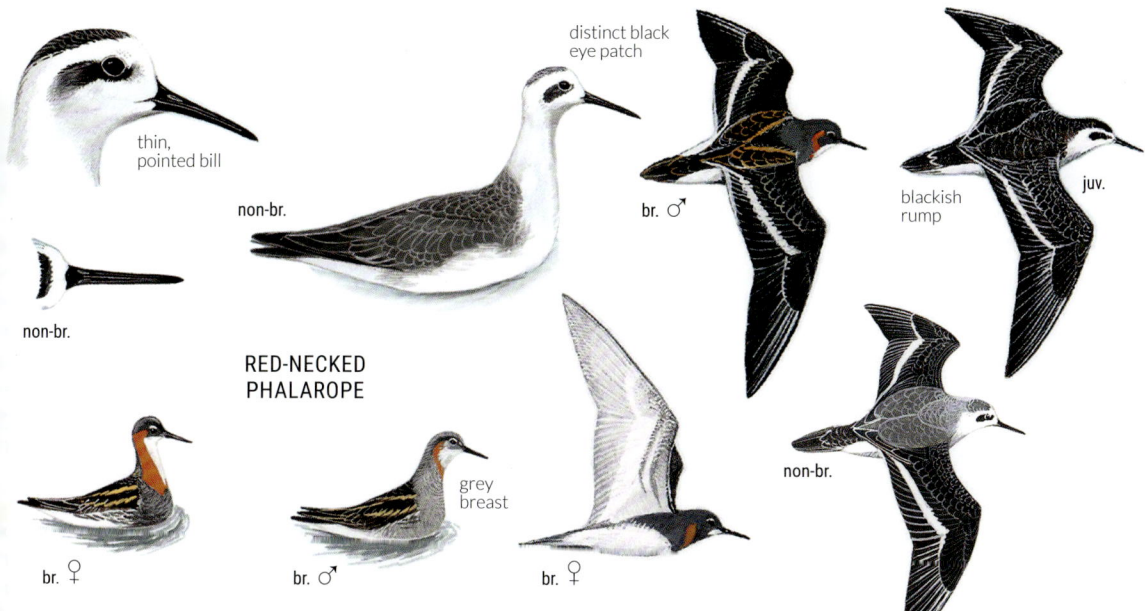

thin, pointed bill

non-br.

non-br.

distinct black eye patch

br. ♂

juv.

blackish rump

non-br.

RED-NECKED PHALAROPE

br. ♀

grey breast

br. ♂

br. ♀

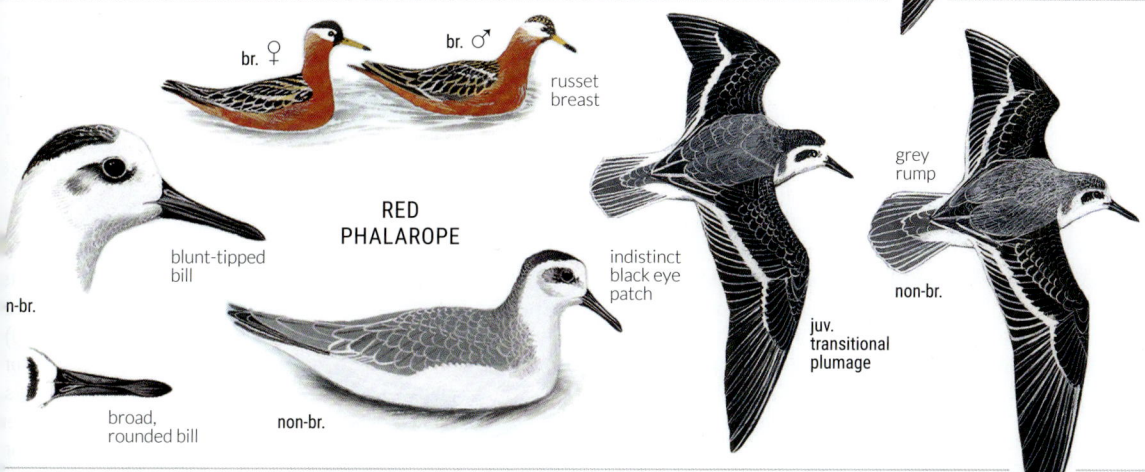

br. ♀

br. ♂

russet breast

n-br.

blunt-tipped bill

broad, rounded bill

non-br.

RED PHALAROPE

indistinct black eye patch

grey rump

non-br.

juv. transitional plumage

WILSON'S PHALAROPE

long, thin bill

n-br.

juv.

♀

yellow legs

projecting toes

juv.

non-br.

SKUAS AND JAEGERS

Medium-sized, gull-like seabirds with mainly brown plumage that often steal food from other birds. Sexes alike, but plumage variable: jaegers have distinct br., non-br. and juv. plumages, and jaegers and some skuas have pale, intermediate and dark morphs. Identification often relies on size and structure. Females slightly larger than males.

BROWN (SUBANTARCTIC) SKUA *Stercorarius antarcticus*

60–66 cm; 1.3–1.8 kg; wingspan 1.4 m A large, heavy-bodied skua with large, white wing flashes (reduced in juv.). Short, broad wings, short rump and tail, and heavy build distinguish it from jaegers. Upperparts are variably streaked and blotched buff; plainer in juv. Larger, chunkier and heavier billed than dark morph of rare South Polar Skua, which has only sparse, fine, pale streaking on its upperparts and black (not dark brown) underwing coverts. Easily told from juv. Kelp Gull (p. 228) by its heavier build and shorter, broader wings with white primary bases; lacks finely streaked plumage. **Voice:** Display call at colonies is a loud '*aah aah-aah-ah-ah-ah-ah-ah*'; generally silent at sea. **Status and biology:** Common in shelf waters, mostly in winter; rarely ventures ashore, but sometimes seen from land. Regularly scavenges at fishing boats. (Bruinroofmeeu)

SOUTH POLAR SKUA *Stercorarius maccormicki*

53–60 cm; 1.2–1.5 kg; wingspan 1.3 m Head, neck and breast colour varies from dark brown (dark morph) to ash grey (pale morph). Pale morph is easily identified by contrast between dark wings and pale head and body. Unfortunately, most records from s Africa are dark morph, which resembles a small, slender juv. Brown Skua with a slightly smaller bill and black (not dark brown) underwing coverts. Body shape recalls Pomarine Jaeger. Plumage rather plain, with only fine, sparse streaks on neck and back; some have paler, greyish feathers at base of bill. Intermediate birds usually show a pale hindneck. Juv. plain brown, lacking paler face or nape. Hybridises with Brown Skua on Antarctic Peninsula; at least some hybrids are indistinguishable from Brown Skua. Chilean Skua *S. chilensis* is a potential vagrant that has been recorded at Tristan and Marion Island and differs from both South Polar and Brown Skua by showing a dark cap that contrasts with paler head and underparts. **Voice:** Silent at sea. **Status and biology:** Rare passage migrant in s Africa. Scavenges at trawlers. (Suidpoolroofmeeu)

POMARINE JAEGER (SKUA) *Stercorarius pomarinus*

50 cm (to 75 cm with streamers); 580–800 g; wingspan 1.2–1.3 m The largest jaeger, characterised by relatively broad wings and large, white wing flashes; can be confused with South Polar Skua. Appears heavily built with a barrel chest and stout, pale-based bill. Flight direct and powerful. Br. ad. has spoon-shaped (not pointed) central tail feathers. Pale morph predominates; more extensive dark cap than other pale-morph jaegers. Non-br. ads barred on vent, flanks and rump. Juv. and imm. boldly barred on belly, vent, rump and underwing. **Voice:** Silent at sea. **Status and biology:** Common Palearctic-br. summer migrant to coastal waters off Namibia, becoming less common further south and scarce off east coast. Rarely ventures far from land; occasionally roosts ashore. Scavenges at trawlers; steals food from gannets, terns and other seabirds. (Knopstertroofmeeu)

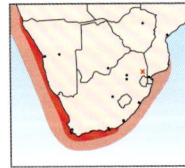

PARASITIC JAEGER (ARCTIC SKUA) *Stercorarius parasiticus*

46 cm (to 65 cm with streamers); 380–500 g; wingspan 1.1–1.2 m A medium-sized jaeger; usually the most abundant in the region. Pale, dark and rare intermediate colour morphs. Larger and more heavily built than Long-tailed Jaeger, with more direct and dashing flight on broader wings with white wing flashes (flashes all but absent in Long-tailed Jaeger); plumage generally darker brown (not pale grey-brown). Can be confused with Pomarine Jaeger, but is smaller and more slender with longer, narrower wings, a smaller bill that appears uniformly dark and a smaller dark hood in pale-phase birds. If protruding, central tail feathers are relatively short and straight. Juv. more rusty-brown than other juv. jaegers, with less strongly barred underparts. **Voice:** Silent at sea. **Status and biology:** Common Palearctic-br. summer migrant to coastal waters; a few overwinter. Occasionally roosts ashore; vagrant inland. Scavenges at trawlers; steals food from terns and gulls. (Arktiese Roofmeeu)

LONG-TAILED JAEGER (SKUA) *Stercorarius longicaudus*

38 cm (to 62 cm with streamers); 250–400 g; wingspan 1–1.1 m The smallest jaeger, with buoyant, tern-like flight on long, slender wings. Almost all birds are pale morphs. Ads are plain, cold greyish above, with only the shafts of outer 2 primaries white; vent and uppertail coverts barred in non-br. plumage. Bill is short; at close range, the nail at the tip of the upper mandible comprises roughly half the bill length (less than a third in other jaegers). In br. plumage, very long central tail feathers are diagnostic, but has shorter streamers when moulting. Juv. colder grey-brown than juv. Parasitic Jaeger, with more boldly barred uppertail coverts and underwing; proportions more tern-like. **Voice:** Silent at sea. **Status and biology:** Fairly common Palearctic-br. summer visitor and passage migrant to oceanic waters off s Africa; scarce close to land; vagrant inland. Scavenges at trawlers, but seldom harries other seabirds. (Langstertroofmeeu)

BROWN SKUA

streaked
upperparts

heavy
build

brown
coverts

**SOUTH POLAR
SKUA**

blackish
underwing
coverts

mostly plain
upperparts

dark
morph

thinner bill than
Brown Skua

often pale at
base of bill

intermediate
morph

slighter
build

blackish
underwing
coverts

pale
morph

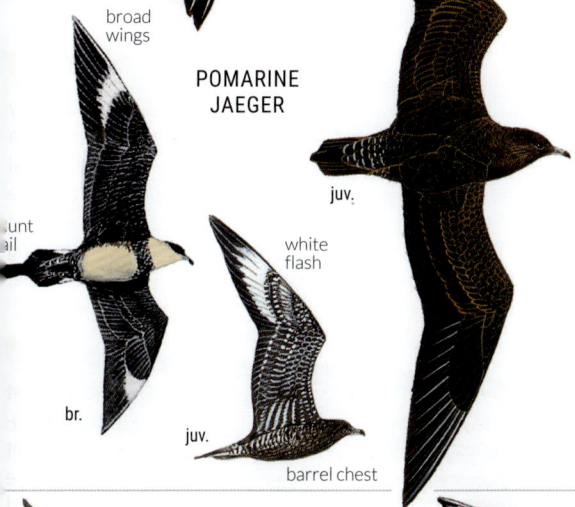
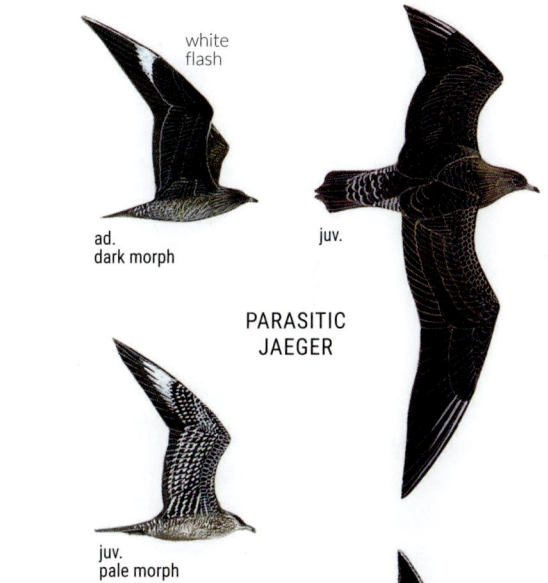

broad
wings

**POMARINE
JAEGER**

juv.

white
flash

blunt
tail

br.

juv.

barrel chest

white
flash

ad.
dark morph

juv.

**PARASITIC
JAEGER**

juv.
pale morph

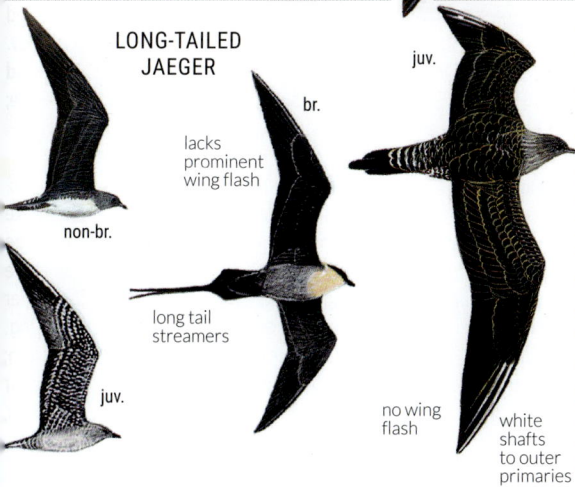
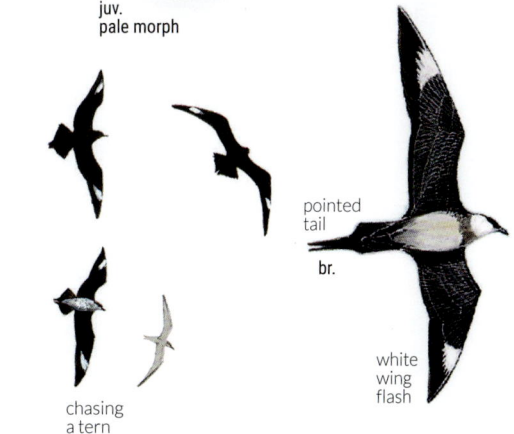

**LONG-TAILED
JAEGER**

juv.

br.

lacks
prominent
wing flash

non-br.

long tail
streamers

juv.

no wing
flash

white
shafts
to outer
primaries

chasing
a tern

pointed
tail

br.

white
wing
flash

GULLS Medium-large sea- and freshwater birds. Sexes alike in plumage; males slightly larger. Juv. body plumage replaced soon after fledging in small gulls (and thus unlikely to be observed in s Africa for migrant and vagrant species). Large *Larus* gulls have 2 or 3 imm. plumages.

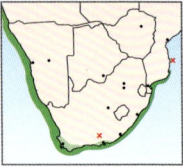

KELP GULL *Larus dominicanus*

55–65 cm; 900–1 100 g; wingspan 1.3–1.4 m The largest gull in the region. More heavily built than Lesser Black-backed Gull, with shorter wings (folded wings do not project well beyond tail tip), a more robust bill, steeper forehead, and typically olive-grey (not yellow) legs, although there is considerable variation in leg colour. Ad. has blackish back and upperwings; eyes dark grey-brown (rarely yellow or silver-grey). Juv. heavily streaked dark brown; bill blackish; legs brownish-pink. Best told from juv. Lesser Black-backed Gull by structure. Could potentially be confused with skuas, but is more slender with finely streaked plumage. Imm. has white body with diffuse brown smudging; mantle slate-grey with some brown feathers; tail tip blackish; bill pinkish at base and tip. Sub-ad. retains some brown in upper- and underwings and has dark tail tip; bill sometimes has a dark subterminal mark. Young ads may retain dark tips to outer-tail feathers and some diffuse, dusky streaking on head and mantle. **Voice:** Display call is long series of *'kee-ah'* notes; alarm calls include single *'kee-ah'*, *'kwok'* and *'yap-yap'*. **Status and biology:** Common resident in coastal habitats and adjacent wetlands; follows trawlers up to 100 km from shore. Increasingly found on fields up to 50 km inland. Endemic race *L. d. vetula* sometimes considered a separate species (Cape or Khoisan Gull) but there is recent gene flow from S America; 20 000 pairs breed in summer. (Kelpmeeu)

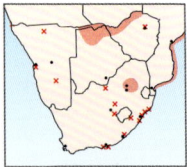

LESSER BLACK-BACKED GULL *Larus fuscus*

52–65 cm; 600–1 150 g; wingspan 1.3–1.4 m Slightly smaller than Kelp Gull with a less robust bill and head, gently sloping forehead and more attenuated appearance; wings project well beyond the tail at rest. Ads have yellow (not olive-grey) legs and feet (but some Kelp Gulls have yellow legs). Three races occur (sometimes treated as separate species): *L. f. fuscus* is smaller (600–800 g), with a blackish back in ad. plumage; 'Heuglin's Gull' *L. f. heuglini* is larger (850–1 150 g), with a longer bill and shallower forehead, giving it a rakish appearance; ad. back dark slate-grey. 'Steppe Gull' *L. f. barabensis* also larger, with a round head, ad.'s bill is tricoloured, tipped ivory. Back of ad. is lighter grey than *L. f. heuglini*. Juvs and imms best differentiated by structure, but legs typically are paler, flesh-coloured (not brownish). Juv. *L. f. heuglini* has pale window in inner primaries. **Voice:** Typical, large-gull *'kow-kow'*; shorter *'kop'*. **Status and biology:** Rare Palearctic-br. summer migrant, mostly Oct–Apr; *heuglini* mostly along east coast; *fuscus* at inland wetlands and along coast; only 1 record of *barabensis* at Mkhombo Dam, Mpumulanga, in 2015. (Kleinswartrugmeeu)

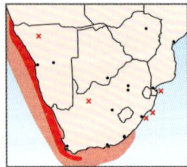

SABINE'S GULL *Xema sabini*

27–32 cm; 160–180 g; wingspan 85–90 cm A small, oceanic gull with a small bill, shallowly forked tail and buoyant, tern-like flight. In flight, boldly tricoloured upperwing is diagnostic. Ad. bill black with yellow tip; legs black. Non-br. ads have white heads with dark smudges on nape; br. ads have dark grey hood. Juv. has all-black bill, grey-brown mid-crown to mantle, back and inner upperwing coverts; appears dark headed; tail tipped black. After first moult, head as non-br. ad.; remainder of upperparts as juv., but saddle dark blue-grey. Smaller than juv. Black-legged Kittiwake, which has back and inner coverts pale grey. **Voice:** Silent in Africa. **Status and biology:** Common Holarctic-br. summer migrant to coastal waters, including large, sheltered bays; rare ashore or at coastal wetlands. Often in flocks; roosts at sea. Regional population 5–10 000 birds. (Mikstertmeeu)

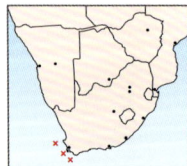

BLACK-LEGGED KITTIWAKE *Rissa tridactyla*

36–40 cm; 320–450 g; wingspan 1.05 m Vagrant. An oceanic gull, slightly larger than Sabine's Gull, with long wings, a shallowly forked tail, yellow bill and black legs. Br. ad. white, with pale grey back; upperwing pale grey with narrow, white trailing edge and black tips to outer primaries (wing tip 'dipped in ink'). Non-br. ad. has grey patch on ear coverts and grey smudging on hindneck. Juv. has black 'M' across upperwing, black tail tip (broadest in centre of tail) and black hind collar; bill black. Juv./imm. could be confused with juv./imm. Sabine's Gull, but is larger, with less deeply forked tail and pale grey (not dark grey-brown or blue-grey) saddle and inner upperwing coverts. **Voice:** Silent at sea. **Status and biology:** VULNERABLE. Palearctic-br. vagrant which seldom crosses the equator: 3 records from the west coast. (Swartpootbrandervoël)

KELP GULL

dark eye

rounded forehead

jet-black back

d.

hort wing rojection

greenish legs

robust bill

juv.

grey legs

imm.

ad.

LESSER BLACK-BACKED GULL

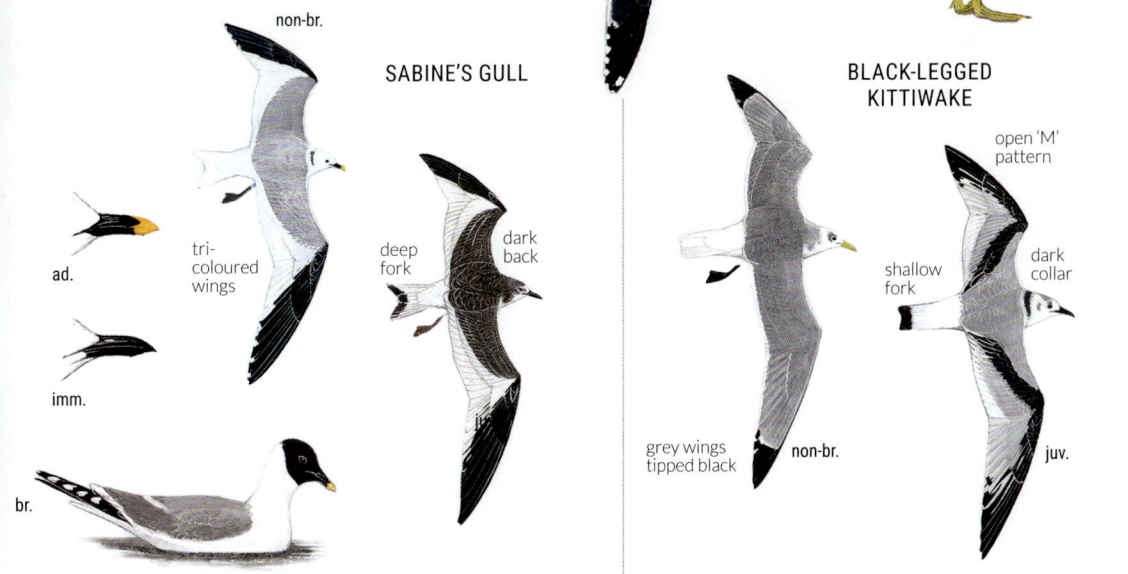

pale eye

sloping forehead

blackish back

fuscus

fuscus

g wing ojection

light grey back

ad.

barabensis

thinner bill

imm.

pink legs

ad.

heuglini

slate-grey back

ad.

heuglini

yellow legs

SABINE'S GULL

non-br.

ad.

tri-coloured wings

imm.

deep fork

dark back

br.

BLACK-LEGGED KITTIWAKE

open 'M' pattern

shallow fork

dark collar

grey wings tipped black

non-br.

juv.

229

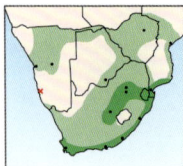

GREY-HEADED GULL   *Chroicocephalus cirrocephalus*

b b b B B B b b b b
J F M A M J J A S O N D

40–42 cm; 220–340 g; wingspan 95–105 cm A medium-sized gull with a pale grey back and upperwings. Br. ad. has diagnostic pale grey head, and bright red bill and legs. Non-br. ad. has largely white head with grey smudges above eye and on cheeks; bill and legs duller. Eyes silver-yellow with narrow, red outer ring. Imm. has dark smudges on ear coverts and a dark-tipped, pink-orange bill. Slightly larger than Hartlaub's Gull, with a longer bill, drooped at the tip. Easily told from vagrant gulls in flight by grey underwing and mostly black outer primaries. Juv. is mottled brown on upperwings; secondaries darker grey than ad., and tail tipped blackish. Compared with juv. Hartlaub's Gull, has more extensive smudges on the head, a dark-tipped, pinkish-orange bill, darker upperwings and more black in tail. **Voice:** Dry 'karrh' and repeated 'pok-pok-pok'. **Status and biology:** Locally common resident along the coast and at coastal and inland wetlands; regional population >2 000 pairs. Occasionally hybridises with Hartlaub's Gull, particularly on central Namibian coast. Readily associates with human activity. (Gryskopmeeu)

HARTLAUB'S GULL   *Chroicocephalus hartlaubii*

b B B B B B b B b b b
J F M A M J J A S O N D

38–40 cm; 220–340 g; wingspan 90–95 cm Endemic to Benguela coast. Slightly smaller than Grey-headed Gull, with a shorter, thinner and darker bill, and deeper red legs; eyes usually dark at all ages, but some birds have whitish eyes. Told from vagrant gulls in flight by grey underwing and mostly black outer primaries. Br. ads have slight, greyish shadow line demarcating pale lavender hood; non-br. ads have plain white heads. Imm. has blackish-red bill (not dark-tipped as Grey-headed Gull) and dull legs; sometimes shows small, dark patches on ear coverts. Juv. has wing coverts mottled brown, but less heavily so than juv. Grey-headed Gull, and brown spots in tail tip. **Voice:** Drawn-out, rattling 'keeerrh' and 'pok-pok-pok'. **Status and biology:** Endemic. Common resident; 13 000 pairs breed on offshore islands, coastal wetlands and on large buildings. Becoming more common along the south coast, now breeding as far east as Port Elizabeth, E Cape. Occasionally hybridises with Grey-headed Gull, particularly on central Namibian coast. An opportunistic scavenger and predator, it does well in urban areas and has spread up to 50 km inland around Cape Town. Often active at night. Large numbers gather to forage at flooded fields. (Hartlaubmeeu)

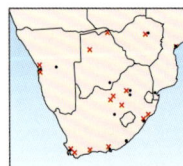

BLACK-HEADED GULL   *Chroicocephalus ridibundus*

X X X X   X X X X
J F M A M J J A S O N D

35–40 cm; 240–360 g; wingspan 90–100 cm Vagrant. Paler grey above than Grey-headed, Hartlaub's and Franklin's Gulls. In flight, has mostly white outer primaries (but aberrant Hartlaub's and Grey-headed Gulls can also show this) and whitish (not grey) underwings. Br. ad. has dark brown hood and partial white eye-ring; bill dark red, tipped black. Non-br. ads have dark smudges on head and paler red bill with dark tip. In flight, shows large, white wedge in outer wing and whitish underwings. These features shared with Slender-billed Gull, but latter has unmarked, white head and pale yellow (not dark brown) eyes. Imm. has a dark brownish subterminal band across its primaries, secondaries and tail feathers. Juv. has mottled brown wing coverts; unlikely to occur in this plumage in s Africa. **Voice:** Typical, small-gull 'kraah'. **Status and biology:** Palearctic-br. summer vagrant, with fewer records in recent years, most often found with other gulls at coastal or inland wetlands. (Swartkopmeeu)

SLENDER-BILLED GULL   *Chroicocephalus genei*

                X
J F M A M J J A S O N D

38–44 cm; 240–320 g; wingspan 1.05 m Vagrant. A pale grey-and-white gull with a white head and distinctive head profile – shallow, sloping forehead and long bill that droops slightly at the tip. Overall, appears long-necked, long-legged and pigeon-breasted. Ad. has pale eye and red bill; underparts washed pink in br. plumage. In flight, wing pattern resembles Black-headed Gull, with extensive white wedge in outer primaries, but body appears longer. Imm. has dark smudge behind eye, but this is less prominent than in imm. Grey-headed and Black-headed Gulls; bill pale orange with a slightly darker tip. **Voice:** 'Ka' or 'kra', deeper than Black-headed Gull. **Status and biology:** Breeds in Palearctic, wintering in ne Africa: 1 record, Durban, Sep 1999. (Dunbekmeeu)

FRANKLIN'S GULL   *Leucophaeus pipixcan*

X X X X X X X X X X X X
J F M A M J J A S O N D

32–36 cm; 240–320 g; wingspan 90 cm Vagrant. A small, black-headed gull with a rather short, stubby bill and at least a partial black hood and white crescents above and below eye at all ages. Darker grey above than other small gulls in the region. In flight, has broad, white trailing edge to secondaries and black subterminal band on outer primaries surrounded by a white band. Underwings whitish. Bill dark red in br. plumage; blackish in non-br. and imm. birds. Imm. has broad, brown subterminal tail band and darker grey wings, but still has broad, white trailing edge to secondaries. Juvs moult before reaching s Africa. **Voice:** A soft 'krrk' or 'wee-ah', seldom heard in the region. **Status and biology:** Nearctic-br. summer vagrant to the coast and adjacent wetlands; rarely inland. Most regular on west coast. (Franklinmeeu)

GREY-HEADED GULL

pale eye, ringed red

juv.

pale-based longish bill

juv.

non-br.

HARTLAUB'S GULL

dark eye & bill

indistinct grey collar

br.

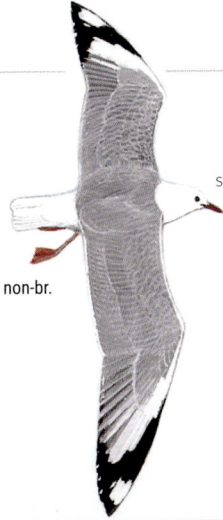

short bill

non-br.

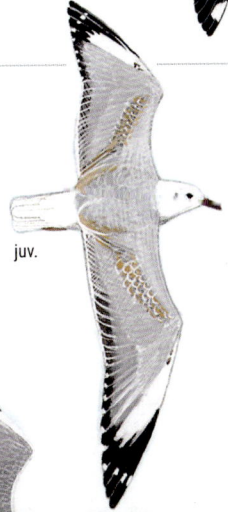

juv.

BLACK-HEADED GULL

white, semi-crescent

imm.

non-br.

br.

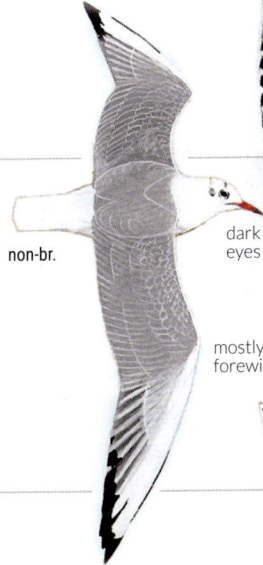

non-br.

dark brown eyes

mostly white forewing

SLENDER-BILLED GULL

shallow forehead

pale yellow eyes

red bill

d.

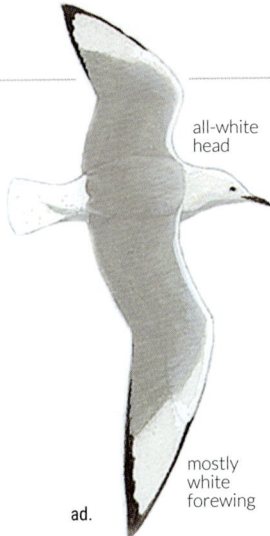

all-white head

mostly white forewing

ad.

FRANKLIN'S GULL

br.

grizzled head

non-br.

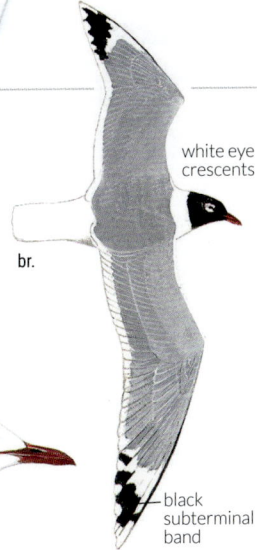

white eye crescents

br.

black subterminal band

231

SKIMMERS

SKIMMERS Large, tern-like birds with long, broad wings. The elongated lower mandible is trailed through the water in flight to locate small fish and other prey. Sexes alike.

AFRICAN SKIMMER *Rhynchops flavirostris*

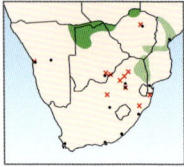

J	F	M	A	M	J	J	A	S	O	N	D
								b	B	B	b	B	b

38–42 cm; 120–220 g; wingspan 1.25–1.35 m Easily identified by its peculiarly shaped, long, red bill with a yellowish tip (but beware vagrant Black Skimmer, most likely encountered on west coast). Blackish-brown upperparts contrast with white underparts, white trailing edge to wing and outer tail. Upperparts blackish in br. ad.; browner and less crisply defined in non-br. ad. Juv. browner above, with pale feather fringes; bill blackish, gradually turning dull yellow, then red from base. **Voice:** Harsh *'rak-rak'*. **Status and biology:** NEAR-THREATENED. Intra-African br. migrant, Apr–Jan, on large rivers, bays and lakes with sandbanks for roosting and breeding; a few remain year-round, especially in drought years. World population some 10 000 birds, with 1 000 in s Africa. Breeds in small colonies. (Waterploeër)

BLACK SKIMMER *Rhynchops niger*

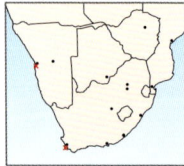

J	F	M	A	M	J	J	A	S	O	N	D
X									X		

40–46 cm; 250–350 g; wingspan 1.4 m Vagrant. Larger than African Skimmer; ad. has well-defined black bill tip (not dusky as in juv. African Skimmer) and blackish upperparts. **Voice:** Mostly silent; occasional harsh, barking call. **Status and biology:** Occurs from the USA to Chile and Argentina. First confirmed record from Cape Town in Oct 2012. Presumably the same bird was seen a week later in Walvis Bay, supporting previous claims from the Namibian coast. (Amerikaanse Waterploeër)

NODDIES

NODDIES Dark brown tropical terns with paler crowns. Flight loose and buoyant; tail long and wedge-shaped with a central notch. Largely oceanic. Lack white underwing coverts and pale belly of imm. Sooty Tern (p. 242); smaller and more slender than dark-morph jaegers (p. 226) with no white in primaries. Sexes alike.

BROWN NODDY *Anous stolidus*

J	F	M	A	M	J	J	A	S	O	N	D
X	X	X									X

36–44 cm; 170–220 g; wingspan 80–85 cm Vagrant. Slightly larger and browner than vagrant Lesser Noddy, with a shorter, heavier bill. Pale crown does not extend onto nape and white frons contrasts sharply with brown lores. Shows greater contrast across wings due to paler brown upperwing coverts contrasting with blackish flight feathers; central underwing coverts paler than flight feathers (underwing not uniformly dark brown). Juv. has pale crown reduced or absent; upperpart feathers pale-edged. **Voice:** Hoarse *'kark'*, seldom heard away from colonies. **Status and biology:** Rare summer visitor to oceanic waters off Mozambique from tropical Indian Ocean colonies; vagrant elsewhere, with only 1 west coast record. Occasionally roosts ashore. (Grootbruinsterretjie)

LESSER NODDY *Anous tenuirostris*

J	F	M	A	M	J	J	A	S	O	N	D
X	X	X	X								

30–34 cm; 90–120 g; wingspan 70–75 cm Vagrant. Smaller than Brown Noddy, with a relatively longer, more slender bill. Whitish forehead merges with brown lores and ash-grey crown extends further back onto nape. Upperwing and underwing are more uniformly dark brown (lacking contrasting coverts). Juv. has pale crown reduced or absent. **Voice:** Short, rattling *'churrr'*, but generally silent at sea. **Status and biology:** Vagrant from tropical Indian Ocean colonies to seas off east coast, usually in summer following cyclones. Occasionally roosts ashore. (Kleinbruinsterretjie)

AFRICAN SKIMMER

ad.

pale tip

BLACK SKIMMER

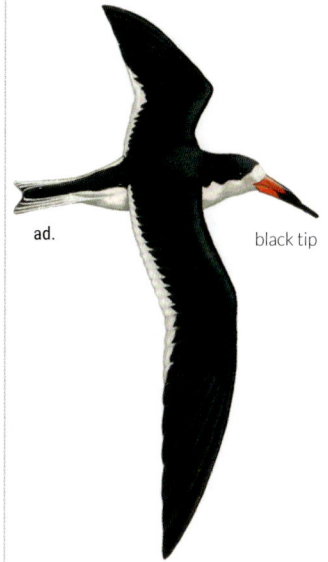

ad.

ad.

black tip

juv.

BROWN NODDY

dark lores

heavy bill

coverts paler than flight
feathers (above & below)

LESSER NODDY

pale lores

long, thin bill

coverts & flight
feathers uniform

233

WEST AFRICAN CRESTED (ROYAL) TERN
Thalasseus albididorsalis

46–48 cm; 340–420 g; wingspan 1.3 m Vagrant. A large, pale tern with a rich orange bill. Smaller than red-billed Caspian Tern (p. 234), with a more slender, relatively longer bill that lacks a dark tip; in flight, undersides of primaries are greyish (not black). Most likely confused with Swift Tern, but bill distinctly orange (not yellow or greenish-yellow); outer primaries darker grey, contrasting with paler coverts and back. Appreciably larger than Lesser Crested and vagrant Elegant Terns, with a heavier bill. Br. ads have full black crown with shaggy crest; non-br. ads have extensive white forehead and crown, with often only a small, black nuchal patch. Juv. has mottled brown wing coverts. Imm. has dark carpal bar. **Voice:** Loud, harsh *'ree-ack'*. **Status and biology:** Breeds in W and central Africa; rare visitor to Kunene River mouth; vagrant further south on Namibian coast. (Wes-Afrikasterretjie)

SWIFT (GREATER CRESTED) TERN   *Thalasseus bergii*

46–49 cm; 320–430 g; wingspan 95–115 cm A large tern with a long, slightly drooped yellow or greenish-yellow bill (not orange, as in West African Crested, Lesser Crested and Elegant Terns). Appreciably larger than Lesser Crested and Elegant Terns. Br. ads have black cap and shaggy crest, but black does not extend to bill. Non-br. ads have white forecrown. Back colour pale grey in nominate race; some darker grey *T. b. velox* may occur in Mozambique. Juv. has dusky, yellow-olive bill; upperparts mottled blackish-brown with blackish margins to flight feathers, appearing dark grey in flight. Imm. retains dark flight feathers. Legs usually black, but some juvs have yellow-orange legs. **Voice:** Harsh *'kree-eck'*; juv. has a thin, vibrating, whistled begging call. **Status and biology:** Common resident and local migrant in coastal waters, rarely out to shelf break; also estuaries and coastal wetlands. Nominate race is endemic to s Africa; population increasing; 8 000–12 000 pairs breed colonially at islands and coastal wetlands from central Namibia to Port Elizabeth, E Cape. Large numbers follow sardine run along the east coast in winter. (Geelbeksterretjie)

ELEGANT TERN   *Thalasseus elegans*

38–41 cm; 200–320 g; wingspan 95–110 cm Vagrant. Slightly larger than Lesser Crested Tern and Sandwich (p. 234) Tern; orange bill longer, with a drooped tip; underwing tips dark grey. Paler grey above than Lesser Crested Tern with a white (not pale grey) rump and a longer, shaggier crest. Appreciably smaller than West African Crested Tern, with a longer, more slender bill. Br. ad. has a full black cap; non-br. ad. has black hind crown (more extensive than West African Crested Tern, extending in front of eye). Juv. paler grey above and slightly scaled; bill yellow. **Voice:** Grating call similar to Sandwich Tern. **Status and biology:** Nearctic-br. vagrant, but a few birds hybridise with Sandwich Terns in Europe. Concerns that European birds may be hybrid Elegant x Sandwich or even Lesser Crested x Sandwich Terns allayed by genetic evidence of 'pure' Elegant Terns in Europe; hybrids typically show some black in bill. (Elegante Sterretjie)

LESSER CRESTED TERN   *Thalasseus bengalensis*

35–37 cm; 180–240 g; wingspan 90–105 cm A medium-sized tern, slightly smaller than Sandwich Tern (p. 234), with a slender, orange-yellow bill. Smaller, more graceful and generally paler above than Swift Tern; bill lacks drooped tip. Much smaller than West African Crested Tern; bill more slender. Main identification risk is vagrant Elegant Tern, but is slightly smaller, with a shorter, straighter bill (not drooped at tip); rump pale grey (not white). Juv. has mottled blackish-brown back and wing coverts, and dark grey flight feathers. Imm. retains dark grey flight feathers. **Voice:** Hoarse *'kreck'*. **Status and biology:** Fairly common non-br. visitor to east coast, mainly in summer; vagrant to W Cape. Forages in inshore waters, bays and estuaries; roosts along coast and at coastal wetlands. (Kuifkopsterretjie)

WEST AFRICAN CRESTED TERN

heavy orange bill

non-br.

imm.

deep fork

br.

yellow or greenish-yellow bill

br.

SWIFT TERN

grey rump

non-br.

juv.

ELEGANT TERN

white rump

non-br.

non-br.

drooping orange bill

LESSER CRESTED TERN

br.

orange bill

imm.

grey rump

237

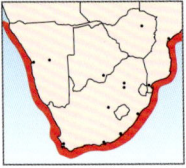

COMMON TERN *Sterna hirundo*

J F M A M J J A S O N D

31–35 cm; 90–150 g; wingspan 72–82 cm Typically the most common small to medium-sized tern, often in large flocks. Differs from Arctic Tern by its slightly longer bill and legs; greyish (not pure white) rump and tail, and patterning of underside of primaries, which have a broad dark outline/wedge to outer primaries (thin, dark outline in Arctic and faintly dark-tipped outline in Antarctic). The dark wedge is also visible in the upperwing in flight. Less dumpy than Antarctic Tern, with a more slender bill; lacks grey wash on underparts in non-br. plumage. Bill shorter and less drooped than Roseate Tern; upperparts darker grey. Rump and tail contrast with darker grey back (uniform grey in vagrant White-cheeked Tern). In br. plumage has black-tipped red bill, light grey wash to breast, and short tail streamers (level with folded wing tips). Rarely has orange or orange-and-black bill in non-br. plumage. Juv. mottled brown above, but has mostly moulted these feathers before arriving in s Africa; imm. retains conspicuous dark carpal bars. **Voice:** '*Kik-kik*' and '*kee-arh*'. **Status and biology:** Abundant Palearctic-br. summer migrant to coastal waters and adjacent wetlands; vagrant inland. At least 300 000 birds visit the region; some imms remain year-round. (Gewone Sterretjie)

ARCTIC TERN *Sterna paradisaea*

J F M A M J J A S O N D

33–35 cm; 80–110 g; wingspan 75–85 cm Similar to Common Tern, but more compact, with a shorter neck, shorter, finer bill and paler outer primaries evident in flight; rump and tail white (not pale grey). At rest, legs distinctly shorter. Not as dumpy as Antarctic Tern, especially at rest; bill smaller; bill and legs usually blackish (not red). Br. ad. has full black cap, pale grey wash on breast, dark red bill and legs; tail streamers extend beyond folded wing (but growing feathers are shorter). Non-br. ad. has extensive white frons, and blackish bill and legs. Imm. has a dark carpal bar, but not as pronounced as imm. Common Tern. **Voice:** '*Kik-kik*' given in flight. **Status and biology:** Fairly common Palearctic-br. passage migrant and summer visitor to coastal and oceanic waters. Most pass offshore en route to and from Antarctic wintering grounds, but some remain over summer; small numbers overwinter. Sometimes roosts ashore, usually with Common Terns; vagrant inland. (Arktiese Sterretjie)

ANTARCTIC TERN *Sterna vittata*

J F M A M J J A S O N D

34–40 cm; 110–170 g; wingspan 72–80 cm A rather dumpy tern, with a heavier bill than either Arctic or Common Terns. Leg length varies among races from short to fairly long. Br. ad. has full black cap, bright red bill and legs, and grey underparts contrasting with white cheek stripe. Underwing similar to Arctic Tern but generally paler, with a very thin outline to outer primaries. Told from vagrant White-cheeked Tern by its white (not grey) rump and heavier build. Non-br. ad. has paler grey underparts and a white forecrown, but retains red legs and some red in bill (black in non-br. Common and Arctic Terns). Juv. has chequered brown, grey and white upperparts and diffuse brown wash on sides of breast; differs from juv. Roseate Tern by its shorter, heavier bill and paler cap, and from juv. Whiskered Tern (p. 242) by its white (not grey) rump. Imm. has black bill and legs; best told by heavy bill and dumpy body; some retain a few barred juv. tertials. **Voice:** High-pitched '*kik-kik*' and harsher '*kreaah*'. **Status and biology:** Fairly common winter migrant to coastal waters of S Africa, mainly in W and E Cape. Global population around 50 000 pairs; regional population >15 000 birds. (Grysborssterretjie)

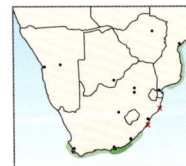

ROSEATE TERN *Sterna dougallii*

J F M A M J J A S O N D

33–38 cm; 100–130 g; wingspan 72–80 cm A sleek tern with very pale grey upperparts and a long, slightly drooped, blackish bill. Legs and wings longer than other similar-sized terns. Underwing appears all-white in flight. Br. ads have full black cap, pink wash to breast, crimson legs, red base to bill and long, white outer-tail feathers that project well beyond wings at rest. Non-br. ads have white forecrown; best identified by long bill and pale colour. Juv. finely barred blackish-brown above; differs from juv. Antarctic Tern by blacker cap, longer bill, greyer wings and more slender body. Imm. retains some juv. tertials and has darker upperwing coverts than non-br. ad. **Voice:** Harsh, grating '*chir-RIK*' when br.; also a grating '*aarh*'. **Status and biology:** Scarce resident; 250 pairs breed May–Oct at islands off south coast, dispersing east and west along the coast. Non-br. vagrants may occur along Mozambique coast from populations br. further north. (Rooiborssterretjie)

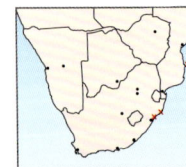

WHITE-CHEEKED TERN *Sterna repressa*

J F M A M J J A S O N D

32–35 cm; 80–100 g; wingspan 80 cm Vagrant. Resembles a dark, slender Common Tern, but told from this (and Arctic and Antarctic Terns) in br. and non-br. plumages in having a uniformly grey back, rump and tail. Br. ad. has grey underparts, darkest on belly; could be confused with br. plumage Whiskered Tern (p. 242), but is larger, with a longer bill, more deeply forked tail and dusky-grey (not white) vent. Imm. white below with mostly black cap and dark carpal bar; best told from imm. Common Tern by grey rump and tail, darker grey upperwings and broad, blackish trailing edge to underwing. **Voice:** Ringing '*kee-leck*'. **Status and biology:** Vagrant to east coast from tropical Indian Ocean; roosts with other terns. (Witwangsterretjie)

COMMON TERN

imm.

br.

non-br.

plain grey rump & tail

imm.

prominent wedge in outer wing

ARCTIC TERN

br.

short legs

non-br.

thin outline to outer wing

white rump & tail

imm.

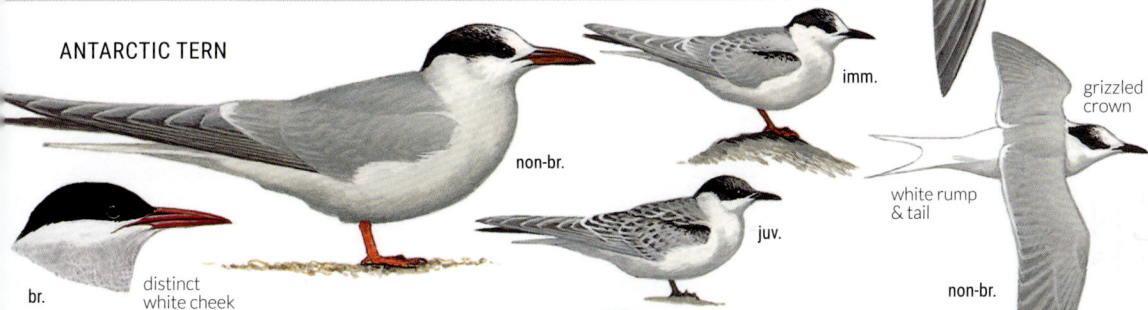

ANTARCTIC TERN

non-br.

imm.

juv.

br.

distinct white cheek

grizzled crown

white rump & tail

non-br.

ROSEATE TERN

long, thin bill

imm.

br.

long tail

pink flush

non-br.

deep fork

uniform dusky-grey above

WHITE-CHEEKED TERN

non-br.

br.

BLACK-NAPED TERN *Sterna sumatrana*

X X X | | | | | | X X X
J F M A M J J A S O N D

30 cm; 90 g; wingspan 60 cm Vagrant. A small, very pale tern, superficially resembling a miniature Sandwich Tern (p. 234), and some have same pink suffusion to underparts. Ad. has black band extending behind eye, broadening across nape; crown pure white. Bill and legs black. Only outer web of outer primary is black (visible at rest and in flight). Juv. has crown feathers tipped black, and is mottled brown above. Imm. has dusky carpal bar. Head and nape patterning can appear similar to non-br. Little, Saunders's and Damara Terns, however, larger size of Black-naped Tern should rule out further confusion. **Voice:** Clipped, repeated *'ki-ki'*. **Status and biology:** Summer vagrant to east coast from tropical Indian Ocean colonies; sometimes roosts at estuaries together with other terns. (Swartneksterretjie)

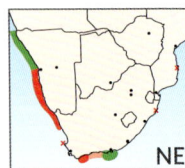

DAMARA TERN *Sternula balaenarum*

B b b b b | | | b B B B
J F M A M J J A S O N D

21–23 cm; 46–60 g; wingspan 50 cm A very small, rather uniform pale grey tern with narrow wings and rapid wing beats. Differs from Little and Saunders's Terns in having a more uniform upperwing, and uniformly grey rump and back; body appears shorter and more dumpy in flight. Br. ad. has diagnostic full black cap and mostly black bill (slight yellowish base to lower mandible). Non-br. ads and imms have mottled grey-and-white frons. Juv. has buffy-brown barring on mantle, and horn-coloured base to the bill. **Voice:** Far-carrying, rapid *'chit-ick'*, higher pitched than Little Tern. **Status and biology:** VULNERABLE. Near-endemic; regional population about 7 000 pairs, most breeding in Namibia; local in S Africa (120 pairs). Favours sheltered coastlines, bays and lagoons. Breeds on coastal dunes and saltpans, singly or in loose colonies. Most migrate to W Africa in winter, but some remain year-round. Recently (Sep 2019) found in small numbers near Vilanculos, Mozambique; origin of these birds is unknown. (Damarasterretjie)

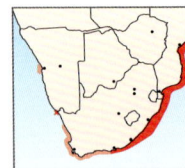

LITTLE TERN *Sternula albifrons*

J F M A M J J A S O N D

22–24 cm; 40–60 g; wingspan 48–55 cm A tiny tern, similar to Damara Tern; rump and tail are paler grey than back (sometimes white) and has greater contrast between dark outer primaries and rest of upperwing. In flight, appears longer-tailed. Br. ad. has a white frons and yellow bill with a small, black tip. Saunders's Tern br. ad. has a broader white frons that does not extend above the eyes. Non-br. ads have varying amounts of yellow in bill, which can appear wholly black; probably inseparable from Damara Tern in the field. Juv. lightly mottled brown above (unlikely to be seen in region). Imm. has indistinct darker carpal bars. **Voice:** Slightly rasping *'ket-ket'*. **Status and biology:** Fairly common Palearctic-br. summer migrant to shallow coastal waters and estuaries; regular along east and south coasts to W Cape. Regional population around 15 000 birds. (Kleinsterretjie)

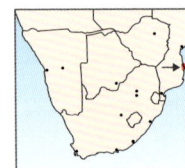

SAUNDERS'S TERN *Sternula saundersi*

J F M A M J J A S O N D

22–24 cm; 40–45 g; wingspan 50–55 cm A small tern, very similar to Little Tern, but br. ad has rounded white frons not extending above eye. The outer primaries (usually 3 or 4) are dark (2 or 3 greyish in Little Tern). Non-br. ad. averages darker above, but is probably not separable from Little Tern in the field. Juv. similar to juv. Little Tern; unlikely to be seen in region. Imm. has indistinct dark carpal bars like imm. Little Tern. **Voice:** Call is similar to Little Tern. **Status and biology:** Breeds from the Arabian Peninsula to Somalia. Non-br. birds migrate south to coastal Tanzania and Madagascar. Small numbers confirmed from near Vilanculos, Mozambique, in Sep 2019, perhaps regular in small numbers. (Arabiese Sterretjie)

juv.

BLACK-NAPED TERN

ad.

pale grey plumage

some with pink below

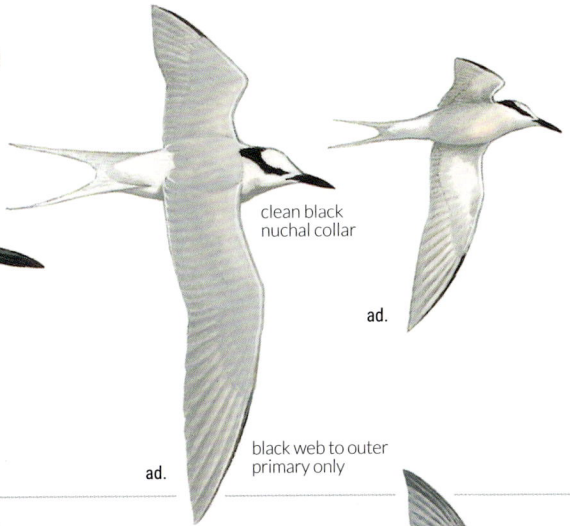

clean black nuchal collar

ad.

black web to outer primary only

ad.

DAMARA TERN

non-br.

br.

black bill & cap

non-br.

non-br.

rump & tail same colour as back (all ages)

br.

LITTLE TERN

br.

white, angular frons extends above eye

mostly yellow bill

black bill & cap

non-br.

non-br.

rump & tail paler than back (all ages)

non-br.

SAUNDERS'S TERN

br.

white, rounded frons does not reach eye

non-br.

non-br.

more extensive black in outer primaries

non-br.

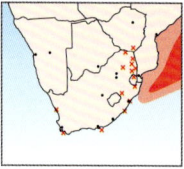

SOOTY TERN   *Onychoprion fuscatus*

40–44 cm; 160–220 g; wingspan 90 cm A fairly large, long-winged, dark-backed tern. Much larger and darker above than Bridled Tern, with a broader white frons that does not extend behind eye; black crown does not contrast with blackish back. In flight, both species have bold, white leading edge to upperwing. Juv. and imm. have blackish throat and breast (juv. and imm. Bridled Tern are white below). Could potentially be confused with noddies (p. 232) or imm. jaegers (p. 226), but neither have largely white underwing coverts. **Voice:** Loud *'wick-a-wick'* or *'wide-awake'* at colonies; also given when attracted to ships' lights at night. Alarm is harsh *'kraark'*. **Status and biology:** Common non-br. summer visitor to oceanic waters off Mozambique and n KZN from tropical Indian Ocean colonies, where it occurs in feeding flocks of several hundred. Typically remains well offshore, but large numbers may be blown inland or wrecked ashore by tropical cyclones. (Roetsterretjie)

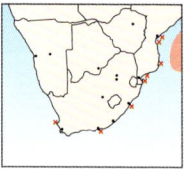

BRIDLED TERN   *Onychoprion anaethetus*

30–32 cm; 110–170 g; wingspan 80 cm Rare visitor or vagrant. An elegant, dark-backed tern. Considerably smaller than Sooty Tern, with paler, grey-brown (not blackish) upperparts. Narrow white frons extends behind eye, and dark crown contrasts with paler back. Non-br. ads have white spotting on crown. Juv. and imm. have wing coverts finely edged buffy-white, and white (not blackish) underparts. **Voice:** *'Wup-wup'* and *'kee-arr'*. **Status and biology:** Rare non-br. visitor to oceanic waters off Mozambique from tropical Indian Ocean colonies; vagrant elsewhere. Mostly pelagic, where often seen perched on floating debris (unlike Sooty Tern), but sometimes roosts ashore. Occurs in much smaller flocks than Sooty Tern. (Brilsterretjie)

LAKE TERNS
Small, mostly freshwater terns with square tails. Generally pick prey from water surface rather than plunge-dive. Sexes alike; br. plumages distinct, but non-br. birds and juvs more difficult to identify.

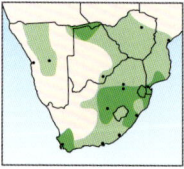

WHISKERED TERN   *Chlidonias hybrida*

24–26 cm; 80–110 g; wingspan 75 cm The largest lake tern, most likely to be confused with a marine tern; relatively long legs and a heavy bill recall much larger Gull-billed Tern (p. 234). Dark grey (not black) underparts in br. plumage are diagnostic; superficially resembles White-cheeked Tern (p. 238), but is smaller and more compact, with white (not dusky grey) vent and less deeply forked tail. Non-br. ads are larger than other lake terns, lacking a dark cheek patch extending below eye. Paler grey above than non-br. Black Tern; rump pale grey (white in White-winged Tern). Juv. is mottled brown on back. **Voice:** Repeated, harsh *'krrkk'*. **Status and biology:** Fairly common resident and intra-African migrant at wetlands and marshes; regional population probably <7 500 pairs. Flight is generally stronger and less erratic than White-winged or Black Terns. Breeds in small colonies, building floating nests of aquatic vegetation. (Witbaardsterretjie)

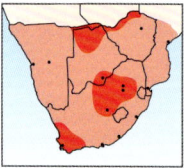

WHITE-WINGED TERN   *Chlidonias leucopterus*

20–22 cm; 45–80 g; wingspan 65 cm The smallest lake tern, with a white rump in all plumages. Br. plumage strikingly black and white, with pale grey upperwings and black underwing coverts; white rump and tail contrast with black back; legs bright red. Non-br. ad. is much paler above than Black Tern, with black confined to rear of crown, and no black shoulder smudge (although birds in transitional plumage can show this feature). Appreciably smaller than Whiskered Tern, with contrasting white rump and different head pattern. Imm. has small brown tips to upperpart feathers. **Voice:** Shrill *'kek-ek-ek'*, higher pitched than Whiskered Tern. **Status and biology:** Common Palearctic-br. summer migrant from central Asia, found at lakes, estuaries and marshes; occasionally in sheltered coastal bays and over open country inland. Picks food from water surface while in flight; hawks insects aerially. (Witvlerksterretjie)

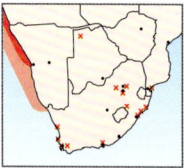

BLACK TERN   *Chlidonias niger*

22–24 cm; 50–75 g; wingspan 65 cm A dark-backed lake tern. In br. plumage, black head, breast and belly merge into dark grey back and wings; lacks contrast between upperwing and back of White-winged Tern. Non-br. ad. has diagnostic dark shoulder smudge (but some White-winged Terns in transitional plumage show a similar smudge), more extensive black on head, and no contrast between dusky-grey back, rump and tail. Darker above than Whiskered Tern, with black cheek spot. Imm. is slightly darker and less uniform above. **Voice:** Usually silent; flight call a quiet *'kik-kik'*. **Status and biology:** Common along n and central Namibian coast; rare elsewhere. Occurs in open ocean, bays and coastal wetlands; usually forages at sea, but many roost ashore; vagrant inland. When krill washes ashore in central Namibia, forages on beaches like a wader. (Swartsterretjie)

black
upperparts

white frons

**SOOTY
TERN**

ad.

ad.

imm.

skua-like; lacks
white wing flashes

BRIDLED TERN

white supercilium

imm.

ad.

ad.

WHISKERED TERN

tail shallowly
forked

grey
rump
& tail

white vent
(all ages)

non-br.

br.

juv.

non-br.

small, black
smudge
behind eye

non-br.

non-br.

br.

white
rump & tail
(all ages)

WHITE-WINGED TERN

br.

large, black
smudge
behind eye

dusky grey
rump & tail
(all ages)

non-br.

black
shoulder
patch

BLACK TERN

non-br.

br.

br.

243

SANDGROUSE

Short-legged terrestrial birds that shuffle on the ground. Wings long, slender and pointed; flight swift. Closest relatives are the mesites of Madagascar, with both more distantly related to the pigeons. Sexes differ in plumage; juvs are similar to females. Eat mainly seeds; low water content of diet requires daily flights to drink, and males carry water to chicks in specially adapted belly feathers All species gather in flocks to drink at certain times of the day, usually early to mid-morning and late afternoon/evening.

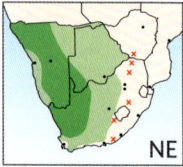

NAMAQUA SANDGROUSE *Pterocles namaqua*

b b b B B B B B b b b
J F M A M J J A S O N D

24–28 cm; 150–230 g The only sandgrouse with a long, pointed tail. Male has double breast band (white above black); face plain buffy-orange, lacking the black and white forehead bands of male Double-banded Sandgrouse; lower breast and belly plain (not barred). Female and juv. differ from female and juv. Double-banded Sandgrouse in being more buffy-yellow on throat and breast, having streaked (not barred) upper breast, pointed (not rounded) tail, and lacking bare, yellow eye-ring. **Voice:** Distinctive, nasal *'kelkie-vein'* call, frequently given in flight. **Status and biology:** Near-endemic. Common nomad and partial migrant in desert, semi-desert, arid savanna and grassland. Moves away immediately after rains when seeds germinate, returning once new growth has set seeds. Mainly drinks 1–4 hours after dawn; sometimes in late afternoon. (Kelkiewyn)

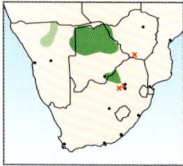

YELLOW-THROATED SANDGROUSE *Pterocles gutturalis*

b b B B b b b
J F M A M J J A S O N D

28–30 cm; 290–400 g The largest African sandgrouse; identified in flight by its short tail, dark brown belly and wholly blackish-brown underwings. Male has a creamy-yellow face and throat, with a broad, black neck collar. Female also has a plain, yellowish face (lacking the black neck collar), but is heavily mottled blackish-brown, buff and white on the neck, breast and upperparts. **Voice:** Flight call a deep, far-carrying bisyllabic *'aw-aw'*, the first note higher pitched; sometimes preceded by *'ipi'*. **Status and biology:** Scarce resident and nomad in grassland, especially on black cotton soils and usually near seasonal wetlands; also arid savanna. Gathers in flocks to drink during early morning; a few drink in late afternoon. (Geelkeelsandpatrys)

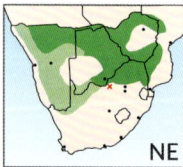

DOUBLE-BANDED SANDGROUSE *Pterocles bicinctus*

b b b B B B B b b
J F M A M J J A S O N D

25–26 cm; 200–270 g A small, short-tailed sandgrouse. Male has a diagnostic head pattern, black and white breast bands and barred lower breast and belly. Female and juv. rather uniformly barred; differ from female and juv. Namaqua Sandgrouse by darker, streaked crown, barred (not streaked) lower breast, short, rounded (not pointed) tail and obvious ring of bare, yellow skin around eye. **Voice:** A whistling *'chwee-chee-chee'* and a soft *'we-we-play-volleyball'* flight call. **Status and biology:** Near-endemic. Fairly common resident and nomad in woodland (especially mopane), savanna, and locally in arid Karoo grassland. Drinks at dusk and often after dark. (Dubbelbandsandpatrys)

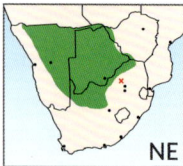

BURCHELL'S SANDGROUSE *Pterocles burchelli*

b B B b b b b
J F M A M J J A S O N D

24–26 cm; 170–230 g A small, compact sandgrouse with a white-spotted, cinnamon breast, belly, back and wing coverts. Female and juv. resemble a drab male, but have buffy (not blue-grey) faces and throats. **Voice:** Call is a soft, mellow *'chup-chup, choop-choop'* given in flight and around waterholes. **Status and biology:** Near-endemic. Scarce to locally common resident and nomad in semi-arid savanna; particularly common on Kalahari sands. Drinks 3–5 hours after dawn, generally later than Namaqua Sandgrouse. (Gevlekte Sandpatrys)

NAMAQUA SANDGROUSE

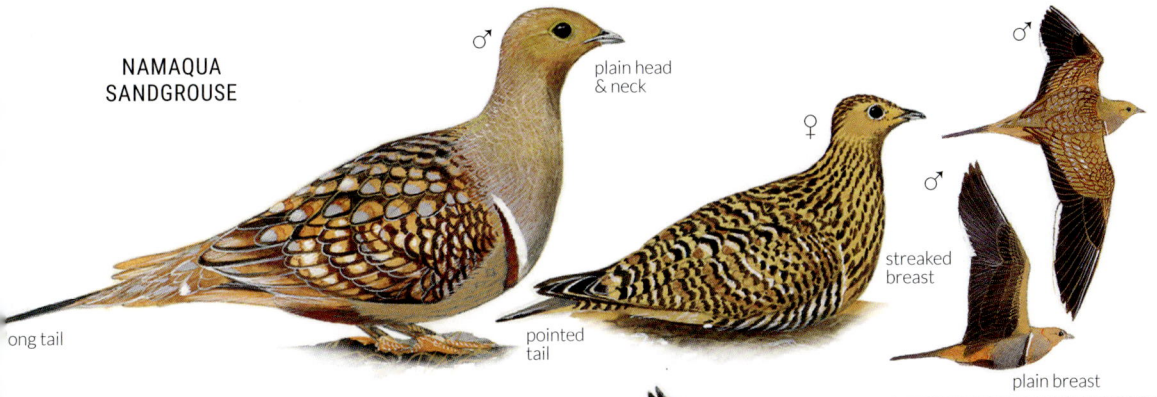

♂ plain head & neck

♀

streaked breast

♂

♂

ong tail

pointed tail

plain breast

YELLOW-THROATED SANDGROUSE

black collar

♂

♂

plain, pale face

grey back & tail

♂

chestnut belly

♀

DOUBLE-BANDED SANDGROUSE

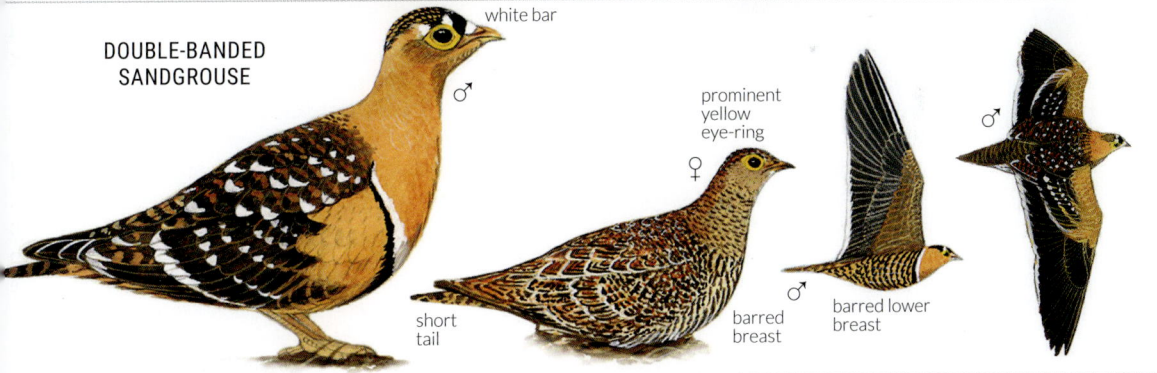

white bar

♂

prominent yellow eye-ring

♀

♂

♂

short tail

barred breast

barred lower breast

BURCHELL'S SANDGROUSE

grey face

♂

yellowish face

white wingbar

♀

♂

♂

spotted body

chestnut coverts

PIGEONS AND DOVES

A distinctive family of birds with rather short legs, rounded heads and chunky bodies. Bills have soft bases with a waxy cere. Sexes alike in most species; juvs duller, often with buff-fringed back feathers.

SPECKLED PIGEON *Columba guinea*

B	b	b	b	b	b	b	b	B	B	B	B
J	F	M	A	M	J	J	A	S	O	N	D

30–34 cm; 280–400 g A large, grey pigeon with a dull brick-red neck, back and inner wing coverts, bare red skin around eyes, red legs and dark grey bill with a whitish cere. Derives its name from its white-spotted wing coverts. Tail broadly tipped blackish with a paler grey subterminal band. Smaller and paler than African Olive Pigeon, with red (not yellow) eye patches and legs, and black (not yellow) bill. Occasionally hybridises with Rock Dove (Feral Pigeon). Juv. browner, with grey-brown facial skin. **Voice:** Deep, booming *'hooo-hooo-hooo'*; softer *'coocoo-coocoo'*. **Status and biology:** Common resident, favouring rocky areas, coastal cliffs and cities, ranging into croplands and grassland to feed. Males clap their wings when taking off on display flights. (Kransduif)

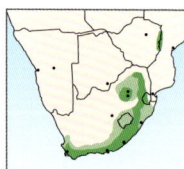

AFRICAN OLIVE PIGEON *Columba arquatrix*

b	b	b	B	B	B	B	b	b	b	b	b
J	F	M	A	M	J	J	A	S	O	N	D

38–40 cm; 300–480 g A large, fairly long-tailed pigeon with dark, purplish plumage, paler grey face and nape, and conspicuous bare, yellow eye patch, bill and legs. Wing coverts and belly are finely spotted white. In flight against a pale sky, appears all-dark with pale legs. Juv. paler, with whitish fringes to feathers; face, bill and legs duller. **Voice:** Soft, deep *'du-du-du-doo-dooo'*; also gives a high-pitched wailing call, especially on landing. **Status and biology:** Fairly common resident and nomad in forest, thickets, gardens and plantations. Movements linked to local fruiting events. (Geelbekbosduif)

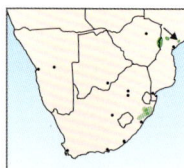

EASTERN BRONZE-NAPED PIGEON *Columba delegorguei*

b	b	b	b							B	B
J	F	M	A	M	J	J	A	S	O	N	D

26–28 cm; 140–170 g A dark, medium-sized forest pigeon, shorter tailed and much smaller than African Olive Pigeon. Male has greyish head and diagnostic whitish, crescent-shaped patch on hindneck collar. Iridescent green neck patches are visible only at close range. Female and juv. lack pale hindneck collar and dark face. Canopy (not understorey) habitat separates it from Lemon Dove. **Voice:** Far-carrying, distinctive *'oo oo oo COO COO COO cu-cu-cu-cu-cu'* with a series of fast notes at the end, descending in pitch. **Status and biology:** Uncommon, localised resident of canopy of tall forest, thick woodland and dense bush. In e Zimbabwe, moves to lower elevations in winter. (Withalsbosduif)

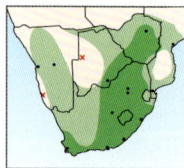

ROCK DOVE (FERAL PIGEON) *Columba livia*

b	b	b	b	b	b	b	b	b	b	b	b
J	F	M	A	M	J	J	A	S	O	N	D

32–34 cm; 320–450 g A feral population, probably derived from domesticated homing pigeons. A large pigeon, closely commensal with humans. Plumage variable; typical form (resembling ancestral Rock Dove of Europe) is bluish-grey with black bars on the wings and tail, a white rump patch, and glossy green and purple on the sides of the neck. Other varieties include black, white and reddish forms. Female and juv. duller. **Voice:** Deep, rolling *'coo-roo-coo'*. **Status and biology:** Locally abundant resident in many urban areas, but also around small, rural towns and villages. (Tuinduif)

LEMON DOVE *Columba larvata*

B	B	B	B	b	b	b	b	b	B	B	B
J	F	M	A	M	J	J	A	S	O	N	D

24–26 cm; 140–190 g A plump, medium-sized dove of the forest floor. Pale forehead and face contrast with darker, iridescent green-bronze hind crown, nape and mantle; underparts rich cinnamon. Eye-ring red. Pale face and habitat separate it from Eastern Bronze-naped Pigeon. In flight, appears all-dark, with slightly paler outer-tail tips. Female duller; juv. darker above, with buff barring on mantle. **Voice:** Deep *'ooop-ooop'*, sometimes repeated 6–10 times. Also gives a series of descending notes similar to Tambourine Dove (p. 250). **Status and biology:** Fairly common resident of forest floor and understorey; secretive and easily overlooked. Flushes to a low branch when disturbed, giving impression of a small, dark dove that lacks the pale, banded rump of Blue-spotted and Emerald-spotted Wood Doves (p. 250). (Kaneelduifie)

heavily spotted
coverts

red legs & feet

SPECKLED
PIGEON

dark tail
band

yellow
bill

EASTERN
BRONZE-NAPED
PIGEON

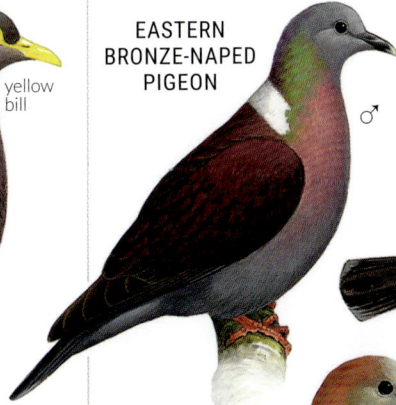

♂

AFRICAN OLIVE
PIGEON

spotted
below

yellow legs
& feet

white
collar

♂

♀

highly variable
plumage

ROCK DOVE

greenish
collar

♂

LEMON DOVE

♂

♀

247

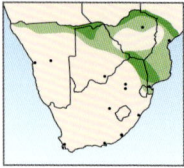

AFRICAN MOURNING (MOURNING COLLARED) DOVE   *Streptopelia decipiens*

b b b b b b b b b b B B
J F M A M J J A S O N D

28–30 cm; 120–200 g A fairly large, collared dove with a plain grey head; wide, red eye-ring contrasts with the pale yellow eye (distinguishing it from all other collared doves of the region). Smaller and paler than Red-eyed Dove, but larger than all other 'collared' doves. In flight, has white in outer tail. Juv. is browner, with wing coverts tipped buff. **Voice:** Loud *'cuck-ook-oooo'*; grating trill *'currrrrrow'*; throaty *'aaooow'* on landing. **Status and biology:** Locally common resident of woodland, riverine forest, thickets and gardens in semi-arid savanna. Particularly common at rest camps in the cental Kruger National Park. Males have a high, towering display flight. (Rooioogtortelduif)

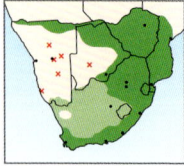

RED-EYED DOVE   *Streptopelia semitorquata*

B b b b b b b B B B B B
J F M A M J J A S O N D

32–34 cm; 190–300 g The largest collared dove; overall dark pinkish-grey with a pale forehead and pinkish head and breast. In flight, has diagnostic broad, buffy band at tip of tail, lacking white outers of all other *Streptopelia* doves in the region. Dull red eye-ring is less prominent than that of African Mourning Dove, and eye is red or orange (not yellow). Juv. is browner, with smaller collar. **Voice:** Typical call, *'coo coo, co-kuk coo coo'* (rendered 'You chew tobacco too.'), is diagnostic; harsh *'chwaa'* alarm call. **Status and biology:** Common resident of woodland, forest, parks and gardens; range has expanded south- and westwards, and numbers have increased in recent decades. Males fly steeply up in towering display, clapping wings, then glide back to perch in a tree. (Grootringduif)

CAPE TURTLE (RING-NECKED) DOVE   *Streptopelia capicola*

b b b b b b b B B B b b
J F M A M J J A S O N D

25–27 cm; 100–160 g A fairly small, pale blue-grey collared dove with a pinkish wash on the neck; grey upperparts average darker in mesic east than in arid west. Smaller and paler than Red-eyed and African Mourning Doves, with dark eyes lacking a red eye-ring. In flight, has conspicuous white tips to all but central tail feathers, contrasting with dark grey bases. Juv. duller, with some buff edgings to feathers. **Voice:** *'Kuk-coorrrr-uk'* (rendered 'How's father?'), middle note descending and rolled; harsh *'kurrrr'* alarm call. **Status and biology:** Abundant resident and nomad in a wide range of habitats, including deserts and semi-deserts where water is available; avoids forests. Males have a high, towering display flight. (Gewone Tortelduif)

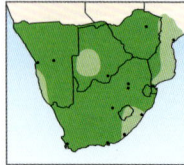

LAUGHING DOVE   *Spilopelia senegalensis*

b b b b b b b B B B B B
J F M A M J J A S O N D

22–24 cm; 80–130 g A fairly small, pinkish-buff dove with no collar. Black mottling on rich rufous upper breast, coupled with blue-grey greater and median coverts and rump are diagnostic. Smaller than vagrant European Turtle Dove, with plain wing coverts and no neck patch. In flight, cinnamon-coloured back contrasts with blue-grey forewings; outer-tail tips white. Female paler. Juv. has pale feather fringes and poorly marked breast. **Voice:** Rising and falling *'uh hu hu huu hu'*, rather like a subdued laugh, hence the common name. **Status and biology:** Abundant resident and nomad in a wide range of habitats, including urban areas; avoids forests. Males have a high, towering display flight. (Rooiborsduifie)

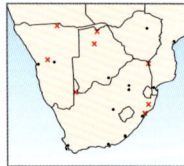

EUROPEAN TURTLE DOVE   *Streptopelia turtur*

X   X X X X   X
J F M A M J J A S O N D

26–28 cm; 110–180 g Vagrant. A richly coloured dove with diagnostic black-and-white neck patch and strongly patterned wing coverts, with dark feather centres contrasting with broad, chestnut margins. Larger than Laughing Dove; lacks rufous upper breast. In flight, has broad, white tips to outer tail, contrasting with blackish subterminal band. Juv. lacks neck patch and has barred and mottled upperparts, distinguishing it from juv. Laughing Dove. **Voice:** Soft, purring *'crrrr roorrrrrrr'*. **Status and biology:** Palearctic-br. vagrant, usually wintering north of equator. 10 records from semi-arid savannas in Namibia, Botswana and S Africa, with a number of other records thought to be escapees, as it is commonly kept in bird collections. (Europese Tortelduif)

AFRICAN MOURNING DOVE

yellow eye, bright red eye-ring

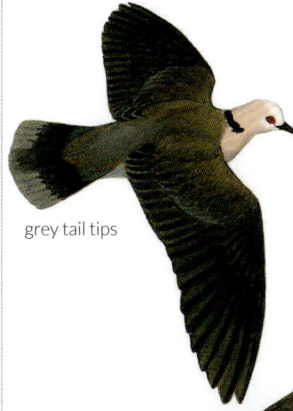

RED-EYED DOVE

grey tail tips

dark eye, purple-red eye-ring

all-black collar

pink flush to chest

CAPE TURTLE DOVE

dark eye, no eye-ring

white border to black collar

pale grey chest

white tail tips

LAUGHING DOVE

black-flecked necklace

EUROPEAN TURTLE DOVE

striped hindneck

heavily scaled coverts

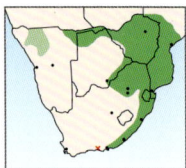

AFRICAN GREEN PIGEON   *Treron calvus*

b b b b b b b b B B B B
J F M A M J J A S O N D

25–28 cm; 210–250 g A distinctive, parrot-like pigeon, often seen in small groups clambering (parrot-like) around canopy of fruiting trees. Green, yellow and grey plumage is unique among local pigeons; bill whitish with red cere; eyes whitish; legs red; thighs yellow. In flight, dark flight feathers and greater coverts contrast with paler green forewing; vent chestnut; tail with broad, grey tip. Ad. has mauve shoulder patches; olive-yellow in juv. The *damarensis* subspecies occurring in the nw of subregion is distinctly paler yellow below, with deep blue (rather than whitish) eyes. Some authorities split the coastal populations (from E Cape to central Mozambique) as a separate species to the inland populations. Coastal populations generally show more grey in plumage compared to the yellow-green of inland populations. **Voice:** A long series of liquid whistles, *'thweeeloo, tleeeoo'*, followed by barking and growling notes and ending with 3 explosive clicks. **Status and biology:** Common resident and local nomad in forest, woodland and savanna. Closely associated with fruiting trees, especially figs; regularly hangs upside-down on branches to obtain fruit. (Papegaaiduif)

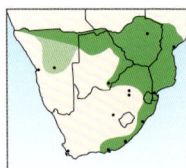

EMERALD-SPOTTED WOOD DOVE   *Turtur chalcospilos*

b b b b b b b b b b b b
J F M A M J J A S O N D

17–20 cm; 55–70 g A small, compact dove with mainly plain, pinkish-brown plumage, relieved by pale grey face and crown, 6–8 iridescent green wing coverts, and with 2 dark grey bands across the back and 1 across the tail. Distinguished from Blue-spotted Wood Dove by its paler plumage, all-dark (not red-and-yellow) bill, and larger, greenish (not blue) wing spots (true colour of wing spots only shows up in good light; otherwise can appear dark). Juv. is browner, barred buff above; wing spots smaller. **Voice:** A long series of muffled notes, starting hesitantly and descending in pitch at the end: *'hu, hu-hu HOO, hu-hu HOO-oo, hu-HOO, hu, hu, hu-hu-hu-hu-hu-hu-hu-hu'*. **Status and biology:** Common resident in woodland, savanna, valley bushveld and gardens; generally in drier habitats than Blue-spotted Wood Dove. (Groenvlekduifie)

BLUE-SPOTTED WOOD DOVE   *Turtur afer*

b b b b b b b b b b b b
J F M A M J J A S O N D

18–21 cm; 55–75 g Similar to Emerald-spotted Wood Dove, but richer brown, with smaller, blue (not green) iridescent wing spots and a yellow-tipped, red (not black) bill (only visible at close range). In flight, back and rump appear more rufous. Juv. has a brown bill and reduced wing spots; told from juv. Emerald-spotted Wood Dove by its more rufous plumage. **Voice:** Series of muffled *'coo'* notes, shorter and deeper in pitch than Emerald-spotted Wood Dove. **Status and biology:** Uncommon to fairly common resident in moist, broadleafed woodland, riparian woodland, forest and thickets. (Blouvlekduifie)

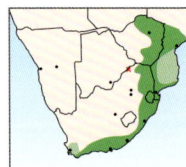

TAMBOURINE DOVE   *Turtur tympanistria*

b b b b b b b B B B B B
J F M A M J J A S O N D

20–22 cm; 60–85 g Darker above than other wood doves; at close range has some metallic green wing coverts, but these are hard to see against dark upperparts. Male has diagnostic white face and underparts. In flight, chestnut underwings contrast strongly with white belly; co-occurring Lemon Dove (p. 246) has dark brown underwings and cinnamon underparts. Female and juv. are greyer below, but still have paler faces and underparts than other wood doves. **Voice:** Series of 20–40 *'du-du-du'* notes, deeper than other wood doves, but not changing intensity or pitch towards the end; Lemon Dove can make a similar call. **Status and biology:** Locally common resident of forest and thickets. Usually flushed before it is seen; flies fast, straight and low. (Witborsduifie)

NAMAQUA DOVE   *Oena capensis*

b b b b b b b b b b b b
J F M A M J J A S O N D

28 cm; 30–45 g A very small dove with a long, pointed tail, rufous outer wings, 3–5 glossy purple wing coverts and 2 blackish bands on its back. Male has diagnostic black face and throat, and yellow-tipped, red bill. Female and juv. lack black face; tails slightly shorter; bills brown. Juv. barred buff and dark brown above. **Voice:** Deep, soft *'hoo huuuu'*, first note sharp, second longer; almost flufftail-like. **Status and biology:** Common resident and nomad in arid and semi-arid savanna, open woodland, Karoo and sparse desert grassland. (Namakwaduifie)

AFRICAN GREEN PIGEON

orientalis and other coastal subspp. are grey-green

damarensis and other inland subspp. are yellow-green

EMERALD-SPOTTED WOOD DOVE

dark bill

green wing spots

BLUE-SPOTTED WOOD DOVE

red & yellow bill

blue wing spots

TAMBOURINE DOVE

♂

♀

dusky breast, paler belly

♂

♂

white breast & belly

NAMAQUA DOVE

black mask

ad. ♂

long tail

juv. ♀

ad. ♀

PARROTS AND LOVEBIRDS

PARROTS AND LOVEBIRDS A familiar family of birds that are the closest living relatives to the passerines. All have stout, hooked bills and short legs with zygodactyl toes (two toes facing forward and two facing back) used for climbing and holding food. Sexes alike in most species. Among the most intelligent of birds, they are popular in the cage-bird trade, and escapees from numerous species are frequent, particularly in urban areas, including budgerigars *Nymphicus hollandicus*, cockatiels *Melopsittacus undulatus*, parakeets (*Psittacula* spp.), conures (*Aratinga*, *Nandayus* and *Cyanoliseus* spp.), parrots (*Amazona* spp.), parrotlets (*Forpus* spp.), extralimital lovebirds (*Agapornis* spp.) and even cockatoos (*Cacatua* spp.).

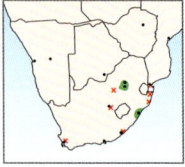

ROSE-RINGED PARAKEET *Psittacula krameri*

J F M A M J J A S O N D

37–43 cm; 115–140 g Introduced. A long-tailed, bright green parakeet with yellowish underwing coverts (contrasting with black flight feathers) and yellow vent. Confusion likely only with escaped parakeets of other species, such as Plum-headed Parakeet *P. cyanocephala*, which has bred in Pretoria, Gauteng, but failed to establish a feral population. Bill dark red (lacks yellow on bill of Plum-headed Parakeet). Ad. male has a distinctive black throat extending into a neck collar. Juv. has a shorter tail. **Voice:** Shrieks and screams. **Status and biology:** Feral populations occur locally in well-wooded parks and gardens in Gauteng (since the 1960s) and KZN (mainly around Durban). Roosts communally at night; particularly vocal when gathering to roost. (Ringnekparkiet)

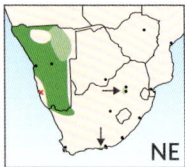

ROSY-FACED LOVEBIRD *Agapornis roseicollis*

J F M A M J J A S O N D

15–18 cm; 46–63 g A small, compact, green parrot with a pink-washed face and breast, and diagnostic blue rump. Flight direct and rapid; blue rump contrasting with green back. Juv. is paler and greyer on the face and upper breast. **Voice:** Short, high-pitched *'shreek'*, sometimes repeated. **Status and biology:** Near-endemic. Locally common resident and nomad in dry, broadleafed woodland, semi-desert and mountainous terrain. Feral populations now occur in a number of S African cities, particularly well established in Johannesburg/Pretoria area and Cape St Francis, E Cape. Often in flocks. Carries nesting material tucked under rump feathers. (Rooiwangparkiet)

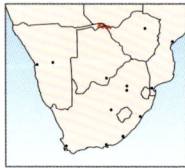

BLACK-CHEEKED LOVEBIRD *Agapornis nigrigenis*

J F M A M J J A S O N D

14–15 cm; 35–46 g Vagrant. Sometimes considered a subspecies of Lilian's Lovebird, but has a dark brown head contrasting with its white eye-ring and red bill. Juv. has a dark head, like ad., but bill more orange, sometimes grey towards base. **Voice:** Shrill chattering at rest; harsh *'shreek'* in alarm. **Status and biology:** VULNERABLE. Locally common in its small range in sw Zambia, but true extent of vagrancy into s Africa unknown; escapees sometimes seen around Victoria Falls and in n central Namibia. Favours tall mopane woodland, riparian woodland and adjacent fields. (Swartwangparkiet)

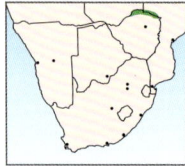

LILIAN'S LOVEBIRD *Agapornis lilianae*

J F M A M J J A S O N D

13–14 cm; 30–40 g The only lovebird in its range; reddish face, prominent, white eye-ring and green (not blue) rump are diagnostic. Female has a paler pink face. Juv. lacks a white eye-ring; cheeks and ear coverts washed blackish; base of upper mandible blackish. **Voice:** Gives shrill chattering at rest; harsh *'shriek'* in alarm. **Status and biology:** NEAR-THREATENED. Locally common resident in broadleafed woodland, especially tall mopane woodland. (Njassaparkiet)

ROSE-RINGED
PARAKEET

♂

♀

♂

♀

ROSY-FACED
LOVEBIRD

pale bill;
pink
face

ad.

blue rump

ad.

BLACK-CHEEKED
LOVEBIRD

juv.

red bill,
chocolate
face

ad.

green rump

ad.

LILIAN'S
LOVEBIRD

dark bill,
reddish face

juv.

ad.

green rump

ad.

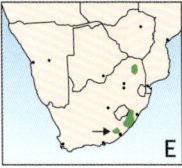

CAPE PARROT   *Poicephalus robustus*

B B b b b b b b b b b
J F M A M J J A S O N D

30–35 cm; 270–320 g A large, green parrot with orange-red wrists and leggings. Told from very similar Grey-headed Parrot by its olive-brown (not silvery-grey) head and deeper bill; ranges do not overlap. Female usually has an orange-red forehead. Juv. lacks orange-red shoulders and leggings, but is readily identified by its large size and massive bill. **Voice:** Various loud, harsh screeches and squawks. **Status and biology:** VULNERABLE. Endemic. Uncommon (500–1 500 birds) and decreasing resident in afromontane forest; commutes to orchards to feed. Usually in pairs or small family groups. Some authorities do not accept species-level taxonomic split from Grey-headed Parrot. (Woudpapegaai)

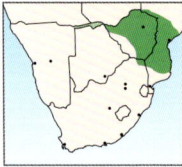

GREY-HEADED (BROWN-NECKED) PARROT
Poicephalus fuscicollis

b B B b b b b b b
J F M A M J J A S O N D

32–36 cm; 310–340 g Slightly larger than Cape Parrot, with a silvery-grey (not brown) head and less massive bill; ranges do not overlap. Larger than Brown-headed Parrot, with a much more massive, pale greyish-white (not grey-and-black) bill. Some birds have an orange-red frons; averages larger in female, with occasionally the entire crown orange-red. **Voice:** Various loud, harsh screeches and squawks, similar to Cape Parrot. **Status and biology:** Locally common resident and nomad in riverine forest and broadleafed woodland with large, emergent trees. Usually in pairs or family groups, but flocks gather at fruiting trees. Some authorities do not accept species-level taxonomic split from Cape Parrot. (Savannepapegaai)

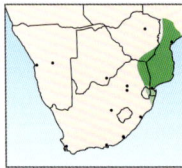

BROWN-HEADED PARROT   *Poicephalus cryptoxanthus*

b b b
J F M A M J J A S O N D

22–25 cm; 125–150 g A predominantly green parrot with a uniformly grey-brown head, diagnostic pale yellow eyes and pale lower mandible. Lacks yellow shoulders and crown of Meyer's Parrot; in flight, rump green (not greenish-blue) and all underwing coverts are yellow (inner coverts not brown); upperwings olive-brown (not brown). Juv. duller, with less vivid yellow underwings and dark eyes. **Voice:** Raucous shrieks. **Status and biology:** Locally common resident and nomad in savanna, riverine forest and open woodland. Hybridises with Meyer's Parrot in narrow overlap zone. (Bruinkoppapegaai)

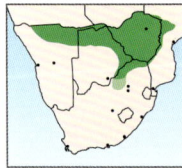

MEYER'S PARROT   *Poicephalus meyeri*

b B B b b b
J F M A M J J A S O N D

21–24 cm; 100–130 g A medium-sized, mainly brown parrot with conspicuous blue-green rump, yellow shoulders and green belly with bluish vent. Bill and eyes dark; some birds have a yellow bar across the crown. Distinguished from Brown-headed Parrot by yellow shoulders and crown, dark eyes and dark lower mandible. Distinguished from Rüppell's Parrot by brown (not grey) head, dark (not red) eyes and green or turquoise (not blue) rump and belly. Differs from both in flight by having brown (not yellow) inner underwing coverts. Juv. duller and lacks yellow on the crown. **Voice:** A loud, piercing *'chee-chee-chee-chee'*, and other screeches and squawks. **Status and biology:** Locally common resident and nomad in broadleafed woodland and savanna. Often in large flocks. Hybridises with Brown-headed Parrot in narrow overlap zone. (Bosveldpapegaai)

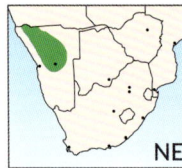

RÜPPELL'S PARROT   *Poicephalus rueppellii*

B B B b b b
J F M A M J J A S O N D

21–24 cm; 100–130 g A grey-brown parrot. Head and throat greyer than Meyer's Parrot, with a grey-brown (not greenish) breast and blue (not green or turquoise) belly; eyes red (not dark). In flight, has blue (not greenish-blue) rump. Female is brighter than male, with more extensive blue on the vent and rump. Juv. duller. **Voice:** Flight call a loud *'shreek'*; contact call a soft *'chip chippi'*. **Status and biology:** Near-endemic. Uncommon to locally common resident and nomad in dry woodland, savanna and dry rivercourses. In n Namibia, favours stands of baobabs and lala palms. (Bloupenspapegaai)

brown
head

♀

♂ **CAPE
PARROT**

♂

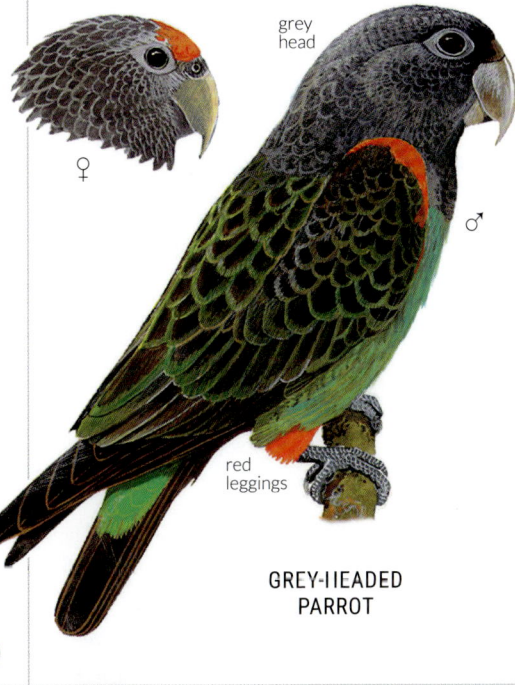

grey
head

♀

♂

red
leggings

**GREY-HEADED
PARROT**

juv.

**BROWN-HEADED
PARROT**

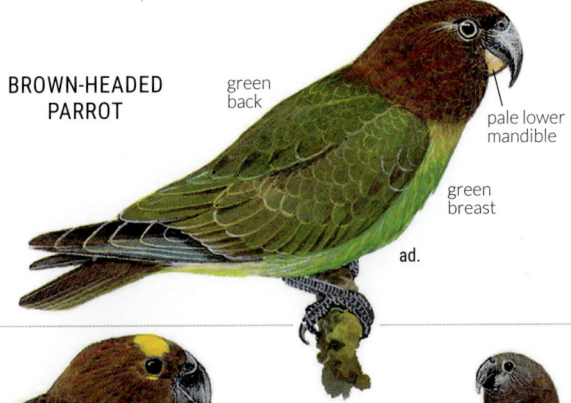

green
back

pale lower
mandible

green
breast

ad.

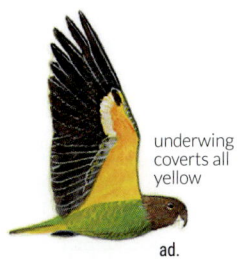

underwing
coverts all
yellow

ad.

brown
back

yellow
shoulder

green
breast

ad.

**MEYER'S
PARROT**

juv.

reduced
yellow on
underwing
coverts

ad.

juv.

grey-brown
back

brown
breast,
blue belly

ad. ♀

**RÜPPELL'S
PARROT**

underwing
coverts all
yellow

ad. ♀

grey head

255

TURACOS AND GO-AWAY-BIRDS (LOURIES)
Endemic to Africa. Fairly large birds with long tails, short, rounded wings and long legs with strong, semi-zygodactyl toes. They bound up through the canopy and glide between trees. Many species have bright red flight feathers. Sexes alike.

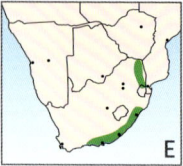

KNYSNA TURACO *Tauraco corythaix*
40–46 cm; 280–350 g A predominantly green turaco with a rounded (not pointed) crest with a narrow all-white fringe. Where range abuts that of Livingstone's Turaco in e Eswatini, occurs at higher elevations; also has a shorter, white-tipped (not black-and-white-tipped) crest. Body is much greener than Purple-crested Turaco, with green (not purple) crest and a white stripe below the eye. Juv. has shorter crest that lacks white tips; red in wing confined to outer primaries. **Voice:** Hoarse *'oop-oop-oop, korr-korr-korr-korr....'*, usually lasting 9–11 notes; quieter *'krrr'* alarm note. **Status and biology:** Endemic. Common resident in afromontane forest. (Knysnaloerie)

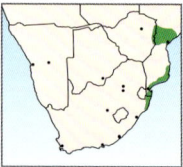

LIVINGSTONE'S TURACO *Tauraco livingstonii*
40–45 cm; 260–370 g A green turaco that occurs at lower elevations than Knysna Turaco where their ranges abut in e Eswatini; differs in having a longer, more pointed, black-and-white-tipped (not white-tipped) crest and slightly darker back. The crest is more rounded and less pointed in front than that of Schalow's Turaco, but their ranges do not overlap; also, tail is dark green (not bluish), and is darker on breast. **Voice:** Hooting calls followed by the typical hoarse *'korr-korr-korr-korr...'* notes, usually lasting 11–15 notes; quieter *'krrr'* alarm note. **Status and biology:** Common resident in dense, riparian woodland, extending into montane forest in e Zimbabwe. (Mosambiekloerie)

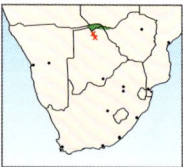

SCHALOW'S TURACO *Tauraco schalowi*
40–45 cm; 210–270 g A green turaco that closely resembles Livingstone's Turaco, but has a longer crest (especially at front of crest), dark blue or purple (not dark green) tail, and slightly paler green underparts; ranges do not overlap. Juv. has shorter crest. **Voice:** A series of 5–7 growling notes uttered at approximately 1-second intervals and becoming increasingly raucous. **Status and biology:** Uncommon, localised resident and nomad in dense riparian woodland along Zambezi River west of Victoria Falls, Zimbabwe, and into e Zambezi Region (Caprivi). (Langkuifloerie)

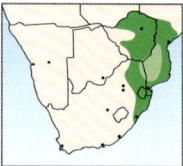

PURPLE-CRESTED TURACO *Tauraco porphyreolophus*
41–43 cm; 220–450 g A rather dark turaco, separated from 'green' turacos by its purple-blue crest (appears black in poor light), bluish wing coverts and tail, green breast washed rose-pink, and lack of white facial markings. Juv. duller, with less extensive red in wings. **Voice:** Loud series of hollow *'kok-kok-kok-kok'*, typically longer and faster than 'green' turacos. **Status and biology:** Common resident in broadleafed woodland and coastal and riverine forests. (Bloukuifloerie)

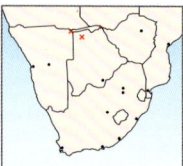

ROSS'S TURACO *Musophaga rossae*
50–52 cm; 300–450 g Vagrant. A dark purple-blue turaco, with crimson primaries, an erect, fez-like red crest and naked yellow face, bill and frontal shield. Juv. duller, with blackish bill. Violet Turaco *M. violacea* occurs as escapees in small numbers in gardens in n Johannesburg, Gauteng; Ross's Turaco distinguished by its all-yellow (not red-tipped) bill, yellow (not red) face and lack of a white cheek stripe. **Voice:** Regular, rather guttural *'ttrer-r-ttrer-rttre-r...'*, often given by several birds at once, producing almost continuous cacophony. **Status and biology:** Vagrant from further north in Africa; only a few records from riverine forest and dense woodland in ne Namibia and n Botswana. (Rooikuifloerie)

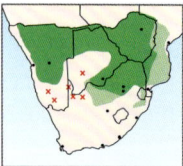

GREY GO-AWAY-BIRD *Corythaixoides concolor*
48–50 cm; 210–300 g An ash-grey turaco with a long tail and loose, pointed crest, similar to that of a mousebird. Juv. paler, more buffy-grey, with a shorter crest. **Voice:** Harsh, nasal *'waaaay'* or *'kay-waaaay'* (rendered 'go-away', hence its common name). **Status and biology:** Common resident of acacia savanna and dry, open woodland; also gardens. Vocal and conspicuous; small groups often perch on top of acacia trees. (Kwêvoël)

KNYSNA
TURACO

rounded
crest
tipped
white

untidy,
pointed crest
tipped black
& white

pointed green
crest tipped
white

LIVINGSTONE'S
TURACO

SCHALOW'S
TURACO

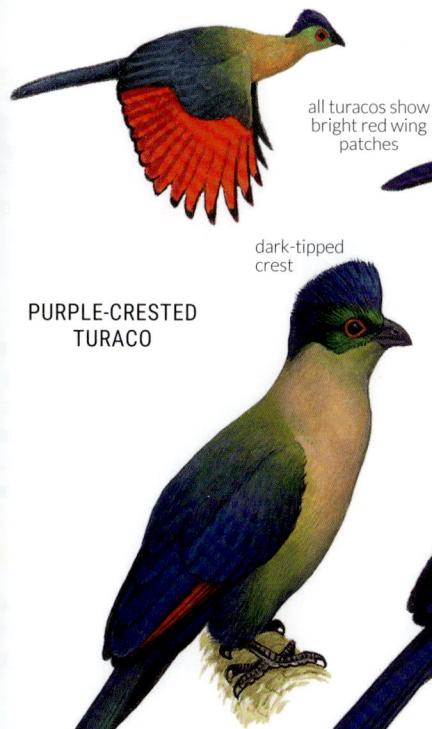

ROSS'S
TURACO

all turacos show
bright red wing
patches

PURPLE-CRESTED
TURACO

dark-tipped
crest

yellow
bill &
facial
skin

GREY GO-AWAY-BIRD

257

CUCKOOS

CUCKOOS Small to medium-sized, slender birds, often with fairly long tails. Flight swift and direct; larger species could be confused with falcons (p. 148) or hawks (p. 138). Sexes alike in some species; differ in others. Most African species are obligate brood parasites. Males call monotonously when br.; easily overlooked if not calling.

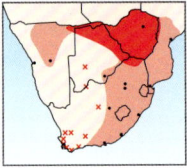

COMMON CUCKOO *Cuculus canorus*

32–34 cm; 105–140 g Difficult to separate from African Cuckoo; ad. typically has less yellow at base of bill; uppertail feathers dark throughout (dark in the uppertail is restricted to a distinct subterminal band in African Cuckoo). Undertail averages darker; mostly black with white feather tips and spots along inner webs of outer-tail feathers; undertail generally more broadly barred in African Cuckoo, with barring across all tail feathers. Belly barred, extending onto undertail coverts. Female of commoner grey morph is similar to male, but has the upper breast barred (not plain grey), often with buff wash. Rare rufous (hepatic) morph female is barred black and rufous above. Juv. may be brown, grey or chestnut; face generally darker than juv. African Cuckoo; upperparts usually barred or tipped rufous; underparts heavily barred; for separation from juv. African Cuckoo see below. **Voice:** Onomatopoeic *'cuck-oo'*, generally silent in Africa. **Status and biology:** Scarce to locally common Palearctic-br. summer migrant in woodland, savanna and riverine forest. (Europese Koekoek)

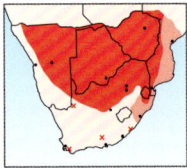

AFRICAN CUCKOO *Cuculus gularis*

30–32 cm; 95–112 g Difficult to separate from Common Cuckoo; ad. typically has a more extensive yellow base to its bill; dark in its uppertail is restricted to distinct subterminal band (uppertail all dark in Common Cuckoo); undertail is more distinctly and broadly barred; much darker undertail feathers in Common Cuckoo with white mostly restricted to outermost tail feathers. Call is most distinctive feature. In Oct/Nov African Cuckoo's plumage is fresh whereas Common Cuckoo's plumage is worn and in heavy moult; reverse true in Mar/Apr. Female is similar to male, but has faint barring on throat and buff wash on lower breast; lacks rufous morph of Common Cuckoo. Juv. has distinct white barring and white feather tips to upperparts, lacks rufous of juv. Common Cuckoo and generally has a plainer face pattern. **Voice:** Far-carrying, two-syllabeled *'coop-coop'*, recalling *'hoop-hoop'* of African Hoopoe, but slower; female gives fast *'kik-kik-kik'*. **Status and biology:** Locally common intra-African br. summer migrant in woodland and savanna. Brood host is Fork-tailed Drongo. (Afrikaanse Koekoek)

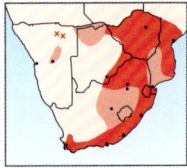

RED-CHESTED CUCKOO *Cuculus solitarius*

28–31 cm; 68–88 g Ad. has a reddish breast and a slate-grey back (darker than African and Common Cuckoos). Female has paler rufous breast, often barred grey. Best located by its characteristic 3-note call. Juv. has blackish head, breast and upperparts (not barred like juv. African and Common Cuckoos), with pale feather edges; lower breast and belly strongly barred black-and-white. **Voice:** Male calls monotonous *'wiet-weet-weeoo'* (rendered 'Piet my vrou' in Afrikaans); female gives shrill *'pipipipipi'*. **Status and biology:** Common intra-African br. summer migrant in forest, woodland and gardens. Brood hosts are mainly robin-chats, but also thrushes, flycatchers and wagtails. (Piet-my-vrou)

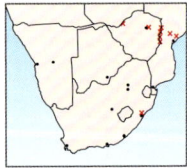

LESSER CUCKOO *Cuculus poliocephalus*

26 cm; 40–58 g Vagrant. A grey cuckoo, smaller than African and Common Cuckoos, with darker upperparts and more heavily barred underparts; in flight, dark tail and rump contrast with paler back. Very similar to Madagascan Cuckoo and probably not safely identifiable in the field, unless calling. In the hand the metacarpal feathering on the underwing can be examined to distinguish the two (see below). Upper breast often with rufous wash, not known in Madagascan Cuckoo. Like Common Cuckoo, there is also a rare rufous morph female (absent in Madagascan Cuckoo). Juv. is barred above and has black face barred with white. **Voice:** Staccato, 5–6 note *'chok chok chi chi chu-chu'*, higher pitched in middle. **Status and biology:** Rare Palearctic-br. summer migrant to savanna and riparian forest. (Kleinkoekoek)

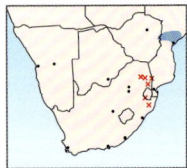

MADAGASCAN CUCKOO *Cuculus rochii*

28 cm; 50–65 g Vagrant. Very similar to Lesser Cuckoo and probably not safely identifiable in the field. Typically has plain or weakly barred undertail coverts (stronger and broader in Lesser Cuckoo) and lacks the rufous wash on upper breast (present in some Lesser Cuckoo); best told by call. In the hand the metacarpal feathering on the underwing can be examined; barred grey and white feathering in Madagascan Cuckoo and mostly white in Lesser Cuckoo. Plumage worn in late summer, when Lesser Cuckoo is in fresh plumage. Lacks a rufous morph, unlike Lesser Cuckoo. Juv. is mottled with rufous on upperparts. **Voice:** Similar to Red-chested Cuckoo, but deeper in tone, and typically has 4 (not 3) notes *'Piet-my-vrou-vrou'*, last note being lower pitched. **Status and biology:** Non-br. vagrant from Madagascar to riverine forest, woodland and dense savanna. Winter visitor to E Africa, but s African records in summer when calling (possibly overlooked in winter). (Madagaskarkoekoek)

rufous morph
(unknown in African Cuckoo)

♂

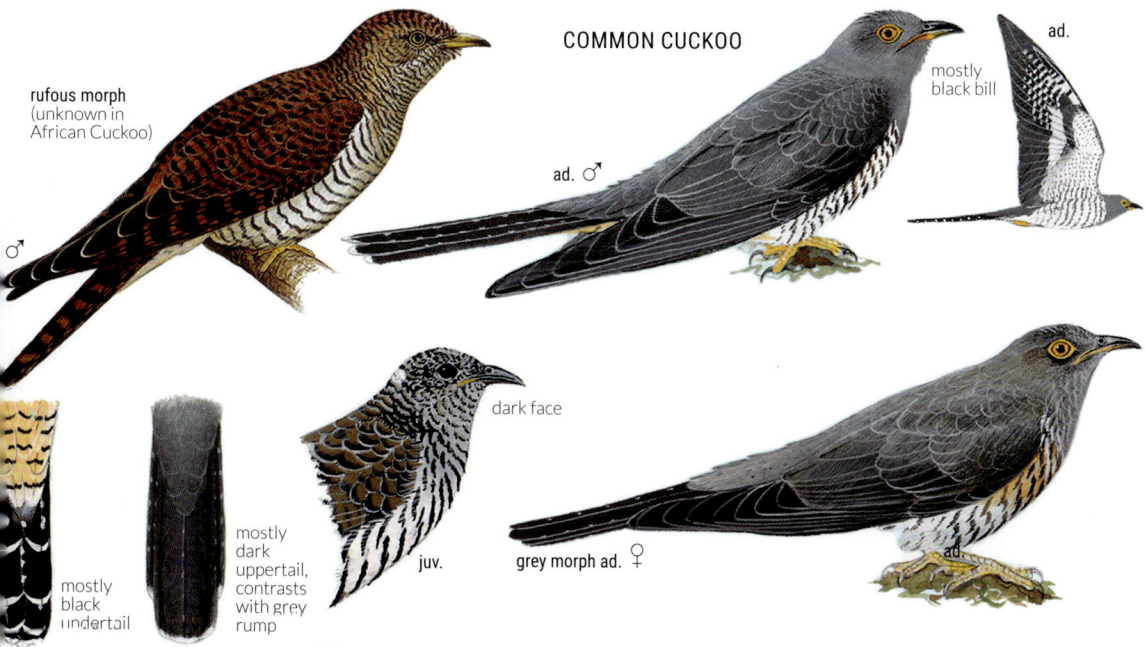

COMMON CUCKOO

ad.

mostly black bill

ad. ♂

mostly black undertail

mostly dark uppertail, contrasts with grey rump

dark face

juv.

grey morph ad. ♀

ad.

broadly barred undertail

pale face

juv.

grey uppertail with dark tip

AFRICAN CUCKOO

clear yellow base to bill

ad.

ad.

grey breast

chestnut breast

RED-CHESTED CUCKOO

black breast & upperparts

ad.

ad.

ad.

juv.

boldly barred undertail coverts

♀

MADAGASCAN CUCKOO

♂

weakly barred undertail coverts

LESSER CUCKOO

ad.

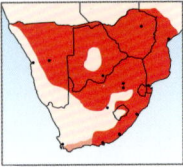

BLACK CUCKOO *Cuculus clamosus*

B | | | | | | | | | b b B B J F M A M J J A S O N D

28–31 cm; 80–100 g The only all-black cuckoo. Lacks crest and white upperwing patches of dark morph Jacobin and Levaillant's Cuckoos (but has white flash in outer underwing). At close range, has narrow, pale tips to tail feathers; female has indistinct paler bars on belly. Juv. duller black; more pointed tail lacks white tips. **Voice:** Male song is mournful *'hoo hooee'* or *'hoo hoo hooeee'* (rendered 'I'm so siiiiick'), with last note rising in pitch, repeated monotonously and sometimes ending in excited, rattling *'whurri whurri whurri'*; female gives fast *'yow-yow-yow-yow'*. **Status and biology:** Common intra-African br. summer migrant in woodland, forest and gardens. Brood hosts are mainly Crimson-breasted Shrike and other boubous. (Swartkoekoek)

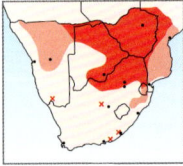

LEVAILLANT'S CUCKOO *Clamator levaillantii*

b B B B B | | | | b B B J F M A M J J A S O N D

38–40 cm; 105–140 g A large, black-and-white cuckoo with a long crest. Larger than Jacobin Cuckoo, with a distinctive striped throat and breast and a larger, shaggier crest. Rare black morph recorded from coastal e Africa has white wing patches and usually has white tail tips; birds with black tails hard to separate from black Jacobin Cuckoo, but this morph not recorded in s Africa. Sexes alike. Juv. is browner above and buff below, but still shows diagnostic throat stripes; best told from juv. Jacobin Cuckoo by its black (not yellow) bill. **Voice:** Loud *'klee-klee-kleeuu'*, followed by descending *'che-che-che-che'*, slowing towards end, this extension of the call distinguishing it from the call of Jacobin Cuckoo. **Status and biology:** Locally common intra-African br. summer migrant in savanna and woodland. Brood hosts are Arrow-marked and other babblers. (Gestreepte Nuwejaarsvoël)

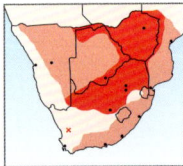

JACOBIN CUCKOO *Clamator jacobinus*

b b b | | | | | | b B B J F M A M J J A S O N D

34 cm; 70–92 g A medium-sized, black-and-white, crested cuckoo. Smaller than Levaillant's Cuckoo, with no stripes on throat and breast. Dark morph birds (mostly coastal) are wholly black, except for white patch at base of primaries; smaller than very rare dark morph Levaillant's Cuckoo, with no white in the tail. Separated from Black Cuckoo by crest, long graduated tail and white on upper primaries. Sexes alike. Juv. is browner above, with creamy-grey underparts; dark morph has dull black underparts. Best told from juv. Levaillant's Cuckoo by its yellow bill and orange gape. **Voice:** Shrill, repeated *'klee-klee-kleeuu-kleeuu'*, similar to start of call of Levaillant's Cuckoo. **Status and biology:** Common intra-African br. summer migrant (*C. j. serratus*) and non-br. migrant from n Africa and s Asia (*C. j. jacobinus* and *C. j. pica*). Occurs in woodland, thicket and acacia savanna. Brood hosts are mainly bulbuls and Southern Fiscal. (Bontnuwejaarsvoël)

THICK-BILLED CUCKOO *Pachycoccyx audeberti*

b | | | | | | | b B B B J F M A M J J A S O N D

36 cm; 92–120 g A large, rather hawk-like cuckoo with a heavy bill. Combination of whitish underparts and lack of a crest is diagnostic. Ad. is dark grey above, with broadly barred undertail. Sexes alike. Juv. has striking white head flecked with black, and white-spotted upperparts; best distinguished in flight from ad. Great Spotted Cuckoo by white (not buffy) throat and mostly white (not grey) cap. Same features, coupled with dark grey (not russet) primaries, distinguish it from juv. Great Spotted Cuckoo. **Voice:** Repeated, ringing *'weee we-wick'*, like a fast, harsh Klaas's Cuckoo (p. 262). **Status and biology:** Uncommon resident and intra-African br. summer migrant in riparian forest and dense woodland. Brood host is Retz's Helmetshrike. (Dikbekkoekoek)

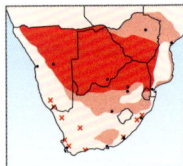

GREAT SPOTTED CUCKOO *Clamator glandarius*

b b b | | | | | | b B B J F M A M J J A S O N D

38–42 cm; 120–165 g A large, distinctively patterned cuckoo with a long, wedge-shaped tail and grey crest. Sexes alike. In flight, best distinguished from juv. Thick-billed Cuckoo by buffy (not white) throat and grey (not whitish) cap. Juv. is also heavily spotted on back, but has small, black crest, buffy underparts and rufous primaries. Latter feature, together with buffy (not white) throat and dark (not whitish) cap, separates it from juv Thick-billed Cuckoo. **Voice:** Loud, far-carrying *'keeow-keeow-keeow'* and shorter, crow-like *'kark'*. **Status and biology:** Common intra-African summer migrant in woodland and savanna, including locally br. *C. g. choragium* and non-br. nominate race from African savannas north of equator. Brood hosts are crows and starlings. (Gevlekte Koekoek)

BLACK
CUCKOO

no crest

ad. ♂

ad. ♀

ad. ♂

ad.

no upperwing
flash

LEVAILLANT'S
CUCKOO

ad.

streaked
throat

juv.

upperwing

ad.

upperwing

ad.

underwing

JACOBIN
CUCKOO

plain throat

underwing

ad. dark
morph

pale
morph

juv.

THICK-BILLED
CUCKOO

juv.

ad.

ad.

juv.

juv.

GREAT SPOTTED
CUCKOO

ad.

spotted
upperparts

ad.

ad.

juv.

rufous
primaries

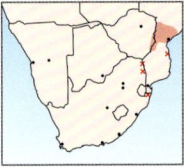

BARRED LONG-TAILED CUCKOO *Cercococcyx montanus*

b | J F M A M J J A S O N D | b b

33 cm; 55–70 g A small, dark cuckoo with a very long tail, brownish, heavily barred upperparts and broadly barred underparts. Sexes alike. Juv. has underparts more dusky, with some streaking. **Voice:** Long series of *'cheee-phweew'* phrases, increasing in intensity, followed by ringing *'whit whew hew hew'*, recalling Red-chested Cuckoo (p. 258) but with 4 or 5 notes; shorter *'hwee-hooa'* or *'hwee-hooo'*; often calls at night. **Status and biology:** Uncommon intra-African br. summer migrant in lowland forest, riparian thicket and mature miombo woodland; decreasing in numbers due to deforestation. Hard to locate even if calling; remains motionless in dense canopy. Brood hosts unknown. (Langstertkoekoek)

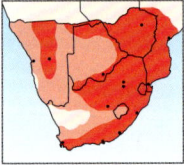

DIEDERIK CUCKOO *Chrysococcyx caprius*

B b b b | J F M A M J J A S O N D | b B B

17–20 cm; 22–42 g A small, glossy green cuckoo with diagnostic white wing spots, red eyes and broadly barred green flanks. Ad. has green malar stripe and white supercilium, which extends in front of reddish eye. Female is duller, with barring extending onto breast; throat is often buffy. Female differs from female Klaas's Cuckoo in having bolder but less extensive barring on flanks, and white spots on forewings. Juv. has conspicuous red bill; occurs in green and rufous morphs. **Voice:** Male's call, uttered from a perch or in flight, is a clear, persistent *'dee-dee-deederick'*; female responds with a plaintive *'deea deea deea'*. **Status and biology:** Common intra-African br. summer migrant in woodland, savanna, grassland and suburban gardens. Brood hosts are mainly weavers, bishops and sparrows. (Diederikkie)

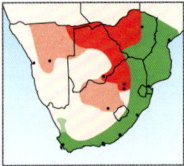

KLAAS'S CUCKOO *Chrysococcyx klaas*

b b b b | J F M A M J J A S O N D | b B B B

16–18 cm; 24–34 g A plain, glossy green cuckoo with a small, white stripe behind the eye. Lacks white wing spots of Diederik Cuckoo, but flight feathers boldly barred. Male is white below, with green spurs extending onto sides of breast, and only a few green bars on thighs. Female is bronzy-brown above and finely barred below; lacks white upperwing markings of female Diederik Cuckoo. Juv. is barred bronze and green above, similar to female African Emerald Cuckoo, but has diagnostic white stripe behind eye. **Voice:** Male's call, uttered from a perch, not in flight, is a far-carrying *'may-i-kie may-i-kie'* (hence the Afrikaans name), repeated 3–6 times. **Status and biology:** Common resident and intra-African br. migrant in forest, woodland, savanna and gardens. Brood hosts are batises, sunbirds, warblers and flycatchers. (Meitjie)

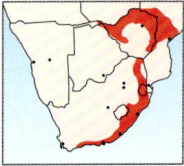

AFRICAN EMERALD CUCKOO *Chrysococcyx cupreus*

b | J F M A M J J A S O N D | b B B

18–20 cm; 33–41 g A forest cuckoo; larger than other glossy green species. Male has sulphur-yellow lower breast and belly contrasting with brilliant emerald-green upper breast, throat and upperparts; vent is barred green and white. Female and juv. are finely barred green and brown above, and green and white below; lacks white eye stripe of juv. Klaas's Cuckoo. **Voice:** Loud, ringing whistle, *'wit-huu, orr-weee'* (rendered 'pretty Georg-eee'). **Status and biology:** Common resident and intra-African br. summer migrant in evergreen forest. Brood host is mainly Green-backed Camaroptera. (Mooimeisie)

MALKOHAS Fairly large, cuckoo-like birds related to coucals (p. 264), with similar long, broad tails and short, rounded wings. Most species found in Asia, but 2 allospecies occur in Africa. Like the coucals, they raise their own chicks. Sexes alike.

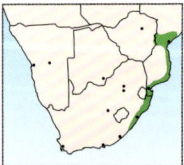

GREEN MALKOHA *Ceuthmochares australis*

J F M A M J J A S O N D | b B b

33 cm; 55–75 g A slender, coucal-like bird with a long, green tail (that appears black in poor light), large, yellow bill, dull green upperparts and grey underparts. Juv. more grey-green above with a duller, greenish bill. **Voice:** A clicking *'kik-kik-kik'*, winding up to a loud *'cher-cher-cher-cher'*; sequence ends with rapid, clicking notes. **Status and biology:** Scarce to locally common resident of forest edge, riparian woodland and thicket, often in creepers. Shy; easily overlooked if not calling. (Groenvleiloerie)

ng,
rrow
l

BARRED LONG-TAILED CUCKOO

white patch
behind & in
front of eye

white
wing
spots
(all ages)

**DIEDERIK
CUCKOO**

red
eye-ring

♂

barred
flanks

♀

juv.

juv.

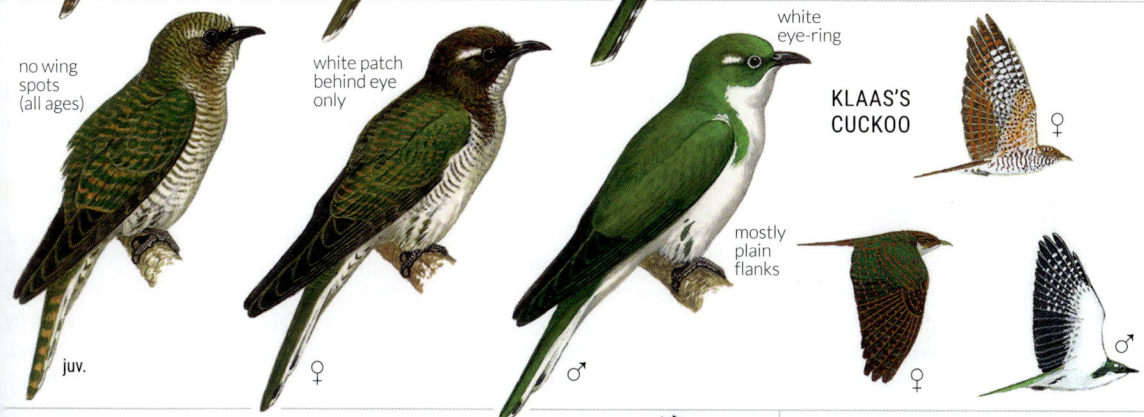

no wing
spots
(all ages)

white patch
behind eye
only

white
eye-ring

**KLAAS'S
CUCKOO**

♀

mostly
plain
flanks

♀

♂

♂

juv.

♀

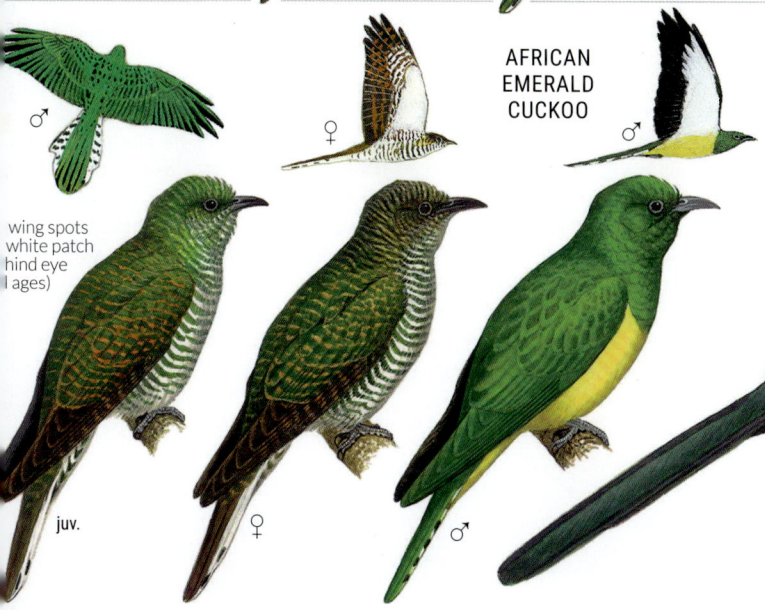

**AFRICAN
EMERALD
CUCKOO**

♀

♂

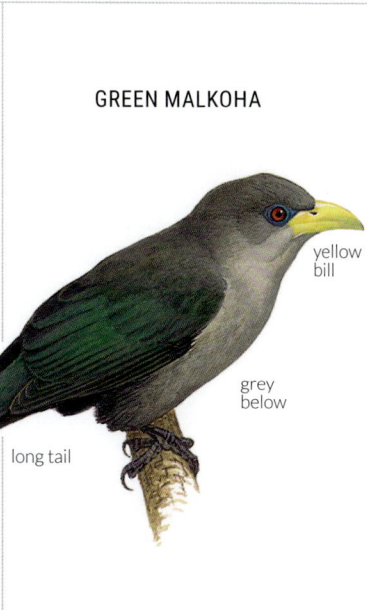

GREEN MALKOHA

yellow
bill

wing spots
white patch
hind eye
l ages)

grey
below

long tail

juv.

♀

♂

263

COUCALS

Fairly large, heavy-set, cuckoo-like birds with long, broad tails and short, rounded wings. Stout, arched bill adapted for tearing flesh. Legs long and robust. Flight cumbersome. Clambers among dense vegetation, but often perches in the open in the early morning or after rain. Sexes alike, but females larger; juvs duller and often heavily barred.

COPPERY-TAILED COUCAL *Centropus cupreicaudus*

b b b J F M A M J J A S O N D

46–52 cm; 240–340 g The largest coucal; glossy blue-black crown extends further onto the mantle than in Senegal Coucal; tail is long and heavy with a violet gloss. Rump finely barred pale brown (not plain brown). In flight, has a distinct dark brown trailing edge to the wing. Juv. duller; barred above and with white streaks on buffy underparts. **Voice:** A deep, loud, resonant series of '*doo*' notes. **Status and biology:** Locally common resident in marshes, thick reed beds and adjoining bush; favours papyrus. (Grootvleiloerie)

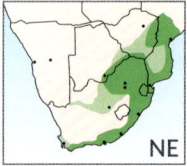

BURCHELL'S COUCAL *Centropus burchelli*

b b b   b B B B J F M A M J J A S O N D

41 cm; 160–210 g Often considered a subspecies of White-browed Coucal, but ad. has a black (not streaked) cap and mantle. Slightly larger than Senegal Coucal (ranges overlap marginally); rump and base of tail with fine, rufous barring (not plain as in Senegal Coucal). Juv. has a dark brown cap with a buffy supercilium and barred mantle; white-streaked, buffy underparts and barred wings separate it from ad. White-browed Coucal, but juvs probably inseparable. **Voice:** Similar to White-browed Coucal; a liquid, bubbling '*doo-doo-doo-doo*', descending in scale then rising towards end. **Status and biology:** Near-endemic. Common resident in riverine scrub, reed beds, thicket and well-wooded gardens. (Gewone Vleiloerie)

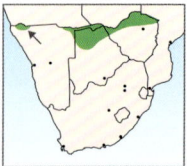

WHITE-BROWED COUCAL *Centropus superciliosus*

b b b   b B B B J F M A M J J A S O N D

40 cm; 145–210 g A fairly large coucal with a diagnostic white supercilium and heavily pale-streaked nape and mantle. Ad. differs from juv. Burchell's Coucal by having plain (not barred) wings and generally is whiter below. Juv. is buffier than ad., with less prominent buff supercilium and barred wings; probably inseparable from juv. Burchell's Coucal. **Voice:** A liquid, bubbling '*doo-doo-doo-doo...*', falling in pitch, then slowing and rising in pitch at the end. **Status and biology:** Common resident in rank grass, riverine scrub, reed beds, thicket and well-wooded gardens. (Gestreepte Vleiloerie)

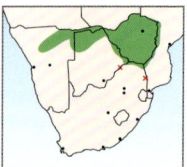

SENEGAL COUCAL *Centropus senegalensis*

b b b   b B B J F M A M J J A S O N D

38–40 cm; 145–180 g Slightly smaller than Burchell's Coucal, with a plain (not barred) rump and uppertail. Much smaller than Coppery-tailed Coucal, with rich chestnut (not black) mantle; lacks a bluish sheen to head. Rare rufous morph, recorded from Okavango Delta, Botswana, has a black head and throat and rufous underparts. Juv. is buffy, heavily barred above (not streaked on head and mantle like Black Coucal). **Voice:** Similar to that of White-browed Coucal; string of bubbling notes, initially falling then rising in pitch. **Status and biology:** Uncommon to locally common, resident or partial migrant found in tangled vegetation and long grass; less tied to water than most other coucals. (Senegalvleiloerie)

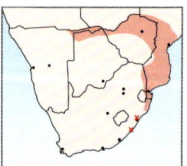

BLACK COUCAL *Centropus grillii*

B B b   b B J F M A M J J A S O N D

35–38 cm; 95–150 g A small, dark coucal. Br. ad. has entirely black head, mantle and underparts, contrasting with plain chestnut wings. Black tail conspicuously barred pale brown. Non-br. ads and juv. have crown and mantle streaked tawny and brown, but lack a clear supercilium. Upperparts, including flight feathers and coverts, mostly barred dark brown and rufous. Dark brown tail has fine, rufous bars. Underparts pale buff and grey, flanks barred black. Darker than juv. Senegal Coucal, with richer chestnut wings and streaked (not barred) head and mantle. **Voice:** Female calls to male with an explosive, repeated '*pop-op*' double note; male responds with '*too-loo-loo*'; also a bubbling call, like White-browed Coucal but faster and higher pitched. **Status and biology:** Uncommon to locally common intra-African br. summer migrant and resident in moist grassland; occurs in drier riverine scrub and rank grass in winter in Zimbabwe. Annual abundance influenced by rainfall. (Swartvleiloerie)

coppery sheen

coppery sheen

COPPERY-TAILED COUCAL

juv.

dull red eye

ad.

barred rump & base to tail

BURCHELL'S COUCAL

white supercilium

streaks

WHITE-BROWED COUCAL

rufous morph

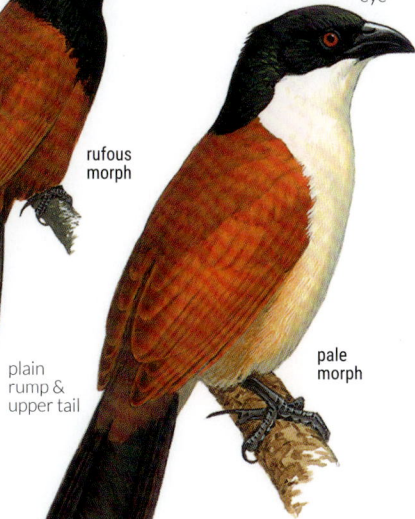

bright red eye

plain rump & upper tail

pale morph

SENEGAL COUCAL

heavily barred wings

non-br.

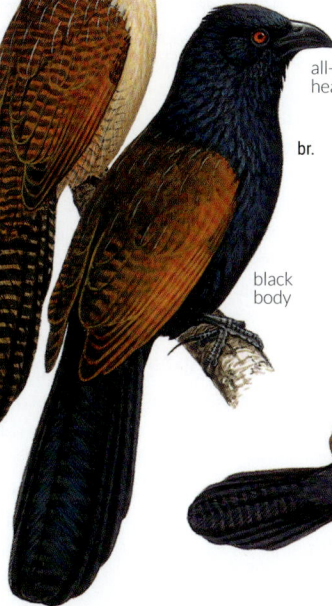

BLACK COUCAL

all-black head

br.

black body

br.

265

OWLS AND BARN OWLS
Small to large predatory birds; mostly nocturnal; remaining concealed during the day, but vocal activity reveals presence at night. Most species have a well-defined facial disc and soft-edged flight feathers that allow near-silent flight. Sexes alike; females typically larger.

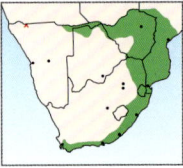

AFRICAN WOOD OWL *Strix woodfordii*
J F M A M J J A S O N D

30–36 cm; 240–350 g A medium-sized, chocolate-brown owl with a rounded head lacking ear tufts, large, dark brown eyes and yellow bill. Facial disc is finely barred and paler than white-spotted brown head and upper breast; belly is barred brown and white. Plumage varies from very dark brown to russet, with dark grey morphs also occurring. **Voice:** Hooting *'hu hu, hu whoo-oo'*; also single hoots and a higher-pitched *'who-uuu'*. Regularly duets; female, with higher-pitched voice, calls first. **Status and biology:** Common resident of forest, mature woodland and alien plantations. (Bosuil)

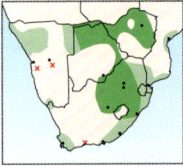

MARSH OWL *Asio capensis*
J F M A M J J A S O N D

35–38 cm; 230–370 g A plain brown, medium-sized owl with a buff-coloured or greyish face, small 'ear' tufts and dark brown eyes. In flight, wings broader than African Grass Owl (p. 268), with much larger buff 'windows' in primaries, less contrast between coverts and flight feathers, and pale trailing edge to secondaries and outer primaries. Underwing has dark carpal marks; tail barred. **Voice:** Harsh, rasping *'krikkk-krikkk'*, likened to sound of material being torn. **Status and biology:** Common resident and nomad in marshes and damp grassland; avoids dense reed beds. Often circles overhead after flushing. Less nocturnal than most other owls; sometimes roosts in flocks. (Vlei-uil)

SOUTHERN WHITE-FACED OWL *Ptilopsis granti*
J F M A M J J A S O N D

25–28 cm; 180–250 g A fairly small, grey-backed owl. Considerably larger than African Scops Owl, with bright orange (not yellow) eyes and prominent white, black-edged facial disc. Juv. browner, less heavily marked and with paler, yellow-orange eyes. **Voice:** A fast, hooting *'doo-doo-doo-doo-hohoo'*, with emphasis on the last note. **Status and biology:** Common resident and nomad in acacia savanna and dry, broadleafed woodland. (Witwanguil)

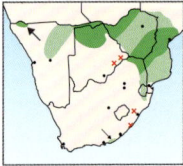

AFRICAN BARRED OWLET *Glaucidium capense*
J F M A M J J A S O N D

20–22 cm; 100–138 g A small, brown-backed owl. Slightly larger than Pearl-spotted Owlet, with faintly barred upperparts (lacking white spots), barred breast and spotted (not streaked) belly; appears larger headed and shorter tailed. Large, white tips to scapulars form conspicuous line of spots on back. Tail buff, with darker brown bars. Juv. is less distinctly spotted below. Birds in the isolated E Cape population are larger, darker backed, with spotted (not barred) heads and more finely barred tails; whether they are specifically distinct awaits DNA evidence. **Voice:** Series of 6–10 notes, starting softly and increasing in volume, *'kerrr-kerrr-kerrr-kerrr'*, often followed by series of purring whistles, *'trru-trrre, trru-trrre'*, second note higher pitched. **Status and biology:** Locally common resident in mature woodland, thicket and forest edge. (Gebande Uil)

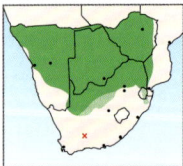

PEARL-SPOTTED OWLET *Glaucidium perlatum*
J F M A M J J A S O N D

17–21 cm; 60–100 g A small, long-tailed owl. Smaller than African Barred Owlet, with small, white spots (not bars) on upperparts and streaked (not barred) breast and flanks; tail dark brown with narrow, white bars. Has 2 black 'false eyes' on nape. Told from African Scops Owl by its rounded head, lack of ear tufts and white spotting on back and tail. Juv. is less spotted on crown and mantle; tail shorter. **Voice:** Series of *'tu-tu-tu-tu-tu-tu-tu tseeu-tseeu-tseeeu-tseeeu'* whistles, rising and then descending in pitch with a brief pause between; regularly calls during the day. **Status and biology:** Common resident in acacia savanna and woodland. Often active by day, when frequently mobbed by small birds. (Witkoluil)

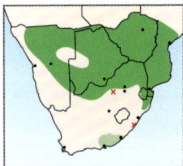

AFRICAN SCOPS OWL *Otus senegalensis*
J F M A M J J A S O N D

14–18 cm; 60–80 g The smallest owl in the region. Most individuals are greyish, but some are more rufous. Smaller than Southern White-faced Owl, with grey (not white) face and yellow (not orange) eyes. Has prominent ear tufts, but these can lie flat on head. **Voice:** Soft, frog-like *'prrrup'*, repeated every 5–8 seconds; often calls for prolonged periods. **Status and biology:** Common resident of savanna and dry, open woodland; avoids forests. Typically roosts by day against a tree trunk; eyes closed to narrow slits and plumage render it nearly invisible against the background. (Skopsuil)

dark 'spectacles', no 'ears'

AFRICAN WOOD OWL

buffy-grey facial disc

plain brown above

dark below

MARSH OWL

SOUTHERN WHITE-FACED OWL

white face; orange eyes

barred breast, head & face

blotched flanks

AFRICAN BARRED OWLET

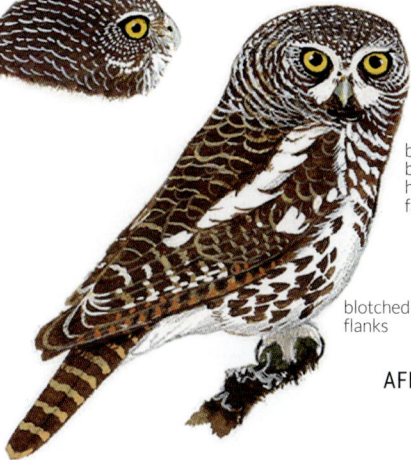

spotted crown

false eyes

grey morph

rufous morph

grey face, yellow eyes

PEARL-SPOTTED OWLET

streaked front

bark-patterned body

AFRICAN SCOPS OWL

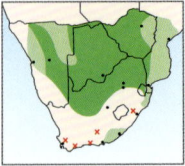

VERREAUX'S (GIANT) EAGLE-OWL *Bubo lacteus*

b B B B b
J F M A M J J A S O N D

58–66 cm; 1.2–3 kg The largest owl in the region; easily identified by its rather short ear tufts and finely vermiculated, pale grey plumage. At close range, pink eyelids and dark brown eyes are distinctive. Other eagle-owls are smaller, with more boldly barred or blotched plumage and yellow or orange eyes. Juv. has grey-barred plumage, lacking white tips to greater coverts; black border to facial disc reduced. **Voice:** A far-carrying, grunting *'unnh-unnh, unnh-unnh'*, pair often call in tandem, female lower pitched; soliciting female and juv. make a shrill wail. **Status and biology:** Uncommon to locally common resident in broadleafed woodland, savanna, thornveld and riverine forest; also desert riverbeds with large trees. (Reuse-ooruil)

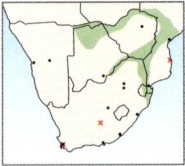

PEL'S FISHING OWL *Scotopelia peli*

b B B B b b
J F M A M J J A S O N D

60–64 cm; 2–2.4 kg A very large, ginger-coloured owl associated with rivers and other waterbodies. Breast spotted or lightly barred; eyes and bill black. Pale, unfeathered legs and feet are difficult to see in the field. Flight rather noisy; lacks soft fringes to flight feathers. When alarmed, it fluffs up its head feathers, giving it a huge, round-headed appearance. Juv. is paler and more buffy coloured, with an almost white head; fledglings still retaining substantial down. **Voice:** Deep, booming *'hoo-huuuum'*, sometimes preceded by series of grunts; juv. gives a jackal-like wailing. **Status and biology:** Uncommon resident that roosts in large trees around lakes and slow-moving rivers. Numbers decreasing due to disturbance and water abstraction from large rivers. (Visuil)

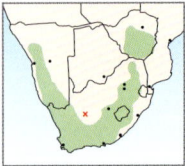

CAPE EAGLE-OWL *Bubo capensis*

b B B b b b
J F M A M J J A S O N D

48–54 cm; 910–1 400 g A large, heavily marked eagle-owl. Easily confused with smaller Spotted Eagle-Owl; has black and chestnut blotching on breast, and bold (not fine) barring on belly and flanks. Feet are much larger, and has orange (not yellow) eyes (but rare rufous morph of Spotted Eagle-Owl also has orange eyes). Upperparts generally richer and more strongly marked than Spotted Eagle-Owl. Juv. paler, with ear tufts absent or shorter; eyes pale orange. *B. c. mackinderi* in Zimbabwe is appreciably larger; darker above and more heavily marked below than nominate race. **Voice:** Deep, far-carrying *'hooooo'* calls, sometimes 2 or 3 notes; or *'hu-hooooo hu'*, first syllable sharp and penetrating; dog-like bark, *'wak-wak-wak'*; seldom duets. **Status and biology:** Generally uncommon resident of rocky and mountainous terrain from sea level to >3 000 metres. (Kaapse Ooruil)

SPOTTED EAGLE-OWL *Bubo africanus*

b b b b b b b B B B b
J F M A M J J A S O N D

43–50 cm; 500–1 100 g The most common large owl. Most birds are greyish; told from Cape Eagle-Owl by smaller size (especially smaller feet), lack of dark breast patches, finely (not boldly) barred belly and flanks, and yellow (not orange) eyes. Rare rufous morph is more heavily blotched and has orange-yellow eyes; best distinguished from Cape Eagle-Owl by its smaller feet and its call. Juv. browner, with shorter ear tufts. Large fledglings still retain fluffy, white down around head. **Voice:** Male gives deep *'hoo-huuu'*, often followed by female's *'huu-ho-huuu'*, with second note higher pitched; softer and less penetrating than Cape Eagle-Owl. Hisses and clicks when alarmed. **Status and biology:** Common resident in all habitats except forest; often in gardens. (Gevlekte Ooruil)

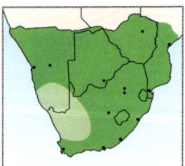

WESTERN BARN OWL *Tyto alba*

b B B B B b b b b b b b
J F M A M J J A S O N D

30–35 cm; 270–500 g A medium-sized, pale owl with golden buff and grey upperparts, a white, heart-shaped facial disc (emphasising small, black eyes) and off-white underparts. Much paler above than African Grass Owl, with less contrast between upperparts and underparts. In flight, has distinctive large head and short tail. Juv. has slightly darker upperparts. **Voice:** Typical call is high-pitched, eerie, screeching *'shreeee'*; also hisses and bill-clicks when disturbed. **Status and biology:** Common resident in most open habitats; avoids dense forest. Roosts in old buildings, caves, hollow trees and mine shafts. (Nonnetjie-uil)

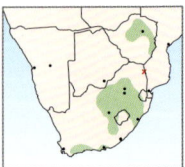

AFRICAN GRASS OWL *Tyto capensis*

B B B B B b b b b
J F M A M J J A S O N D

34–37 cm; 355–520 g Slightly larger than Western Barn Owl, with much darker brown upperparts that contrast strongly with pale buff underparts; also differs in habitat. Paler below than Marsh Owl (p. 266), with pale (not dark) face; lacks ear tufts. In flight has longer, more slender wings, with darker coverts contrasting with paler flight feathers and only small, buffy bases to primaries; outer tail white. Juv. has rufous facial disc and darker underparts; upperparts lack white spotting. **Voice:** Soft, cricket-like *'tk-tk-tk-tk'* given in flight; also screeching *'shree'*, shorter than similar call of Western Barn Owl. **Status and biology:** Uncommon resident and nomad at marshes and tall grassland; roosts on the ground. (Grasuil)

VERREAUX'S
EAGLE-OWL

pink
eyelids

overall
grey

PEL'S
FISHING
OWL

plain face;
dark eyes

overall
rufous

CAPE
EAGLE-OWL

orange
eyes

blotched
underparts

large
talons

SPOTTED
EAGLE-OWL

yellow
eyes
(rarely
orange)

barred
underparts

WESTERN BARN
OWL

pale back
& wings

AFRICAN GRASS
OWL

white
outer tail

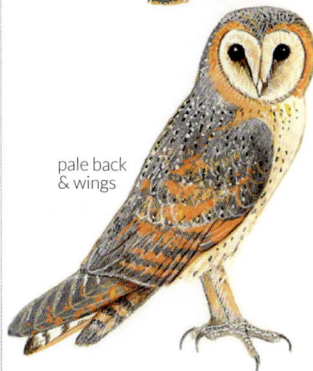

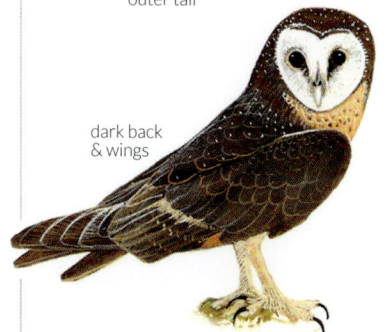

dark back
& wings

NIGHTJARS
NIGHTJARS Nocturnal insectivores that roost on the ground or along branches during the day; most prey taken in the air. Identification based largely on wing and tail patterns, and voice; separating roosting birds is often tricky if they are not calling. Sexes differ.

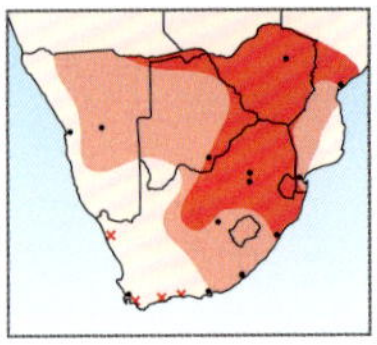

EUROPEAN NIGHTJAR *Caprimulgus europaeus*

J F M A M J J A S O N D

26–28 cm; 55–80 g A large, long-winged, long-tailed nightjar. At rest, appears rather pale grey, similar to Rufous-cheeked Nightjar, but lacks a rufous nuchal collar. Plumage more patterned than Freckled Nightjar (p. 274), with a buff bar on the folded wing and a pale moustachial streak. Overall coloration variable from buff-grey to silver-grey, depending on its origins in the Palearctic. In flight, male has white tips to outer tail and white outer primary window. Female has tail and wings all dark; appears greyer than female Pennant-winged Nightjar (p. 274); tail more rounded. **Voice:** Churring song, similar to Rufous-cheeked Nightjar, but rarely sings in Africa; occasionally utters a nasal grunt on flushing. **Status and biology:** Locally common Palearctic-br. summer migrant in woodland, savanna and plantations. Usually roosts lengthwise on branches, less often on the ground. (Europese Naguil)

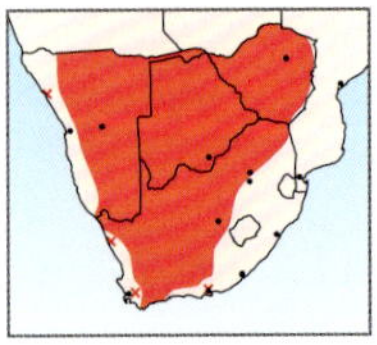

RUFOUS-CHEEKED NIGHTJAR *Caprimulgus rufigena*

J F M A M J J A S O N D

22–24 cm; 45–65 g A fairly small, variegated nightjar; paler grey than Fiery-necked Nightjar (p. 272), with narrower, orange-buff (not rufous) collar. Male has more prominent white patches in primaries but less white in tail than Fiery-necked Nightjar; female has small, buff tail tips and buff primary patches. Smaller than European Nightjar, which lacks a rufous collar. **Voice:** Prolonged churring, usually preceded by choking *'chukoo, chukoo'*. Unlike Square-tailed Nightjar (p. 272), does not change pitch while calling. **Status and biology:** Locally common intra-African br. summer migrant in arid savanna, woodland and desert scrub. Roosts on the ground, usually in areas shaded by vegetation. (Rooiwangnaguil)

NIGHTJAR IDENTIFICATION

Identification of nightjars is rarely straightforward and is often further confounded by brief views in bad light. Even with in-hand examination, identification can be tricky. In-hand features to concentrate on include patterning of the second outermost primary feather (P9) and the outer-tail feathers. The size of the pale wing spot/s on the emargination of P9 is usually distinct enough to help eliminate other species, however, if not, further examination of the patterning of the outer-tail feathers will exclude other species.

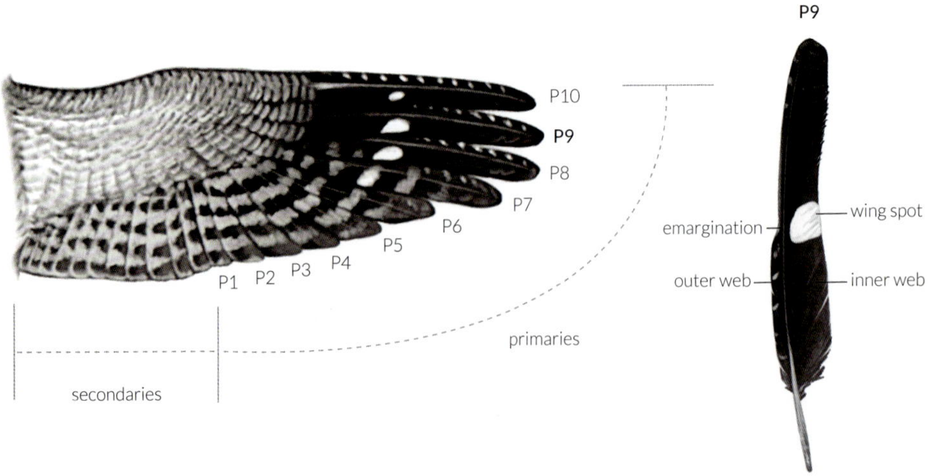

♂ ♀

P9 P9

EUROPEAN NIGHTJAR

grey crown; thinly streaked

often roosts lengthwise on horizontal branches

prominent white bar under eye

unwini

small white wing spots

♂

long primary projection

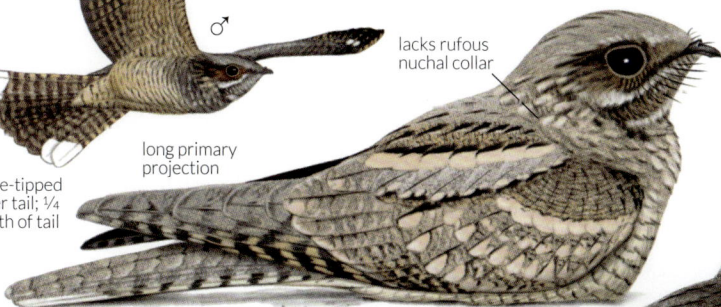

lacks rufous nuchal collar

plumipes

white-tipped outer tail; ¼ length of tail

wings & tail finely barred no pale spots)

♀

distinct pale wingbar

europaeus

♂ ♀

large white spot on emargination of primaries (cf. Fiery-necked Nightjar)

P9 P9

white-tipped outer tail; ⅓ length of tail

♀

RUFOUS-CHEEKED NIGHTJAR

♂

indistinct pale streak above eye

♂

sandy-tipped covert feathers, often with a central black spot

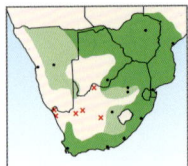

FIERY-NECKED NIGHTJAR   *Caprimulgus pectoralis*

b B B B b
J F M A M J J A S O N D

23–25 cm; 48–65 g A dark brown, heavily marked nightjar with a rich rufous collar (not orange-buff as in Rufous-cheeked Nightjar, p. 270), and a white moustache and throat patch. Rictal bristles (visible with good views of sitting birds) are white-based, differentiating it from all other local nightjar species. In flight, male has broad, white tips to outer tail and small, white primary patches; creamy in female, which also has smaller white tail tips. Amount of rufous on face and breast varies considerably. Northern races average paler above with a more prominent collar. **Voice:** Characteristic night sound of Africa: plaintive, whistled *'good lord, deliiiiiver us'*, descending in pitch, first note often repeated. Also 2-note whistle. **Status and biology:** Common resident and intra-African br. summer migrant in woodland, savanna, fynbos shrubland and plantations. Roosts on the ground under bushes, but often perches in trees at night. (Afrikaanse Naguil)

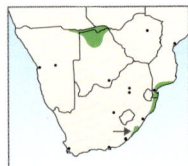

SWAMP NIGHTJAR   *Caprimulgus natalensis*

b B B B b
J F M A M J J A S O N D

20–24 cm; 60–70 g A distinctive nightjar with scaly upperparts formed by buff fringes to mantle feathers; at rest, lacks prominent pale wingbar of most other nightjars, including Square-tailed Nightjar. In flight, male has white trailing edge to wings, white spots in outer primaries and extensive white tips to the 2 outer-tail feathers; outer webs white. Primary spots and outer tail buff in female. White/buff in outer tail more extensive than in Square-tailed Nightjar. **Voice:** A slow *'chow-chow-chow'* or *'chop-chop-chop'*. **Status and biology:** Fairly common, localised resident and local nomad in open grassland and palm savanna, often near damp areas. Roosts on the ground. (Natalse Naguil)

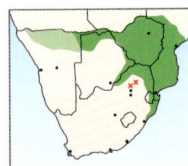

SQUARE-TAILED NIGHTJAR   *Caprimulgus fossii*

b B B b
J F M A M J J A S O N D

22–25 cm; 45–75 g At rest, recalls Fiery-necked Nightjar, but male has white outer tail extending to base of tail and more extensive white primary panel. White outer tail narrower than male Swamp Nightjar; differs at rest by having white lines on the wing coverts and back, and cinnamon nuchal collar; lacks scaly mantle. Female has buff outer tail and primary patch. **Voice:** Prolonged churring, changing in frequency ('changing gears', unlike constant tone of Rufous-cheeked Nightjar's (p. 270) churring call). **Status and biology:** Common resident and local migrant in coastal dune scrub and sandy woodland, often near water. Resident in eastern lowlands; moves to higher elevations and more arid areas in summer. Roosts on the ground, usually under vegetation. (Laeveldnaguil)

FIERY-NECKED NIGHTJAR

small white spot on emargination of primaries

dark crown feathers with grey fringe

pale-based rictal bristles

extensive white tail tips

♂

rufous nuchal collar & ear coverts

sandy-tipped black scapular feathers form a thin pale bar

♀

♂

♂ P9

♀ P9

SWAMP NIGHTJAR

pale spot midway along primaries

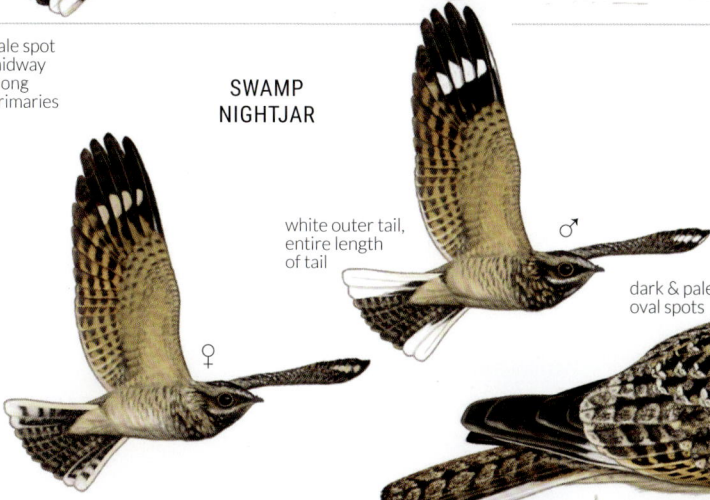

white outer tail, entire length of tail

♂

very dark facial markings below/ behind eye

dark & pale oval spots

♀

♂

♂ P9

♀ P9

natalensis

SQUARE-TAILED NIGHTJAR

complete pale bar across primaries

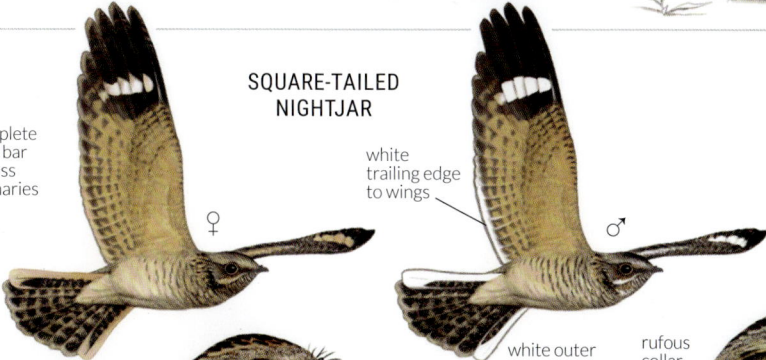

white trailing edge to wings

♀

♂

buff outer tail, entire length of tail

white outer tail, entire length of tail

rufous collar

♂ P9

♀ P9

♀

♂

pale line on wing coverts

FRECKLED NIGHTJAR   *Caprimulgus tristigma*

b B B B b
J F M A M J J A S O N D

26–28 cm; 70–90 g A large, dark grey nightjar; differs from European (p. 270) and Pennant-winged Nightjars by its plain greyish upperparts finely freckled and vermiculated with greyish-buff, lacking wingbars at rest; blends well with its rocky habitat. In flight, male has white outer-tail tips and tiny, white spots in primary bases (smaller than European Nightjar). Female and juv. lack white tail patches but have small, creamy primary spots (absent in female European and Pennant-winged Nightjars). **Voice:** Yapping, double-noted *'cow-wow'*, sometimes extending to 3 or 4 syllables. **Status and biology:** Locally common resident and nomad at rocky outcrops in woodland and hilly terrain; roosts in the open on boulders; also on buildings in cities. (Donkernaguil)

PENNANT-WINGED NIGHTJAR   *Caprimulgus vexillarius*

b   b B B b
J F M A M J J A S O N D

23–26 cm (excl. br. male's 'pennants'); 50–95 g Male has a diagnostic broad, white band across primaries, with longer inner primaries giving the wing a peculiar shape. Br. male has long, white inner primaries (especially primary 2), which trail behind the bird. Female lacks any white on wings or tail; more rufous than European Nightjar (p. 270), with boldly barred flight feathers and squarer tail. Juv. like female but duller. **Voice:** Male makes a soft, high-pitched, insect-like twittering lasting 10 seconds or more during display flight or from a perch. **Status and biology:** Locally common intra-African br. summer migrant (Sep–Apr) in broadleafed woodland. Roosts on ground, but often perches on branches after flushing. (Wimpelvlerknaguil)

no white in outer web of primaries

♂

♀

P9

P9

FRECKLED NIGHTJAR

thin white wingbar, further reduced in female

♀

♂

♀
predominantly grey plumage

♂
plain upperparts

PENNANT-WINGED NIGHTJAR

br. ♂

♂

♀

P9

P9

broadly barred tail & wings

♀

black mottled crown

♀
prominent pale bar on scapulars

small-headed appearance

long primary projection

♂

white-tipped secondaries (visible at rest)

SWIFTS AND SPINETAILS
The most aerial of birds, feeding and even roosting in flight. Legs and toes short, with strong claws. Could be confused with swallows (p. 330), but have scythe-shaped wings and mainly black or brown plumage. Sexes alike.

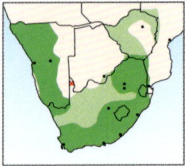

ALPINE SWIFT *Tachymarptis melba*

	b		b					b	b	b	b	b
J	F	M	A	M	J	J	A	S	O	N	D	

20–22 cm; 68–90 g A very large, brown swift with a diagnostic white throat and lower breast and belly. Flight swift and direct, with deep beats of long, scythe-like wings. Juv. has pale fringes to brown feathers. **Voice:** A drawn-out canary-like trill, lasting about 8 seconds and varying in pitch. **Status and biology:** Common resident and intra-African br. summer migrant. Often in large, mixed flocks with other swifts. (Witpenswindswael)

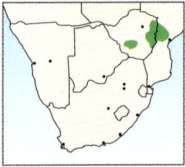

MOTTLED SWIFT *Tachymarptis aequatorialis*

	b					b	b	B	B	b	b	b
J	F	M	A	M	J	J	A	S	O	N	D	

22 cm; 84–92 g A very large, brown swift. Size, structure and flight action are similar to Alpine Swift, but underparts wholly scaled and mottled greyish-brown (not brown and white) and with a slightly longer, more deeply forked tail; throat pale. **Voice:** Dry, deep, insect-like trill. **Status and biology:** Uncommon resident, usually around rocky areas; often solitary. (Bontwindswael)

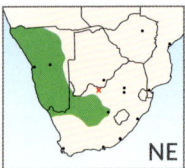

BRADFIELD'S SWIFT *Apus bradfieldi*

b	b	b	b	b			b	b	b	b	b
J	F	M	A	M	J	J	A	S	O	N	D

17 cm; 33–50 g A grey-brown or warm brown swift with scaled underparts; usually paler than Common and African Black Swifts. Lacks strongly contrasting pale uppersides to secondaries of African Black Swift; flight feathers tend to be darker than body and wing coverts. Bulkier than Common Swift, appearing similar to vagrant Pallid Swift. Unlikely to be distinguishable from latter unless they are seen together: undersides of flight feathers of Bradfield's Swift are uniform in colour; outer primaries of Pallid Swift appear darker than inner primaries and secondaries. **Voice:** High-pitched 'sweer' screams and titters around br. sites. **Status and biology:** Locally common near-endemic resident and nomad. (Muiskleurwindswael)

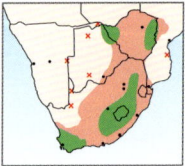

AFRICAN BLACK SWIFT *Apus barbatus*

b	b								B	B	b	b
J	F	M	A	M	J	J	A	S	O	N	D	

18 cm; 35–50 g A large, robust, blackish-brown swift with a small, pale throat patch; belly often scaled. Slightly larger and bulkier than Common Swift, with a shorter, less deeply forked tail and less sharply pointed wings. In good light, the dark back contrasts with paler secondaries from above. **Voice:** High-pitched 'sweeer' screams in air around br. sites. **Status and biology:** Common resident and partial migrant. Breeds in small colonies, sometimes with Alpine Swifts. (Swartwindswael)

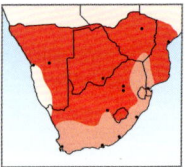

COMMON SWIFT *Apus apus*

J	F	M	A	M	J	J	A	S	O	N	D	

17 cm; 30–44 g A fairly large, blackish-brown (*A. a. apus*) or grey-brown (*A. a. pekinensis*) swift with a pale throat patch. Appears more sleek and rakish than African Black and Pallid Swifts because of relatively longer wings and tail; tail sharply pointed and rather deeply forked (when not moulting). From above, secondaries are the same colour as its back (not paler). Safely separable from Pallid Swift only when the two species are together: Common Swift has less white on the throat, a more slender body and a darker forehead (but eastern race *pekinensis* in worn plumage is very similar to Pallid Swift). Juv. has a pale forehead. **Voice:** Shrill screaming 'shreeee', seldom heard in Africa. **Status and biology:** Common Palearctic-br. migrant, often in large flocks; roosts on the wing. Range has expanded south and west since the 1960s. (Europese Windswael)

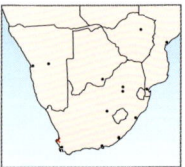

PALLID SWIFT *Apus pallidus*

X												
J	F	M	A	M	J	J	A	S	O	N	D	

16 cm; 32–48 g Vagrant. Difficult to separate from other all-brown swifts. In flight, appears rather robust, with broad wings (especially broad primaries) and slow, leisurely flight action. Typically paler than African Black and Common Swifts, with more extensive white throat patch and paler forehead (but juvs of other species often have paler frons). Tail is less pointed and more shallowly forked than Common Swift, closer to African Black Swift. Shows some contrast between paler secondaries and back, but less than in African Black Swift. Hardest to separate from Bradfield's Swift unless they are seen together, although underparts typically are less scaled. Outer primaries appear darker than inner primaries and secondaries (uniformly coloured in Bradfield's Swift). **Voice:** High-pitched, screaming 'shree-er'. **Status and biology:** Rare Palearctic-br. vagrant that usually winters north of the equator; 1 record from W Cape. (Bruinwindswael)

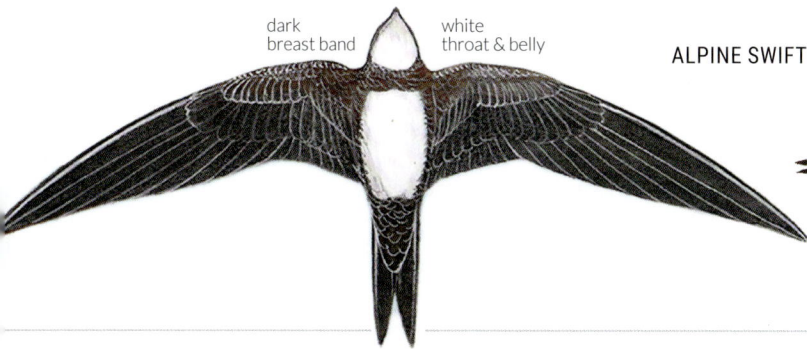

dark breast band

white throat & belly

ALPINE SWIFT

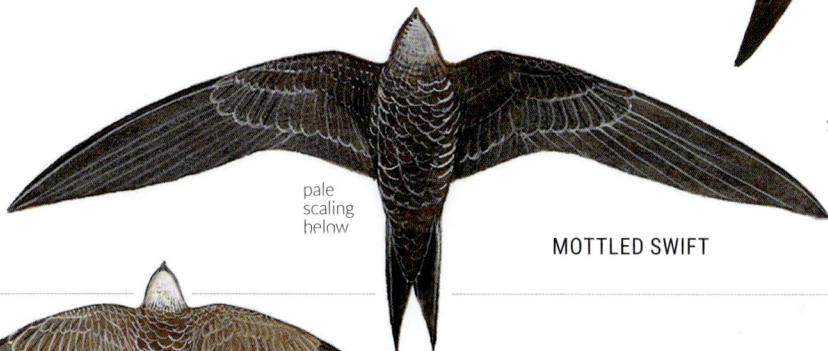

pale scaling below

MOTTLED SWIFT

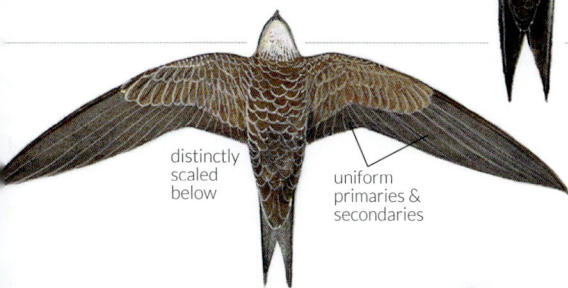

distinctly scaled below

uniform primaries & secondaries

BRADFIELD'S SWIFT

AFRICAN BLACK SWIFT

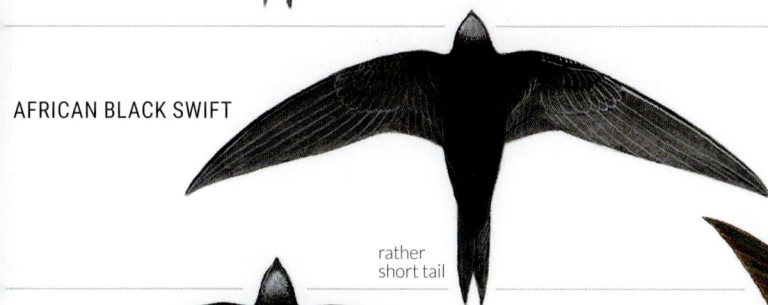

rather short tail

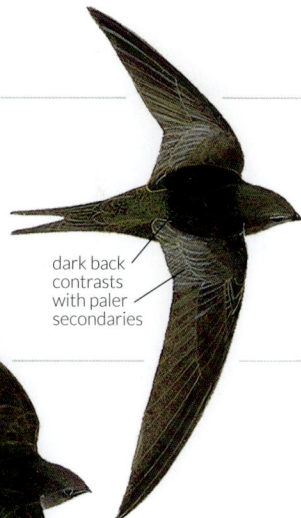

dark back contrasts with paler secondaries

COMMON SWIFT

long, rakish tail

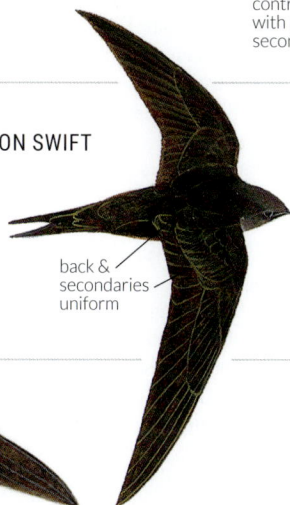

back & secondaries uniform

PALLID SWIFT

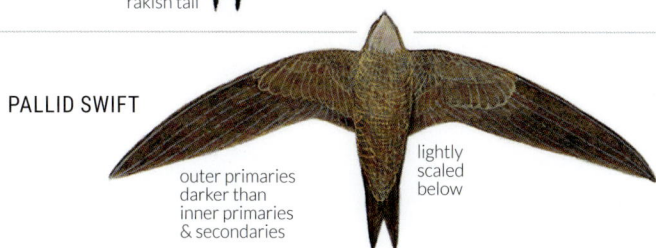

outer primaries darker than inner primaries & secondaries

lightly scaled below

277

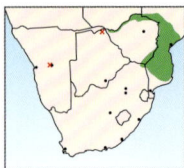

BÖHM'S SPINETAIL   *Neafrapus boehmi*

B B B b | b b B B B
J F M A M J J A S O N D

10 cm; 12–16 g A tiny, very short-tailed swift with a white belly and rump; appears nearly tailless in flight. Flight is erratic, almost bat-like, with short glides interspersed with rapid bursts of flapping. 'Tailless' appearance distinguishes it from all other swifts in the region. **Voice:** High-pitched *'tsit-tsit-tsee-tseeuu'*, but seldom calls. **Status and biology:** Uncommon localised resident and local migrant over savanna and open, broadleafed woodland, often near baobab trees. (Witpensstekelstert)

MOTTLED SPINETAIL   *Telacanthura ussheri*

b b b b b | | | | b b
J F M A M J J A S O N D

14 cm; 28–36 g A spinetail with a fairly long, square tail, broad, white rump, mottled throat and diagnostic white line across the vent. Larger than Little Swift, with narrow secondaries giving the wings a pinched look near body; pale throat colour extends to upper breast. **Voice:** A rasping *'kak-k-k-k'* and a softer *'tt-rrit tt-rrit'*. **Status and biology:** Uncommon localised resident over woodland and savanna; often along forested rivers, frequently near baobab trees. (Gevlekte Stekelstert)

LITTLE SWIFT   *Apus affinis*

b b b | | | b B B B B
J F M A M J J A S O N D

12 cm; 20–30 g A small, square-tailed swift with a broad, white rump patch that wraps around the flanks (visible from below), and a large, white throat patch. In flight, appears squat and dumpy, with rather short, rounded wings. Smaller than Mottled Spinetail, with no white stripe across belly, plain white throat (not extending to upper breast) and less erratic flight action. **Voice:** High-pitched, screeching *'tit-trrrrrrrrrr'*, especially while wheeling in tight flocks during display flights. **Status and biology:** Common resident and partial migrant; often in towns. (Kleinwindswael)

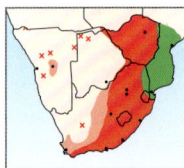

HORUS SWIFT   *Apus horus*

b b b b b b b b b b b b
J F M A M J J A S O N D

14 cm; 20–34 g A bulky version of Little Swift, with a large, white rump and throat, but with a shallowly forked tail; extent of white rump and depth of tail fork intermediate between Little and White-rumped Swifts. Brown-rumped 'Loanda' Swift of Angola apparently is a colour morph of Horus Swift; possible vagrant to nw Namibia. **Voice:** Normally silent; screams *'drreep'* and *'whi-whi-whi'* at br. sites; also buzzy, trilled *'prreeoooo'*. **Status and biology:** Uncommon resident and intra-African br. summer migrant. Forages over grassland, montane shrubland, woodland and semi-desert. (Horuswindswael)

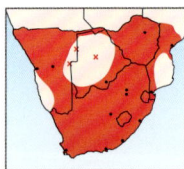

WHITE-RUMPED SWIFT   *Apus caffer*

B b b b | | | | B B B B
J F M A M J J A S O N D

14–16 cm; 18–20 g A slender, black swift with a prominent white throat, a fairly long, deeply forked tail and diagnostic narrow, white 'U' on the rump, much narrower than in Little or Horus Swifts. Tail is frequently held closed, appearing pointed. Secondary feathers are white-tipped, forming a thin white line; dark in Horus Swift. **Voice:** Deeper screams than Little Swift; generally less vocal; scream often preceded by 5 or 6 *'sip'* notes. **Status and biology:** Common intra-African br. summer migrant over open country, often near water; regular in towns and cities. Usually breeds and roosts in swallow or Little Swift nests. (Witkruiswindswael)

SCARCE SWIFT   *Schoutedenapus myoptilus*

| | | | | | | | | | | |
J F M A M J J A S O N D

17 cm; 25–30 g A grey-brown swift with a long, deeply forked tail and dark rump. Shape intermediate between White-rumped Swift and African Palm Swift. Flight action rapid and slightly jerky; glides with wings angled down. **Voice:** A series of 3 or 4 *'tic'* notes, followed by short trills. **Status and biology:** Localised but fairly common br. visitor (mostly Aug–Feb) to eastern highlands of Zimbabwe and adjacent Mozambique, usually over cliffs and rocky bluffs in forested mountain areas. (Skaarswindswael)

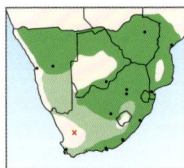

AFRICAN PALM SWIFT   *Cypsiurus parvus*

B B b b b b b B B B B B
J F M A M J J A S O N D

16 cm; 12–16 g A pale grey-brown, very slender, streamlined swift with long, thin, sickle-shaped wings and a very long, deeply forked tail. Has the longest tail of any swift, but beware birds in moult and juvs, which have shorter, less streamer-like tails, and could be confused with Scarce Swift. **Voice:** Soft, high-pitched, twittering screams; gives quiet *'tseep'* while foraging. **Status and biology:** Common resident and local summer migrant to E Cape, usually in vicinity of palm trees, including those in towns. Range has expanded westward in recent decades. Often in small groups or in flocks with other swifts. (Palmwindswael)

'lless'

'normal'
tail shape

square
tail

LITTLE
SWIFT

slightly
forked
tail

HORUS
SWIFT

ÖHM'S SPINETAIL

ostly
hite
elow

MOTTLED
SPINETAIL

white
vent bar

mottled
throat

plain
throat

more extensive
white 'thighs'

ail often
held
closed

white-tipped
secondaries

WHITE-RUMPED
SWIFT

slight white
on 'thighs'

eeply
orked tail
cf. Horus
wift)

SCARCE SWIFT

grey-
brown
below

broader
forked
tail (cf.
African
Palm
Swift)

bulkier body

AFRICAN
PALM SWIFT

slender
body

very deeply
forked tail usually
held closed

MOUSEBIRDS
An endemic African order. Fairly small birds with long, stiff tails, short crests and short legs. Flight direct. Clamber in vegetation and sun their bellies after feeding to accelerate digestion. Occur in small flocks; roost huddled to conserve heat. Sexes alike.

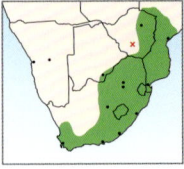

SPECKLED MOUSEBIRD   *Colius striatus*

b b b b b b b b B B B b
J F M A M J J A S O N D

32 cm; 35–65 g A warm brown mousebird with a buffy belly and rather broad tail. Face, eyes and upper mandible blackish; lower mandible bluish-white. Mantle, rump and breast finely barred. Legs black or purple-brown in most of region, but dull red or pink in e Zimbabwe and central Mozambique. Darker and browner than Red-faced Mousebird, with weaker flight. Juv. has shorter tail and lacks black face patch. **Voice:** A sharp *'chee chee chik chik'*, and harsh *'zhrrik-zhrrik'* alarm call. **Status and biology:** Common resident and nomad in thicket, riparian woodland and forest edge; also at fruiting trees in gardens. Red-backed Mousebird *Colius castanotus* may occur occasionally in nw Namibia; similar to Speckled Mousebird, but has pale eyes and a reddish back. (Gevlekte Muisvoël)

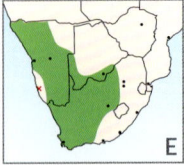

WHITE-BACKED MOUSEBIRD   *Colius colius*

b b b b b b b b B B B B
J F M A M J J A S O N D

31 cm; 30–55 g Paler and greyer than Speckled Mousebird, with bluish-white bill, tipped black, and coral-pink legs. In flight, central back white, bordered by glossy violet stripes (appears black in the field). Juv. has greyish-brown rump and blue-green bill with dark lower mandible. **Voice:** Whistling *'zwee we-wit'*. **Status and biology:** Endemic. Common resident of strandveld, coastal fynbos, arid savanna and scrubby areas in semi-desert. (Witkruismuisvoël)

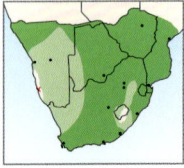

RED-FACED MOUSEBIRD   *Urocolius indicus*

B b b b b b b b B B B B
J F M A M J J A S O N D

34 cm; 40–70 g A blue-grey mousebird with red facial skin and a buffy-cinnamon wash on its face, breast and underwings; legs reddish. Tail longer than *Colius* mousebirds; flight faster and more sustained. In flight, pale rump contrasts with darker greenish-grey back, wings and tail. Juv. has greenish facial skin. **Voice:** A whistled *'chi-vu-vu'*, first note highest pitched. **Status and biology:** Common resident and nomad in savanna, woodland, lowland fynbos and gardens. (Rooiwangmuisvoël)

TROGONS
Brightly coloured, squat, long-tailed forest birds with short, broad bills and short legs. Sexes differ.

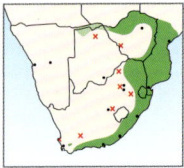

NARINA TROGON   *Apaloderma narina*

b b b B B
J F M A M J J A S O N D

30–34 cm; 55–70 g Male bright green with crimson belly, waxy-yellow bill and turquoise or greenish skin patches on the face and throat (this colour varies seasonally and regionally). Female duller with rufous-brown face merging into greyer breast and dull crimson belly. Juv. paler than female. **Voice:** Deep, hoarse *'hoo hook'*, with emphasis on second syllable, repeated 6–10 times and increasing in volume; wags tail slightly downwards when calling. **Status and biology:** Fairly common resident and local migrant in forest, dense woodland and thicket. Often sits quietly in mid-storey and can be hard to locate if not calling; normally sits with back to observer, green upperparts well camouflaged by foliage. (Bosloerie)

PITTAS
Brightly coloured passerines, related to broadbills (p. 414) with long, robust legs and very short tails. Mainly feed on the ground. Sexes alike; juvs duller.

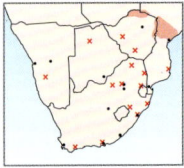

AFRICAN PITTA   *Pitta angolensis*

b B B
J F M A M J J A S O N D

18–22 cm; 80–95 g A brilliantly coloured, but unobtrusive forest-floor bird. In flight, has white panel in primaries. Juv. duller, with bi-coloured bill. When disturbed, flies into mid-storey and freezes for long periods. **Voice:** Displaying birds give frog-like *'preert'* while jumping up from a low perch; after calling, they parachute back to perch, then slowly raise their tail, displaying crimson vent. Call is a short, ascending croak. **Status and biology:** Uncommon intra-African br. summer migrant to riverine thicket and forest. Vagrants occasionally stray well outside usual range. Best located when displaying. (Angolapitta)

SPECKLED MOUSEBIRD

black mask, black & white bill

purple-brown legs & feet

tail slightly fanned in flight

WHITE-BACKED MOUSEBIRD

blue-white bill, tipped black

red legs & feet

black underwing stripe

white back with black border

RED-FACED MOUSEBIRD

red mask; mostly red bill

ad.

juv.

grey-green above

thin tail in flight

NARINA TROGON

Not to scale

♀

♂

AFRICAN PITTA

Not to scale

KINGFISHERS

Strikingly patterned or coloured birds with long, dagger-like bills and short legs. The family comprises three distinct subfamilies: cerylin and alcedinin species on this page and halcyonin species on p. 284. Most species hunt from a perch, but Pied Kingfishers regularly hunt by hovering. Cerylin and some alcedinin species feed on aquatic prey, plunging into water, but other alcedinins are terrestrial. Sexes similar in most species, but differ in cerylin species.

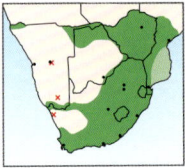

GIANT KINGFISHER *Megaceryle maxima*

b							b	B	B	B	b
J	F	M	A	M	J	J	A	S	O	N	D

40–45 cm; 320–440 g A huge kingfisher with a massive, black bill, dark grey upperparts (with fine, white spots) and some chestnut on its underparts. Male has chestnut breast, white belly with blackish spots and white underwing coverts. Female has white breast with blackish spots, chestnut belly and chestnut underwing coverts. Juv. male has black-speckled, chestnut breast and white underwings; juv. female has whitish breast and chestnut underwings. **Voice:** Loud, harsh *'kahk-kah-kahk'*. **Status and biology:** Common resident at wooded streams and dams, fast-flowing rivers and coastal lagoons; on open coast in s Cape. Usually hunts from exposed perch, but occasionally hovers. (Reusevisvanger)

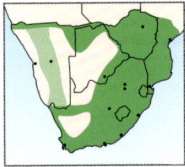

PIED KINGFISHER *Ceryle rudis*

b	b	b	b	B	B	b	B	B	B	B	b
J	F	M	A	M	J	J	A	S	O	N	D

23–25 cm; 70–110 g A large, black-and-white kingfisher, with a long, black bill and short crest. Male has double breast band; female has a single, incomplete breast band. Juv. like female, but feathers from chin to breast edged buff. **Voice:** Rattling twitter; sharp, high-pitched *'chik-chik'*. **Status and biology:** Common resident at freshwater wetlands, coastal lagoons and tidal pools. Frequently hovers over water. Often seen in small groups; breeds cooperatively. (Bontvisvanger)

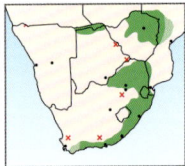

HALF-COLLARED KINGFISHER *Alcedo semitorquata*

b	b	b				b	b	B	B	b	b
J	F	M	A	M	J	J	A	S	O	N	D

18 cm; 35–40 g A black-billed, aquatic kingfisher with a brilliant blue back and rump. Combination of black bill and blue cheeks diagnostic. Larger than Malachite Kingfisher, with more subdued crest, blue (not white) cheeks and black (not red) bill (although juv. Malachite Kingfisher also has dark bill). Juv. duller with black-tipped breast feathers. Blue (not orange) ear coverts separate it from juv. Malachite and African Pygmy Kingfishers. **Voice:** High-pitched *'chreep'* or squeaking *'tsip-ip-ip-ip-eep'*; also a softer *'peeek-peek'*. **Status and biology:** Uncommon resident and nomad at wooded streams, channels in large reed beds and coastal lagoons. (Blouvisvanger)

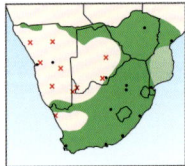

MALACHITE KINGFISHER *Alcedo cristata*

b	b	b	b	b	b	b	b	b	B	B	b
J	F	M	A	M	J	J	A	S	O	N	D

14 cm; 13–19 g A small, aquatic kingfisher with a turquoise-and-black-barred crown. Slightly larger and in a different habitat from African Pygmy Kingfisher (which has orange supercilium and lilac ear coverts). Some birds in north have pale bellies and paler blue back and crest. Ad. has red bill; juv. bill black and is blackish on back, but is much smaller than Half-collared Kingfisher with rufous (not blue) ear coverts and diagnostic barred crown. **Voice:** High-pitched *'peep-peep'* in flight. **Status and biology:** Common resident and local migrant at lakes and dams, and along streams and lagoons. (Kuifkopvisvanger)

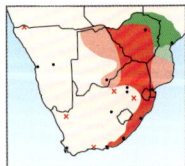

AFRICAN PYGMY KINGFISHER *Ispidina picta*

b	b	b					b	B	B	B
J	F	M	A	M	J	J	A	S	O	N	D

12 cm; 12–17 g A tiny, richly coloured, terrestrial kingfisher. Smaller than aquatic Malachite Kingfisher, with an orange supercilium, mainly rufous face and lilac-washed ear coverts. Juv. has blackish bill and diffuse, dark moustachial stripe. **Voice:** High-pitched *'chip-chip'* in flight. **Status and biology:** Common intra-African br. summer migrant in woodland, savanna and coastal forest. (Dwergvisvanger)

juv. ♂

♂

♀

♂

♀

GIANT KINGFISHER

PIED KINGFISHER

♂

♀

♀

white collar, bordered blue

black bill, blue cheeks

HALF-COLLARED KINGFISHER

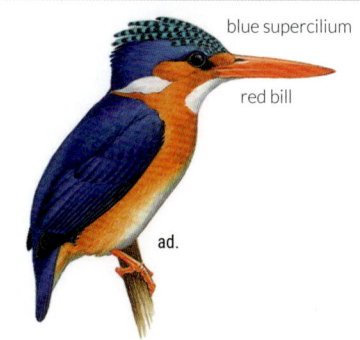

brilliant blue back

blue supercilium

red bill

ad.

MALACHITE KINGFISHER

juv.

black bill, orange cheeks

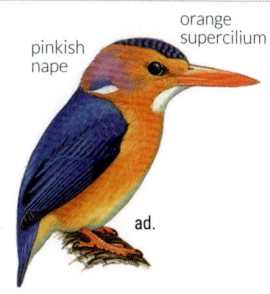

orange supercilium

pinkish nape

ad.

AFRICAN PYGMY KINGFISHER

juv.

HALCYON KINGFISHERS

The so-called tree kingfishers in the subfamily Halcyoninae are largely terrestrial species, belying their common name 'kingfisher'. Of the species in the region, only the Mangrove Kingfisher feeds in aquatic habitats, and mainly only in winter.

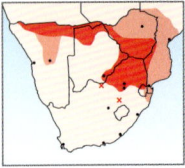

WOODLAND KINGFISHER *Halcyon senegalensis*

22–24 cm; 55–80 g A striking, blue-backed kingfisher with a red-and-black bill (rarely all-red). Told from Mangrove Kingfisher by its two-tone bill, black (not reddish) legs, black eye stripe extending behind its eye and paler head with blue wash extending onto the hind crown; underwing all white. Juv. has dusky, reddish-brown bill and is lightly barred grey on sides of breast. **Voice:** Loud, piercing *'chip-cherrrrrrrrrr'*, latter part a descending trill. Often spreads wings while calling. **Status and biology:** Common intra-African br. summer migrant in woodland and savanna with tall trees. Tracking studies show that birds br. in s Africa spend the austral winter in the Sahel region in Chad, the Central African Republic and South Sudan. (Bosveldvisvanger)

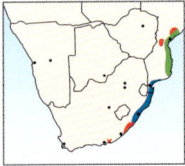

MANGROVE KINGFISHER *Halcyon senegaloides*

22–24 cm; 75–90 g Slightly heavier than Woodland Kingfisher with an all-red (not red-and-black) bill, reddish (not black) legs, black carpal patch on underwing, and darker grey head. Pale blue back separates it from other *Halcyon* kingfishers. Juv. has brownish bill and dark scaling on its breast; best distinguished from juv. Woodland Kingfisher by black facial mark not extending behind eye. **Voice:** Noisy; loud, ringing and accelerating *'cheet cheet cheet cheet chroo-choo-chroo-chroo'*, slower and deeper than Woodland Kingfisher. **Status and biology:** Uncommon local migrant; breeds in summer in coastal forest, lowland forest and riparian thicket; winters in mangrove swamps. (Manglietvisvanger)

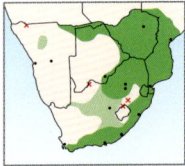

BROWN-HOODED KINGFISHER *Halcyon albiventris*

20–22 cm; 50–75 g Differs from other red-billed *Halcyon* kingfishers by its brownish head streaked with black, rufous patches on sides of breast, and streaked flanks. Larger than Striped Kingfisher, with all-red (not black-and-red) bill, and lacks white in upper- and underwing. Male has black back; female and juv. have brown back; juv. underparts whiter. **Voice:** Whistled *'tyi-tu-tu-tu-tu'*; harsher *'klee-klee-klee'* alarm note. **Status and biology:** Common resident in woodland, coastal forest, parks and gardens. (Bruinkopvisvanger)

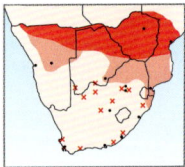

GREY-HEADED KINGFISHER *Halcyon leucocephala*

20–22 cm; 38–48 g A striking kingfisher with a pale grey head and chestnut belly and underwing coverts (more extensive in male than female). Juv. has a blackish bill and buffy, grey-barred head and underparts; lacks streaking on head and flanks of Brown-hooded Kingfisher. **Voice:** High-pitched, rapid *'chee-chi-chi-chi-chi'*; also a high-pitched trill. **Status and biology:** Locally common intra-African br. summer migrant in broadleafed woodland and savanna. (Gryskopvisvanger)

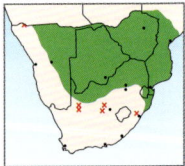

STRIPED KINGFISHER *Halcyon chelicuti*

16–18 cm; 30–50 g A small, dryland kingfisher with a grey-streaked crown, black-and-red bill and white collar. Smaller than Brown-hooded Kingfisher, with a darker cap and white collar. In flight, shows extensive white on underwing, white flash in upperwing, and blue rump. Male has darker crown than female and more prominent black carpal patch on underwing. Juv. has dusky bill and blackish, scaled breast and flanks. **Voice:** High-pitched, piercing *'cheer-cherrrrrrrrr'*, rising and falling in pitch and ending in a trill. **Status and biology:** Common resident in acacia woodland, savanna and edges of riverine forest. (Gestreepte Visvanger)

blue head

red & black bill

black lores & eye stripe

WOODLAND KINGFISHER

ad.

juv.

grey head

red bill

blue back

pale belly

MANGROVE KINGFISHER

ad.

black lores only

juv.

brown hood

black mantle

♂

dark brown mantle

♀

BROWN-HOODED KINGFISHER

♀

cinnamon underwing coverts

♂

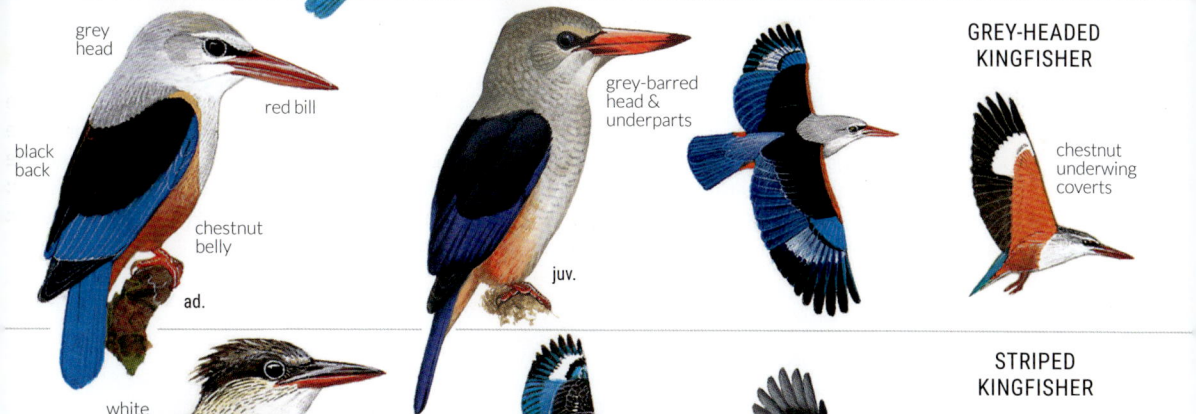

grey head

red bill

black back

chestnut belly

ad.

grey-barred head & underparts

juv.

GREY-HEADED KINGFISHER

chestnut underwing coverts

white collar

striped underparts

♂

♂

STRIPED KINGFISHER

♀

BEE-EATERS
Colourful, aerial insectivores with long, decurved bills. Sally from perches or during longer flights. Legs small and weak; shuffle on ground. Sexes alike, but juvs and ads in worn plumage duller.

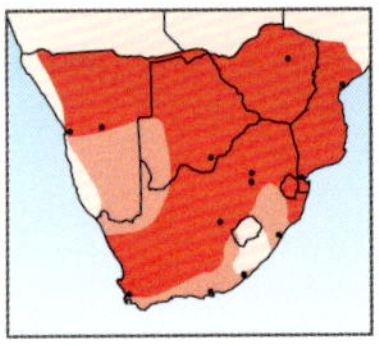

EUROPEAN BEE-EATER *Merops apiaster*

25 cm (28 cm incl. streamers); 40–66 g A large, distinctively coloured bee-eater with a chestnut crown and back, golden mantle, yellow throat and blue breast and belly. Local br. birds are more intensely coloured than Palearctic-br. migrants. Juv. lacks elongated central tail feathers, and has a green back and pale blue underparts. **Voice:** Far-carrying, frog-like flight call, *'prrrup'* or *'krroop-krroop'*. **Status and biology:** Common Palearctic-br. migrant Oct–Mar and intra-African br. migrant in sw South Africa Aug–Feb, where it breeds in small colonies. Occurs in savanna, broadleafed woodland, fynbos and adjacent grassy areas, with passage birds crossing the Karoo to reach br. grounds in southwest. (Europese Byvreter)

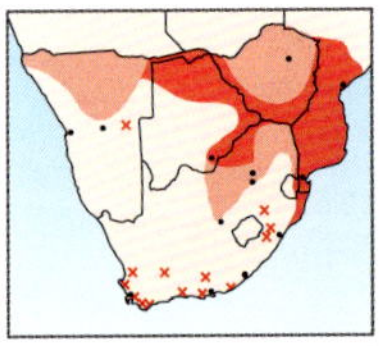

BLUE-CHEEKED BEE-EATER *Merops persicus*

25 cm (33 cm incl. streamers); 40–54 g A large, green bee-eater. Slightly larger than Olive Bee-eater, with green (not brown) crown, blue (not white) frons, supercilium and cheek stripe, and yellow (not whitish) chin; rufous throat patch smaller. In worn plumage, blue facial stripes appear white; best told from Olive Bee-eater by green (not brown) crown. Juv. duller, with blue tips to feathers giving scaled appearance; lacks tail streamers. **Voice:** Liquid *'prrrup'* and *'prrreo'*, slightly higher pitched than European Bee-eater. **Status and biology:** Common Palearctic-br. summer migrant to flood plains and adjacent woodland. (Blouwangbyvreter)

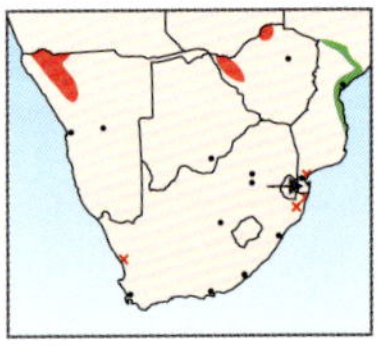

OLIVE BEE-EATER *Merops superciliosus*

24 cm (30 cm incl. streamers); 38–48 g Slightly smaller than Blue-cheeked Bee-eater, with a brown (not green) crown, white (not blue) supercilium and cheek stripe, whitish (not yellow) chin, more extensive rufous-brown throat and paler green underparts. *M. s. alternans* of nw Namibia has crown washed green; best told by brownish throat. Much larger than Böhm's Bee-eater (p. 288), with a dull brown (not chestnut) cap and brown (not green) mantle. Juv. duller; feathers blue-tipped, giving scaly appearance; chin yellowish; lacks tail streamers. **Voice:** High-pitched *'pit-ilup'*, higher than Blue-cheeked Bee-eater. **Status and biology:** Locally common intra-African br. summer migrant in broadleafed woodland near lakes, rivers and swamps; easiest seen in nw Namibia. Uncommon resident in Mozambique. (Olyfbyvreter)

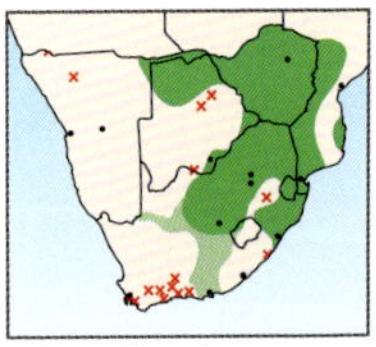

WHITE-FRONTED BEE-EATER *Merops bullockoides*

22–24 cm; 30–40 g A mainly green bee-eater with a white and crimson throat, white cheek stripe and frons, and striking blue rump, thighs and vent; lacks elongated central tail streamers. Juv. duller, with a green crown. **Voice:** Querulous *'qerrr'* or *'querry'*, like that of Greater Blue-eared Starling (p. 426); twittering noises when roosting. **Status and biology:** Common resident in grassland and savanna, usually near rivers and wetlands. Occasionally takes prey from ground. (Rooikeelbyvreter)

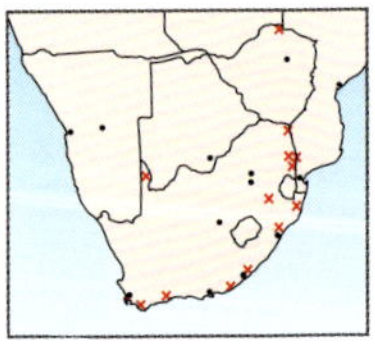

WHITE-THROATED BEE-EATER *Merops albicollis*

20 cm (32 cm incl. streamers); 22–30 g Vagrant. A slender bee-eater with a distinctive black-and-white-striped head, black breast band and blue-green upperparts. Ad. has very long central tail streamers. Juv. has buffy throat and is duller green above, lightly scalloped; lacks tail streamers. **Voice:** Mellow *'terruw-uw'* or *'tsip-tsip-terruwuw'*. **Status and biology:** Vagrant; intra-African migrant in semi-arid and mesic savanna further north in Africa; scattered records throughout s Africa, mostly Dec–Apr. (Witkeelbyvreter)

chestnut
back &
cap

yellow
throat

pale blue
breast & belly

**EUROPEAN
BEE-EATER**

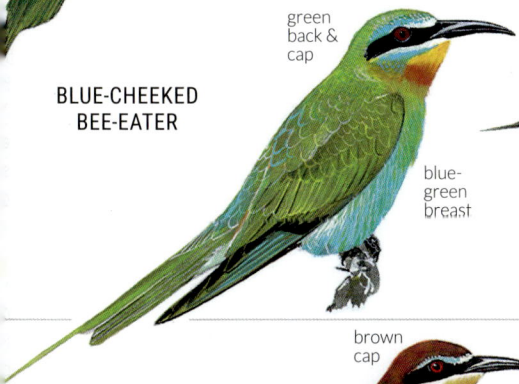

green
back &
cap

**BLUE-CHEEKED
BEE-EATER**

blue-
green
breast

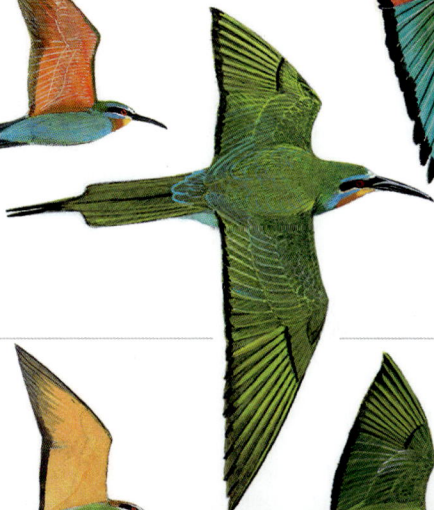

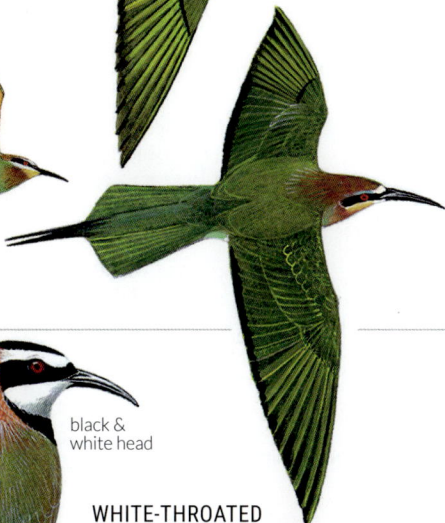

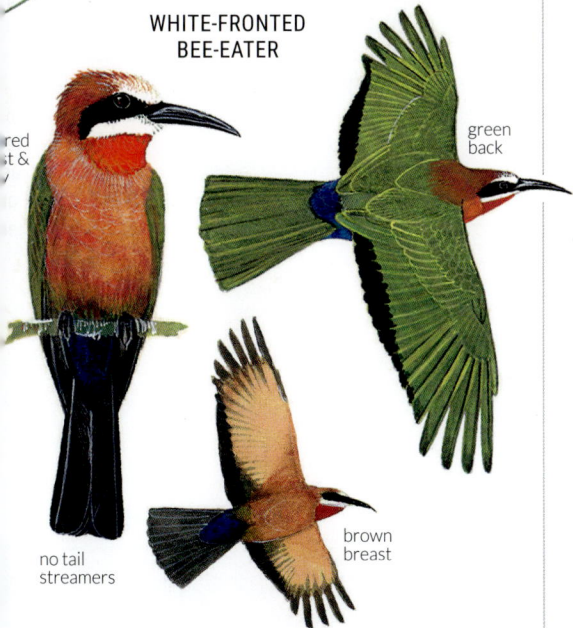

brown
cap

**OLIVE
BEE-EATER**

green
back

blue-
green
breast

**WHITE-FRONTED
BEE-EATER**

green
back

red
t &

black &
white head

**WHITE-THROATED
BEE-EATER**

blue tail
with
streamers

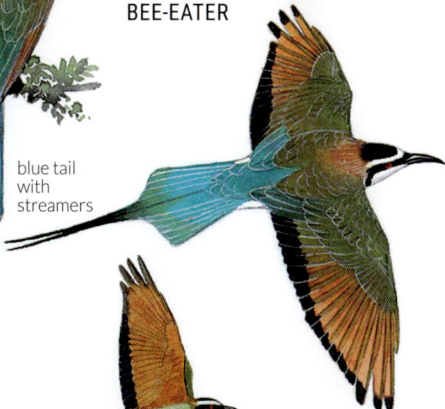

no tail
streamers

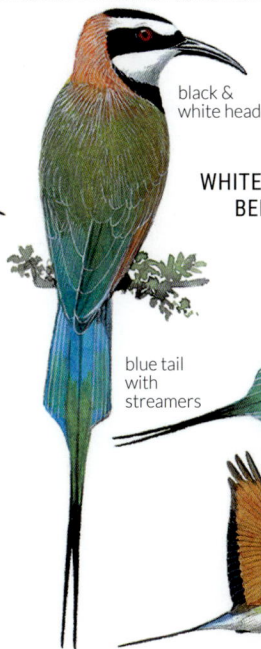

brown
breast

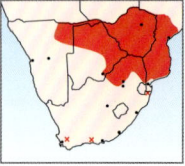

SOUTHERN CARMINE BEE-EATER *Merops nubicoides*

J F M A M J J A S O N D

26 cm (38 cm incl. streamers); 52–75 g A bright pinkish-red bee-eater with a black eye stripe, blue crown, rump and belly, and elongated, blackish central tail feathers. Juv. duller, with browner back; lacks elongated central tail streamers. Rarely, juvs have pale blue chin, throat and cheeks that could cause confusion with Northern Carmine Bee-eater *M. nubicus* (vagrant to Zambia; not yet recorded from s Africa). **Voice:** Rather nasal, deep *'gra-gra-gra'*. **Status and biology:** Common intra-African br. summer migrant in woodland, savanna and flood plains. Usually breeds in large colonies in sandbanks. Often attends grass fires. (Rooiborsbyvreter)

ROSY BEE-EATER *Merops malimbicus*

J F M A M J J A S O N D

24 cm (28 cm incl. streamers); 45 g Vagrant. A fairly large bee-eater, with slate-grey upperparts, rosy pink underparts and elongated central tail feathers. Told from Southern Carmine Bee-eater by its uniform, slate-grey (not reddish) upperparts and pale cheek stripe. Juv. lacks tail streamers and is duller, with an indistinct buffy cheek stripe. **Voice:** Trilling *'prrp-prrp'*, higher pitched than European Bee-eater. **Status and biology:** A tropical forest species that occurs from the Congo Basin to Ghana; 1 record from s Africa: Cape Recife, E Cape, Apr 2003. (Pienkborsbyvreter)

SWALLOW-TAILED BEE-EATER *Merops hirundineus*

J F M A M J J A S O N D

20–22 cm; 15–28 g The only bee-eater with a deeply forked tail. Blue tail and belly are distinctive among small, green bee-eaters. Juv. duller with buffy throat and green underparts; lacks a blue collar; tail less deeply forked. Larger than Little Bee-eater with a blue (not black) collar, blue-green (not buffy-yellow) underparts and blue (not green and chestnut) tail; lacks russet in upper wings. **Voice:** Soft, twittering *'kwit-kwit-kwit'* and a high-pitched *'kweep kweepi bzzz, kweep'*. **Status and biology:** Common nomad and local migrant in a wide range of habitats from semi-desert scrub to forest margins. (Swaelstertbyvreter)

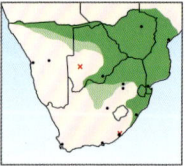

LITTLE BEE-EATER *Merops pusillus*

J F M A M J J A S O N D

15–17 cm; 11–18 g A tiny, green-backed bee-eater with a yellow throat, black collar and rufous underparts. In flight, shows conspicuous rufous flight feathers with dark trailing edge; underwings are entirely rufous. Tail square or slightly notched. Central tail green; rest of tail is rufous with black tip. Juv. has paler, buffy throat and lacks a black collar; breast tinged green. **Voice:** *'Zeet-zeet'* or *'chip-chip'*. **Status and biology:** Common resident in savanna, woodland, forest edges and around wetlands. (Kleinbyvreter)

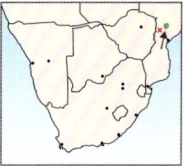

BÖHM'S BEE-EATER *Merops boehmi*

J F M A M J J A S O N D

18–22 cm; 15–20 g A small, slender bee-eater with a chestnut crown and throat, neat blue cheek stripe and very long central tail streamers. Much smaller than Olive Bee-eater (p. 286), with dark tail tip and neater head pattern, lacking white stripes. Juv. duller with yellowy-green throat; central tail feathers shorter. **Voice:** Soft high-pitched *'sip'*; song a whistled *'swee-deedle-ee-dee jeep'*. **Status and biology:** A small population recently discovered on the south bank of the Zambezi River, near Sena, central Mozambique, where it is considered a rare resident. (Roeskopbyvreter)

SOUTHERN CARMINE BEE-EATER

reddish back

pink throat

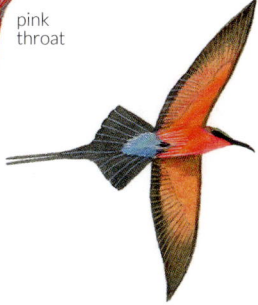

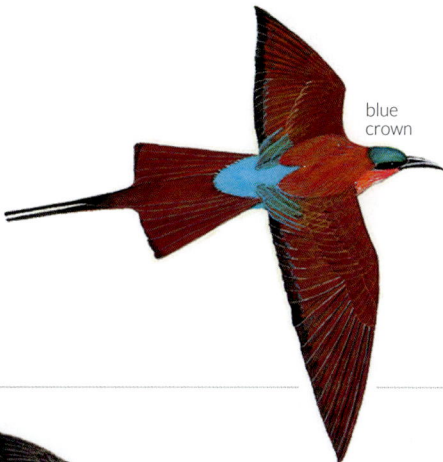

blue crown

ROSY BEE-EATER

grey back

pink breast

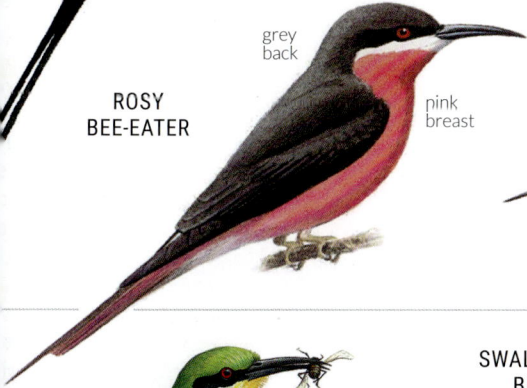

pale cheek stripe

SWALLOW-TAILED BEE-EATER

blue rump & forked tail

LITTLE BEE-EATER

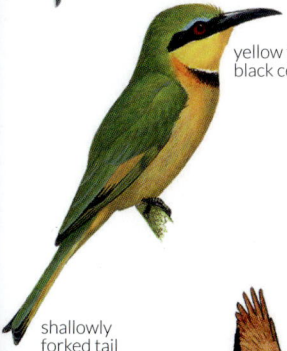

yellow throat; black collar

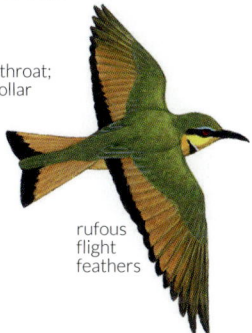

rufous flight feathers

shallowly forked tail

mostly chestnut head

BÖHM'S BEE-EATER

green upperparts

tail streamers

green lower breast & belly

289

ROLLERS

Brightly coloured, medium-sized birds with robust bills and short legs. Hunt from perches, dropping to the ground or taking prey in the air. Sexes alike.

LILAC-BREASTED ROLLER   *Coracias caudatus*

29 cm (37 cm incl. streamers); 90–135 g Ad. distinctive with pointed tail streamers, green crown, nape and mantle, and lilac breast. Streamers lost during winter moult, but crown and breast distinctive. Juv. duller; lacks elongated tail feathers; best told from juv. European Roller by smaller head and bill and blue (not black) flight feathers. Told from juv. Racket-tailed Roller by blue (not brown) median wing coverts and uniformly dark blue flight feathers (lacking pale blue bases). **Voice:** Harsh *'chak'*, often repeated when agitated; flight call *'grak grak grarrak'*. **Status and biology:** Common resident in savanna. Perches conspicuously, often along telephone lines. (Gewone Troupant)

RACKET-TAILED ROLLER   *Coracias spatulatus*

29 cm (36 cm incl. streamers); 80–115 g Ad. tail is diagnostic, but moulting ad. (winter) and juv. lack diagnostic spatulate tail streamers. At all ages has brown back colour extending onto median coverts (blue in European and Lilac-breasted Rollers) and pale blue wingbar formed by pale bases to flight feathers (uniformly black in European Roller, dark blue in Lilac-breasted Roller). Also lacks sharp demarcation between brownish back and greenish hindneck. **Voice:** *'Chaw'* and *'cheer'*; less harsh than other rollers; *'kaairsh kaairsh'* in display flight. **Status and biology:** Uncommon resident and nomad in tall woodland, especially miombo. Unobtrusive; often perches just below the canopy unlike other roller species, which mostly perch in prominent positions. (Knopsterttroupant)

EUROPEAN ROLLER   *Coracias garrulus*

30–32 cm; 105–150 g A large, stocky roller with a square tail lacking tail streamers. Slightly larger than Lilac-breasted and Racket-tailed Rollers, with a bigger head and bill; flight feathers blackish (not blue) from above. Often in rather scruffy plumage, but ad. has bluish head separated from brown back. Juv. is more olive-green with pinkish wash on throat; told from juv. Lilac-breasted Roller by larger head and black (not dark blue) flight feathers. **Voice:** Normally silent in Africa; dry *'krask-kraak'* when alarmed. **Status and biology:** Common Palearctic-br. summer migrant to savanna; rare in more open habitats. Range apparently expanding southwards; becoming increasingly common in south coast fynbos. (Europese Troupant)

PURPLE ROLLER   *Coracias naevius*

35–40 cm; 130–210 g A large, distinctive roller with a broad, whitish supercilium, greenish crown and purple underparts strongly streaked with white. Juv. duller, underparts washed greenish-brown. **Voice:** Harsh, repeated *'karaa-karaa'* in display flight, accompanied by exaggerated, side-to-side rocking motion. **Status and biology:** Fairly common resident and partial migrant in dry thornveld and open, broadleafed woodland. (Groottroupant)

BROAD-BILLED ROLLER   *Eurystomus glaucurus*

27–29 cm; 85–135 g A compact, dark cinnamon and purple roller with a bright, waxy yellow bill. In flight, blue tail with dark central stripe contrasts with purple uppersides of flight feathers; underwing has purple coverts and blue flight feathers. Juv. duller, with browner upperparts and dull, blue-grey underparts mottled greyish; bill duller yellow. **Voice:** Harsh screams *'garrr'* and cackles *'kek-kek-kek'*; flight call a rattling *'g-r-r-r-r-r-g'*. **Status and biology:** Locally common intra-African br. summer migrant to riverine forest and adjacent savanna. Often perches and breeds in dead trees. (Geelbektroupant)

green crown & nape

lilac breast

LILAC-BREASTED ROLLER

pointed tail streamers

brown crown & nape

blue breast

RACKET-TAILED ROLLER

bulbous-tipped tail streamers

blue head & nape

EUROPEAN ROLLER

no tail streamers

pale supercilium

heavy streaking

BROAD-BILLED ROLLER

yellow bill

short tail

PURPLE ROLLER

GROUND HORNBILLS
Huge, terrestrial hornbills in a separate family from other hornbills. Sexes differ in facial ornamentation; males larger than females. Occur in small family groups; breed cooperatively. Female not sealed into nest like other hornbills.

SOUTHERN GROUND HORNBILL   *Bucorvus leadbeateri*

90–130 cm; 2.5–6 kg A very large, black hornbill with a long, decurved bill, and conspicuous red face and throat; white primaries obvious in flight. Ad. female has small violet-blue throat patch and extensive red neck pouches. Juv. sooty-brown, with dull yellowish face and throat. **Voice:** Loud, booming *'ooomph ooomph'* duet given early in the morning. **Status and biology:** VULNERABLE. Scarce resident and nomad in savanna, woodland and grassland with adjoining forest. Range and numbers decreased by about two-thirds during the 20th century; now mainly confined to large reserves. (Bromvoël)

HORNBILLS
Medium to large, long-tailed birds with characteristic long, decurved bills, often with casques. Sexes similar in most species, but females have smaller casques. Female moults all flight feathers while sealed into nest during incubation and early chick-rearing stages.

SILVERY-CHEEKED HORNBILL   *Bycanistes brevis*

60–80 cm; 1.1–1.4 kg A large, black-and-white hornbill; larger than Trumpeter Hornbill, with mostly black wings (lacking white trailing edge), more extensive black underparts, white (not black) back and a paler bill. Male has huge, creamy casque. Lower belly and outer-tail tips white. Cheek feathers have silvery tips, giving pale-faced appearance. Female and juv. have much reduced casques; juv. lacks paler cheek. Flight heavy and noisy compared to Trumpeter Hornbill. Black-and-white-casqued Hornbill *B. subcylindricus*, a much larger species from W Africa, occasionally reported in KZN after escaping from captivity. **Voice:** Deep wail, harsh *'quark-quark'* and nasal *'wheeer-eer'* calls. **Status and biology:** Fairly common resident and local nomad in montane and coastal forests. (Kuifkopboskraai)

TRUMPETER HORNBILL   *Bycanistes bucinator*

50–65 cm; 480–900 g A medium-large, black-and-white hornbill. Smaller than Silvery-cheeked Hornbill with a white belly and lower breast, a black (not white) back and a smaller, darker bill. In flight, has white trailing edges to its wings and white underwing coverts (wings of Silvery-cheeked Hornbill have a black trailing edge). Uppertail coverts white. At close range, has pinkish-red (not blue) eye skin. Female has smaller bill and casque. Juv. has almost no casque. **Voice:** Wailing, plaintive *'waaaaa-weeeee-waaaaa'*, similar to a crying baby. **Status and biology:** Common resident and local nomad in lowland, coastal and riverine evergreen forests; also well-wooded suburbs. (Gewone Boskraai)

AFRICAN GREY HORNBILL   *Lophoceros nasutus*

44–50 cm; 130–220 g A drab, grey-brown hornbill with a creamy supercilium and narrow, whitish back, visible in flight. Wings brown, with narrow, pale feather edges; outer tail tips white. Male has dark grey bill with creamy stripe at base of upper mandible. Upper part of female's bill (including small casque) is pale yellow; tip is maroon. Juv. lacks a casque; black-and-white patch at base of upper mandible (both sexes). **Voice:** Highly vocal; plaintive, whistling *'pee pee pee pee phee pheeoo phee pheeoo'*, with bill held vertically and wings flicked open on each note. **Status and biology:** Common resident in acacia savanna and dry, broadleafed woodland. Often in flocks outside breeding season. Flight extremely undulating. (Grysneushoringvoël)

SOUTHERN YELLOW-BILLED HORNBILL   *Tockus leucomelas*

48–60 cm; 145–230 g Combination of yellow bill, spotted wing coverts and pale underparts is diagnostic. In flight, has white inner secondaries and outer tail like Southern and Damara Red-billed Hornbills (p. 294). Female and juv. have smaller bills and shorter casques; juv. has dark head and throat. **Voice:** Rapid, hollow-sounding *'tok tok tok tok tok tokatokatoka'*, given with head lowered and wings fanned. **Status and biology:** Common near-endemic resident in thornveld and dry, broadleafed woodland. (Geelbekneushoringvoël)

SOUTHERN GROUND HORNBILL

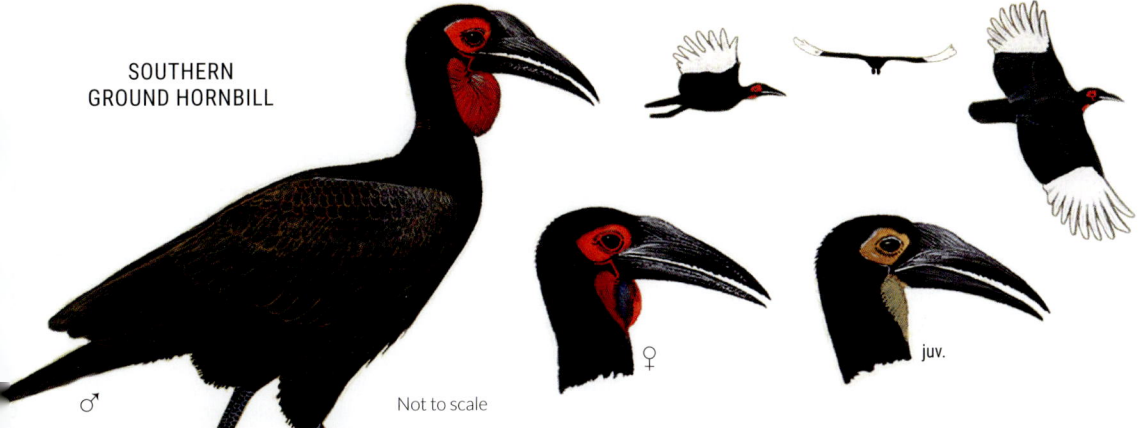

♂

♀

juv.

Not to scale

SILVERY-CHEEKED HORNBILL

cream bill & casque

black trailing edge

♂

♀

♂

TRUMPETER HORNBILL

red bare skin around eye

grey bill & casque

♂

white trailing edge

♂

♂

♀

AFRICAN GREY HORNBILL

♀

blackish bill

♂

♂

white back

SOUTHERN YELLOW-BILLED HORNBILL

♂

large yellow bill

♂

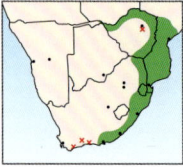

CROWNED HORNBILL   *Lophoceros alboterminatus*

b | | | | | | | | | | b B B b
J F M A M J J A S O N D

50–54 cm; 180–330 g A dark brown hornbill with a white belly and a long, red bill. Darker brown than slightly larger Bradfield's Hornbill (small range overlap in e Zambezi Region (Caprivi) and nw Zimbabwe), with a yellow (not red) eye, a shorter, deeper red bill with a yellow line at base, and more extensive white tips to outer tail. Male has a large casque and black throat; female has a smaller casque and turquoise throat. Juv. has an orange bill that lacks a casque; wing coverts with some white flecks. **Voice:** Whistling *'chleeoo chleeoo'*. **Status and biology:** Common resident and local nomad in dry season. Occurs in coastal and riverine forests, often in flocks. Flight extremely undulating. (Gekroonde Neushoringvoël)

BRADFIELD'S HORNBILL   *Lophoceros bradfieldi*

b | b | | | | | | | b b B b
J F M A M J J A S O N D

50–57 cm; 180–340 g Paler brown than Crowned Hornbill (small range overlap in e Zambezi Region (Caprivi) and nw Zimbabwe), with a longer, orange-red bill without a distinct casque; lacks yellow line at base of bill and has orange (not pale yellow) eyes. Easily told from Monteiro's Hornbill by lacking white in wings and having mainly brown (not white) outer-tail feathers. Female has a smaller bill, with turquoise (not black) facial skin. Juv. has smaller bill. **Voice:** Rapidly repeated, whistling *'chleeoo'* note, with bill raised vertically. **Status and biology:** Near-endemic. Fairly common resident in open mopane woodland and thornveld. (Bradfieldneushoringvoël)

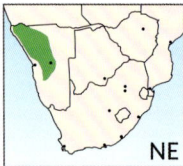

MONTEIRO'S HORNBILL   *Tockus monteiri*

b B B b | | | | | | | b b b
J F M A M J J A S O N D

50–58 cm; 280–420 g A brown hornbill with a large, red bill. White-spotted (not uniform) wing coverts, and white (not dark) secondaries and outer tail separate it from Bradfield's Hornbill. Female has smaller bill and turquoise (not blackish) facial skin. Juv. has smaller, dull orange bill and pink facial skin. **Voice:** Croaking series of *'tooaak tooaak'* notes, given with head lowered and wings closed. **Status and biology:** Near-endemic. Common resident and local nomad in dry thornveld and broadleafed woodland, often in stony, hilly country. Winter flocks undertake local movements, often moving into open, flat country. (Monteironeushoringvoël)

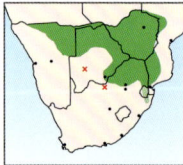

SOUTHERN RED-BILLED HORNBILL   *Tockus rufirostris*

b b b | | | | | | b B B B
J F M A M J J A S O N D

38–45 cm; 100–170 g A fairly small hornbill with a rather short, red bill. White-spotted wing coverts, pale face and white (not brown) throat are diagnostic, apart from very similar Damara Red-billed Hornbill in n Namibia. Told from Damara Red-billed Hornbill by its brown-streaked (not white) facial feathers and pale yellow (not brown) eyes. Male has black patch at base of lower mandible; female has smaller, all-red bill. Juv. has grey eyes, a smaller, orange bill with a black patch at the base, and buff (not white) spotting on the back and wing coverts. **Voice:** A long series of *'kuk kuk kuk'* calls, becoming faster and louder, ending with double notes *'kuk-we kuk-we'*. Does not raise its wings during its display. **Status and biology:** Locally common in savanna and semi-arid woodland. (Rooibekneushoringvoël)

DAMARA RED-BILLED HORNBILL   *Tockus damarensis*

B B | | | | | | | | | |
J F M A M J J A S O N D

40–50 cm; 180–230 g Slightly larger than Southern Red-billed Hornbill, with a rather plain, white head and neck with a narrow, dark crown stripe extending to nape; typically has brown (not yellow) eyes. White (not dark) face and breast, more boldly spotted upperwing coverts and relatively small bill separate it from Monteiro's Hornbill. Female and juv. have less black on lower mandible; juv. has grey eyes. **Voice:** Staccato *'kwa kwa kwa kokkok kokkok kokkok'*; wings of male partly opened in display. **Status and biology:** Near-endemic. Locally common resident in savanna and semi-arid woodland, often in hilly areas; also along dry watercourses with large trees. (Damararooibekneushoringvoël)

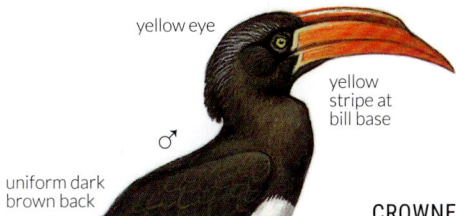

yellow eye

♂

yellow stripe at bill base

uniform dark brown back

♀

CROWNED HORNBILL

♂

orange eye

scalloped coverts

BRADFIELD'S HORNBILL

dark outer tail

dark eye

spotted coverts

MONTEIRO'S HORNBILL

white outer tail

pale eye

variably grey cheeks & neck

SOUTHERN RED-BILLED HORNBILL

brown eye

white face & neck

DAMARA RED-BILLED HORNBILL

WOOD HOOPOES, SCIMITARBILLS AND HOOPOES

Three families with long, decurved beaks for probing out invertebrates. Hoopoes strikingly plumaged black, white and buff with an erectile crest; wood hoopoes and scimitarbills more sombre, although plumage often glossy. All produce foul-smelling secretions from preen glands that deter predators. Sexes similar, but bill length differs in wood hoopoes.

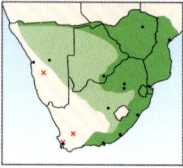

GREEN WOOD HOOPOE *Phoeniculus purpureus*

b b B B B B b b b B B B b
J F M A M J J A S O N D

32–36 cm; 60–90 g A glossy green wood hoopoe, with a long, decurved, red bill, red legs, white wingbars and long tail with white subterminal bars in outer feathers. In good light, bottle-green head and back distinguish it from Violet Wood Hoopoe. Male has longer and more decurved bill than female. Juv. has a black bill and lacks gloss on plumage; juv. male has a brown throat patch (black in female). Red legs and feet and white primary coverts separate juvs from Common Scimitarbill, but probably inseparable in field from juv. Violet Wood Hoopoe. **Voice:** Garrulous chattering and cackling, usually by groups during territorial encounters. **Status and biology:** Common resident in woodland, thicket and forest edges, usually in groups. Occasionally in mixed groups with Violet Wood Hoopoes. (Rooibekkakelaar)

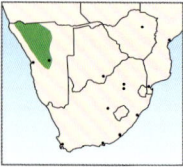

VIOLET WOOD HOOPOE *Phoeniculus damarensis*

b b b b
J F M A M J J A S O N D

34–40 cm; 76–94 g Very similar to Green Wood Hoopoe, but slightly larger with a longer tail that flops around more in flight; the violet and coppery (not bottle-green) head, mantle and back are only evident in good light. Female has a shorter, less decurved bill than male. Juv. has a black bill and lacks gloss on plumage; probably inseparable from juv. Green Wood Hoopoe. Juv. male has a brown throat patch (black in female). **Voice:** Harsh cackling, slightly slower and deeper than Green Wood Hoopoe. **Status and biology:** Fairly common resident in dry thornveld and mopane woodland, also along dry watercourses with large trees; usually in groups. Occasionally in mixed groups with Green Wood Hoopoes. (Perskakelaar)

COMMON SCIMITARBILL *Rhinopomastus cyanomelas*

b b b B B B b
J F M A M J J A S O N D

24–28 cm; 25–40 g Resembles a small, slender wood hoopoe with a black, sickle-shaped bill and black (not red) legs and feet. White bars on primaries and white tips to outer-tail feathers visible in flight. Male glossy black; female and juv. have brownish heads and shorter bills. Some individuals, probably juvs, lack white primary bar. **Voice:** Fairly high-pitched, whistling *'sweep-sweep-sweep'*. Also a harsh chattering alarm call. **Status and biology:** Common resident of dry savanna and open, broadleafed woodland. (Swartbekkakelaar)

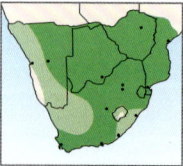

AFRICAN HOOPOE *Upupa africana*

b b b B B B b
J F M A M J J A S O N D

25–28 cm; 40–60 g Unmistakable, with its long, decurved bill, cinnamon, black and white plumage, and long, black-tipped crest that is raised when the bird is alarmed. Flight buoyant on broad, rounded wings. Sexes distinguishable in flight by the amount of white in the secondaries: male has a broad white central panel; female has three narrow white bars. Female duller with face and breast washed grey; less white at the base of the secondaries. Juv. dull buff below, with a shorter bill and crest; white wingbars tinged buff. **Voice:** *'Hoop-oop'* or *'hoop-oop-oop'*, typically all notes at the same pitch. **Status and biology:** Common resident and local nomad in savanna, broadleafed woodland, parks and gardens. Probes in ground for prey. (Hoephoep)

GREEN WOOD HOOPOE

♂

juv.

♀

green head & mantle

♂

♂

red feet

VIOLET WOOD HOOPOE

♀

juv.

violet & coppery head & mantle

♂

black, sharply decurved bill

black feet

ad.

♀

♂

COMMON SCIMITARBILL

juv.

erects crest when alarmed

AFRICAN HOOPOE

♀

♀

♂

297

BARBETS AND TINKERBIRDS
Mainly frugivorous birds with large, robust bills used to dig cavities in dead wood for roosting and breeding. Flight direct, often with noisy wing beats. Sexes alike in most species.

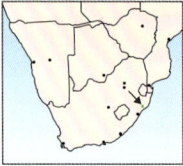

GREEN BARBET   *Stactolaema olivacea*

B										b	B
J	F	M	A	M	J	J	A	S	O	N	D

17 cm; 40–60 g A chunky, olive-green barbet confined to Ngoye Forest, inland from Mtunzini, KZN; other races occur in montane forests of E Africa. Birds in s Africa are slightly paler below than birds to the north, with a darker head and yellow-green ear coverts. Underwing coverts yellowish-white. In flight, shows pale primary bases. Larger, darker and more heavily built than Green Tinkerbird (p. 300). Juv. duller, with bare yellow skin on face; lacks yellow ear coverts. **Voice:** Hollow, repetitive *'tjop tjop tjop'*; regularly duets. **Status and biology:** Fairly common within its very restricted range, but easily overlooked if not calling. (Groenhoutkapper)

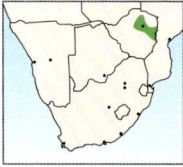

WHYTE'S BARBET   *Stactolaema whytii*

B									B	B	b
J	F	M	A	M	J	J	A	S	O	N	D

16 cm; 40–55 g A brown barbet with a prominent white wing panel, pale belly and diagnostic yellowish frons and lores, extending below and behind eye. Bases of outer primaries are white, giving a small, second white panel in folded wing. Paler than White-eared Barbet; lacks white ear patches. Juv. has pale base to bill and less extensive yellow on face. **Voice:** Deep hooting *'coo'*, repeated 1 or 2 times per second; alarm call *'skreeek'*. **Status and biology:** Locally common resident and nomad in miombo woodland and riverine forest, usually near fig trees. (Geelbleshoutkapper)

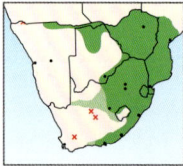

BLACK-COLLARED BARBET   *Lybius torquatus*

b	b	b	b			b	b	B	B	B	
J	F	M	A	M	J	J	A	S	O	N	D

18–20 cm; 48–62 g A fairly large barbet with a bright red face and throat, broadly bordered with black. Rare yellow morph has yellow face and throat. Juv. has a dark brown head and throat, streaked with orange and red; bill horn-coloured (not black). **Voice:** Duet that starts with harsh *'krrr krrrr'*, followed by ringing *'tooo puudly tooo puudly'*, the *'tooo'* being higher pitched. **Status and biology:** Common resident in forest, woodland, savanna and gardens; often in groups. (Rooikophoutkapper)

WHITE-EARED BARBET   *Stactolaema leucotis*

B	b	b	b			b	b	b	B	B	
J	F	M	A	M	J	J	A	S	O	N	D

18 cm; 48–60 g A dark brown barbet with a blackish head and prominent white ear stripes and belly. Darker than Whyte's Barbet, lacking yellow frons and lores and white wing patches. Juv. has paler base to bill. **Voice:** Loud, twittering *'treee treeetee teeetree'*; harsher *'waa waa'* notes. **Status and biology:** Common resident in coastal forest and bush, often near wetlands. Moves to lower altitudes in winter in e Zimbabwe. Usually in groups. (Witoorhoutkapper)

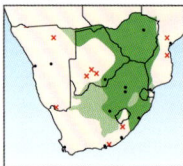

CRESTED BARBET   *Trachyphonus vaillantii*

B	B	b	b	b	b	b	b	b	B	B	B
J	F	M	A	M	J	J	A	S	O	N	D

24 cm; 60–80 g The only *Trachyphonus* barbet in s Africa, with a distinctive shaggy crest and black, yellow and red plumage. Juv. is browner and duller; breast band lacks white spots. **Voice:** Male gives sustained, trilling *'trrrrrrrr'*; female responds with repeated *'puka-puka'*. **Status and biology:** Common resident in woodland, savanna, riverine forest and gardens. Feeds on the ground more than other barbets. (Kuifkophoutkapper)

GREEN BARBET

yellow ear
coverts

WHYTE'S BARBET

yellow
forehead;
brown head

**BLACK-COLLARED
BARBET**

are yellow-
eaded
norph

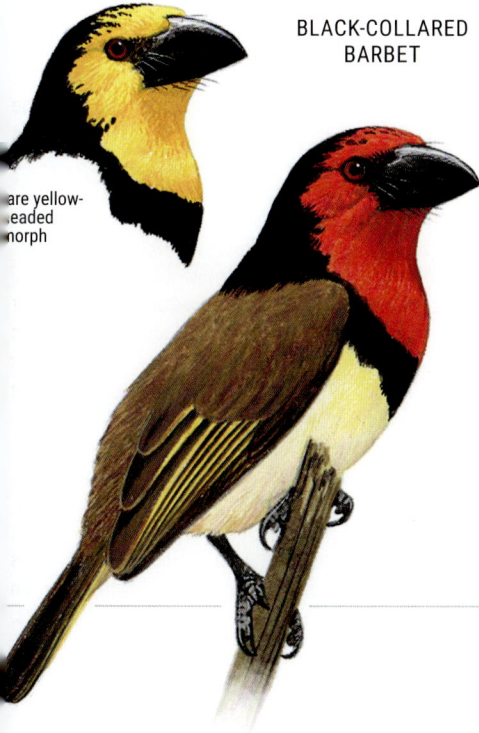

**WHITE-EARED
BARBET**

white ear stripe;
blackish head

CRESTED BARBET

299

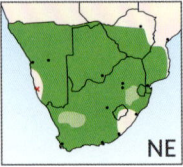

ACACIA PIED BARBET *Tricholaema leucomelas*

b b b b   b B B B b
J F M A M J J A S O N D

16–18 cm; 25–36 g A chunky barbet with a yellow-spotted black back, red-and-yellow forehead, and black-and-white-striped head; underparts white, apart from black throat. Lacks black moustachial stripes of much smaller Red-fronted Tinkerbird; wings less yellow; underparts white (not pale yellow). Juv. lacks red forehead; bill tipped yellowish. SW race *leucomelas* has darker, more streaked underparts than northern subspecies. **Voice:** Soft, low-pitched *'poop-oop-oop-oop'* and nasal *'nehh, nehh, nehh'* (toy trumpet) call. **Status and biology:** Common, near-endemic resident in woodland and savanna, especially arid acacia woodland; also stands of alien trees and gardens. (Bonthoutkapper)

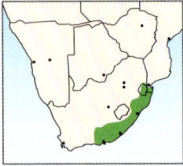

RED-FRONTED TINKERBIRD *Pogoniulus pusillus*

b   b b B B B
J F M A M J J A S O N D

13 cm; 13–20 g A small tinkerbird with white-streaked black upperparts and yellow-washed underparts. Darker above than Yellow-fronted Tinkerbird with a bright red (not yellow or orange) forehead and more extensive yellow feather edges on folded wing. Typically has a neater facial pattern. Separated from larger Acacia Pied Barbet by pale yellow (not black) throat and pale yellow (not white) underparts. Juv. has black (not red) forehead, but wing pattern differs from juv. Yellow-fronted Tinkerbird. **Voice:** Continuous, monotonous *'pop-pop-pop...'*, averaging slightly faster and higher pitched than Yellow-fronted Tinkerbird's call. **Status and biology:** Common resident in coastal forest. (Rooiblestinker)

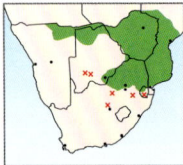

YELLOW-FRONTED TINKERBIRD *Pogoniulus chrysoconus*

b b b b b b B B B B
J F M A M J J A S O N D

11 cm; 11–16 g A small tinkerbird, best told from similar Red-fronted Tinkerbird by its yellow or orange (rarely red) forehead and less extensive yellow feather edges on folded wing. Forehead colour varies from pale yellow to bright orange. Juv. has black (not yellow) forehead, but wing pattern differs from juv. Red-fronted Tinkerbird. **Voice:** Continuous *'pop-pop-pop ...'* or *'tink tink tink ...'* that may continue unabated for several minutes, averaging slightly slower and deeper than Red-fronted Tinkerbird. **Status and biology:** Common resident in woodland and savanna. (Geelblestinker)

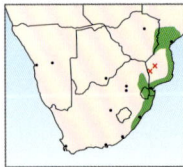

YELLOW-RUMPED TINKERBIRD *Pogoniulus bilineatus*

B b b   b b B B
J F M A M J J A S O N D

12 cm; 12–18 g A small tinkerbird with a plain, black back and crown, and 2 white stripes on sides of head. Small, yellow rump patch is not easy to see in the field. Juv. has black upperparts narrowly barred and spotted yellow-green; base of bill is paler. Lacks white stripe from base of bill, below eye to lower neck. **Voice:** *'Pop pop pop pop'*; a lower-pitched, more ringing note than that of Red- or Yellow-fronted Tinkerbirds, repeated in phrases of 3–9 notes, not continuously; also short trills. **Status and biology:** Common resident of forest, forest edge and dense woodland. (Swartblestinker)

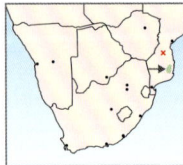

GREEN TINKERBIRD *Pogoniulus simplex*

b b   b
J F M A M J J A S O N D

10 cm; 8–10 g Rare, localised resident. A dull green tinkerbird with pale yellow wingbars and a yellow rump. Much smaller than Green Barbet (p. 298); ranges do not overlap. Juv. more olive-green above and greyer below than ad.; basal half of bill yellow. **Voice:** Very fast series of 8–10 *'pop'* notes, accelerating into a trill. **Status and biology:** Forages in the canopy of woodland and coastal forest; often in secondary bush. Poorly known in the subregion, but several singing males were discovered in Jan 2013 west of Unguana, 30–40 km inland in Inhambane, central Mozambique, close to where a specimen was collected in the 1950s. Likely to be a rare resident. (Groentinker)

ACACIA PIED
BARBET

large red
forehead

black
throat

RED-FRONTED
TINKERBIRD

small red
forehead

broad,
yellow-gold
wing patch

YELLOW-FRONTED
TINKERBIRD

yellow or
orange
forehead

speckled
shoulder

black
crown

striped
face

yellow
rump

YELLOW-RUMPED
TINKERBIRD

plain olive-green
upperparts

yellow
rump

GREEN
TINKERBIRD

WOODPECKERS
Attractively patterned birds with stiffened tails; mostly feed on tree branches, hammering off bark and probing for prey. Many species advertise territories by drumming. Flight undulating, often with noisy wing beats. Sexes usually differ in facial pattern; key identification features include pattern on face, back and underparts.

GOLDEN-TAILED WOODPECKER *Campethera abingoni*

J F M A M J J A S O N D

19–22 cm; 62–75 g Male has full red crown and red moustachial stripes; told from Bennett's Woodpecker by its streaked (not spotted) underparts and ear coverts. Back is greenish, barred pale yellow. Female has white-spotted black crown, red nape, and black-flecked moustachial stripes. *C. a. anderssoni* of semi-arid savanna in the west of its range is more heavily streaked below, sometimes appearing black throated. Paler than Knysna Woodpecker, with streaked (not spotted) underparts and more prominent barring on the back; ranges barely overlap. Juvs duller. **Voice:** Loud, nasal shriek 'wheeeeeaa', likened to a rusty nail being pulled from a plank of wood. Seldom drums. **Status and biology:** Common resident of woodland, thicket and coastal forest. (Goudstertspeg)

KNYSNA WOODPECKER *Campethera notata*

J F M A M J J A S O N D

19–21 cm; 60–70 g Both sexes are darker than Golden-tailed Woodpecker, with dense, blackish spotting (not streaks) on face and underparts; back darker green, less prominently barred golden; ranges barely overlap. Male's dark red forehead and moustachial stripes are heavily marked with black. Female has indistinct, black moustachial stripes and red of head restricted to hind crown. **Voice:** Loud, nasal shriek, 'wheeeeeaa' slightly softer than Golden-tailed Woodpecker's. Does not drum. **Status and biology:** NEAR-THREATENED by virtue of small range. Locally common, endemic resident in forest, riparian woodland, euphorbia scrub, milkwood thickets and alien acacias and eucalypts; sometimes moves into shrublands to forage. Easily overlooked if not calling. (Knysnaspeg)

BENNETT'S WOODPECKER *Campethera bennettii*

J F M A M J J A S O N D

22–24 cm; 65–80 g Male has full red crown and moustachial stripes contrasting with plain face and throat. Best told from male Golden-tailed Woodpecker by white throat and spotted (not streaked) breast. Female has diagnostic brown throat and cheek stripe; forecrown blackish with white spots. Sides of neck, breast and flanks are spotted in nominate race, but *C. b. capricorni* (n Namibia and nw Botswana) has little spotting, and is paler, with a yellow wash on underparts and rump. **Voice:** High-pitched, chattering 'wrrrrr, whirrr-itt, whrrr-itt...', often given in duet; does not drum. **Status and biology:** Fairly common resident of broadleafed woodland and savanna; often feeds on ground. (Bennettspeg)

BEARDED WOODPECKER *Chloropicus namaquus*

J F M A M J J A S O N D

23–25 cm; 70–90 g The largest arboreal woodpecker with a bold black-and-white face pattern and faintly barred back, wings and tail. Breast and belly lead-grey, diffusely barred with white. Male has red hind crown; black in female. Juvs of both sexes have more red in crown than ad. male. **Voice:** Loud, rapid 'wik-wik-wik-wik'; drums very loudly. **Status and biology:** Common resident in woodland, riverine forest and thicket, favouring areas with large, dead trees. (Baardspeg)

SPECKLE-THROATED WOODPECKER
Campethera scriptoricauda

J F M A M J J A S O N D

20–20 cm; 60–70 g Sometimes considered a subspecies of Bennett's Woodpecker, but is slightly smaller with a distinct facial pattern. Male has fine speckling on throat (not plain white); female has a lightly speckled (not brown) throat, densely speckled (not white) moustachial stripes and white (not brown) cheek stripe. Told from smaller Green-backed Woodpecker (p. 306) by its prominent malar stripes. **Voice:** High-pitched, chattering 'whirrr-itt-whirrr-itt', similar to Bennett's Woodpecker. **Status and biology:** Fairly common resident of dense woodland and thicket between Beira and the lower Zambezi River in Mozambique. (Tanzaniese Speg)

streaked
cheeks

♀

♂

KNYSNA
WOODPECKER

♀

blotched
underparts
(all)

♂

capricorni

♂

BENNETT'S
WOODPECKER

plain
underparts

♂

plain throat,
red malar stripe

spotted
necklace
& upper
breast

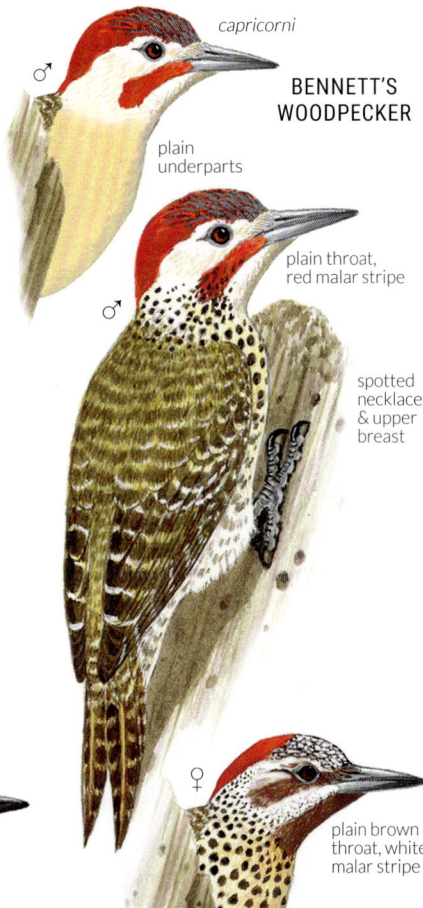

streaked
underparts
(all)

GOLDEN-TAILED
WOODPECKER

very dark
throat

anderssoni

juv.

♀

plain brown
throat, white
malar stripe

spotted
underparts

black & white
face pattern

♂

BEARDED
WOODPECKER

grey
underparts,
finely barred
white

♀

yellowish lower
mandible

speckled
throat

♀

SPECKLE-THROATED
WOODPECKER

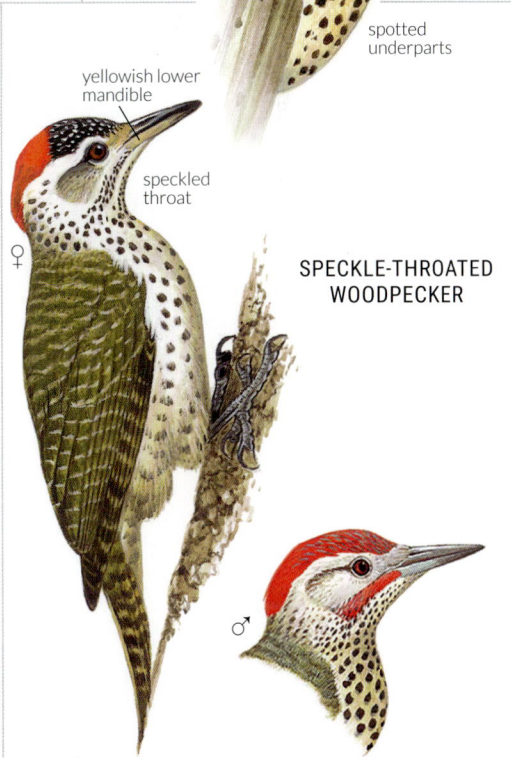

♂

305

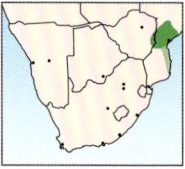

GREEN-BACKED WOODPECKER *Campethera cailliautii*

b b | | | | | | b b b b b
J F M A M J J A S O N D

15–18 cm; 35–46 g A small woodpecker with a rather short bill and plain face, lacking moustachial stripes. Male has a full red crown; female has a red nape. Back green, spotted pale yellow; rump and wings finely barred. Underparts creamy, spotted blackish. Both sexes distinguished from Golden-tailed (p. 304) and Cardinal Woodpeckers by having spotted (not streaked) underparts. Distinguished from larger Bennett's and Speckle-throated Woodpeckers (p. 304) by lack of malar stripes. Juv. resembles female, but red on nape is reduced or absent. **Voice:** High-pitched, whining 'whleeee'; also drums softly. **Status and biology:** Uncommon to locally common resident of forest edge, riparian forest, broadleafed woodland, and thicket. (Gevlekte Speg)

CARDINAL WOODPECKER *Dendropicos fuscescens*

| | | | | | b B B B b b
J F M A M J J A S O N D

14–16 cm; 20–36 g A small woodpecker with a streaked breast and belly. Back boldly barred in south, becoming plainer in northeast. Both sexes have a plain, whitish face with a bold, black moustachial stripe. Forecrown brownish, merging into red hind crown in male; black in female. Slightly smaller than Green-backed Woodpecker: both sexes distinguished by barred (not spotted) upperparts, streaked (not spotted) underparts, and presence of bold, dark malar stripe. Juv. duller and greyer with red on crown of both sexes. **Voice:** High-pitched, rather dry, rattling 'krrrek krrrek krrrek' or 'kik-ik-ik krrrek krrrek...'; also drums softly. **Status and biology:** Common resident in most wooded habitats, from forest edge to dry thornveld. (Kardinaalspeg)

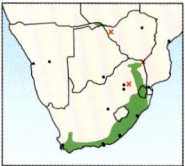

OLIVE WOODPECKER *Dendropicos griseocephalus*

| | | | | | b B B b
J F M A M J J A S O N D

17–19 cm; 40–50 g A plain woodpecker with a grey head, olive body and red rump. Male has a red crown; female grey. Isolated population of *D. g. ruwenzori* in e Zambezi Region (Caprivi) has red belly patch. Juvs of both sexes resemble male. **Voice:** Loud 'weet' or 'weet-er', repeated at intervals; seldom drums. **Status and biology:** Common resident in forest and dense woodland; often in small forest patches and near forest edge. (Gryskopspeg)

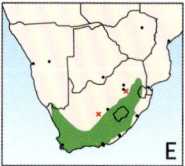

GROUND WOODPECKER *Geocolaptes olivaceus*

| | | | | | b B B b b b
J F M A M J J A S O N D

24–30 cm; 110–130 g A large, olive-grey woodpecker of open country. Pinkish-red belly and rump, pale eyes and cream-barred wings and tail are diagnostic. Ad. male has red moustachial stripes; female has dark grey. Female and juv. have less pink on belly and rump. **Voice:** Far-carrying 'dwerr' or 'tik-werr'; ringing 'ree-chick'; alarm call a harsh 'pee-aargh'. **Status and biology:** NEAR-THREATENED. Endemic. Locally common resident on rocky hill slopes in fynbos, Karoo and grassland; not associated with trees. Usually in small family parties. Has become scarce in many areas (Grondspeg)

WRYNECKS Cryptically patterned birds related to woodpeckers but with soft (not stiffened) tail feathers. Sexes alike.

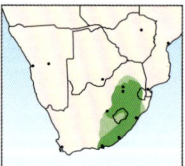

RED-THROATED WRYNECK *Jynx ruficollis*

b b | | | | | b b b B b b
J F M A M J J A S O N D

19 cm; 46–58 g A peculiar thrush- or warbler-like bird with a wedge-shaped bill, rufous throat and breast, and mottled, nightjar-like plumage. Juv. paler, with less prominent rufous wash on breast. Upperparts and belly darker, more heavily barred. **Voice:** Series of 2–10 squeaky 'kweek' notes, lower pitched in male; also repeated, scolding 'peegh'. **Status and biology:** Locally common resident in grassland and open savanna, woodland, forest edge, plantations and gardens. Feeds on ground, moving jerkily like a woodpecker. (Draaihals)

GREEN-BACKED WOODPECKER

stumpy dark bill

no malar stripe

♂

♀

speckled throat

spotted underparts (all)

juv.

CARDINAL WOODPECKER

brown frons (all)

♀

juv. ♂

♂

streaked underparts (all)

OLIVE WOODPECKER

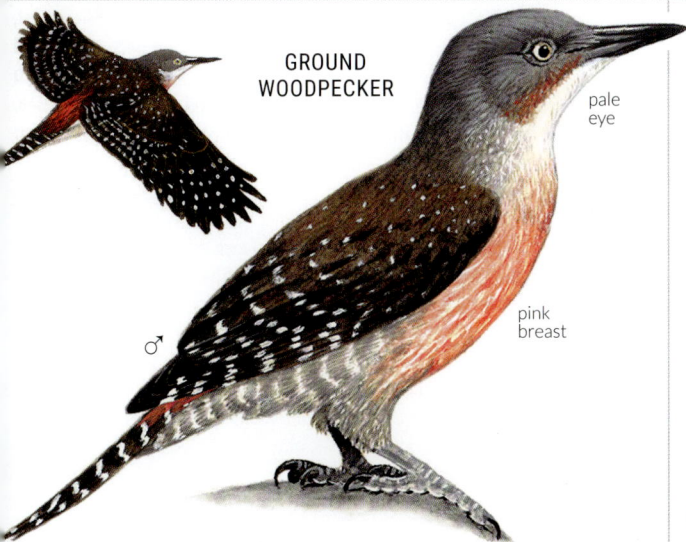

♀

♂

plain body

juv. ♂

GROUND WOODPECKER

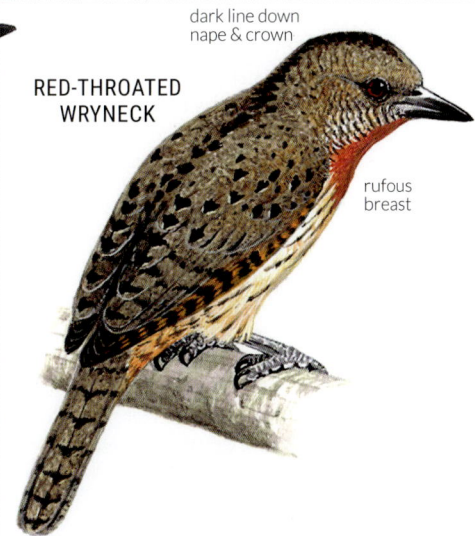

♂

pale eye

pink breast

RED-THROATED WRYNECK

dark line down nape & crown

rufous breast

307

LARKS

Cryptic, ground-dwelling birds, most likely to be confused with pipits (p. 320), which are easily overlooked when not calling or displaying. Sexes alike, except in sparrow-larks. Short-lived juv. plumages are typically characterised by pale fringes to upperpart feathers and diffuse breast spots. Many species show marked geographic variation in plumage colour.

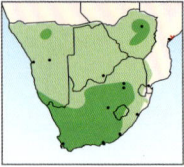

RED-CAPPED LARK *Calandrella cinerea*

14–15 cm; 20–26 g A long-winged, slender lark with a short, blackish bill and plain white underparts. Ad. has distinctive rufous cap and epaulettes on the sides of the breast; these are larger and more prominent in males. In flight, dark brown wings are more pointed than other larks, with broad bases. Juv. dark brown above, feathers edged white; breast heavily spotted. Widespread *C. c. cinerea* is intermediate in coloration between dark, richly coloured *C. c. saturatior* (ne Namibia to nw Zimbabwe) and much paler *C. c. spleniata* (nw Namibia). Juv. lacks rufous cap and epaulettes of ad. Upperparts dark brown, spotted with white; pale rufous breast heavily spotted with dark brown. **Voice:** Flight call is a sparrow-like *'tchweerp'*. Male song is a sustained jumble of melodious, high-pitched phrases, given during high, undulating display flight. **Status and biology:** Common resident, local nomad and intra-African migrant, found in short-grassy areas and croplands. Often in flocks. (Rooikoplewerik)

LARGE-BILLED LARK *Galerida magnirostris*

18 cm; 35–48 g A robust, heavily built lark with a thick-based bill that has a yellow base to the lower mandible. Compared to large-billed forms of Sabota Lark, it has a broad buff (not white) supercilium and more heavily streaked underparts. Upperparts sandy grey-brown, streaked dark brown; wings dark brown, lacking rufous panels in flight. It has a small crest that is raised when the bird is alarmed or singing. Juv. broad, buff tips to upperpart feathers; breast streaking more diffuse. **Voice:** A far-carrying, ascending *'troo-lee-liiii'*, like a rusty gate being opened. Highly vocal; mimics other species. **Status and biology:** Endemic. Common resident in grassland, arid scrubland and agricultural lands. (Dikbeklewerik)

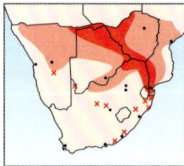

DUSKY LARK *Pinarocorys nigricans*

19–20 cm; 30–46 g A large, thrush-like lark, with a striking black-and-white facial pattern, heavily streaked breast and strange wing-flicking behaviour. Superficially similar to Groundscraper Thrush (p. 352), but is smaller, with a whitish ring around the eyes (not a vertical dark line through the eyes), no wingbars and shorter, whitish (not brown) legs. Forages on the ground, but often perches on trees or bushes. Male is slightly darker above and whiter below than female, with stronger facial markings. Juv. has broader buff margins to upperpart feathers, appearing scaly in fresh plumage; breast buffy with faint streaking. **Voice:** A soft *'chrrp, chrrp'*, uttered when flushed. **Status and biology:** Locally common intra-African migrant that breeds in central African woodlands; occurs in open grassy areas in woodland and savanna, often in recently burnt areas; regularly in loose flocks. (Donkerlewerik)

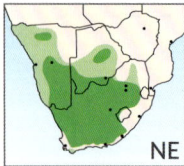

SPIKE-HEELED LARK *Chersomanes albofasciata*

13–15 cm; 20–34 g A fairly small lark with a characteristic upright stance and short, white-tipped tail (tips conspicuous in flight). Bill long and slightly decurved. Coloration variable, but the white throat contrasts with a rufous or buffy breast and belly; upperparts scaled due to buff fringes to feathers. Almost invariably in small groups; 1 bird often stands sentry on a low bush while the others forage. Juv. darker above with whitish feather margins; breast diffusely spotted. **Voice:** A trilling *'trrrep, trrrep, trrrep'*. **Status and biology:** Near-endemic. Common resident in a wide range of habitats, from moist grassland through Karoo shrubland and semi-desert grassland to gravel plains. (Vlaktelewerik)

SABOTA LARK *Calendulauda sabota*

15 cm; 21–31 g A medium-sized lark; more compact and heavily streaked than Fawn-coloured Lark (p. 310); differs in having malar stripes and lacking rufous wing panel. The prominent white supercilium extends from frons to nape, giving it a capped appearance. Breast boldly streaked, contrasting with pale throat and belly. Upperpart colour varies regionally, but always lacks rufous in the wing. Bill size varies considerably, with large-billed forms ('Bradfield's Lark') typically in more arid areas in the west of its range. Juv. is darker above with white feather tips, appearing spotted. **Voice:** A jumbled song of rich, melodious *'chip'* notes and twitterings; mimics other birds. Often calls from an elevated perch such as a treetop or telephone wire. **Status and biology:** Near-endemic. Common resident in arid savanna and Nama-Karoo. (Sabotalewerik)

spleniata

generally pale appearance

LARGE-BILLED LARK

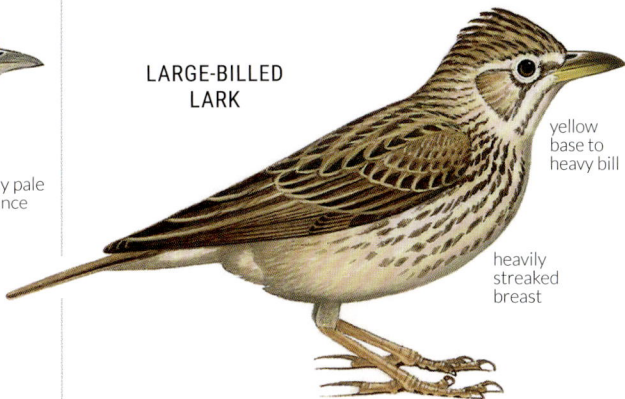

yellow base to heavy bill

heavily streaked breast

RED-CAPPED LARK

juv.

wing-lifting behaviour

red cap & shoulder

whitish underparts

saturatior

ad.

dark back

DUSKY LARK

bold facial pattern

pale legs

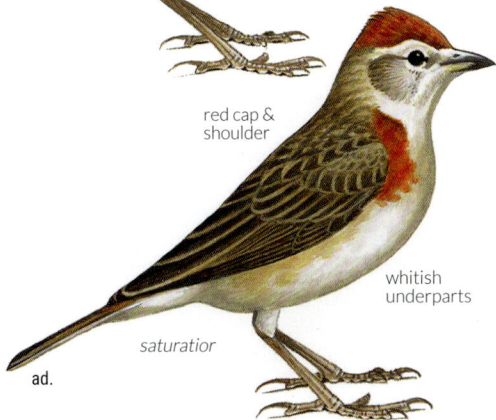

SPIKE-HEELED LARK

long, decurved bill

pale throat

ale races

rufous races

short white-tipped tail

bold, whitish supercilium

malar stripe

SABOTA LARK

thin-billed race

faintly streaked breast

thick-billed race

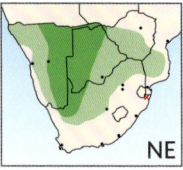

FAWN-COLOURED LARK *Calendulauda africanoides*

B B B b | | | | | b B B B
J F M A M J J A S O N D

14–16 cm; 21–30 g A medium-sized lark with a fairly long tail, broad, white supercilium and white throat; lacks malar stripes of Sabota (p. 308) and Red Larks; usually found on sandy soils. Upperpart feathers dark brown with broad, buff margins; wing with rufous panel. Underparts mainly white; breast buffy with brown streaks. Smaller than Red Lark, with paler, more streaked upperparts. Juv. is darker above with white feather tips, appearing spotted. **Voice:** A jumble of harsh *'chip'* notes and twitterings lasting 3–4 seconds and ending in a buzzy slur, given from a treetop or during short aerial display flight. **Status and biology:** Near-endemic. Fairly common resident in Kalahari scrub, broadleafed woodland, savanna and thornveld. (Vaalbruinlewerik)

THE KAROO LARK COMPLEX A group of 4 larks related to Fawn-coloured and Sabota (p. 308)

Larks with rather long, heavy tails and lack rufous in the wing; bills range from slender to fairly stout. Sexes alike, but males larger, with longer bills. Males sing from the ground, low perches or in a fluttering aerial display. Songs are stereotyped, but with up to 5 song types in any area. All occur on sandy soils; plumage colour varies in relation to soil colour. Identification depends on size, shape and plumage streaking; ranges typically do not overlap; Barlow's Lark sometimes lumped with Dune Lark. Juvs of all species have buff tips and darker brown subterminal bars to upperpart feathers, as well as diffuse brown breast spots.

KAROO LARK *Calendulauda albescens*

| | | | | | b b b b b
J F M A M J J A S O N D

16–17 cm; 25–33 g The smallest and most heavily streaked lark in the Karoo Lark complex; the only species that has streaking on the flanks; bill small and slender. White supercilium and throat contrast with dark ear coverts and boldly streaked breast. Upperparts range from sandy brown (coastal *C. a. albescens* and *codea*), through reddish (w Karoo *C. a. guttata*), to dark brown (e Karoo *C. a. karruensis*). Identification is most problematic around contact zone with Barlow's Lark. **Voice:** Male song *'chleeep-chleeep-trrr-trrrrrrr'*, shorter and higher pitched than Barlow's Lark, with only 1 or 2 lead-in notes; much higher pitched than Red Lark. Often located by guttural alarm calls or strident contact calls. **Status and biology:** Endemic. Common resident in Karoo and coastal shrublands. Hybridises with Barlow's Lark in a narrow contact zone. (Karoolewerik)

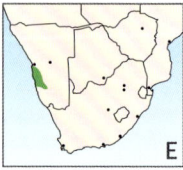

DUNE LARK *Calendulauda erythrochlamys*

b b b | | | | | b b b b b
J F M A M J J A S O N D

17–18 cm; 26–33 g The plainest lark in the Karoo Lark complex; slightly longer legged than other species. Differs from Barlow's Lark by its unstreaked upperparts and fine, rufous breast streaks (streaking darker and generally heavier in Barlow's Lark), and longer song. Some Barlow's Larks inland from Lüderitz are virtually indistinguishable from Dune Larks, but ranges not known to overlap. Plain, sandy-coloured upperparts could lead to confusion with much smaller Gray's Lark (p. 318), but habitat differs (dunes, not gravel plains). **Voice:** Male song a series of 10+ *'tip-ip-ip-ip'* lead-in notes followed by a whistle and long, uniform trill. Also has strident contact calls and a chittering alarm call. **Status and biology:** Endemic. Fairly common resident in sparsely vegetated areas among dunes in the Namib Desert between the Koichab and Kuiseb rivers. (Duinlewerik)

BARLOW'S LARK *Calendulauda barlowi*

| | | | | | b b b | |
J F M A M J J A S O N D

17–18 cm; 26–36 g Slightly larger and longer billed than Dune and Karoo Larks. Differs from Karoo Lark in having plain (not streaked) flanks, and often appears 'bull-necked'. Coastal *C. b. patae* is pale sand-brown above; inland *cavei* and *barlowi* reddish. Nominate *barlowi* closely resembles Dune Lark in the north, but usually has brown (not rufous) breast streaks and some dark brown streaks in tertials and central tail feathers. **Voice:** Male song is similar to Dune Lark, but with fewer (5–8) lead-in notes and a shorter trill. Also has strident contact calls and a chittering alarm call. **Status and biology:** Endemic. Locally common resident in arid scrubland and vegetated dunes. Hybridises in a narrow contact zone with Karoo Lark between Port Nolloth and the Orange River, N Cape. (Barlowlewerik)

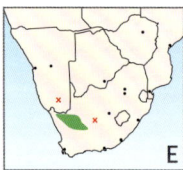

RED LARK *Calendulauda burra*

b b b b b | | | b B b b
J F M A M J J A S O N D

18–19 cm; 32–43 g The largest lark in the Karoo Lark complex, with a shorter, deeper bill than other species. Upperpart colour varies from plain red on dunes to brown with darker streaks on the plains. Range is not known to overlap with Karoo Lark. Red Lark is appreciably larger with a much heavier bill and plain (not streaked) flanks. Overlaps with Fawn-coloured Lark in north, but is much larger, with a plainer, darker red-brown (not buffy) back. Juv. has upperparts brighter than ad.; crown and back feathers tipped whitish. **Voice:** Song starts with 4 or 5 repeated notes and ends with a rattled trill *'chee-chee-chee-chee, chrrrrrreee'*, slower and much lower pitched than Karoo Lark; sings in the air or from a bush. Short *'chrrk'* given when flushed. Also has strident contact calls and a chittering alarm call. **Status and biology:** Endemic. VULNERABLE due to habitat degradation through overgrazing. Locally fairly common in scrub-covered sand dunes and on Nama-Karoo plains. (Rooilewerik)

FAWN-COLOURED LARK

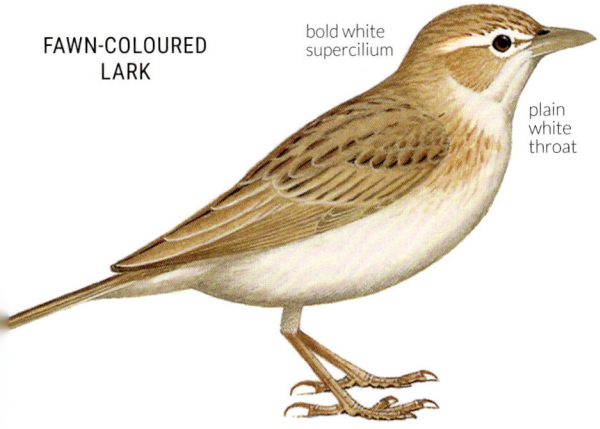

bold white supercilium

plain white throat

KAROO LARK

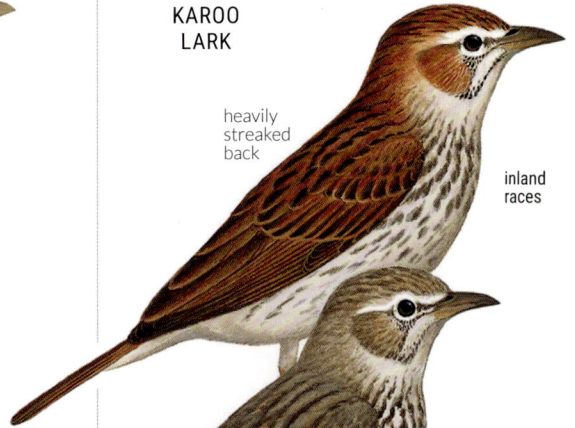

heavily streaked back

inland races

coastal races

DUNE LARK

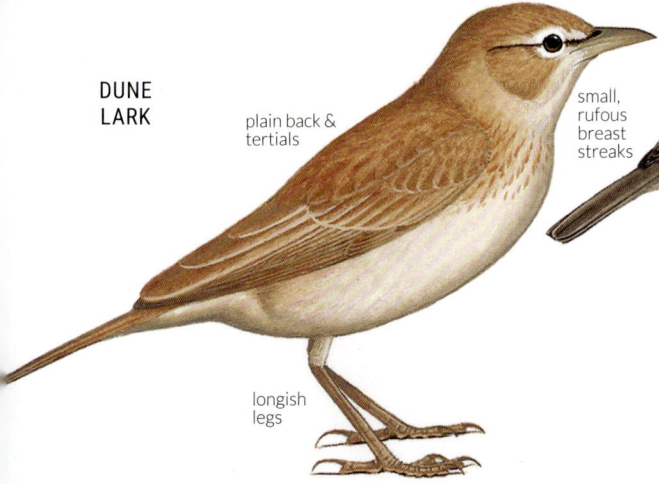

plain back & tertials

longish legs

small, rufous breast streaks

streaked flanks

BARLOW'S LARK

finely streaked or plain back

RED LARK

stout bill

plain or streaked back

cavei

311

MONOTONOUS LARK *Mirafra passerina*

B b B | | | | | | | b b b
J F M A M J J A S O N D

14 cm; 21–28 g A medium to small, compact, stout-billed lark. Plumage similar to Melodious Lark; it is also slightly larger, with a shorter tail, white (not buff) belly and less distinct facial markings, lacking a strong supercilium; best distinguished by habitat and call. White throat contrasts with relatively plain, buff breast, and is prominent when singing. In flight, distinguished from paler Stark's Lark (p. 316) by its chestnut wing patches. **Voice:** Usually sings from a bush or tree; a monotonous *'trrp-chup-chip-choop'* (rendered 'purple jeep') day and night. Short display flight from a perch, rising 15–20 m, calling all the time, with plumage fluffed. **Status and biology:** Near-endemic. Common resident and nomad in thornveld and mopane woodland with sparse grass cover. Summer visitor to south of range; irruptive in high-rainfall years. Inconspicuous when not calling. (Bosveldlewerik)

MELODIOUS LARK *Mirafra cheniana*

B b b | | | | | | | b b B B
J F M A M J J A S O N D

12 cm; 13–15 g A small, compact grassland lark with a broad, whitish supercilium. Plumage similar to Monotonous Lark (see above for differences); best distinguished by habitat and call. Much smaller than Fawn-coloured Lark (p. 310), with darker centres to upperpart feathers and buffy underparts. Shares white throat with Pink-billed Lark (p. 316), but has a longer, brown (not short, pink) bill, longer tail, and prominent white supercilium. Juv. darker above with broad, buff margins to feathers. **Voice:** Song jumbled and melodious, mainly mimicking other birds; usually given during high (up to 50 m), protracted (sometimes >40 min) aerial display. Alarm call *'chuk chuk chucker chuk'*. **Status and biology:** NEAR-THREATENED. Endemic. Locally fairly common resident in grassland and pastures. Inconspicuous when not displaying. (Spotlewerik)

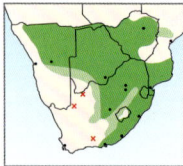

RUFOUS-NAPED LARK *Mirafra africana*

B b b | | | | b b b B B B
J F M A M J J A S O N D

15–18 cm; 34–50 g The most common large lark of savanna and grassland, characterised by a small, erectile crest, rufous nape and a fairly long, decurved bill. Males sing year-round from the tops of bushes, termite mounds and poles. In flight, has extensive rufous wing panels like Flappet and Eastern Clapper Larks, but is larger and longer billed. Plumage varies regionally; generally paler in more arid areas. Juv. darker above than ad., with pale tips to feathers; breast has diffuse dark spots. **Voice:** Male song a repetitive, trisyllabic *'tree tree-leeooo'*, with numerous variations; sometimes preceded by rapid wing vibrating. Fluttering display flight accompanied by a jumbled song often including mimicry of other species' calls. **Status and biology:** Common resident in grassland, savanna and old fields, typically where there are suitable song perches. (Rooineklewerik)

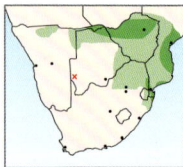

FLAPPET LARK *Mirafra rufocinnamomea*

B B b | | | | | | b B B
J F M A M J J A S O N D

14–15 cm; 21–32 g A fairly small, compact, dark-backed lark, recalling a small, dark Rufous-naped Lark, but with a shorter bill and much less rufous in the wing in flight. Smaller than Eastern Clapper Lark, with a slightly longer tail, darker back and different aerial display. Tertials are streaked (not barred); in area of overlap usually occurs in more wooded habitats. Juv. is darker above than ad., with broad, buffy fringes to feathers; breast diffusely spotted dark brown. **Voice:** Characteristic display flight includes brief bursts of rapid wing clapping (2 or 3 phrases, separated by several seconds), but no song. Gives a short *'tuee'* when perched. **Status and biology:** Common resident in grassland, savanna and open woodland. (Laeveldklappertjie)

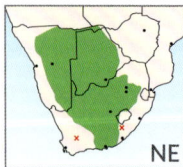

EASTERN CLAPPER LARK *Mirafra fasciolata*

b b b | | | | | b B B B
J F M A M J J A S O N D

15 cm; 26–44 g A medium-sized, compact lark; larger than Cape Clapper Lark, with less heavily barred, more rufous upperparts and paler underparts, more extensive rufous in the wing in flight, a paler, heavier bill, and different display. Southern grassland birds are rufous above and buff below, resembling a small Rufous-naped Lark, but with a shorter bill and no crest; song distinctive. Similar to Flappet Lark, but occurs in more open habitats and has a slightly shorter tail and barred (not streaked) tertials. **Voice:** Climbs steeply with trailing legs in aerial display, clapping its wings, then parachutes down, giving a long, ascending whistle *'pooooeeee'*, sometimes followed by other notes. Wing-clapping is slow: only 12–14 claps per second. Often mimics other birds. **Status and biology:** Near-endemic. Common resident in grassland and open savanna. (Hoëveldklappertjie)

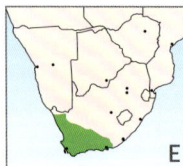

CAPE CLAPPER LARK *Mirafra apiata*

| | | | | | | | b B B b
J F M A M J J A S O N D

15 cm; 23–38 g Smaller and darker than Eastern Clapper Lark, with richly barred black-and-rufous upperparts, appearing dark from a distance; underparts buffy-rufous, with darker breast streaks; bill more slender and darker grey-brown (not pale horn-coloured). In flight, rufous in the wings is much less prominent. South coast *M. a. marjoriae* is greyer above; sometimes treated as a separate species, Agulhas Clapper Lark. Unobtrusive when not displaying; reluctant to flush. **Voice:** A long, ascending whistle *'pooooeeee'* (*M. a. apiata*) or 2 descending whistles *'tseeoo tseeuuuu'* (*M. a. marjoriae*), preceded by fast wing-clapping at 25–28 claps per second. Rarely mimics other birds. **Status and biology:** Endemic. Common resident in Karoo scrub, coastal fynbos and rank old fields; favours areas rich in restios. Rarely observed, unless displaying. (Kaapse Klappertjie)

MONOTONOUS LARK

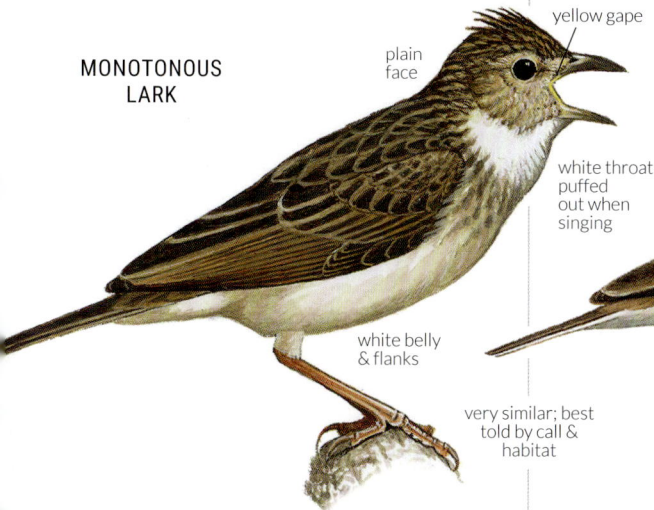

plain face

yellow gape

white throat puffed out when singing

white belly & flanks

very similar; best told by call & habitat

MELODIOUS LARK

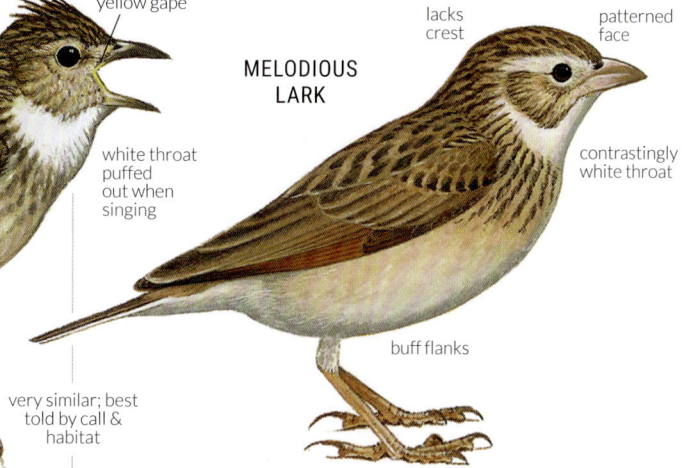

lacks crest

patterned face

contrastingly white throat

buff flanks

RUFOUS-NAPED LARK

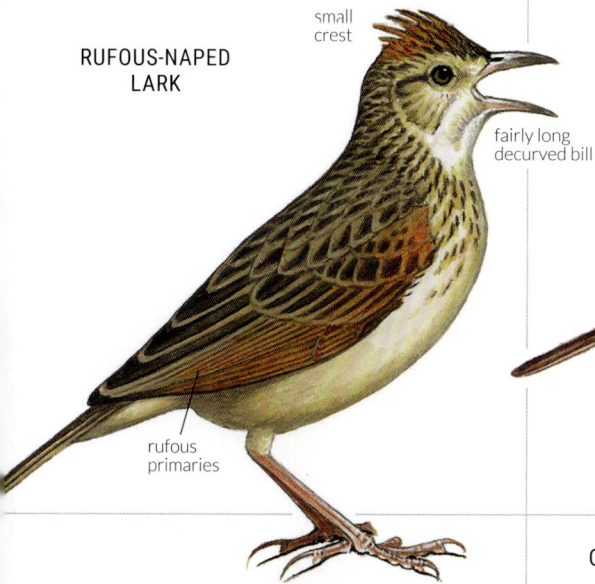

small crest

fairly long decurved bill

rufous primaries

FLAPPET LARK

dark brown above

tertials streaked

EASTERN CLAPPER LARK

pale heavy bill

rather plain rufous back

ntly red tials

CAPE CLAPPER LARK

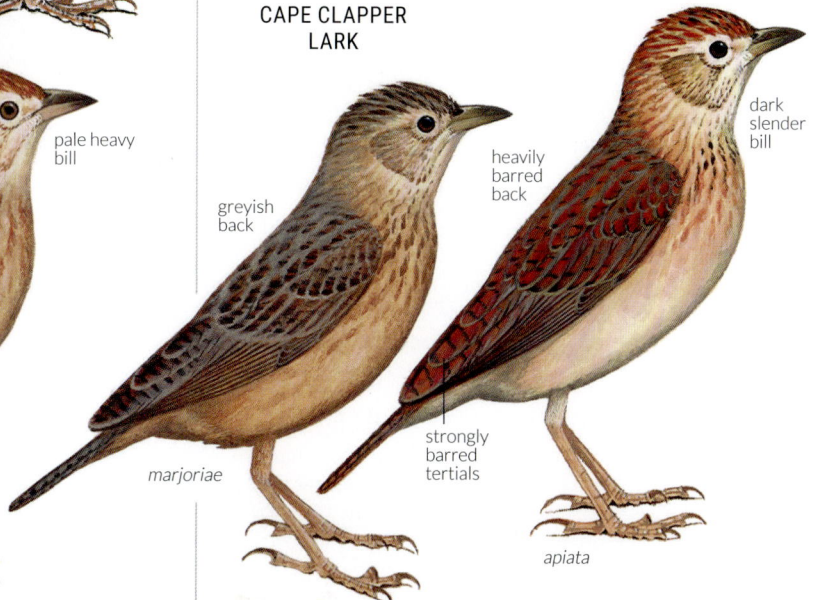

greyish back

heavily barred back

dark slender bill

strongly barred tertials

marjoriae

apiata

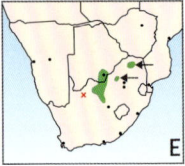

SHORT-CLAWED LARK *Certhilauda chuana*

B B b | | | | | b b B B B
J F M A M J J A S O N D

17–19 cm; 28–44 g A fairly large, slender lark; recalls a long-billed lark, but with distinctively patterned upperparts: feathers dark brown with broad, buff fringes; range barely overlaps with that of Eastern Long-billed Lark. The long, buff-white supercilium imparts a capped appearance. Lacks the rufous wings and nape, and erectile crest of Rufous-naped Lark (p. 312), but rufous rump visible in flight. Juv. has pale tips to upperpart feathers, appearing spotted above. **Voice:** Display flight with descending *'peeeeuuuuu'* whistle given on descent, similar to long-billed larks. Also a short *'chreep-chuu-chree'*, given when perched. **Status and biology:** Endemic. Uncommon to locally common resident in semi-arid thornveld and dry acacia savanna. (Kortkloulewerik)

THE LONG-BILLED LARK COMPLEX
Five large, slender larks that all have long tails and lack rufous in the wing. Bills are slender and decurved. Males sing from the ground, low perches or in a spectacular aerial display in which they fly close to the ground, then rise vertically 10–15 m, close their wings just before the top of the climb, call, and drop, opening their wings just above the ground. Identification depends on size, plumage colour and extent of streaking; ranges typically do not overlap. Males are appreciably larger and longer billed than females. Juvs of all species are darker above, with buff tips to upperpart feathers and diffuse brown breast spots.

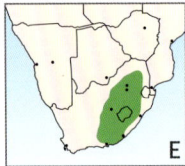

EASTERN LONG-BILLED LARK *Certhilauda semitorquata*

b | | | | | | | b B B b
J F M A M J J A S O N D

16–20 cm; 30–48 g The smallest, least-streaked and shortest-billed long-billed lark. Upperparts reddish, lightly streaked darker brown in west, virtually unstreaked in east. Buffy below, with light streaking confined to breast. Elongated shape and relatively short, straight bill (especially females) can result in confusion with pipits (p. 320), but different gait, reddish plumage and lack of pale outer-tail feathers diagnostic. **Voice:** A long, descending whistle *'seeeooooo'*. **Status and biology:** Endemic. Fairly common resident in grassland, generally on rocky hill slopes. (Grasveldlangbeklewerik)

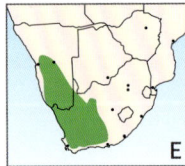

KAROO LONG-BILLED LARK *Certhilauda subcoronata*

b b b b | | | b b b b b
J F M A M J J A S O N D

18–22 cm; 31–55 g The most widespread and abundant long-billed lark. Upperparts vary from dark chocolate-brown in south to reddish in north; hindneck greyer. Streaking decreases in north, but belly and flanks always largely unstreaked (unlike Cape Long-billed Lark). In Breede River Valley, W Cape, differs from Agulhas Long-billed Lark in being rufous-brown (not buffy-brown) above, longer billed and longer tailed, and with plainer flanks, belly and undertail coverts. Not known to overlap with Eastern Long-billed Lark, which is less streaked. **Voice:** At close range a soft *'inhalation'* can be heard before the long descending whistle *'uh-seeeooooo'*. **Status and biology:** Endemic. Common resident in Karoo scrub and grasslands west of 25˚E; typically in rocky areas. (Karoolangbeklewerik)

BENGUELA LONG-BILLED LARK *Certhilauda benguelensis*

| | | b b | | | | | |
J F M A M J J A S O N D

18–20 cm; 35–53 g Replaces Karoo Long-billed Lark in n Namibia, but exact range uncertain because recognition as a distinct species is based on its large genetic difference; identification criteria require clarification. Resembles the northern form of Karoo Long-billed Lark, but tends to be slightly more heavily streaked on the crown, back and breast. **Voice:** A slightly quavering, long, descending whistle *'seeeoeeooooo'* given in aerial display, also from ground or perch. **Status and biology:** Near-endemic. Common resident on arid hill slopes and plains, apparently north of Brandberg. (Kaokolangbeklewerik)

CAPE LONG-BILLED LARK *Certhilauda curvirostris*

| | | | | | b b | | |
J F M A M J J A S O N D

20–24 cm; 40–60 g The largest lark in the region; bill long to very long. Upperparts are grey-brown with well-marked, dark streaking; underparts white, densely streaked blackish-brown, streaking extending onto flanks and belly. Northern *C. c. falcirostris* is larger and greyer, with truly impressive bill. **Voice:** Song a far-carrying, descending whistle *'seeeooooo'* in the north, 2-note *'whit seeeooooo'* in the south. Irritable *'whir-irry'* contact call. **Status and biology:** Endemic. Fairly common resident in coastal scrub, vegetated dunes and croplands, especially fallow and recently planted fields. (Weskuslangbeklewerik)

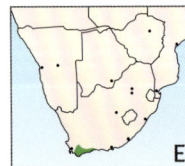

AGULHAS LONG-BILLED LARK *Certhilauda brevirostris*

| | | | | | b b | | |
J F M A M J J A S O N D

18–20 cm; 35–48 g Similar to the southern (nominate) race of Cape Long-billed Lark, but plumage is buffier and bill and tail shorter; their ranges are not known to overlap. Range abuts that of Karoo Long-billed Lark in the Breede River Valley, W Cape; differs by being buffy-brown (not rufous-brown) above and by having streaking extending onto the flanks and belly. **Voice:** Display song *'seeoo seeooo'*; first note longer than that of the nominate race of Cape Long-billed Lark. **Status and biology:** NEAR-THREATENED. Endemic. Fairly common resident in fallow fields, croplands, coastal fynbos and semi-arid Karoo scrub. (Overberglangbeklewerik)

SHORT-CLAWED
LARK

large buff
supercilium

heavily scaled
back

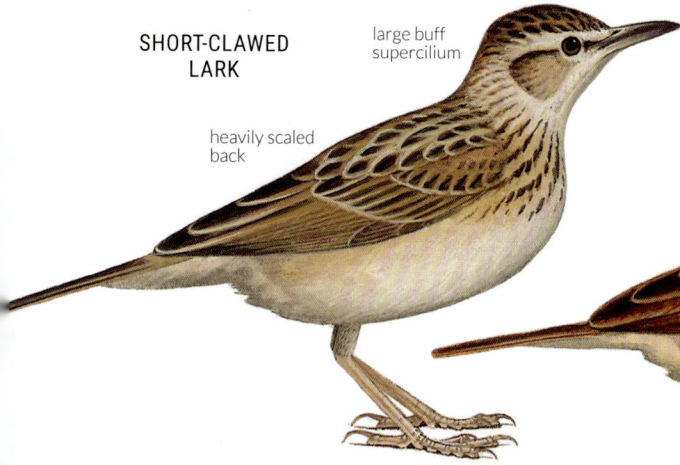

EASTERN
LONG-BILLED LARK

lightly streaked
back

plain
flanks

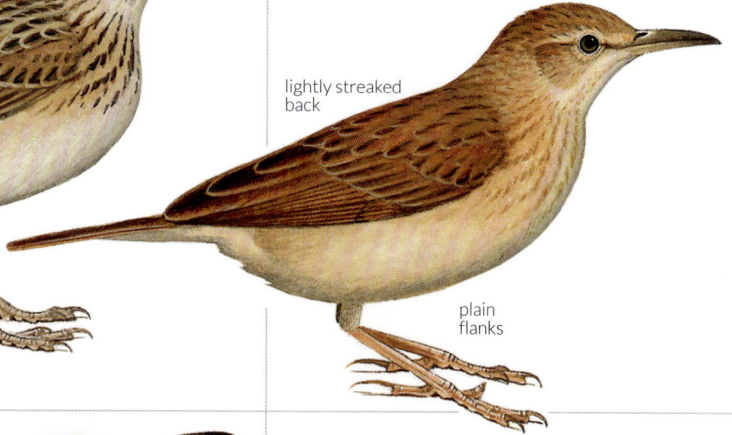

KAROO LONG-BILLED
LARK

grey nape
contrasts
with reddish-
brown back

plain
flanks

southern races

BENGUELA
LONG-BILLED
LARK

streaked
crown
& back

CAPE LONG-BILLED
LARK

very long bill

streaked
flanks

long tail

curvirostris

AGULHAS
LONG-BILLED
LARK

streaked
flanks

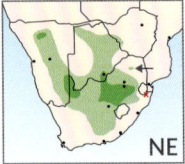

PINK-BILLED LARK   *Spizocorys conirostris*

12–13 cm; 12–17 g A small, compact lark; ad. has a distinctive short, conical, pink bill. Differs from Botha's Lark by its rich rufous-buff (not white) underparts, which contrast boldly with the white throat and unstreaked flanks; breast lightly streaked. In flight, outer tail edged buff (not white). Western races are paler below and can be confused with Stark's Lark, but are less grey and lack a crest; bill shorter and more conical; lacks white eye-ring. Overlaps marginally with Sclater's Lark, but has shorter, pink (not horn-coloured) bill and lacks dark 'teardrop' mark on face. Juv. has a blackish bill; upperparts darker brown with buffy spots; breast with diffuse brown spots. **Voice:** Gives a soft *'si-si-si'* when flushed. Male song mixes this call with short, sweet whistles, usually given during a short, jerky aerial display. **Status and biology:** Near-endemic. Common resident and nomad in grassland, pasture and desert scrub. (Pienkbeklewerik)

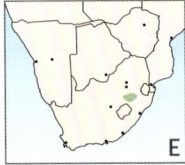

BOTHA'S LARK   *Spizocorys fringillaris*

12–13 cm; 16–21 g A small lark with a short, fairly stout bill and heavily streaked upperparts. Ad. bill orange-pink, slightly less stout than that of Pink-billed Lark; also differs by its more prominent white supercilium, heavier breast streaking, streaked flanks, white (not rufous) belly and white (not buff) outer-tail feathers. Juv. has white spots on crown, buff margins to flight feathers and diffuse brown breast spots; bill dull pinkish-brown. **Voice:** A cheerful, repeated *'chiree'*; a *'chuk, chuk'* given in flight. **Status and biology:** Endemic. ENDANGERED due to loss of habitat. Uncommon resident and local nomad in heavily grazed, upland grassland, often in small flocks of 3–6 birds. (Vaalrivierlewerik)

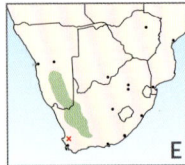

SCLATER'S LARK   *Spizocorys sclateri*

13–14 cm; 17–21 g A small, compact lark with a remarkably large, long, horn-coloured bill. Ground colour is buffy-brown, distinctly warmer than grey-brown Stark's Lark. At close range the dark brown 'teardrop' mark below the eye is diagnostic. In flight, has a dark triangle on the tail (broader at the tip), with the white outer tail broadening towards the tail base. Large, horn-coloured (not short, conical, pink) bill (often tilted skywards) and white (not buff) outer-tail feathers distinguish it from Pink-billed Lark. Juv. has pale spotted upperparts and diffuse breast spots; teardrop on face less distinct. **Voice:** A repeated *'tchweet-tchweet'*, given in flight. Displaying males give similar calls in a short aerial display. **Status and biology:** Endemic. NEAR-THREATENED. Uncommon resident and partial nomad in arid Nama-Karoo shrubland, favouring stony plains. Strongly associated with Kalkgras *Enneapogon desvauxii*. Regularly visits water to drink. (Namakwalewerik)

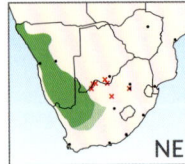

STARK'S LARK   *Spizocorys starki*

13–14 cm; 16–22 g A small, compact lark with an erectile crest, stubby, pale bill and whitish eye-ring, often with a slight squint to the eye. Much paler above and below than Sclater's and Pink-billed Larks. Could possibly be confused with Gray's Lark (p. 318), but has streaked (not plain) upperparts. Juv. spotted white above. **Voice:** A short *'chree-chree'*, given in flight. Male song is a melodious jumble of notes, given in a high, extended display flight. **Status and biology:** Near-endemic. Common resident and nomad in stony desert scrub, gravel plains and arid grassland. Large flocks follow rains, often in association with Grey-backed Sparrow-Larks. (Woestynlewerik)

RUDD'S LARK   *Heteromirafra ruddi*

14 cm; 25–28 g A small, large-headed lark with a short, thin tail. Legs long; gait often more erect than most larks, recalling Spike-heeled Lark (p. 308), but bill is short and broad, underparts are whitish, and has white edges to outer tail (not white tail tips). It has an erectile crest with a buff median stripe, but this can be hard to see in the field. Unobtrusive when not displaying (displays mainly Oct–Feb). Juv. spotted buff above. **Voice:** Males give a bubbling, whistled song, with 3 or 4-note *'pee-witt-weerr'* phrases repeated several times, before switching to a new variation; typically sung in high, protracted circular display flight that can last up to 30 min. **Status and biology:** Endemic. ENDANGERED due to habitat loss and degradation. Uncommon, localised resident in short, upland grassland. (Drakensberglewerik)

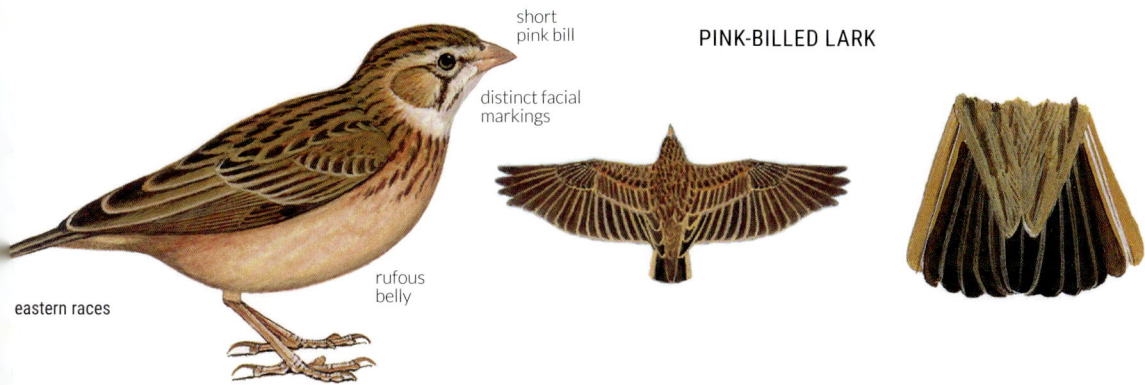

short pink bill

PINK-BILLED LARK

distinct facial markings

rufous belly

eastern races

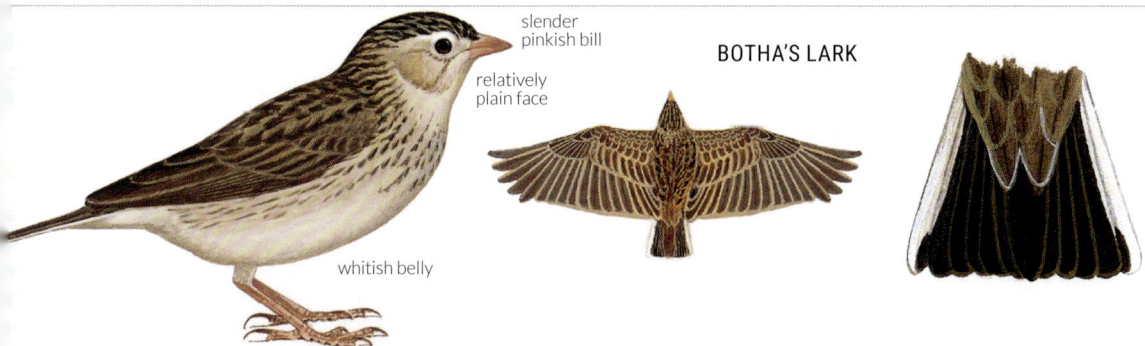

slender pinkish bill

BOTHA'S LARK

relatively plain face

whitish belly

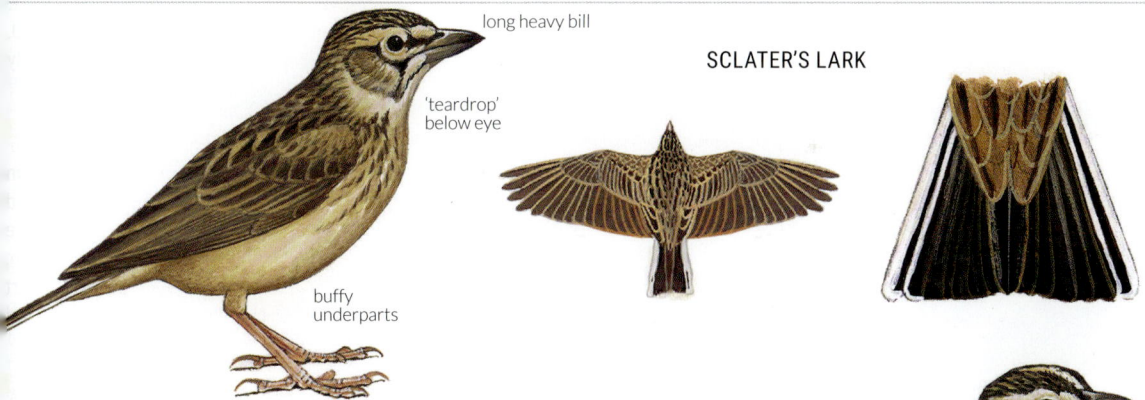

long heavy bill

SCLATER'S LARK

'teardrop' below eye

buffy underparts

erectile crest

pale eye-ring

large 'lollipop-like' head with buff crown stripe

STARK'S LARK

whitish underparts

RUDD'S LARK

short tail

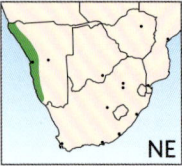

GRAY'S LARK *Ammomanopsis grayi*

b b b b b b b
J F M A M J J A S O N D

13–14 cm; 17–27 g A small, plain lark of desert gravel plains, with a fairly long, stout, pale bill with a dark tip; usually in small groups year-round. Does not co-occur with larger, pale-backed Dune Lark (p. 310). Distinguished from Stark's Lark (p. 316) by its unstreaked back, dark tip to the outer tail and rounded head (lacking a crest). Upperparts darker brown in n Namibia (*A. g. hoeschi*) than further south. Female Grey-backed Sparrow-Lark has more mottled upperparts and a shorter, more conical bill. The desert form of Tractrac Chat (p. 358) has a more upright stance, longer black (not whitish) legs, a dark, slender bill and a white rump and base to its tail. Juv. lightly mottled above. **Voice:** A short *'tseet'* or *'tew-tew'*, given in flight. Song is a high-pitched mix of descending, metallic-sounding *'tink'* notes and whistles, normally given before dawn. **Status and biology:** Fairly common, near-endemic resident and local nomad on coastal gravel plains; further inland occurs in more sandy habitats. Easily seen immediately north of Swakopmund, Namibia. (Namiblewerik)

SPARROW-LARKS (FINCH-LARKS) Small, compact larks with conical bills and marked sexual dimorphism in plumage. Boldly patterned males are easily identified, but females and juvs could be confused with other small larks; differ in having blackish underwing coverts. Some juv. males have a partial moult into ad. plumage, resulting in a transitional imm. plumage.

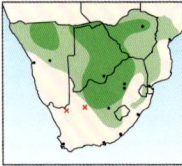

CHESTNUT-BACKED SPARROW-LARK *Eremopterix leucotis*

B B B B B B B b b b b b
J F M A M J J A S O N D

12–13 cm; 12–21 g Male's rich chestnut back and wings and contrasting black-and-white head are diagnostic; lacks white crown patch of male Grey-backed Sparrow-Lark; vent white (not black). Female is mottled buff and brown above. Darker than female Grey-backed Sparrow-Lark, with chestnut upperwing coverts, with black belly merging into heavily streaked chest (not defined black belly patch) and a more obvious white nuchal collar. Told from female Black-eared Sparrow-Lark by its black lower breast and belly, and pale rump. Juv. like female, but with pale-spotted upperparts; dark belly patch smaller or absent. **Voice:** A short *'chip-chwep'* given in flight; song a soft *'kree, kree, kree, hu-hu'*. **Status and biology:** Common resident and nomad in sparsely grassed savanna, cultivated lands and road verges, especially recently burnt areas; usually in flocks. (Rooiruglewerik)

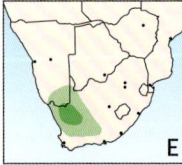

BLACK-EARED SPARROW-LARK *Eremopterix australis*

b b b b b b b b b b b b
J F M A M J J A S O N D

12–13 cm; 12–16 g Male black with broad, chestnut margins to back, wing and tail feathers; entirely black underwings are conspicuous in flight. Female is streaked chestnut and dark brown above and heavily streaked with black below; lacks the dark belly patch of other female sparrow-larks and has black secondaries visible in flight. Juv. like female, but spotted buff above. **Voice:** A short *'preep'* or *'chip-chip'*, given in flight; male has a butterfly-like aerial display accompanied by a buzzy song. **Status and biology:** Endemic. Locally common, nomad in Karoo shrubland and grassland, Kalahari sandveld, gravel plains and, occasionally, cultivated lands. Usually in flocks. (Swartoorlewerik)

GREY-BACKED SPARROW-LARK *Eremopterix verticalis*

b b B B B B b b b b b b
J F M A M J J A S O N D

12–13 cm; 13–21 g Greyish-brown (not chestnut) back and upperwings separate both sexes from Chestnut-backed Sparrow-Lark. Male also has a white patch on the hind crown and a black (not white) vent. Separated from female Chestnut-backed Sparrow-Lark by smaller, more defined, black belly patch, paler upperparts and less distinct (or absent) nuchal collar. Female's black belly patch and unstreaked flanks separate it from female Black-eared Sparrow-Lark. Juv. like female, but warmer brown with buff spots above. **Voice:** A sharp *'chruk, chruk'*, given in flight. **Status and biology:** Common, near-endemic nomad and local migrant in Karoo shrubland, semi-desert, grassland, arid savanna and cultivated land. Usually in groups; often congregates in huge flocks following good rains; regularly mixes with Chestnut-backed Sparrow-Larks in north of range. (Grysruglewerik)

GRAY'S LARK

faint shoulder smudge

oeschi

plain head & back

dark-tipped pale bill

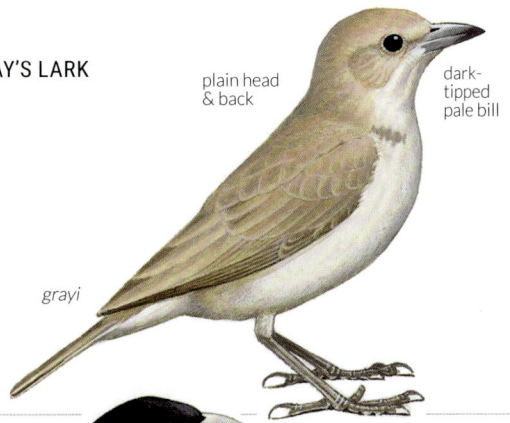

grayi

CHESTNUT-BACKED SPARROW-LARK

warm brown back

black crown

white cheek patch

♀

♂

♀

BLACK-EARED SPARROW-LARK

fully black head

♀

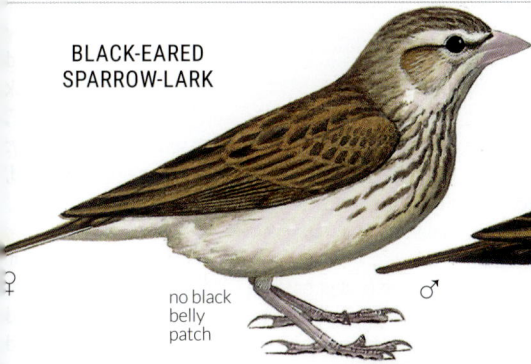

♀

no black belly patch

♂

♀

♂

GREY-BACKED SPARROW-LARK

pale crown

cheek & neck patches connected

pale upperparts

♀

♀

dark races

♂

♂

pale races

♂

PIPITS Pipits are ground-dwelling birds in the same family as longclaws and wagtails. Most are cryptically marked and superficially resemble larks, but are more slender, with longer legs and tails, more slender bills and a more horizontal stance. All are insectivorous and live and nest on the ground; a few species also perch in trees. Sexes alike. Many species look confusingly similar and within-species geographic variation often exceeds differences between species. Identification of larger species is particularly problematic, with considerable individual and regional variation within species. Range, habitat and differences in vocalisations aid identification. The species on this page all have mottled backs. African Pipit is the commonest, most widespread in the region and is the 'standard' against which other pipits should be compared. Nicholson's and Wood Pipits are segregated mainly by habitat.

AFRICAN PIPIT *Anthus cinnamomeus*

B	b	b	b			B	B	B	B	B
J	F	M	A	M	J	J	A	S	O	N	D

17 cm; 25 g The commonest pipit in most areas but plumage varies considerably across the region, from dark, heavily streaked birds in the northeast to pale, lightly streaked birds in the arid west. Smaller and more slender than other large, mottled-backed pipits on this page. Diagnostic features are its crisp white outer-tail feathers (buff in the other similarly large species), the yellow (not pink) base to the lower mandible, and its song and calls. Juv. is darker above, with more heavily streaked underparts. **Voice:** Song a repeated 3–5-note *'pli-pli-pli-pli'*, usually given in an undulating display flight, followed by 15 or more rapid notes as it dives to the ground; alarm call on being flushed is a *'tshisik'*. **Status and biology:** Common, widespread resident in much of its range, but moves from high-lying ground to lower elevations in winter. Occurs in open, short grassland, especially where heavily grazed, recently burnt or mowed, including playing fields and road verges. (Gewone Koester)

MOUNTAIN PIPIT *Anthus hoeschi*

b										b	b
J	F	M	A	M	J	J	A	S	O	N	D

BE

19 cm; 27 g Larger than African Pipit, with pink (not yellow) base to lower mandible and, in flight, shows buff (not white) outer-tail feathers. Its restricted br. range (>2 000 m in alpine grassland of Maluti Mountains) overlaps only marginally with Nicholson's Pipit (distinguished by display song and less marked face and breast) and African Rock Pipit (p. 322, distinguished by display song and unstreaked upperparts). Juv. is blackish-brown above and heavily mottled below. **Voice:** Display song, given in flight, is similar to that of African Pipit, but is deeper in pitch and slower in tempo. **Status and biology:** Br. endemic. Common visitor, Oct–Apr, in montane grassland, non-br. range presumed to be in central Africa, based on old specimens collected on passage in n Namibia and Botswana. Could be encountered anywhere north of the Drakensberg in winter, but no recent data. (Bergkoester)

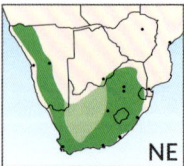

NICHOLSON'S PIPIT *Anthus nicholsoni*

B	b	b	b				b	b	B	B	B
J	F	M	A	M	J	J	A	S	O	N	D

NE

19 cm; 30 g Nicholson's Pipit of s Africa recently split from E African populations which retain the name Long-billed Pipit (*A. similis*). Larger than African Pipit, with a longer bill, plainer face, less streaked breast and buff (not white) outer tail. Colouring very variable across its range, with palest forms in the northwest, darkest birds in the southeast. Where it overlaps with Mountain Pipit, it is distinguished by its plainer face, less heavily marked breast and different call and display: Nicholson's Pipit glides to ground at the end of its display, with wings and tail spread; Mountain Pipit dives steeply at end of display. Juv. is scaled buff above and more heavily mottled on breast. Recent studies suggest that the so-called 'Kimberley Pipit' *A. pseudosimilis*, described in 2002, was misidentified Nicholson's Pipits. **Voice:** Repeated, monotonous sparrow-like notes, *'tchreep-tritit-churup'*, given from a perch or during aerial display; also a sharp *'wheet'* and a *'killink'* when flushed. **Status and biology:** Near-endemic. Locally common resident, mainly found on open, rocky hillsides in grassland and savanna, but also in well-grazed pastures and burnt lands. Occurs singly or in pairs; frequently perches on trees and shrubs where present. (Nicholsonkoester)

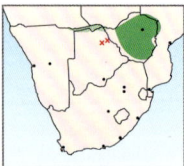

WOOD PIPIT *Anthus nyassae*

b	b					b	b	B	B	B	b
J	F	M	A	M	J	J	A	S	O	N	D

18 cm; 24 g Sometimes treated as a race of Nicholson's Pipit, but restricted to woodland; no evidence for overlapping ranges. Tail and bill shorter than Nicholson's Pipit; supercilium whitish (not buffy), and tail darker, with more extensive pale areas in base of outer-tail feathers. Juv. is spotted above, and more heavily streaked below. **Voice:** Sparrow-like song similar to that of Nicholson's Pipit, but sharper and more variable in pitch. **Status and biology:** Locally common resident of miombo, teak and other broadleafed woodland. Solitary or in pairs; forages on the ground, but flushes into trees when disturbed. (Boskoester)

AFRICAN PIPIT

white outer-tail feathers

undertail

typical races

yellow base to bill

rather small & slender

paler races

MOUNTAIN PIPIT

pink base to bill

buff outer-tail feathers

undertail

NICHOLSON'S PIPIT

pinkish-yellow base to bill

diffuse breast markings

buff outer-tail feathers

undertail

WOOD PIPIT

whitish supercilium

buff outer-tail feathers

undertail

walks along branches

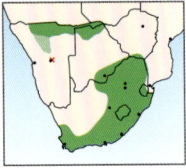

PLAIN-BACKED PIPIT   *Anthus leucophrys*

bbBB
J F M A M J J A S O N D

17 cm; 26 g A large pipit with an unmarked back. Confusion is most likely with Buffy Pipit; it is typically more compact and less slender than Buffy Pipit, with shorter bill, legs and tail; averages colder and darker above, with more uniform underparts (lacking buffy flanks). Face more strongly marked; diagnostic differences are a yellowish (not pink) lower mandible and a long (11–14 mm vs. 8–11 mm) hind claw, but close views are needed for this. Tail-wagging is less pronounced; typically doesn't raise tail above horizontal (see further under Buffy Pipit). Juv. is heavily scaled above and mottled below. **Voice:** A monotonous, sparrow-like *'chrrrup-chereeoo'*, sung from a low perch or during a brief aerial display. A brief *'tsip-tsip'* or *'tsissik'* alarm note on being flushed. **Status and biology:** Locally common resident and nomad, favouring short grass on heavily grazed or recently burnt ground; in small flocks at times. Some birds move to lower elevations in winter. (Donkerkoester)

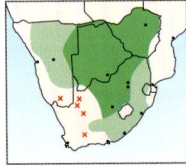

BUFFY PIPIT   *Anthus vaalensis*

bb   bb BBBB
J F M A M J J A S O N D

18 cm; 30 g A large pipit with an unmarked back. Confusion is mainly with the slightly smaller Plain-backed Pipit; Buffy Pipit typically (but not invariably) has warmer brown upperparts and a less striking facial pattern. Consistent differences between the 2 species are the colour of the base of the lower mandible (pinkish in Buffy Pipit, yellow in Plain-backed Pipit) and the length of the hind claw (8–11 mm in Buffy Pipit, 11–14 mm in Plain-backed Pipit). Morphologically, Buffy Pipit has a 'big-chested' profile, whereas Plain-backed Pipit has a 'big-bellied' profile. Both species wag their tails up and down in an exaggerated manner, but this is more pronounced in Buffy Pipit, with the rear half of the body included in the wagging, and the upward wag extending above the horizontal. Juv. is scaled above and mottled below. **Voice:** Sparrow-like chipping notes uttered at a slower frequency than Plain-backed Pipit, *'tchreep-churup'*, usually given from a low perch; alarm call when flushed a brief *'sshik'* or *'ship-ip'*. **Status and biology:** Uncommon to locally common resident, nomad and partial migrant in shortly grazed or burnt grassland and scrub, fallow lands, airfields and other open, bare areas. May gather in loose groups when not br. Non-br. birds from central Africa thought to overwinter in s Africa. (Vaalkoester)

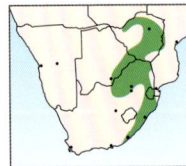

STRIPED PIPIT   *Anthus lineiventris*

b   bBBB
J F M A M J J A S O N D

18 cm; 34 g A large, dark pipit with a heavily striped chest and belly; tail dark brown, with conspicuous white outer-tail feathers. At close range, yellow-edged wing coverts are diagnostic. Larger than Tree Pipit (p. 324), with a more marked supercilium and strongest streaking on belly (not breast); lacks white edges to wing coverts. Juv. is paler above and less heavily streaked below. **Voice:** Loud, penetrating, thrush-like song, given from a rock or tree perch. **Status and biology:** Locally common resident on boulder-strewn hill slopes in woodland. Flies up into trees when disturbed; easily overlooked when not singing. (Gestreepte Koester)

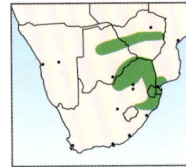

BUSHVELD PIPIT   *Anthus caffer*

BBb   bBB
J F M A M J J A S O N D

13 cm; 18 g A tiny, short-tailed pipit, with heavily marked upperparts and a well-streaked breast. Possibly confused in shared woodland habitat with Tree Pipit (p. 324), but much smaller; face appears unmarked (no dark malar stripe, supercilium very short and narrow) and breast markings more diffuse. Short-tailed Pipit (p. 324) is darker, even shorter tailed, blunter billed, more heavily streaked below (extending to flanks) and occupies different habitat (woodland, not grassland). Juv. is paler, scaled buff above. **Voice:** Characteristic *'bzeeent'* as it flies from ground to tree; song a treble-note *'zrrrt-zrree-chreee'* from perch in tree. **Status and biology:** Locally fairly common resident and nomad in open woodland, especially in rocky areas where there is good grass cover. Typically flies up to an open perch in a tree when flushed. (Bosveldkoester)

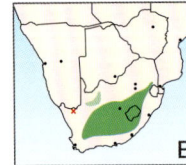

AFRICAN ROCK PIPIT   *Anthus crenatus*

b   bBB
J F M A M J J A S O N D

18 cm; 31 g A dark pipit of rocky hill slopes most closely related to Striped Pipit, but plumage is closer to Plain-backed and Buffy Pipits. Upperparts are uniformly olive-brown, underparts buff with suffused streaks on the upper breast. Rocky habitat and whistled song are diagnostic. Distinctive morphological features are yellow-green edges to wing coverts, dark ear coverts and a creamy supercilium extending well behind the eye; has shorter legs and tail than other large pipits. Juv. is mottled above. **Voice:** Male sings for extended periods from a low vantage point on a hillside, bill pointed upwards. A drawn-out, far-carrying *'wheeu, prrreeeuuuu'*, second note descending and quavering; usually located by this distinctive song and easily overlooked when not calling. **Status and biology:** Endemic. Locally common resident in higher-lying areas (mainly >1 000 m) on boulder-strewn, steep, grassy hillsides and Karoo koppies. (Klipkoester)

PLAIN-BACKED PIPIT

grey-brown above

buff outer-tail feathers

undertail

long hind claw

yellowish base to bill

compact with heavy belly

BUFFY PIPIT

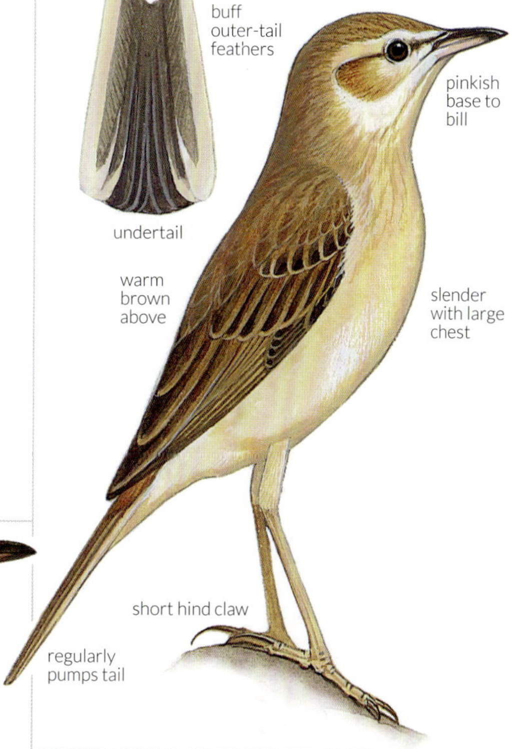

buff outer-tail feathers

undertail

warm brown above

pinkish base to bill

slender with large chest

short hind claw

regularly pumps tail

STRIPED PIPIT

yellow-edged wing coverts

heavily streaked underparts

white outer tail

AFRICAN ROCK PIPIT

points head up when calling

bold facial pattern

dark brown above

yellowish edges to wing coverts

sings in bushes/ trees

BUSHVELD PIPIT

heavily streaked above

diffuse breast streaks

short tail

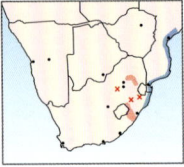

SHORT-TAILED PIPIT *Anthus brachyurus*

b b | | | | | | | b B B
J F M A M J J A S O N D

12 cm; 17 g A tiny, dark-coloured pipit with a short tail, dark face, no pale supercilium and heavily streaked underparts. When flushed, it could be mistaken for a large cisticola or female bishop, but the short, thin tail with prominent white outer feathers is distinctive. Most similar to Bushveld Pipit (p. 322), but has a shorter tail and bill, colder brown upperparts and more extensive streaking below, extending onto the flanks; habitats differ. Male has yellowish wash to plumage; female whiter beneath. Juv. has warm rufous-buff margins to upperpart feathers; breast streaking broader and more diffuse. **Voice:** A buzzy, bubbling *'chrrrrt-zhrrrreet-zzeeep'*, similar to Bushveld Pipit; sings from a perch or during a low, circling, pre-dawn display flight. **Status and biology:** Uncommon resident and local migrant (leaving high-lying areas in winter); on grassy hillsides; prefers areas of short, dense grass, including recently burnt areas. Unlike Bushveld Pipit, does not flush into trees. (Kortstertkoester)

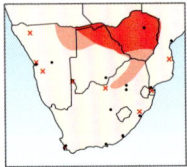

TREE PIPIT *Anthus trivialis*

■ ■ ■ ■ | | | | | | | | |
J F M A M J J A S O N D

14 cm; 22 g A smallish pipit with a heavily streaked breast. Smaller than Striped Pipit (p. 322) with a mostly plain (not streaked) belly and white edges to wing coverts forming 2 short, white bars; supercilium less marked. Larger and longer tailed than Bushveld (p. 322) or Short-tailed Pipits, showing more contrast between pale throat and dark upperparts. Bill rather short and weak. Easily confused with non-br. Red-throated Pipit, but has less clearly streaked underparts and lacks dark brown streaking on rump; at close range, pale eye-ring visible. Juv. is more buffy than ad. **Voice:** Soft, nasal *'teeez'* in flight or when flushed. Melodic, canary-like song is seldom given in s Africa. **Status and biology:** Palearctic-br. visitor, Nov–Mar; locally common in Zimbabwe, rare elsewhere. Singly, or in loosely associated parties in woodland glades. Feeds on the ground, but usually flushes into a tree when disturbed, walking along branches like a Wood Pipit. (Boomkoester)

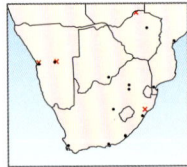

RED-THROATED PIPIT *Anthus cervinus*

X X X | | | | | | | | |
J F M A M J J A S O N D

14 cm; 20 g Vagrant. Structure similar to Tree Pipit, but slightly dumpier, with a heavier belly. In br. plumage, easily recognised by its dull red face, throat and breast. Non-br. birds lack the red throat; average darker and richer than Tree Pipit, with no pale eye-ring, less prominent buffy (not whitish) wingbars and bolder streaking: flank streaks broader and rump streaked dark brown (not uniform). Mantle often shows 2 creamy stripes on either side. Larger than Short-tailed Pipit, with a longer tail and more strongly marked supercilium. **Voice:** Distinctive, buzzy *'skeeeaz'* given in flight or on flushing; also a clear, penetrating *'chup'* call. **Status and biology:** Palearctic-br. vagrant to damp grassland and fields, usually near water; only 5 records. (Rooikeelkoester)

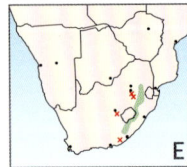

YELLOW-BREASTED PIPIT *Anthus chloris*

B B B | | | | | | | b B B
J F M A M J J A S O N D

17 cm; 25 g A bright yellow, unmistakable pipit in summer; less obvious in winter. Br. birds have diagnostic yellow throat and breast; upperparts have a scaled appearance from dark back and covert feathers being edged with shades of buff. Sexes virtually alike; yellow in female less extensive. In non-br. plumage, yellow is mostly lost from underparts, which become whitish; then distinguishable from other pipits by its distinctly scaled upperparts. Juv. is more crisply scaled above and buff below. **Voice:** Song is a soft *'se-chik'*, repeated 10–20 times from a perch or while performing a lazy display flight; alarm call is a rapid *'chip, chip, chip'*. **Status and biology:** VULNERABLE. Endemic, localised summer visitor (Sep–Apr) to short montane grassland above 1 500 m; most birds move to lower elevations in winter. Genetic evidence suggests it is actually a longclaw. (Geelborskoester)

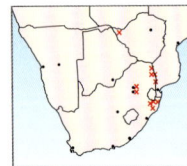

GOLDEN PIPIT *Tmetothylacus tenellus*

X X X | | | | | | | X X
J F M A M J J A S O N D

15 cm; 20 g Vagrant. Male resembles a small, bright golden Yellow-throated Longclaw (p. 326), but has mainly yellow wings and tail with black tips to the outer primaries and tail feathers; black breast band does not extend up sides of neck. Female is duller, with whitish underparts variably washed yellow and lacking a black breast band; wing and tail feathers edged yellow. Female could be confused with Western Yellow Wagtail (p. 326), but is smaller, with a shorter tail, patterned (not plain) mantle, and different gait. Juv. like female but duller, with whitish margins to upperpart feathers and diffuse brown breast spots. **Voice:** Song is a short burst of scratchy, whistled notes with a weaver-like quality; vagrants usually silent. **Status and biology:** Vagrant from e Africa to open, dry woodland and savanna; all records have been of males, mostly in summer. May be more common than records indicate, with females and juvs overlooked. When not foraging, commonly perches in trees. Genetic evidence suggests it is actually a longclaw. (Goudkoester)

SHORT-TAILED PIPIT

heavily streaked above & below

tail narrow in flight, white outer tail

TREE PIPIT

plain rump

lightly streaked flanks

heavily streaked breast

RED-THROATED PIPIT

br.

streaked rump

heavily streaked breast & flanks

non-br.

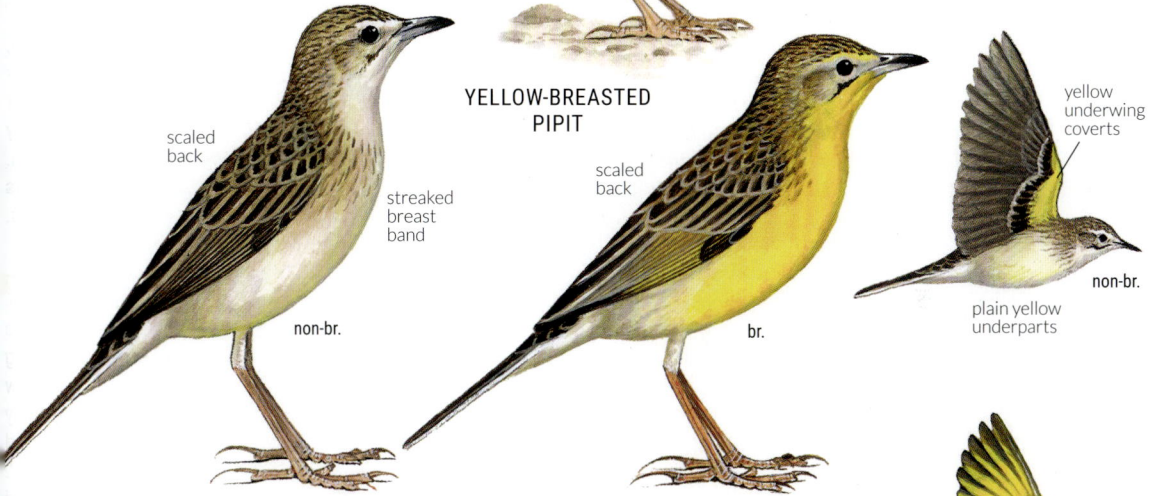

YELLOW-BREASTED PIPIT

scaled back

streaked breast band

non-br.

scaled back

br.

yellow underwing coverts

non-br.

plain yellow underparts

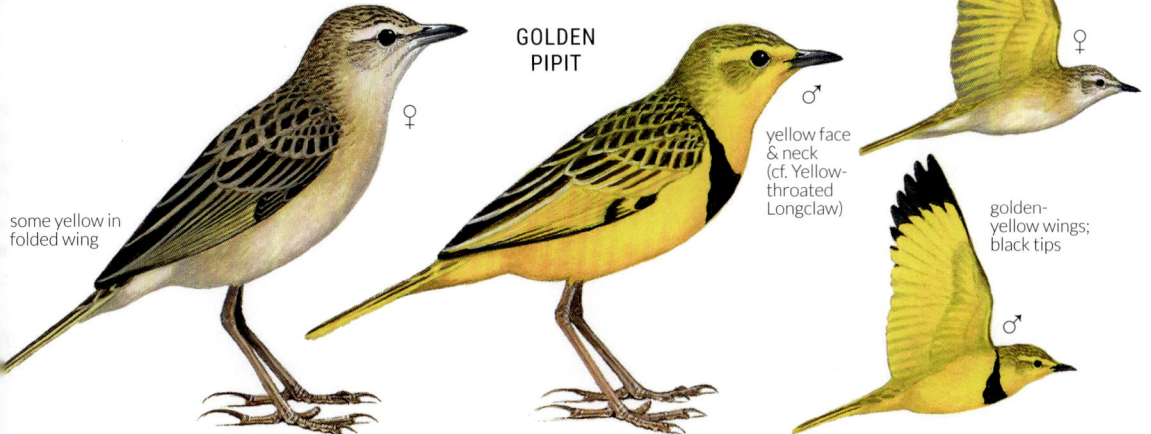

GOLDEN PIPIT

some yellow in folded wing

♀

♂

yellow face & neck (cf. Yellow-throated Longclaw)

♀

golden-yellow wings; black tips

♂

LONGCLAWS
Robust, pipit-like birds with cryptically coloured upperparts and brightly coloured underparts with a black breast band. Flight jerky, with bursts of stiff-winged flaps interspersed with short glides on downward-angled wings. Females similar to males, but with duller underparts. Juvs rather drab, with reduced breast bands. Occur singly or in pairs, foraging for invertebrates on the ground; sometimes perch on rocks or bushes when singing.

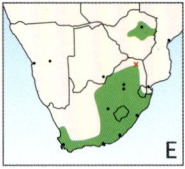

CAPE LONGCLAW *Macronyx capensis*

B	b	b	b			b	B	B	b	B	B
J	F	M	A	M	J	J	A	S	O	N	D

20 cm; 45 g The only longclaw in the region with an orange throat. In flight, shows white tail tips, broader towards edges, like Yellow-throated Longclaw. Female duller, with narrower breast band. Juv. has a warm buff throat, an indistinct, brown-streaked breast band and buffy belly; wing feathers edged with buff. **Voice:** A loud, cat-like *'meew'*, a single note usually given from a vantage point; in flight, a repeated *'cheewit-cheewit'*; also a far-carrying, high-pitched *'tsweet'* whistle and a nasal *'skeeaaa'* alarm call. **Status and biology:** Endemic. Common resident in moist grasslands from sea level to montane uplands. (Oranjekeelkalkoentjie)

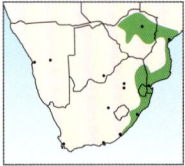

YELLOW-THROATED LONGCLAW *Macronyx croceus*

B	b	b			b	b	b	b	B	B	
J	F	M	A	M	J	J	A	S	O	N	D

21 cm; 49 g A large longclaw with bright yellow underparts and a black necklace broadening into a breast band. Told from Cape Longclaw by its yellow (not orange) throat. Much larger than male Golden Pipit (p. 324); breast band extends up the sides of the neck; wings brown (not mostly yellow). Female duller, washed buffy below. Juv. lacks black breast band; told from juv. Cape Longclaw by yellow (not buff-orange) belly and less streaked upperparts. **Voice:** Loud, whistled *'phooooeeet'* or series of whistles, frequently given from the top of a bush or small tree; also calls in flight. Alarm call is a loud *'whip-ip-ip-ip-ip'*. **Status and biology:** Common resident in grassy savanna, open woodland and grassland. (Geelkeelkalkoentjie)

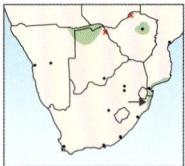

ROSY-THROATED LONGCLAW *Macronyx ameliae*

B	B	b	b					b	B	B	
J	F	M	A	M	J	J	A	S	O	N	D

19 cm; 34 g A slender longclaw, with dark, heavily streaked upperparts and pinkish underparts. Male has a bright pink throat and broad, black breast band. In flight, dark back and all-white outer-tail feathers distinguish it from other longclaws. Female has duller underparts and a narrower, sometimes incomplete, breast band. Juv. has a brown-streaked breast band with a creamy throat and only a faint rosy wash on the belly. **Voice:** Melodious, rather deep *'cheet errr'* or *'cheet eeet eet eet eer'*; plaintive *'chewit'* alarm call. **Status and biology:** Uncommon resident in moist grassland surrounding open areas of fresh water. (Rooskeelkalkoentjie)

WAGTAILS
Slim, long-tailed birds related to longclaws and pipits, that frequently wag their tails up and down; mostly associated with aquatic habitats. The 3 resident species have no seasonal or sex-related plumage differences, whereas the 4 migrant/vagrant species are sexually dimorphic with distinct br. and non-br. plumages. Residents defend br. or permanent territories. All are insectivorous, preying on flies and other small insects, which are pursued on foot on the ground or hawked in the air.

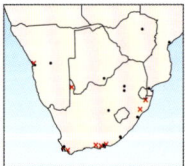

CITRINE WAGTAIL *Motacilla citreola*

X		X	X	X		X					
J	F	M	A	M	J	J	A	S	O	N	D

17 cm; 20 g Vagrant. Slightly larger than Western Yellow Wagtail, with a greyish back (lacking olive tones); appears slimmer due to longer tail and legs. Br. male has a diagnostic all-yellow head and black nape. Non-br. ads easily confused with Western Yellow Wagtail, but have a paler yellow vent, a more prominent supercilium (extending in a complete loop around the ear coverts) and typically bolder white wingbars. Juv. drab grey, with only a buff-yellow wash on its breast. **Voice:** Similar to Western Yellow Wagtail, but shorter and shriller *'trsiiip'*. **Status and biology:** Palearctic-br. vagrant; with few records from KZN, E Cape, W Cape and Namibia, mainly in autumn, suggesting reverse migration. (Sitrienkwikkie)

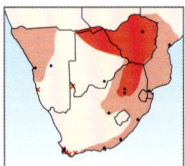

WESTERN YELLOW WAGTAIL *Motacilla flava*

■	■	■	■					■	■	■	■
J	F	M	A	M	J	J	A	S	O	N	D

17 cm; 14–21 g A highly variable migrant wagtail. Smaller than rare Grey Wagtail (p. 328), with a shorter tail and greenish (not blue-grey) back. Slightly smaller than vagrant Citrine Wagtail, with typically greenish (not greyish) back and a shorter supercilium. Five races occur in the region; br. males vary in head colour: *M. f. lutea* (head yellow, sometimes washed olive-green on crown and cheeks); *M. f. beema* (pale grey crown and cheeks; broad white supercilium); *M. f. flava* (dark grey crown, black cheeks and narrow, white supercilium); *M. f. thunbergi* (dark grey crown, cheeks black with whitish stripe below; breast washed olive-green); and *M. f. feldegg* (black crown and cheeks; underparts entirely yellow). Juv. is yellowish-brown above and pale buff below, with a narrow, blackish breast band; told from Cape Wagtail by its yellow underparts. **Voice:** Weak, thin *'tseeep'*. **Status and biology:** Uncommon to locally common Palearctic-br. visitor. Usually in small parties in moist, short-grass areas close to wetlands. (Geelkwikkie)

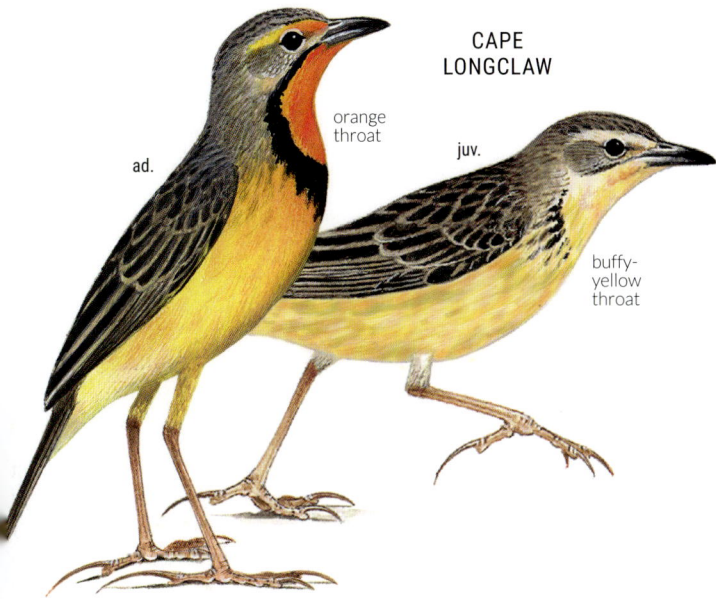

CAPE LONGCLAW

orange throat

ad.

juv.

buffy-yellow throat

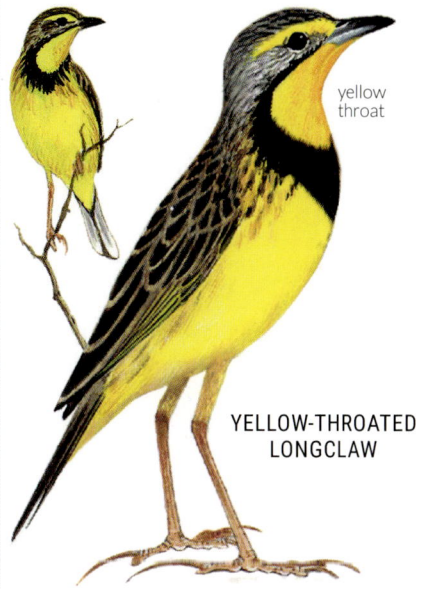

yellow throat

YELLOW-THROATED LONGCLAW

ROSY-THROATED LONGCLAW

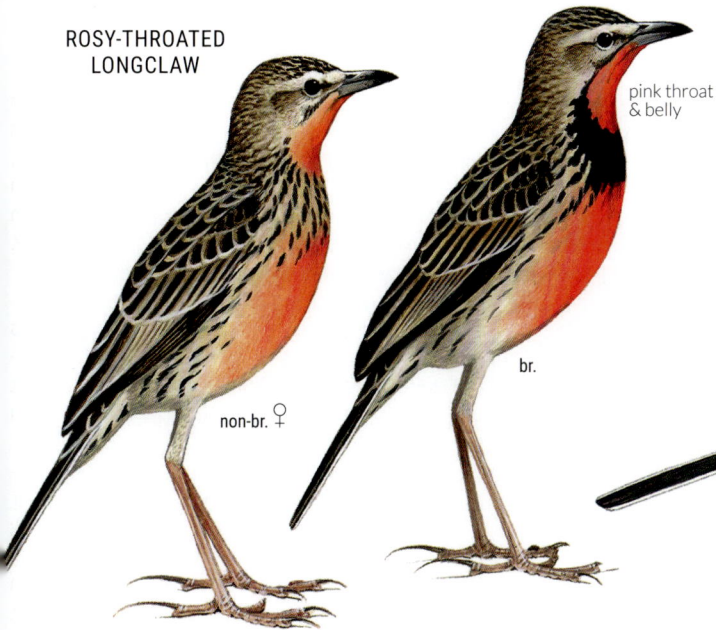

pink throat & belly

non-br. ♀

br.

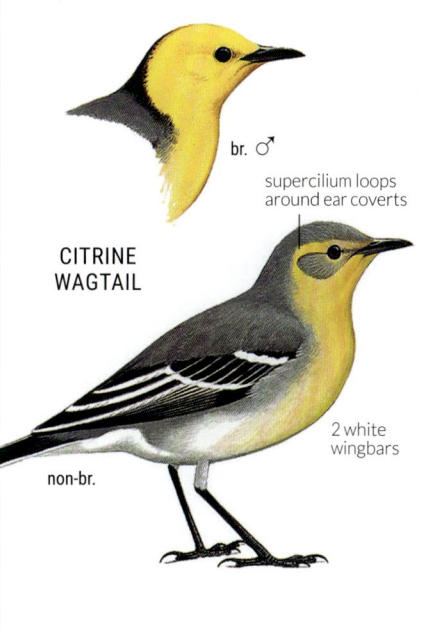

br. ♂

supercilium loops around ear coverts

CITRINE WAGTAIL

2 white wingbars

non-br.

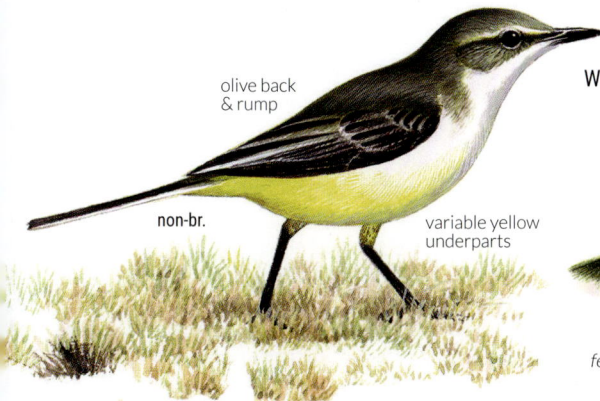

olive back & rump

non-br.

variable yellow underparts

WESTERN YELLOW WAGTAIL

lutea

feldegg

thunbergi

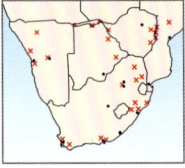

GREY WAGTAIL *Motacilla cinerea*

X X X X X   X X X
J F M A M J J A S O N D

19 cm; 18 g Vagrant. A very long-tailed wagtail; structure similar to Mountain Wagtail, but has at least some yellow on its vent, buff (not whitish) flanks and less white in its wing and outer tail. Appreciably longer tailed than Western Yellow Wagtail (p. 326), with a blue-grey (not green) back that contrasts with the greenish-yellow rump. Tail-wagging is more exaggerated. Br. male has a black throat; speckled in non-br. plumage. Female has a white throat. Juv. similar to female, but duller. **Voice:** Single, sharp *'tit'*. **Status and biology:** Rare, Palearctic-br. migrant, mostly Nov–Apr, with a few winter records. Occurs singly along wooded streams. (Gryskwikkie)

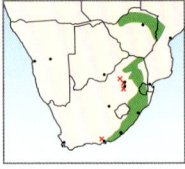

MOUNTAIN WAGTAIL *Motacilla clara*

b b b b b   b B B B B
J F M A M J J A S O N D

20 cm; 20 g Closely related to Grey Wagtail, but has white underparts (lacking any yellow) and more white in its wing and outer tail. Told from Cape Wagtail by its much longer tail, blue-grey (not grey-brown) upperparts, clean, white underparts, and more white in wings and outer tail. Juv. is browner. **Voice:** Sharp, high-pitched *'cheeerip'* or *'chissik'* call. **Status and biology:** Locally common resident along streams in hilly, forested country; favours streams with emergent rocks, rapids and small waterfalls. (Bergkwikkie)

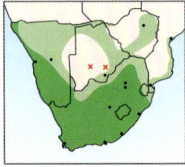

CAPE WAGTAIL *Motacilla capensis*

b b b b b b B B B B B
J F M A M J J A S O N D

20 cm; 21 g A familiar garden bird, usually with a prominent breast band. Told from Mountain Wagtail by its shorter tail, grey-brown (not bluish-grey) upperparts and greyish (not crisp white) underparts. *M. c. simplicissima* in n Botswana has breast band reduced to a small spot. Juv. is browner above, with buffy tips to feathers and a smaller breast band; belly washed buff-yellow; could be confused with juv. Western Yellow Wagtail (p. 326). **Voice:** Clear, ringing *'tseee-chee-chee'* and *'tseep'* calls; whistled, trilling song. **Status and biology:** Common resident around wetlands and along the coast; also on lawns and in urban parks. Some move away from high-lying areas in winter. May roost communally outside br. season. (Gewone Kwikkie)

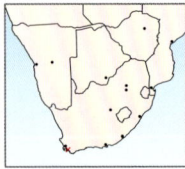

WHITE WAGTAIL *Motacilla alba*

X
J F M A M J J A S O N D

16–19 cm; 17–25 g Vagrant. A grey-backed wagtail, superficially recalling Cape Wagtail but with boldly patterned wings (formed by blackish secondaries and greater coverts with broad white margins) and a paler face in most plumages. Slightly smaller and more slender than Cape Wagtail with a long, thin black-and-white tail, which is pumped vigorously as it forages. Ad. readily identified by large black bib, with extensive white facial markings onto the forehead. Imm. has an indistinct black bib, ending under the eye in a diagnostic horizontal black sickle mark. **Voice:** In undulating flight utters *'zeep'* calls. **Status and biology:** Palearctic migrant, regularly reaching central Africa. 1 record in Jan 2018 of an imm. bird at Bot River, W Cape. (Witkwikkie)

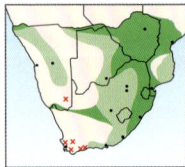

AFRICAN PIED WAGTAIL *Motacilla aguimp*

b B B B b b
J F M A M J J A S O N D

20 cm; 27 g The common black-and-white wagtail in s Africa; white in wing more extensive than other wagtails. Female and non-br. male have duller black upperparts. Juv. is grey-brown above; breast band narrow; wing coverts edged pale buff, but much more broadly than in Cape Wagtail. **Voice:** Loud, shrill *'chee-chee-cheree-cheeroo'*. **Status and biology:** Locally common resident along large rivers and wetlands, including coastal lagoons; also in parks and gardens, especially in north of its range. May roost communally outside br. season. (Bontkwikkie)

GREY WAGTAIL

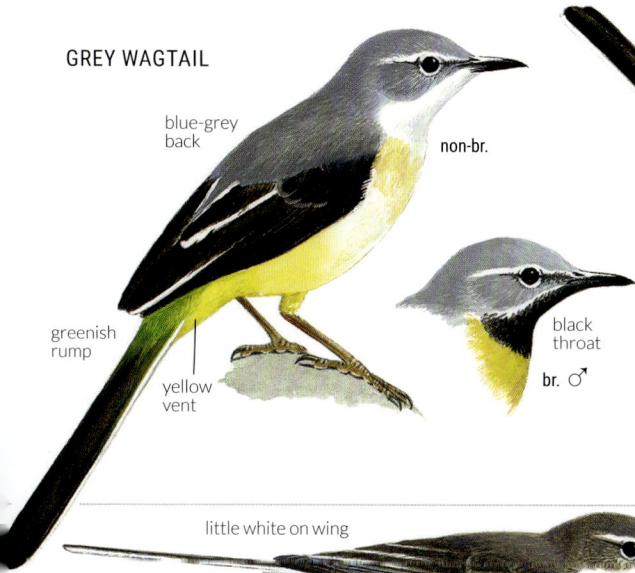

blue-grey back

non-br.

greenish rump

yellow vent

black throat

br. ♂

MOUNTAIN WAGTAIL

long tail

blue-grey upperparts

CAPE WAGTAIL

reduced breast band

simplicissima

little white on wing

juv.

ad.

shorter tail

WHITE WAGTAIL

pale grey mantle

boldly patterned wings

ad. ♀

AFRICAN PIED WAGTAIL

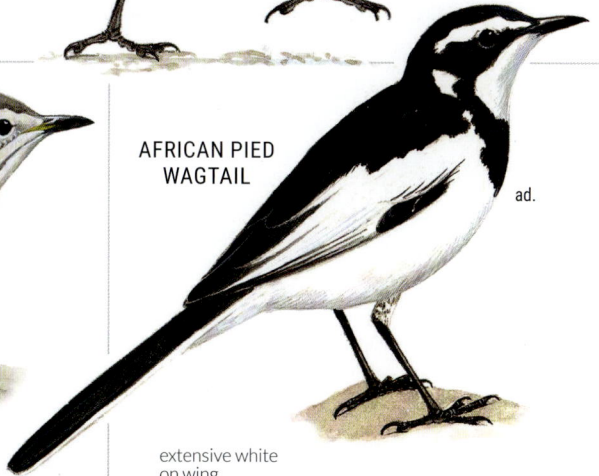

ad.

ad. ♂

white face, black throat

extensive white on wing

juv.

SWALLOWS

Aerial feeders; told from swifts (p. 276) by angled wings and more flapping flight; also frequently perch. Swallows typically have pale underparts, but some martins are brown below. Sexes alike; juvs duller; in species with tail streamers, those of females and juvs are shorter. Some species breed colonially.

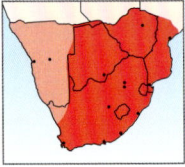

BARN SWALLOW *Hirundo rustica*

J F M A M J J A S O N D

15–20 cm; 16–24 g Dark blue or brown above, with white panels in the outer tail; breast and belly vary from off-white to rich buff. Ad. striking in fresh plumage (usually Mar–May), with long tail streamers and a reddish frons and throat, but streamers often short or missing and throat dull brown when worn. Slightly larger than rare Angola Swallow; red largely confined to throat (not onto breast), bordered by a complete blackish breast band of relatively uniform thickness; remainder of underparts creamy-buff (not grey). Juv. duller, with browner frons and throat; outer-tail feathers short. **Voice:** Soft, high-pitched twittering. **Status and biology:** Abundant Palearctic-br. migrant found in all habitats; often in loose flocks. Roosts communally, usually in reeds; occasionally overwinters. (Europese Swael)

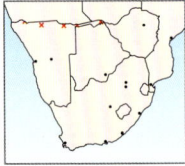

ANGOLAN SWALLOW *Hirundo angolensis*

X | | | | | | | X X X
J F M A M J J A S O N D

14–15 cm; 12–20 g Vagrant. Slightly smaller and more compact than Barn Swallow; red throat extends onto breast and is narrowly bordered by an incomplete black band. Upperparts are darker blue and appear more iridescent; underparts grey (not creamy-buff). Outer-tail streamers are much shorter than br. ad. Barn Swallow's, and tail is less deeply forked. Juv. has red of ad. replaced by pale rufous, and is less glossy above. **Voice:** Weak, twittering song; loud *'tsip'* call. **Status and biology:** Rare visitor to grasslands and over woodlands of extreme n Namibia. (Angolaswael)

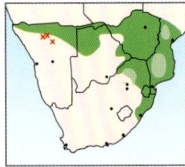

WIRE-TAILED SWALLOW *Hirundo smithii*

B B B B b b B B B B B b
J F M A M J J A S O N D

12–14 cm; 10–15 g A tiny, blue-backed swallow with a chestnut crown and white underparts. Long, very thin tail streamers often difficult to see. Appreciably smaller than White-throated Swallow, with only a partial breast band (but beware confusion with juv. White-throated Swallow which also may have an incomplete breast band); entire crown (not just forehead) rufous, dark vent band and longer, narrower tail streamers. Flight extremely rapid. Juv. is less glossy blue above, with a brown crown; underparts washed buff; could be confused with Grey-rumped Swallow (p. 334) but still shows the dark vent band. **Voice:** Call is a sharp, metallic *'tchik'*; song is twittering *'chirrik-weet, chirrik-weet'*. **Status and biology:** Common resident and intra-African br. migrant, usually near water. (Draadstertswael)

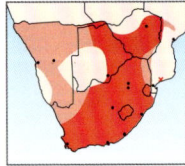

WHITE-THROATED SWALLOW *Hirundo albigularis*

b b b b | | | b B B B b
J F M A M J J A S O N D

15 cm; 18–28 g A blue-backed swallow with white underparts and a complete, blue-black breast band. Larger than Wire-tailed Swallow, with rufous confined to the frons (not entire cap), a complete breast band, no vent band, and short, relatively broad tail streamers. Juv. is less glossy above, and has a brownish forehead; breast band dull brown, narrower than ad.'s (sometimes incomplete). **Voice:** Harsh twittering and nasal notes. **Status and biology:** Common intra-African br. migrant, closely associated with water; often nests under bridges over water. (Witkeelswael)

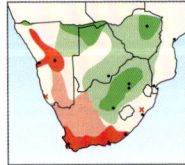

PEARL-BREASTED SWALLOW *Hirundo dimidiata*

b b b | | | | b B B B B
J F M A M J J A S O N D

13 cm; 10–15 g A small, rather plain blue-and-white swallow, lacking red on head and white in tail. Tail has a shallow fork, but lacks streamers. Best told from Grey-rumped Swallow (p. 334) and Common House Martin (p. 334) by its plain blue-black (not pale grey or white) rump. From below it differs by its white (not dark) underwing coverts; has shorter tail than Grey-rumped Swallow, and is more slender than Common House Martin. Juv. less glossy above and has shorter outer-tail feathers; innermost secondaries tipped white. **Voice:** A subdued and repeated *'chip chip chip'* given in flight. **Status and biology:** Locally common resident and intra-African br. summer migrant in grassland, savanna, strandveld, open woodland and farmland. (Pêrelborsswael)

BARN SWALLOW

nm.

ad.

br.

red throat, broad breast band

pale underwing coverts

ANGOLAN SWALLOW

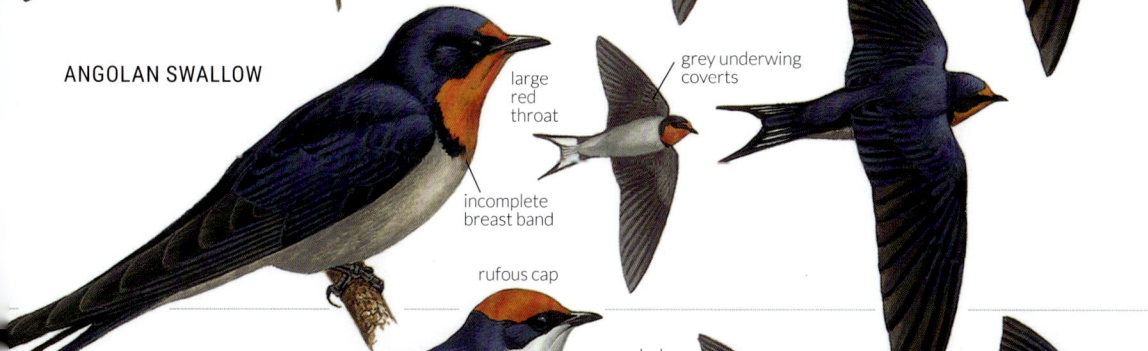

large red throat

incomplete breast band

rufous cap

grey underwing coverts

WIRE-TAILED SWALLOW

dark vent band

grey-brown cap

ender tail
creamers

juv.

WHITE-THROATED SWALLOW

rufous frons

black breast band

PEARL-BREASTED SWALLOW

no breast or thigh bands

unfeathered legs (cf. Common House Martin)

uniform upperparts

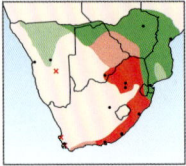

LESSER STRIPED SWALLOW *Cecropis abyssinica*

B b b b b b b | | B B B
J F M A M J J A S O N D

15–19 cm; 16–22 g Smaller and darker than Greater Striped Swallow, with more prominent blackish striping contrasting strongly with whiter underparts; rump darker rufous; ear coverts rufous (not whitish). Female has shorter tail streamers than male. Juv. duller, lacking blue-black gloss above, and has a brown (not rufous) crown; breast and flanks washed buff; tail streamers reduced. **Voice:** Descending series of squeaky, nasal *'zeh-zeh-zeh-zeh'* notes. **Status and biology:** Common resident and intra-African br. summer migrant in diversity of mesic habitats, often near water. (Kleinstreepswael)

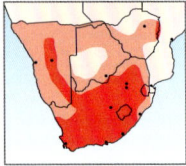

GREATER STRIPED SWALLOW *Cecropis cucullata*

J F M A M J J A S O N D

16–20 cm; 20–35 g Larger and paler than Lesser Striped Swallow; striping on the buffy underparts is paler and less well defined; orange crown slightly paler; rump paler rufous; ear coverts whitish (not rufous). Paler overall than vagrant Red-rumped Swallow, with much more pronounced streaking on underparts; crown rufous-orange (not dark blue). Juv. duller with a reddish-brown crown and more heavily streaked underparts; tail streamers reduced. **Voice:** Twittering *'chissick'* and querulous, nasal notes. **Status and biology:** Common intra-African br. summer migrant in grassland, shrubland, savanna, agricultural lands and suburbia; often around wetlands. (Grootstreepswael)

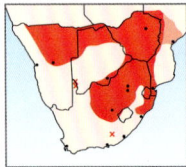

RED-BREASTED SWALLOW *Cecropis semirufa*

B B b b | | b B B B B
J F M A M J J A S O N D

20–24 cm; 24–38 g A large, dark, red-rumped swallow; slightly smaller than Mosque Swallow. Ad. has a red (not white) throat and breast, dark blue (not pale rufous) ear coverts and dull buff (not white) underwing coverts. Larger than vagrant Red-rumped Swallow, with a longer tail, darker rufous underparts and dark (not rufous) nape and cheeks. Juv. has a creamy-white throat and breast and short tail streamers; differs from Mosque Swallow by its darker underwing coverts. **Voice:** Soft, warbling song; twittering notes in flight. **Status and biology:** Common intra-African br. summer migrant in grassland and savanna. (Rooiborsswael)

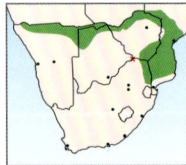

MOSQUE SWALLOW *Cecropis senegalensis*

b b b b | | B B b b b
J F M A M J J A S O N D

22–26 cm; 38–50 g A large, red-rumped swallow; slightly larger than Red-breasted Swallow, with a white (not red) throat (but beware confusion with juv. Red-breasted Swallow which has a creamy throat), face and upper breast, pale rufous (not dark blue) ear coverts and white (not buffy) underwing coverts. Juv. duller, with short tail streamers and pale buff underwing coverts, but coverts always paler than those of Red-breasted Swallow. **Voice:** Nasal *'harrrrp'*, guttural chuckling and a piping *'peeuuu'*. **Status and biology:** Locally common resident and partial summer migrant (in north and northwest of region) in open woodland, often near rivers, and especially near baobabs. (Moskeeswael)

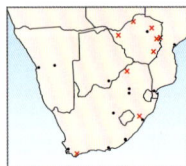

RED-RUMPED SWALLOW *Cecropis daurica*

X X X | | | | | | | | X
J F M A M J J A S O N D

17–19 cm; 23–30 g Vagrant. Paler than ad. Red-breasted Swallow with pale rufous (not dark blue) ear coverts and nape. Smaller than Mosque Swallow; paler below with a rufous (not dark blue) nape, blackish (not rufous) undertail coverts and pale grey (not blackish) undersides to flight feathers. Differs from Greater Striped Swallow by its dark blue (not reddish) cap and blackish undertail coverts and tail; underparts plain or only faintly streaked. Juv. duller above; lacks tail streamers. **Voice:** Single-note *'djuit'*; also a soft, twittering song. **Status and biology:** Vagrant, presumably from resident E and central African population, although Palearctic-br. migrants may occasionally reach the region; 7 confirmed records, most from Zimbabwe but 1 ringed individual in Limpopo and 1 in central KZN. (Rooinekswael)

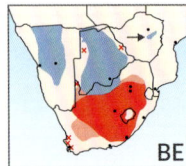

SOUTH AFRICAN CLIFF SWALLOW *Petrochelidon spilodera*

b b b b | | b b B B B
J F M A M J J A S O N D

14–15 cm; 16–24 g A martin-like swallow with broad wings and an almost square tail, but with a buffy-rufous rump and creamy underparts washed rufous on breast and vent; breast diffusely streaked dark brown. Crown dark brown, slightly glossed in front. Lacks forked tail and well-defined breast streaks of Lesser and Greater Striped Swallows. Non-br. Barn Swallow (p. 330) often shows similar scruffy breast markings, but lacks a pale rump. Juv. duller, lacking gloss on upperparts; underparts less boldly marked. **Voice:** Twittering *'chooerp-chooerp'*. **Status and biology:** Br. endemic. Locally common intra-African migrant in upland grassland, open savanna and Karoo scrub. Breeds colonially (20–900 pairs), usually in culverts and under road bridges. Red-throated Cliff Swallow *P. rufigula* is a potential vagrant in the north (unconfirmed sighting from Victoria Falls); smaller and more slender than South African Cliff Swallow with a longer tail, plain rufous throat and diagnostic small, white tail spots. (Familieswael)

LESSER STRIPED SWALLOW

orange ear coverts

bold streaking

GREATER STRIPED SWALLOW

pale striped ear coverts

diffuse streaking

RED-BREASTED SWALLOW

rufous throat

buffy underwing coverts

MOSQUE SWALLOW

pale throat

white underwing coverts

RED-RUMPED SWALLOW

dark cap & red nape

dark vent

SOUTH AFRICAN CLIFF SWALLOW

rufous/off-white vent

square tail

diffuse breast band

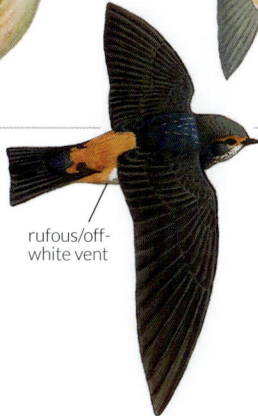

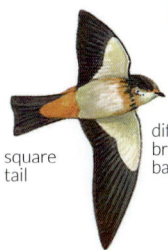

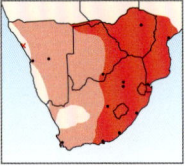

COMMON HOUSE MARTIN   *Delichon urbicum*

J F M A M J J A S O N D

14 cm; 10–16 g A rather plump, swallow-like martin with pure white underparts and a diagnostic white rump. In flight, told from Grey-rumped Swallow by its shorter, less deeply forked tail, broader-based, shorter wings, and white (not off-white) underparts. Pearl-breasted Swallow (p. 330) lacks a pale rump. Perched birds can be separated from other swallows by white feathering on the legs down to the toes. Juv. is less glossy above, with a narrower, pale grey rump band; breast washed grey-brown, inner secondaries tipped white. **Voice:** Hard *'chirrp'* or *'prt-prt'*. **Status and biology:** Fairly common Palearctic-br. migrant over most open habitats; often feeds higher in the sky than other swallows. Thought to roost mainly on the wing, but also recorded roosting on cliffs, in reed beds and trees and on buildings; flocks often gather on wires in early morning, especially prior to northward migration in Mar. Occasional br. attempts recorded in S Africa and Namibia. (Huisswael)

GREY-RUMPED SWALLOW   *Pseudhirundo griseopyga*

J F M A M J J A S O N D

14 cm; 8–12 g A small, slender swallow with glossy blue upperparts, a long, deeply forked tail, pale grey rump and grey-brown crown; underparts off-white. Juv. Common House Martin also has a pale grey rump, but is much plumper, with a shorter tail. Could be mistaken for juv. Wire-tailed Swallow (p. 330), which also has a grey-brown crown, but Grey-Rumped Swallow lacks dark vent bar. Juv. is duller above with buffy tips to feathers, appearing scaly in fresh plumage; breast washed grey. **Voice:** A soft *'chraa'* in flight, audible only at close range. **Status and biology:** Locally common resident and intra-African migrant, moving out of wet areas in the rainy season. Occurs in open woodland and grassland, often near water; breeds in holes in the ground. (Gryskruisswael)

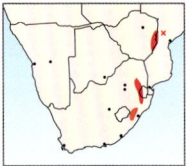

BLUE SWALLOW   *Hirundo atrocaerulea*

J F M A M J J A S O N D

18–25 cm; 13 g A dark blue swallow with variable amounts of white mottling on the neck, flanks and rump. Most likely to be confused with a saw-wing, but flight action is less erratic. In good light, the glossy blue plumage is diagnostic. Ad. male has much longer outer-tail feathers than a saw-wing; tail streamers shorter in female, but still more slender than in a saw-wing. Juv. has a dull brown throat and less glossy upperparts; belly blackish. **Voice:** Song in courtship display is a musical *'bee-bee-bee-bee'*; soft *'chip chip'* contact calls while foraging. Typical flight call is a soft *'chip'*. **Status and biology:** VULNERABLE due to habitat loss. Uncommon and very localised intra-African br. migrant in upland grassland, often bordering forests. (Blouswael)

SAW-WINGS   Slender, mainly black swallows of forest and dense woodland. Males have barbs of outer web of outer primary recurved, giving a saw-edged effect. Taxonomy contentious, with Black and Eastern Saw-wings recently lumped in a widespread *P. pristoptera*. Breed in burrows, usually self-excavated.

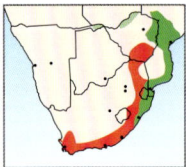

BLACK SAW-WING   *Psalidoprocne pristoptera*

J F M A M J J A S O N D

14 cm; 11–13 g A small, slender, glossy black swallow with a fairly long, deeply forked tail. Two distinct forms in the region, formerly split but now generally lumped due to hybridisation among other forms further north in Africa: 'Black Saw-wing' *P. p. holomelas* (breeds from W Cape to s Zimbabwe and winters along the Mozambique coastal plain) has black underwing coverts; 'Eastern Saw-wing' *P. p. orientalis* (highlands of central and ne Zimbabwe and adjacent Mozambique) and *P. p. reichenowi* (e Zambezi Region (Caprivi) and occasionally in central Zambezi Valley) has crisp white underwing coverts and has a slightly longer tail. Told from Blue Swallow by its glossy, greenish-black (not dark blue) plumage and shorter tail streamers. Juv. sooty-brown; lacks gloss and has shorter tail streamers than ad., but tail longer and more deeply forked than juv. White-headed Saw-wing; throat sooty-brown (not grey). **Voice:** Soft *'chrrp'* and weak *'skre-aa'* alarm call. **Status and biology:** Locally common intra-African br. migrant favouring fringes and clearings in forests and plantations, often near water; mainly winters on the coastal plain of KZN and Mozambique. (Swartsaagvlerkswael)

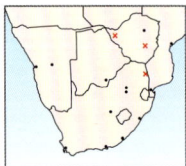

WHITE-HEADED SAW-WING   *Psalidoprocne albiceps*

J F M A M J J A S O N D

13 cm; 7–14 g Vagrant. A dark brown saw-wing; ad. male has a distinctive, snowy-white head, bisected by a diagnostic dark brown line running through the eye to the nape; upperparts have a slight greenish sheen. Female has crown grizzled black; throat pale grey. Juv. duller than female, often with brown crown and grey throat. Some juv. females may be entirely brown; told from other juv. saw-wings by shorter, less deeply forked tail. **Voice:** Soft, weak twittering, with harsher chatters. **Status and biology:** Rare vagrant from central Africa; 3 records from Zimbabwe and Kruger National Park. (Witkopsaagvlerkswael)

COMMON HOUSE MARTIN

blackish cap

shallowly forked tail

white-feathered legs

whitish rump

GREY-RUMPED SWALLOW

grey cap, dark mask

off-white below

long, deeply forked tail

pale grey rump

BLUE SWALLOW

very long streamers

♂

♀

glossy blue plumage

♂

BLACK SAW-WING

orientalis

glossy black plumage

holomelas

orientalis

holomelas

variably white coverts

WHITE-HEADED SAW-WING

white crown

white throat

MARTINS

A miscellaneous group of swallows with predominantly brown plumage. The sand martins (*Riparia*) breed in holes in banks, but the Rock Martin builds a mud cup nest like most swallows. Sexes alike.

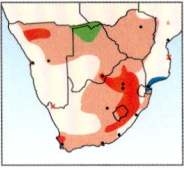

BANDED MARTIN   *Riparia cincta*

17 cm; 23–29 g A large, broad-winged martin with mainly white underparts and a brown breast band. Larger than other martins, with white (not dark) underwing coverts, a small, white supercilium and a square-ended (not forked) tail. Often has a thin, brown line across vent. Juv. has upperparts scaled with pale buff. **Voice:** Flight call is *'che-che-che'*; song is jumble of harsh *'chip'*, *'choop'* and *'chiree'* phrases. **Status and biology:** Locally common intra-African br. migrant to grassland, scrub and other low vegetation; some overwinter in tropical north and east. Roosts communally with other swallows. (Gebande Oewerswael)

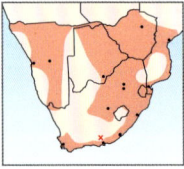

SAND MARTIN   *Riparia riparia*

12 cm; 11–16 g A small, slender martin with mainly white underparts and a brown breast band. Much smaller than Banded Martin, with dark underwings (not white underwing coverts), a shallow-forked (not square) tail, and no white supercilium. Differs from Brown-throated Martin by its white (not brown) throat and brown breast band. Juv. resembles ad. by the time birds reach the region. **Voice:** Grating *'chrrr-chrrr'*. **Status and biology:** Locally common Palearctic-br. migrant, usually near fresh water; often feeds in mixed flocks with Brown-throated Martins, but tends to roost with Barn Swallows. (Europese Oewerswael)

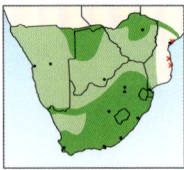

BROWN-THROATED MARTIN   *Riparia paludicola*

12 cm; 10–16 g A small, grey-brown martin that lacks the white throat and dark breast band of Sand and Banded Martins. Belly usually white, but brown in some individuals. Dark form (mainly in south of range) is smaller and slimmer than Rock Martin, with colder grey-brown plumage and no creamy tail spots; underwing coverts brown (not pale russet); flight more erratic. Juv. is warmer brown than ad. below, with buffy fringes to upperpart feathers, appearing scaly in fresh plumage. **Voice:** Soft twittering. **Status and biology:** Common resident and local migrant in open areas, usually near water. Roosts communally in reeds and other vegetation, usually over water. (Afrikaanse Oewerswael)

MASCARENE MARTIN   *Phedina borbonica*

15 cm; 18–23 g A fairly large, robust martin, most similar in structure to Banded Martin. Easily identified by its white underparts that are heavily streaked dark brown. Could be confused with juv. Lesser Striped Swallow (p. 332), but lacks chestnut crown and pale rump. Juv. has pale fringes to secondaries and tertials. **Voice:** Soft *'siri-wiri siri wiri'*, but usually silent in Africa. **Status and biology:** Uncommon non-br. winter visitor from Madagascar to coastal lowlands of Mozambique, mainly over dense, lowland miombo woodland and forests, often in flocks with other swallows. (Gestreepte Kransswael)

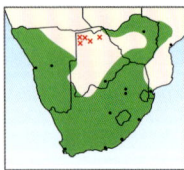

ROCK MARTIN   *Ptyonoprogne fuligula*

14 cm; 14–16 g A brown martin with a warm, buffy wash on the throat and breast. Tail slightly forked, with 8 creamy-white spots near the tail tip, visible when the tail is fanned. Larger and bulkier than brown form of Brown-throated Martin, with warmer brown plumage, paler throat, breast and underwing coverts, and has diagnostic white tail spots; flight action less erratic, more swallow-like. Juv. has pale edges to upperwing coverts and secondaries. **Voice:** Soft, indistinct twitterings; short *'wik'* given in flight. **Status and biology:** Common resident, usually in rocky areas, although also around farms and old buildings. Some move out of high-altitude areas in winter. Often roosts in small groups on cliff faces or buildings when not br. (Kransswael)

BANDED MARTIN

small white supercilium

broad breast band

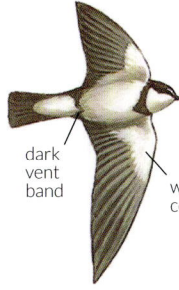

dark vent band

whitish coverts

square tail

SAND MARTIN

narrow breast band

dark coverts

forked tail

BROWN-THROATED MARTIN

brown throat

uniform brown underparts

variable white belly

MASCARENE MARTIN

streaked underparts

wings darker than body

ROCK MARTIN

buffy underparts

stocky build

body paler than wings

creamy tail spots

DRONGOS
Noisy, conspicuous, black birds with erect posture. Hawk from perches, taking prey in air or from the ground. Insectivorous; frequently join bird parties. Sexes alike.

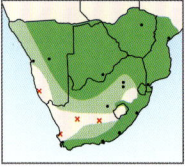

FORK-TAILED DRONGO *Dicrurus adsimilis*

b | | | | | | | | **b B B B b**
J F M A M J J A S O N D

23–26 cm; 38–55 g A large drongo with a deeply forked tail (but moult can affect tail shape). Ad. is uniformly glossy black, with a dark red eye; in flight, primaries noticeably paler than coverts. Larger than Common Square-tailed Drongo; tail more deeply forked, and absent from forest habitats. Juv. has buff-tipped wing coverts, mottled underparts, dark brown eye and yellow gape. Smaller *D. a. apivorus* has paler primaries – considered a separate race by some authorities, 'Clancey's Drongo'. **Voice:** A variety of loud, grating and shrill notes; mimics birds of prey, especially Pearl-spotted Owlet, also cats and small predatory mammals. **Status and biology:** Common resident in woodland, savanna and plantations. Occurs singly, in pairs and, when attending bush fires, in dozens, occasionally hundreds. Pirates food from other birds, sometimes small mammals, often mimicking alarm calls to make them drop their prey. Frequently harasses raptors. (Mikstertbyvanger)

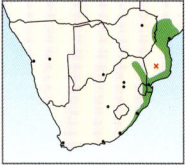

COMMON SQUARE-TAILED DRONGO *Dicrurus ludwigii*

b | | | | | | | | **b b B B b**
J F M A M J J A S O N D

18–19 cm; 20–40 g Smaller than Fork-tailed Drongo, with a shorter and more shallowly forked tail; in flight, primaries dark (not pale). Heavy bill, forked tail and bright red eye distinguish it from Southern Black Flycatcher (p. 406). Vertical posture, habits, black (not yellow) gape and slightly forked (not rounded) tail distinguish it from Black Cuckooshrike. Juv. has pale tips to wing coverts, mantle and underpart feathers. **Voice:** Strident *'cheweet-weet-weet'* and whistled phrases; often very vocal in bird parties; commonly mimics raptors, especially African Goshawk. **Status and biology:** Locally common resident in forests, dense riparian woodland and thickets. Usually in pairs. (Kleinbyvanger)

CUCKOOSHRIKES
Thrush-sized, unobtrusive birds that forage in woodland and forest canopy. Sexes differ in most species; juvs resemble females.

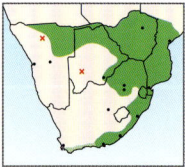

BLACK CUCKOOSHRIKE *Campephaga flava*

b | | | | | | | | **b B B B**
J F M A M J J A S O N D

18–21 cm; 29–36 g Ad. male is all black, with a prominent yellow-orange gape; some individuals have a yellow shoulder patch. Told from Southern Black Flycatcher (p. 406) and Common Square-tailed Drongo by its habits, yellow (not black) gape and rounded (not square or shallowly forked) tail. Ad. female is distinctively barred below, with bold, yellow edges to wing feathers. Juv. is more heavily barred than female, including on crown. **Voice:** High-pitched, prolonged *'trrrrrrrr'*. **Status and biology:** A fairly common resident and local migrant in mature woodland and forest margins. Many birds leave the plateau in winter, apparently moving to lower elevations. Commonly joins mixed-species bird parties. Presence often revealed by its call. (Swartkatakoeroe)

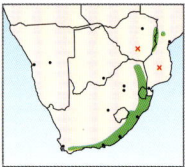

GREY CUCKOOSHRIKE *Ceblepyris caesius*

b | | | | | | | | | **b B B**
J F M A M J J A S O N D

25–27 cm; 53–68 g The only all-grey cuckooshrike; has a narrow, white eye-ring. Male has a dark loral patch, female has grey lores. Juv. has buff barring below and white edges to flight feathers and outer-tail feathers. **Voice:** Soft, thin *'tseeeeep'*. **Status and biology:** Uncommon resident and local altitudinal migrant; confined to evergreen forests and adjacent woodland, usually in canopy. Commonly joins mixed-species bird parties. (Bloukatakoeroe)

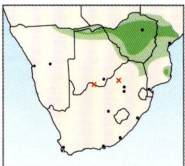

WHITE-BREASTED CUCKOOSHRIKE *Ceblepyris pectoralis*

| | | | | | | | | **b B B B b**
J F M A M J J A S O N D

27 cm; 50–58 g A grey-and-white cuckooshrike with a contrasting black eye and pale eye-ring; distinguished from Grey Cuckooshrike by white underparts and habitat. Male has grey throat and upper breast, contrasting with white belly; female and juv. are paler, with whitish throats. Juv. barred black-and-white above; underparts spotted blackish; flight feathers tipped white, tail black. **Voice:** *'Duid-duid'* by male; *'tchee-ee-ee-ee'* by female. **Status and biology:** Uncommon to locally common resident of tall woodland, especially miombo, mopane and riverine forests. Singly or in pairs; presence usually given away by its wispy call. (Witborskatakoeroe)

FORK-TAILED DRONGO

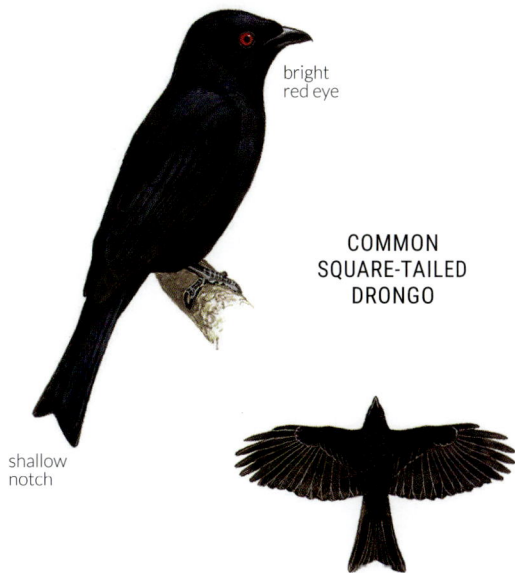

♂

dull red eye

heavy bill (cf. Southern Black Flycatcher)

tail during moult

juv.

deep notch

COMMON SQUARE-TAILED DRONGO

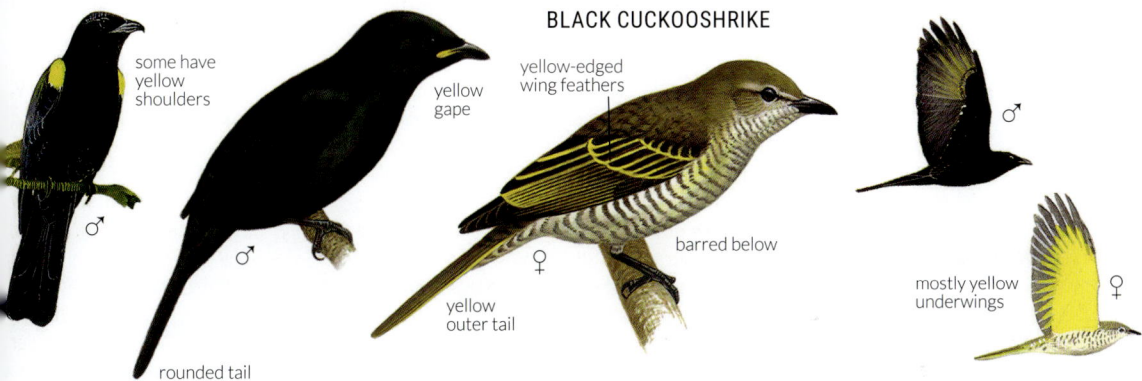

bright red eye

shallow notch

BLACK CUCKOOSHRIKE

some have yellow shoulders

♂

yellow gape

♂

yellow-edged wing feathers

barred below

♂

mostly yellow underwings

♀

yellow outer tail

rounded tail

♀

GREY CUCKOOSHRIKE

grey underwing coverts

grey breast & belly

♂

WHITE-BREASTED CUCKOOSHRIKE

white underwing coverts

♂

♀

white breast & belly

♂

339

ORIOLES

Large, bright yellow birds of woodland and forest canopy; often located by their loud, ringing songs. Sexes alike in dark-headed species, differ in golden orioles.

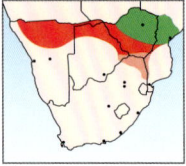

AFRICAN GOLDEN ORIOLE *Oriolus auratus*

| | | | | | | | | | | | |
|b|b|b| | | | | |b|b|B|B|
J F M A M J J A S O N D

20–24 cm; 70–80 g Ad. male distinguished from male Eurasian Golden Oriole by mainly yellow (not black) wings, all-yellow outer-tail feathers, and black eye stripe extending well behind the eye. Female less intense yellow than male, green-tinged above, and faintly streaked on breast. Female is brighter yellow than female Eurasian Golden Oriole, and is less streaked below, with the black eye stripe extending further behind the eye. Juv. duller and greener than female, more streaked below; eye and bill brown. **Voice:** Liquid whistle, *'feeyoofeeyoo'*; also mewling, up-slurred call. **Status and biology:** Locally common intra-African br. migrant, with occasional birds overwintering. In tall woodland, especially miombo, mopane and riverine forest. Usually in pairs. Presence often given away by its call. (Afrikaanse Wielewaal)

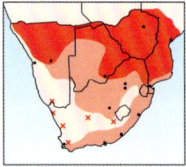

EURASIAN GOLDEN ORIOLE *Oriolus oriolus*

J F M A M J J A S O N D

22–25 cm; 50–80 g Ad. male is vivid yellow and black; differs from male African Golden Oriole by mainly black (not yellow) wings, partly black outer tail and truncated black eye stripe which extends only marginally behind eye. Female is similar to female African Golden Oriole, but underparts are paler and more heavily streaked, with plain wing coverts (lacking yellow edges), and less extensive dark line behind eye. Juv. duller and greener above than female, with dark eye and bill. **Voice:** Mainly silent in the region; song a liquid *'chleeooop'*, with chattering subsong and grating *'naaah'* calls. **Status and biology:** Fairly common Palearctic migrant to mature woodland, savanna and alien plantations. Occurs singly, in pairs or sometimes in small groups. (Europese Wielewaal)

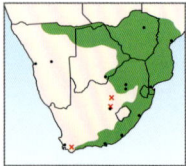

BLACK-HEADED ORIOLE *Oriolus larvatus*

| | | | | | | | | | | | |
|b|b| | | | | | |b|B|B|b|
J F M A M J J A S O N D

20–24 cm; 60–80 g The only black-headed oriole in the region. Ad. has plain black head, throat and central breast; mantle, upperwing coverts and tail more olive-green than in golden orioles. Juv. is duller, with dark brown, slightly mottled head, and streaked throat and breast; bill darker, dull red. **Voice:** Most typical call is an explosive, whistled *'pooodleeoo'*; at onset of br. has a melodious, rambling song mixed with mimicked notes of other birds; alarm note a harsher *'kweeer'*. **Status and biology:** Common resident in mature woodland, forest edge and alien plantations. Some evidence of movement to coastal areas in winter. In pairs, sometimes aggregating at fruiting trees or flowering aloes. (Swartkopwielewaal)

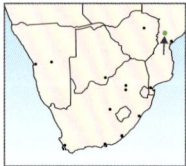

GREEN-HEADED ORIOLE *Oriolus chlorocephalus*

| | | | | | | | | | | | |
| | | | | | | | | | |b|b|
J F M A M J J A S O N D

20–24 cm; 65 g A montane forest oriole, confined in s Africa to montane forest on Mt Gorongoza, central Mozambique. Distinguished from superficially similar Black-headed Oriole by moss-green head, yellow collar and green back. Juv. has dull yellow underparts, breast slightly streaked with olive, and pale olive wash on head and throat. **Voice:** Explosive, liquid song, typical of orioles; distinctive nasal mewing, *'waaaarrr'*. **Status and biology:** Locally common resident within its restricted range; keeps to the upper canopy of tall forest trees. (Groenkopwielewaal)

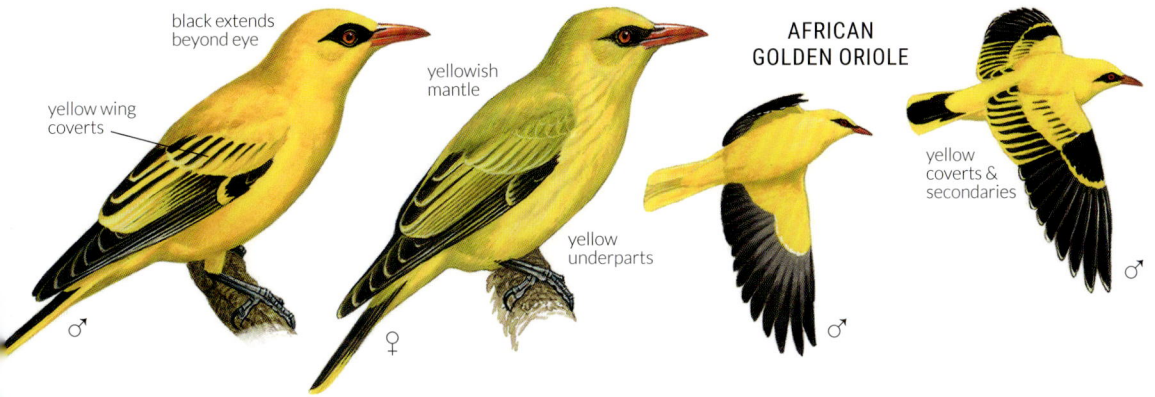

black extends beyond eye

yellow wing coverts

♂

yellowish mantle

yellow underparts

♀

AFRICAN GOLDEN ORIOLE

♂

yellow coverts & secondaries

♂

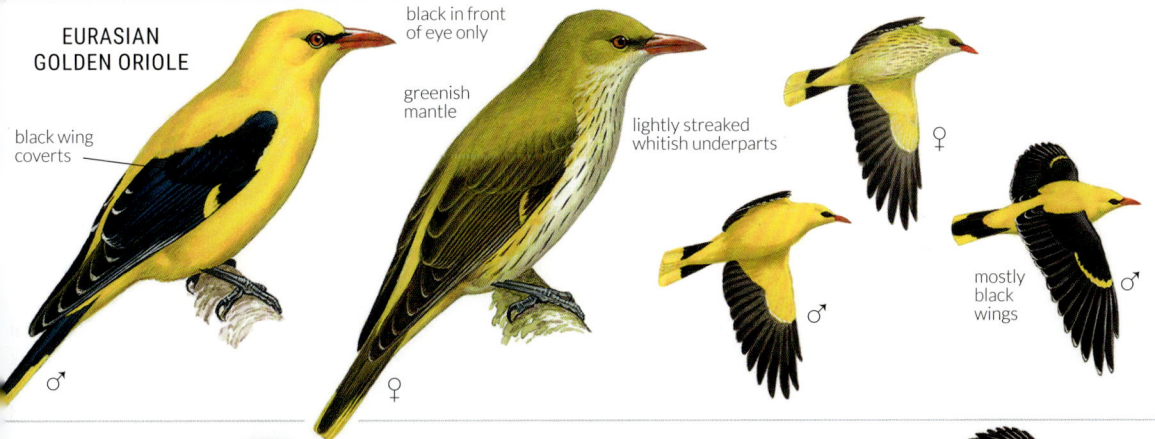

EURASIAN GOLDEN ORIOLE

black wing coverts

♂

black in front of eye only

greenish mantle

lightly streaked whitish underparts

♀

♂

♀

mostly black wings

♂

♀

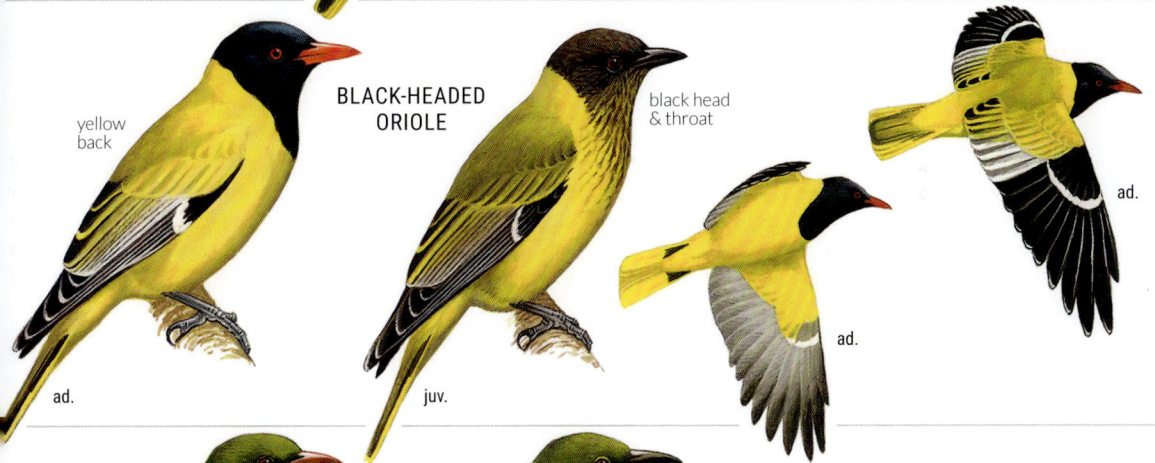

yellow back

ad.

BLACK-HEADED ORIOLE

black head & throat

juv.

ad.

ad.

ad.

green back & wings

ad.

green head & throat

juv.

GREEN-HEADED ORIOLE

ad.

ad.

plain green upperwing & rump

CROWS AND RAVENS
Large, vaguely raptor-like passerines, with either all-black or mainly black-and-white plumage and stout legs and feet. Some species hoard food, demonstrating remarkable memories; others fashion and use tools. Sexes alike.

WHITE-NECKED RAVEN *Corvus albicollis*

50−56 cm; 750−880 g A large crow, with a white crescent on the back of the neck and a heavy, white-tipped bill. In flight, has broader wings and a shorter, broader tail than other crows. Female slightly larger than male. Juv. is brownish-black, especially on head and neck; may have some white feathers on its breast. **Voice:** Deep, throaty *'kwaak'* or *'kraak'*. **Status and biology:** Confined to mountainous and hilly areas with cliffs, but becoming increasingly common in urban environments in W Cape. Locally common resident; in pairs year-round but sometimes flocking (5−150 birds) in non-br. season. Scavenges much of its food from carcasses, roadkills and refuse dumps, also takes live prey, fruit and seeds. (Withalskraai)

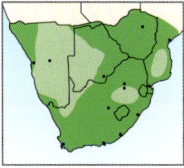

PIED CROW *Corvus albus*

46−50 cm; 410−610 g The only white-breasted crow in the region. From above, has a longer tail, more slender wings and smaller head and bill than White-necked Raven. Juv. is slightly duller, with dusky tips to white feathers, and less crisp margin between black throat and white breast. **Voice:** Croaking, cawing *'krow'* or *'kwooork'*. **Status and biology:** Widespread, common and increasing resident in virtually all habitats except driest desert areas. Often occurs in flocks; roosts communally, sometimes in thousands. A familiar roadside bird, often feeding on roadkills and using telephone poles as perches and as nesting sites. (Witborskraai)

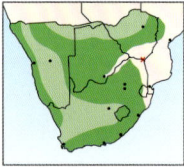

CAPE CROW *Corvus capensis*

45−50 cm; 410−630 g An all-black crow, with a longer, more slender bill than other crows in the region. Larger than House Crow, with uniformly black (not grey-and-black) plumage. Juv. is duller, brownish-black. **Voice:** Deep, cawing *'kaah-kaah'*, and a bizarre, liquid *'kwollop, kwollop'* and other gargling sounds. **Status and biology:** Common resident in grassland, open country, cultivated fields and dry, desert regions. Usually in pairs, but sometimes in flocks (up to 50 birds), and hundreds may roost communally. Forages on the ground, probing with its long bill; less dependent on carrion than other crows in the region. (Swartkraai)

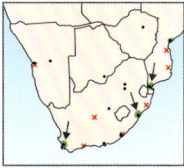

HOUSE CROW *Corvus splendens*

38−43 cm; 260−400 g Introduced; a ship-assisted colonist down the east coast of Africa from Asia that established itself in several coastal cities since the early 1970s. Smaller than Cape Crow, with a diagnostic grey body and proportionally heavier bill. Juv. is duller, with a more uniform blackish body. **Voice:** Hurried, high-pitched *'kah, kah'*. **Status and biology:** Locally common resident, largely restricted to urban environments. Found in Cape Town (still common, despite eradication efforts), Durban, Richards Bay, Maputo and a recent colonist to a few other coastal towns and cities. Sightings away from established populations should be reported to BirdLife. Scavenges food scraps but also takes small birds, eggs and nestlings, often raiding heronries and weaver colonies. Often in flocks of up to 50 birds. (Huiskraai)

WHITE-NECKED RAVEN

white nape

huge bill, white tip

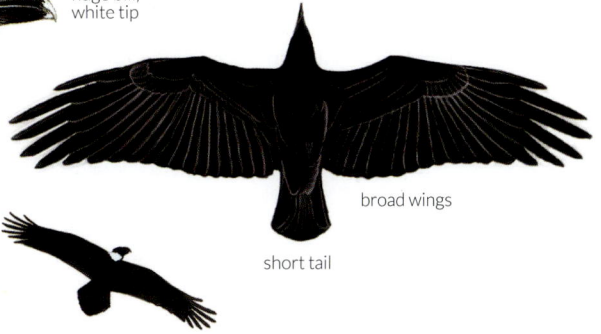

broad wings

short tail

PIED CROW

white collar

all-black bill

white lower breast & belly

CAPE CROW

long slender bill

uniformly black

HOUSE CROW

greyish breast, nape & mantle

TITS
Small, sparrow-sized passerines, most with some black-and-white plumage. Social and noisy; they occur in pairs or small family parties (3–6 birds), frequently join bird parties, and forage in the foliage of trees and bushes. Sexes alike in most species. Nest in holes, usually in trees, often with helpers at the nest.

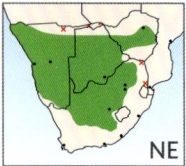

ASHY TIT   *Melaniparus cinerascens*

b	b	b							B	B	B	b
J	F	M	A	M	J	J	A	S	O	N	D	

15 cm; 19–22 g The arid savanna member of the Grey Tit complex, which are all characterised by a striking black-and-white head and black throat extending as a stripe down the central breast and belly. Differs from Grey Tit by its blue-grey (not brownish-grey) back, grey (not buffy) flanks and belly, and white (not buffy) nape spot. Told from Miombo Tit by its grey (not whitish) flanks, white (not buffy) cheeks, narrower white edges to its wing coverts and heavier bill. Juv. is duller, with a sooty-brown cap and throat; back washed brown. **Voice:** Song is ringing trill, *'tlu-tlu-tlu-tlu-tlu'*, *'tu-tu-tu-tu'* or *'tuweee-tuweee-tuweee'*; also a harsh alarm call. **Status and biology:** Near-endemic. Common resident in thornveld and arid savanna. (Akasiagrysmees)

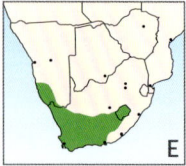

GREY TIT   *Melaniparus afer*

b	b	b				b	B	B	B	b	b
J	F	M	A	M	J	J	A	S	O	N	D

15 cm; 18–22 g Distinguished from Ashy Tit by its brownish-grey (not blue-grey) back, buffy (not grey) belly and flanks, and buff (not white) nape spot. Juv. browner above with pale-edged wing coverts and duller black head and throat. Range does not overlap with Miombo Tit. **Voice:** Song is ringing, whistled *'chiree-wuu-wuu'* or *'swit-weeuu-weeuuz'*. Alarm call is a series of rasping notes, sometimes preceded by 2 or 3 high-pitched whistles; typically less harsh than alarm calls of Ashy Tit. **Status and biology:** Endemic. Locally common resident in fynbos and Karoo scrub, often near rocky outcrops and old buildings. (Piet-tjou-tjougrysmees)

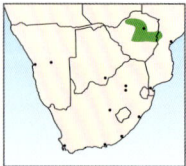

CINNAMON-BREASTED TIT   *Melaniparus pallidiventris*

									b	b	b	b
J	F	M	A	M	J	J	A	S	O	N	D	

14 cm; 16–20 g Sometimes treated as a subspecies of Rufous-bellied Tit, but is slightly smaller and much paler, with a more sharply defined divide between the blackish head and greyish-buff (not rufous) underparts; eyes brown (not yellow); ranges do not overlap in s Africa. Juv. duller with pale edges to wing feathers. **Voice:** Whistles and churrs similar to those of Rufous-bellied Tit. **Status and biology:** Local and generally uncommon resident in miombo woodland; sometimes also in climax mopane woodland. (Swartkopmees)

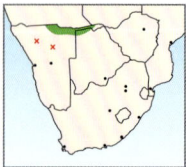

RUFOUS-BELLIED TIT   *Melaniparus rufiventris*

									b	b	b	b
J	F	M	A	M	J	J	A	S	O	N	D	

15 cm; 18–22 g Slightly larger and darker above than Cinnamon-breasted Tit, with a rich rufous (not greyish-buff) belly. Ad. has bright yellow eyes, conspicuous at close range. Juv. duller, with brown eyes and buffy edges to wing feathers. **Voice:** Harsh, tit-like *'chrrr chrrr'*; clear *'chick-weeu, chick-weeu'* song. **Status and biology:** Locally common resident in broadleafed woodland. (Rooipensmees)

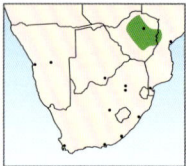

MIOMBO TIT   *Melaniparus griseiventris*

									b	B	B	b	b
J	F	M	A	M	J	J	A	S	O	N	D	

13 cm; 14–17 g Paler than Ashy Tit, with whitish (not grey) flanks, creamy-buff (not white) cheek patches, broader white edges to wing coverts and thinner bill; habitat also differs. Range does not overlap with Grey Tit and it has blue-grey upperparts (not brownish-grey) and pale blue-grey (not buffy) flanks. Juv. duller, with smaller bib reaching only onto its breast; wing feathers edged buff (not white). **Voice:** Scolding *'tjou-tjou-tjou-tjou'* and churring notes. **Status and biology:** Common resident in miombo and other broadleafed woodland. (Miombogrysmees)

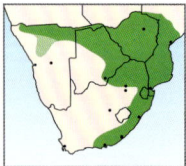

SOUTHERN BLACK TIT   *Melaniparus niger*

b									b	b	B	B	b
J	F	M	A	M	J	J	A	S	O	N	D	

16 cm; 18–25 g A mainly black tit, with white edges to wing feathers. Told from Carp's Tit by its barred grey (not black) vent and less white in wings. Male head and body uniformly black; female has greyish underparts, paler than those of female Carp's Tit. Juv. like female. **Voice:** Lively mix of harsh, chattering *'chrr-chrr-chrr'* and musical *'phee-cher-phee-cher'* notes. **Status and biology:** Common resident in forest and broadleafed woodland. (Gewone Swartmees)

CARP'S TIT   *Melaniparus carpi*

b	B	B	B	b	b						
J	F	M	A	M	J	J	A	S	O	N	D

14 cm; 14–21 g Similar to Southern Black Tit, but has a black (not grey-barred) vent and more extensive white in wings. Male has all-black body; female is sooty-grey, with a darker belly than female Southern Black Tit. Juv. duller than female, with yellowish fringes to flight feathers. **Voice:** Similar to that of Southern Black Tit. **Status and biology:** Near-endemic. Fairly common resident in semi-arid savanna woodland. (Ovamboswartmees)

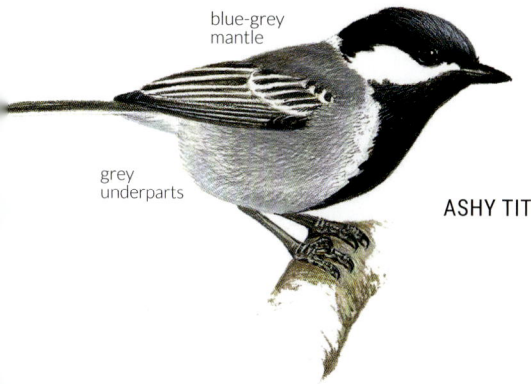

blue-grey mantle

grey underparts

ASHY TIT

grey-brown mantle

buff-grey underparts

GREY TIT

CINNAMON-BREASTED TIT

greyish-buff underparts

RUFOUS-BELLIED TIT

pale eye

rich rufous underparts

MIOMBO TIT

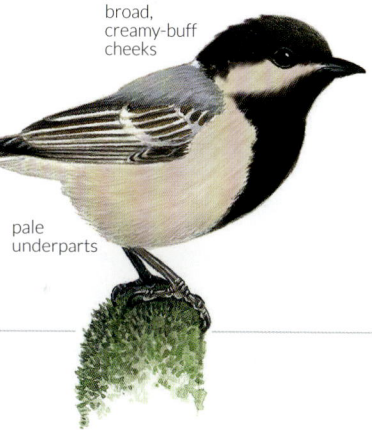

broad, creamy-buff cheeks

pale underparts

♂

barred undertail

mostly black primaries

CARP'S TIT

SOUTHERN BLACK TIT

♀

grey vent

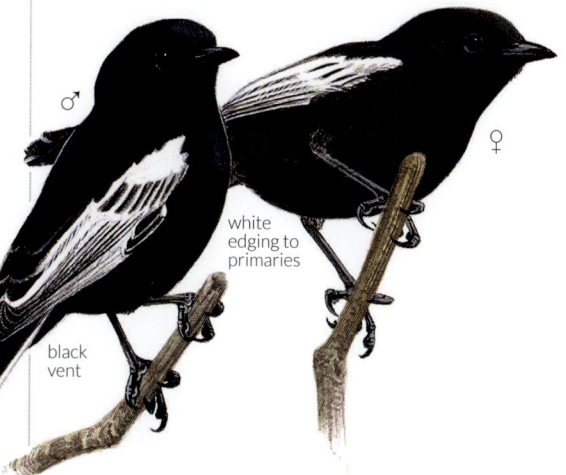

♂

♀

white edging to primaries

black vent

BABBLERS Large, thrush-like birds that live in cohesive family groups of up to 15 birds year-round. Noisy; group members maintain contact with loud babbling and cackling calls. Seldom fly >2 m above the ground, and with long glides between flaps. Mostly forage on the ground, especially among twigs and leaf litter. May form mixed-species foraging groups in areas of overlap. Sexes alike; juvs duller. All s African species, except perhaps Black-faced Babbler, are brood hosts for Levaillant's Cuckoo (p. 260).

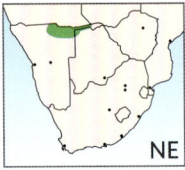

BLACK-FACED BABBLER *Turdoides melanops*

J F M A M J J A S O N D

24–28 cm; 75–80 g A localised, grey-brown babbler; black lores encircling bright cream eye diagnostic. Pale, grey-edged feathers make head and breast appear flecked and scalloped; belly buff. Most similar to Hartlaub's Babbler, but has cream (not reddish) eyes and brown (not white) rump and vent. Juv. is paler, with less scalloping below; eyes brown. **Voice:** Nasal *'wha-wha-wha'*, and harsh, fast *'papapapa'*. **Status and biology:** Uncommon to locally common, near-endemic resident in semi-arid woodland; year-round in family groups of 4–7 birds (3–15). (Swartwangkatlagter)

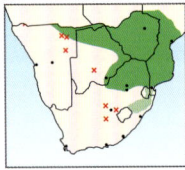

ARROW-MARKED BABBLER *Turdoides jardineii*

J F M A M J J A S O N D

22–25 cm; 60–82 g The most widespread brown babbler in the region; bright yellow eyes with red rims and white arrow-marks on neck, throat and breast are diagnostic. Juv. has brown eyes; arrow-marks less contrasting with brown body. **Voice:** Raucous *'chow-chow-chow-chow...'*, with several birds calling together. **Status and biology:** Common resident of thickets and bush clumps in wooded savanna; territorial groups usually number 6 or 7 birds (3–15); noisy, especially during territorial interactions. (Pylvlekkatlagter)

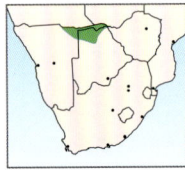

HARTLAUB'S BABBLER *Turdoides hartlaubii*

J F M A M J J A S O N D

24–26 cm; 70–85 g The only brown babbler in the region with a white rump. At rest, told from Black-faced and Arrow-marked Babblers by its whitish lower belly and vent. Eyes orange-red to crimson. Juv. is paler, especially on the throat and breast. **Voice:** Noisy; loud *'kwek-kwek-kwek'* or *'papapapapapa'*. **Status and biology:** Common resident, usually close to water; in woodland and thickets fringing rivers and flood plains; ranges into papyrus and reed beds. Occurs in groups of 5–15 birds. (Witkruiskatlagter)

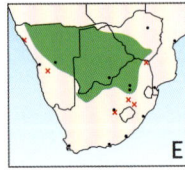

SOUTHERN PIED BABBLER *Turdoides bicolor*

J F M A M J J A S O N D

23–26 cm; 70–84 g A striking, black-and-white babbler. Ad. told from Bare-cheeked Babbler by its blackish wings and unmarked white head and back. In flight, white wing coverts contrast with blackish flight feathers. Juv. is pale brown, whitening with age and passing through series of mottled plumages over 4–6-month period; usually paler than juv. Bare-cheeked Babbler, and always accompanied by ads. **Voice:** High-pitched *'kwee kwee kwee kweer'* babbling. **Status and biology:** Locally common, endemic resident in arid savanna, especially thornveld. Usually in groups of 6–10 birds (3–15). (Witkatlagter)

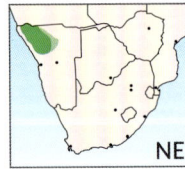

BARE-CHEEKED BABBLER *Turdoides gymnogenys*

J F M A M J J A S O N D

24–26 cm; 70–90 g A distinctive babbler with a mainly white body. Differs from Southern Pied Babbler by its brown (not white) back, rufous nape, brown (not black) wings, and small patches of bare, black skin below and behind its eyes. Juv. has crown and nape pale grey-brown; edges of wing feathers edged buff; appears darker than juv. Southern Pied Babbler. **Voice:** Typical babbler *'kerrrakerrra-kek-kek-kek'*. **Status and biology:** Near-endemic. Uncommon to locally common resident in arid savanna, favouring wooded hills and dense vegetation along rivers; usually in groups of 5 or 6 birds (2–11). (Kaalwangkatlagter)

yellow eyes,
black lores

**BLACK-FACED
BABBLER**

lightly scalloped
underparts

**ARROW-MARKED
BABBLER**

red &
yellow eye

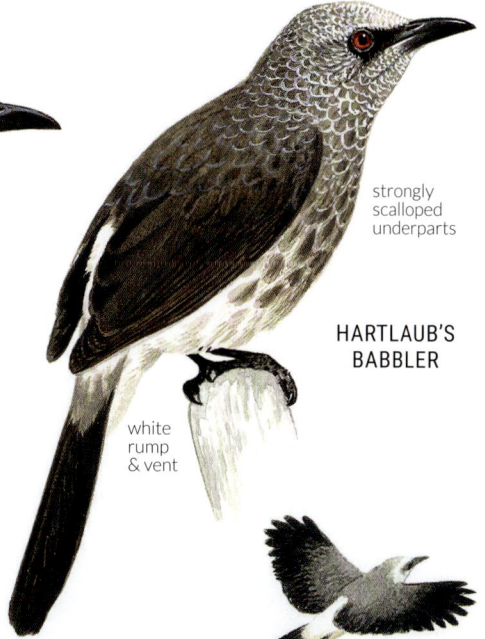

strongly
scalloped
underparts

**HARTLAUB'S
BABBLER**

pale arrow-
marks on
underparts

ad.

white
rump
& vent

juv.

**SOUTHERN
PIED BABBLER**

**BARE-CHEEKED
BABBLER**

entirely
white
head

rufous
nape

white
back

dark brown
back

imm.

ad.

347

BUSH BLACKCAP

BUSH BLACKCAP A forest-living, bulbul-like bird in its own genus, now known to be related to the sylviid warblers (p. 378). Sexes alike.

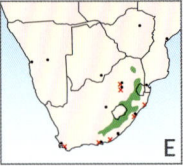

BUSH BLACKCAP *Lioptilus nigricapillus*

16–18 cm; 26–33 g A brown-backed warbler whose structure and rich song recall a bulbul. Black cap, bluish-grey neck and underparts and pinkish-red bill and legs are diagnostic. Juv. duller, with a pink bill. **Voice:** Song a rich, melodious warbling *'plik plik toodley-oodley-oodley-ooo'*. **Status and biology:** VULNERABLE. Uncommon, endemic resident and altitudinal migrant; juvs occasionally recorded well out of range. Lives singly or in pairs in montane forest and forest edge, also gardens and coastal forest in winter. Mostly in leafy canopy of trees, presence usually given away by its song. (Rooibektiptol)

BULBULS AND GREENBULS

BULBULS AND GREENBULS *Pycnonotus* bulbuls are common and conspicuous medium-sized passerines known locally as 'toppies' and characterised, in s Africa, by yellow vents and dark brown caps. Greenbuls and brownbuls are a large, diverse group of drab bulbuls of forest, woodland and thickets. They are notoriously difficult to identify; knowing their calls helps greatly with detection and identification. Sexes alike.

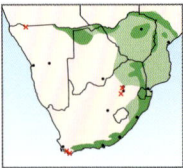

TERRESTRIAL BROWNBUL *Phyllastrephus terrestris*

18–22 cm; 24–44 g An all-brown forest bulbul with a white throat, paler brown underparts and white eye-ring. Juv. is paler with a more rufous rump and tail, and yellow wash on underparts. **Voice:** Soft, chattering *'chrrt-chrrt-chrrt, cherrup trrup trrup'*. **Status and biology:** Common resident in forest and forest-edge thickets; occurs in pairs or family groups of 3–6. Secretive, gleaning among low, tangled vegetation and on the forest floor; presence often revealed by its frequent contact calling between group members. Joins mixed-species bird parties. (Boskrapper)

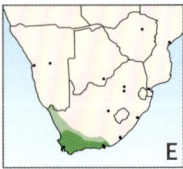

CAPE BULBUL *Pycnonotus capensis*

21 cm; 30–46 g A chocolate-brown bulbul, larger than African Red-eyed and Dark-capped Bulbuls; white eye-ring and darker underparts diagnostic. In E Cape, hybrids with Dark-capped Bulbul (along the Sundays River) and with African Red-eyed Bulbul (between Prince Albert and Somerset East) have intermediate characters between parents. Juv. has a purplish-grey eye-ring. **Voice:** Song is a lively, liquid whistle, *'chip chee woodely'*, higher pitched and sharper than Dark-capped Bulbul. **Status and biology:** Common, endemic resident in fynbos, coastal scrub and gardens. Occurs singly, in pairs, or in flocks; sometimes hawks insects. (Kaapse Tiptol)

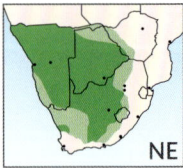

AFRICAN RED-EYED BULBUL *Pycnonotus nigricans*

19 cm; 22–36 g Red eye-ring diagnostic; head blacker than Dark-capped Bulbul, contrasting with a greyish neck collar and breast. Juv. has whitish eye-ring, turning pale pink after 2–3 months. Hybrids with Dark-capped and Cape Bulbuls have intermediate characters between parents. **Voice:** Liquid whistles, slightly more fluty than Dark-capped Bulbul. **Status and biology:** Common, near-endemic resident and local nomad in arid savanna, riverine bush and gardens. Occurs singly, in pairs or in flocks; sometimes hawks insects. (Rooioogtiptol)

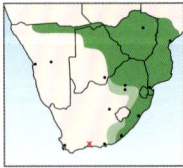

DARK-CAPPED BULBUL *Pycnonotus tricolor*

21 cm; 30–48 g The most common bulbul in the east of the region; told from other bulbuls by its black (not red or white) eye-ring. Juv. paler. Hybrids with African Red-eyed and Cape Bulbuls have intermediate characters between parents. **Voice:** Harsh *'kwit, kwit, kwit'* alarm call; song is a liquid *'sweet sweet sweet-potato'*. **Status and biology:** Abundant resident in a wide range of wooded habitats from savanna to forest edge and gardens. Vocal and conspicuous; occurs singly, in pairs or in flocks (especially in winter). Sometimes hawks insects, especially termite alates. (Swartoogtiptol)

pink bill

BUSH
BLACKCAP

ad.

juv.

brown
upperparts

white
throat

TERRESTRIAL BROWNBUL

white
eye-ring

CAPE BULBUL

red
eye-ring

AFRICAN RED-EYED
BULBUL

black
eye-ring

DARK-CAPPED
BULBUL

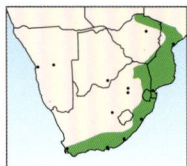

SOMBRE GREENBUL *Andropadus importunus*

b b B B
J F M A M J J A S O N D

18 cm; 30–39 g A medium-sized, plain olive-green greenbul with a diagnostic pale eye and distinctive song. *A. i. hypoxanthus* (central Mozambique) is greener above and more yellow below; could be confused with Yellow-bellied Greenbul, but has white (not red) eye and lacks narrow, white eye crescents. Juv. duller, with dark eyes, a narrow, pale eye-ring and prominent yellow gape. **Voice:** Song is a piercing *'willie'*, often followed by a rapid warble, ending in a nasal, querulous note (*'willie, come-and-have-a-fight, scaaared'*). Alarm call is a loud *'plee plee plee'*. **Status and biology:** Common resident in canopy and mid-strata of forest and thicket; solitary or in pairs. (Gewone Willie)

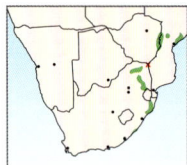

YELLOW-STREAKED GREENBUL *Phyllastrephus flavostriatus*

b b b b b B B
J F M A M J J A S O N D

18–20 cm; 22–38 g A slender greenbul with pale underparts and a long, slender bill; appreciably larger than Lowland Tiny Greenbul. Best identified by its foraging action: gleans insects by creeping up branches, frequently flicking open one wing at a time. The faint yellow breast streaks are only visible at close range and in good light. Juv. has breast washed olive. **Voice:** Song is a joyful *'klip, klip-ip-ip, klip-ip-ip, weet-weet-weet-weaat'*; also sharp *'kleet kleet kleeat'* and dry *'trl-rl-rl-rl'* calls. **Status and biology:** Locally common resident of forest mid-stratum and canopy. Usually in groups of 5 or 6 birds (max. 15). (Geelstreepboskruiper)

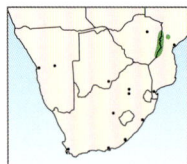

STRIPE-CHEEKED GREENBUL *Arizelocichla milanjensis*

b b b b B B
J F M A M J J A S O N D

20 cm; 31–47 g A localised, montane forest greenbul; combination of dark olive-green body, grey head, dark eye and white eye crescent is diagnostic; faint white cheek streaks are only visible at close range. Juv. has a green-washed crown. **Voice:** Throaty *'chrrup-chip-chrup-chrup'*. **Status and biology:** Locally common resident in montane forest and forest edge. Occurs singly or in pairs. Typically remains concealed in foliage in mid- and upper forest strata; presence usually given away by its loud, guttural call. (Streepwangwillie)

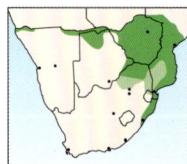

YELLOW-BELLIED GREENBUL *Chlorocichla flaviventris*

b b b B B B B
J F M A M J J A S O N D

20–23 cm; 32–50 g A thickset greenbul with a heavy bill, olive upperparts, yellow-washed underparts and a diagnostic white crescent above its eye. Larger than Sombre Greenbul, with much brighter yellow underparts and dark red (not white) eyes. Bright yellow underwing coverts are conspicuous in flight. Juv. duller and paler, with grey eyes and a prominent yellow gape. **Voice:** A querulous, nasal *'nehr-nehr-nehr-nehr'*, often repeated, becoming faster and higher pitched when alarmed, and interspersed by short, guttural calls. Contact call is a hoarse *'kwoar-tooarr'*. **Status and biology:** Common resident of coastal and riverine forest, dense woodland and thickets; also gardens and mangroves. Usually in groups of 5 or 6 birds. Sometimes hawks insects and occasionally gleans parasites from antelopes. (Geelborswillie)

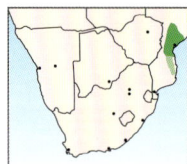

LOWLAND TINY GREENBUL *Phyllastrephus debilis*

b b b b
J F M A M J J A S O N D

14 cm; 13–16 g A small, warbler-like greenbul. A diminutive version of Yellow-streaked Greenbul, sharing its pale grey head, whitish throat and variably yellow-washed underparts, but lacking its distinctive wing-flicking foraging action. Eyes vary from white to grey-brown; bill slender with a pale grey base. Juv. has a greener crown. **Voice:** Nasal, bubbling song *'kwerr kerr ker ker kr-r-rrrr'*, increasing in pace; shrill *'shriiip'* call. **Status and biology:** Uncommon resident in low- and mid-elevation forest, forest edge and adjacent thickets. Occurs in pairs or small groups. (Kleinboskruiper)

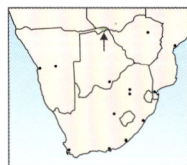

YELLOW-THROATED LEAFLOVE *Atimastillas flavicollis*

b b b b
J F M A M J J A S O N D

22 cm; 40–60 g A large greenbul of riparian woodland and thicket that reaches its southern limit along the Zambezi River in Katima Mulilo, Namibia. Drab olive-brown above and slightly paler below with a striking sulphur-yellow throat. Flight and outer-tail feathers edged greenish-olive. Eyes are dull yellow. Juv. is darker overall, with a paler throat and dark eyes. **Voice:** Babbler-like chattering. **Status and biology:** Found in thick undergrowth, wooded streams and rivers, thickets and overgrown cultivated areas. Occurs in pairs or small groups. (Geelkeelwillie)

SOMBRE GREENBUL

uniform olive plumage

white eye

YELLOW-STREAKED GREENBUL

grey crown

dark eye

pale throat

STRIPE-CHEEKED GREENBUL

dark eye, pale eye-ring

faintly streaked cheeks

YELLOW-BELLIED GREENBUL

dark red eye

bright yellow underparts

LOWLAND TINY GREENBUL

grey cap

pale eye

pale throat

yellowish underparts

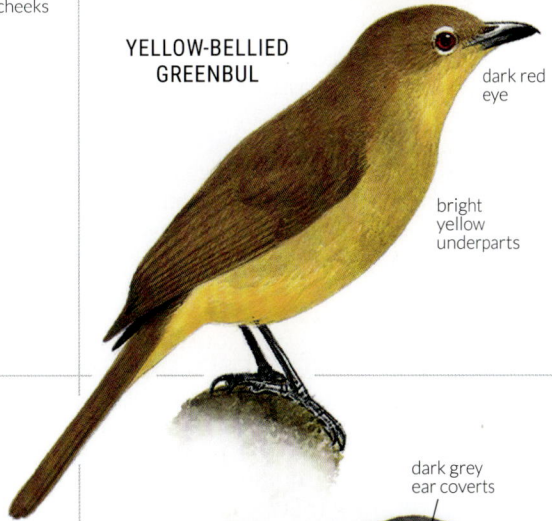

dark grey ear coverts

striking yellow throat

YELLOW-THROATED LEAFLOVE

THRUSHES

THRUSHES Fairly large passerines that live solitarily or in pairs in well-wooded habitats. Mainly forage on the ground, turning over leaves in search of food. All sing melodiously, often before dawn. Ground thrushes have distinctive white wingbars. Sexes alike.

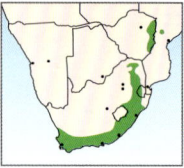

OLIVE THRUSH *Turdus olivaceus*

b	b	b	b	b	b	b	b	B	B	B	B
J	F	M	A	M	J	J	A	S	O	N	D

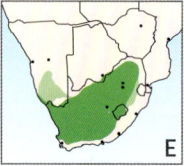

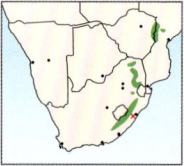

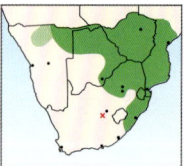

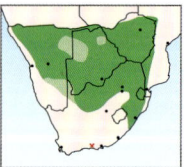

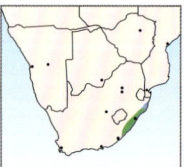

22–24 cm; 55–78 g A forest thrush with an orange-washed belly and yellow legs and bill. Told from very similar Karoo Thrush by its brighter underparts with orange (not whitish) central belly, a generally paler vent; at close range has dark brown (not yellow) eye-ring and brown smudge around nostril. Juv. has mottled underparts and buff edging to upperwing coverts; occasionally hybridises with Karoo and Kurrichane Thrushes. **Voice:** Sharp *'chink'* or thin *'tseeep'* call; song is a rich, melodic whistle, *'wheeet-tooo-wheeet'*, usually given before dawn. **Status and biology:** Common resident and altitudinal migrant in evergreen forest; in s and W Cape habitat extends to parks, gardens and plantations, where it is often confiding. Solitary or in pairs. (Olyflyster)

KAROO THRUSH *Turdus smithi*

b	b	b	b						b	B	B
J	F	M	A	M	J	J	A	S	O	N	D

22–24 cm; 60–86 g A woodland thrush, very similar to Olive Thrush but typically found in drier habitats, told by duller underparts, a generally darker vent, yellow (not dark) eye-ring and richer orange bill that lacks a dark saddle around the nares. Ranges seldom overlap; intermediates at contact zones suggest some hybridisation. Juv. has mottled underparts. **Voice:** Soft *'tseeeep'* contact call; song similar to Olive Thrush. **Status and biology:** Endemic. Locally common resident, mostly in riparian woodland in semi-arid Karoo and introduced woodland on the highveld; common garden bird. Occurs singly or in pairs, but semi-gregarious outside br. season. (Geelbeklyster)

ORANGE GROUND THRUSH *Geokichla gurneyi*

b								b	B	B
J	F	M	A	M	J	J	A	S	O	N	D

20–22 cm; 60–74 g A secretive forest thrush, distinguished from other thrushes with orange underparts by its diagnostic white bars on its upperwing coverts, blackish (not yellow-orange) bill and white eye-ring. From below, told from Olive Thrush by its white belly. Juv. has pale, spotted upperparts and mottled underparts; throat whitish. **Voice:** Sibilant *'tseeep'*; song is a series of rich, melodic whistles. **Status and biology:** Uncommon resident in montane forests, favouring stream edges with an open understorey. Solitary outside br. season. (Oranjelyster)

KURRICHANE THRUSH *Turdus libonyanus*

b	b	b					b	B	B	B	b
J	F	M	A	M	J	J	A	S	O	N	D

21–22 cm; 50–70 g A woodland thrush with a white belly and orange-buff flanks; upperparts and breast are paler and greyer than Olive or Karoo Thrushes. Black speckling on throat is concentrated into diagnostic malar stripes; orange (not yellow or dark) eye-ring; bill is brighter orange than other thrushes. Occasionally hybridises with Olive and Karoo Thrushes. Juv. has mottled underparts and buffy tips to upperwing coverts. **Voice:** Loud, whistled *'peet-peeoo'*. **Status and biology:** Common resident and local nomad in woodland with an open understorey; also parks and gardens. (Rooibeklyster)

GROUNDSCRAPER THRUSH *Turdus litsitsirupa*

b	b	b					b	B	B	B	b
J	F	M	A	M	J	J	A	S	O	N	D

21–23 cm; 70–85 g A short-tailed savanna thrush; bold, black spotting on white underparts and face, and lack of wingbar are diagnostic. Has a characteristic upright stance and habit of running forward, stopping and tipping back, sometimes partly lifting one wing. In flight, shows prominent chestnut bars across primaries. Superficially similar to Dusky Lark (p. 308), but is larger with yellow (not grey) legs, plain (not scalloped) back and an orange lower mandible. Juv. like ad. **Voice:** Song is less melodic than other thrushes; series of slow notes *'lit-sit-si-rupa'*. **Status and biology:** Common resident in open woodland with a short-grass or bare understorey. (Gevlekte Lyster)

SPOTTED GROUND THRUSH *Geokichla guttata*

b	b							B	B	B	b
J	F	M	A	M	J	J	A	S	O	N	D

20–22 cm; 62–76 g A secretive forest thrush with a spotted breast; white wingbars and habitat distinguish it from similar-looking Groundscraper Thrush; pale pink legs are conspicuous. Juv. has head and mantle spotted buff; underparts buff with darker streaks. **Voice:** Quiet *'tseeeep'* call; song is whistled and fluty, with short phrases of 4 or 5 notes. **Status and biology:** ENDANGERED. Rare resident and partial migrant in forest understorey. Breeds in coastal forests in E Cape and slightly further inland in KZN; both populations winter in coastal forests along the KZN coast. (Natallyster)

OLIVE THRUSH

brown eye-ring

brown top to upper mandible

white vent

orange belly

ad.

juv.

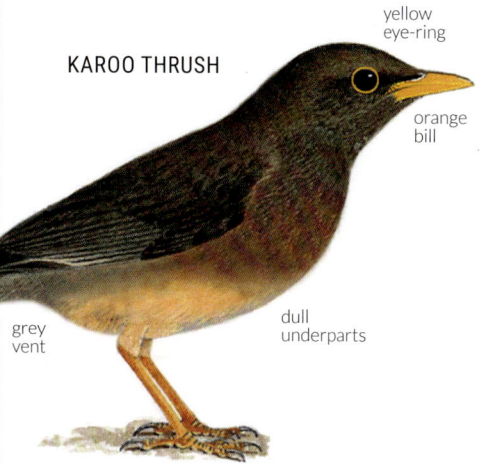

KAROO THRUSH

yellow eye-ring

orange bill

grey vent

dull underparts

white eye crescents

white wingbars

black bill

ORANGE GROUND THRUSH

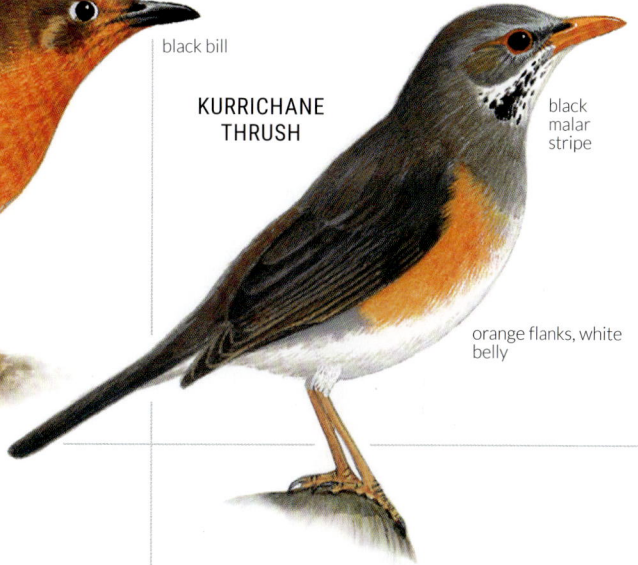

orange-red eye-ring & bill

KURRICHANE THRUSH

black malar stripe

orange flanks, white belly

GROUNDSCRAPER THRUSH

upright stance

plain grey wings

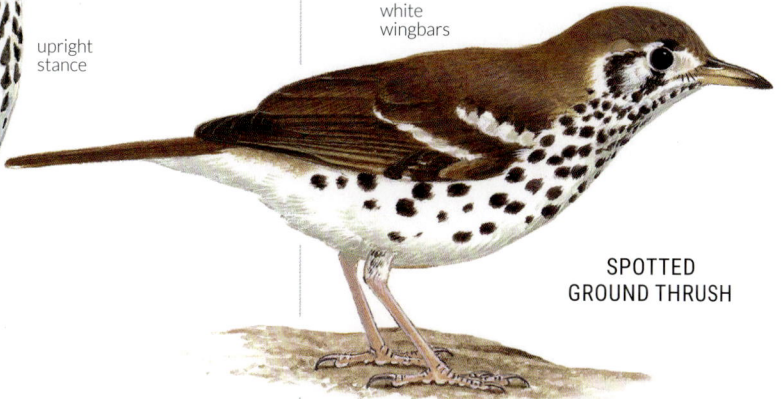

white crescent behind eye

white wingbars

SPOTTED GROUND THRUSH

ROCK THRUSHES

Distinctive-looking, thrush-like birds related to robins and chats. Males have slate-blue heads and rufous or buff underparts; females duller, typically lacking blue on head. Terrestrial, open-country birds, most restricted to rocky habitats.

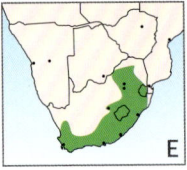

CAPE ROCK THRUSH *Monticola rupestris*

20–22 cm; 50–65 g Largest of the rock thrushes in the region, with a relatively long tail and more horizontal stance. Male has diagnostic brown back. Female distinguished from other species by darker rufous underparts extending to upper breast. Juv. like female, but spotted buff above, scaled blackish below. **Voice:** Song is a far-carrying, rather stereotyped *'tsee-tseu-tseet tsee-tseu-tseet chweeeoo'* whistle; harsh, grating alarm calls. **Status and biology:** Endemic. Locally common resident in rocky areas, favouring steep hillsides and ravines in grassland and heaths. Some birds leave high elevations in winter. (Kaapse Kliplyster)

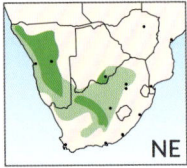

SHORT-TOED ROCK THRUSH *Monticola brevipes*

16–18 cm; 28–35 g A rather small rock thrush of semi-arid areas. Nominate male (*M. b. brevipes*) has diagnostic whitish crown which distinguishes it from other male rock thrushes. Male of 'Pretoria' race (*M. b. pretoriae*) lacks white crown (head, back and shoulders all slate-blue) and is distinguished from very similar male Sentinel Rock Thrush by its smaller size, shorter legs (tarsus 25 mm versus 34 mm) and deeper rufous underparts. Female has more extensively white throat than female Cape Rock Thrush and more rufous underparts than female Sentinel Rock Thrush. Juv. is spotted with buff on upperparts and with black below. Juv. of *M. b. pretoriae* is duller, with a browner crown. **Voice:** Thin *'tseeep'*; song of whistled phrases is like those of other rock thrushes, including some mimicry of other birds' calls. **Status and biology:** Near-endemic. Common resident and local nomad on bushed and wooded rocky hillsides and outcrops. (Korttoonkliplyster)

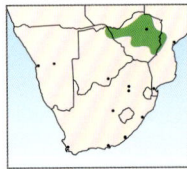

MIOMBO ROCK THRUSH *Monticola angolensis*

16–18 cm; 40–48 g The only woodland rock thrush in the region. A rather small species, easily distinguished from all other rock thrushes in the region by distinctive black mottling on the hind crown, nape, back and shoulders on both sexes. Juv. is more heavily mottled below. **Voice:** A 2-note whistle; song is high-pitched variety of melodic phrases. **Status and biology:** Locally common resident and local nomad in miombo woodland, often in hilly, rocky country or among granite outcrops. Range does not overlap with any other local rock thrushes. Unobtrusive and easily overlooked; often perches motionless on low branches for long periods. (Angolakliplyster)

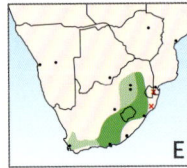

SENTINEL ROCK THRUSH *Monticola explorator*

18–20 cm; 44–51 g A long-legged rock thrush. Male distinguished from other male rock thrushes by the blue-grey head colour extending onto mid-breast, back and shoulders. Slightly larger than male of 'Pretoria' race of Short-toed Rock Thrush (*M. b. pretoriae*), with longer legs (tarsus 34 mm versus 25 mm) and paler rufous-orange underparts. Some females have bluish heads and orange bellies, but most are dull brown above with a pale ginger-buff belly grading into a mottled white-and-buff upper breast and throat; lack rufous underparts of female Cape Rock Thrush and white throat of female Short-toed Rock Thrush. Juv. has pale-spotted upperparts and brown-scaled underparts. **Voice:** Whistled song is more varied and softer than that of Cape Rock Thrush. **Status and biology:** NEAR-THREATENED. Endemic. Locally common resident and altitudinal migrant, moving in winter to treeless, open grasslands, especially where burnt. (Langtoonkliplyster)

CAPE ROCK THRUSH

dark rufous underparts
(cf. female Sentinel
Rock Thrush)

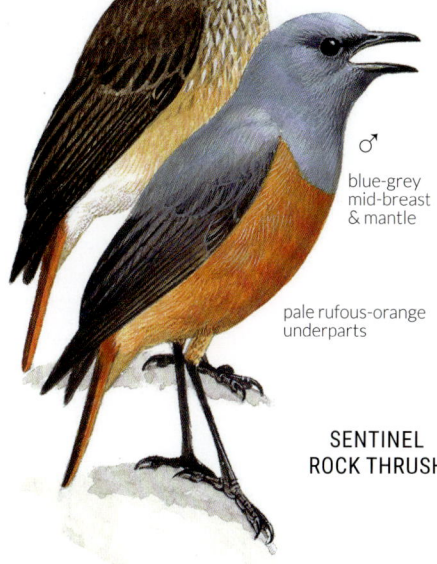

♀

brown-red
back

dark orange
breast &
underparts

♂

variable
white cap

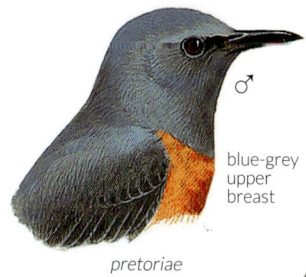

blue-grey
back

♀

brevipes

♂

brevipes

blue-grey
upper
breast

♂

pretoriae

♀

pretoriae

SHORT-TOED ROCK THRUSH

malar
stripes

mottled
back

♀

♂

MIOMBO ROCK THRUSH

♀

♂

blue-grey
mid-breast
& mantle

pale rufous-orange
underparts

SENTINEL ROCK THRUSH

ALETHES, ROCKJUMPERS AND CHATS
Alethes are thrush-like birds in the Muscicapidae family (robins, chats and flycatchers) found in the understorey of tropical African forests; the genus has been split and most species (including the only 1 found in our region) are now in the genus *Chamaetylus*, not *Alethe*. Rockjumpers are an endemic family of strikingly coloured, terrestrial, babbler-like birds of open, rocky mountain slopes, allied to picathartes from W and central Africa and the enigmatic Rail-babbler *Eupetes macrocerus* of SE Asia. The 2 large, long-tailed chats here, although not closely related, share a preference for wooded, boulder-strewn hills.

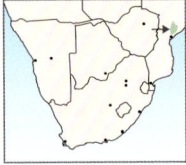

WHITE-CHESTED ALETHE *Chamaetylus fuelleborni*
b b b
J F M A M J J A S O N D

18–20 cm; 52 g A large, rather chunky alethe with crisp white underparts contrasting sharply with its dark grey face and crown, and rich chestnut back, wings and tail; flanks washed grey. Sexes alike. Juv. is blackish-brown above, spotted with buff; underparts creamy-buff, variably mottled and scaled dark grey. **Voice:** Lively, slightly mournful and vibrato *'fweer-her-hee-her-hee-her'* series of whistles. **Status and biology:** Uncommon resident of coastal forest in central Mozambique. Unobtrusive, and very difficult to obtain views; solitary or in pairs. Forages on forest floor, flying to a perch if disturbed. (Witborswoudlyster)

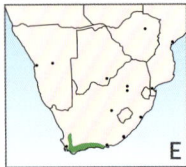

CAPE ROCKJUMPER *Chaetops frenatus*
b b b B b b
J F M A M J J A S O N D

23–25 cm; 50–72 g An unmistakable mountain fynbos endemic; male striking with red eye, black face and throat, white moustachial stripe and dark rufous underparts and rump. Female and juv. less colourful, with buff-and-grey-streaked head, back and breast; they differ from female and juv. Drakensberg Rockjumper by their brownish (not buff) underparts. Ranges of the 2 do not overlap, but come within 100–150 km of each other in E Cape. **Voice:** Series of loud, high-pitched whistles, *'pee-pee-pee-pee-pee-pee'*, falling in pitch. **Status and biology:** NEAR-THREATENED. Endemic. Locally common resident and partial altitudinal migrant on rocky slopes. Usually in small groups that forage on the ground, occasionally hopping onto rocks, then disappearing. (Kaapse Berglyster)

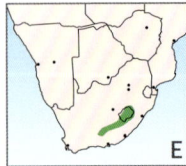

DRAKENSBERG ROCKJUMPER *Chaetops aurantius*
b b b b B B b
J F M A M J J A S O N D

21–23 cm; 45–56 g Paler than Cape Rockjumper; ranges of the 2 do not overlap, but come within 100–150 km of each other in E Cape. Male has an orange (not rufous) belly and rump. Female and juv. are pale buff (not rufous) below. **Voice:** Repeated, piping whistles, like Cape Rockjumper, including a loud, staccato *'prreee prreee prreee'*. **Status and biology:** NEAR-THREATENED. Locally common, endemic resident on rocky slopes, typically above 2 000 m. Habits similar to Cape Rockjumper. (Oranjeborsberglyster)

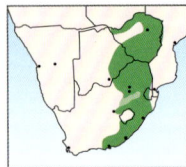

MOCKING CLIFF CHAT *Thamnolaea cinnamomeiventris*
b B B B b
J F M A M J J A S O N D

22 cm; 46–53 g A large, strikingly coloured, rock-dwelling chat. Male is glossy black, with a bright chestnut belly, vent and rump, and white shoulder patches. Female is dark grey above and chestnut below. Juv. like female but duller, with buff tips to its contour feathers. **Voice:** Loud, melodious, whistled song, often mimicking other birds; also shrill, piercing calls. Both sexes sing. **Status and biology:** Common but localised resident on cliffs and boulder-strewn hillsides and ravines, usually in pairs or family parties; becomes confiding around human dwellings. Forages on the ground. (Dassievoël)

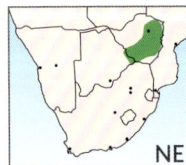

BOULDER CHAT *Pinarornis plumosus*
b B B B b
J F M A M J J A S O N D

25 cm; 48–72 g A large, blackish-brown, babbler-like chat; told from Mocking Cliff Chat by its dark, blackish-brown (not chestnut) rump and belly. In flight, white outer-tail feathers and a line of white spots across each wing (tips of greater coverts) are conspicuous. Sexes alike. Juv. duller than ad., with paler underparts. **Voice:** Clear, sharp whistle; softer *'wink, wink'* call, like a squeaky wheel. **Status and biology:** Near-endemic. Locally common resident confined to large boulder outcrops in woodland. Active and vocal, running and bounding over large boulders, occasionally raising tail well over back when landing. Forages in pairs or family parties on the ground in leaf litter. (Swartberglyster)

russet wings & tail

WHITE-CHESTED ALETHE

ad.

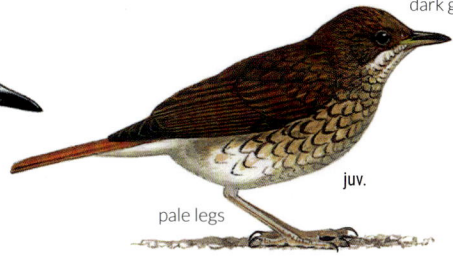

dark grey cap

juv.

pale legs

♂

dark rufous underparts

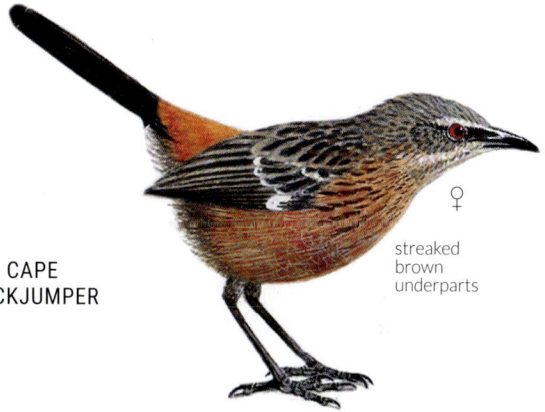

♀

CAPE ROCKJUMPER

streaked brown underparts

♂

white outer-tail tips

orange underparts

♀

DRAKENSBERG ROCKJUMPER

pale buff underparts

♀

MOCKING CLIFF CHAT

well-defined chestnut belly & rump

♂

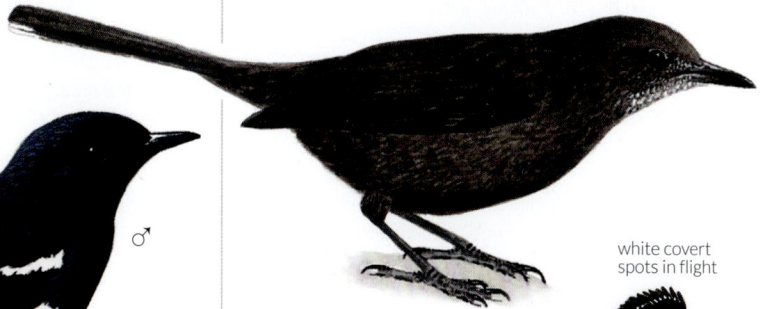

white covert spots in flight

white outer-tail tips

BOULDER CHAT

CHATS AND WHEATEARS

Small to medium-sized passerines that spend most of their time on the ground or perched on low vantage points; usually in pairs or family groups. Sexes differ in some species, similar in others. The species on this page were traditionally placed in a single genus, *Cercomela*, but genetic evidence shows that the Familiar Chat is a wheatear (*Oenanthe*), not closely related to the other 3 species, which are now placed in a new genus, *Emarginata*. They are still kept together here because of their similar appearance.

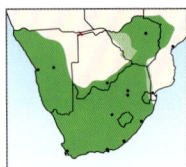

FAMILIAR CHAT   *Oenanthe familiaris*

b b b b | b B B B B b
J F M A M J J A S O N D

15 cm; 14–26 g A pale grey-brown chat with a rufous rump and tail; invariably flicks its wings after landing; this behaviour and the rufous-and-black tail pattern are diagnostic. At close range, narrow, pale eye-ring and rufous wash on the ear coverts evident. Juv. is spotted buff above and mottled on the throat and breast. **Voice:** Male song intersperses whistles among harsh *'chak'* notes; harsh, scolding *'shek-shek'* alarm call. **Status and biology:** Common resident in semi-arid areas, more localised elsewhere. Usually in open, treeless country, often in hilly or rocky terrain and around farm buildings; confiding around humans. Occurs singly, in pairs or, after br., in family groups. Perches on rocks or low shrubs, dropping to the ground to catch insect prey. (Gewone Spekvreter)

SICKLE-WINGED CHAT   *Emarginata sinuata*

B b b | b b B B B
J F M A M J J A S O N D

15 cm; 17–24 g More slender than Familiar Chat, with a more conspicuous, pale eye-ring, and greater contrast between the grey-brown upperparts and paler underparts; flight feathers edged buff. Combination of pale, orange-buff rump and the tail pattern is diagnostic: black in tail forms broad triangle, not reaching base of tail, rather than a 'T' as in Familiar Chat. Tail pattern similar to Tractrac Chat, but orange-buff (not white or creamy) rump and tail edges distinctive. Has a more 'leggy' appearance than Familiar Chat, due to longer tarsi. Common name derives from its extremely attenuated outer primary, but this is only visible in the hand. Juv. is spotted buff above and mottled on the throat and breast. **Voice:** Soft *'tree-tree'* or buzzy *'brrr-brrr'* call; male gives a warbled song prior to br. **Status and biology:** Endemic. Locally common resident in semi-arid Karoo scrub, cultivated lands and pastures; also in Lesotho highlands (summer only). Occurs singly or in pairs; drops from low perches to catch prey. (Vlaktespekvreter)

KAROO CHAT   *Emarginata schlegelii*

b b b | b b B B b
J F M A M J J A S O N D

17 cm; 25–35 g Larger and longer tailed than other *Emarginata* chats, with a more horizontal posture and diagnostic blackish, triangular central tail, framed by grey rump and white outer tail. Upperparts vary from pale grey-brown in arid north to darker grey in south. Much paler than female Mountain Wheatear (p. 362), with pale grey (not white) rump and completely white outer-tail feathers. Juv. is spotted buff above and mottled on the throat and breast. **Voice:** Fairly harsh *'chak-chak'* or *'trrat-trrat-trrat'*. **Status and biology:** Near-endemic. Common resident and local nomad in Karoo and semi-desert scrub. Occurs singly or in pairs, perching prominently on bushes; forages on the ground. (Karoospekvreter)

TRACTRAC CHAT   *Emarginata tractrac*

b b b b | B B B b b
J F M A M J J A S O N D

14–15 cm; 20–25 g A small, pale, short-tailed chat; upperparts vary from very pale grey-brown (some almost white) on gravel plains in the Namib Desert to darker sandy-brown in the south. Most similar to Sickle-winged Chat; shares same tail pattern, but has white (not orange-buff) rump and base to outer tail. Very pale Namib form could be confused with Gray's Lark (p. 318), but bill much finer and legs longer; tail pattern differs. Juv. is spotted buff above and mottled on the throat and breast. **Voice:** Soft, fast *'tactac'*; song is quiet, musical bubbling; territorial defence call is a loud chattering. **Status and biology:** Near-endemic. Common resident and local nomad in Karoo and desert scrub, hummock dunes and gravel plains. Occurs singly or in pairs, perching on stones or low bushes; drops to the ground to catch prey. (Woestynspekvreter)

wing
flicking

rufous rump

FAMILIAR CHAT

dark race

juv.

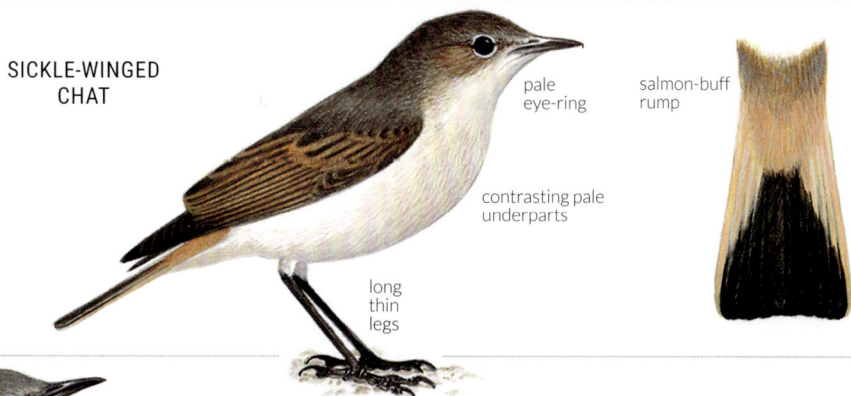

SICKLE-WINGED
CHAT

pale
eye-ring

salmon-buff
rump

contrasting pale
underparts

long
thin
legs

KAROO CHAT

grey rump,
white outer tail

dark race
(Karoo)

pale race
(Namibia)

TRACTRAC CHAT

very pale
overall

whitish
rump

Namib race

359

WHEATEARS Only 3 wheatears breed in the region: Capped Wheatear, Mountain Wheatear (p. 362) and Familiar Chat (p. 358). The other 3 wheatear species are rare, Palearctic-breeding vagrants that can be confused with juv. and imm. Capped Wheatears. Sightings of vagrants need supporting photographic evidence.

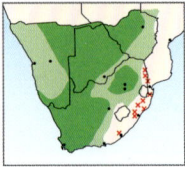

CAPPED WHEATEAR *Oenanthe pileata*

17 cm; 23–32 g A large, striking wheatear with dark brown upperparts, a black crown, face and breast band, and contrasting white throat and supercilium. Sexes alike. Juv. initially dark buff above and pale buff below, with variable dark mottling on the breast and a faint, pale supercilium behind the eye. Upperparts typically darker than in Isabelline and Northern Wheatears; coverts and flight feathers buffy-brown with pale edging. **Voice:** Song is loud warbling with slurred chattering; *'chik-chik'* alarm note. **Status and biology:** Common resident in parts of its range, a seasonal visitor in others. Favours bare, open, flat ground, including grasslands after fires, cultivated and recently harvested croplands. Nests below ground in rodent holes and occurrence partly dependent on availability of these. Occurs singly or in pairs, foraging on the ground. (Hoëveldskaapwagter)

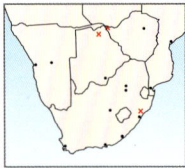

PIED WHEATEAR *Oenanthe pleschanka*

14 cm; 14–20 g Vagrant. Tail has black T-pattern, but black tip is broader on outer feathers in all plumages (diagnostic); underwing coverts blackish, darker than other wheatears. Male in fresh plumage (Sep) has buffy crown, brown margins to back and wing feathers and buff wash across its breast. As plumage wears, buff feather tips are lost, revealing a silvery crown contrasting with black back, wings and throat; remainder of underparts white. Smaller than male Mountain Wheatear (p. 362), with black (not white) shoulders and more extensive white in outer tail. Female and juv. are dusky grey-brown above, lores not appreciably darker than face; smaller and darker than female Northern and Isabelline Wheatears. Could be confused with female Buff-streaked Chat (p. 362), but has darker throat and different tail pattern. **Voice:** Soft *'zack'* call; song mixes this note with soft whistles. **Status and biology:** Palearctic-br. summer vagrant; 3 s African records (all males); Mtunzini, KZN, Jan 1984; Chobe National Park, Botswana, Dec 2014; Victoria Falls, Zimbabwe, Feb 2017. (Bontskaapwagter)

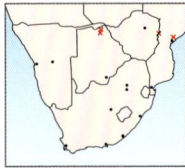

ISABELLINE WHEATEAR *Oenanthe isabellina*

16–17 cm; 21–27 g Vagrant. A uniform, sandy-brown wheatear; sexes alike. Closely resembles non-br. Northern Wheatear, but is slightly larger and paler, has plainer upperwing coverts similar in colour to mantle (upperwing coverts in Northern Wheatear darker than mantle and more similar to flight feather coloration) and looks more upright due to its longer legs and shorter, more rounded tail. At close range, blackish alula contrasts with sandy-brown coverts. Underwing coverts white or whitish-buff. Black tail tip is generally broader than in Northern Wheatear; combined with the shorter tail this results in a shorter stem to the black T-pattern. Appears larger headed, and tends to have more direct flight. **Voice:** Call a high-pitched *'wheet-whit'*. **Status and biology:** Palearctic-br. summer vagrant; 4 s African records; n Botswana, Dec 1972; Rio Savane, Mozambique, Dec 2012; Chobe National Park, Botswana, Jan 2013; Honde Valley, Zimbabwe, Mar 2013; many claims unconfirmed due to similarity with juv. Capped Wheatears. (Isabellaskaapwagter)

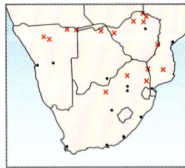

NORTHERN WHEATEAR *Oenanthe oenanthe*

14–16 cm; 21–27 g Vagrant. Br. male has a diagnostic blue-grey crown and back, and black face; underparts white with buff wash on breast. Other plumages are nondescript buffy-brown with a paler supercilium. Easily confused with vagrant Isabelline Wheatear, but has slightly darker wings, shorter legs, a more horizontal stance and narrower black tail tip; dark brown alula shows less contrast with wing coverts. Underwing coverts sooty-grey (not whitish). Smaller than juv. Capped Wheatear or female Buff-streaked Chat (p. 362), with more white on rump and outer tail, and is usually paler, with a plain (not mottled) breast. **Voice:** Harsh *'chak-chak'* or *'wee-chak'*; song is jumbled mix of high whistles and squeaks. **Status and biology:** Palearctic-br. summer vagrant, with <20 records, Sep–Mar; solitary, in open grassland and dry plains. (Europese Skaapwagter)

CAPPED WHEATEAR

black breast band

ad.

diffusely scalloped

juv.

diffuse breast band

imm.

PIED WHEATEAR

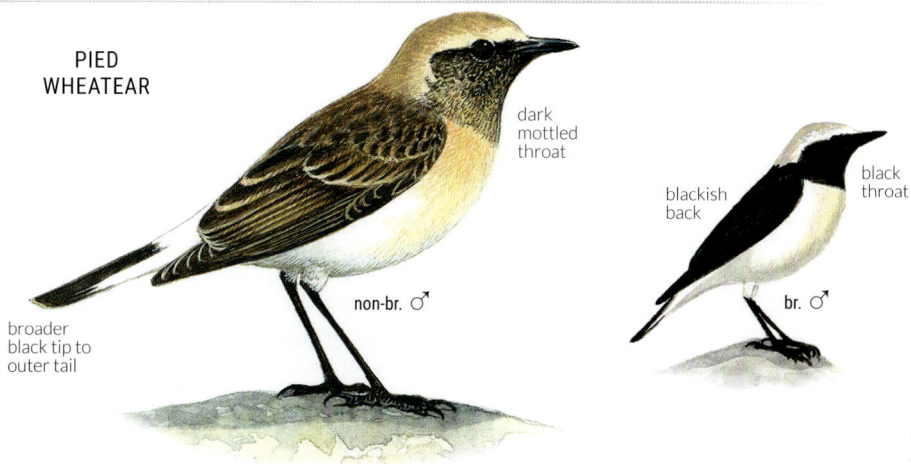

dark mottled throat

non-br. ♂

blackish back

black throat

br. ♂

broader black tip to outer tail

ISABELLINE WHEATEAR

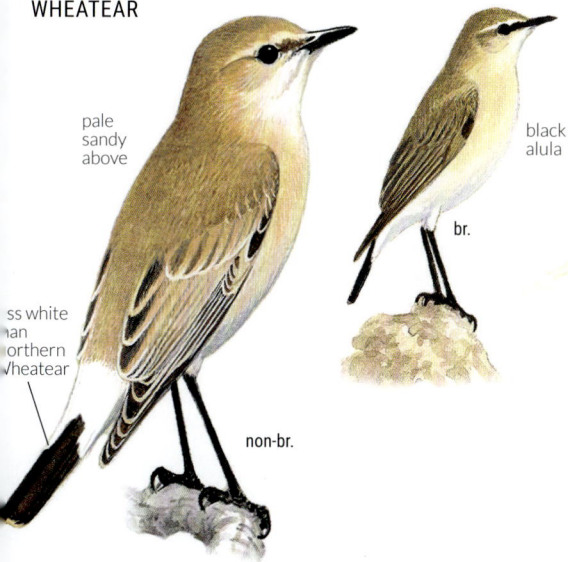

pale sandy above

black alula

br.

ss white an orthern Vheatear

non-br.

blue-grey crown to mantle

black ear coverts

NORTHERN WHEATEAR

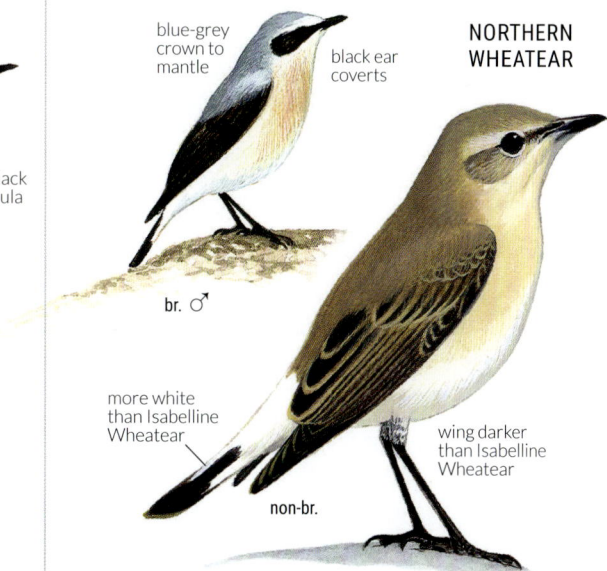

br. ♂

more white than Isabelline Wheatear

wing darker than Isabelline Wheatear

non-br.

MOUNTAIN WHEATEAR *Oenanthe monticola*

19 cm; 32–45 g A variable wheatear of rocky or hilly areas. Males polymorphic: the more common black morph is all-black, with white shoulders, vent, rump and outer tail; may or may not have a white cap, and in the west has a white belly. Grey-morph male has a pale grey body, white rump, white outer tail and may or may not have white shoulder patches and a white cap. Female is uniform dark sooty-brown, except for the white rump and outer-tail feathers. Juv. like female; juv. males start to show ad. feathering after 2–3 months. **Voice:** Clear, thrush-like, whistling song, interspersed with harsh chatters. **Status and biology:** Near-endemic. Common resident and local nomad. Favours rocky hillsides, erosion gulleys, road cuttings, old mine workings and other broken ground. Occurs in pairs or small family groups, usually seen flying between rocks or perching prominently; forages on the ground. (Bergwagter)

OTHER CHAT-LIKE BIRDS
Angola Cave Chat is a *Cossypha* robin-chat with unusual black and white plumage, which is most likely to be confused with the Mountain Wheatear, so is placed here rather than with the robin-chats. The 2 *Myrmecocichla* are compact chats with mostly blackish plumage, and Buff-streaked Chat is a South African endemic placed in its own genus, most closely related to the stonechats.

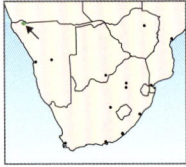

ANGOLA CAVE CHAT *Cossypha ansorgei*

18–19 cm; c. 30 g A striking, black-and-white robin-chat recently discovered in southern Africa, just south of the Kunene River. Superficially recalls White-throated Robin-Chat (p. 368), but has white (not orange) flanks, vent and outer tail; ranges do not overlap. Tail much longer than Mountain Wheatear, and has white (not black) throat and breast. **Voice:** A warbling *'plu-ee uuee-e-e-e'* increasing in strength, then fading away; also a harsh, 2-note call. **Status and biology:** Formerly considered endemic to the Angolan scarp, but a small population was discovered in the Zebra Mountains in 2012. Occurs singly or in pairs on steep boulder slopes with patches of dense thicket. Creeps in crevices among boulders, but sings from exposed perches. Apparently fairly common within its restricted range. (Angolajanfrederik)

ANT-EATING CHAT *Myrmecocichla formicivora*

17–18 cm; 40–60 g An all-brown chat with paler feather margins, appearing scaly at close range. In flight, silvery-white primary bases form prominent pale wing panels. Male has a small, white carpal patch; female is paler brown. Juv. has more prominent pale feather margins; white in primaries reduced. **Voice:** Short, sharp *'pee-ik'* or *'piek'* call; song a mix of whistles and grating notes. **Status and biology:** Endemic. Common resident in open, flat landscapes, especially short grassland or short scrub; favours areas dotted with termite mounds (often used as perches). Conspicuous; usually in small groups. Roost and nest in holes in banks or an aardvark burrow. Sooty Chat *M. nigra* may occur occasionally in nw Namibia; male is blacker than male Ant-eating Chat, and lacks white bases to primaries in all plumages. (Swartpiek)

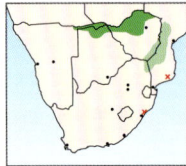

ARNOT'S CHAT *Myrmecocichla arnoti*

17 cm; 36–44 g A striking, black-and-white chat, most similar to Mountain Wheatear (but ranges non-overlapping). Male all-black with white cap and shoulders; female lacks white cap, but has conspicuous white throat and upper breast, often scaled with black. Juv. is duller, with a mainly black head and throat, but has white shoulder patches. **Voice:** Song is a varied series of canary-like, warbling whistles; call is a quiet, whistled *'fick'* or *'feee'*. **Status and biology:** Locally common resident, restricted to mature miombo, mopane and teak woodlands with an open understorey. Occurs in pairs or family groups; lively and conspicuous birds, moving frequently from tree to tree, dropping onto the ground at intervals to forage. (Bontpiek)

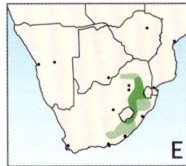

BUFF-STREAKED CHAT *Campicoloides bifasciata*

16–17 cm; 31–38 g A buff-and-black chat. Male is easily identified by its distinctive black face and throat and long, broad, whitish supercilium. Female is warm rufous-brown above with buffy underparts; darker than juv. Capped Wheatear (p. 360), with a buff (not white) rump and all-black (not black-and-white) tail. Juv. has buff-spotted upperparts and mottled underparts. **Voice:** A short series of rich, warbling notes, given by both sexes. **Status and biology:** Endemic. Locally common resident, found on rock-strewn, grassy slopes. Occurs in pairs or small family groups. Perches conspicuously on large rocks but feeds on the ground; occasionally hawks flying insects. (Bergklipwagter)

ANGOLA
CAVE CHAT

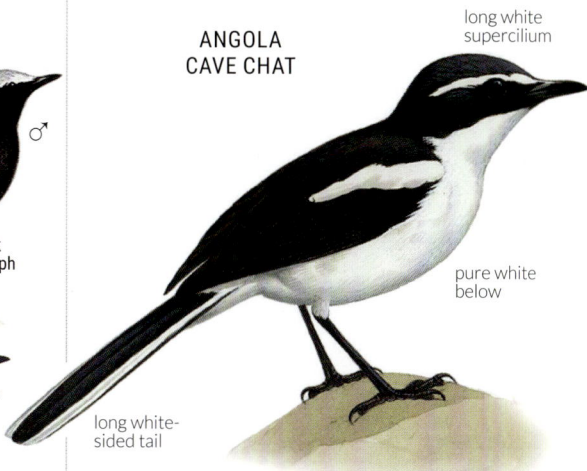

long white
supercilium

pure white
below

long white-
sided tail

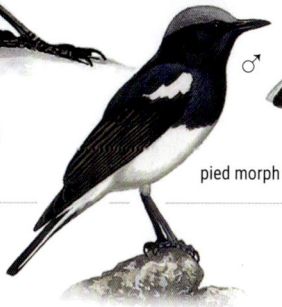

white
rump &
outer tail

grey morph

dark
morph

pied morph

MOUNTAIN WHEATEAR

ANT-EATING
CHAT

black rump
& tail

white
carpal
patch

white
primary
patch

hovering

ARNOT'S
CHAT

broad pale
supercilium

BUFF-STREAKED
CHAT

buffy
supercilium

extensive
black
throat

buff
rump

plain buff
underparts

363

CHATS AND ALLIES
Five chat-like species: Herero Chat, endemic to the escarpment of central and n Namibia and sw Angola, is a peculiar chat that behaves somewhat like a flycatcher or robin-chat. African Stonechat is a widespread resident, whereas Whinchat, Common Redstart and White-throated Robin are Palearctic-br. vagrants that overshoot their usual wintering areas further north in Africa.

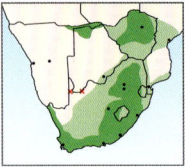

AFRICAN STONECHAT *Saxicola torquatus*

J F M A M J J A S O N D

13 cm; 12–17 g A small, plump-looking chat; ad. male has a striking black head, chestnut breast and flanks, and white neck patch, belly, rump and wingbar. Female duller, with a mottled brown head and orange-buff breast; uppertail coverts whitish. Often shows a pale supercilium, but not as broad or prominent as in vagrant Whinchat. Juv. is spotted buff above, and is paler below than female, with mottled breast. **Voice:** Song is a fast series of high-pitched warbled whistles; calls *'weet-weet'* and harsh *'chak'*. **Status and biology:** Common resident, with winter movement out of high-lying areas; frequents grassland, fynbos and scrub, often on fringes of wetlands or forest edge. Occurs singly or in pairs. Active, vocal birds, perching prominently on twigs or fences, dropping periodically into the grass to forage. (Gewone Bontrokkie)

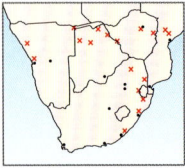

WHINCHAT *Saxicola rubetra*

J F M A M J J A S O N D

14 cm; 13–24 g Vagrant. Much like female African Stonechat in appearance, but brown (not white) rump and uppertail coverts diagnostic in both sexes and at all ages; also has a more prominent supercilium that broadens behind the eye and extends around the top of the ear coverts; outer-tail feathers have white bases (black in African Stonechat). Br. male has dark cheeks, white supercilium and moustachial stripe and warm buff underparts, but resembles female in winter quarters. **Voice:** Scolding *'tick-tick'*. **Status and biology:** Palearctic-br. summer vagrant, c. 20 records for s Africa. Solitary, in open grassland with patches of stunted scrub. (Europese Bontrokkie)

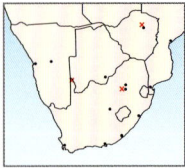

COMMON REDSTART *Phoenicurus phoenicurus*

J F M A M J J A S O N D

13–14 cm; 11–22 g Vagrant. Br. male is striking, with its black face and throat, white forecrown, slate-grey upperparts and red breast. Smaller than male White-throated Robin, with black throat, and red rump and outer tail. Female and non-br. male are much duller; could be confused with Familiar Chat (p. 358), but have more extensive red rumps and brighter red outer tails lacking dark tip; tail is continuously 'trembled' up and down; doesn't wing-flick. Juv. is spotted buff above and barred darker brown below; unlikely to be encountered in s Africa. **Voice:** Loud *'hooeeet'*, similar to that of Willow Warbler, often with harsher notes *'hooeet-tucc-tucc'*. **Status and biology:** Palearctic-br. summer vagrant in semi-arid savanna and woodland; only 2 records from s Africa. (Europese Rooistert)

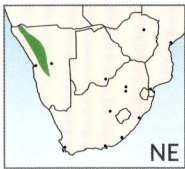

HERERO CHAT *Namibornis herero*

J F M A M J J A S O N D

17 cm; 23–30 g An unobtrusive, brown, chat-like bird that has the posture of a *Malaenornis* flycatcher (p. 406) and could be mistaken for Red-backed Shrike (p. 418). Pale rufous rump and tail (latter with a brown centre), blackish eye-mask and whitish, streaked underparts are diagnostic. Sexes alike. Juv. is buff-spotted above and has a mottled breast. **Voice:** Mostly silent; melodious, warbling *'twi-tedeelee-doo'* song when br. **Status and biology:** Near-endemic. Uncommon resident in arid, sparsely wooded, hilly, rocky country of central and n Namibia. Occurs singly or in pairs; hunts from low perches in trees. Frequently flicks wings when foraging, apparently to flush prey. (Hererospekvreter)

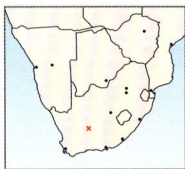

WHITE-THROATED ROBIN *Irania gutturalis*

J F M A M J J A S O N D

16–17 cm; 16–27 g Vagrant. A skulking robin that frequently cocks its diagnostic black tail; structure and behaviour recall Thrush Nightingale (p. 372). Ad. male is striking; told from male Common Redstart by the narrow, white centre to throat (not entirely black), smaller black face (not extending onto frons), clean white vent and black tail. Female has plain grey upperparts and head, with only a faint, pale stripe in front of its eye; ear coverts washed rufous. Underparts buffy with faint, darker barring and richer rufous flanks; throat paler; vent white. Juv. even duller than female, with buff-spotted upperparts and a more heavily mottled breast. **Voice:** Contact call is a grating *'krrrk'*; song is a mixture of musical whistles and scratchy notes. **Status and biology:** Palearctic-br. vagrant, favouring dry thickets and scrub that mainly winters in E Africa. Only 1 record, near Williston, N Cape, in Jul 2006. (Irania)

AFRICAN STONECHAT

dusky supercilium

♀

plain head

♂

dark base to outer tail

WHINCHAT

white supercilium

black ear coverts

tawny throat

♂

white base to outer tail

heavily marked mantle

pale supercilium

♀

COMMON REDSTART

white forehead

black throat

white eye-ring

br. ♂

chestnut tail

rufous rump & tail

♀

HERERO CHAT

dark ear coverts

white supercilium

lightly streaked breast

russet rump & outer tail

WHITE-THROATED ROBIN

white contrasting throat

grey back & rump

black tail

♀

white supercilium

pale central throat

♂

SCRUB ROBINS
Mostly rather drab robins with long, white-tipped tails that are often held cocked. Tails lack dark central feathers of robin-chats; most species have white wingbars. All have well-marked heads with a dark eye stripe and white supercilium. Sexes alike. Most live in pairs on year-round territories.

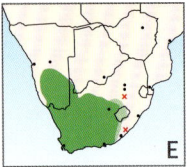

KAROO SCRUB ROBIN *Cercotrichas coryphoeus*

J F M A M J J A S O N D

14–16 cm; 18–23 g A dark, grey-brown scrub robin with a blackish tail and contrasting white tail tips; no white in wing. Pale throat contrasts with greyish or grey-brown underparts. Much darker and greyer-brown than Kalahari Scrub Robin, with no rufous in the tail, rump or wing. Juv. is browner above, with buff spots on head and mantle, and buffy margins to wing coverts; underparts mottled darker grey. **Voice:** Harsh, chittering *'tchik, tchik, tcheet'*; song is a mixture of whistles and harsh, grating notes. **Status and biology:** Endemic. Common resident in Karoo scrub and strandveld. Usually in pairs or family groups; some pairs have helpers at the nest. (Slangverklikker)

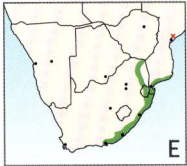

BROWN SCRUB ROBIN *Cercotrichas signata*

J F M A M J J A S O N D

16–18 cm; 33–42 g A dark grey-brown forest scrub robin; most resembles Bearded Scrub Robin, but in most of its range lacks buffy orange on chest and flanks, and rufous on the rump and uppertail coverts; *C. s. tongensis* (n KZN and s Mozambique) has buffier breast and rump; told from Bearded Scrub Robin by narrower white tail tips. Juv. is warmer brown above, with black tips to feathers; underparts creamy buff, with grey mottling or scaling on breast. **Voice:** Sings year-round, especially at dawn; a series of high-pitched, melodious phrases introduced by a *'twee-choo-sree-sree'*; alarm note is sibilant *'zeeeeet'*. **Status and biology:** Endemic. Locally common resident in thick tangles of coastal and evergreen forests. Shy and skulking, but becomes tame in camps. Mostly solitary, spending much of its time on the ground foraging; flies to a branch if disturbed. (Bruinwipstert)

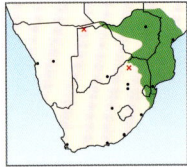

BEARDED SCRUB ROBIN *Cercotrichas quadrivirgata*

J F M A M J J A S O N D

14–16 cm; 22–31 g Much like a warmly coloured version of Brown Scrub Robin; unstreaked underparts with buffy-orange breast band and flanks diagnostic; lacks double white wingbars and rufous base to tail of White-browed Scrub Robin. Juv. has dark tips to mantle and breast feathers, appearing mottled; wing coverts tipped buff. **Voice:** Clear, penetrating song of often-repeated, mixed phrases punctuated with short pauses; alarm call comprises 1 or 2 sharp notes followed by *'churr, chek-chek kwezzzzz'*. **Status and biology:** Common resident of sand and riparian forests and termitaria thickets in mature broadleafed woodland. Occurs singly or in pairs, generally shy. Forages in leaf litter under thickets and dense tangles; flies to a perch if disturbed. (Baardwipstert)

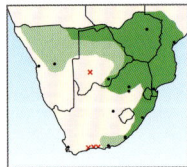

WHITE-BROWED SCRUB ROBIN *Cercotrichas leucophrys*

J F M A M J J A S O N D

14–16 cm; 14–22 g The common scrub robin of savanna and woodland, easily identified by its 2 white wingbars and, in most of its range, streaked underparts; streaking confined to sides of rufous-washed breast in *C. l. ovamboensis* (Namibia to nw Zimbabwe), which could be confused with Bearded Scrub Robin. Rump and base of tail rufous; remainder of tail dark brown, feathers with narrow, white tips. Juv. paler above, with buff spots; breast diffusely mottled (not streaked). **Voice:** Fluty, repetitive song; characteristic call at dawn and dusk is whistled *'seeep po go'*; alarm note a harsh *'trrrrrr'*. **Status and biology:** Common resident in woodland and savanna where there is ample grass cover; usually in pairs. Easily detected by its song, but mostly keeps concealed, especially when singing. (Gestreepte Wipstert)

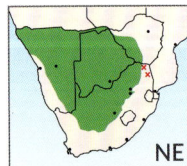

KALAHARI SCRUB ROBIN *Cercotrichas paena*

J F M A M J J A S O N D

14–16 cm; 17–23 g A pale, sandy brown scrub robin. Told from White-browed Scrub Robin by its unmarked breast and absence of white wingbars; dark subterminal tail band narrower, so tail appears mostly orange. Juv. has black tips to head, mantle and breast feathers, appearing mottled. **Voice:** Musical song of whistles and chirps, more varied than those of Karoo or White-browed Scrub Robins; frequently mimics other birds. Contact call is a whistled *'seeeup'*; alarm call is a harsh *'zzeee'*. **Status and biology:** Near-endemic. Common resident in dry acacia savanna, favouring thickets and bushy areas. (Kalahariwipstert)

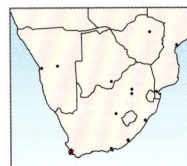

RUFOUS-TAILED SCRUB ROBIN *Cercotrichas galactotes*

J F M A M J J A S O N D

15 cm; 20–28 g Vagrant. Similar to Kalahari Scrub Robin with rufous rump and tail, but blackish subterminal tail band usually narrower, often appearing as a series of black spots on the spread tail; upperparts browner (not greyish on crown) and wing feather margins paler (whitish, not buffy); bill slightly heavier with a pale brown base (not mostly dark with yellowish base) and usually with more dusky underparts. Juv. is paler with reduced black and white tail markings. **Voice:** Song is repetitive warbling, less musical than Kalahari Scrub Robin. Contact call is a whistled *'si-sip'*; alarm call is *'zi-zi-zi'* or *'piu'*. **Status and biology:** Palearctic-br. migrant that winters in the Sahel and E Africa; 1 record from Zeekoeivlei, W Cape, in Jul–Aug 2016 presumably was a reverse migrant. (Rooistertwipstert)

KAROO SCRUB ROBIN

plain wings

white-tipped blackish-brown tail

BROWN SCRUB ROBIN

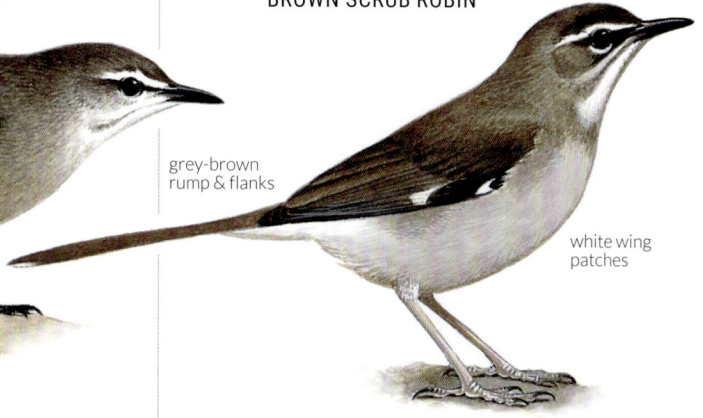

grey-brown rump & flanks

white wing patches

BEARDED SCRUB ROBIN

buffy-orange rump & flanks

WHITE-BROWED SCRUB ROBIN

distinct white wingbars

streaked breast

ovamboensis

buff, lightly streaked chest

'cleaner' facial markings

slender, black bill

no white wingbars

plain whitish breast

russet rump & upper tail

KALAHARI SCRUB ROBIN

RUFOUS-TAILED SCRUB ROBIN

'dirty' facial markings

heavy, pale-based bill

dusky brown underparts

ROBIN-CHATS

Large robins with rufous rumps and outer tails. Sexes alike. Juv. plumage, retained for only a few months, is spotted buff above and mottled buffy-brown below. Typically found close to the ground in dense vegetation. Most are accomplished songsters and mimics, males singing through the year from concealed perches in the foliage of bushes and trees. Some defend year-round territories, others territorial only while breeding, moving to lower altitudes in winter.

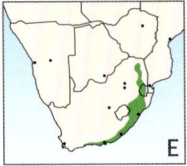

CHORISTER ROBIN-CHAT *Cossypha dichroa*

J F M A M J J A S O N D

20 cm; 37–56 g A large, dark-backed robin with orange underparts and nuchal collar; head blackish, with no white eye stripe. Occasionally hybridises with Red-capped Robin-Chat; hybrids have orange supercilia and dusky cheeks. Juv. is sooty, mottled tawny-buff above and below. **Voice:** Contact call is plaintive *'toy-toy, toy-toy'*; song is loud and bubbly, including much mimicry of other birds and even frogs, humans whistling, dogs barking, etc. Unlike other robin-chats, male mostly sings from high in a forest tree. **Status and biology:** Endemic. Common resident in forest and coastal thickets; some local movements from interior to coastal forests in winter. Occasionally attends driver ant columns and gleans parasites from forest antelopes. (Lawaaimakerjanfrederik)

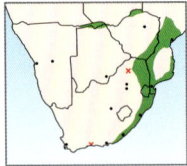

RED-CAPPED ROBIN-CHAT *Cossypha natalensis*

J F M A M J J A S O N D

17 cm; 25–36 g The only robin-chat with a plain rufous face and a slate-blue back and wings; crown and nape reddish-brown. When alarmed, frequently flicks up its tail, then lowers it slowly. Imm. like ad., but has dusky ear coverts. Juv. dark brown above, with buff spots; underparts duller, mottled on breast. Occasionally hybridises with Chorister Robin-Chat; hybrids have orange supercilium and dusky cheeks. **Voice:** Call is soft, slightly trilled *'seee-saw'*, often repeated; song is rambling series of melodious whistles, including much mimicry. **Status and biology:** Common resident and local migrant in thickets, forest and dense woodland. (Nataljanfrederik)

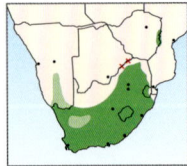

CAPE ROBIN-CHAT *Cossypha caffra*

J F M A M J J A S O N D

17 cm; 23–38 g A dark-backed robin-chat with a fairly short, white supercilium and pale orange throat, upper breast, vent and rump; remainder of underparts pale grey. Juv. is browner above, with buff spotting; underparts dull buffy-brown, mottled darker brown on breast. Imm. retains some buff-fringed wing coverts. **Voice:** Song is a series of melodious phrases, usually starting *'cherooo-weet-weet-weeeet'*; often mimics other birds. Alarm call is a guttural *'wur-der-durrr'*. **Status and biology:** Common resident and altitudinal migrant, found in a wide range of habitats including forest edge, thickets, bracken, heath and scrub; also in gardens, often becoming very confiding. Confined to higher elevations in north, but common down to sea level in S Africa. (Gewone Janfrederik)

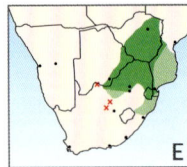

WHITE-THROATED ROBIN-CHAT *Cossypha humeralis*

J F M A M J J A S O N D

16 cm; 19–26 g The only robin-chat with a white wingbar; combination of black-and-white patterned head and orange-rufous underparts (with white throat and upper breast) is diagnostic. Juv. lacks white wingbar and supercilium; upperparts brown, spotted with buff; underparts buffy, mottled black. **Voice:** Alarm call is a repeated *'seet-cher, seet-cher'*; song, a rather short series of rich whistles, less robust than songs of other robin-chats, but often including mimicry of other birds, animals, and other sounds. **Status and biology:** Endemic. Common resident of thickets and riverine scrub in woodland and savanna; in more arid areas, confined to riparian vegetation. (Witkeeljanfrederik)

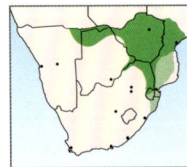

WHITE-BROWED ROBIN-CHAT *Cossypha heuglini*

J F M A M J J A S O N D

19 cm; 29–44 g A large robin-chat, recalling Chorister Robin-Chat, but with paler upperparts and a prominent white supercilium; also occurs at lower elevations and in a different habitat. Juv. is dark brown above, with fine buff spots; underparts buffy, mottled blackish. **Voice:** Characteristic, loud, crescendo song of repeated phrases; also an accomplished mimic. Alarm call is a harsh *'tserrrk-tserrrk'*. **Status and biology:** Common resident of dense thickets and tangles, also in gardens and parks. (Heuglinjanfrederik)

no white
supercilium

CHORISTER
ROBIN-CHAT

ad.

juv.

RED-CAPPED
ROBIN-CHAT

juv.

blue-grey
back

orange
face

ad.

CAPE
ROBIN-CHAT

juv.

no white
wingbar

ad.

pale
belly

juv.

WHITE-THROATED
ROBIN-CHAT

white wingbar

ad.

white
supercilium

no
wingbar

WHITE-BROWED
ROBIN-CHAT

juv.

ad.

369

SMALL FOREST ROBINS
Small, ground-dwelling forest robins, mainly yellow or orange in colour; eyes unusually large as an adaptation to forest living; sexes alike or nearly so. Skulking, presence mostly given away by their song.

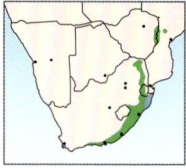

WHITE-STARRED ROBIN *Pogonocichla stellata*

16 cm; 18–26 g A fairly small, compact forest robin. Lemon-yellow breast and glossy blue-black head diagnostic in ad.; white 'stars' on lower throat and forehead usually not visible. Juv. is sooty, mottled with yellowish-buff above and below. Imm. dull olive above, greyish-yellow below; tail pattern of both juv. and imm. like ad., but duller yellow. **Voice:** Soft 'chuk' or 'zit'; whistled 'too-twee' contact call; frequently repeated song is quiet and warbling. **Status and biology:** Common resident of forest understorey, subject to some altitudinal movements in south of range. Solitary or in pairs; often confiding. (Witkoljanfrederik)

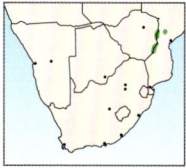

SWYNNERTON'S ROBIN *Swynnertonia swynnertoni*

14 cm; 14–20 g Smaller than White-starred Robin; white bib on breast above thin black collar diagnostic in ad.; tail uniformly dark grey (lacking yellow windows). Female is duller, with greenish crown and face. Juv. is brown above, spotted yellow-buff; underparts duller, with dark mottling on breast; throat crescent pale grey with brown band below. Imm. like pale female but retains some buff-tipped juv. wing feathers. **Voice:** Song, given by male, is subdued, trisyllabic 'zitt, zitt, slurr', last syllable lower pitched; alarm call is a monotonous, quiet purring. **Status and biology:** VULNERABLE. Uncommon resident of montane forest understorey. (Bandkeeljanfrederik)

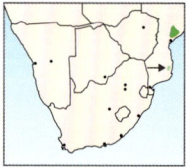

EAST COAST AKALAT *Sheppardia gunningi*

12 cm; 16–19 g A small, lowland forest robin; combination of bronze-brown upperparts, orange throat and breast, and white belly is diagnostic. At close range, shows blue-grey face and shoulders. Head has a faint, paler supercilium; white pre-orbital spots only exposed when displaying. Juv. darker above with buff edging on upperwing coverts; underparts dull yellow-buff, mottled brown. **Voice:** Song is a soft, fast series of high-pitched notes in oft-repeated, short phrases. **Status and biology:** NEAR-THREATENED. Uncommon, localised resident, found singly or in pairs in understorey of lowland forest and thickets in dense woodland. Unobtrusive and easily overlooked. (Gunningjanfrederik)

PALM THRUSHES
Medium-sized, robin-like birds with plain, rufous-brown wings, rump and tail. Usually found near palm trees, which provide their preferred nest sites. Sexes alike. Juvs have mottled underparts. Very vocal, especially in early morning. Forage on the ground or in thickets; occasionally hawk insects.

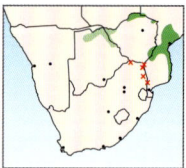

COLLARED PALM THRUSH *Cichladusa arquata*

20 cm; 29–38 g A strongly patterned palm thrush; incomplete narrow, black necklace, which divides pale buff throat and chest from greyish head, is diagnostic. Pale blue-grey eye and contrasting black pupil is striking. Juv. is streaked on crown and nape, and mottled dark brown below; necklace indistinct. **Voice:** Explosive, varied song, often including 'weet-chuk' or 'cur-lee chuk-chuk' phrases and imitations of other birds' songs. **Status and biology:** Locally common resident in thickets in *Hyphaene* and *Borassus* palm savanna. (Palmmôrelyster)

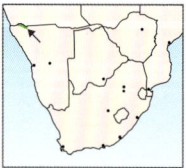

RUFOUS-TAILED PALM THRUSH *Cichladusa ruficauda*

18 cm; 22–37 g A richly coloured palm thrush that lacks the black necklace of Collared Palm Thrush and has dull red (not pale blue-grey) eyes. Juv. has dark grey mottling on the breast. **Voice:** Loud and protracted, melodious whistling, including imitations of other birds' songs. **Status and biology:** Locally common resident in riverine thickets, usually close to *Hyphaene* palms. Easily seen along the Kunene River between Ruacana and Epupa falls. (Rooistertmôrelyster)

slate-grey head

ad.

extensive yellow in tail

WHITE-STARRED ROBIN

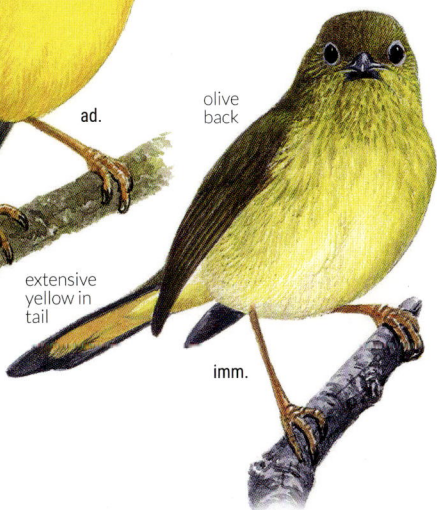

olive back

imm.

SWYNNERTON'S ROBIN

juv.

small, white bib bordered black

ad.

dark tail

EAST COAST AKALAT

brown head & back

indistinct white spot

dark tail

COLLARED PALM THRUSH

imm.

pale eye

RUFOUS-TAILED PALM THRUSH

red eye

plain breast

rufous tail

necklace

ad.

NIGHTINGALES
Skulking, warbler-like birds related to akalats and robins. Renowned songsters when br., but easily overlooked on their wintering grounds. Sexes alike.

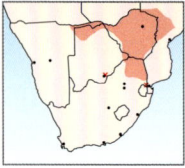

THRUSH NIGHTINGALE *Luscinia luscinia*

J F M A M J J A S O N D

16 cm; 18–33 g A skulking, warbler-like bird, easily overlooked if not calling. Best identified by its warm olive-brown upperparts and rufous-washed rump and tail; underparts whitish, with diffuse grey-brown streaking on the breast, flanks and vent. Tail often cocked or waved from side to side. Remains low down in thick tangles, feeding mainly on the ground; could be confused with Terrestrial Brownbul (p. 348), but solitary, not in groups. Juv. has buff tips to tertials and greater upperwing coverts. **Voice:** Rich, warbling song interspersed with harsh, grating notes. **Status and biology:** Uncommon to fairly common Palearctic-br. migrant, in thickets in savanna and woodland, often along rivers. (Lysternagtegaal)

MACROSPHENID WARBLERS
A small endemic family of African warblers, containing the extra-limital longbills, crombecs (p. 374) and 4 monotypic genera of rather skulking warblers, all found in s Africa, and best located by their distinctive songs. Sexes alike.

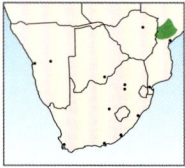

MOUSTACHED GRASS WARBLER *Melocichla mentalis*

J F M A M J J A S O N D

20 cm; 33–38 g A large, plain grass warbler with a broad, blackish, rounded tail. Much larger than Fan-tailed Grassbird (p. 380), with a black malar stripe, pale (not dark) eye, more strongly marked face and plain undertail (lacking buff tips to feathers). Juv. duller, lacking chestnut forehead and dark malar stripe; breast mottled brown. **Voice:** Rich, bubbling *'tip-tiptwiddle-iddle-see'*. **Status and biology:** Uncommon resident in rank grass adjoining forests and in open glades, often near water. Occurs singly or in pairs. (Breëstertgrasvoël)

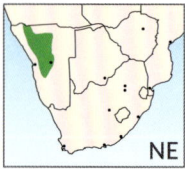

ROCKRUNNER *Achaetops pycnopygius*

J F M A M J J A S O N D

20 cm; 26–34 g A large, striking warbler with a heavily streaked, dark back, white breast spotted with black, and a bright rufous belly and undertail. Long tail, horizontal stance and distinctive behaviour separate it from chats; could possibly be confused with a scrub robin (p. 366) or robin-chat (p. 368). Juv. is less distinctly marked and duller. **Voice:** Rich, liquid, melodious song, more varied than Cape Grassbird; sings from exposed perches. **Status and biology:** Locally common, near-endemic resident on boulder-strewn, grassy hillsides and bases of small hills; bounds across rocks on long legs, often perches in low bushes. Occurs singly or in pairs; easily overlooked if not calling. (Rotsvoël)

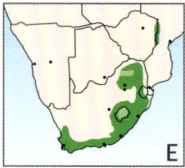

CAPE GRASSBIRD *Sphenoeacus afer*

J F M A M J J A S O N D

17–19 cm; 26–34 g A large, richly coloured warbler with a long, ratty tail, chestnut cap and black malar stripes; flight weak. Heavily streaked plumage separates it from Moustached Grass Warbler; much larger than cisticolas. Juv. duller, with a more streaked cap. Northeastern race *S. a. natalensis* has underparts washed with rufous. **Voice:** Song is a stereotyped crescendo, starting as a soft series of warbling notes, building in volume, pace and pitch, and ending in a descending trill; also nasal *'where'* call. **Status and biology:** Endemic. Common resident in fynbos and rank grass and ferns. Mostly remains in dense vegetation, but easily located by its song. Occurs singly or in pairs. (Grasvoël)

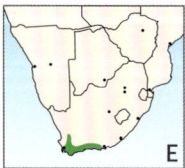

VICTORIN'S WARBLER *Cryptillas victorini*

J F M A M J J A S O N D

16 cm; 20 g A colourful warbler with cinnamon-orange underparts, blue-grey face and striking, yellow eyes. Female duller, with grey-brown upperparts. Juv. is even greyer above and paler below. **Voice:** Male sings a rollicking, repetitive *'whit-itty-weeo, wit-itty weeo'*, accelerating towards the end (could be confused with Cape Grassbird, but lacks the descending trill at the end); female gives loud *'weet-weet-weeeo'*; both sexes give a harsh, pishing alarm call. When aroused, alarm calls run onto the song. **Status and biology:** Endemic. Common resident in damp, montane fynbos, typically in thick tangles alongside streams and in gullies. Occurs singly or in pairs. Until recently placed in *Bradypterus*, but forms part of an African warbler radiation that includes Cape Grassbird and crombecs. (Rooiborsruigtesanger)

THRUSH NIGHTINGALE

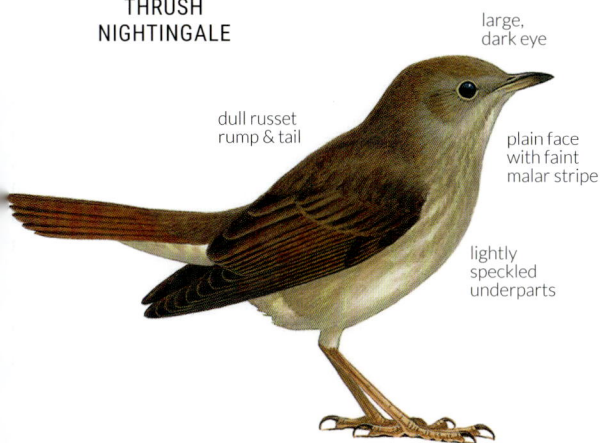

large, dark eye

dull russet rump & tail

plain face with faint malar stripe

lightly speckled underparts

MOUSTACHED GRASS WARBLER

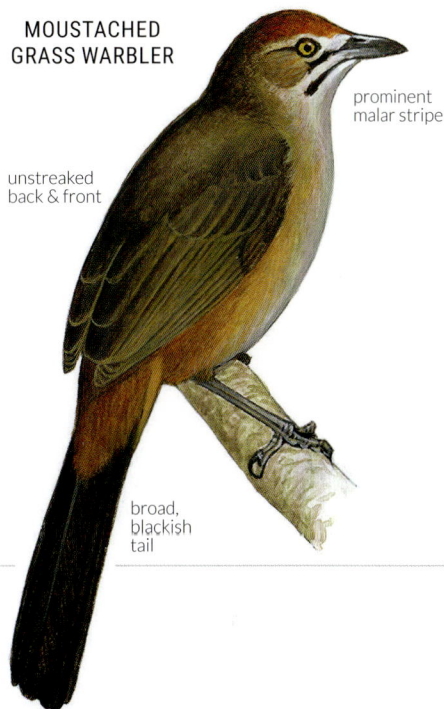

prominent malar stripe

unstreaked back & front

broad, blackish tail

ROCKRUNNER

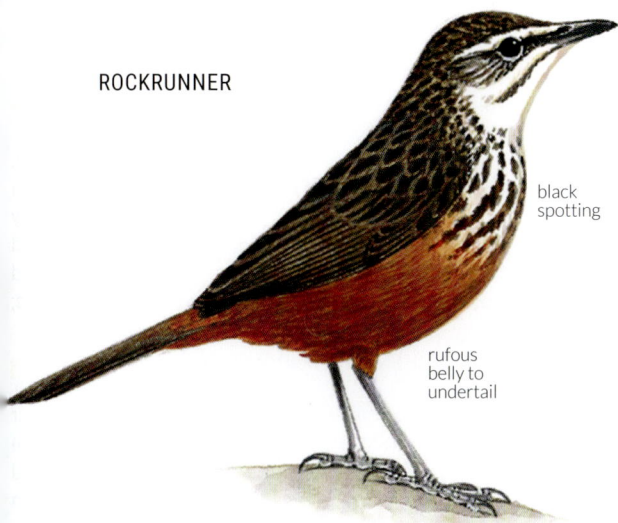

black spotting

rufous belly to undertail

CAPE GRASSBIRD

rufous crown

heavily patterned back

long tail (appears 'tatty')

VICTORIN'S WARBLER

pale eye

grey cheeks

373

CROMBECS
Small warblers that appear almost tailless; bills slender and slightly decurved. Superficially similar to eremomelas (p. 376), but with shorter tails and longer bills. Sexes alike; juvs duller. Occur singly or in pairs in wooded or shrubland habitats; commonly join mixed-species bird parties. Primarily insectivorous, gleaning caterpillars, spiders and other invertebrates from branches and twigs.

LONG-BILLED CROMBEC *Sylvietta rufescens*

B b b | | | | b B B B B
J F M A M J J A S O N D

11 cm; 9–14 g Longer billed than other crombecs in the region. Face paler than Red-faced Crombec, with a diagnostic darker grey eye stripe; upperparts brownish-grey (not ash-grey). Lacks chestnut breast band and ear patches of vagrant Red-capped Crombec. Juv. paler below. **Voice:** Song is a loud, whistled and repeated *'trree-rriit, trree-rriit'* or *'trree reee rit'*; dry *'prrit'* contact call given regularly. **Status and biology:** Widespread resident in woodland, savanna and arid scrublands; the most common crombec in the region. Never still when foraging, hopping along branches and twigs, moving frequently between trees; in woodlands, forages in lower strata rather than canopy. (Bosveldstompstert)

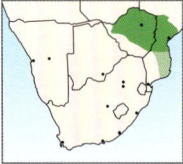

RED-FACED CROMBEC *Sylvietta whytii*

b b b b | | | b B B B b
J F M A M J J A S O N D

9 cm; 8–11 g Slightly smaller than Long-billed Crombec, with plain rufous face and underparts; lacks a dark eye stripe. Juv. brownish-grey above. **Voice:** Trilled, far-carrying and repeated *'wit-wit-wit-wit'*; thin *'si-si-si-see'*. **Status and biology:** Locally common resident in miombo and teak woodland and riparian forests; side-by-side with Long-billed Crombec in places. Typically forages in the canopy, whereas lower strata used by Long-billed Crombec. (Rooiwangstompstert)

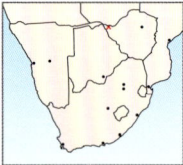

RED-CAPPED CROMBEC *Sylvietta ruficapilla*

| | | | X | | | | | | | |
J F M A M J J A S O N D

10 cm; 10–12 g Vagrant. Rare, localised crombec with light grey underparts, and with diagnostic chestnut ear patches and breast band. Juv. more buffy above and paler below. **Voice:** Ringing, repeated *'richi-chichi-chichir'*. **Status and biology:** One specimen collected west of Victoria Falls, Zimbabwe, in Apr 1967; may be resident in small numbers in Tete Province, n Mozambique; found in tall miombo woodland in adjacent Zambia. (Rooikroonstompstert)

PENDULINE TITS
Tiny, warbler-like birds – these are the smallest birds in s Africa; pale coloured, with short, pointed, conical bills. Sexes alike. Found in pairs or small family groups, often joining mixed-species bird parties in woodlands. They make densely woven, ball-like nests in trees or shrubs, and use them both for breeding and as a winter roost.

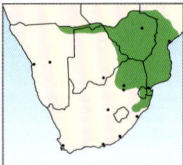

GREY PENDULINE TIT *Anthoscopus caroli*

b b | | | | | b B B B b
J F M A M J J A S O N D

8 cm; 6–7 g A plain penduline tit, grey above, with a buffy face, flanks and belly; lacks black facial markings of Cape Penduline Tit and has buff (not yellowish) underparts. Bill much shorter and tail longer than Long-billed Crombec. Juv. paler buff on belly. **Voice:** Song a high-pitched *'tsi-tsi-tsweeee'*, repeated 3 or 4 times; also buzzy *'chi-zeeee'* and *'tsee-wee'* calls. **Status and biology:** Common resident and local migrant in broadleafed and miombo woodlands. Usually occurs in small groups. (Gryskapokvoël)

CAPE PENDULINE TIT *Anthoscopus minutus*

B b b b | b B B B B B B
J F M A M J J A S O N D

8–10 cm; 7–9 g Differs from Grey Penduline Tit in having a black-speckled forehead extending as eye stripe, yellowish (not buff) belly and flanks, and speckled throat. Black-mottled forehead differentiates it from Yellow-bellied Eremomela (p. 376). Juv. is paler below. **Voice:** Song a high-pitched *'tseeep tseeep tseeep'*; plaintive *'chweee'* and bell-like contact calls. **Status and biology:** Near-endemic. Common resident in fynbos, Karoo scrub, semi-desert and arid savanna. Usually occurs in small groups. (Kaapse Kapokvoël)

LONG-BILLED CROMBEC

dark eye stripe

RED-FACED CROMBEC

plain rufous face

RED-CAPPED CROMBEC

indistinct chestnut breast band

tailless appearance (cf. Burnt-necked Eremomela)

GREY PENDULINE TIT

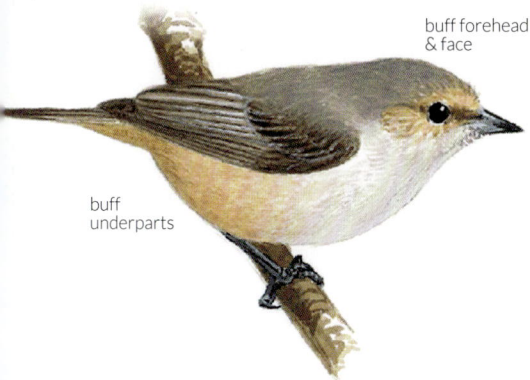

buff forehead & face

buff underparts

yellowish underparts

black-speckled forehead & eye stripe

CAPE PENDULINE TIT

EREMOMELAS

Small, resident warblers, reminiscent of a Willow Warbler (p. 382) in size and shape, and frequenting woodland, savanna and arid scrub. Usually in small groups; regularly join bird parties. Sexes alike; juvs duller. Insectivorous, foraging in outer twigs and leaves of trees and shrubs.

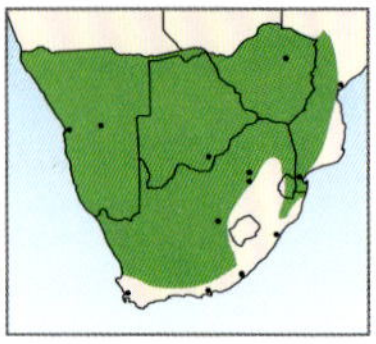

YELLOW-BELLIED EREMOMELA *Eremomela icteropygialis*

J	F	M	A	M	J	J	A	S	O	N	D
b	b	b	b	b			b	B	B	B	b

10 cm; 6–9 g A pale, grey-backed eremomela with a yellow-washed belly and red-brown (not pale) eyes; yellow belly most extensive and brightest in moist areas. Juv. duller. **Voice:** Song is a crombec-like, high-pitched, frequently repeated *'tchee-tchee-tchee'*. **Status and biology:** Common resident in semi-arid savanna, woodland and scrub. Usually solitary or in pairs, moving restlessly when foraging, typically keeping to shrubs and the lower strata of trees. (Geelpensbossanger)

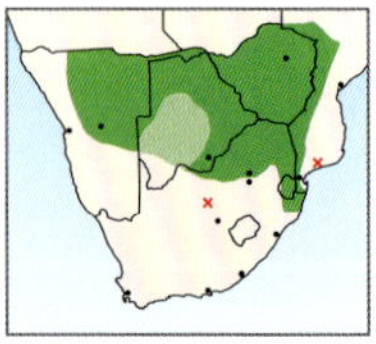

BURNT-NECKED EREMOMELA *Eremomela usticollis*

J	F	M	A	M	J	J	A	S	O	N	D
b	b	b	b					b	b	B	b

10 cm; 7–10 g A pale buff-and-grey eremomela with a striking white eye; in br. season, small, rusty throat band and malar stripe present, vestigial or absent at other times. Most like Grey Penduline Tit (p. 374), but is larger, has white (not blackish) eye, and a longer bill and tail. Juv. like non-br. ad., with brown eyes. **Voice:** High-pitched *'chii-cheee-cheee-cheee'*, often followed by *'trrrrrrrrr'*. Calls include *'tee-up tee-up tee-up'*. **Status and biology:** Common resident in acacia savanna; in pairs or small family groups, maintaining frequent contact with their trilling call. Usually encountered moving through or between tree canopies; often in mixed-species bird parties. (Bruinkeelbossanger)

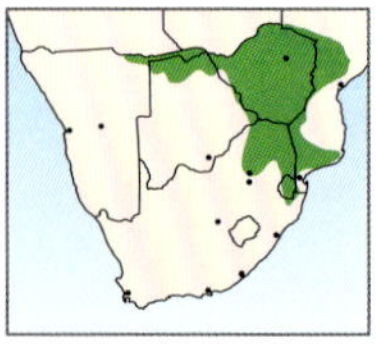

GREEN-CAPPED EREMOMELA *Eremomela scotops*

J	F	M	A	M	J	J	A	S	O	N	D
b	b	b	b				b	B	B	B	b

11–12 cm; 8–11 g A handsome eremomela with a greenish crown and grey back; underparts pale yellow, brightest on the breast. Belly paler in n Namibia, Botswana and Zambezi Valley. Most likely to be confused with a white-eye (p. 404), but eyes whitish with a red (not white) eye-ring. Juv. paler above; off-white below. **Voice:** Song is a monotonous *'twip-twip-twip-twip'*; alarm call a rasping or chattering *'tweeer'*. **Status and biology:** Fairly common resident in broadleafed woodland, especially miombo; also in mopane and riverine forests, usually in small groups. Forages mainly in the canopy, and frequently joins mixed-species bird parties, especially with white-eyes. (Donkerwangbossanger)

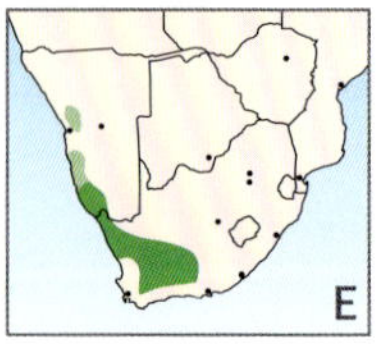

KAROO EREMOMELA *Eremomela gregalis*

J	F	M	A	M	J	J	A	S	O	N	D
b	b	b		b	b	b	B	B	B	b	b

11–12 cm; 7–9 g A fairly long-tailed eremomela with olive-green upperparts and grey face contrasting sharply with its silvery-white underparts. Conspicuous pale eyes, yellow flanks and undertail coverts distinguish it from camaropteras (p. 400); habitat differs. Juv. browner above. **Voice:** Monotonous, frog-like *'swee-er swee-er swee-er'* song, usually given before dawn; also a soft *'pink pink'* contact call. **Status and biology:** Endemic. Uncommon to locally common resident in Karoo and semi-desert scrub, usually in flat areas with taller than average shrubs (especially euphorbias); avoids watercourses with trees. Usually in small groups, commonly detected by contact calling between party members. (Groenbossanger)

CREEPERS

Small passerines related to wrens that creep up trunks and large branches, gleaning invertebrates with their long, slender bills. Sexes alike.

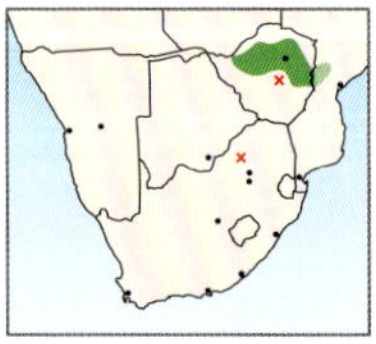

AFRICAN SPOTTED CREEPER *Salpornis salvadori*

J	F	M	A	M	J	J	A	S	O	N	D
								b	B	b	b

14–15 cm; 14–16 g A distinctive, small passerine with distinctive spotted and barred plumage. Much smaller than woodpeckers (p. 304) or wrynecks (p. 306). Juv. duller, with whiter underparts. **Voice:** A series of very high-pitched, rather soft notes, *'sweepy-swip-swip-swip'*; also a harsher *'keck-keck'*. **Status and biology:** Uncommon to locally common resident in miombo woodland; often joins mixed-species bird parties. (Boomkruiper)

yellow-washed belly

western races

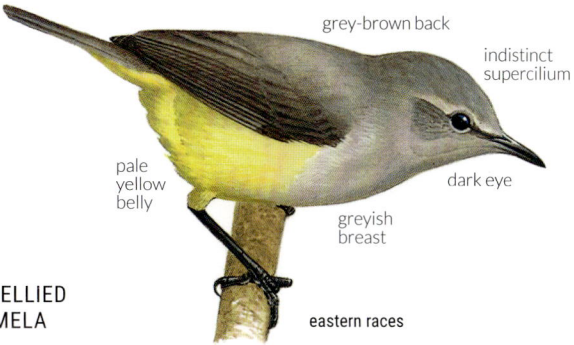

grey-brown back

indistinct supercilium

pale yellow belly

greyish breast

dark eye

YELLOW-BELLIED EREMOMELA

eastern races

pale eye

chestnut cheeks & breast bar

br.

non-br.

BURNT-NECKED EREMOMELA

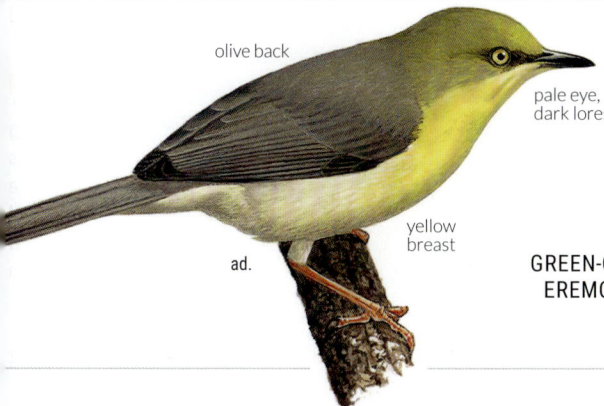

olive back

pale eye, dark lores

yellow breast

ad.

greenish crown

ad.

GREEN-CAPPED EREMOMELA

KAROO EREMOMELA

pale eye

greyish-green back

white throat

ad.

AFRICAN SPOTTED CREEPER

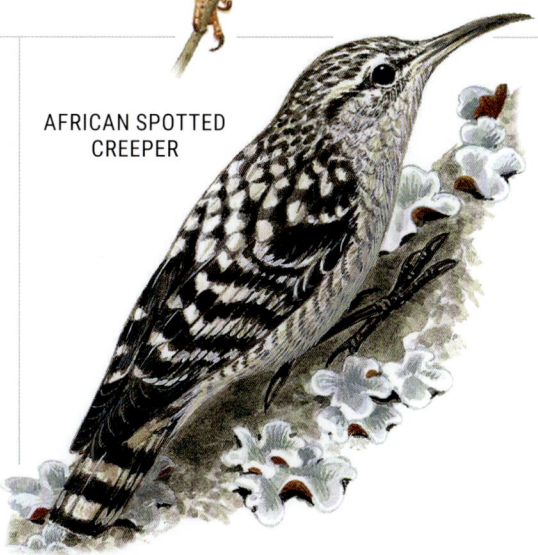

SYLVIA WARBLERS

SYLVIA WARBLERS Despite their distinct appearance, Chestnut-vented and Layard's Warblers (formerly known as tit-babblers in s Africa and currently known as parisomas in E Africa) are now known to belong to the mainly Palearctic genus of *Sylvia* warblers, of which 3 species overwinter in Africa. These resident sylviids are mainly grey-coloured with black-and-white tails and strikingly pale eyes. Sexes alike or nearly so. Occur singly or in pairs. The migrant sylviids are mainly brown, drab-plumaged birds frequenting woodland and dense undergrowth; solitary and unobtrusive in their winter range.

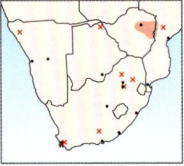

EURASIAN BLACKCAP *Sylvia atricapilla*

X	X	X		X	X				X	X	
J	F	M	A	M	J	J	A	S	O	N	D

13 cm; 14–20 g Vagrant. A grey-brown warbler, distinguished by its black (male) or reddish-brown (female) crown. Juv. like female, but duller. If head not seen, could be confused with Garden Warbler. **Voice:** Call hard *'tac'*. Song (Jan–Mar) is a series of varied warbles, initially subdued, but becoming more persistent; very similar to Garden Warbler. **Status and biology:** A rare Palearctic-br. migrant, perhaps under-reported because of inconspicuous behaviour. Frequents dense undergrowth. (Swartkroonsanger)

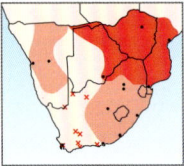

GARDEN WARBLER *Sylvia borin*

J	F	M	A	M	J	J	A	S	O	N	D

14 cm; 16–24 g A nondescript, drab, grey-brown warbler, paler below. Only distinctive feature is its rather large, dark eye with an indistinct eye-ring in its plain face; lacks eye stripe or supercilium found in most other warblers. Occasionally shows a slightly paler nuchal collar. Juv. duller, with slightly warmer plumage. **Voice:** Subdued, monotonous song, interspersed with soft, grating phrases; very similar to Eurasian Blackcap; call a harsh *'tec'*. **Status and biology:** Fairly common, Palearctic-br. visitor. Occurs singly in thick tangles in forest, bush and riverine habitats; skulks and is easily overlooked. (Tuinsanger)

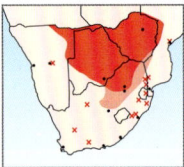

COMMON WHITETHROAT *Sylvia communis*

J	F	M	A	M	J	J	A	S	O	N	D

14 cm; 12–20 g A slender, long-tailed, mainly brown warbler; white throat contrasts with pinkish-grey underparts; rufous-edged wing feathers are diagnostic. Male has grey head; female head browner. Head often slightly crested, especially when alarmed. Juv. duller, with buffy (not white) outer-tail feathers. **Voice:** Soft *'whit'* and grating *'tchack'* and *'tchurr'* alarm calls; song a harsh, snappy mixture of grating and melodious notes. **Status and biology:** Locally common, Palearctic-br. migrant. Solitary; in dry woodland, especially thornveld in arid savanna. (Witkeelsanger)

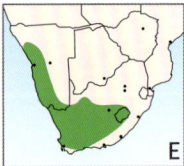

LAYARD'S WARBLER (TIT-BABBLER) *Sylvia layardi*

J	F	M	A	M	J	J	A	S	O	N	D
					b	b	B	B	B	B	b

14 cm; 14–16 g Differs from darker Chestnut-vented Warbler by having a white (not chestnut) vent; throat streaking often less pronounced. Pale eyes contrast with dark head. Female slightly browner than male; vent tinged buff. Juv. is washed buff; lacks throat streaking. **Voice:** Song is a short, attractive warble; alarm call a ratchet-like *'tit-trt-trt-trt-trt'*. **Status and biology:** Endemic. Common resident in Karoo scrub, arid savanna, coastal thicket, and montane scrub in Lesotho; often found in rocky, hilly areas. Occurs singly or in pairs, foraging in low bushes and shrubs. (Grystjeriktik)

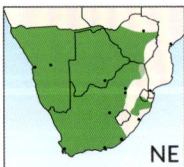

CHESTNUT-VENTED WARBLER (TIT-BABBLER)
Sylvia subcaeruleum

b	b	b	b			b	b	B	B	B	B
J	F	M	A	M	J	J	A	S	O	N	D

15 cm; 12–18 g Differs from Layard's Warbler in having chestnut (not pale grey) vent; throat streaking often bolder and more extensive. Juv. lacks throat streaking; vent paler rufous. **Voice:** A cheerful, varied song, usually initiated by a characteristic *'cher-eeetr tik-tik-tik'* and continuing with a medley of mimicked notes of other birds. **Status and biology:** Near-endemic. Common resident in savanna, especially thornveld and coastal thicket; confined to dry watercourses in arid shrubland. Occurs singly or in pairs, foraging from tree canopy to shrubby understorey; presence often given away by its song. (Bosveldtjeriktik)

red-brown
crown

EURASIAN
BLACKCAP

black crown

♂

♀

GARDEN
WARBLER

pale nuchal collar

plain head
with large
dark eye

shortish
bill

♂

COMMON
WHITETHROAT

♀

rufous
wing
panel

long tail

chestnut
vent

pale eye

pale eye

LAYARD'S
WARBLER

white
vent

streaked
throat

boldly
streaked
throat

CHESTNUT-VENTED
WARBLER

LOCUSTELLID WARBLERS

Five rather similar, thickset, longish-tailed, drab-coloured warblers belong to this main Eurasian family. All skulk in dense undergrowth, and are most easily detected by their calls. Sexes alike.

FAN-TAILED GRASSBIRD (BROAD-TAILED WARBLER)
Catriscus brevirostris

B	b										b	B
J	F	M	A	M	J	J	A	S	O	N	D	

15 cm; 15–17 g A small warbler with a long, broad, black tail that is conspicuous in flight. Tail is strongly graduated; tail feathers tipped buff, visible as scaling on the underside. Flattened head shape is distinctive, with forehead and culmen forming an almost straight line. Juv. has warmer brown upperparts and yellowish underparts; tail shorter and narrower. **Voice:** Soft, metallic *'zeenk'*, repeated at intervals of a few seconds; clear, high-pitched *'peee, peee'*. **Status and biology:** Locally common resident and altitudinal migrant, skulking in long, rank grass, usually in damp areas. Often perches in the open after heavy rains to dry its tail and wings. (Breëstertsanger)

LITTLE RUSH WARBLER   *Bradypterus baboecala*

B	b	b							b	B	B	B
J	F	M	A	M	J	J	A	S	O	N	D	

15–17 cm; 12–18 g Predominantly found in dense reeds and riparian vegetation; told from reed warblers (p. 384) by its distinctive song, dark brown upperparts, dappled throat and breast, and long, rounded tail; dark bill is small and slender. Extent of breast streaking varies from well-defined streaks to diffuse mottling. Juv. has underparts washed pale yellow. **Voice:** Harsh, ratchet-like song accelerates towards the end *'krrak krrak krrak krrak krk-krk-rk-rk-rk-rk'*; often accompanied by a wing rattle; also nasal *'wheeaaa'*. **Status and biology:** Locally common resident and nomad, typically associated with tangled vegetation around wetlands; not usually over open water. (Kaapse Vleisanger)

KNYSNA WARBLER   *Bradypterus sylvaticus*

										b	B	B
J	F	M	A	M	J	J	A	S	O	N	D	

14–15 cm; 15–20 g A small, dark olive-brown warbler confined to dense undergrowth in and around afromontane forest and coastal thickets. Best told by its distinctive song. Tail is broad and graduated, but is shorter and less graduated than in Barratt's Warbler; breast is plain or finely streaked (less marked than Barratt's Warbler). Smaller and darker below than Little Rush Warbler. Male has a white loral spot and dark brown eye stripe; female has paler throat and less marked face. Juv. has underparts washed pale yellow; breast streaked. **Voice:** Song is an accelerating series of sharp whistles, ending in a dry trill, *'wit it it it it-it-it-ititititrrrrrrr'*; both sexes make deep *'chack'* and dry *'prrt'* calls. **Status and biology:** VULNERABLE. Endemic. Uncommon and localised in dense forest understorey, bracken and briar thickets, often along watercourses; very secretive. (Knysnaruigtesanger)

BARRATT'S WARBLER   *Bradypterus barratti*

										b	B	B	b
J	F	M	A	M	J	J	A	S	O	N	D	

15 cm; 17–20 g Slightly larger and longer tailed than Knysna Warbler, with paler underparts and a more heavily streaked throat and breast, but best identified by song where ranges overlap in E Cape. Also resembles Little Rush Warbler, but occurs in forest (not reed beds) and has a different song. Juv. is warmer olive above and more yellowish below than ad. **Voice:** Song, delivered in startling bursts, begins with repeated high-pitched *'tik'* notes, then accelerates to a trill; alarm note a sparrow-like *'chrrrrr'*. **Status and biology:** Endemic. Locally common resident and altitudinal migrant found in brambles and other low, thick, tangled growth on edges of evergreen forests and plantations. (Ruigtesanger)

RIVER WARBLER   *Locustella fluviatilis*

J	F	M	A	M	J	J	A	S	O	N	D	

14 cm; 14–20 g A secretive, dark brown warbler with slightly paler underparts; tail graduated, undertail coverts brown, with broad, white tips. At close range, shows diffuse streaking on the throat and breast. Could be confused with Thrush Nightingale (p. 372), which also skulks in thickets. Juv. is warmer rufous above and creamy buff below. **Voice:** Buzzy, insect-like *'derr-derr-zerr-zerr'* song, given mainly at dawn during Mar–Apr; contact call is a series of 6–8 *'chick'* notes lasting 1 second, given before singing. Alarm call is a sharp *'pwit'*, accompanied by wing- and tail-flicking. **Status and biology:** Scarce, Palearctic-br. migrant in dense thickets in woodland and riverine scrub; secretive, best located by its song. (Sprinkaansanger)

FAN-TAILED GRASSBIRD

flattened crown

heavy bill

dark undertail coverts with buff feather tips

LITTLE RUSH WARBLER

dark brown upperparts

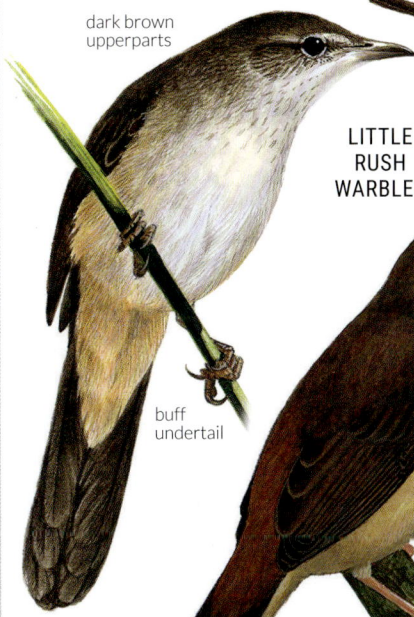

rattles wings in display flight

faint breast streaking

buff undertail

KNYSNA WARBLER

dark underparts

shortish tail

BARRATT'S WARBLER

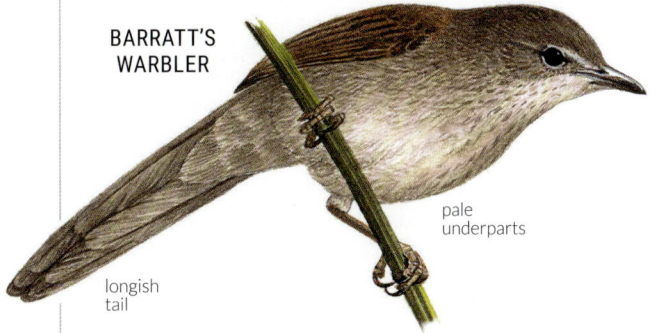

pale underparts

longish tail

streaked throat

long scalloped undertail coverts

RIVER WARBLER

LEAF WARBLERS
Small, slender warblers, with small, thin bills, that forage in trees and shrubs. Sexes alike.

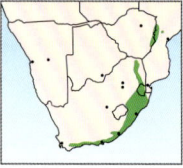

YELLOW-THROATED WOODLAND WARBLER
Phylloscopus ruficapilla

11 cm; 6–9 g A small, forest canopy warbler; recalls Willow Warbler in shape but easily distinguished by ginger crown and bright yellow throat, breast and supercilium. Juv. has a green wash on the breast. **Voice:** Song a loud *'seee suuu seee suuu'*. **Status and biology:** Fairly common resident, inland populations move to lower elevations in winter. Occurs in the canopy and mid-strata of evergreen forest, singly, in pairs or small groups; often joins bird parties, especially with white-eyes. (Geelkeelsanger)

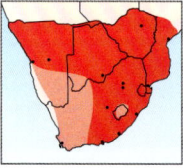

WILLOW WARBLER
Phylloscopus trochilus

11 cm; 7–12 g A small, slender warbler with a dark eye stripe and a pale, yellowish supercilium. Most birds are from w Europe and have olive-washed upperparts and yellow-washed throat and breast (*P. t. trochilus*), but some are larger, duller birds from Asia (*P. t. yakutensis*) that have the yellow-green wash confined to the face. Bill smaller, darker and more slender than Icterine Warbler; yellow usually restricted to throat and breast. Juv. has more extensive yellow underparts. **Voice:** Soft, 2-note contact call *'foo-eet'*; short, melodious song, descending in scale (often heard before northward migration). **Status and biology:** Abundant, Palearctic-br. migrant in woodland and savanna. Forages in leafy tree canopies, constantly on the move and frequently giving its characteristic contact call. Mostly solitary, but several may forage in same tree. Joins mixed-species bird parties. (Hofsanger)

HIPPOLAIS WARBLERS
Heavily built warblers related to reed warblers with more stout bills and peaked crowns than leaf warblers. Sexes alike.

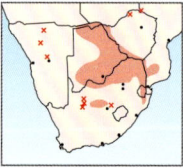

OLIVE-TREE WARBLER
Hippolais olivetorum

15–18 cm; 14–22 g A large, pale grey warbler with a long bill and distinctly angled crown; combination of size, head shape, absence of yellow or green in plumage, white edges to outer tail and habitat separates it from all species except for the vagrant Upcher's Warbler (see below). Unworn secondaries are white edged, creating a pale panel in the folded wing; has a distinct white eye-ring; legs and feet are robust. **Voice:** A rambling song that can continue for minutes, made up of harsh, grating, metallic notes, recalling Great Reed Warbler (p. 384). **Status and biology:** Scarce, possibly decreasing. Palearctic-br. migrant found in semi-arid thornveld, particularly in thickets of *Vachellia tortilis* (= *Acacia tortilis*) and *Senegalia mellifera* (=*A. mellifera*) <6 m high. Skulks in tree crowns; best located by its song (sings from late Dec). (Olyfboomsanger)

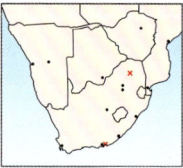

UPCHER'S WARBLER
Hippolais languida

14–15 cm 10–20 g Vagrant. Slightly smaller and browner than Olive-tree Warbler with a more rounded head and shorter bill, with a pinkish-yellow (not orange-yellow) lower mandible and typically a longer supercilium, extending behind the eye. Tertials and secondaries are white-edged, creating a pale wing panel like Olive-tree Warbler, but greater coverts are edged warmer brown (not pale grey-brown). Primary projection shorter ($^2/_3$ tertial length, not $^3/_4$ in Olive-tree). Uppertail and tips of primaries dark grey, appearing black in low light. Habitually swings tail pendulum-like, side-to-side; also fans and depresses tail. **Voice:** In the region, full song not heard, harsh *'tchack'* calls observed. **Status and biology:** Two records, one in Aug 2017 near Port Elizabeth, and one in Apr 2018, in Limpopo. Possibly overlooked due to confusion with Olive-tree Warbler. (Vaalspotsanger)

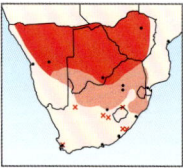

ICTERINE WARBLER
Hippolais icterina

13–14 cm; 10–18 g A yellowish warbler, larger than Willow Warbler with a more robust bill and an angular, sloping forehead. It has a short, yellowish supercilium and narrow, pale eye-ring. Secondaries are yellow-edged, creating a pale panel in the folded wing. Ad. has yellow-washed underparts and olive-grey upperparts. Juv. and imm. duller, with yellow wash confined to face and throat. Very drab birds could be confused with Olive-tree and Upcher's Warblers, but are smaller, with a shorter bill; lack white in outer tail. **Voice:** Song is a mix of harsh and melodious notes, including *'tree-tjuu'* and *'di de roy'*; alarm a harsh *'tac, tac'*. Sings regularly, especially before northward migration. **Status and biology:** Fairly common, Palearctic-br. migrant to thickets in savanna and arid woodland, dry riverbeds, plantations and gardens. Mostly solitary. (Spotsanger)

brown cap, yellow supercilium

yellow vent

yellow throat

YELLOW-THROATED WOODLAND WARBLER

long pale supercilium

WILLOW WARBLER

short, thin bill

typical ad.

brownish legs

yakutensis

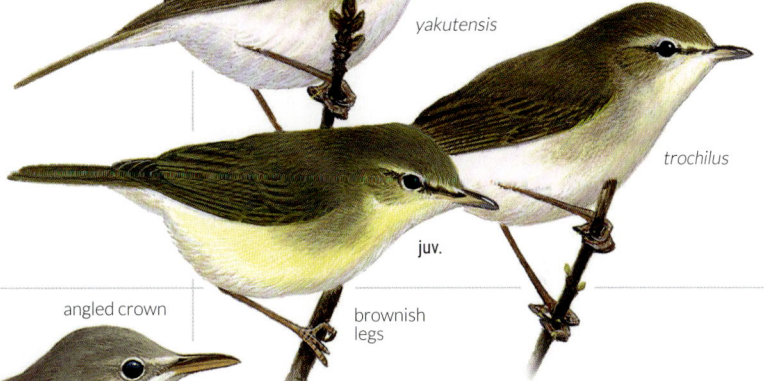

trochilus

juv.

brownish legs

OLIVE-TREE WARBLER

angled crown

uniform pale grey upperparts

long, heavy bill

pale wing panel

long primary projection

robust legs & feet

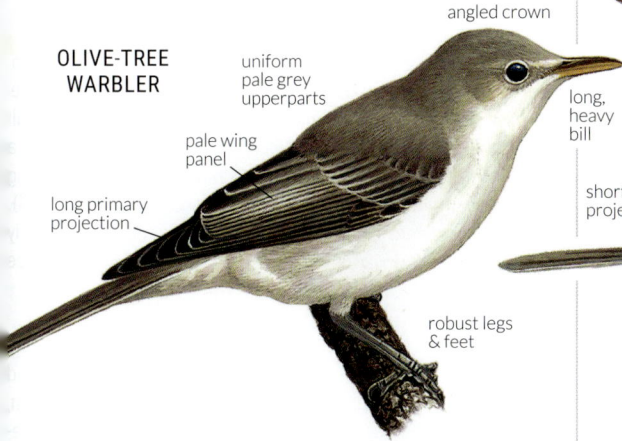

very faint wing panel

shorter primary projection

slightly shorter bill

ICTERINE WARBLER

short supercilium

pale wing panel

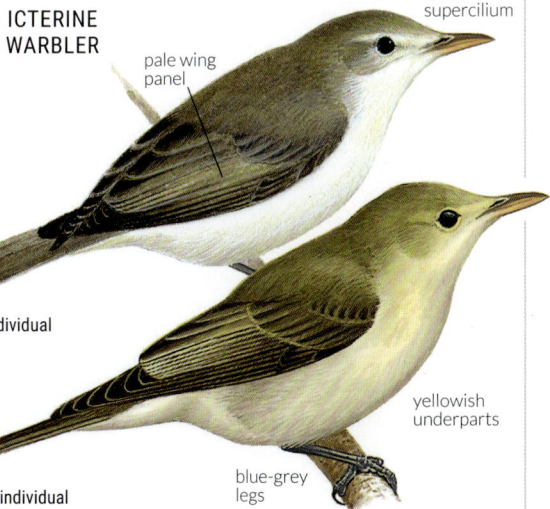

rab individual

yellow individual

blue-grey legs

yellowish underparts

UPCHER'S WARBLER

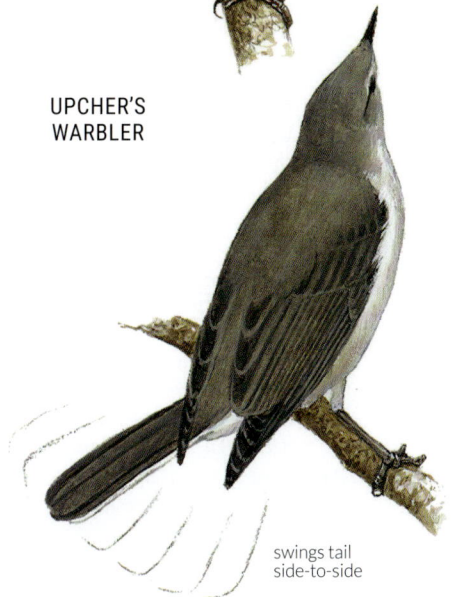

swings tail side-to-side

383

LARGE REED WARBLERS
Largish warblers that live in reeds and other rank vegetation, mostly keeping under cover. Their presence is often given away by their distinctive songs. Sexes alike.

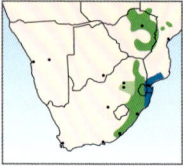

DARK-CAPPED (AFRICAN) YELLOW WARBLER
Iduna natalensis

14–15 cm; 9–14 g A distinctive reed warbler of rank vegetation with rich yellow underparts and underwing coverts. Bill yellowish with broad, flycatcher-like base. Female duller, with less contrast between back and underparts. Darker above than Icterine Warbler (p. 382) with a dark brownish-olive cap and yellowish rump; tail longer; unlikely to occur in the same habitat. Juv. has buff wash above; paler below. **Voice:** A rich, throaty medley of chipping and bubbling notes repeated in various combinations 'chip chip chip, tititi, churrup, chuurp cheezee'; reminiscent of Lesser Swamp Warbler. **Status and biology:** Locally common resident and partial migrant in bracken, sedges and tangled vegetation at forest edges and along streams. Mostly inconspicuous but sometimes hawks insects from exposed perches. (Geelsanger)

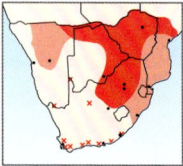

SEDGE WARBLER *Acrocephalus schoenobaenus*

12–13 cm; 9–16 g The only *Acrocephalus* warbler in the region with a streaked back; head has a striking broad, creamy supercilium and a streaked crown. The relatively short, dark brown tail contrasts with its plain rufous rump. Larger than cisticolas (p. 388); tail lacks black-and-white tips. Juv. is yellower and more distinctly marked than ad. **Voice:** Wheezy, buzzy, rapidly repeated refrain of loud and soft phrases, sometimes mingled with snatches of mimicry; alarm a sharp 'tuk'. **Status and biology:** Locally common, Palearctic-br. migrant in marshy habitats, including reed beds over water and rank, weedy tangles bordering wetlands; sometimes in thickets far from water. (Europese Vleisanger)

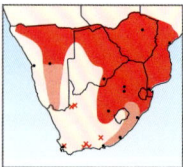

GREAT REED WARBLER *Acrocephalus arundinaceus*

18–20 cm; 20–42 g A very large, robust reed warbler with a fairly prominent buffy supercilium. Large size readily distinguishes it from most other reed warblers. Appreciably larger than Lesser Swamp Warbler (p. 386), with a heavier bill, more extensive rufous wash on flanks, and longer wings. Paler and warmer brown than Greater Swamp Warbler. Confusion most likely with rare Basra Reed Warbler; averages larger and warmer brown above, with a heavier bill and shorter supercilium; at close range has dark shafts to throat feathers. Juv. warmer brown above, underparts with a buffy-orange wash. **Voice:** Very vocal; a sustained, low-pitched grating wheezing and prolonged, rambling 'chee-chee-chaak-chaak-chuk-chuk' song comprises harsh elements repeated 2–3 times; delivery slower than smaller reed warblers. **Status and biology:** Fairly common Palearctic-br. migrant in reed beds and thickets, often near water. Usually located by its harsh, guttural song (Grootrietsanger)

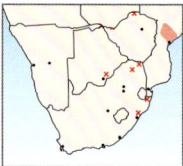

BASRA REED WARBLER *Acrocephalus griseldis*

15–17 cm; 16–20 g A large, slim reed warbler with a dark tail and a long, slender bill accentuated by its flat forehead. Slightly smaller than Great Reed Warbler, with colder, olive-grey (not rufous-brown) upperparts and whiter underparts (washed pale yellow in fresh plumage, Mar–Apr); lacks dark throat streaks. Narrow white supercilium extends well behind eye and contrasts with dark lores and eye stripe. Build is less robust, closer to that of smaller reed warblers. In the hand, wing 75–88 mm (85–105 mm in Great Reed Warbler). **Voice:** Slow, rather weak nasal churring and chattering, reminiscent of blue-eared starling call, 'chuc-chac-churruc-charra, tchuktchuk weea churruc-chuc', weaker and more monotonous than Great Reed Warbler; calls continuously for long periods, even in the heat of the day. **Status and biology:** ENDANGERED. Status in the region uncertain; observed fairly regularly in central Mozambique, but possibly overlooked elsewhere. Occurs in reed beds, thickets and rank vegetation near water. (Basrarietsanger)

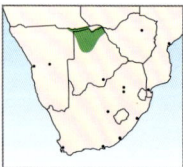

GREATER SWAMP WARBLER *Acrocephalus rufescens*

16–18 cm; 18–28 g A large, long-billed warbler confined to papyrus swamps; habitat, distinctive song, dark grey-brown upperparts and virtual absence of supercilium are diagnostic. Often raises short crest when singing from top of papyrus. Slightly smaller than Great Reed Warbler and lacks pale rufous-brown in upperparts; larger and darker than Lesser Swamp Warbler (p. 386), with no white supercilium. Structure bulbul-like; could be confused with Terrestrial Brownbul (p. 348), but habitat and song distinctive. **Voice:** Loud, gurgling 'churrup, churr-churr, churrup, chuckle', interspersed with harsher notes. **Status and biology:** Locally common resident in papyrus swamps. Mostly remains hidden, its presence given away by its far-carrying song. (Rooibruinrietsanger)

DARK-CAPPED
YELLOW WARBLER

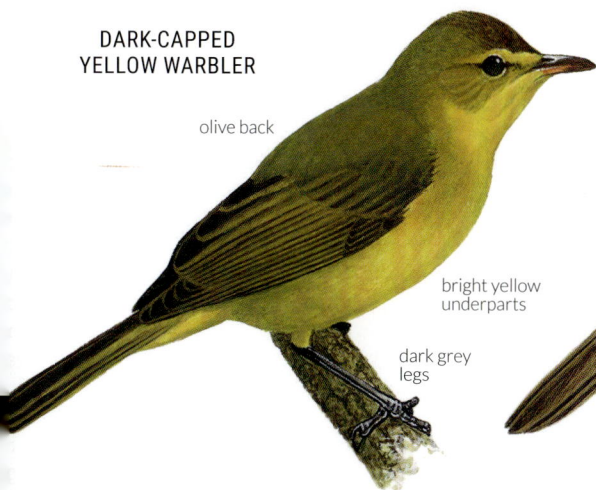

olive back

bright yellow
underparts

dark grey
legs

broad supercilium

streaked
back

SEDGE
WARBLER

streaked crown,
broad white
supercilium

GREAT
REED WARBLER

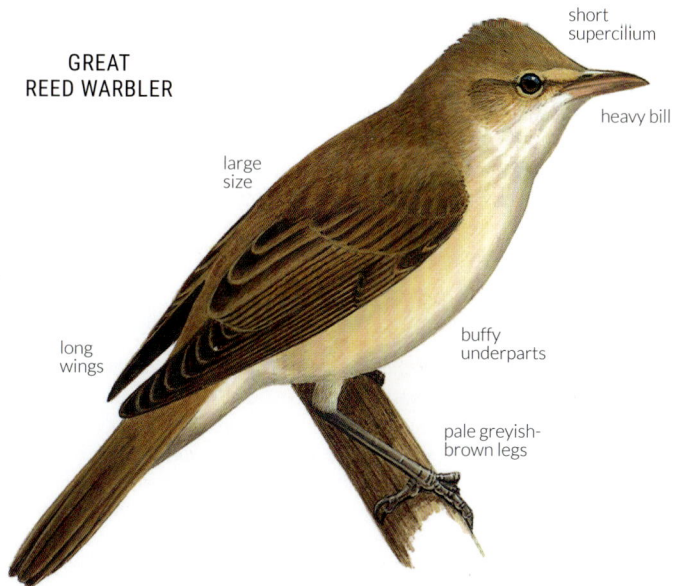

short
supercilium

heavy bill

large
size

long
wings

buffy
underparts

pale greyish-
brown legs

BASRA REED
WARBLER

flat
forehead

long
supercilium

long
slender
bill

whitish
underparts

GREATER
SWAMP WARBLER

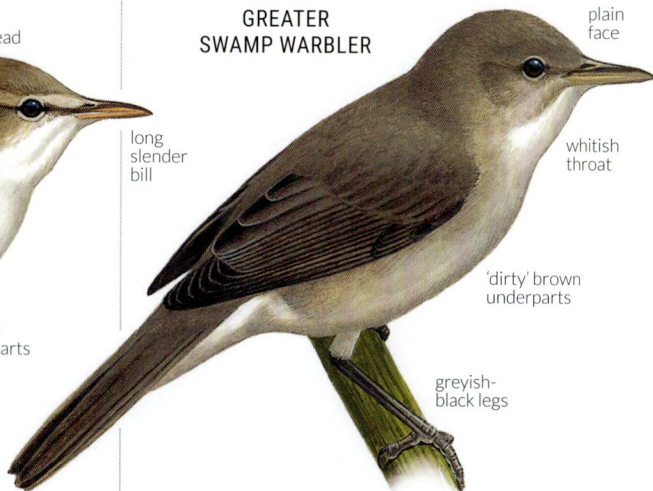

plain
face

whitish
throat

'dirty' brown
underparts

greyish-
black legs

385

SMALL REED WARBLERS

Uniformly sandy-brown warblers mainly found in reeds and adjacent vegetation. Widespread, resident Lesser Swamp Warbler presents no identification problems, but remaining 3 notoriously difficult to identify. African Reed Warbler is a resident br. species in parts of s Africa and migratory in others; the other 2 are Palearctic-br. migrants. With experience, Marsh Warbler can be distinguished on slight shape and colour differences, but separating Eurasian and African Reed Warblers requires examination in the hand. Sexes alike, but plumage varies with amount of wear.

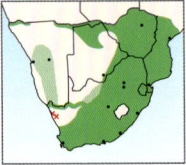

LESSER SWAMP WARBLER *Acrocephalus gracilirostris*

B	B	B	b	b		b	B	B	B	B	B
J	F	M	A	M	J	J	A	S	O	N	D

14–16 cm; 12–20 g A rather large, slender reed warbler, usually with a fairly distinct, whitish supercilium, long bill, and dark brown legs that appear blackish in the field. Underparts mainly white, with rufous wash confined to thighs and lower flanks. Larger than African Reed Warbler, with a more distinct supercilium, warmer brown upperparts, white breast and darker legs. Wings short and rounded, showing less primary projection than migrant Great and Basra Reed Warblers (p. 384); bill more slender. Slightly smaller than Greater Swamp Warbler (p. 384), with warmer rufous upperparts, whiter underparts and a white supercilium. **Voice:** Rich, fluty *'cheerup-chee-chiree-chiree'* song is distinctive; lacks harsh notes of other reed warblers. **Status and biology:** Common resident in *Phragmites* and *Typha* reed beds over water; some seasonal movement. (Kaapse Rietsanger)

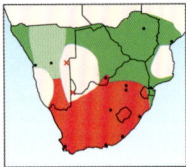

AFRICAN REED WARBLER *Acrocephalus baeticatus*

B							b	B	B	B	
J	F	M	A	M	J	J	A	S	O	N	D

12–13 cm; 8–14 g A small, brown reed warbler with a dull white throat; breast and flanks washed rufous. Told from Marsh Warbler by its song and by its slightly longer, more slender bill, warmer brown upperparts, buffy (not greyish) underparts, darker legs and shorter wings (folded wings do not extend beyond rump). Forehead profile is less steep. Separating it from Eurasian Reed Warbler is easy if it is found nesting (Eurasian Reed Warbler does not breed in s Africa), otherwise in-the-hand examination is needed: moults Apr–Jun (Dec–Mar in Eurasian Reed Warbler) and has a different wing structure (see that species). **Voice:** Song is harsher, more churring and repetitive than that of Marsh Warbler, typically repeating notes 2–4 times. Sometimes mimics other birds. **Status and biology:** A common br. migrant in the south; resident further north. In arid areas, occurs in scrub along dry watercourses and in gardens, but in eastern areas, where overlap with Marsh Warbler more likely, usually confined to wetland habitats (but seldom over water like Lesser Swamp Warbler). (Kleinrietsanger)

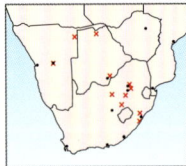

EURASIAN REED WARBLER *Acrocephalus scirpaceus*

J	F	M	A	M	J	J	A	S	O	N	D

12–13 cm; 9–18 g Closely related to African Reed Warbler and indistinguishable on plumage characters alone; all records from s Africa identified in the hand: wing longer (60–76 mm vs 58–62 mm); emargination on P9 less extensive, level with tip of P3 (extends beyond tips of secondaries in African Reed Warbler). Typically appears colder grey-brown above and whiter below, with a more prominent white throat. Moults Dec–Mar (not Apr–Jun). Differs from Marsh Warbler by its longer, more slender bill, less rounded forehead and typically warmer, olive-brown upperparts, often with a rufous wash on the rump. **Voice:** Song similar to African Reed Warbler; call notes include low *'churrr'*. **Status and biology:** Rare, Palearctic-br. migrant; possibly overlooked. Favours reed beds and rank vegetation close to water. (Hermanrietsanger)

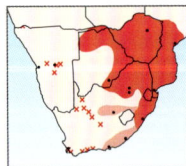

MARSH WARBLER *Acrocephalus palustris*

J	F	M	A	M	J	J	A	S	O	N	D

12–13 cm; 8–15 g Similar in appearance to African and Eurasian Reed Warblers, but bill is slightly shorter and stouter; upperparts more olive-grey (less rufous) and it has a steeper, more rounded forehead profile. Wings average longer than African Reed Warbler (primary tips extend beyond rump to uppertail coverts). In the hand, claws are all dark (lack pale undersides). Juv. is warmer brown above, with rufous-washed rump; averages paler and more olive than juv. Eurasian Reed Warbler. **Voice:** Sings mostly in late summer; clear, melodious phrases are more varied and less scratchy than African and Eurasian Reed Warblers, with little repetition; rich in mimicked phrases from other birds. Frequently gives a *'chuck'* or *'tushr'* call note while foraging. **Status and biology:** Common, Palearctic-br. migrant; frequents rank undergrowth, bracken and briar, forest edge, riverine thickets and dense gardens. Unobtrusive; sings from a concealed perch. (Europese Rietsanger)

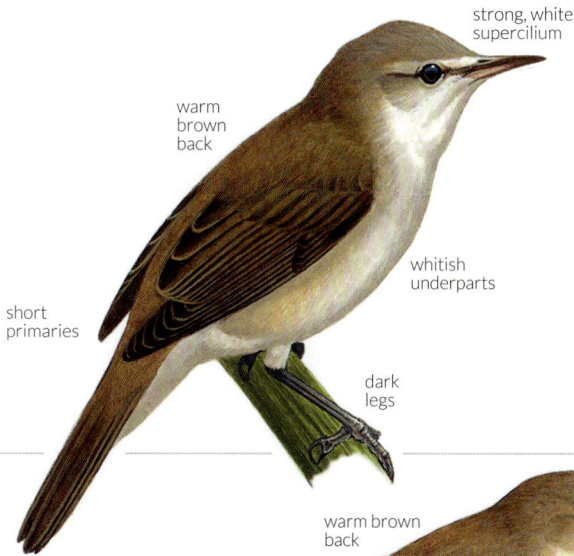

LESSER SWAMP WARBLER

strong, white supercilium

warm brown back

whitish underparts

short primaries

dark legs

AFRICAN REED WARBLER

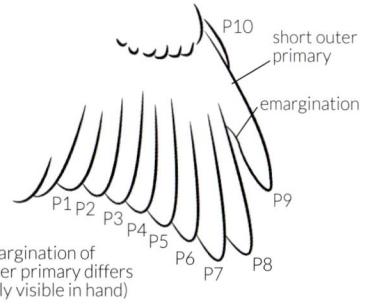

warm brown back

short primaries

dark brown legs

P10

short outer primary

emargination

P1 P2 P3 P4 P5 P6 P7 P8 P9

emargination of outer primary differs (only visible in hand)

These two species are probably indistinguishable in the field

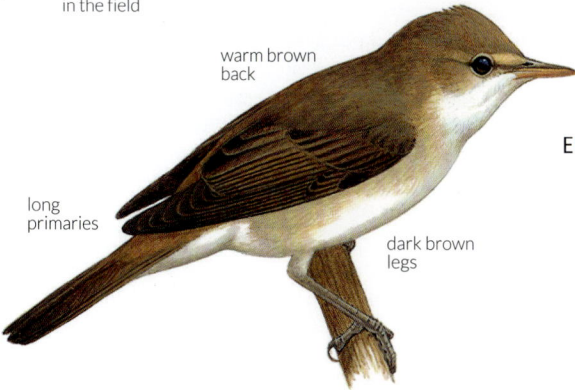

warm brown back

long primaries

dark brown legs

EURASIAN REED WARBLER

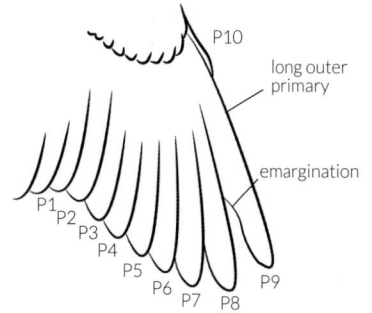

P10

long outer primary

emargination

P1 P2 P3 P4 P5 P6 P7 P8 P9

MARSH WARBLER

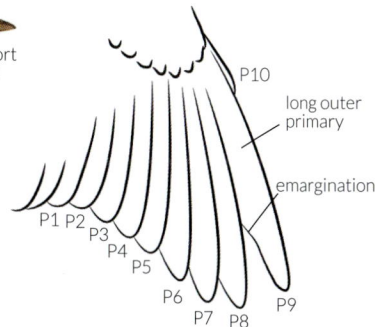

greyish back

rounded head

short bill

long primaries

pinkish-brown legs

P10

long outer primary

emargination

P1 P2 P3 P4 P5 P6 P7 P8 P9

SHORT-TAILED CISTICOLAS

Cisticolas are small, insectivorous warblers with mainly brown plumage. The many species, combined with seasonal and age-, sex- and regional plumage differences, makes them notoriously difficult to identify. Most can be identified by their calls and habitat preferences, but distinguishing them in winter when they are silent and have different plumage and shape (altered tail length) is a challenge. Most species occur singly, in pairs or small family groups. The 5 short-tailed species comprise 2 with very short tails (26–28 mm; Wing-snapping and Cloud) and 3 with slightly longer tails (33–44 mm; Zitting, Desert and Pale-crowned).

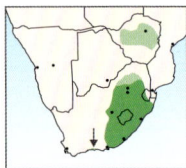

WING-SNAPPING CISTICOLA *Cisticola ayresii*

9 cm; 9 g Very short tail distinguishes this species from all other cisticolas, except Cloud Cisticola. Wing-snapping Cisticola is slightly smaller and shorter winged, and mantle feathers are distinctly darker; br. male has dark lores and dull rufous crown. Non-br. male and female have longer tails, pale lores and well-streaked crowns and napes. Juv. is duller; breast washed yellow. **Voice:** Males spend long periods displaying in summer, cruising 30–50 m above the ground, repeatedly uttering 3 or 4 even-pitched notes, *'si, siee sie si'* or *'sooey sooey sooey'*, sometimes preceded by 1 or 2 higher-pitched *'chik'* notes; returns to the ground in a rapid vertical descent making sharp clicks and wing snaps while diving. **Status and biology:** Common resident in short, moist grasslands, with some local movement to lower elevations in winter. Unobtrusive outside br. season, flying far when flushed. (Kleinste Klopkloppie)

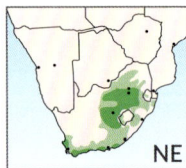

CLOUD CISTICOLA *Cisticola textrix*

10 cm; 11 g Very short tail distinguishes this species from all other cisticolas, except Wing-snapping Cisticola; they occur together widely, but Cloud Cisticola's range extends further into more arid grasslands. *C. t. textrix* (W Cape) has diagnostic breast streaking. In other races, br. male is slightly larger and longer winged than Wing-snapping Cisticola and has less dark mottling on mantle. Non-br. ad. has more streaking on crown and upperparts and longer tail. Juv. is duller, with breast washed yellow. **Voice:** Displaying male easily distinguished by its song. Spends long periods in the air in summer, cruising 30–50 m up, every few seconds giving a short refrain of 3–5 soft notes followed by 3–5 tongue clicks, *'so-si-si-si, clik-clik-clik-cli'*; returns to the ground in a rapid vertical descent, making sharp clicks. **Status and biology:** Near-endemic. Common resident in short grassland, grazed fields and lowland fynbos. Unobtrusive outside the br. season, flying far when flushed. (Gevlekte Klopkloppie)

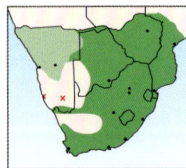

ZITTING CISTICOLA *Cisticola juncidis*

11 cm; 9 g A widespread grassland cisticola. Similar in structure to Desert Cisticola, but appears warmer brown; both sexes (in br. and non-br. plumages) have a rufous (not greyish-brown) rump and tail has a more prominent dark brown subterminal bar. Female like male, but upperparts have a more rufous wash. Juv. more rufous and more streaked above than ad., yellower below. **Voice:** Male performs an undulating display flight 5–20 m up, uttering a single note at 1-second intervals, *'zit'* or *'klink'*, on each dip; may also call from a low perch. Alarm a rapid *'chik-chik-chik-chik...'*. **Status and biology:** Common resident in grassland and fields, especially in low-lying damp areas where there are dense grass tufts. (Landeryklopkloppie)

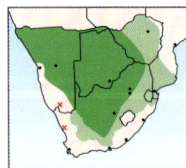

DESERT CISTICOLA *Cisticola aridulus*

11 cm; 9 g The common cisticola of semi-arid and arid grasslands. Structure similar to Zitting Cisticola, but usually is paler and greyer looking; rump greyish-brown (not rufous); upper tail lacks a prominent dark brown subterminal bar. Non-br. birds paler above and more buffy below. Juv. more rufous, with yellow wash below; easily confused with Zitting Cisticola. **Voice:** Male performs a fast, jinking display flight 2–10 m up, calling while doing so, a series of short notes *'zink zink zink'*, *'sii sii sii'* or *'su-ink su-ink su-ink'*, 3 or 4 notes/second, interspersed with wing-snapping. **Status and biology:** Common resident in arid grassland and old fields; where alongside Zitting Cisticola always in sparser, drier grassland. Solitary, in pairs or small family groups. (Woestynklopkloppie)

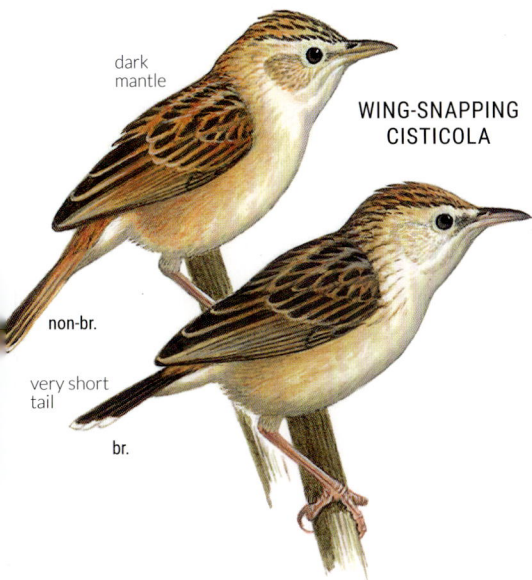

dark mantle

WING-SNAPPING CISTICOLA

non-br.

very short tail

br.

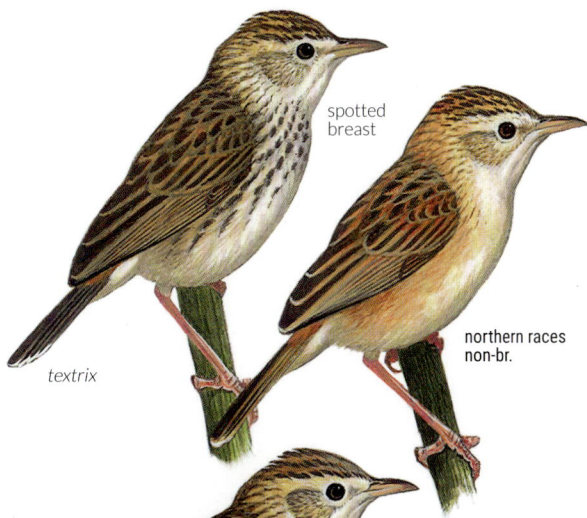

spotted breast

textrix

northern races non-br.

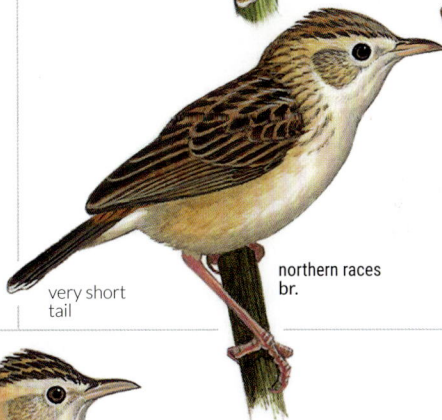

CLOUD CISTICOLA

very short tail

northern races br.

non-br.

rufous flanks

mid-length tail with dark subterminal bar

warm brown

br.

rufous unstreaked rump

ZITTING CISTICOLA

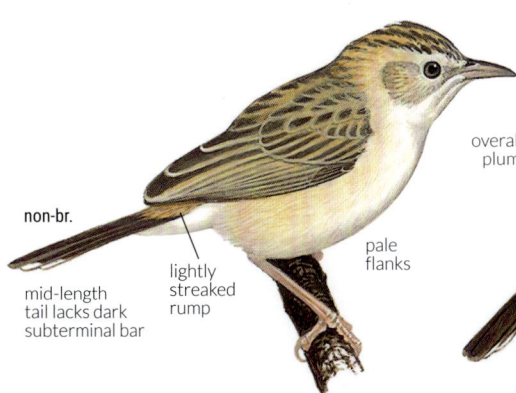

non-br.

lightly streaked rump

mid-length tail lacks dark subterminal bar

pale flanks

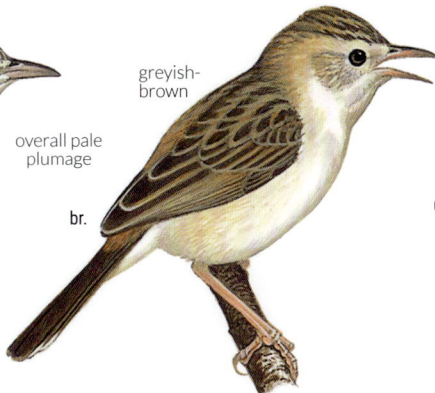

greyish-brown

overall pale plumage

br.

DESERT CISTICOLA

389

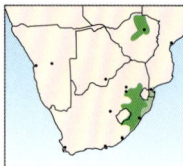

PALE-CROWNED CISTICOLA   *Cisticola cinnamomeus*

B b b | | | | | | b b B
J F M A M J J A S O N D

10 cm; 10 g A localised cisticola of moist grasslands; overlaps widely with Cloud, Wing-snapping and Zitting Cisticolas (p. 388) and hard to identify unless displaying or br. male seen. In br. plumage, male has a diagnostic, plain buffy crown and black lores. Non-br. male and female have pale lores, well-streaked buff-and-black crowns and backs and a rufous rump. Lacks Zitting Cisticola's dark brown subterminal tail bar, and is longer tailed than Cloud and Wing-snapping Cisticolas. Juv. has underparts washed yellow. **Voice:** Male sings during rapid undulating display flight. Song softer and much higher pitched than other cisticolas; 3–7 plaintive thin notes, *'tsee tsee tsee'*, followed by several more shriller *'chree chree-chree...'* notes or *'tsee-tsee-tsee-itititititi'*. Alarm call a high-pitched repeated *'teee'* accompanied by wing snaps. **Status and biology:** Localised resident in moist grasslands, often around vleis and depressions; possibly some movement to lower elevations in winter. Unobtrusive outside the br. season, flying far when flushed. (Bleekkopklopkloppie)

PLAIN-BACKED CISTICOLAS   Five rather plain-coloured cisticolas with mainly unmarked backs and longish tails. Sexes alike; br. and non-br. plumages similar. During the br. season males sing from open perches; at other times they are silent and unobtrusive.

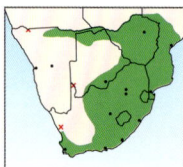

NEDDICKY   *Cisticola fulvicapilla*

B b b | | | | | b b B B
J F M A M J J A S O N D

11 cm; 9 g A widespread, plain-coloured cisticola with a rufous cap and fawn-coloured upperparts; underparts dark grey (Cape to KZN) or off-white (elsewhere in range). In small area of overlap with Short-winged Cisticola, Neddicky is distinguished by its rufous (not brown) cap, plain (not lightly mottled) upperparts, longer and uniformly plain tail (lacking pale tips), and different song. Non-br. ad. has rufous-washed upperparts. Juv. more rufous above, whiter below than ad. **Voice:** Br. male calls from top of bush, tree or rock, giving a monotonous, frog-like ticking note, 3–4x per second *'wiep-wiep-wiep-wiep...'* or *'tsiek-tsiek-tsiek-tsiek...'*, continuing for a minute or longer. Alarm call is a fast, ratchet-like *'tic-tic-tic-tic'*. **Status and biology:** Common resident in grassy understorey of woodland and savanna; also mountain fynbos and plantations, especially where there are rocks or dead trees. (Neddikie)

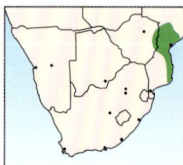

SHORT-WINGED CISTICOLA   *Cisticola brachypterus*

b b b | | | | | | b b
J F M A M J J A S O N D

11 cm; 9 g A plain-coloured cisticola of lowland woodlands in central Mozambique and adjacent Zimbabwe. Told from Neddicky by its brown (not rufous) crown, mottled back, and marginally shorter tail (37 mm vs 41 mm for br. males) that has a pale (not all-brown) tip. Non-br. ad. is more heavily streaked above, especially on crown; underparts more buffy. Juv. has underparts washed yellow. **Voice:** Male calls from treetop during summer, giving a series of soft, unevenly pitched notes, *'tsip tsip seu'*, *'see-see-sippi-ippi'*; also slurred warble, *'tsip tsiddle'*. **Status and biology:** Locally common resident in rank grass in open woodland and savanna. (Kortvlerktinktinkie)

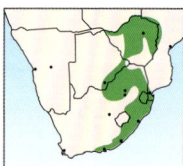

LAZY CISTICOLA   *Cisticola aberrans*

B b b b | | | | b b B B
J F M A M J J A S O N D

14 cm; 13–15 g A plain-backed cisticola typically found in rocky areas; long tail is often cocked like that of a prinia. Has a dull rufous cap and an indistinct, pale supercilium that extends well behind eye. Non-br. ad. has a longer tail; upperparts streaked, underparts greyish. Juv. more rufous than non-br. ad. **Voice:** Most characteristic call is a long, plaintive *'zweeee'* or *'tweeee'*, repeated at intervals; also a loud, clear *'twink twink twink'* or more plaintive *'tjwert tjwert tjwert'*, interspersed by short trills and stutters, sometimes followed by *'peet-peet-peet'*. **Status and biology:** Locally common resident on rocky hill slopes in wooded country. Often rather skulking, remaining in dense cover. (Luitinktinkie)

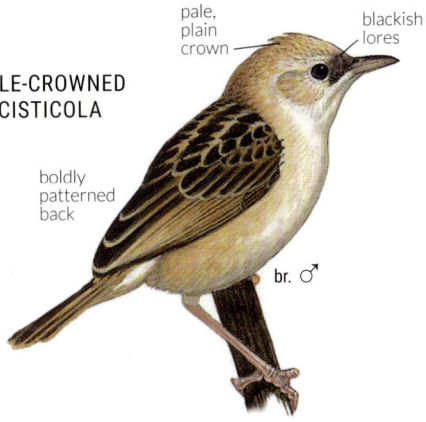

PALE-CROWNED CISTICOLA

pale, plain crown

blackish lores

boldly patterned back

non-br.

medium-length tail

br. ♂

NEDDICKY

rufous cap

plain back

uniform short tail

grey breast

southwestern races

plain back

northern races

buffy breast

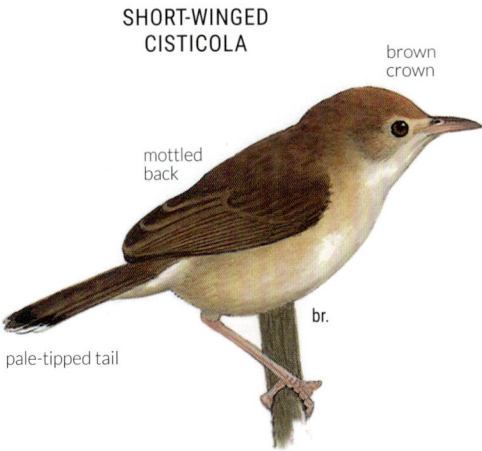

SHORT-WINGED CISTICOLA

brown crown

mottled back

br.

pale-tipped tail

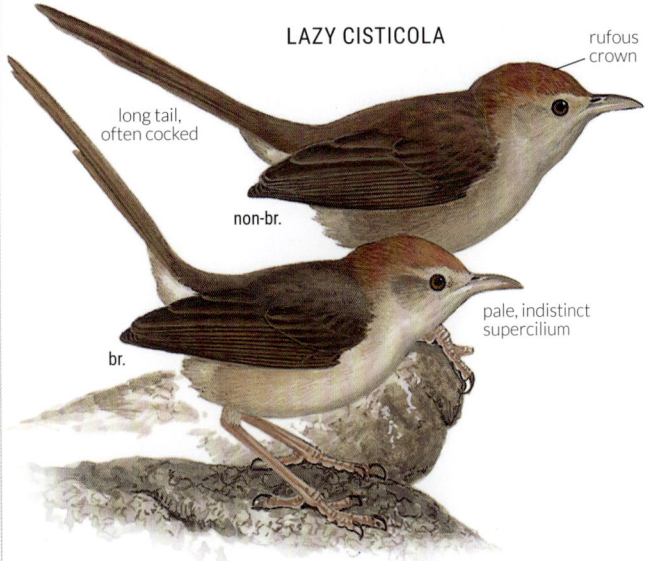

LAZY CISTICOLA

rufous crown

long tail, often cocked

non-br.

br.

pale, indistinct supercilium

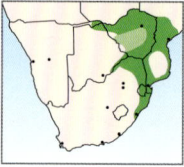

RED-FACED CISTICOLA *Cisticola erythrops*

B b b | b B B
J F M A M J J A S O N D

14 cm; 14 g A plain, greyish-backed cisticola with a diagnostic brown crown and rufous face, warm buff flanks and white-tipped tail year-round. In area of overlap with Singing Cisticola, distinguished by its grey-brown (not rufous) crown, pale (not black) lores, rufous (not pale grey) cheeks and uniformly greyish-brown wings (lacking rufous wing panel). Non-br. ad. has browner crown and richer-coloured underparts. Juv. is browner than ad. **Voice:** Male gives a series of loud, ringing notes descending in scale and building to a crescendo *'tee-tee-tee-TAY-TAY-TAY-TOY-TOY-TOY'*; female usually joins in with a lower-pitched *'zidit'*. May also simply repeat a single, loud *'weet'*, rapidly or slowly. Alarm call is a thin, high-pitched *'tseeeep'*. **Status and biology:** Common resident of rank, weedy growth and other low vegetation fringing reed beds, rivers and streams. (Rooiwangtinktinkie)

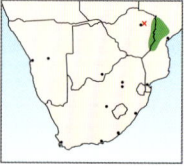

SINGING CISTICOLA *Cisticola cantans*

B b b b | b B B
J F M A M J J A S O N D

13 cm; 12 g A plain-backed cisticola of upland forest edge in e Zimbabwe and adjacent Mozambique. Told from Red-faced Cisticola by its rufous (not brown) crown, pale greyish (not rufous) cheeks and rufous (not brown) wing panel. Br. ad. has black lores, a short, white supercilium and a longish, grey tail with a dark subterminal bar and white tips. Non-br. ad. lacks black lores and has streaked back and wing coverts. Juv. has paler underparts, washed yellow. **Voice:** Loud, disyllabic *'jhu-weee'* or *'whee-cho'* and other variations at same pitch; alarm a plaintive chittering. **Status and biology:** Common resident in rank vegetation, bracken and tall grass with scattered bushes, usually on forest verges. (Singende Tinktinkie)

STREAK-BACKED CISTICOLAS
A mixed group of long-tailed cisticolas with streaked backs. The 2 species on this page are both savanna species that exhibit marked sexual dimorphism, with males 20–30% larger than females. The next 4 species are all associated with wetlands and marshy habitats, and have relatively bright rufous wings and caps, and black-striped backs. They mostly have non-overlapping ranges. The remaining 3 species are found away from wetlands, in arid woodland (Tinkling Cisticola), grassland (Wailing Cisticola) or heathland (Grey-backed Cisticola). Sexes similar in plumage; summer and winter plumages differ.

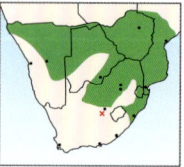

RATTLING CISTICOLA *Cisticola chiniana*

B b b b | b B B
J F M A M J J A S O N D

15 cm; male 18 g, female 14 g A common, widespread, savanna cisticola, and one of the most common birds of the bushveld. A rather large, but nondescript cisticola with a dull rufous cap and tail and indistinctly streaked back. Larger than Tinkling Cisticola; lacks its conspicuous white supercilium and rich rufous cap, tail and wing panels. Birds in semi-arid northwest (*C. c. frater*) are paler and less heavily streaked than eastern races (incl. *C. c. chiniana*). Female smaller than male. Non-br. ad. is more buffy above and below. Juv. has underparts washed pale yellow. **Voice:** Br. male calls loudly from tops of trees and shrubs, 2–4 high-pitched notes followed by a lower-pitched trill *'che che che CHRRRRRRRRR'* or *'chi chi chi CHIRRREEEEE'*, the trill (rattle) variable in length and pitch. Alarm call a shrill, persistent *'chree chree chree'*. **Status and biology:** Common resident in woodland, savanna and scrub. (Bosveldtinktinkie)

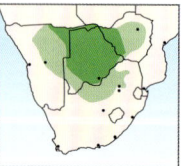

TINKLING CISTICOLA *Cisticola rufilatus*

B b b | b b B B
J F M A M J J A S O N D

14 cm; 11 g The only cisticola with a Kalahari-centred range, found in semi-arid woodland and scrub. Combination of bright rufous crown and ear coverts separated by a pale supercilium is diagnostic; its long tail is warm rufous. Non-br. ad. has a less heavily streaked back and is more rufous above; tail longer. Juv. is more buffy above. **Voice:** Br. male sings from tops of trees and shrubs, a loud *'weeet weeet weeet...'*; this may or may not be followed by high-pitched *'trrrrrrrrrrrrr'* (the tinkle). Alarm call is harsher *'chirrrrr'* or *'chit-it-it-rrrrrrrr'*. **Status and biology:** Uncommon to locally common resident in open broadleafed woodland or second-growth scrub and savanna in semi-arid areas. Shy and easily overlooked; often runs mouse-like along the ground. (Rooitinktinkie)

RED-FACED CISTICOLA

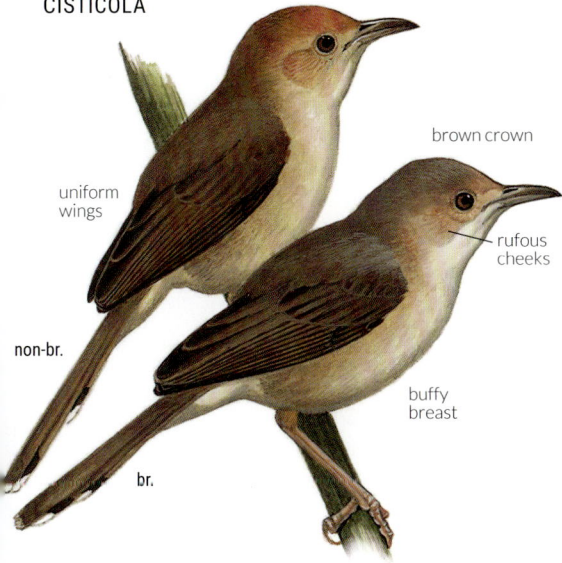

brown crown

uniform wings

rufous cheeks

non-br.

buffy breast

br.

SINGING CISTICOLA

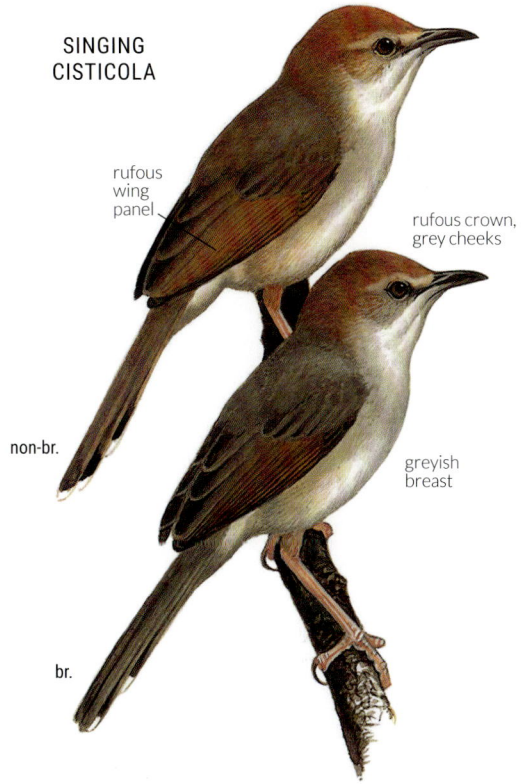

rufous wing panel

rufous crown, grey cheeks

non-br.

greyish breast

br.

RATTLING CISTICOLA

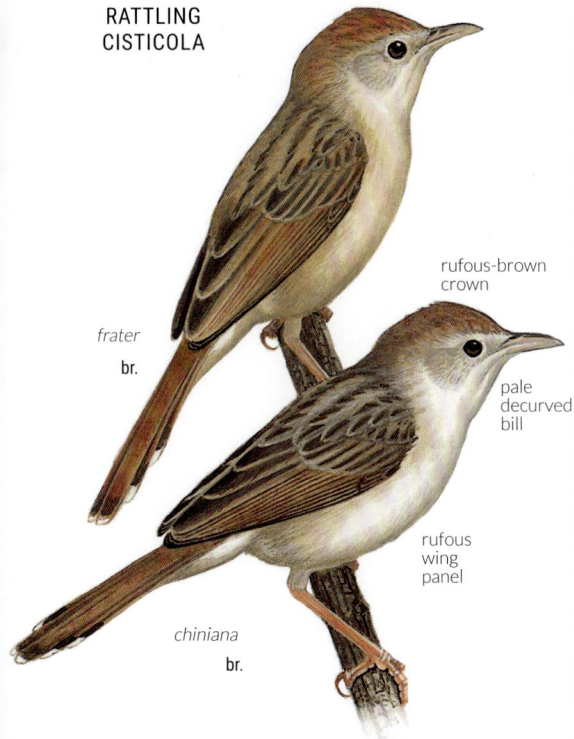

rufous-brown crown

frater
br.

pale decurved bill

chiniana
br.

rufous wing panel

TINKLING CISTICOLA

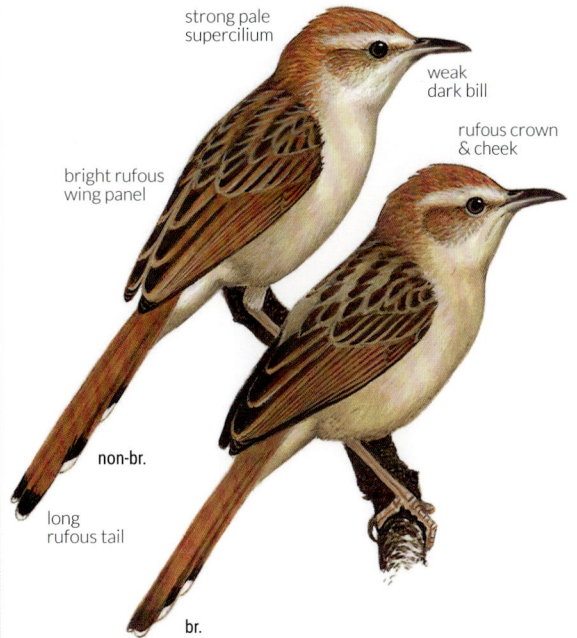

strong pale supercilium

weak dark bill

rufous crown & cheek

bright rufous wing panel

non-br.

long rufous tail

br.

393

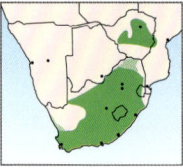

LEVAILLANT'S CISTICOLA *Cisticola tinniens*

13 cm; 11 g A common, widespread, wetland cisticola in temperate/upland areas; range overlaps only narrowly with Rufous-winged Cisticola and not at all with Luapula and Chirping Cisticolas. Distinguished from Rufous-winged Cisticola by its rufous (not grey-brown) tail, streaked (not plain) rump and diagnostic song. Non-br. ad. has buff-brown (not greyish) margins to the mantle feathers and a streaked crown. Juv. is duller above, with less prominent streaking; breast washed pale lemon. **Voice:** Male song, given from the top of a plant stem or in short display flight, is a cheerful, warbling *'tsi tsiororee'*, sometimes followed by a jumble of disyllabic notes; alarm a harsh *'tee tee tee'*. **Status and biology:** Common resident in marshy vegetation and areas of long, rank grass in open grassland. (Vleitinktinkie)

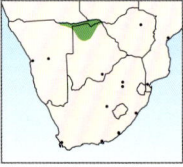

CHIRPING CISTICOLA *Cisticola pipiens*

14 cm; 13 g The least strikingly marked of the wetland cisticolas; overlaps widely in the Okavango and elsewhere with Luapula Cisticola (but not with Rufous-winged or Levaillant's Cisticolas). Br. male distinguished by its song, duller rufous cap, rufous ear coverts, pale rufous (not rich rufous) wing panel, grey-and-black-striped (not buff-and-black) back, and more buffy underparts. Non-br. male is slightly paler above, with buffy margins to mantle and back feathers. Female is paler, less buffy below in fresh plumage. Juv. has crown streaked dark brown; base of tail reddish. **Voice:** Song loud: 2 or 3 sharp notes followed by a dry, buzzy trill *'chit chit-it trrrrrrrrrrrrr'*; also plaintive *'chwer-chwer-chwer'* and sharp *'chit chit'* calls. Sometimes snaps its wings while calling. **Status and biology:** Locally common resident in reed beds and papyrus swamps. Favours taller vegetation and wetter areas than Luapula Cisticola. (Piepende Tinktinkie)

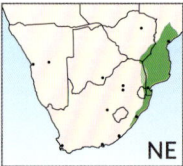

RUFOUS-WINGED CISTICOLA *Cisticola galactotes*

13 cm; 13 g Restricted to marshy areas along the coastal plain and lowlands from s KZN to Mozambique; overlaps narrowly with Levaillant's Cisticola. Br. male distinguished from Levaillant's Cisticola by its grey (not brown) tail and plain (not streaked) rump. Br. ads of some subspecies show dark lores. Non-br. ad. has more rufous streaking above and is more buffy below, with a browner tail. Juv. has crown streaked dark brown, underparts washed yellow. **Voice:** Various piping and rasping notes; an accelerating *'tschick tshcick tschick'*, higher pitched towards the end. Also a repeated, ascending *'phweeep'*; a buzzy *'tzeet'* or *'trrrt trrrt'*. **Status and biology:** Near-endemic. Common resident in rank grass, sedges and reed beds in marshy ground. (Swartrugtinktinkie)

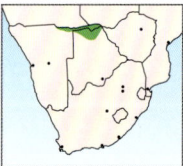

LUAPULA CISTICOLA *Cisticola luapula*

14 cm; 13 g Recently split from the very similar Rufous-winged Cisticola; ranges do not overlap with it or Levaillant's Cisticola, but overlaps widely with Chirping Cisticola. Distinguished from this species by its richer rufous crown and wing panel, more boldly marked back and greyer tail; underparts paler. Non-br. ad. is more rufous above and more buffy below. Juv. has crown streaked dark brown, underparts washed yellow. **Voice:** Loud 2-note display song, *'tid-ick'*; also *'tic tic'*, *'we-we-we'* and *'zrrtttt'* calls, sometimes given in display flight. **Status and biology:** Locally common resident in swamps and marshes, preferring *Phragmites* reeds to the dense papyrus favoured by Chirping Cisticola. (Luapulatinktinkie)

LEVAILLANT'S CISTICOLA

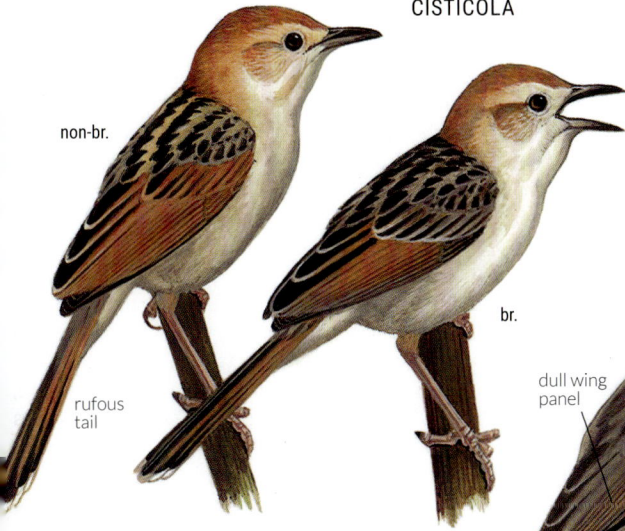

non-br.

rufous
tail

br.

CHIRPING CISTICOLA

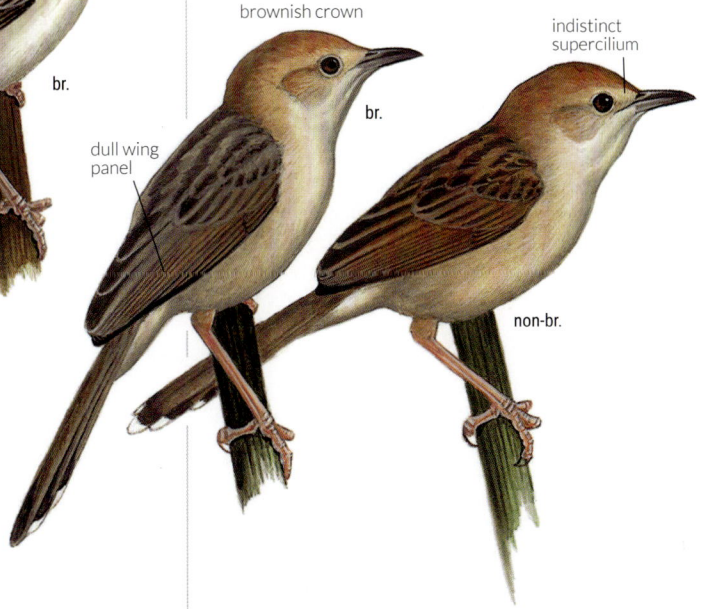

brownish crown

dull wing
panel

indistinct
supercilium

br.

non-br.

RUFOUS-WINGED CISTICOLA

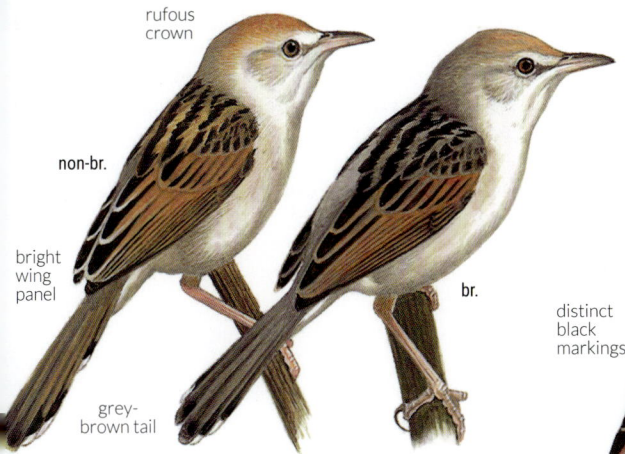

rufous
crown

non-br.

bright
wing
panel

grey-
brown tail

br.

LUAPULA CISTICOLA

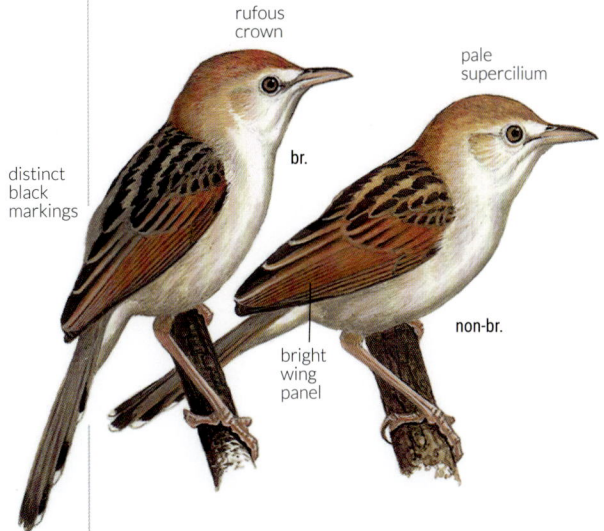

rufous
crown

pale
supercilium

distinct
black
markings

br.

bright
wing
panel

non-br.

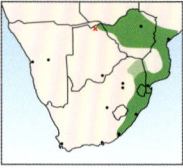

CROAKING CISTICOLA   *Cisticola natalensis*

Bbb ‖ bBb
J F M A M J J A S O N D

15 cm; male 25 g, female 18 g The largest cisticola in the region, confined to moist grasslands. More robustly built than other cisticolas, with a heavier bill that is blackish in br. male and yellowish-pink in female and non-br. male. Female much smaller than male; could be confused with Rattling Cisticola, but proportionately shorter tailed, is heavier billed and cap paler and clearly streaked. Juv. has underparts washed yellow. **Voice:** Br. male very vocal and conspicuous, giving a loud, croaking *'q-RRRRRRP'* repeated at 2–3-second intervals when perched, or more frequently in display flight, when croaks are higher pitched. Alarm a sharp, frog-like *'pRRRp'*. **Status and biology:** Locally common resident in grassland and grassy savanna. (Groottinktinkie)

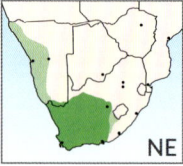

GREY-BACKED CISTICOLA   *Cisticola subruficapilla*

bb ‖ bBBBBb
J F M A M J J A S O N D

12 cm; 10–12 g A Karoo-fynbos shrubland cisticola with greyish underparts, a greyish, streaked back and contrasting rufous crown, tail and wing panel. Northern races are browner above and more buffy below; most like Wailing Cisticola, but smaller and finer billed, with colder, greyish-buff underparts. Breast lightly streaked, especially in the southwest. Sexes alike, and little seasonal plumage variation. Juv. is duller, with yellow-washed underparts. **Voice:** Muffled *'tr-r-rrrrrt'* and loud, plaintive *'hu-weeeee'*, slower and deeper than Wailing Cisticola, given from perch or in air; also harsher *'chee chee'* call. **Status and biology:** Near-endemic. Common resident in lowland fynbos, Karoo scrub and arid, grassy hillsides; typically in drier habitats than Wailing Cisticola. Disjunct nw Namibian population with paler and warmer brown underparts is less common; may be a distinct species. (Grysrugtinktinkie)

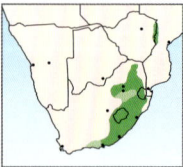

WAILING CISTICOLA   *Cisticola lais*

Bbb ‖ bbBB
J F M A M J J A S O N D

14 cm; male 15 g, female 12 g A hill slope and grassland cisticola with a streaked, rufous head and rufous-brown tail and wing panels contrasting with greyer back. Slightly larger than Grey-backed Cisticola, with a heavier bill, heavier back streaking, warmer buff (not cold grey-buff) belly and brighter rufous head. E Cape *C. l. maculatus* has a variably streaked breast. Sexes alike. Non-br. ad. has more rufous and heavily streaked upperparts. Juv. is duller, washed pale yellow on face and throat. **Voice:** Rattled *'t-trrrrrreee'*, often followed by 3–10 high-pitched piping notes, *'t-pee t-pee t-pee'*, or a slightly plaintive, drawn-out *'phweeeep'*; more rapid than Grey-backed Cisticola (4 vs 3 peeps/second). **Status and biology:** Common resident in rank grass and bracken on rocky hill slopes; usually in moister areas than Grey-backed Cisticola. (Huiltinktinkie)

PRINIAS AND PRINIA-LIKE WARBLERS   Long-tailed cisticolid warblers of scrubby undergrowth in woodland, savanna and semi-arid shrubland. Sexes alike, or females slightly duller; some species have distinct non-br. plumages. Sedentary; occur singly or in pairs.

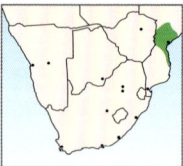

RED-WINGED PRINIA (WARBLER)   *Prinia erythropterus*

b ‖ bb
J F M A M J J A S O N D

14 cm; 12 g A long-tailed, prinia-like warbler with rich rufous wings. Restricted to lowland woodland in central Mozambique and lowland e Zimbabwe. The only possible confusion within this range is Tawny-flanked Prinia (p. 398); distinguished from it by its uniformly greyish head and ear coverts and lack of a white supercilium. Sexes alike. Non-br. birds have brown (not grey) crown. Juv. is paler. **Voice:** Monotonous, rather prinia-like *'pseep-pseep-pseep'*; also a chattering alarm call. **Status and biology:** Localised resident in rank grass in miombo woodland and forest edge. (Rooivlerksanger)

CROAKING CISTICOLA

heavy bill

♂

dull rufous wing panel

pale brown crown

on-br.

br. ♂

shorter tail

br. ♀

browner back

br. n Namibia

paler underparts

GREY-BACKED CISTICOLA

br. South Africa

lightly streaked greyish underparts

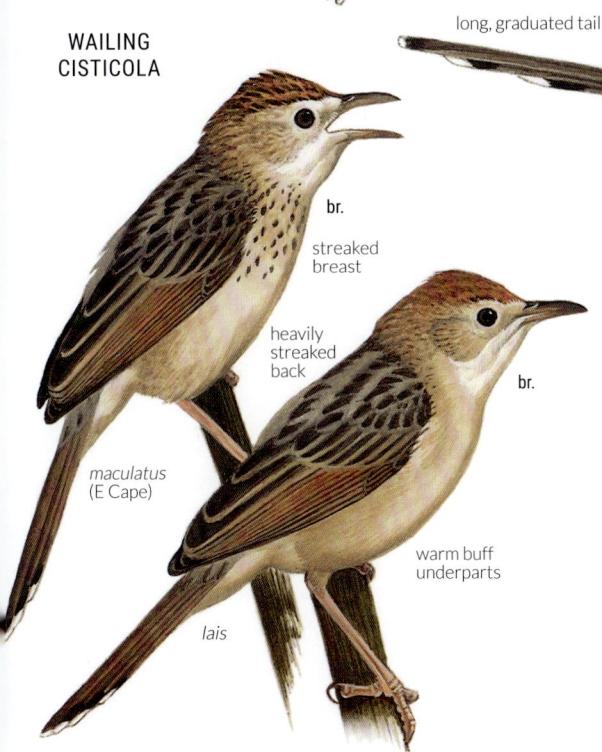

WAILING CISTICOLA

br.

streaked breast

heavily streaked back

br.

maculatus (E Cape)

warm buff underparts

lais

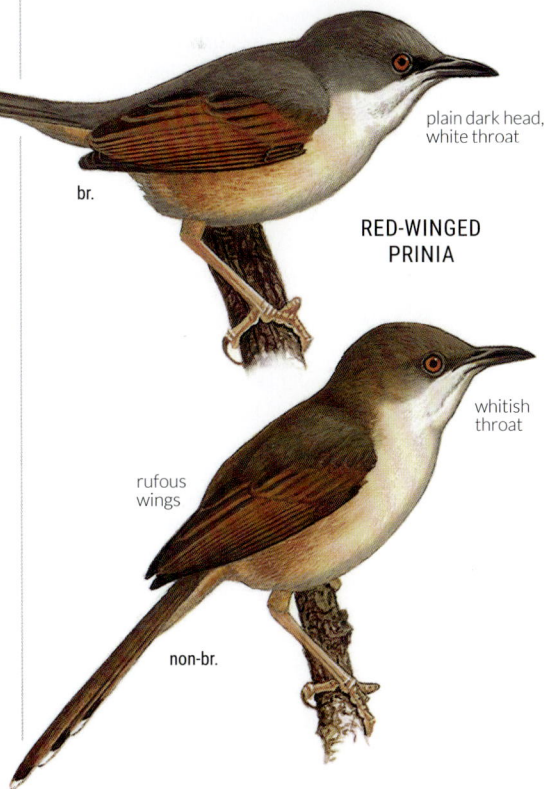

long, graduated tail

plain dark head, white throat

br.

RED-WINGED PRINIA

whitish throat

rufous wings

non-br.

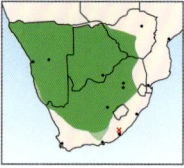

BLACK-CHESTED PRINIA *Prinia flavicans*

B b b | | | | | b b B B
J F M A M J J A S O N D

14 cm; 9 g The only prinia with a broad, black breast band (narrower in female). In non-br. plumage, when breast band is much reduced or absent, could be mistaken for Tawny-flanked Prinia, but lacks russet edges to its wing and tail feathers. Underpart colour varies across range from a pale yellow wash in southwest to rich yellow in northeast. Occasionally hybridises with Karoo Prinia. Juv. resembles non-br. ad., but is yellower below. **Voice:** Loud, repetitive *'prrt-prrt-prrt...'* or *'dzrrt-dzrrt-dzrrt'*; ratchet-like *'zrrrrrrt'* alarm call. **Status and biology:** Common resident in arid savanna, thornveld and scrub; also orchards, plantations and gardens. (Swartbandlangstertjie)

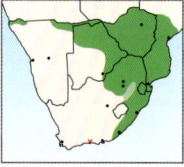

TAWNY-FLANKED PRINIA *Prinia subflava*

B B b b | | b b b b B B
J F M A M J J A S O N D

13 cm; 9 g A nondescript prinia with unstreaked, whitish underparts and brown upperparts; long, broad supercilium and russet flight feathers are diagnostic. Br. male has black bill and mouth; bill brown in non-br. male and female. Smaller than localised Red-winged Prinia (p. 396), with a more strongly marked head and less prominent rufous wing panel. Told from non-br. Black-chested Prinia by its buff flanks and russet edges to wing and tail feathers. Juv. is washed yellow below. **Voice:** Rapidly repeated *'przzt-przzt-przzt...'* or *'tsip-tsip-tsip...'*; mewing *'zbee'* or *'cheeee'* alarm call. **Status and biology:** Common, widespread resident in broadleafed woodlands and especially along watercourses in thick, rank vegetation. A vocal and conspicuous bird, solitary or in pairs. (Bruinsylangstertjie)

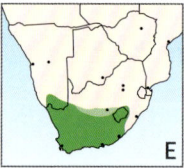

KAROO PRINIA *Prinia maculosa*

b b b | | | b b B B B b
J F M A M J J A S O N D

14 cm; 9 g Sometimes lumped with Drakensberg Prinia, but is more heavily streaked below, with streaking usually extending onto its throat and flanks (throat and flanks unstreaked in Drakensberg Prinia); ranges overlap slightly in Drakensberg. Also more heavily streaked than Namaqua Warbler, and is greyer brown above with a broader tail, with buff-tipped (not plain) tail feathers. Female is less heavily spotted. Occasionally hybridises with Black-chested Prinia. Juv. is washed yellow below. **Voice:** Wide range of scolding calls, including buzzy *'dzeeep dzeeep dzeeep...'*, harder *'chip-chip-chip...'* and faster *'kli-kli-kli...'*. **Status and biology:** Endemic. Common resident in fynbos, thickets and taller Karoo scrub. Solitary or in pairs which are territorial year-round. (Karoolangstertjie)

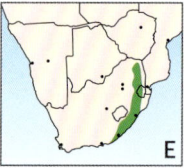

DRAKENSBERG PRINIA *Prinia hypoxantha*

B b | | | | | b b B
J F M A M J J A S O N D

14 cm; 10 g Sometimes lumped with Karoo Prinia, but warmer brown above and with yellow-washed underparts; streaking usually confined to its breast (not onto throat and only occasionally onto flanks); ranges overlap slightly in Drakensberg. Juv. is paler and less streaked below; upperparts redder brown. **Voice:** Similar to Karoo Prinia. **Status and biology:** Endemic. Common resident of forest edges, wooded gullies and bracken tangles, normally at higher elevations than Karoo Prinia. Solitary or in pairs. (Drakensberglangstertjie)

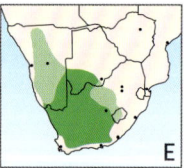

RUFOUS-EARED WARBLER *Malcorus pectoralis*

b b b | | | b b b b b b
J F M A M J J A S O N D

15 cm; 10 g An attractive, prinia-like warbler of arid scrub. Bright rufous face, white throat, black breast band and long, slender tail often held cocked are diagnostic. Breast band reduced or absent in winter. Female has narrower breast band than male. Juv. lacks breast band and has a brown face. Isolated Etosha subspecies is considerably paler than southern populations. **Voice:** Scolding, high-pitched *'tzee tzee tzee tzee...'* repeated 6–20 times; plaintive *'peeee'* alarm call. **Status and biology:** Endemic. Common resident and local nomad in arid scrub. Restless, keeping to low shrubs, briefly up on bushes, then on the ground, moving rodent-like between bushes. Solitary or in pairs. (Rooioorlangstertjie)

NAMAQUA WARBLER *Phragmacia substriata*

b b b | | | b B B B
J F M A M J J A S O N D

14 cm; 11 g A slender, long-tailed warbler, recalling a delicate Karoo Prinia, but with a more russet-coloured back, narrow streaking confined to the breast, plain face and thinner, more graduated tail that lacks buff tips. Female has less streaking. Juv. is duller, and streaked on the back. **Voice:** Both sexes give a high-pitched, rattling *'trip-trip-trrrrrrrrrrrr'* song; also *'chit'* contact call and *'chewy chewy chewy'* alarm call. **Status and biology:** Endemic. Common resident, mainly confined to vegetation along drainage lines in arid Karoo, especially *Phragmites* reed beds, acacia woodland and tangled thickets. Forages on or close to the ground. Pairs defend territories year-round. (Namakwalangstertjie)

BLACK-CHESTED PRINIA

non-br.

br.

no russet on wing

juv.

yellow wash below

russet wings

buff vent

TAWNY-FLANKED PRINIA

KAROO PRINIA

ad.

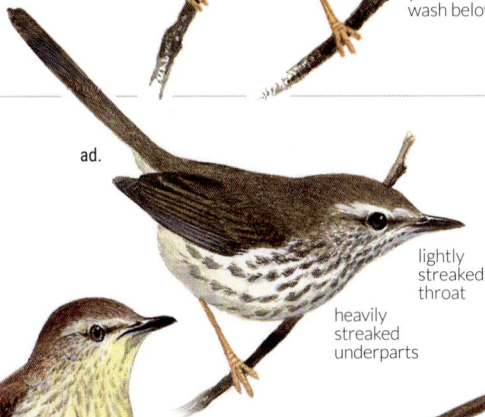

lightly streaked throat

heavily streaked underparts

juv.

pale plain throat

lightly streaked yellow-washed breast

ad.

DRAKENSBERG PRINIA

RUFOUS-EARED WARBLER

juv.

NAMAQUA WARBLER

rufous cheeks

ad.

variably grey underparts

russet back

pale face

very faintly streaked breast

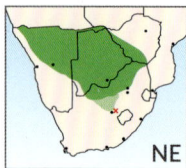

BARRED WREN-WARBLER *Calamonastes fasciolatus*

14 cm; 13 g Arid savanna replacement of Stierling's Wren-Warbler; br. male differs by its brown breast markings, which largely obscure the darker barring. Females and non-br. male**s** have buff underparts with dusky bars, less well defined than those of Stierling's Wren-Warbler and do not extend to vent; eyes and legs usually darker. Juv. is more rufous above, with yellowish wash on breast. **Voice:** Soft, tremulous *'trrrreee trrrreee trrrreee'* and rather mechanical *'pleelip-pleelip'*; also chattering alarm call. **Status and biology:** Near-endemic. Common resident in dry acacia savanna. (Gebande Sanger)

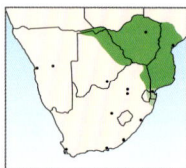

STIERLING'S WREN-WARBLER *Calamonastes stierlingi*

14 cm; 13 g A small, skulking, wren-like warbler with whitish underparts, barred from throat to vent. Upperparts are uniformly rufous-brown, with white spots on tips of upperwing coverts visible at close range. Resembles non-br. Barred Wren-Warbler, but has barring extending to vent (not with plain, pale buff belly and undertail coverts); legs and eyes usually paler. Br. and non-br. plumages alike (whereas male Barred Wren-Warbler has mottled brown throat and chest in br. plumage). Juv. is washed yellow below and more rufous above, with dark brown eyes. **Voice:** Song is a fast, repeated *'tlip-tlip-tlip'* (bicycle pump); also a soft *'preep'* and sharp *'tsik'* alarm call. **Status and biology:** Fairly common resident in broadleafed woodland, especially miombo. Solitary or in pairs; keeps to lower canopy and understorey. (Stierlingsanger)

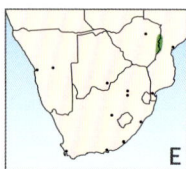

ROBERTS'S WARBLER *Oreophilais robertsi*

14 cm; 10 g A greyish, prinia-like warbler confined to e Zimbabwe highlands and adjacent Mozambique. Within its small range, most similar to Chirinda Apalis (p. 402), but has off-white (not grey) underparts, dull brownish wings, red-rimmed yellow eyes, and occurs in forest edge (not canopy). Told from Tawny-flanked Prinia (p. 398) by its uniformly grey head and cheeks (not patterned with a dark eye stripe and white supercilium); lacks rufous edging to flight feathers. Non-br. birds are paler below. Juv. has dark eye. **Voice:** Raucous chorus of high-pitched, babbler-like chattering. **Status and biology:** Endemic. Common resident in scrub and bracken undergrowth on montane forest edge. In pairs or small groups; noisy birds, all members of the group often calling simultaneously. (Woudlangstertjie)

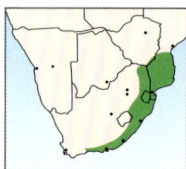

GREEN-BACKED CAMAROPTERA *Camaroptera brachyura*

12 cm; 10 g A small, wren-like warbler, with olive-green upperparts, greyish-white underparts and a greyish forehead and face; often cocks its short tail. Often lumped with Grey-backed Camaroptera, but mantle, back and tail are olive-green (not grey). Lacks distinct br. and non-br. plumages. Intergrades with Grey-backed Camaroptera in Mozambique. Juv. is washed yellow below. **Voice:** Male song is a loud, snapping *'bidup-bidup-bidup'*. Nasal *'neeehhh'* alarm call accounts for its alternative common name of Bleating Warbler. **Status and biology:** Common resident in dense undergrowth and tangles in montane, riparian and lowland forest. Males may call from high in the forest canopy, but spend most of their time within a few metres of the ground. (Groenrugkwêkwêvoël)

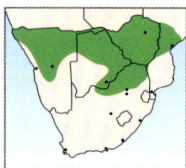

GREY-BACKED CAMAROPTERA *Camaroptera brevicaudata*

12 cm; 10 g A small, wren-like warbler very similar to Green-backed Camaroptera; wings olive-green, but mantle, rump and tail grey. Intergrades with Green-backed Camaroptera in Mozambique. Juv. is washed yellow below. **Voice:** Similar to Green-backed Camaroptera. **Status and biology:** Common resident in understorey of deciduous woodland, especially in tangled thickets. In areas of overlap with Green-backed Camaroptera, typically occurs in drier habitats. (Grysrugkwêkwêvoël)

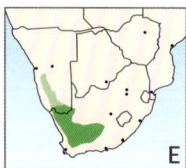

CINNAMON-BREASTED WARBLER *Euryptila subcinnamomea*

12 cm; 10–13 g A small, dark brown, rock-loving warbler of rocky outcrops and steep valleys in arid regions. At close range the rufous breast band, flanks, rump and forehead are diagnostic. Juv. is more uniform rufous-brown. **Voice:** Song is loud, piercing series of 1–3 high-pitched whistles: ascending *'peeeee'* notes are often followed by several short, rapid whistles. **Status and biology:** Endemic. Uncommon, localised resident in scrub-covered, rocky hillsides and gorges. Secretive; creeps among jumbled boulders, often disappearing into crevices; usually detected by its song. (Kaneelborssanger)

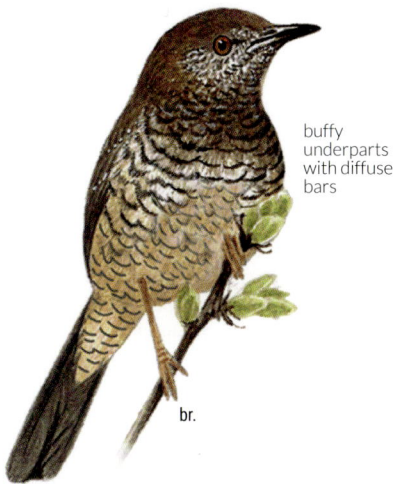

buffy underparts with diffuse bars

br.

BARRED WREN-WARBLER

plain vent

non-br.

STIERLING'S WREN-WARBLER

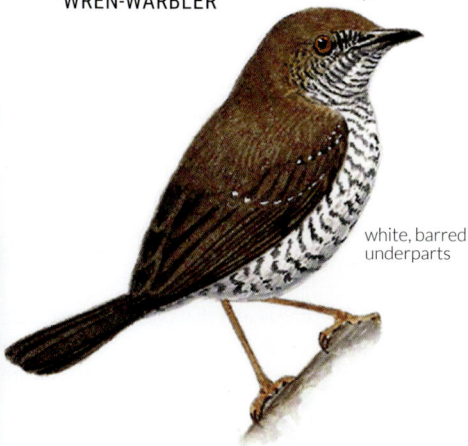

slightly paler eye

white, barred underparts

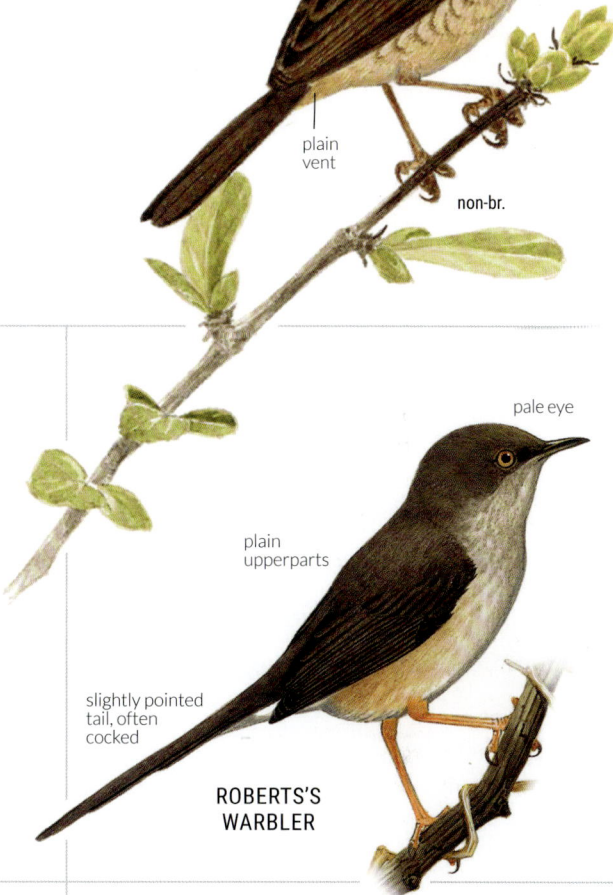

pale eye

plain upperparts

slightly pointed tail, often cocked

ROBERTS'S WARBLER

green back & nape

reen ail

GREEN-BACKED CAMAROPTERA

grey back

grey tail

GREY-BACKED CAMAROPTERA

CINNAMON-BREASTED WARBLER

black tail

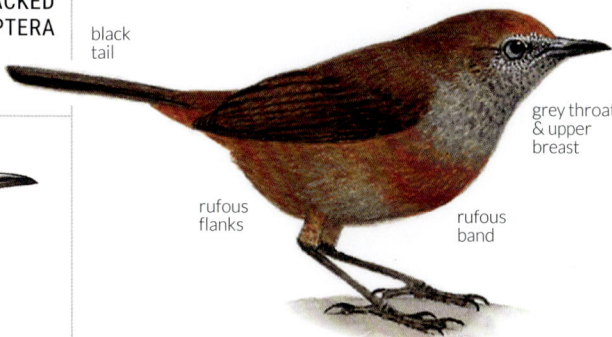

grey throat & upper breast

rufous flanks

rufous band

401

APALISES
Long-tailed cisticolid warblers of forest, woodland or thicket. Active birds that forage in tree canopies, often joining mixed-species bird parties; call frequently, male and female duetting in many species. Sexes similar in most species, but males brighter or more strongly marked in some species. Juvs duller. All are resident, occurring singly, in pairs or small family groups.

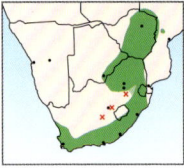

BAR-THROATED APALIS *Apalis thoracica*
13 cm; 11 g A common, widespread apalis with a narrow, black breast band (broader in male). Coloration variable across its range; upperparts vary from grey to green, underparts from white to greyish and yellow. Pale eyes and white outer tail distinguish it from Rudd's Apalis. Juv. collar sometimes incomplete. **Voice:** Male gives a harsh *'prrup-prrup-prrup'* or *'tilllup-tilllup-tilllup'*, female responds with a high-pitched *'ti-ti-ti-ti'*. Snaps bill and wings when agitated. **Status and biology:** Common resident of forest, dense woodland and coastal thicket; confined to montane forest in the north. (Bandkeelkleinjantjie)

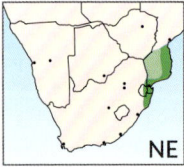

RUDD'S APALIS *Apalis ruddi*
13 cm; 10 g Superficially similar to Bar-throated Apalis, but dark eye, and green tail lacking white in outer feathers distinctive. Dark green back contrasts with grey crown. Female has narrower breast band. Juv. breast band indistinct or lacking. **Voice:** Male calls a fast *'trrt trrt trrt...'*, answered by female with a slower *'prp prp prp...'*. **Status and biology:** Near-endemic. Locally common, resident of low-lying thornveld and sand forest thickets along eastern coastal plain. (Ruddkleinjantjie)

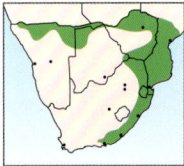

YELLOW-BREASTED APALIS *Apalis flavida*
12 cm; 8 g The only apalis with a bright yellow breast and no breast band. Males of northern races have a small, black bar beneath the centre of the yellow breast. Crown grey, contrasting with olive-green back. Female duller; lacks black breast bar. Juv. like female, with a greenish crown and a paler yellow breast. **Voice:** Male song is a fast, buzzy *'chirip chirip chirip...'* or *'kri-krrik kri-krrik, kri-krrik...'*; female responds with a series of strident shrieks, *'jee jee jee'*. **Status and biology:** Common resident of moist woodland and riparian forest, but absent from montane forest. Often forages in tree canopies alongside white-eyes, hyliotas and other small warblers. (Geelborskleinjantjie)

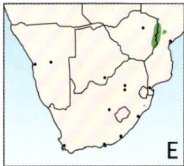

CHIRINDA APALIS *Apalis chirindensis*
13 cm; 9 g A rather plain, grey apalis with a white outer tail; confined to e Zimbabwe and adjacent Mozambique. In poor light could be confused with Black-headed Apalis, but has grey (not blackish) upperparts and orange-brown (not yellow) eyes. Very similar to Roberts's Warbler (p. 400) but has grey (not off-white) underparts, grey (not brownish) wings, orange-brown (not red-rimmed yellow) eyes and occurs in forest canopy, rather than forest edge. Creeps through the canopy; does not fan or flirt its tail like White-tailed Crested Flycatcher (p. 410). Juv. is washed olive above, yellow below, with buffy flanks; eyes grey. **Voice:** Male song is a strident *'swik swik swik...'*, female responds with a lower-pitched *'chip chip chip...'*; also a quavering alarm trill. **Status and biology:** Endemic. Locally common resident of montane forest; in pairs or small family groups. Forages at all levels in the forest; often associated with mixed-species bird parties. (Gryskleinjantjie)

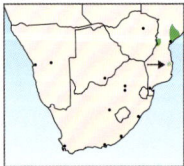

BLACK-HEADED APALIS *Apalis melanocephala*
13 cm; 9 g A two-toned apalis with blackish upperparts and creamy-white underparts. Closely related to Chirinda Apalis, but has whitish (not grey) underparts that show more contrast with the upperparts, and yellow (not orange-brown) eyes; ranges don't overlap. Juv. paler olive-brown above, with yellowish wash on underparts. **Voice:** Both sexes give a piercing, repeated *'wiii-tiiit-wiii-tiiit'* or *'sweep sweep sweep...'*; alarm call a soft *'puit'*. **Status and biology:** Locally common resident in the canopy of lowland and riverine forests, especially where there are tangled creepers. Usually high in canopy; often associated with mixed-species bird parties. (Swartkopkleinjantjie)

dark face, pale eye

RUDD'S
APALIS

dark eye

white outer tail

no white on outer tail

♂

BAR-THROATED
APALIS

dark face, pale eye

pale race

♀

very long tail

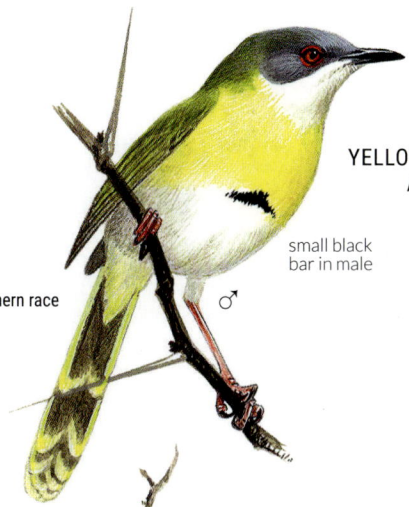

grey ear coverts

red eye

YELLOW-BREASTED
APALIS

broad yellow breast band

small black bar in male

northern race

♂

♀

CHIRINDA
APALIS

BLACK-HEADED
APALIS

pale eye

wholly black upperparts

uniformly grey overall

403

HYLIOTAS

HYLIOTAS A small family of warbler-like birds, the Hyliotidae, confined to Africa. Their black-and-white upperparts could be confused with *Ficedula* flycatchers (p. 408) or perhaps batises. They are insectivores, gleaning from twigs and foliage in the canopies of tall woodland or forest. Occur singly or in pairs; commonly join mixed-species bird parties. Males brighter than females; juvs barred buffy above.

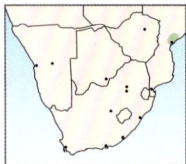

YELLOW-BELLIED HYLIOTA *Hyliota flavigaster*

J F M A M J J A S O N D

12 cm; 12 g Very similar to Southern Hyliota. Male distinguished from male Southern Hyliota by white in the wings extending onto the folded secondaries and tertials (not confined to upperwing coverts); upperparts are glossy blue-black (not velvety black) and underparts are richer yellow. Female is grey-brown above (not warm brown as female Southern Hyliota). Juv. has buff margins to upperpart feathers; underparts creamy. **Voice:** A soft, rather querulous *'tueet tueet'*, similar to Southern Hyliota, but deeper and slower. **Status and biology:** Perhaps a rare resident in tall miombo woodland in lowland Mozambique, however, no recently confirmed sightings. May occur side-by-side with Southern Hyliota. (Geelborshyliota)

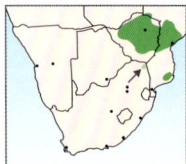

SOUTHERN HYLIOTA *Hyliota australis*

J F M A M J J A S O N D

12 cm; 11 g Very similar to Yellow-bellied Hyliota. Male distinguished from male Yellow-bellied Hyliota by white shoulder patch restricted to coverts (not extending onto secondaries and tertials); also sooty-black above (not glossy blue-black) and underparts paler yellow. Female is brown above with a creamy breast; differs from female Yellow-bellied Hyliota by having warmer brown upperparts. Juv. has buff margins to upperpart feathers; underparts washed buff. **Voice:** A soft, high-pitched, rather querulous *'tueet tueet'*; also a chittering call. **Status and biology:** Locally common resident in miombo and other broadleafed woodland. May occur side-by-side with Yellow-bellied Hyliota. (Mashonahyliota)

WHITE-EYES

WHITE-EYES Small, greenish-yellow, warbler-like birds allied to babblers with prominent white eye-rings. Common and familiar garden birds wherever they occur. They are all rather similar, and taxonomy probably is not fully resolved; southern and northern forms of African Yellow White-eye recently split. Usually in flocks of 5–20 birds, except when br.; individuals in groups maintain contact with frequent twittering calls. They have a varied diet that includes small insects, especially aphids, nectar and fruit. Sexes alike; juvs duller, with smaller eye-rings. *Zosterops* is the largest genus in the family Zosteropidae, which is centred in Asia.

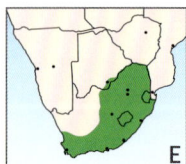

CAPE WHITE-EYE *Zosterops virens*

J F M A M J J A S O N D

11 cm; 11 g A variable white-eye; lower breast and belly vary from grey (*Z. v. capensis* in southwest) to green (*Z. v. virens* in north and east), contrasting with yellow throat and vent. Black lores split white eye-ring, unlike Southern Yellow White-eye, which has complete white eye-rings. Darker green above than Southern Yellow White-eye and has at least some green wash on flanks. Also darker above than Orange River White-eye; flanks green or grey (not buffy-peach). **Voice:** Loud, warbling song given mostly at dawn; also constant chittering contact calls. **Status and biology:** Endemic. Common resident in woodland, forests, thickets, plantations and gardens. (Kaapse Glasogie)

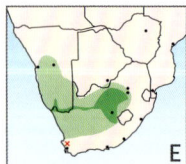

ORANGE RIVER WHITE-EYE *Zosterops pallidus*

J F M A M J J A S O N D

11 cm; 9 g Once lumped with Cape White-eye, but is paler above with distinctive, peach-coloured flanks and a more prominent yellow supra-loral stripe. Hybridises with both grey- and green-flanked forms of Cape White-eye in narrow overlap zone in Free State. **Voice:** Higher pitched and more trilling than Cape White-eye (hence alternative name of Warbling White-eye). **Status and biology:** Endemic. Common resident and local nomad in riparian woodland and thickets in the Karoo and semi-arid savanna, extending down wooded watercourses to bushed areas of the Namibian coast; regularly wanders outside normal range in W Cape. (Gariepglasogie)

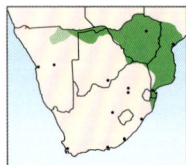

SOUTHERN YELLOW WHITE-EYE *Zosterops anderssoni*

J F M A M J J A S O N D

11 cm; 10 g Bright yellow; lacks any green tones in underparts and has yellower upperparts than Cape White-eye; ranges largely discrete. Black lores do not extend into the eye-ring, unlike Cape White-eye which has a split eye-ring. Juv. duller, with greener upperparts. **Voice:** Loud, melodious, whistled *'tweee-tuuu-twee-twee'*. **Status and biology:** Common resident in woodland, scrub, forest and gardens. Recently split from northern races of Yellow White-eye. (Geelglasogie)

glossy
upperparts

extended
wingbar

♂

**YELLOW-BELLIED
HYLIOTA**

♀

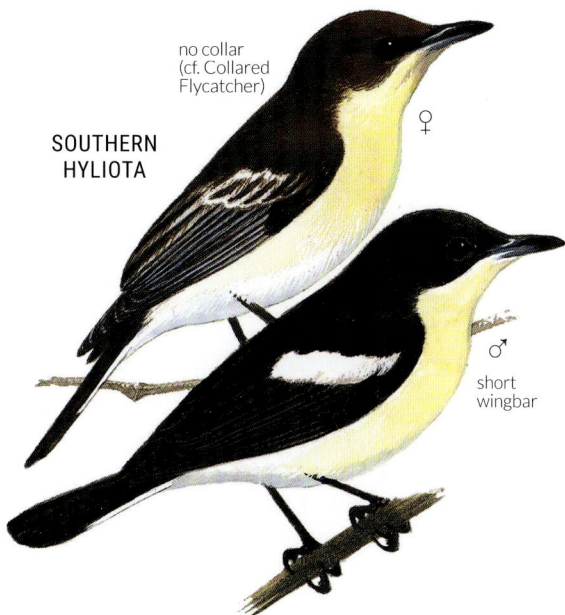

no collar
(cf. Collared
Flycatcher)

**SOUTHERN
HYLIOTA**

♀

♂

short
wingbar

**CAPE
WHITE-EYE**

contrast
between
upper- &
underparts

yellow form

grey belly   *grey form*

broken white
eye-ring

olive form

yellow frons

broken
white
eye-ring

peachy-
buff flanks

**ORANGE RIVER
WHITE-EYE**

solid
white
eye-ring

**SOUTHERN YELLOW
WHITE-EYE**

little difference
between upper-
& underparts

LARGE FLYCATCHERS
Mainly insectivorous passerines, either hawking prey in the air or on the ground. The species on this page are medium-sized birds, rather chat-like in demeanour, that perch erect on low branches and scan the ground below for prey. Bills slender; tails rather long and square-ended or slightly notched. Sexes alike, except for Fiscal Flycatcher. Occur singly, in pairs or family groups.

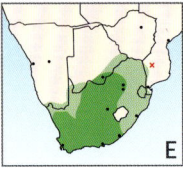

FISCAL FLYCATCHER *Melaenornis silens*
J F M A M J J A S O N D

18–20 cm; 22–36 g A striking, pied flycatcher with white panels in otherwise black wings and tail. Resembles Southern Fiscal (p. 418) in broad colour pattern, but bill more slender, tail shorter, legs longer, white in wings confined to secondaries (not wing coverts) and white windows in the tail; lacks white outer-tail feathers. Larger and longer tailed than Collared Flycatcher (p. 408) and nape black (not white). Female is browner above and washed darker grey on breast and belly. Juv. is much browner above, with buff spots; underparts mottled grey-brown. **Voice:** A string of high-pitched, weedy notes, sometimes extended by mimicking calls of other birds; *'tssisk'* alarm call. **Status and biology:** Endemic. Common resident in south and east of range, a non-br. winter visitor (Mar–Sep) in north of range. Found in woodland and thickets, scrub, gardens and plantations. (Fiskaalvlieëvanger)

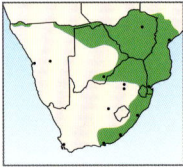

SOUTHERN BLACK FLYCATCHER *Melaenornis pammelaina*
J F M A M J J A S O N D

18–20 cm; 24–32 g An entirely black, somewhat glossy flycatcher that is most easily confused with a Common Square-tailed Drongo (p. 338), but has blackish-brown (not red) eyes, a taller stance and more slender body. Easily told from Fork-tailed Drongo (p. 338) by its smaller size and square or slightly forked (not deeply forked) tail. Juv. is sooty-brown, heavily spotted above and below with buff. **Voice:** High-pitched, wheezy 2 or 3-note song, often preceded by a shrill note, *'tziiii tsooo-tsoo'*. **Status and biology:** Common resident in broadleafed woodland, savanna and forest edges. (Swartvlieëvanger)

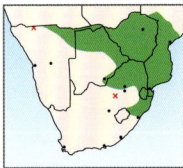

PALE FLYCATCHER *Melaenornis pallidus*
J F M A M J J A S O N D

17 cm; 20–24 g A rather uniformly buff-coloured flycatcher, darkest on the wings and palest on the throat, with a thin white eye-ring bisected by a dark eye stripe. Told from Marico Flycatcher by its pale buff (not white) underparts. Range overlaps slightly with Chat Flycatcher, but Pale Flycatcher is smaller and shorter tailed; typically shows more contrast between upper- and underparts; Chat Flycatcher shows a slight pale panel on folded secondaries; habitats differ. Juv. has fine buff spots above, with broad, buff margins to flight feathers; underparts streaked dark brown. **Voice:** Song is a melodious warbling interspersed with harsh chitters; alarm call is a soft *'churr'*. **Status and biology:** Locally common resident in moist, broadleafed woodland. (Muiskleurvlieëvanger)

CHAT FLYCATCHER *Melaenornis infuscatus*
J F M A M J J A S O N D

20 cm; 39 g A large, chat-like flycatcher with long wings and tail. Rather nondescript, with paler edges to wing feathers forming slight panel on folded secondaries. Best told from chats by its long, uniformly brown tail. Larger than Marico and Pale Flycatchers, with darker brown underparts that contrast less with the upperparts. Spends more time on ground than other flycatchers; often raises its wings. Juv. is spotted buff above; flight feathers edged rufous, breast mottled dark brown. **Voice:** Song is a deep, warbled *'cher-cher-cherrip'*, interspersed with harsh, grating notes. **Status and biology:** Near-endemic. Common resident or local nomad in semi-arid and arid shrublands. (Grootvlieëvanger)

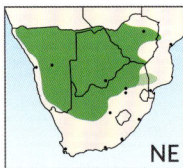

MARICO FLYCATCHER *Melaenornis mariquensis*
J F M A M J J A S O N D

18 cm; 22–28 g A sandy brown-and-white flycatcher, easily told from Pale and Chat Flycatchers by the sharp divide between brown upperparts and white underparts; wing feathers have pale edges. Juv. is dull brown above, spotted with buff, with rufous edges to flight feathers. Underparts off-white, with dark brown streaking; probably indistinguishable from juv. Pale Flycatcher. **Voice:** Song is a series of sparrow-like *'tsii-cheruk-tukk'* chirps; soft *'tsee tsee'* alarm call. **Status and biology:** Near-endemic. Common resident in semi-arid acacia savanna and sparse woodland. Where its range overlaps with Pale Flycatcher, Marico Flycatcher is confined to thornveld, Pale Flycatcher to broadleafed woodland. (Maricovlieëvanger)

FISCAL
FLYCATCHER

white-edged
secondaries

♂

white tail
patches

juv.

♀

dark eye

slim bill
(cf. drongos)

SOUTHERN
BLACK
FLYCATCHER

ad.

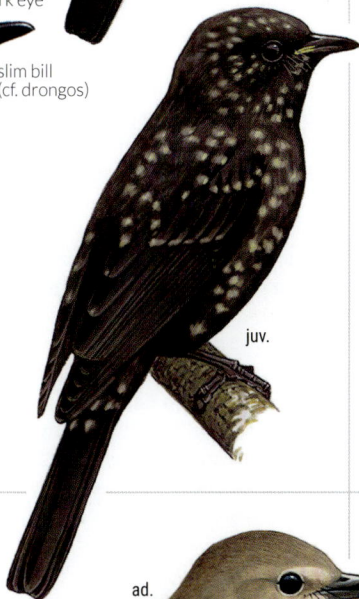

juv.

PALE
FLYCATCHER

grey-brown
below

square tail

juv.

ad.

ad.

juv.

pale
secondary
panel

pale
brown
below

strong
contrasting
plumage

CHAT
FLYCATCHER

MARICO
FLYCATCHER

SMALL FLYCATCHERS

SMALL FLYCATCHERS Smaller and more compact flycatchers than the species on the preceding page; bills slender, but often broad-based, with rictal bristles that enhance gape width for catching insects in flight. Sexes alike in *Muscicapa* and *Myioparus*, but differ in *Ficedula*. Juvs spotted buff above and mottled below. Occur singly or, for resident species, in pairs or family groups.

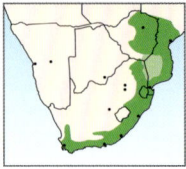

AFRICAN DUSKY FLYCATCHER *Muscicapa adusta*

J F M A M J J A S O N D

12 cm; 10–14 g A small, dumpy, grey-brown flycatcher. Smaller than Spotted Flycatcher, with proportionally shorter wings and tail; lacks streaking on the crown and has a more conspicuous white eye-ring. Juv. is darker brown, with buff-spotted upperparts and mottled breast. **Voice:** High-pitched, descending *'tseeeuu'*; also a sharper *'ti-ti-ti-trrrrr'* or *'tsirit'*; alarm call is a series of sharp clicks. **Status and biology:** Common resident in forest edges and glades, riverine forest and well-wooded gardens. Easily overlooked when not calling until it flits from its perch after prey. (Donkervlieëvanger)

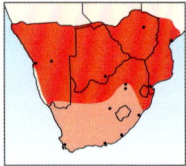

SPOTTED FLYCATCHER *Muscicapa striata*

J F M A M J J A S O N D

13–14 cm; 15–20 g A small, grey-brown flycatcher with pale underparts and a streaked crown and breast. Larger and longer winged than African Dusky Flycatcher, with a streaked crown and diffusely streaked breast; wings extend almost halfway down its tail. Hind crown is slightly peaked. Often flicks its wings on landing. Juv. body plumage is moulted before reaching s Africa, but some tertials and wing coverts with broad, buff margins may remain. Mediterranean Flycatcher *M. tyrrhenica* (breeds Balearic Islands, Corsica, Sardinia and adjacent Italian coast) sometimes split and may occur in the region, but doubtfully distinct in the field. **Voice:** Occasionally gives soft *'tzee'* and *'zeck, chick-chick'* calls. **Status and biology:** Common, Palearctic-br. summer visitor in virtually all wooded habitats, from forest edges to semi-arid savanna. Perches on low branches. (Europese Vlieëvanger)

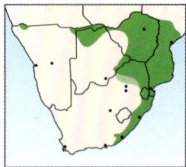

ASHY FLYCATCHER *Muscicapa caerulescens*

J F M A M J J A S O N D

14–15 cm; 15–17 g A blue-grey flycatcher with a conspicuous white eye-ring bisected by a blackish eye stripe. Told from Grey Tit-Flycatcher by its erect stance, different foraging behaviour (doesn't flit its tail while gleaning) and plain grey (not white-edged) tail. Blue-grey plumage distinguishes it from Pale (p. 406), Spotted and African Dusky Flycatchers. Juv. is darker above with buff spotting; breast mottled blackish. **Voice:** Song is a short, descending warble, *'sszzit-sszzit-sreee-sree'*, lasting 1–2 seconds; alarm call is a piercing hiss. **Status and biology:** Common resident in riverine forest and moist, broadleafed woodland. Some birds move to lower elevations in winter. Solitary or in pairs. (Blougrysvlieëvanger)

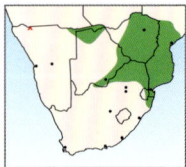

GREY TIT-FLYCATCHER *Myioparus plumbeus*

J F M A M J J A S O N D

14 cm; 12 g A slender, apalis-like, grey bird with a black-and-white tail which is fanned and rotated continuously as it forages; this behaviour and tail pattern distinguishes it from similar-looking Ashy Flycatcher. Juv. is spotted brown and buff above and below. **Voice:** Soft, tremulous, whistled *'treeoo'* or *'treee-trooo'*. **Status and biology:** Locally common resident in dense woodland and savanna thickets, riparian woodland and forest edge. Solitary or in pairs. (Waaierstertvlieëvanger)

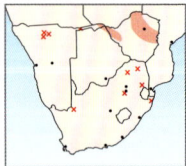

COLLARED FLYCATCHER *Ficedula albicollis*

J F M A M J J A S O N D

13 cm; 12 g A small, compact flycatcher with prominent white wingbars. Told from hyliotas (p. 404) by its longer wings, rounder head and hawking behaviour (not warbler-like gleaning). For separation from vagrant European Pied Flycatcher, see that species. Br. male blackish above with a complete white collar, large, white forecrown and extensive white primary bases. Female and imm. birds are grey-brown with partial collar and reduced white primary patches (can be almost invisible in some imm. birds). Both sexes have a small, pale rump patch and narrow, white edges to base of outer tail. Imm. similar to female but with reduced white margins to the outer tertials. **Voice:** Drawn-out *'seep'* or soft *'whit-whit-whit'*. **Status and biology:** Rare, Palearctic-br. migrant to woodland, especially miombo. Solitary, perching within tree canopy, darting out after prey and returning to perch. (Withalsvlieëvanger)

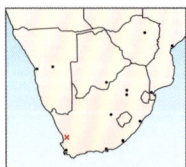

EUROPEAN PIED FLYCATCHER *Ficedula hypoleuca*

J F M A M J J A S O N D

12–13 cm 9–13 g Vagrant. Very similar to Collared Flycatcher; br. male differs by having a darker rump, smaller white forehead spot, less white in the wings (only a small mark on the base of the primaries) and lacks the white collar. Other plumages duller with brown upperparts, closely resembling Collared Flycatchers at the same stages but with white restricted to inner primary coverts, not extending close to the edge of the wing. Outer primary (P9) shorter than P6 (length similar in Collared Flycatcher). Often flicks one wing at a time and bobs its tail. **Voice:** Loud *'pik pik pik'*, sharper and harder than Collared Flycatcher. **Status and biology:** Only 1 record of an imm. bird from near Clanwilliam in Dec 2016. (Bontvlieëvanger)

plain
crown

AFRICAN DUSKY FLYCATCHER

indistinct
streaks/
blotches

juv.

streaked
crown

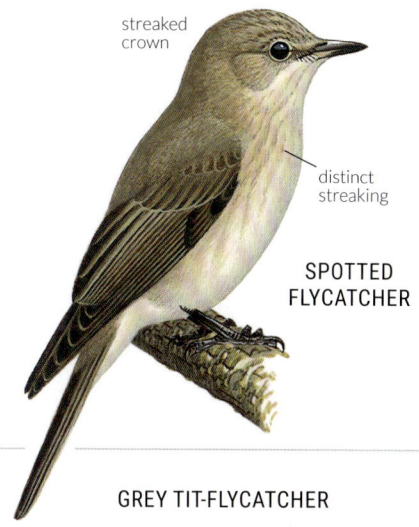

distinct
streaking

SPOTTED FLYCATCHER

distinct white
eye-ring

ASHY FLYCATCHER

black
lores

GREY TIT-FLYCATCHER

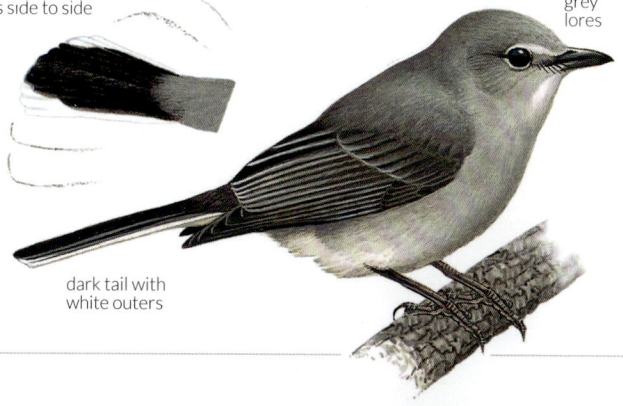

fans cocked tail and
flicks side to side

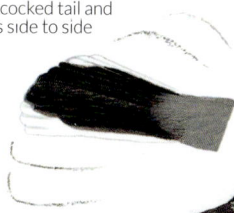

grey
lores

dark tail with
white outers

grey tail

partial collar

white
collar

large
white
frons

COLLARED FLYCATCHER

br. ♂

non-br.

large white patch
extends to outer
primaries

♀

small
white frons

no collar

slightly
darker
rump

EUROPEAN PIED FLYCATCHER

small
white mark
restricted
to inner
primaries

br. ♂

♀

CRESTED FLYCATCHERS AND FLYCATCHER WARBLERS

The taxonomy of flycatchers has undergone a major shake-up since the molecular revolution. *Trochocercus* and *Terpsiphone* are monarch flycatchers (Monarchidae), which form part of the Australasian radiation of basal passerines that are very different from the muscicapid flycatchers of the preceding two plates. Livingstone's Flycatcher is 1 of only 3 species in the Erythrocercidae, an endemic African family allied to the cettiid warblers, and *Elminia* and *Stenostira* are in another small family, the Stenostiridae, mostly found in Africa and SE Asia. All 5 forage for insects in the canopies of trees and forest undergrowth; some fan their tails while foraging, but this is independent of their evolutionary relationships. Sexes differ in *Trochocercus* and *Terpsiphone*; alike in other species. All lack the buff-spotted juv. plumage typical of muscicapids. Most occur singly or in pairs.

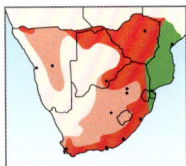

AFRICAN PARADISE FLYCATCHER   *Terpsiphone viridis*

17 cm (br. male up to 34 cm); 14 g A striking chestnut-and-grey, crested flycatcher with a blue bill and blue eye-ring. Most males acquire long central tail feathers while br. female and juv. have short tails and less vividly blue bills and eye-rings. **Voice:** Harsh *'zway-ter'* call; song is loud, whistled *'tswee-tswitty-tswee-tswitty-ter'*. **Status and biology:** Common br. summer visitor, Oct–Apr, but resident in eastern lowlands. Frequents forest, woodland, plantations and well-wooded gardens; often along streams. (Paradysvlieëvanger)

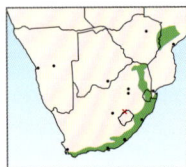

BLUE-MANTLED CRESTED FLYCATCHER
Trochocercus cyanomelas

15 cm; 10 g A black-and-white, crested flycatcher with a prominent white wingbar; tail dark grey, lacking white tip of White-tailed Crested Flycatcher. Male has glossy black head, crest and throat, contrasting with its white lower breast and belly. Female is paler grey, with fine streaking on throat, and less white in wing. Juv. resembles female, but has shorter crest and buffy wingbar. **Voice:** Frequently calls a rasping *'zweet-zwa'*, similar to African Paradise Flycatcher; song is a short series of whistles. **Status and biology:** Common resident in forests, mostly in the lower strata; some may move to lower elevations in winter. Frequently fans its wings while foraging, and often joins mixed-species bird parties. (Bloukuifvlieëvanger)

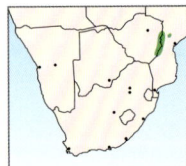

WHITE-TAILED CRESTED FLYCATCHER   *Elminia albonotata*

14 cm; 8 g A small, sombre-looking forest flycatcher-like warbler. Most like Blue-mantled Crested Flycatcher, but lacks a white wingbar, has white tail tips, and dusky throat and breast merge into paler belly. Sexes alike. Juv. is duller. **Voice:** Song is a series of very high-pitched notes, *'tsee-tsee-teuu-choo'*; rapid *'zitt-zitt-zitt'* alarm call. **Status and biology:** Restricted. A common but localised resident, restricted in the region to montane forest in e Zimbabwe and adjacent Mozambique; some may move to lower elevations in winter. Forages at all levels within the forest, often in mixed-species bird parties; characteristically fans its tail and droops its wings while zigzagging from twig to twig. (Witstertvlieëvanger)

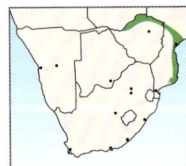

LIVINGSTONE'S FLYCATCHER   *Erythrocercus livingstonei*

12 cm; 7 g A small, flycatcher-like warbler. Lemon-yellow underparts and orange tail with an indistinct subterminal black band are diagnostic. Juv. has olive back colour extending onto hind crown; tail lacks dark subterminal band. **Voice:** Melodic, warbled song, often preceded by a series of fast, chittering notes; sharp *'chip-chip'* call. **Status and biology:** Locally common resident in riverine and lowland forests; in pairs or small groups. While foraging, keeps to tree canopies, moving position constantly and maintaining contact with brief snatches of song. Between foraging, often sit huddled together on a branch. (Rooistertvlieëvanger)

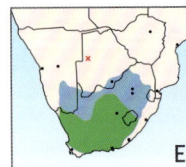

FAIRY FLYCATCHER   *Stenostira scita*

11 cm; 6 g A very small, flycatcher-like warbler with distinctive grey, black and white plumage and a peachy wash on lower breast and belly. Juv. is browner above and lacks peach-coloured flanks. **Voice:** Wispy, high-pitched *'tisee-tchee-tchee'* and descending *'cher cher cher'*. **Status and biology:** Endemic. Common resident in the Karoo, but seasonal elsewhere, leaving high-lying br. areas to winter in acacia savanna in the north. Forages in twiggy branches of trees and shrubs. Restless, constantly moving position from twig to twig. (Feevlieëvanger)

AFRICAN PARADISE FLYCATCHER

♂

♀

black band in orange tail

BLUE-MANTLED CRESTED FLYCATCHER

prominent white wingbar

♀

blue-grey mantle

strongly contrasting underparts

♂

WHITE-TAILED CRESTED FLYCATCHER

plain upperparts, no wingbar

LIVINGSTONE'S FLYCATCHER

FAIRY FLYCATCHER

black mask, white supercilium

BATISES

BATISES An African genus of small, flycatcher-like forest and savanna birds which, together with wattle-eyes, form another endemic African family, most closely related to helmetshrikes and bushshrikes (p. 424). Thickset, with large heads, a broad, black eye-mask and contrasting eye. All are sexually dimorphic, with males typically having black collars and females chestnut or buff collars; females often more distinctive than males. Found singly, in pairs or family groups; often join mixed-species bird parties. Forage for insects in the canopies of trees and bushes, hopping between branches and making short flights after prey. Most have a bouncing flight, with their wings making a characteristic 'fripping' sound. Although some species pairs are rather similar, there is little or no overlap between the ranges and preferred habitats of the 5 s African species.

CAPE BATIS *Batis capensis*

J F M A M J J A S O N D

13 cm; 12 g A richly coloured batis; rich chestnut wingbar and feather edges in both sexes distinguish it from all other batises in the region (female Woodwards' Batis has paler, buffy wingbar). Male has a broad, black breast band and golden eyes; female has a narrower, chestnut breast band, diffuse chin spot and reddish eyes. Juv. like female, but duller, with buff spots above. **Voice:** Song is a series of hooting whistles, *'foo-foo-foo'*; both sexes give a bubbling alarm call, *'kshee kshee kshee ksh ksh ksha'*. **Status and biology:** Near-endemic. Common resident in forests, riparian thickets and well-wooded gardens. Pairs maintain year-round territories over most of their range, with localised movements from high-lying areas to lower elevations in winter. (Kaapse Bosbontrokkie)

WOODWARDS' BATIS *Batis fratrum*

J F M A M J J A S O N D

11 cm; 12 g A localised batis restricted to lowland and coastal forest. Both sexes have a diffuse, buff-washed breast, similar to female Pririt Batis; female told by buffy (not white) wingbar; male by white (not buffy) throat and richer eye colour; ranges do not overlap. Female has paler buffy wingbar than female Cape Batis; narrow, white supercilium diagnostic. Buff breast is paler and more diffuse than chestnut breast band of female Pale Batis. Juv. like female, but duller, with buff spots above. **Voice:** Clear, penetrating whistle, *'tch-tch-phoooo'*, often repeated; also clicks bill. **Status and biology:** Near-endemic. Localised resident in low-lying evergreen forests along eastern coastal plain. (Woodwardbosbontrokkie)

CHINSPOT BATIS *Batis molitor*

J F M A M J J A S O N D

13 cm; 12 g The common batis across moist savanna, narrowly overlapping with Pririt Batis in the west and Pale Batis in the east. Male distinguished from male Pririt Batis by its white (not dark-mottled) flanks and from male Pale Batis by its broader breast band and plain grey (not mottled) back. Female distinguished from female Pale Batis by its mottled back. The call-notes of the 3 species are quite different. Juv. like female, but upperparts spotted buff and wingbars buffy. **Voice:** Male song is 3 clear, descending whistles, *'teuu-teuu-teuu'* ('three blind mice'); female gives harsh *'chrr-chrr'* notes. **Status and biology:** Common resident in broadleafed woodland and acacia savanna. (Witliesbosbontrokkie)

PALE BATIS *Batis soror*

J F M A M J J A S O N D

11 cm; 10 g The common batis of the savannas of Mozambique's coastal plain. Slightly smaller than Chinspot Batis, with a dappled grey and white (not uniform grey) back in both sexes; male has a narrower black breast band. Juv. like female, but spotted buff above. **Voice:** Ten or more piping notes repeated in quick succession, *'wit-wit-wit-wit ...'* or *'tlop-tlop-tlop-tlop ...'*. **Status and biology:** Fairly common resident in lowland woodland, especially miombo. (Mosambiekbosbontrokkie)

PRIRIT BATIS *Batis pririt*

J F M A M J J A S O N D

12 cm; 10 g The common batis of the arid savanna and Karoo. Female has distinctive buff-washed breast and throat, but male closely resembles male Chinspot Batis; distinguished by having dark-speckled (not all-white) flanks; calls differ and ranges overlap only marginally. Juv. like female, but buff-spotted above; breast mottled dark brown and buff. **Voice:** A long series of sharp, whistled notes *'teuu, teuu, teuu, teuu...'*, descending in scale, often with sharp, clicking calls. **Status and biology:** Near-endemic. Common resident in thornveld and scrub, extending into desert areas along wooded, dry watercourses. (Priritbosbontrokkie)

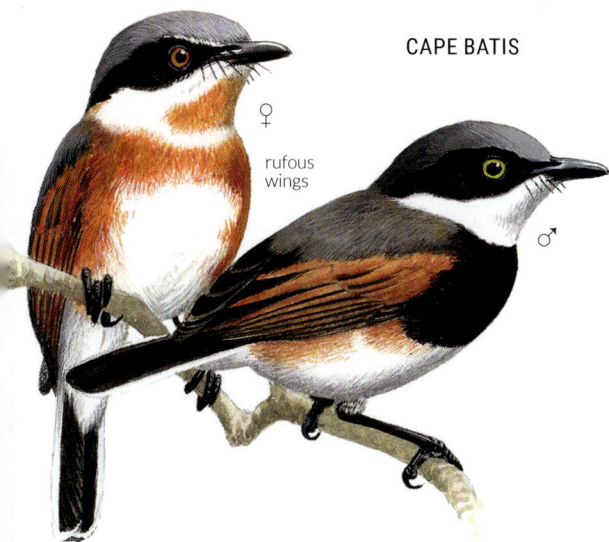

CAPE BATIS

♀

rufous wings

♂

WOODWARDS' BATIS

rufous wing panel

♀

no black breast band

♂

CHINSPOT BATIS

chestnut throat spot

♀

plain back

broad breast band

♂

plain white flanks

PALE BATIS

long supercilium

mottled back

♀

narrow, black breast band

♂

PRIRIT BATIS

♀

buff-yellow wash

♂

flanks speckled grey

WATTLE-EYES AND ALLIES

The White-tailed 'Shrike' is a peculiar, ground-dwelling batis confined to arid and semi-arid savannas in nw Namibia and sw Angola. It is in the same family, Platysteiridae, as the wattle-eyes, which are compact forest flycatchers characterised by their bright fleshy eye wattles. Sexes differ in eye-wattles. Occur singly, in pairs or small family groups.

WHITE-TAILED SHRIKE *Lanioturdus torquatus*

J F M A M J J A S O N D

15 cm; 27 g A striking black, white and grey bird, with yellow eyes, long legs and a short, white tail. Its upright posture contributes to its very short-tailed appearance. Sexes alike. Juv. has a mottled black-and-white crown and dark eyes. **Voice:** A variety of clear, drawn-out whistles, churrs and mews; also a harsh cackling. **Status and biology:** Near-endemic. Common resident of dry savanna, especially mopane, acacia or bush-willow. Often forages on the ground, bounding with long hops, sometimes flying up to tree perches. (Kortstertlaksman)

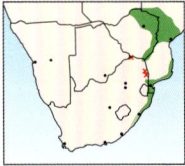

BLACK-THROATED WATTLE-EYE *Platysteira peltata*

J F M A M J J A S O N D

13 cm; 13 g A small, batis-like, black-and-white bird with distinctive red eye-wattles. Male has a narrow, black breast band; female has throat and upper breast black. Imm. grey above; throat mottled brown. Juv. has pale brown upperparts with rufous spotting; underparts off-white with a diffuse breast band; eye-wattles small and orange. **Voice:** Dry 'wichee-wichee-wichee-wichee...' song; also sings 'ptec ptec ptec' in aerial display. Alarm call is a harsh 'chit chit chit'. **Status and biology:** Locally common resident in coastal, lowland and riparian forest, favouring areas with sub-canopy creepers and dense undergrowth, often along watercourses; also in coastal mangroves. (Beloogbosbontrokkie)

SHRIKE-FLYCATCHERS

Formerly Vanga Flycatcher, genetic evidence confirms that this species belongs in the Vangidae, a family of mainly African birds known for their radiation in Madagascar, but also with some SE Asian representatives. Helmetshrikes (p. 416) also belong to this family. Sexes differ.

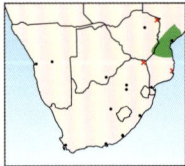

BLACK-AND-WHITE SHRIKE-FLYCATCHER (VANGA FLYCATCHER) *Bias musicus*

J F M A M J J A S O N D

14 cm; 19 g A squat flycatcher, larger than a batis or wattle-eye, with an obvious crest. Legs and eyes yellow. Male is black and white with small, white patches in the primary bases. Imm. male has blackish breast and rufous back. Female has a bright chestnut back and tail, recalling a female African Paradise Flycatcher (p. 410), but is stockier, with all-white underparts and short tail. Juv. resembles female, but is duller and streaked on head. **Voice:** Song is loud, whistled 'whitu-whitu-whitu'; alarm note is sharp 'wee-chip'. **Status and biology:** Uncommon and local resident of lowland and riparian forest, mostly found in canopies of the tallest forest trees. Male frequently performs a characteristic butterfly display flight at onset of br. season. (Witpensvlieëvanger)

BROADBILLS

Small to medium-sized, forest-living insectivores that together with the pittas and asities are the only Old World representatives of the sub-oscine passerines; they are placed here because of the superficial resemblance of *Smithornis* to a flycatcher. Broadbills are characterised by having a large, flat head and a short, unusually wide bill. Most species occur in Asia, with only 4 species in Africa. Sexes similar.

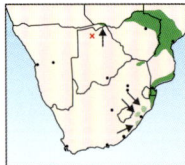

AFRICAN BROADBILL *Smithornis capensis*

J F M A M J J A S O N D

12–14 cm; 20–25 g An unobtrusive, small, flycatcher-like bird with heavily streaked underparts. Male has blacker forehead and crown than female. Juv. is buffy and less streaked above. **Voice:** During display, male gives a distinctive, frog-like trilling call, 'prrrrurrrr', usually dipping in pitch in the middle of the song; the sound is made during a short, horizontal, circular flight from a low perch when the white 'puffball' on its back is exposed. **Status and biology:** Uncommon to locally common resident in lowland, riparian and sand forest. Solitary or in pairs; easily overlooked if not calling. (Breëbek)

WHITE-TAILED
SHRIKE

BLACK-THROATED
WATTLE-EYE

♀

lacks
wingbar

♂

♂

BLACK-
AND-WHITE
SHRIKE-
FLYCATCHER

♂

AFRICAN
BROADBILL

broad-based bill

♀

♂

♀

HELMETSHRIKES
Highly sociable, group-living bushshrikes, previously placed in their own family Prionopidae, but recently reduced to a genus in the family Vangidae; characterised by stiff, bristle-like feathers protruding from the forehead, colourful, fleshy eye-wattles and bright yellow eyes. Sexes alike. Groups remain in close contact by calling while foraging in woodland or forest canopy.

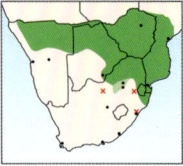

WHITE-CRESTED HELMETSHRIKE *Prionops plumatus*

J F M A M J J A S O N D

19 cm; 34 g A striking and unmistakable black-and-white helmetshrike with grey hind crown, broad, white collar, yellow eye-wattles and orange-pink legs. Juv. is duller, with brown-washed upperparts, feathers tipped buff; eyes brown, lacking yellow wattles. **Voice:** Repeated, bubbling *'cherow'* and various clicks and chattering calls, often chorused by a group. **Status and biology:** Common resident in mature woodland, but may range widely beyond br. habitat in winter, especially after drought periods. Occurs in groups of 2–10 birds, up to 22 in winter; often associates with mixed-species bird parties. Forages in trees in summer but often on the ground in winter. (Withelmlaksman)

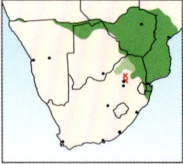

RETZ'S HELMETSHRIKE *Prionops retzii*

J F M A M J J A S O N D

22 cm; 44 g A large, black helmetshrike; larger and darker than Chestnut-fronted Helmetshrike with black (not chestnut) forehead bristles, brown (not slate-grey) back, black (not grey) belly and red (not blue-grey) eye-wattles. Juv. is paler grey-brown, with pale tips to mantle and wing feathers; dull pink bill and legs; lacks eye-wattles and is less easily separated from juv. Chestnut-fronted Helmetshrike, but juvs remain with ads. **Voice:** Harsh bubbling and winding *'tweeerr-r-r-r'* call. **Status and biology:** Common resident in woodland, including miombo, and riparian forest. Occurs in groups of 5 or 6 (2–15) birds. They form mixed groups with other helmetshrikes and join mixed-species bird parties. (Swarthelmlaksman)

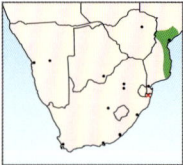

CHESTNUT-FRONTED HELMETSHRIKE *Prionops scopifrons*

J F M A M J J A S O N D

18 cm; 30 g Slightly smaller and paler than Retz's Helmetshrike, with a bristly chestnut (not black) forehead, blue-grey (not red) eye-wattles and a small, white loral spot. When viewed in canopy, appears paler due to grey (not black) underparts; mantle slate-grey (not dark brown). Juv. is grey-brown with white tips to upperpart feathers; lacks chestnut forehead and eye-wattles; eyes brown (not yellow), bill and legs dull pink. **Voice:** Repeated, trilling *'churee'*, with bill-snapping and other whirring and chattering notes. **Status and biology:** Locally common resident in lowland and riparian forest. Occurs in groups of 4–12 birds (up to 30 in winter). Sometimes joins groups of Retz's Helmetshrikes. (Stekelkophelmlaksman)

SMALL BUSHSHRIKES
Bushshrikes superficially resemble true shrikes, but are related to batises, vangas and cuckooshrikes. Skulking insectivores that range from forest undergrowth to forest canopy and are highly vocal. Sexes differ slightly.

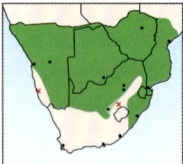

BRUBRU *Nilaus afer*

J F M A M J J A S O N D

14 cm; 24 g A small, black-and-white, arboreal bushshrike with chestnut flanks. Black above, with chequered back and prominent white wingbar and supercilium. Batis-like (p. 412), but larger, with a thicker bill. Female is sooty-brown above with light streaking on the chin, face and breast. Juv. mottled and barred with buff and brown above; wingbar buffy; underparts creamy, barred brown. **Voice:** Trilling *'prrrrr'* or *'tip-ip-ip prrrrrrrr'* given by male, sometimes answered *'eeeu'* by female. **Status and biology:** Common resident of acacia savanna and open, broadleafed woodland. (Bontroklaksman)

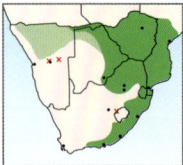

BLACK-BACKED PUFFBACK *Dryoscopus cubla*

J F M A M J J A S O N D

17 cm; 26 g A smallish, black-and-white bushshrike with white scapular bars and white-edged wing feathers. Male has black crown, red eyes and silky white, plume-like feathers on the rump which are raised during display to form a half-sphere. Female has a prominent white supercilium (shorter than Brubru), slate-grey upperparts and grey-washed underparts; eyes orange-yellow. Juv. like female, but browner above and washed buffy below; eyes brown. **Voice:** Sharp, repeated *'chick, weeo'* by male, answered with a scolding note by female. In flight, male gives loud *'chok chok chok'*. **Status and biology:** Common resident in woodland, thickets and forest canopy. (Sneeubal)

WHITE-CRESTED HELMETSHRIKE

ad.

juv.

red eye-ring

black plumage

ad.

RETZ'S HELMETSHRIKE

juv.

CHESTNUT-FRONTED HELMETSHRIKE

blue eye-ring

slate-grey plumage

ad.

juv.

BRUBRU

♀

♂

♂ display

BLACK-BACKED PUFFBACK

♀

bright red eye

♂

juv.

417

TRUE SHRIKES

Medium-sized passerines with an upright stance and stout, hook-tipped bills; most have whitish underparts and darker backs. Sexes alike or differ. Juvs duller and often finely barred. All are primarily insectivorous, hunting for prey by scanning the ground from elevated perches.

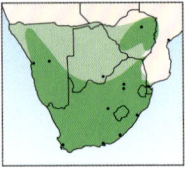

SOUTHERN FISCAL *Lanius collaris*

21–23 cm; 25–50 g A common black-and-white shrike of open country; often goes by the local name of 'jackie hangman' from its habit of cacheing prey on thorn 'larders'. Superficially similar to boubous (p. 420), but its erect stance, long tail and perch-hunting behaviour distinguish it. Also resembles Fiscal Flycatcher (p. 406), but tail is longer and is white-edged (not white in panels); bill heavier and stance more squat. In most races, female has chestnut flanks. Juv. is greyish-brown, with fine dark bars on its head, back and flanks. Desert and arid savanna races have a white supercilium. **Voice:** Harsh, grating call *'dwrrr-wrrrr'*; song is melodious and jumbled, often with harsher notes and mimicry of other birds. **Status and biology:** Common resident in most open habitats, including parks and gardens. Solitary or in pairs, in permanent territories; often hunts from roadside fences and telephone wires. (Fiskaallaksman)

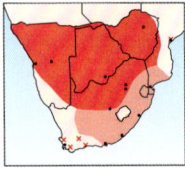

LESSER GREY SHRIKE *Lanius minor*

20–22 cm; 40–65 g A crisp-looking, grey-backed shrike with a bold, black face mask; tail shorter than Southern Fiscal. Lacks the white scapular bar of Southern Fiscal and Souza's Shrike. Male has pinkish wash to underparts in fresh plumage. Female is duller, with a slightly smaller black mask. Imm. is like female, but paler with a smaller mask that does not extend onto forecrown. Often retains some fine juv. barring on crown and flanks; base of bill pink. **Voice:** Soft, warbled, chattering song sometimes heard in autumn before northward migration; alarm call is a harsh, grating *'grrr-grrrr'*. **Status and biology:** Common Palearctic-br. visitor. Found in arid savanna, especially open thornveld; also semi-desert scrub. Solitary. (Gryslaksman)

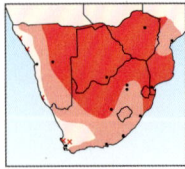

RED-BACKED SHRIKE *Lanius collurio*

18 cm; 22–44 g A smallish, compact shrike. Male easily identified by its white underparts, warm rufous back and wings, grey head and broad, black face mask. Female is brownish above and off-white below; breast and flanks with pale brown chevrons. Female has a partial brown mask; superficially similar to juv. Southern Fiscal and Souza's Shrike, but lacks their white or pale scapulars. Imm. is like female, with upperparts finely barred darker brown; see below for separation from vagrant Red-tailed Shrike. **Voice:** Soft, warbled song is given in autumn before northward migration; alarm call is a harsh *'chak, chak'*. **Status and biology:** Common and widespread, Palearctic-br. visitor to savanna and open woodland. Solitary. (Rooiruglaksman)

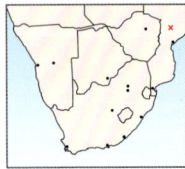

RED-TAILED SHRIKE *Lanius phoenicuroides*

16–18 cm; 19–32 g Vagrant. Similar to Red-backed Shrike; male easily identified by its reddish tail (not black with white sides); upperparts grey-brown (not reddish) but warmer brown on the crown; white bases to inner primaries form a white spot on folded wing; flanks washed peach (not pink). Female and imm. very similar to Red-backed Shrike, but rufous tail contrasts with grey-brown (not rufous-brown) back and lacks contrasting grey-washed nape. **Voice:** Harsh *'chack'* or quiet *'ch-ch-ch'*. **Status and biology:** Palearctic-br. migrant from e Europe and Asia that winters in E Africa. 1 record near Villa de Sena in Mozambique in Jan 2014. (Rooistertlaksman)

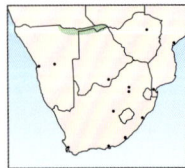

SOUZA'S SHRIKE *Lanius souzae*

17 cm; 22–30 g An unobtrusive, slender shrike with a long, thin tail. Most similar to female Red-backed Shrike; white scapular bars diagnostic (although aberrant Red-backed Shrikes may show this feature); also has a longer tail, with outer feathers all-white (not black-tipped). Female has duller black face mask and a rufous wash on its flanks. Juv. is finely barred above and below; could be confused with juv. Southern Fiscal, but is smaller and warmer brown above. **Voice:** A long, descending *'beeeeerrrr'* song; also grating alarm calls. **Status and biology:** Rare resident of broadleafed woodland. Solitary, in pairs or family groups. Hunts from low branches, retreating into tree canopy if disturbed; very easily overlooked. (Souzalaksman)

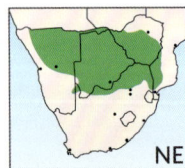

SOUTHERN WHITE-CROWNED SHRIKE
Eurocephalus anguitimens

23–25 cm; 56–82 g A large, heavy-bodied shrike with a distinctive white crown and dark brown eye stripe extending to ear coverts and nape. Sexes alike. Juv. is paler, with crown mottled brown, smaller dark face patch and yellow bill. **Voice:** Shrill, buzzy *'kree, kree, kree'*, often chorused by group members; also bleating and harsh chattering calls. **Status and biology:** Near-endemic. Common resident of dry woodland and acacia savanna, favouring areas with tall trees and a short-grass understorey. Occurs in groups of up to 10 birds. In flight, often glides, recalling a flock of babblers. (Kremetartlaksman)

SOUTHERN FISCAL

juv.

western races

♂

LESSER GREY SHRIKE

black forehead

juv.

ad.

grey forehead

plain wings back

♂

juv.

RED-BACKED SHRIKE

♀

RED-TAILED SHRIKE

grey-brown back

juv.

white wing patch

ad.

rufous-red tail

SOUZA'S SHRIKE

white bar

dark ear coverts

juv.

SOUTHERN WHITE-CROWNED SHRIKE

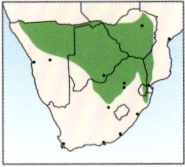

MAGPIE SHRIKE   *Urolestes melanoleucus*

b b b | | | | b b b b b
J F M A M J J A S O N D

40–50 cm (incl. tail); 60–95 g A large, very long-tailed, black shrike with white scapular bars, white tips to flight feathers and a whitish rump. Female has slightly shorter tail and whitish flanks. Juv. is shorter tailed, dull sooty-brown, with buff tips to underpart feathers. **Voice:** Liquid, whistled *'pee-leeo'* or *'pur-leeoo'*; also scolding *'tzeeaa'* alarm call. **Status and biology:** Common resident of acacia savanna with a short-grass understorey. In pairs or, more often, groups of 4–12 birds. (Langstertlaksman)

BOUBOUS   The genus *Laniarius* (boubous and gonoleks, such as the Crimson-breasted Shrike) is a distinctive group of large bushshrikes with glossy black upperparts (often with a white wingbar) and either white or coloured underparts. Sexes alike. The 3 boubous have narrowly overlapping ranges; male calls are largely indistinguishable, but female responses readily identify the species.

CRIMSON-BREASTED SHRIKE   *Laniarius atrococcineus*

b b b | b | | b b B B b
J F M A M J J A S O N D

25 cm; 53 g A striking, crimson-and-black boubou with a white wingbar. Rare yellow morph has lemon-yellow underparts. Juv. is finely barred black-and-buff above and barred greyish-brown below, with crimson initially confined to vent; crimson spreads patchily from 1 month after fledging. **Voice:** Pair sing in duet year-round, but most intensely in early summer. Various duets are performed, e.g. male initiates with a boubou-like whistled *'qwip-qwip'*, followed by female's harsher *'qwee-er, qwee-er'*; also various grating alarm calls. **Status and biology:** Near-endemic. Common resident of acacia thickets in arid savanna. Solitary or found in pairs, which defend year-round territories. (Rooiborslaksman)

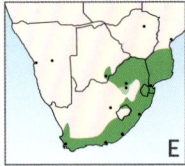

SOUTHERN BOUBOU   *Laniarius ferrugineus*

b b b b b b b b B B B b
J F M A M J J A S O N D

22 cm; 59 g A black-and-white bushshrike with a white wingbar and buff wash on the belly. Underparts more richly coloured than Tropical and Swamp Boubous, with cinnamon (not pinkish) wash concentrated on the belly (not breast). Female has less glossy black upperparts and more extensive buff on underparts. Juv. is mottled buff-brown above and vermiculated below; wingbar buffy; base of bill paler. **Voice:** Variable duet, started by male with one or several whistled notes followed by synchronised whistled response from the female, e.g. *'whooo'* (male), *'wieee'* (female); or *'wiet-weeo'* (male), *'wot-wot-wot'* (female). Both give harsh, grating alarm calls. **Status and biology:** Endemic. Common resident of forest edges, thickets and dense coastal scrub. Solitary or found in pairs, which defend year-round territories. (Suidelike Waterfiskaal)

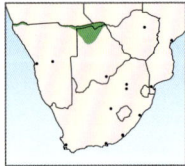

SWAMP BOUBOU   *Laniarius bicolor*

b b b | | b b b B B B b
J F M A M J J A S O N D

23 cm; 50 g Separated from Southern and Tropical Boubous by its pure white underparts. Sexes alike. Juv. is spotted with buff above and barred grey-brown below. **Voice:** A variable duet, initiated by male's hollow whistle, followed synchronously by a ratchet-like cackle from the female, e.g. *'cherhaaw'* (male), *'... k-k-k-k-k-k-'* (female). **Status and biology:** Locally common in thickets alongside rivers and papyrus swamps. Occurs singly or in pairs. (Moeraswaterfiskaal)

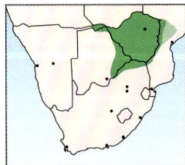

TROPICAL BOUBOU   *Laniarius major*

b b | b | | b b B B b
J F M A M J J A S O N D

20–25 cm; 45–65 g Similar to Southern Boubou, but with pink or cream-tinged breast and belly; underparts are not pure white, as in Swamp Boubou. Sexes alike. Juv. is duller, spotted with buff above; underparts whitish, with brown wash and darker barring on breast and flanks. **Voice:** A variable duet, initiated by male, followed synchronously by harsh, scolding note from female, e.g. *'haaw'* (male), *'... wheehaw'* (female); or *'chrrr'* (male), *'... weerrhooo'* (female); also various harsh, grating alarm calls. **Status and biology:** Common resident in thickets, riverine and evergreen forests, and gardens. Solitary or found in pairs, which defend year-round territories. (Tropiese Waterfiskaal)

MAGPIE SHRIKE

juv.

♂

CRIMSON-BREASTED SHRIKE

rare yellow morph

juv.

ad.

SOUTHERN BOUBOU

♀

♂

rufous flanks

SWAMP BOUBOU

pure white flanks

TROPICAL BOUBOU

pink wash on flanks

TCHAGRAS
Bushshrikes characterised by having rufous wings and generally brownish upperparts and pale underparts. Sexes alike. Males often sing in short aerial displays, flying up, then descending with fanned tail and 'fripping' wings.

MARSH TCHAGRA *Bocagia minuta*

b	b	b								b	b
J	F	M	A	M	J	J	A	S	O	N	D

17 cm; 33 g A small, compact tchagra restricted to marshes. Male has plain chestnut upperparts and creamy-buff underparts, a red eye, black tail and black cap. Female differs in having a broad, white supercilium. Juv. like female, but duller, with a buffy supercilium; crown sooty-brown, mottled buff. **Voice:** Song is a short, melodic 2 or 3-note whistle, *'whit-whee-wheeeu'*; also grating contact and alarm calls. **Status and biology:** Uncommon resident in rank bracken, sedges and reeds on edges of streams and swamps. (Vleitjagra)

BROWN-CROWNED TCHAGRA *Tchagra australis*

b	b	b						b	B	B	B
J	F	M	A	M	J	J	A	S	O	N	D

18 cm; 33 g Smaller than Black-crowned Tchagra; the brown central crown is diagnostic but it is bordered by black lateral crown stripes which can be mistaken for an all-back crown. Calls of the 2 species are entirely different. Smaller and shorter billed than Southern Tchagra, with buff (not grey) underparts. Juv. is duller, with mottled breast. **Voice:** Song is a series of 8–15 notes, descending in pitch and increasing in duration: *'chi-chi-che-chee-cheeyu-cheeyu-cheeeyuu...'*; also nasal and grating alarm calls. **Status and biology:** Common resident in thickets and understorey of savanna and woodland. (Rooivlerktjagra)

SOUTHERN TCHAGRA *Tchagra tchagra*

		b					b	B	B	B	b
J	F	M	A	M	J	J	A	S	O	N	D

21 cm; 47 g A large, long-billed tchagra with pale grey breast and belly, and grey central tail feathers. Long bill and grey (not buff) underparts distinguish it from smaller Brown-crowned Tchagra, and brown (not black) crown from Black-crowned Tchagra. Juv. is duller, with buff tail tips. **Voice:** Aerial display and song similar to Brown-crowned Tchagra, but averages deeper and slower. **Status and biology:** Endemic. Common resident of coastal scrub, forest edges and thickets, but easily overlooked if not calling. (Grysborstjagra)

BLACK-CROWNED TCHAGRA *Tchagra senegalus*

b	b	b	b					b	b	B	B
J	F	M	A	M	J	J	A	S	O	N	D

22 cm; 54 g The largest tchagra; very similar to, but larger than, Brown-crowned Tchagra; size, all-black crown and voice are diagnostic. Juv. is duller, with crown mottled brown; bill paler at base. **Voice:** Song is a series of mournful, whistled notes, *'whee-cheree, cherooo, cheree-cherooo'*, rising and descending in pitch; female sometimes responds in duet with a trilling note. Alarm calls include various rattling and tearing sounds. **Status and biology:** Common resident of savanna, thickets and riverine scrub. (Swartkroontjagra)

NICATORS
A small African family of 3 species; habits and behaviour recall a blend of greenbuls and bushshrikes. Sexes alike.

EASTERN NICATOR *Nicator gularis*

b										b	b
J	F	M	A	M	J	J	A	S	O	N	D

20–23 cm; 34–62 g Bright yellow spotting on the wing coverts and tertials is diagnostic. Superficially resembles Grey-headed Bushshrike (p. 424), but is smaller, has breast buffy-olive (not yellow-orange), throat whitish (not yellow) and eye dark (not yellow). Yellow tips to outer tail are obvious in flight. Juv. duller, with yellow-tipped primaries. **Voice:** A short, rich, explosive, liquid jumble of notes that includes snatches of mimicry. **Status and biology:** Common resident in dense riverine and coastal forests and scrub. Occurs singly or in pairs. Unobtrusive and easily overlooked; presence given away by distinctive song. (Geelvleknikator)

GREEN BUSHSHRIKES
Brightly coloured bushshrikes. Most are woodland or forest-dwelling species usually detected (and identified) by their whistled calls. Sexes similar in some, different in others. Several species are polymorphic.

BOKMAKIERIE *Telophorus zeylonus*

b	b	b	b	b	b	B	B	B	b	b	b
J	F	M	A	M	J	J	A	S	O	N	D

23 cm; 60–66 g A striking olive, black and yellow bushshrike. Sexes alike. In flight, green and black tail with yellow tips is conspicuous. Juv. duller; lacks black breast band and bold head markings. **Voice:** Varied whistled calls, mostly sung in duets initiated by male, e.g. *'wit, witawit'* (male), *'... toowip toowip'* (female); a trilling *'trrrreeeee'* and onomatopoeic *'bok-bok-kik-ik'*; also harsher calls and clicking notes. **Status and biology:** Endemic. Common resident in shrublands, strandveld and grassland with scattered bushes. Conspicuous, lively birds, spending much time on the ground; perches prominently while duetting; in territorial pairs year-round. (Bokmakierie)

solid
black
cap

♂

MARSH TCHAGRA

♀

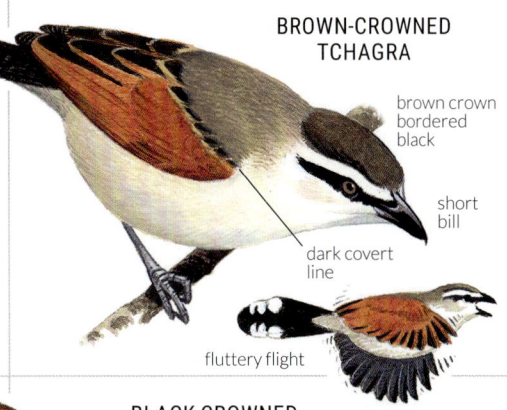

BROWN-CROWNED TCHAGRA

brown crown
bordered
black

short
bill

dark covert
line

fluttery flight

lacks black
scapular
bar

brown crown

long
bill

BLACK-CROWNED
TCHAGRA

black crown

SOUTHERN
TCHAGRA

EASTERN
NICATOR

shrike-like
bill

yellow spots
on wing

ad.

juv.

yellow tail tip

ad.

BOKMAKIERIE

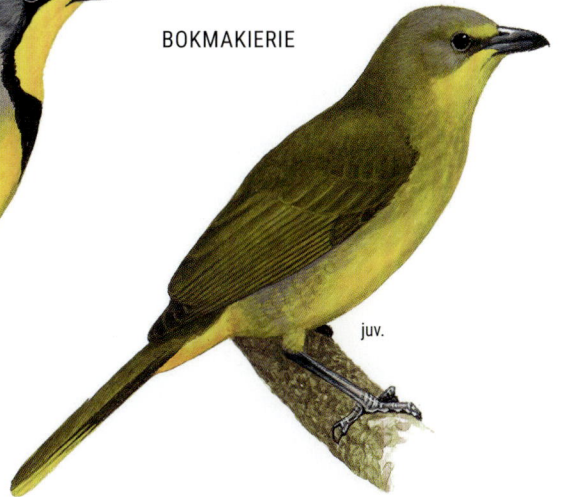

juv.

bright yellow
tail tip

423

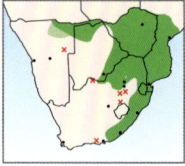

ORANGE-BREASTED BUSHSHRIKE
Chlorophoneus sulfureopectus

b	b	b						b	b	B	B	B
J	F	M	A	M	J	J	A	S	O	N	D	

19 cm; 27 g A bright yellow bushshrike of acacia savanna; its dark eyes, yellow forehead and supercilium are diagnostic, distinguishing it from the similar-looking but much larger Grey-headed Bushshrike (with all-grey head and yellow eyes). Male has black lores, a black line through the eye and black malar stripe. Female has plainer grey face. Juv. has drab grey head; underparts finely barred black and whitish-yellow, variably barred grey-brown. **Voice:** Male song is a frequently repeated, variably pitched whistle, *'poo-poo-poo-pooooo'* ('coffee, tea, or meee'), female sometimes responds in a short *'skzzz'*; alarm a harsher *'titit-eeezz'*, with bill-clicking. **Status and biology:** Common but often inconspicuous resident in acacia savanna and riverine forests. (Oranjeborsboslaksman)

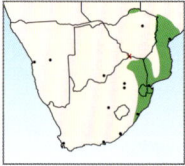

GORGEOUS BUSHSHRIKE   *Telophorus viridis*

b	b									B	B	B
J	F	M	A	M	J	J	A	S	O	N	D	

19 cm; 37 g A stunning, unmistakable, crimson-throated bushshrike. Male has black lores, a black line through eyes and black malar stripe joining broad, black breast band. Female lacks black facial pattern and has a narrow, black breast band and olive-green underparts. Juv. has dull yellow underparts, no breast band and a plain yellowish-green crown and face; breast band and red throat develop from 3 months. Olive-green head distinguishes it from juvs of other bushshrikes except juv. olive-morph Olive Bushshrike (from which it is inseparable in the field). **Voice:** Male song is a penetrating 2–5-syllabled *'kong-kon-kooit'* or *'kong-kooit-koit'*; also rasping calls. Sings year-round. **Status and biology:** Common resident in dense, tangled thickets, especially among low undergrowth on fringes of coastal and lowland forest. A skulking bird, more often heard than seen. (Konkoit)

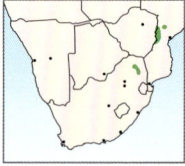

BLACK-FRONTED BUSHSHRIKE   *Chlorophoneus nigrifrons*

b	b								b	B	B
J	F	M	A	M	J	J	A	S	O	N	D

19 cm; 35 g A colourful bushshrike of tall, evergreen forest. Orange underparts, dark grey crown, broad, black face mask and lack of pale supercilium are diagnostic. Female has smaller black mask and orange wash mainly on sides of breast. Juv. duller, with buffy-yellow underparts barred olive-grey; wing feathers with pale tips. **Voice:** Male song is a ringing double note, the first a mellow whistle, followed by a harsh rasping sound, *'wooo-haaaa'*, *'wooo-hoooee'* or *'ooop eeerrk'*, female sometimes responding with a scolding *'skaaaa'*; also other harsh calls and bill-clicking when alarmed. **Status and biology:** Uncommon resident of creeper- and liana-draped forest canopy and mid-stratum. (Swartoogboslaksman)

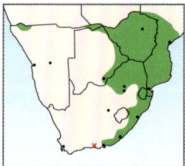

GREY-HEADED BUSHSHRIKE   *Malaconotus blanchoti*

b	b		b			b	b	B	B	B	b
J	F	M	A	M	J	J	A	S	O	N	D

26 cm; 77 g The largest bushshrike in the region; large size, all-grey head, massive bill and yellow eyes are diagnostic. The olive-green upperparts and pale spotting on the wing coverts resemble those of Eastern Nicator (p. 422), but larger size, grey head and yellow underparts distinguish the bushshrike. Sexes alike. Juv. has dark eyes and brown mottling on head; underparts pale yellow. **Voice:** A mournful, drawn-out whistle lasting up to 1 second, *'ooooooooooop'* (hence colloquial name 'Ghostbird'); also makes a call starting with bill-clicking, followed by a descending whistle, *'click-click-clik-click ...whayoooo'*, and a harsh *'skeeerrrr'*. **Status and biology:** Locally common resident of mature broadleafed and acacia woodland, riverine forest, suburbia and, less frequently, forest edge; far-carrying call usually gives away its presence. (Spookvoël)

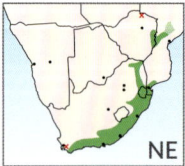

OLIVE BUSHSHRIKE   *Chlorophoneus olivaceus*

b	b							b	b	B	B
J	F	M	A	M	J	J	A	S	O	N	D

19 cm; 34 g A forest bushshrike with 2 colour forms: a widespread grey (ruddy-fronted) morph, and a more range-restricted olive (yellow-fronted) morph. Male of grey morph easily identified by its pinkish-buff throat and breast merging into a whitish vent. Female has pink-washed underparts and plain grey crown and face. Olive morph has olive-green upperparts and yellow underparts; distinguished from Black-fronted and Orange-breasted Bushshrikes by its olive (not grey) crown. Female like male, but lacks mask; ear coverts grey (not black). Juvs are duller than respective females, with pale tips to wing feathers and faint barring below. Olive-morph juv. inseparable from juv. Gorgeous Bushshrike. **Voice:** A very vocal species, calling year-round, occasionally in duet. Varied series of whistles, including *'wheee hoo hoo hoo hoo'* or descending *'whee-whee-wheoo-wheoo-whoo-whooo'*; also harsh, rattling alarm call, *'krrrr krrrr krrrrr'*. **Status and biology:** Near-endemic. Common resident of evergreen and riverine forests. (Olyfboslaksman)

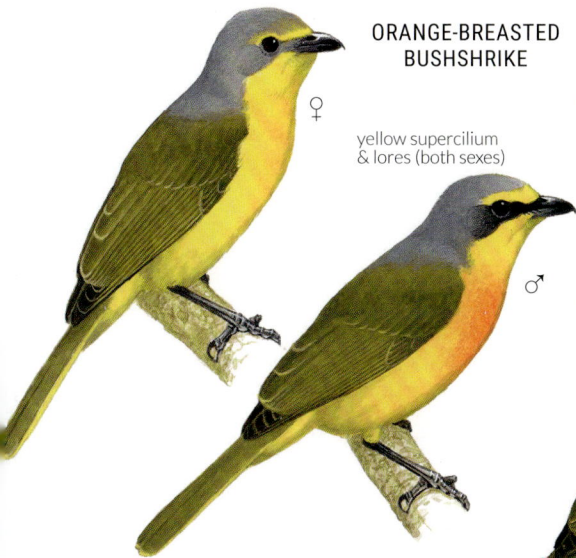

ORANGE-BREASTED BUSHSHRIKE

♀

yellow supercilium & lores (both sexes)

♂

GORGEOUS BUSHSHRIKE

♂

♀

juv.

BLACK-FRONTED BUSHSHRIKE

♀

black forehead

♂

rich orange chest

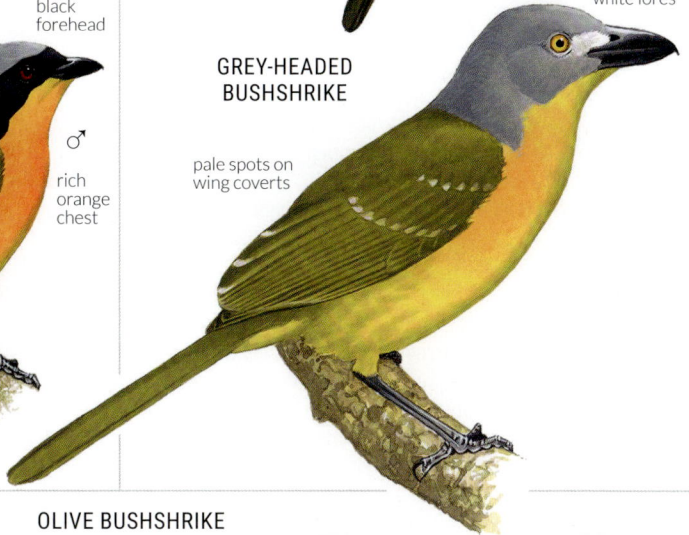

pale eye, white lores

GREY-HEADED BUSHSHRIKE

pale spots on wing coverts

olive forehead

♂

OLIVE BUSHSHRIKE

grey forehead

♂

♀

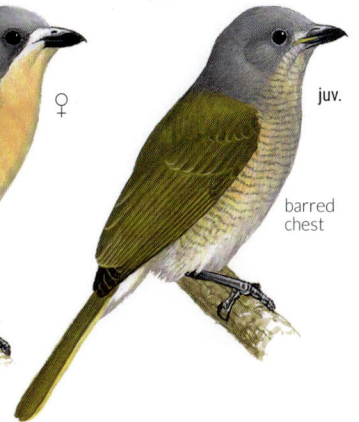

juv.

barred chest

olive morph

buff morph

STARLINGS AND MYNAS
Fairly large passerines with medium-sized bills indicative of their omnivorous diet. Most species have some gloss on their plumage. The glossy starlings are difficult to identify because plumage colour is dependent on lighting and angle. Sexes alike in most species; juvs duller. Only a few species have distinct br. plumages. Often roost communally (sometimes in huge flocks) and forage in small groups when not br.; some species breed colonially. Most breed in cavities in trees, including old woodpecker or barbet holes, or holes in cliffs or buildings.

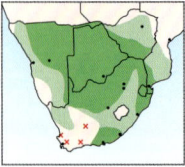

CAPE (GLOSSY) STARLING *Lamprotornis nitens*

b	b								b	B	B	B
J	F	M	A	M	J	J	A	S	O	N	D

25 cm; 65–112 g A fairly large, short-tailed, glossy starling that differs from the blue-eared starlings by its uniformly glossy, green head (lacking darker blue ear coverts), reddish-bronze shoulder patch, and glossy, green belly and flanks (lacking blue or magenta on flanks). Larger than Black-bellied Starling, with much brighter, glossier plumage; belly glossy green (not dull black); spends more time foraging on ground than Black-bellied Starling, which is mostly arboreal. Juv. is duller, with straw-yellow (not bright orange-yellow) eyes. **Voice:** Song a slurred warble, *'trrr-chree-chrrrr'*; flight call *'turreeeu'*. **Status and biology:** Common resident in savanna, mixed woodland and gardens; often in quite arid regions. (Kleinglansspreeu)

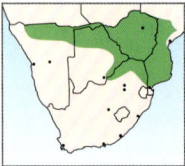

GREATER BLUE-EARED STARLING *Lamprotornis chalybaeus*

b								b	b	B	B	b
J	F	M	A	M	J	J	A	S	O	N	D

24 cm; 68–105 g A fairly large, short-tailed, glossy starling that differs from Cape Starling by its dark blue (not green) ear coverts, more prominent black tips to greater coverts, and blue (not green) belly and flanks. Larger than Miombo Blue-eared Starling, with a broader ear patch and longer, heavier bill. Belly and flanks are blue (not magenta), but this is often hard to distinguish as it depends on lighting and viewing angle. Juv. is duller and less glossy, with dark brown tones to its underparts (not chestnut-brown as in juv. Miombo Blue-eared Starling). **Voice:** Distinctive, nasal *'squee-aar'*, often incorporated in its warbled song. **Status and biology:** Common resident in savanna and mopane woodland. (Groot-blouoorglansspreeu)

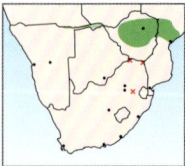

MIOMBO BLUE-EARED STARLING *Lamprotornis elisabeth*

b	b							b	B	B	b	b
J	F	M	A	M	J	J	A	S	O	N	D

19 cm; 52–66 g Smaller than Greater Blue-eared Starling, with a more compact head and finer bill. The blackish-blue ear patch is narrow, appearing as a black line through the eye. Belly and flanks are magenta (not blue), but this is often hard to distinguish as it depends on lighting and viewing angle. Juv. has distinctive, chestnut-brown underparts, becoming mottled blue and brown in imm. **Voice:** A series of short, staccato notes, *'chip chirroo krip kreuup krip'*, higher pitched and more musical than Greater Blue-eared Starling; *'wirri-gwirri'* flight call. **Status and biology:** Fairly common resident in miombo woodland. Often in large flocks in winter. (Klein-blouoorglansspreeu)

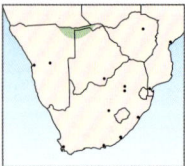

SHARP-TAILED STARLING *Lamprotornis acuticaudus*

								b	b	b	b	b
J	F	M	A	M	J	J	A	S	O	N	D

26 cm; 61–76 g Similar to Cape and Greater Blue-eared Starlings, but has a distinctive, wedge-shaped tail (not square-tipped). In flight, undersides of primaries appear pale (not black as in other glossy starlings). Eyes red (male) or orange (female). Juv. is duller, with a matt grey body, scaled buffy; wings and tail slightly glossy; eyes brown. **Voice:** Reedy *'chwee-chwee-chwee'*, higher pitched and less varied than Cape Starling. **Status and biology:** Rare resident and local nomad in broadleafed woodland and dry riverbeds. Often in small flocks. (Spitsstertglansspreeu)

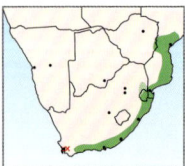

BLACK-BELLIED STARLING *Notopholia corrusca*

b									b	B	B
J	F	M	A	M	J	J	A	S	O	N	D

19 cm; 46–68 g The smallest and least glossy of the glossy starlings, with a black (not blue or green) belly and flanks. Male has a bronze gloss on its belly that is visible at close range, and has yellow-orange eyes (red for a short period during the br. season). Female and juv. are duller than male, often appearing black in the field. **Voice:** Song is a short, rather fast, warble of whistles and chipping notes, sometimes including mimicry of other bird songs. **Status and biology:** Locally common resident in coastal and riverine forests; irregular visitor in extreme south of range. Often seen foraging in the canopy of trees, unlike other glossy starlings, which mostly forage on the ground. (Swartpensglansspreeu)

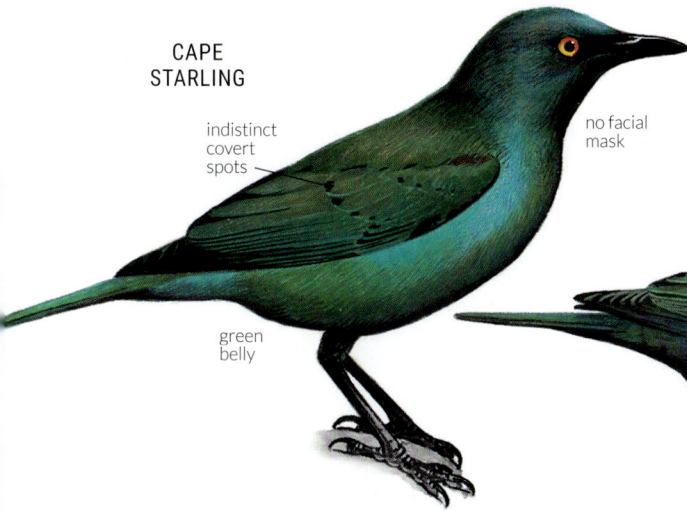

CAPE STARLING

indistinct covert spots

no facial mask

green belly

GREATER BLUE-EARED STARLING

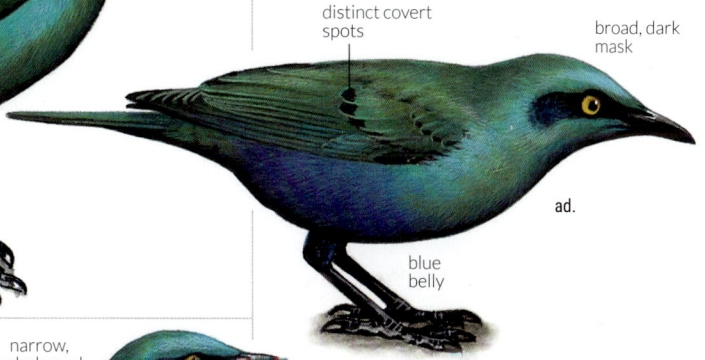

distinct covert spots

broad, dark mask

ad.

blue belly

MIOMBO BLUE-EARED STARLING

rich chestnut-brown

juv.

narrow, dark mask

mauve belly

ad.

dull brown

juv.

SHARP-TAILED STARLING

ad.

edge-shaped tail

dark red eye

juv.

BLACK-BELLIED STARLING

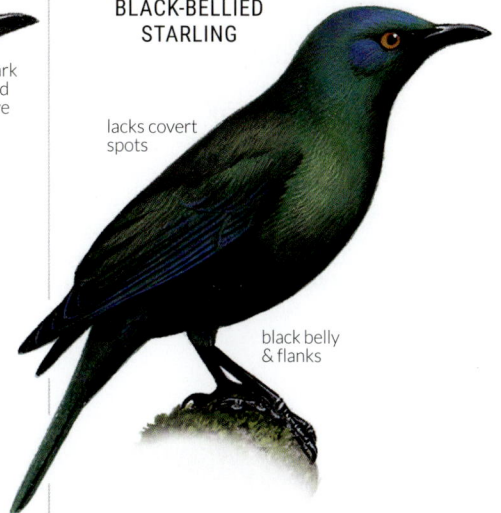

lacks covert spots

black belly & flanks

pale underwing, dark coverts

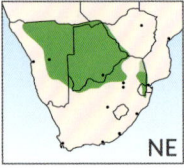

BURCHELL'S STARLING   *Lamprotornis australis*

J F M A M J J A S O N D

32 cm; 112–128 g The largest glossy starling. More heavily built than Meves's Starling, with broader, more rounded wings and a shorter, broader tail, which, together with its more erect stance, make it appear longer legged. Juv. is duller above; matt black below. **Voice:** Querulous, throaty calls, *'mreeow'* and *'chirow-eeow'*; song is a jumble of chortles and chuckles. **Status and biology:** Near-endemic. Common resident in savanna and dry, broadleafed woodland. (Grootglansspreeu)

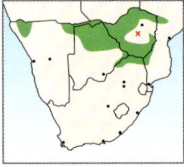

MEVES'S STARLING   *Lamprotornis mevesii*

J F M A M J J A S O N D

34 cm; 65–95 g A medium-sized, glossy starling with a very long, graduated tail. Smaller than Burchell's Starling, with a much longer tail. Juv. is duller, with brownish head and matt black underparts; tail shorter, but still distinctly long and graduated. *L. m. violacior* 'Cunene Long-tailed Starling' from n Namibia is split by some authorities – differs by having a more purple breast, reddish-copper rump and uppertail coverts, belly is darker and it is longer winged. **Voice:** Querulous, rather harsh *'keeeaaaa'*, higher pitched than Burchell's Starling; also various squawks. **Status and biology:** Locally common resident in tall mopane woodland and riverine forest. Sometimes breeds in loose groups, with up to 3 pairs in the same tree. (Langstertglansspreeu)

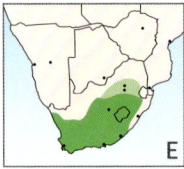

PIED STARLING   *Lamprotornis bicolor*

J F M A M J J A S O N D

27 cm; 94–112 g A large, blackish-brown starling with a conspicuous white vent. At close range, exaggerated yellow gape, creamy-white eyes and glossy wing coverts are visible. Paler primary bases are visible in flight. Juv. is matt black with white vent; gape white; eyes dark. **Voice:** Nasal *'skeer-kerrra-kerrra'*, often given in flight; also a warbling song. **Status and biology:** Endemic. Common resident in grassland and Karoo scrub. Often in flocks, frequently with Wattled Starlings (p. 430). Usually roosts and breeds colonially in holes in earth banks; noisy and conspicuous when commuting to and from its colonies at dusk and dawn. Sometimes gleans ectoparasites from livestock. (Witgatspreeu)

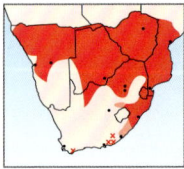

VIOLET-BACKED STARLING   *Cinnyricinclus leucogaster*

J F M A M J J A S O N D

17 cm; 39–56 g Male unmistakable, with its glossy amethyst head, upperparts and upper breast, and plain white belly. Upperparts may appear bluish or coppery at certain angles. Female and juv. are brown above and white below, both densely streaked dark brown. **Voice:** Song is a short series of buzzy whistles, *'cheez-i-werrr'*; also soft, sharp *'tip, tip'* notes. **Status and biology:** Common intra-African br. migrant in woodland, riverine forests and gardens; a few remain year-round in north of region. (Witborsspreeu)

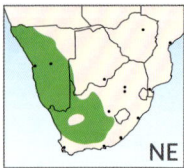

PALE-WINGED STARLING   *Onychognathus nabouroup*

J F M A M J J A S O N D

27 cm; 80–120 g Smaller and shorter tailed than Red-winged Starling, with whitish (not rufous) patches in its primaries, visible in flight, and bright, pale orange (not dark red) eyes. Sexes alike. Juv. is duller than ad. **Voice:** Ringing *'preeoo'* in flight; song is a rambling series of musical whistles; alarm call is a harsh *'garrrr'*. **Status and biology:** Near-endemic. Common resident and local nomad in rocky ravines and cliffs in dry and desert regions. Flocks may move long distances to exploit specific food sources. Often breeds in loose colonies. Gleans ectoparasites from zebras and antelope. (Bleekvlerkspreeu)

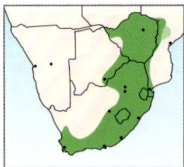

RED-WINGED STARLING   *Onychognathus morio*

J F M A M J J A S O N D

28 cm; 120–155 g A large, black starling with a long, graduated tail and rufous primaries that are conspicuous in flight. Larger and longer tailed than Pale-winged Starling, with dark red (not pale orange) eyes and obvious sexual dimorphism: male all glossy black; ad. female has head and upper breast lead-grey. Juv. is dull black, resembling male. **Voice:** Clear, whistled *'cher-leeeeoo'* or *'who-tuleooo'* call; song is a rambling series of musical whistles; alarm call is a harsh *'garrrr'*. **Status and biology:** Common resident in rocky ravines, cliffs and suburbia, where it may nest on sheltered ledges on large buildings. Wanders more widely in winter, when often in flocks. Sometimes gleans ectoparasites from cattle and game. (Rooivlerkspreeu)

BURCHELL'S STARLING

strong, bulky appearance

MEVES'S STARLING

mostly violet plumage

bluer plumage

violacior

mevesii

PIED STARLING

white eye, yellow gape

VIOLET-BACKED STARLING

♂

♀

PALE-WINGED STARLING

orange eye

shorter tail

RED-WINGED STARLING

dark eye

♂

thin graduated tail

♀

429

COMMON MYNA *Acridotheres tristis*

B b b b b b b b B B B
J F M A M J J A S O N D

25 cm; 65–135 g Introduced from Asia. Easily identified by its chestnut plumage, white wing patches, white tail tips and yellow facial skin. Moulting ads sometimes lose most of their head feathers, when bare head appears all yellow. Juv. is paler, with a grey-brown head and buffy tail tips that are narrower than ad's. **Voice:** Varied song comprises jumbled titters and chattering as well as mimicry of other sounds. **Status and biology:** Locally common resident in urban and suburban areas; also around farms. Range continues to expand from its original point of introduction in Durban, KZN, with a subsequent introduction in Gauteng, expanding north into Botswana and Zimbabwe. Usually roosts communally, often in very large flocks. Sometimes gleans ectoparasites from livestock and game. (Indiese Spreeu)

ROSY STARLING *Pastor roseus*

X X X
J F M A M J J A S O N D

21 cm; 62–88 g Vagrant. Structure similar to Common Starling, but ad. has striking pink body contrasting with black head, wings and tail; bill pink. Juv. has grey-brown head and body, contrasting with darker brown wings and tail; bill yellow-pink (not blackish, as juv. Common Starling). Lacks white rump of juv. Wattled Starling. Imm. has blackish head, breast, wings and tail contrasting with pinkish, grey-brown body, palest on belly. **Voice:** Harsh, chittering calls. **Status and biology:** Rare, Palearctic-br. vagrant that usually winters in s Asia; only a few records: Kgalagadi Transfrontier Park, Jul 2005 and Sep 2019, Kasane, n Botswana, Sep 2007, and s Kruger National Park, Oct 2017. (Roosspreeu)

COMMON STARLING *Sturnus vulgaris*

b B B b
J F M A M J J A S O N D

21 cm; 65–95 g Introduced from Europe. A compact starling with a pointed bill. Ad. has glossy black plumage with buff-tipped feathers above, white-tipped feathers below; pale tips gradually wear off, leaving plain black plumage. Bill yellow during br. season; otherwise blackish. Female paler and less glossy. Juv. is drab brown, with a paler throat and black bill; lacks pale rump of juv. Wattled Starling. **Voice:** Song is a chattering, high-pitched warble, often including mimicry. **Status and biology:** Common resident in towns and open farmland. Range continues to expand from its original point of introduction in Cape Town. Often joins flocks of Wattled and Pied (p. 428) Starlings. (Europese Spreeu)

WATTLED STARLING *Creatophora cinerea*

b b b b b b b
J F M A M J J A S O N D

21 cm; 52–84 g A compact, pale grey starling with a pale, slender bill and whitish rump that is conspicuous in flight. Male has black flight feathers; female has a darker grey body and browner flight feathers. Br. male has a bare black-and-yellow head with variable black wattles on its crown and throat. Some br. females also develop bare heads and small wattles. Juv. resembles a dull, brown female. **Voice:** High-pitched, wheezy hisses and cackles, similar to Common Starling; *'sreeeeo'* flight call. **Status and biology:** Common nomad in grassland, farmland, savanna and open woodland. Often in large flocks, regularly with other starlings. Breeds erratically, usually in vast colonies. (Lelspreeu)

OXPECKERS
Elongate, slender passerines adapted for grooming ectoparasites from large mammals. Tails long and stiffened, legs short with strong feet, and bills flattened, enabling scissoring through fur. Also eat blood and mucus, and hawk insects. Usually in small groups. Nest in tree cavities, often br. cooperatively.

YELLOW-BILLED OXPECKER *Buphagus africanus*

b b b b B B b
J F M A M J J A S O N D

20 cm; 53–71 g Slightly larger than Red-billed Oxpecker, with a pale rump that contrasts with its brown back at all ages. Bill red, with an extensive yellow base; lacks an eye-ring. Juv. has a pale brown (not black-tipped) bill that is deeper than that of juv. Red-billed Oxpecker. Hybrids of the 2 species, having intermediate bill and eye colouring, occasionally recorded. **Voice:** Rasping *'kriss'* or *'churr'*; also short, high-pitched notes. **Status and biology:** Locally common resident of savanna and broadleafed woodland, often near water. Usually associated with buffalo, rhino, hippo and livestock. (Geelbekrenostervoël)

RED-BILLED OXPECKER *Buphagus erythrorynchus*

b b b b b B B
J F M A M J J A S O N D

19 cm; 42–59 g Slightly smaller than Yellow-billed Oxpecker, with uniform brown upperparts at all ages (rump not paler than back); bill more slender. At close range, has prominent yellow eye-ring and an all-red bill. Juv. has a black bill with a pale base (not uniform pale brown). **Voice:** Scolding *'churrrr'*, hissing *'zzzzzzist'* and short, high-pitched *'zit'* notes. **Status and biology:** Locally common resident and local nomad in savanna and woodland; regularly wanders beyond normal range. Usually associated with game and cattle. (Rooibekrenostervoël)

COMMON
MYNA

prominent
wing
patches

ROSY STARLING

non-br.

br.

dark
bill

pale
bill

juv.

juv.

non-br.

COMMON
STARLING

pale
rump

br. ♂

WATTLED
STARLING

br.

YELLOW-BILLED
OXPECKER

pale
rump

ad.

pale
rump

juv.

stout
bill

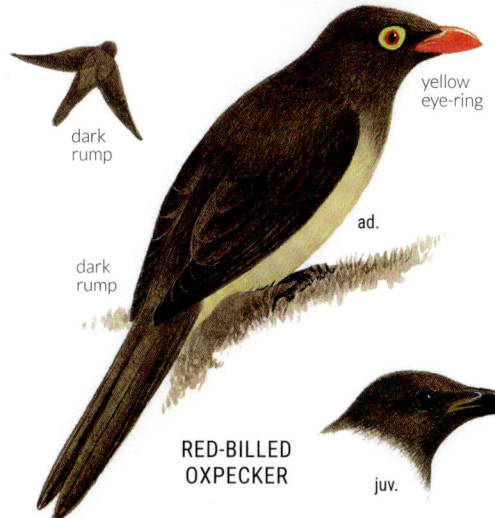

dark
rump

dark
rump

yellow
eye-ring

ad.

RED-BILLED
OXPECKER

juv.

SUGARBIRDS
Large, long-tailed nectarivores closely associated with proteas. Bizarrely, they are most closely related to the enigmatic Modulatricidae, 3 robin-like species found in central and E African forest understorey. Together with rockjumpers, sugarbirds are the only families endemic to s Africa. Often located by their raucous, rattling calls. Males make fripping sounds with wings and clatter tails in aerial display. Females and juvs have shorter tails than ad. males.

CAPE SUGARBIRD *Promerops cafer*

B	B	B	B	B	b	b	b
J F M A M J J A S O N D

28–44 cm; 25–50 g Slightly larger than Gurney's Sugarbird, with a brown (not rufous) crown and breast; lacks a supercilium but has distinct malar stripes. Often has yellow, pollen-stained crown. Tail of ad. male varies considerably in length (1.5–3 x body length; female and juv. tail usually 1–1.5 x body length). Juv. is duller, with a pale brown vent. **Voice:** Rattling song includes starling-like chirps and whistles; harsh, grating noises. **Status and biology:** Endemic. Common resident and local nomad in fynbos; breeds in winter; visits gardens and coastal scrub, mainly in summer. (Kaapse Suikervoël)

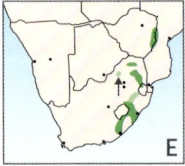

GURNEY'S SUGARBIRD *Promerops gurneyi*

B	b	b	b			b	B	B	B
J F M A M J J A S O N D

23–29 cm; 24–46 g Slightly smaller than Cape Sugarbird, with a conspicuous rufous breast and crown, bright yellow vent, greyish (not brown) cheeks and a whitish supercilium; lacks malar stripes. Tails of ad. males are not as long as the longest tails of Cape Sugarbirds. Ranges only rarely overlap in E Cape. Juv. has brown breast and dull yellow vent. **Voice:** Rattling song, higher pitched and more melodious than Cape Sugarbird. **Status and biology:** NEAR-THREATENED. Endemic. Locally common resident and partial migrant in stands of flowering proteas and aloes; some move to lower altitudes in winter. (Rooiborssuikervoël)

SUNBIRDS
Small, active, warbler-like birds with decurved bills that enables them to take nectar from flowers. Often pugnacious, chasing other birds from flowers. Sexes differ in most species. Males usually have some iridescent plumage and in many species have brightly coloured pectoral tufts that are fluffed out when displaying. Some species have a female-like eclipse plumage. Juvs resemble females.

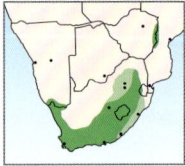

MALACHITE SUNBIRD *Nectarinia famosa*

B	b	b	b	b	b	B	B	B	B	B	B
J F M A M J J A S O N D

14–25 cm; 11–25 g A large, long-billed sunbird. Br. male has metallic green plumage with yellow pectoral tufts (only visible when displaying) and elongated central tail feathers. Female is rather pale, with diffuse mottling on the breast, a small supercilium, pale yellow moustachial stripe and white outer tail; fairly strong contrast between pale underparts and darker upperparts. Eclipse male resembles female, but often retains a few green feathers. **Voice:** Piercing *'tseep-tseep'* calls; song is a series of twittering notes, usually including loud *'tseep'* notes. **Status and biology:** Common resident, nomad and local migrant in fynbos, grassland and mountain scrub. (Jangroentjie)

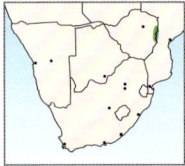

BRONZY SUNBIRD *Nectarinia kilimensis*

b	b	b						b	B	B	B
J F M A M J J A S O N D

14–22 cm; 12–17 g A large sunbird with a strongly decurved bill. Male is metallic bronze-green; much darker than Malachite Sunbird, appearing black in poor light. Female is smaller than female Malachite Sunbird, with a shorter, more strongly curved bill, paler throat lacking clear moustachial stripes, and more clearly streaked, yellowish breast; tail tip more pointed. Male has no eclipse plumage. **Voice:** Loud, piercing *'chee-wit'*, repeated every half-second; also high-pitched twittering song. **Status and biology:** Locally common resident and nomad of forest edge, bracken and adjoining grassland. (Bronssuikerbekkie)

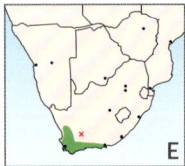

ORANGE-BREASTED SUNBIRD *Anthobaphes violacea*

b	b	b	B	B	B	B	B	B	b	b	b
J F M A M J J A S O N D

12–15 cm; 7–13 g Male easily identified by its orange belly and purple breast band; central tail feathers elongated. Yellow pectoral tufts only visible in display. Female is olive-green above and yellow-green below (not grey like smaller female Southern Double-collared Sunbird, p. 436); tail slightly pointed. Juv. is like female, with yellow gape; imm. male has patchy plumage. **Voice:** Typical call is a metallic, nasal twang. Song is a soft, high-pitched warble, sometimes mimicking other birds. Also gives rapidly repeated *'tick'* in pursuit flight. **Status and biology:** Endemic. Common resident and local nomad in fynbos and adjacent gardens. (Oranjeborssuikerbekkie)

malar stripe

♀

brown crown

♂

CAPE SUGARBIRD

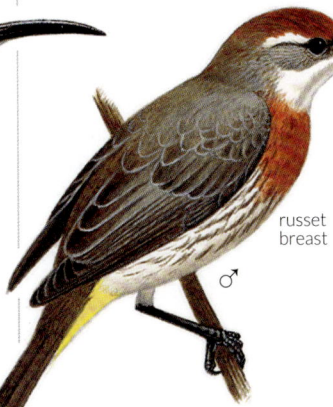

russet crown

russet breast

♂

russet breast

♀

GURNEY'S SUGARBIRD

MALACHITE SUNBIRD

contrasting pale underparts

very long bill

♀

white outer tail

br. ♂

eclipse ♂

imm. ♂

ORANGE-BREASTED SUNBIRD

BRONZY SUNBIRD

♂

dark cheek contrasts with pale, streaked underparts

♀

olive-green plumage

♀

♂

slightly pointed tail

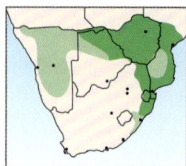

SCARLET-CHESTED SUNBIRD *Chalcomitra senegalensis*

B b b b b b b b B B B B
J F M A M J J A S O N D

14 cm; 10–17 g Male easily identified by its mainly black body and scarlet breast; iridescent green crown and throat visible in good light. Lacks pectoral tufts; no eclipse plumage. Female is dark grey-olive above; throat and breast darker and more heavily mottled than female Amethyst Sunbird; pale supercilium starts above eye (not at base of bill). At close range, whitish edges to wrist feathers/primary coverts visible (lacking in Amethyst Sunbird). **Voice:** Loud, whistled *'cheeup, chup, toop, toop, toop'* song; *'chak chak'* alarm call. **Status and biology:** Common resident, nomad and local migrant in woodland, savanna and suburban gardens; some birds move out of arid areas in winter. (Rooiborssuikerbekkie)

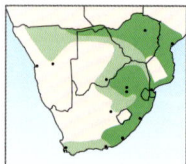

AMETHYST SUNBIRD *Chalcomitra amethystina*

B B b b b b b b B B B b
J F M A M J J A S O N D

15 cm; 11–19 g Male appears all-black, but in good light shows an iridescent green forecrown and violet-purple throat, shoulders and rump. Lacks pectoral tufts; no eclipse plumage. Female similar to female Scarlet-chested Sunbird, but throat and breast paler and less heavily streaked; supercilium starts at base of bill (not above eye). **Voice:** Quite deep, fast, twittering song; loud *'chip chip'* calls. **Status and biology:** Common resident, nomad and local migrant in woodland, forest edge and gardens; some birds move out of arid areas in winter. Range has expanded westward in recent years. (Swartsuikerbekkie)

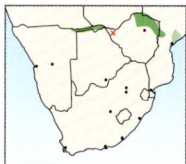

COPPER SUNBIRD *Cinnyris cupreus*

b b b   b
J F M A M J J A S O N D

13 cm; 8–12 g A slender, short-billed sunbird with a fairly long tail. Male is blackish, with coppery iridescence; smaller than Bronzy Sunbird (p. 432) and lacks elongated central tail feathers. Female grey-brown above, creamy below, with a pale supercilium contrasting with its dark face. Tail blue-black, with narrow white tips and outer edges. Eclipse male brown, but retains coppery wing coverts and rump, and black flight feathers. **Voice:** Harsh *'chit-chat'* call; high-pitched warbling song. **Status and biology:** Uncommon to locally common resident and local nomad in woodland and forest edge. (Kopersuikerbekkie)

GREY SUNBIRD *Cyanomitra veroxii*

B b b   b B B B
J F M A M J J A S O N D

14 cm; 11–15 g A nondescript sunbird; sexes alike. Dark grey face and upperparts contrast with paler grey underparts (unlike Olive Sunbird and female Greater Double-collared Sunbird, p. 436); pectoral tufts red. At close range, faint blue-green wash to crown, back and wing coverts is visible. Juv. is more yellow-olive below, more closely resembling Olive Sunbird. **Voice:** Song is a loud series of well-spaced *'chreep chreep choop'* phrases; call harsh, grating *'tzzik, tzzik'*. **Status and biology:** Locally common resident in coastal and riverine forests; often feeds at *Strelitzia* trees. (Gryssuikerbekkie)

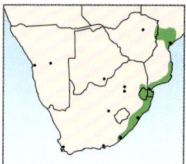

OLIVE SUNBIRD *Cyanomitra olivacea*

B b b   B B B B
J F M A M J J A S O N D

15 cm; 8–15 g A dull, olive sunbird. Much greener than Grey Sunbird, with less contrast between the dark upperparts and paler underparts. Both sexes have yellow pectoral tufts in lowland populations; males only in e Zimbabwe highlands (previously split as Eastern and Western Olive Sunbirds, but not supported by subsequent genetic studies). Juv. has a rust-coloured throat. **Voice:** Call is a sharp *'tuk, tuk, tuk'*; song is a series of descending, piping notes, accelerating in pace, and sometimes rising in pitch at the end. Male also makes a monotonous insect-like *'teep teep teep'* advertisement call. **Status and biology:** Common resident in coastal, riverine and montane forests; usually the most abundant forest sunbird within its range. (Olyfsuikerbekkie)

SCARLET-CHESTED SUNBIRD

dark, mottled underparts

pale primary covert patch

♀

♂

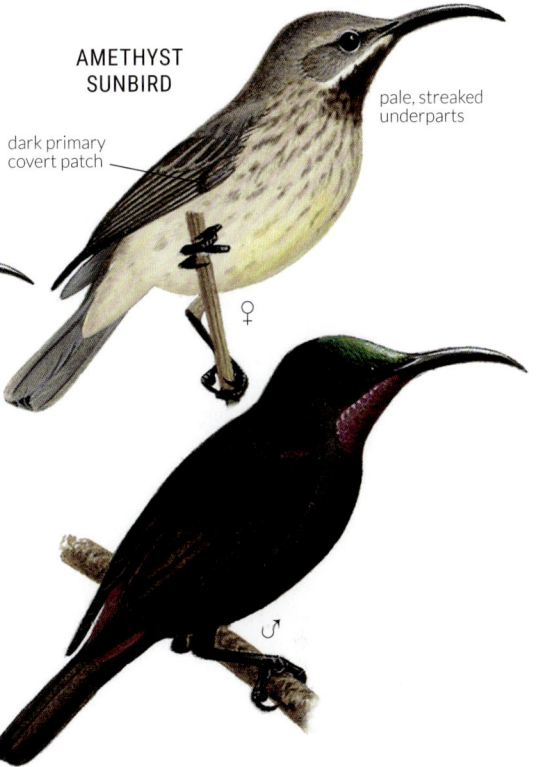

AMETHYST SUNBIRD

pale, streaked underparts

dark primary covert patch

♀

♂

COPPER SUNBIRD

♂

plain pale underparts

♀

GREY SUNBIRD

red pectoral tufts when displaying

uniform pale grey underparts

yellow pectoral tufts

OLIVE SUNBIRD

unstreaked, dull olive underparts

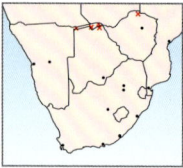

SHELLEY'S SUNBIRD *Cinnyris shelleyi*

X X X X X X X X X X X X
J F M A M J J A S O N D

11 cm; 9–11 g Vagrant. Male has a broad, red breast band and sooty-black belly. Larger and longer billed than male Neergaard's Sunbird, with a broader red breast band, but ranges do not overlap. Eclipse male has a paler belly; could be confused with Eastern Miombo Sunbird, but rump iridescent green (not blue). Female has lightly streaked underparts (not plain as female Eastern Miombo Sunbird); shorter billed, greyer above and less yellow below than female Marico Sunbird. **Voice:** Call is a *'didi-didi'*; song is a nasal *'chibbee-cheeu-cheeu'*. **Status and biology:** Vagrant or possibly rare resident and nomad in broadleafed woodland; recorded with any regularity only in e Zambezi Region (Caprivi), Namibia, and at Mana Pools, n Zimbabwe. (Swartpensuikerbekkie)

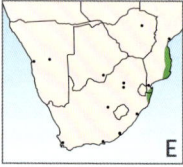

NEERGAARD'S SUNBIRD *Cinnyris neergaardi*

b   b B B B
J F M A M J J A S O N D

10 cm; 5–7 g A tiny, compact sunbird with a short, thin bill. Male has a red breast band and dark grey-black belly; smaller and shorter billed than rare Shelley's Sunbird; ranges do not overlap. Pectoral tufts yellow. Female is grey-brown above and unstreaked, pale grey below (female Purple-banded Sunbird has light streaking on breast). **Voice:** Thin, wispy *'weesi-weesi-weesi'* and short, chippy song. **Status and biology:** NEAR-THREATENED. Endemic. Uncommon, localised resident in acacia savanna and sand forest. (Bloukruissuikerbekkie)

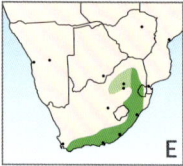

GREATER DOUBLE-COLLARED SUNBIRD *Cinnyris afer*

b b b b b b B B B B B b
J F M A M J J A S O N D

14 cm; 10–18 g Larger than Southern Double-collared Sunbird, with a longer, heavier bill. Male has a broader, red breast band bordered above by a violet-blue band. Pectoral tufts yellow. Female grey-brown above, paler below; best separated from female Southern Double-collared Sunbird by its larger size and longer, heavier bill. Imm. male has only a few red breast feathers, sometimes a narrow breast band. **Voice:** Harsh *'tchut-tchut-tchut'* calls; song is fast and twittering, but deeper and slower than Southern Double-collared Sunbird. **Status and biology:** Endemic. Common resident in tall shrubs, forest fringes, clearings and gardens. (Groot-rooibandsuikerbekkie)

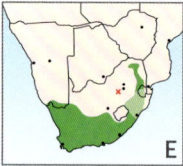

SOUTHERN DOUBLE-COLLARED SUNBIRD *Cinnyris chalybeus*

b b b b b B B B B B b b
J F M A M J J A S O N D

12 cm; 6–10 g Smaller than Greater Double-collared Sunbird, with a shorter, more slender bill; male has a narrower, red breast band (but beware, northern subspecies has a broader breast band). Pectoral tufts yellow. Imm. male has patchy plumage. Female is grey-brown (not olive, as female Orange-breasted Sunbird, p. 432), darker above than below, but not as pale below as female Dusky Sunbird (p. 438). Juv. is like female, with yellow gape. **Voice:** Harsh *'chee-chee'* call; song is a very rapid, high-pitched twitter, rising and falling in pitch. **Status and biology:** Endemic. Common resident in coastal and Karoo scrub, fynbos and forests (in northeast and extreme west). (Klein-rooibandsuikerbekkie)

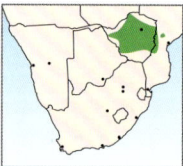

EASTERN MIOMBO SUNBIRD *Cinnyris manoensis*

b b b b b b B B B B B
J F M A M J J A S O N D

13 cm; 8–11 g Similar to Southern Double-collared Sunbird, but male has dull blue rump (not iridescent green); ranges do not overlap. Told from Shelley's Sunbird by pale olive (not blackish) belly and blue (not green) rump. Female is grey-brown above, pale grey below, often with a yellowish wash on the central belly. *C. m. amicorum* on Mt Gorongoza, central Mozambique, is larger, recalling Greater Double-collared Sunbird (and was once classified as that species). **Voice:** Song is a rapid warble, similar to Southern Double-collared Sunbird. **Status and biology:** Common resident in miombo woodland, gardens and edges of montane forest. (Miombo-rooibandsuikerbekkie)

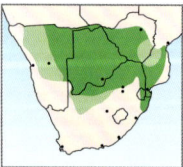

MARICO SUNBIRD *Cinnyris mariquensis*

b b   b B B B b
J F M A M J J A S O N D

12 cm; 8–14 g Markedly larger than Purple-banded Sunbird, with a longer, thicker and more decurved bill. Male has a black or sooty-grey belly and a narrow, violet-blue breast band above a broader purple-maroon band. Purple breast band is broader than that of male Purple-banded Sunbird; rump green (not blue). Female has grey-brown upperparts and pale yellow, dusky-streaked underparts; tail dark grey with white outer edges. Eclipse and juv. male resemble female, but have dark throats. **Voice:** Long series of rapid *'tsip-tsip-tsip'* phrases; fast, warbling song. **Status and biology:** Common resident in woodland and savanna, often in fairly arid areas. (Maricosuikerbekkie)

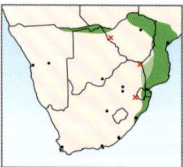

PURPLE-BANDED SUNBIRD *Cinnyris bifasciatus*

b b b   b B B B b
J F M A M J J A S O N D

10 cm; 6–9 g Smaller than Marico Sunbird, with a thinner, shorter and less decurved bill. Male has a narrower purple breast band and blue (not green) rump. Female is pale yellow below with less streaking than female Marico Sunbird; lacks white outer edge to tail. Eclipse and juv. male resemble female, but have dark throats. **Voice:** High-pitched *'teeet-teeet-tit-tit'* song, accelerating at end, not sustained as song of Marico Sunbird. **Status and biology:** Fairly common resident in woodland, moist savanna and coastal scrub. (Purperbandsuikerbekkie)

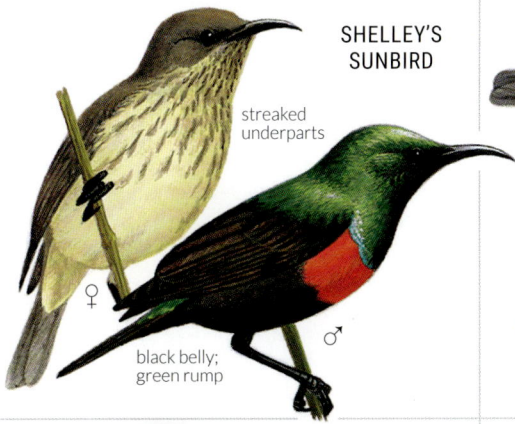

SHELLEY'S
SUNBIRD

streaked
underparts

black belly;
green rump

♀

♂

NEERGAARD'S
SUNBIRD

♀

black belly,
blue rump

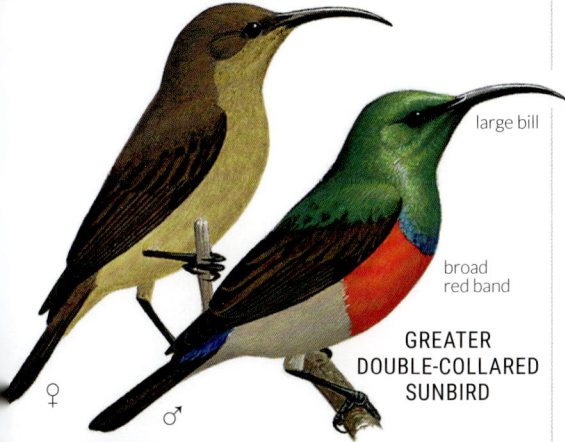

♂

GREATER
DOUBLE-COLLARED
SUNBIRD

large bill

broad
red band

♀

♂

SOUTHERN
DOUBLE-COLLARED
SUNBIRD

narrow
red band

♀

♂

EASTERN MIOMBO
SUNBIRD

♀

blue
rump

♂

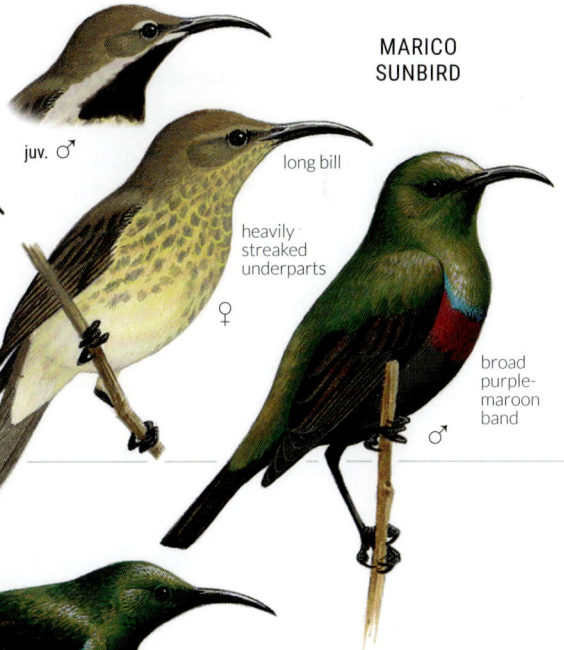

MARICO
SUNBIRD

juv. ♂

long bill

heavily
streaked
underparts

♀

broad
purple-
maroon
band

♂

juv. ♂

shortish
bill

lightly
streaked
breast

♀

narrow
purple
band

♂

PURPLE-BANDED
SUNBIRD

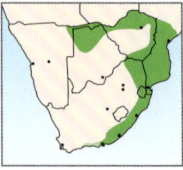

COLLARED SUNBIRD *Hedydipna collaris*

B b b | | b b b B B B B
J F M A M J J A S O N D

10 cm; 6–11 g A small, very short-billed sunbird with metallic green upperparts. Male has an iridescent green throat and upper breast, separated from the yellow belly by a narrow, iridescent purple band. Pectoral tufts yellow. Smaller and much shorter billed than Variable Sunbird; throat green (not blue) and extends less far onto breast. Female slightly duller, with yellow underparts; lacks white eye-ring of white-eyes (p. 404). **Voice:** Soft *'tswee'* and sharp *'tic'* calls; song a series of rather harsh, chirpy notes. **Status and biology:** Common resident in forest, dense woodland and gardens. (Kortbeksuikerbekkie)

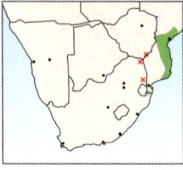

PLAIN-BACKED SUNBIRD *Anthreptes reichenowi*

| | | | | | | | | | b b b
J F M A M J J A S O N D

10 cm; 7–9 g A slender, warbler-like sunbird, with olive-brown upperparts and a short bill. Male is pale yellow below, with an iridescent blue-black forehead and narrow throat patch, bordered by a yellow moustachial stripe. Imm. male Variable and White-bellied Sunbirds can have small, blackish throats, but have longer bills, lack iridescent frons and usually have a few iridescent back or wing feathers. Female is duller, with a grey-brown throat and breast; told from female Collared Sunbird by its olive-brown (not iridescent green) upperparts. Greener above than female Variable Sunbird; lacks white eye-ring of white-eyes (p. 404). **Voice:** A warbler-like *'tsweep tsweep twit tweet tweet'* song; soft *'tik-tik'* call. **Status and biology:** NEAR-THREATENED. Uncommon resident in moist woodland and coastal forests. (Bloukeelsuikerbekkie)

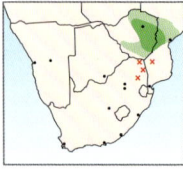

VARIABLE SUNBIRD *Cinnyris venustus*

| | | B | B B B B B B b
J F M A M J J A S O N D

11 cm; 6–10 g Male is iridescent blue-green above, with a violet-blue crown and throat, and purple breast band. Told from male White-bellied Sunbird by its yellow belly (in s Africa); has a longer, more decurved bill and more extensive dark breast than Collared Sunbird. Female is grey-olive above and plain yellow-olive below; has blackish tail with more extensive white tail tip than female White-bellied Sunbird. Juv. male has blackish throat; told from Plain-backed Sunbird by its longer bill and some iridescence on wing coverts. **Voice:** Trilling song often starts with a rapid *'tsui-tse-tse...'*. **Status and biology:** Common resident in woodland, forest edge, gardens and plantations. (Geelpenssuikerbekkie)

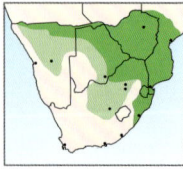

WHITE-BELLIED SUNBIRD *Cinnyris talatala*

b b b b | b b b B B B b
J F M A M J J A S O N D

11 cm; 6–9 g Male has an iridescent green back, head and breast, a purple breast band and a white belly. Female is grey-brown above and off-white (rarely yellowish) below; told from female Dusky Sunbird by its indistinctly streaked (not plain) breast, and only outer-tail feathers tipped whitish (not whole tail). Juv. male has a blackish throat. **Voice:** Song is a rapid series of trilled notes, often preceded by *'pitchee'* call. **Status and biology:** Common resident in dry woodland, savanna and gardens. (Witpenssuikerbekkie)

WESTERN VIOLET-BACKED SUNBIRD *Anthreptes longuemarei*

| | | | | | | B B B b b
J F M A M J J A S O N D

13 cm; 10–13 g A short-billed sunbird with a glossy, purple-black tail. Male has a violet head, throat, mantle and rump, appearing blackish in poor light; underparts white. Female is grey-brown above with a conspicuous white eye stripe; underparts white, with belly washed yellow. **Voice:** Sharp *'chit-chit'* or *'skee'*; song is a soft, high-pitched warble. **Status and biology:** Uncommon to locally common resident in miombo woodland. Often joins bird parties. Pairs remain together year-round. (Blousuikerbekkie)

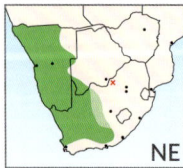

DUSKY SUNBIRD *Cinnyris fuscus*

b B B B b b b b b b b b
J F M A M J J A S O N D

11 cm; 6–11 g Male has a glossy, black head, throat and back, merging into a white belly; pectoral tufts orange. Female is pale grey-brown above and off-white below; paler below than female Southern Double-collared Sunbird (p. 436). Breast plain, not lightly streaked as female White-bellied Sunbird. Juv. male has a blackish throat. **Voice:** Short, canary-like song usually ends in a trill; sharp *'tsik'* and *'ji-dit'* calls. **Status and biology:** Near-endemic. Common resident and nomad in arid savanna, acacia thickets and Karoo scrub; irrupts into more mesic areas in drought years. (Namakwasuikerbekkie)

COLLARED SUNBIRD

shiny green above

short bill

narrow, purple band

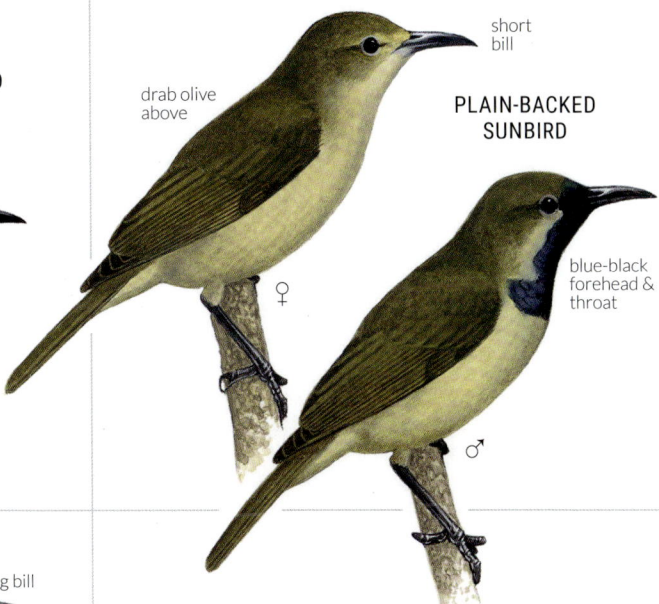

♀

♂

PLAIN-BACKED SUNBIRD

short bill

drab olive above

♀

blue-black forehead & throat

♂

VARIABLE SUNBIRD

imm. ♂

long bill

dull yellow belly & flanks

♀

broad, purple band

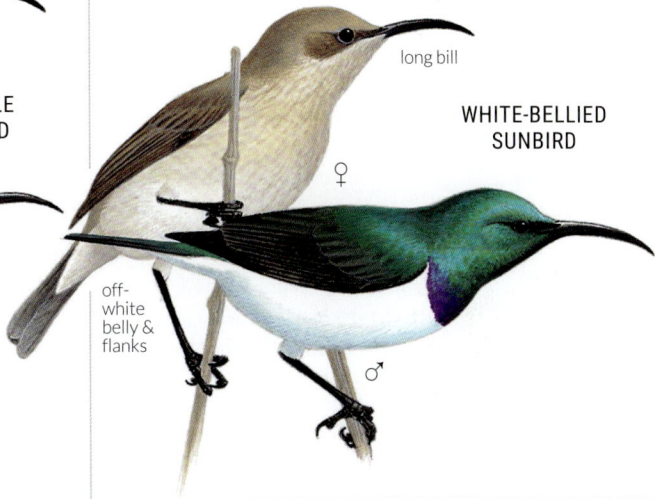

♂

WHITE-BELLIED SUNBIRD

long bill

♀

off-white belly & flanks

♂

WESTERN VIOLET-BACKED SUNBIRD

white eyebrow

♀

short bill

violet tail

♂

DUSKY SUNBIRD

♀

whitish vent

♂

SPARROWS

SPARROWS Finch-like passerines with stout, conical bills. Most species have rufous or chestnut backs. Sexes differ in most species; juvs like dull females. Bill often black when br. Nests are untidy, ball-like structures, built in cavities by some species.

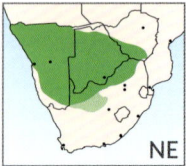

GREAT SPARROW *Passer motitensis*

B	B	B	b					b	B	B	B
J	F	M	A	M	J	J	A	S	O	N	D

15 cm; 26–35 g Larger than House Sparrow, with a heavier bill. Male has paler chestnut nape and back, and chestnut (not grey) rump. Female has chestnut back and shoulders (not sandy brown as in female House Sparrow); crown pale grey (not brown). Told from female Cape Sparrow by its dark eye stripe, chestnut supercilium and dark-streaked (not plain) chestnut mantle. **Voice:** Slow, measured *'cheereep, cheereeu'*, slower and slightly deeper than House Sparrow. **Status and biology:** Near-endemic. Locally common resident and nomad in semi-arid acacia woodland. Nests in a thorn tree or bush. (Grootmossie)

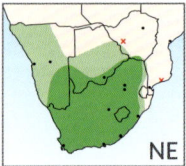

CAPE SPARROW *Passer melanurus*

B	B	b	b	b	b	b	b	B	B	B	B
J	F	M	A	M	J	J	A	S	O	N	D

15 cm; 22–36 g Male has a striking black-and-white head. Female duller, with pale grey head, whitish supercilium that almost links with whitish throat (unlike plain grey head of grey-headed sparrows). Mantle unstreaked, unlike female House and Great Sparrows. **Voice:** Series of musical cheeps, *'chirp chroop, chirp chroop'*; also a short, rattled alarm call. **Status and biology:** Near-endemic. Common resident in grassland, fields and large gardens; also in some urban areas. Nests in bushes or small trees (often thorn trees), or in a wall or roof cavity. (Gewone Mossie)

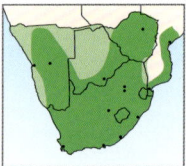

HOUSE SPARROW *Passer domesticus*

b	b	b	b	b	b	b	b	B	B	B	B
J	F	M	A	M	J	J	A	S	O	N	D

14 cm; 22–30 g Introduced from Europe and India. Smaller than Great Sparrow, with darker, reddish-brown upperparts and grey (not chestnut) rump. Black bib concealed by buff feather tips in fresh plumage; gradually exposed by wear to be most prominent in br. season; size signals social status. Female is sandy grey-brown, with a pale buff supercilium that is much smaller than that of Yellow-throated Bush Sparrow; also smaller, with a shorter, stouter bill. **Voice:** Song is a medley of chirps, chips and *'chissick'* notes. **Status and biology:** Locally common resident in and around human habitation. Small flocks may occur away from buildings. Nests singly or in loose groups, usually in a cavity in a building (rarely in a tree). (Huismossie)

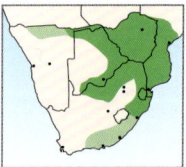

YELLOW-THROATED BUSH SPARROW (PETRONIA)
Gymnoris superciliaris

b	b	b						b	B	B	b
J	F	M	A	M	J	J	A	S	O	N	D

15 cm; 21–29 g A grey-brown sparrow with a broad, creamy-white supercilium. Sexes alike. The small, yellow spot at the base of the white throat is seldom visible; absent in juv. May be confused with Streaky-headed Seedeater (p. 466), but latter has a streaked crown and lacks pale wingbars. **Voice:** 3 or 4 fast, chipping notes, either all the same, or first or last notes higher or lower pitched. **Status and biology:** Locally common resident in dry woodland, savanna and riverine bush. Forages on the ground and along larger branches of trees. Breeds in a tree cavity or old woodpecker or barbet hole. (Geelvlekmossie)

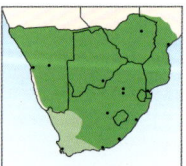

SOUTHERN GREY-HEADED SPARROW *Passer diffusus*

b	b	b	b	b	b	b	b	b	b	b	b
J	F	M	A	M	J	J	A	S	O	N	D

15 cm; 20–30 g A plain sparrow with a warm brown back and dove-grey head; sexes alike. Told from all other species, except localised Northern Grey-headed Sparrow, by its plain grey head. **Voice:** Rather slow series of loud chirping notes, *'tchep tchierp tchep'*; also an alarm rattle. **Status and biology:** Common resident in woodland and other areas with trees; occurs in gardens, but generally avoids urban areas. Range has expanded westwards in recent years. Breeds in a hole in a tree, pole or building; also in old nests of other birds. (Gryskopmossie)

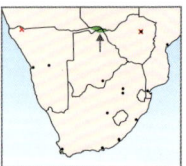

NORTHERN GREY-HEADED SPARROW *Passer griseus*

B	B	b	b								
J	F	M	A	M	J	J	A	S	O	N	D

16 cm; 34–43 g Slightly larger and darker than Southern Grey-headed Sparrow. Dark grey head contrasts more with rufous mantle and whitish throat; white shoulder stripe reduced or sometimes absent. Bill is heavier and is black year-round (lacks pale bill of Southern Grey-headed Sparrow outside br. season). **Voice:** More varied than Southern Grey-headed Sparrow, with some higher-pitched notes. **Status and biology:** Locally common resident, mainly around settlements. Confined to a few areas in north of region, especially between Chobe, Botswana, and Mana Pools, Zimbabwe. (Witkeelmossie)

GREAT SPARROW

broad chestnut
supercilium

heavy
bill

small
bib

chestnut
rump

♀

♂

CAPE
SPARROW

HOUSE
SPARROW

large
bib

♂

pale
supercilium

plain rufous
back

grey
rump

thin, buff
supercilium

mottled
back

♂

♀

♀

YELLOW-THROATED
BUSH SPARROW

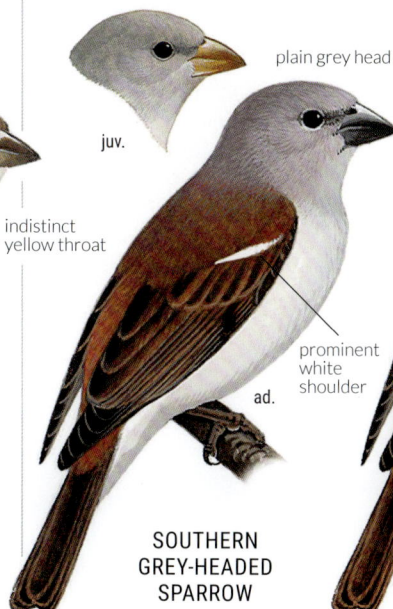

broad white
supercilium

plain grey head

large
bill

juv.

weak white
shoulder

indistinct
yellow throat

ad.

prominent
white
shoulder

prominent
white
throat

juv.

ad.

SOUTHERN
GREY-HEADED
SPARROW

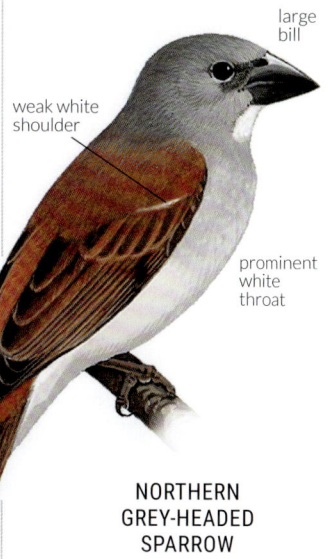

NORTHERN
GREY-HEADED
SPARROW

WEAVERS
A diverse group of sparrow-like finches renowned for weaving complex, ball-like nests. Sexes usually differ. Males of open-country species have an eclipse plumage and are often hard to identify outside the br. season, whereas forest and woodland species lack an eclipse plumage and have more distinctive females. Juvs resemble dull females.

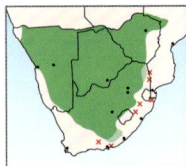

WHITE-BROWED SPARROW-WEAVER *Plocepasser mahali*

b	b	b	b	b	b	b	b	b	B	B	B
J	F	M	A	M	J	J	A	S	O	N	D

17 cm; 41–54 g A chunky, sparrow-like weaver, with blackish, brown and white plumage. Larger and much more boldly plumaged than Yellow-throated Bush Sparrow (p. 440). Bill black (male), horn-coloured (female) or pinkish-brown (juv.). In northeast, *P. m. pectoralis* has brown breast spots. **Voice:** Loud, liquid *'cheeooop-preeoo-chop'* song; also harsh *'chik-chik'* call. **Status and biology:** Common resident in acacia savanna and dry woodland. Usually in small groups; breeds cooperatively. Nests are untidy bundles of grass built at the edge of the canopy of a thorn tree. Most nests for roosting; only the dominant pair breeds. (Koringvoël)

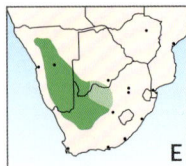

SOCIABLE WEAVER *Philetairus socius*

b	b	B	B	b	b	b	b	b	b	b	b
J	F	M	A	M	J	J	A	S	O	N	D

14 cm; 24–32 g A distinctive, sparrow-like weaver with a pale bill, black face mask and scaly back (but see Scaly-feathered Weaver and Red-headed Finch, p. 456). Sexes alike. Ads have scaly, black flanks. Juv. is duller; lacks black chin and has a finely streaked crown. **Voice:** Metallic, chattering *'chicker-chicker'* song; *'chip'* contact call. **Status and biology:** Endemic. Common resident and local nomad in semi-arid savanna and Karoo scrub. Gregarious, cooperative breeder in a communally thatched nest, built in a tree or on a pole or windmill. The largest nest of any bird, with up to 500 chambers. Provision of telephone poles has allowed westward range expansion. (Versamelvoël)

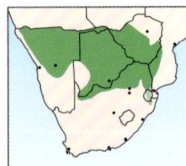

RED-BILLED BUFFALO WEAVER *Bubalornis niger*

B	B	B	b	b	b			b	B	B	B
J	F	M	A	M	J	J	A	S	O	N	D

22 cm; 64–92 g A large, heavy-billed weaver. Male black, with white wing panel and an orange-red bill. Female is browner and slightly scaled. Juv. is paler below with more prominent scaling. **Voice:** Chattering medley of slightly scratchy, nasal notes; also harsh calls. **Status and biology:** Common resident and local migrant in dry woodland and savanna. Breeds in large, untidy stick nests, usually thorny, each containing 1–10 br. chambers; built in a large tree or on an electricity pylon. Some males polygynous, building multiple nests to attract females; other males form coalitions and assist with raising chicks. (Buffelwewer)

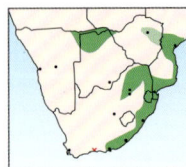

THICK-BILLED WEAVER *Amblyospiza albifrons*

B	B	b							b	B	B
J	F	M	A	M	J	J	A	S	O	N	D

18 cm; 35–56 g A large, dark brown weaver with a massive bill and large head; appears long tailed in flight. Male warm brown, with a white frons and wing patches; bill blackish. Female is whitish below, heavily streaked brown, with a paler bill. **Voice:** High-pitched, chattering song; high-pitched *'tweek tweek'* or *'pink pink'* flight call. **Status and biology:** Common resident and local nomad of forest edge and rank vegetation, often near water. Usually breeds in small colonies in neat, finely woven nests with a side entrance attached to reeds; polygynous. Roosting nests have larger entrances than br. nests. (Dikbekwewer)

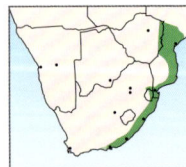

DARK-BACKED WEAVER *Ploceus bicolor*

b	b								b	B	B
J	F	M	A	M	J	J	A	S	O	N	D

15 cm; 28–43 g A forest weaver with blackish-brown upperparts that contrast with golden-yellow breast and belly. Sexes alike, with no eclipse plumage; throat variably grizzled grey. Could be confused with the localised Olive-headed Weaver (p. 444) in Mozambique, but has dark brown (not olive) upperparts and dark (not pale) eyes. **Voice:** Song is a squeaky, musical duet, interspersed with shrill, buzzing notes; also a soft, high-pitched *'tseep'* call. **Status and biology:** Common resident in forest and dense woodland; often joins bird parties. Territorial; pairs remain together year-round. Pairs nest solitarily; builds a roughly woven nest of vine stems with a long entrance tube; monogamous. (Bosmusikant)

WHITE-BROWED
SPARROW-WEAVER

♀

♂

SOCIABLE
WEAVER

juv.

ad.

RED-BILLED
BUFFALO WEAVER

♀

white in
wings

♂

THICK-BILLED
WEAVER

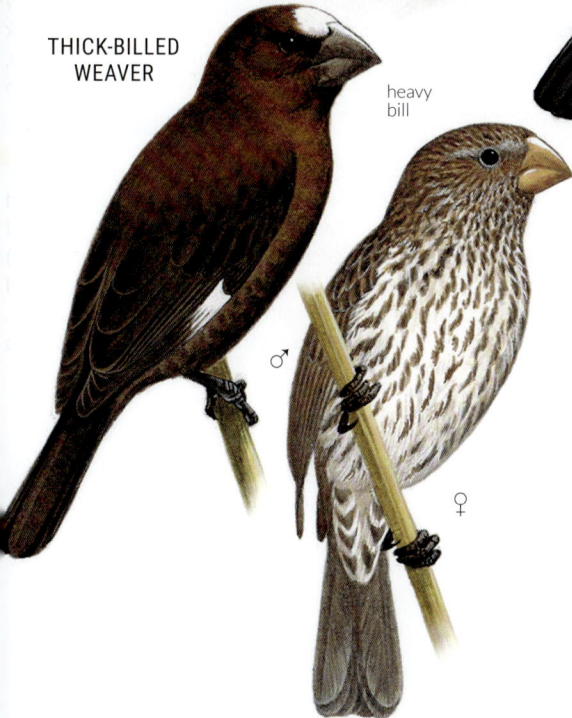

heavy
bill

♂

♀

DARK-BACKED
WEAVER

♀

♂

443

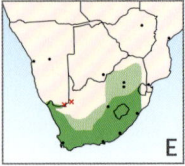

CAPE WEAVER *Ploceus capensis*

b b | | | | | b b B B B B
J F M A M J J A S O N D

17 cm; 28–54 g A large weaver with a long, pointed bill. Br. male is less yellow above than 'golden' weavers, with chestnut wash on face and throat; lacks well-defined bib of Southern Brown-throated Weaver and has whitish (not red-brown) eyes. Female has brown eyes. Female and non-br. male are olive above and dull yellow below; lack pale belly of female Southern Masked Weaver (p. 446). Bill is longer and dagger-like and is a useful identification feature in all plumages. **Voice:** Harsh, hysterical swizzling song; *'chack'* contact call. **Status and biology:** Endemic. Common resident in grassland, scrub and gardens. Breeds in small or large colonies in reeds or tall trees; builds a large, neatly woven nest with short entrance tube; polygynous. (Kaapse Wewer)

SOUTHERN BROWN-THROATED WEAVER
Ploceus xanthopterus

b | | | | | | | b B B
J F M A M J J A S O N D

14 cm; 18–30 g Br. male is bright yellow, with a distinctive chestnut (not black) bib confined to face and throat, not extending onto forecrown as in Cape Weaver; eyes red-brown (not whitish). Br. female has yellow underparts and brown eyes. Non-br. birds duller, with horn-coloured bills, buff breasts and flanks, and white bellies; rumps and backs washed rufous. **Voice:** Swizzling song often ends with 2 buzzy notes, *'zweek, zweek'*. **Status and biology:** Locally common resident in forest and scrub close to marshy br. sites. Breeds in small or large colonies, mostly in reeds over water; builds a small nest with no entrance tube; polygynous. (Bruinkeelwewer)

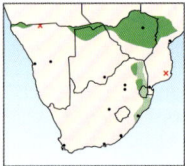

HOLUB'S (AFRICAN) GOLDEN WEAVER *Ploceus xanthops*

B B b b | | | b b B B B
J F M A M J J A S O N D

17 cm; 35–50 g A large, golden weaver with pale yellow eyes (often appearing white) and a heavy bill; lacks an eclipse plumage. Male's bright yellow head and breast lack richer tones of male Eastern Golden Weaver; eyes yellow, not red as in Eastern Golden Weaver. Female duller, with an olive wash on the head and underparts; lacks whitish belly of most other female weavers. Juv. has brown eyes. **Voice:** Typical swizzling weaver song, deeper than Eastern Golden Weaver; also *'chuk'* call. **Status and biology:** Locally common resident in woodland and savanna. Breeds in small, usually single-male, colonies; builds an untidy, roughly woven nest that lacks an entrance tube; polygynous. (Goudwewer)

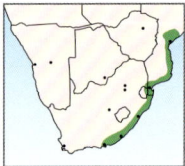

EASTERN GOLDEN (YELLOW) WEAVER *Ploceus subaureus*

b b | | | | | | B B B B
J F M A M J J A S O N D

15 cm; 22–39 g Br. male is smaller than Holub's Golden Weaver, with a smaller, shorter bill and orange wash to head and breast; eyes red (not yellow). Non-br. male duller, with olive crown and cheeks contrasting with yellow supercilium. Female has heavily streaked back and wings, olive-yellow head and breast and a whitish belly; bill pale pink; eyes brown. **Voice:** Typical swizzling weaver song, higher pitched than Holub's Golden Weaver; soft *'chuk'* contact call. **Status and biology:** Locally common resident in woodland, savanna and gardens. Breeds in small or large colonies, usually in reeds over water; builds a neatly woven nest with no entrance tube; polygynous. (Geelwewer)

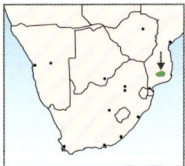

OLIVE-HEADED WEAVER *Ploceus olivaceiceps*

b b | | | | | b b b b
J F M A M J J A S O N D

14 cm; 18–24 g A woodland weaver that differs from Dark-backed Weaver (p. 442) by its olive (not blackish-brown) upperparts and pale (not dark) eyes. Male has an olive-yellow crown and chestnut breast. Female has an olive head and small, chestnut breast patch. Juv. is paler. **Voice:** Soft, whistled contact calls and short song. **Status and biology:** NEAR-THREATENED. Uncommon resident in miombo woodland with abundant epiphytic lichens; some 100 pairs around Panda in s Mozambique, but deforestation threatens the regional population with extinction. Often joins bird parties, gleaning along branches. Breeds singly in a well-camouflaged nest of lichens with a longish entrance tube slung under a branch in the canopy of a tall tree. (Olyfkopwewer)

CAPE WEAVER

long pointed bill

white eye

♀

br. ♂

SOUTHERN BROWN-THROATED WEAVER

brown eye

chestnut wash

♀

br. ♂

HOLUB'S GOLDEN WEAVER

♀

dull olive upperparts

pale eye

large, black bill

♂

EASTERN GOLDEN WEAVER

♀

red eye

br. ♂

OLIVE-HEADED WEAVER

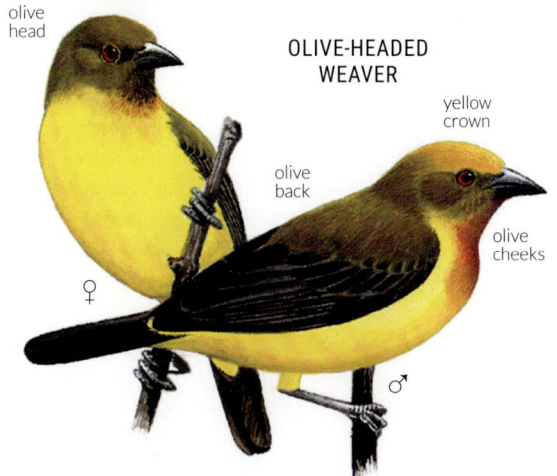

olive head

yellow crown

olive back

olive cheeks

♀

♂

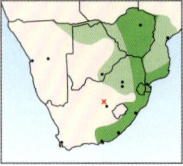

VILLAGE WEAVER *Ploceus cucullatus*

B | b | b | | | b B B B B
J F M A M J J A S O N D

16 cm; 30–46 g A large, heavy-billed, masked weaver with a mottled black-and-yellow back; eyes dark red. Extent of black on br. male's head varies: crown mostly yellow in *P. c. spilonotus* (S Africa and adjacent Mozambique); entire head and nape black in *P. c. nigriceps* further north. Female and non-br. male have prominent yellow supercilium, throat and breast, and whitish belly. Back dull, mottled, greyish; eyes red-brown. **Voice:** Buzzy, swizzling song; throaty *'chuck-chuck'* call. **Status and biology:** Common resident in savanna, fields and gardens; often in large flocks. Breeds in large, multiple-male colonies in reeds or in tall trees; builds a large, coarsely woven nest with short entrance tube; polygynous. (Bontrugwewer)

SOUTHERN MASKED WEAVER *Ploceus velatus*

B B b b | | b b B B B B
J F M A M J J A S O N D

16 cm; 30–44 g Br. male told from Village Weaver by its mostly plain, greenish back (not mottled black-and-yellow). Larger than Lesser Masked Weaver, with a heavier bill and red (not whitish) eyes; legs pinkish (not blue-grey). Female and non-br. male much duller, with buffy-yellow breast and white belly; bill paler and eyes duller red-brown. **Voice:** Typical swizzling weaver song; sharp *'zik, zik'* call. **Status and biology:** Common resident in savanna, grassland, watercourses in arid areas, fields and gardens. Breeds in single-male colonies (rarely multiple-male) in trees or reeds; builds a neatly woven nest with short entrance tube; polygynous. (Swartkeelgeelvink)

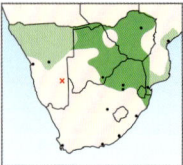

LESSER MASKED WEAVER *Ploceus intermedius*

B b b | | | b b B B B
J F M A M J J A S O N D

13 cm; 15–32 g A small masked weaver with a slender bill and whitish eyes. Br. male has a black mask extending onto its forecrown and a notch above its eye. Eye whitish (not red as in Village Weaver and Southern Masked Weaver). Female and non-br. male washed yellow below; told by pale eyes and blue-grey (not reddish) legs. Juv. has brown eyes. **Voice:** Typical swizzling weaver song. **Status and biology:** Locally common resident in savanna and woodland. Breeds colonially in trees and reeds near water, building a rather untidy nest with a short to medium-length entrance tube; polygynous. (Kleingeelvink)

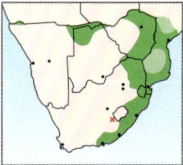

SPECTACLED WEAVER *Ploceus ocularis*

B b b | | | | b B B B
J F M A M J J A S O N D

15 cm; 27–32 g A bright yellow weaver with a plain, olive-green back, slender bill and black eye stripe; eyes whitish. Male has black throat and chestnut-brown wash on face and breast. Lacks an eclipse plumage. Juv. lacks eye stripe. **Voice:** Descending, whistled *'dee-dee-dee-dee-dee'* call given by both sexes; male has short, swizzling song. **Status and biology:** Common resident of forest edge, moist woodland, thickets and gardens; usually in pairs. Pairs nest solitarily, building a neat, finely woven nest with a long entrance tube; monogamous. (Brilwewer)

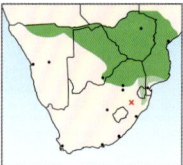

RED-HEADED WEAVER *Anaplectes rubriceps*

b b | | | b b B B B B
J F M A M J J A S O N D

14 cm; 17–26 g A grey-backed weaver with yellow-edged wing feathers and a long, slender, orange-red bill; belly white. Head and breast of br. male scarlet; lemon-yellow in female and non-br. male. **Voice:** High-pitched swizzling song; squeaky *'cherra-cherra'* calls. **Status and biology:** Locally common resident and nomad in woodland, savanna and gardens. Breeds in small or large colonies in trees; builds a large, roughly woven nest of fine twigs with a long entrance tube; monogmous or polygynous. (Rooikopwewer)

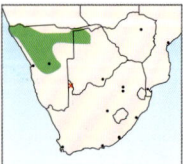

CHESTNUT WEAVER *Ploceus rubiginosus*

B B B b | | | | | | | b
J F M A M J J A S O N D

15 cm; 25–37 g Br. male has distinctive chestnut body and black head. Female and non-br. male sandy brown, heavily streaked dark brown above, with whitish-buff edges to wing feathers; lack yellow-green in plumage of other female and non-br. weavers. Recall non-br. widowbirds; best distinguished by reddish eyes and heavy bills. Juv. has streaked breast. **Voice:** Swizzling song at colonies. Typical call is a melodic *'whut whut'*, also harsh *'chuk'*. **Status and biology:** Locally common resident and partial migrant in arid savanna and grassy woodland. Entered and bred in S Africa for the first time in summer of 2010/11, following heavy La Niña-associated rains. Breeds in large multiple-male colonies in trees; builds a bulky, roughly woven nest with a short entrance tube; polygynous. (Bruinwewer)

mottled
back

yellow frons

VILLAGE
WEAVER

black head,
red eye

br. ♂
spilonotus

red-
brown
legs

br. ♂ *nigriceps*

♀

SOUTHERN
MASKED
WEAVER

reddish
eye

black
frons

br. ♂

black frons,
whitish eye

LESSER MASKED
WEAVER

whitish
eye

delicate
bill

br. ♂

grey
legs

♀

♀

red-
brown
legs

SPECTACLED
WEAVER

♀

black eye
stripe

♂

grey
back

sharp
orange-
red bill

RED-HEADED
WEAVER

♀

br. ♂

chestnut
eye

large,
pale bill

brownish
breast

♀

CHESTNUT
WEAVER

br. ♂

QUELEAS

QUELEAS Small, compact weavers, often in large flocks. Males have striking br. plumages; females, non-br. males and juvs nondescript. Typically more slender than bishops, with plain (not streaked) breasts. All build an oval nest with a side entrance.

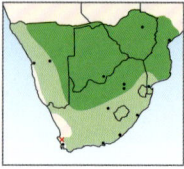

RED-BILLED QUELEA *Quelea quelea*

B B B b | | | | | | | b B
J F M A M J J A S O N D

12 cm; 15–26 g Most br. males have a black face and a buff or pinkish crown and breast; some have creamy or pinkish-purple faces. Bill red (male) or yellow (female). Non-br. male and female have a strong, whitish supercilium and paler underparts than other queleas. Juv. has a pale brown bill. **Voice:** Song is jumble of harsh, weaver-like calls interspersed with more melodious notes; chittering contact calls. **Status and biology:** Abundant nomad in savanna and cereal fields, often in drier areas than other queleas. Range has expanded westwards in recent years. Feeds, roosts and breeds in large flocks, sometimes numbering millions. Breeds colonially, usually in thorn trees; sometimes hundreds or thousands of nests per tree. (Rooibekkwelea)

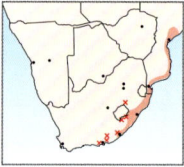

RED-HEADED QUELEA *Quelea erythrops*

B b b | | | | | | | b B
J F M A M J J A S O N D

11 cm; 15–25 g Br. male has a neat, red head and a blackish (not red) bill; most likely confused with rare Cardinal Quelea, but see also Red-headed Finch (p. 456). Br. female has a pale orange wash to the face and throat. Non-br. birds have brown bills (not red or yellow as Red-billed Quelea); supercilium and throat yellowish; breast buff. Juv. is more buffy on head, with broad, buff margins to back feathers. **Voice:** Sizzling chatter; also a sharp 'chip' call. **Status and biology:** Uncommon intra-African br. migrant in damp grassland and adjoining woodland. Breeds in small colonies, usually in a reed bed. (Rooikopkwelea)

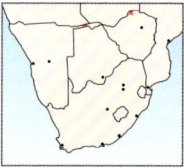

CARDINAL QUELEA *Quelea cardinalis*

| | | | X | | | | | | | |
J F M A M J J A S O N D

10–11 cm; 11–15 g Vagrant. Slightly smaller than other queleas. Br. male has a red head that extends further onto its breast than in male Red-headed Quelea; throat plain red (not finely barred blackish). Non-br. male and female have a streaked crown and face and a strong buff supercilium; throat and breast yellowish; bill horn-coloured. Juv. is buff below, with dusky breast streaking. **Voice:** Sizzling chatter; also a soft 'zeet' call. **Status and biology:** Rare vagrant to wooded grassland and cultivated areas. Only 1 confirmed record: Mana Pools, Zimbabwe, Mar 1999. (Kardinaalkwelea)

BISHOPS AND WIDOWBIRDS

BISHOPS AND WIDOWBIRDS Strongly dimorphic, polygamous birds in the weaver family. Br. males predominantly black; females, non-br. males and juvs are nondescript. More compact than queleas, with streaked breasts. Frequently in flocks when not br. All build an oval nest with a side entrance.

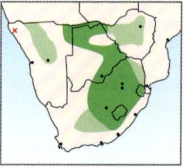

YELLOW-CROWNED BISHOP *Euplectes afer*

B B b b | | | | | b b b
J F M A M J J A S O N D

10 cm; 13–19 g The smallest bishop. Br. male has a diagnostic, yellow crown. Displaying birds appear almost spherical, fluffing out their golden back and rump feathers. Non-br. male and female are smaller than other bishops, with whiter (less buffy) underparts and a prominent buffy-yellow supercilium contrasting with darker ear coverts. **Voice:** High-pitched buzzing and chirping notes. **Status and biology:** Locally common resident and local nomad in grassland. Breeds in small colonies, building nests low down in flooded grass or reeds. (Goudgeelvink)

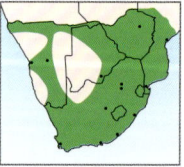

SOUTHERN RED BISHOP *Euplectes orix*

B B b b b b b b b B B B
J F M A M J J A S O N D

12 cm; 18–29 g Br. male is black and red. Told from male Black-winged Bishop by its black (not red) forecrown and brown (not black) flight feathers; latter difference retained year-round. Non-br. male and female have a broad, buffy supercilium, finely streaked breast and brownish bill. **Voice:** Advertising males give buzzing, chirping song; also 'cheet-cheet' flight call and nasal 'wheet' contact call. **Status and biology:** Common, flock-feeding resident and local nomad in grassland, savanna and fields. Usually roosts and breeds in reed beds. (Rooivink)

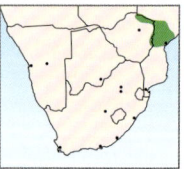

BLACK-WINGED BISHOP *Euplectes hordeaceus*

b B B b | | | | | | | |
J F M A M J J A S O N D

12 cm; 18–28 g Br. male told from Southern Red Bishop by its full red crown and blackish (not brown) flight feathers; latter retained in non-br. plumage. Female similar to non-br. Southern Red Bishop, but has a slightly heavier bill; probably indistinguishable in the field. **Voice:** Buzzing chatter, faster and higher pitched than Southern Red Bishop. **Status and biology:** Locally common resident in damp grassy areas and fields; usually breeds in rank vegetation. (Vuurkopvink)

variable head
pattern

**RED-BILLED
QUELEA**

br. ♂

♀

non-br. ♂

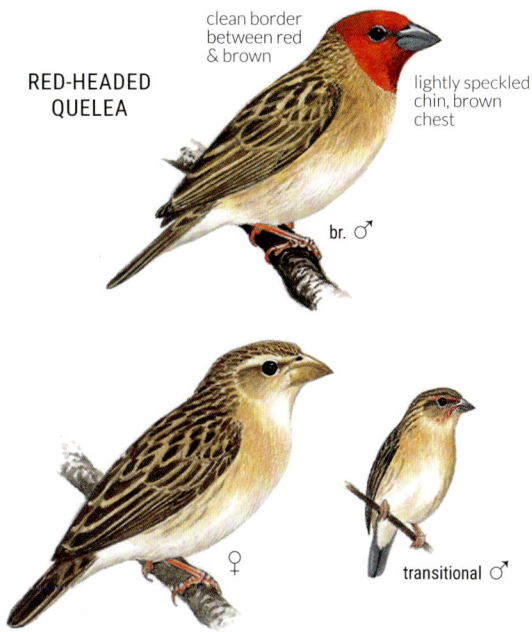

clean border
between red
& brown

**RED-HEADED
QUELEA**

lightly speckled
chin, brown
chest

br. ♂

♀

transitional ♂

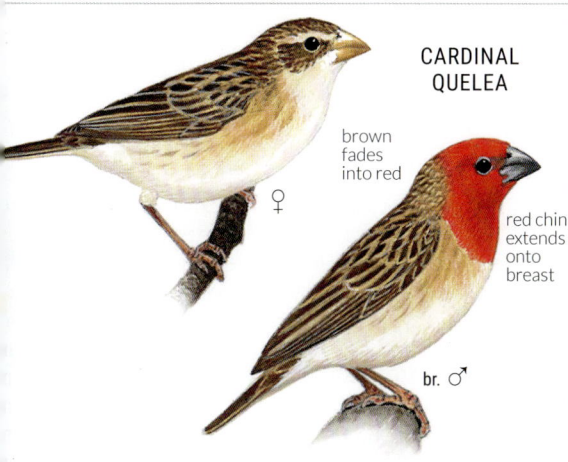

**CARDINAL
QUELEA**

brown
fades
into red

♀

red chin
extends
onto
breast

br. ♂

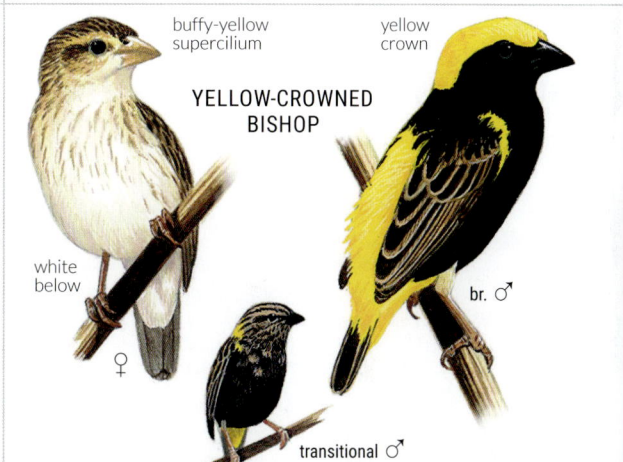

buffy-yellow
supercilium

yellow
crown

**YELLOW-CROWNED
BISHOP**

white
below

♀

br. ♂

transitional ♂

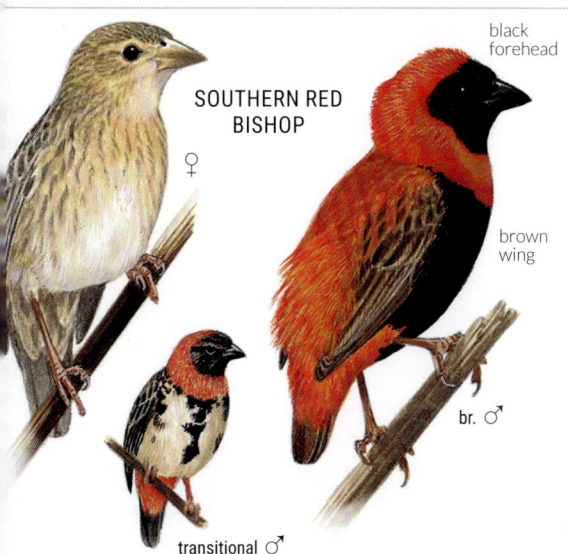

black
forehead

**SOUTHERN RED
BISHOP**

♀

brown
wing

br. ♂

transitional ♂

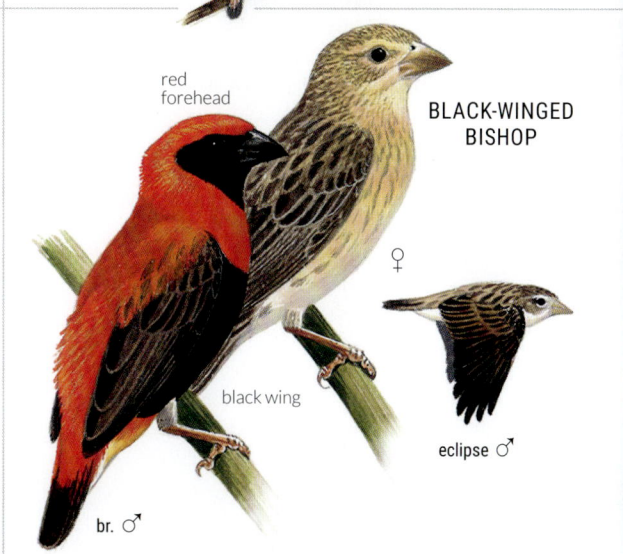

red
forehead

**BLACK-WINGED
BISHOP**

♀

black wing

eclipse ♂

br. ♂

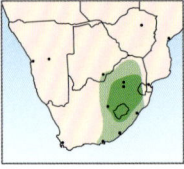

LONG-TAILED WIDOWBIRD *Euplectes progne*

B B b b b b | | | b B B
J F M A M J J A S O N D

16–20 cm (br. male 60 cm); 25–46 g A large widowbird. Br. male has a very long tail and broad, blackish wings with red shoulders bordered whitish-buff. In display, male flies slowly, with trailing tail laterally compressed. Non-br. male retains black flight feathers and red shoulders; larger than Fan-tailed Widowbird, and wings are disproportionately broad. Female is larger than other female widowbirds; underparts buffy-white, lightly streaked on breast and flanks. **Voice:** Male gives a buzzy *'chi-chi-cheet, cheet'* and harsher *'zzit, zzit'* in display. **Status and biology:** Common resident in open grassland, especially in valleys and damp areas, nesting in dense grass. (Langstertflap)

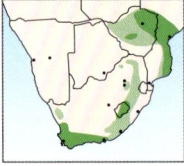

YELLOW BISHOP *Euplectes capensis*

B B b b | | b B B B b b
J F M A M J J A S O N D

15–18 cm; 25–50 g Br. male is black and yellow. Much larger than Yellow-crowned Bishop (p. 448), with a black (not yellow) crown. Shorter tailed than Yellow-mantled Widowbird, with a yellow rump and lower back (not a yellow mantle). Non-br. male is heavily streaked buffy-brown, but retains the bright yellow rump and shoulder. Female is more heavily streaked below than female Yellow-mantled Widowbird, with an olive-yellow (not brown-streaked) rump; appreciably larger and more streaked than other female bishops. Juv. has buff back and less streaking below. Bill size increases along a gradient from northeast to southwest. **Voice:** Displaying males give high-pitched *'zeet, zeet, zeet'* or buzzy *'zzzzzzzzt'* calls. **Status and biology:** Common resident in damp, grassy areas and heathlands. (Kaapse Flap)

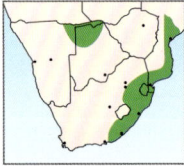

FAN-TAILED WIDOWBIRD *Euplectes axillaris*

B b b | | | | | b b B
J F M A M J J A S O N D

15 cm; 20–32 g Br. male has red shoulders and buff-orange greater coverts. Tail much shorter than Long-tailed Widowbird, but could be confused with males moulting into br. plumage. Broad, rounded tail is fanned in display. Non-br. male retains black primaries and red shoulders. Female is heavily streaked above with reddish-brown shoulders and a large, buffy supercilium; underparts buffy-white, lightly streaked on breast and flanks. Juv. resembles female. **Voice:** High-pitched trilling and hissing calls given in display. **Status and biology:** Common resident in reed beds, rank grassland and sugar cane fields. (Kortstertflap)

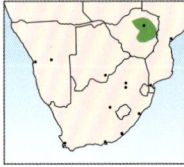

YELLOW-MANTLED WIDOWBIRD *Euplectes macroura*

B b b | | | | | | | | b
J F M A M J J A S O N D

14–20 cm; 15–28 g Br. male is black and yellow with a longer tail than Yellow Bishop; yellow confined to mantle (not lower back and rump). Non-br. male retains yellow shoulders and blackish wings; has longer tail and is less heavily streaked than male Yellow Bishop. Female is less heavily streaked on breast; lacks olive-yellow rump of female Yellow Bishop. Juv. resembles female. **Voice:** Buzzing song *'buzz-itty buzz-itty...'*, typically varying in pitch; also dry *'tseep'* call. **Status and biology:** Locally common resident in damp grassland and marshy areas. (Geelrugflap)

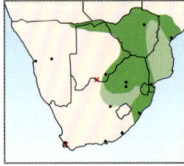

WHITE-WINGED WIDOWBIRD *Euplectes albonotatus*

B B B b b | | | | | | b
J F M A M J J A S O N D

16 cm; 16–26 g Male has a diagnostic, white wingbar formed by white bases to its black flight feathers; coverts also edged white. Br. male black, with yellow shoulders; lacks yellow on back or rump. Non-br. male buffy-brown, but retains white wingbars and yellow shoulders. Female has brown wings; lesser coverts edged yellow; underparts whiter and usually less streaked than other female widowbirds. Juv. resembles female. **Voice:** Dry, high-pitched rattle *'zeh-zeh-zeh'*, sometimes with squeaky or buzzy notes. **Status and biology:** Common resident in tall grass in savanna; also marsh edges and rank vegetation. (Witvlerkflap)

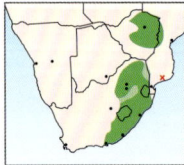

RED-COLLARED WIDOWBIRD *Euplectes ardens*

B B b b | | | | | b B B
J F M A M J J A S O N D

12 cm (br. male 25 cm); 14–25 g Br. male black, with a red breast band and a long, slender tail. Much smaller than Long-tailed Widowbird, with uniform black wings. Female and non-br. male have boldly streaked upperparts with a large, buffy supercilium and unstreaked underparts with a chestnut-brown wash to the breast; male retains black flight feathers. **Voice:** Fast, high-pitched *'tzee-tzee-tzee'* or *'tee-tee-tee'* by displaying male. **Status and biology:** Locally common resident in rank grassland and fields. (Rooikeelflap)

LONG-TAILED WIDOWBIRD

elongated shape

♀

non-br. ♂

white bar below red shoulder

br. ♂

YELLOW BISHOP

non-br. ♂

♀

yellow shoulder & lower back

br. ♂

heavily streaked below

FAN-TAILED WIDOWBIRD

red shoulder lacks white wingbar

br. ♂

♀

br. ♂

yellow shoulder

♀

non-br. ♂

YELLOW-MANTLED WIDOWBIRD

non-br. ♂

black shoulder

br. ♂

RED-COLLARED WIDOWBIRD

non-br. ♂

no streaking

♀

WHITE-WINGED WIDOWBIRD

yellow shoulder & white wingbar

non-br. ♂

br. ♂

♀

pale, lightly streaked below

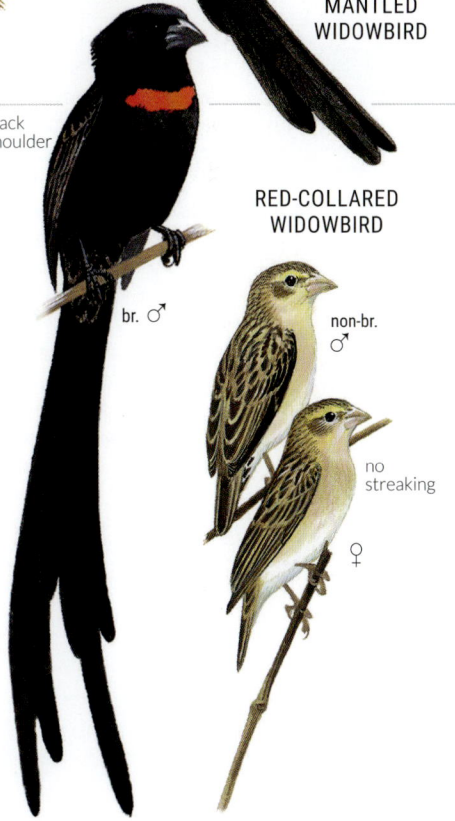

WHYDAHS AND INDIGOBIRDS
Brood-parasitic finches that typically parasitise estrildid finches. Br. male whydahs are distinctive, but indigobirds are difficult to identify; song, soft-part coloration and host range are important. Males establish display sites in summer and, with the exception of Pin-tailed Whydah and Cuckoo-Finch (p. 454), they sing from these to attract females, mimicking the songs of their specific hosts. Sexes differ; non-br. males resemble females. Juvs are plain brown, with a paler supercilium and belly. Unlike cuckoos, the chicks do not eject host eggs or chicks, and the two species are reared together; chicks have mouth-spot patterns that mimic those of their host species.

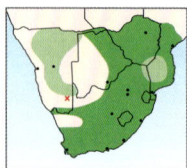

PIN-TAILED WHYDAH *Vidua macroura*

B	B	B	B	b		b	B	B	B	b
J	F	M	A	M	J	J	A	S	O	N	D

12 cm (br. male 34 cm); 13–17 g Br. male is readily identified by its black-and-white plumage and long, wispy tail. Non-br. males have boldly black-and-buff-striped heads, with diagnostic moustachial stripes and red bills. Underparts typically less buffy than other non-br. whydahs and indigobirds. Br. female has a blackish bill, becoming pinker when not br. Juv. is plain brown, with a buffy supercilium; throat whitish, remainder of underparts pale buff; bill black with a red base. May hybridise with Dusky Indigobird (p. 454), resulting in all-black males with shorter tails than typical br. male Pin-tailed Whydah. **Voice:** Sharp *'chip-chip-chip'*, often while hovering jerkily. **Status and biology:** Common resident and local nomad in savanna, grassland, scrub, parks and gardens. Br. males are very aggressive, chasing other species. Main host is Common Waxbill, but also parasitises Orange-breasted and Swee Waxbills. (Koningrooibekkie)

SHAFT-TAILED WHYDAH *Vidua regia*

B	B	B	B	b						b
J	F	M	A	M	J	J	A	S	O	N	D

12 cm (br. male 34 cm); 12–17 g Br. male easily told from male Pin-tailed Whydah by its buff-and-black plumage, all-dark wings, black rump and diagnostic spatulate tail tips. At a distance, the shaft-like central tail is almost invisible, leaving the tips 'suspended' behind the bird. Female and non-br. male have short tails and less boldly streaked heads lacking moustachial stripes; bills orange-pink or brownish; legs red (dark in other female/non-br. male whydahs). Juv. is dull brown, with faint, darker streaking on its back; throat and breast buff-brown, merging into a white belly; bill and legs dark. Occasionally hybridises with indigobirds, producing long-tailed, black males. **Voice:** High-pitched, squeaky and slurred whistles; mimics brood host, Violet-eared Waxbill. **Status and biology:** Near-endemic. Common resident and local nomad in grassy areas in dry acacia savanna. (Pylstertrooibekkie)

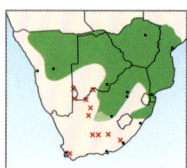

LONG-TAILED PARADISE WHYDAH *Vidua paradisaea*

b	B	B	B	b						
J	F	M	A	M	J	J	A	S	O	N	D

15 cm (br. male 36 cm); 15–28 g Br. male has a long, 'humped-backed' tail. Told from Broad-tailed Paradise Whydah by its more slender tail with long, pointed (not rounded) tips, as well as by its golden-yellow (not chestnut-orange) nape patch and paler chestnut breast. Female and non-br. male have heads striped black and whitish-buff; remainder of upperparts streaked grey-brown. Breast pale buffy-grey, lightly streaked; bill dark grey, typically darker than Broad-tailed Paradise Whydah. Juv. is plain brown, with a paler belly. **Voice:** Male has a high-pitched song including mimicry of the chittering calls and high-pitched whistles of brood host, Green-winged Pytilia; also sharp *'chip-chip'* call. **Status and biology:** Common resident and local nomad in woodland and acacia savanna. (Gewone Paradysvink)

BROAD-TAILED PARADISE WHYDAH *Vidua obtusa*

b	b	b								
J	F	M	A	M	J	J	A	S	O	N	D

15 cm (br. male 27 cm); 18–26 g Br. male has a shorter, much broader tail than male Long-tailed Paradise Whydah, with a darker, chestnut-orange (not golden-yellow) nape patch and breast. Female and non-br. male typically are less heavily streaked above than Long-tailed Paradise Whydah and have a paler, pinkish-grey bill; face less strongly marked (blackish eye stripe doesn't wrap around the ear coverts to the same extent). Juv. is plain brown, with a paler belly. **Voice:** Mimics piping whistles and sharp *'chip'* calls of brood host, Orange-winged Pytilia. **Status and biology:** Uncommon to locally common resident and nomad in miombo and other broadleafed woodland; abundance varies between years. (Breëstertparadysvink)

PIN-TAILED WHYDAH

red bill

br. ♂

♀

non-br. ♂

dark legs
(all ages/sexes)

juv.

br. ♂ displaying

SHAFT-TAILED WHYDAH

juv.

LONG-TAILED PARADISE WHYDAH

dark 'C'
on cheeks

♂ transitional

♀

red legs
(all ages/
sexes)

♀

br. ♂

br. ♂

BROAD-TAILED
PARADISE WHYDAH

br. ♂

♂ displaying

lacks dark
'C' on
cheeks

♀

slender- vs
broad-tipped
tails

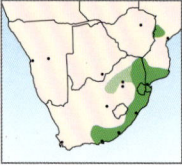

DUSKY INDIGOBIRD *Vidua funerea*

b B B B
J F M A M J J A S O N D

12 cm; 12–18 g Br. male is black, with a slight bluish gloss (typically less pronounced than in Purple Indigobird); wings dark brown. In S African *V. f. funerea*, combination of whitish bill and red or orange-red legs is diagnostic, but northern *V. f. nigerrima* has pinkish-white legs, similar to Purple Indigobird. Female and non-br. male resemble a female whydah, but have whitish bills and red (in the south) or whitish (in the north) legs. Juv. plainer than female, with more buffy underparts; bill and legs grey. **Voice:** Short, canary-like jingle and scolding *'chit-chit-chit'* mimicry of alarm call of brood host, African Firefinch. **Status and biology:** Common resident of forest edge, woodland and moist savanna. (Gewone Blouvinkie)

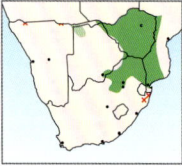

PURPLE INDIGOBIRD *Vidua purpurascens*

b B B B b
J F M A M J J A S O N D

11 cm; 12–15 g The combination of white or pinkish-white bill, legs and feet distinguish both male and female from other indigobirds in S Africa. In Zimbabwe and central Mozambique, could be confused with northern race of Dusky Indigobird, which also has pinkish-white legs; in good light, male typically shows more purple gloss on plumage and calls differ. Female and non-br. male resemble Dusky Indigobird; only separable in S Africa, where leg colour differs. Juv. indistinguishable from juv. Dusky Indigobird. **Voice:** Mimics purring calls, whistles and trills of brood host, Jameson's Firefinch. **Status and biology:** Common resident and local nomad in savanna, especially acacias. (Witpootblouvinkie)

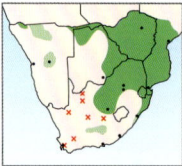

VILLAGE INDIGOBIRD *Vidua chalybeata*

b b B B b b   b
J F M A M J J A S O N D

11 cm; 11–15 g In most of its range, the red bill, legs and feet differentiate both male and female from other indigobirds. West of Victoria Falls, *V. c. okavangensis* has a whitish bill, but it is the only indigobird in the area with red legs. Br. male has blackish wings, darker than most other male indigobirds, with narrow white edges; body plumage has a bluish sheen. Female and non-br. male are told from non-br. Pin-tailed Whydah (p. 452) by pink (not dark) legs and less boldly striped head. Told from non-br. Shaft-tailed Whydah (p. 452) by its smaller bill and paler head. Juv. indistinguishable from juv. Dusky Indigobird. **Voice:** Mimics dry rattling calls and clear, whistled *'wheeetoo-wheeetoo-wheet'* song of main brood host, Red-billed Firefinch. **Status and biology:** Common resident and local nomad in woodland and savanna. Also parasitises Brown Firefinch. (Staalblouvinkie)

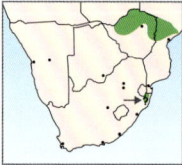

ZAMBEZI (TWINSPOT) INDIGOBIRD *Vidua codringtoni*

b b
J F M A M J J A S O N D

11 cm; 12–14 g Br. male is glossy greenish (rarely bluish), with black (not dark brown) wings; lacks pale primary edges of other male indigobirds and, in good light, shows a matt, dark grey breast. Bill whitish; told from northern race of Dusky Indigobird by its orange-red (not pinkish-white) legs and feet. Female and non-br. male have greyer breasts than other indigobirds; leg colour distinctive. Juv. indistinguishable from juv. Dusky Indigobird. **Voice:** Mimics trilled call and song of brood host, Red-throated Twinspot. **Status and biology:** Little known; apparently scarce resident in moist woodland and forest edge. Birds mimicking Pink-throated Twinspot in ne KZN and s Mozambique may represent an undescribed species of indigobird. (Groenblouvinkie)

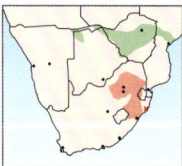

CUCKOO-FINCH *Anomalospiza imberbis*

B B b   b b b B
J F M A M J J A S O N D

13 cm; 20–23 g A weaver-like finch related to indigobirds which resembles a plump, short-tailed weaver with a stubby bill. Male could be confused with a weaver, but is smaller, brighter yellow (especially below) and has a heavy, black bill. Female and juv. are buffy, with heavy dark brown streaks above; told from female bishops and queleas by relatively plain, buffy face and short, heavy bill. Juv. resembles female, but face and underparts more buffy. **Voice:** Swizzling song and soft, chipping flight calls. **Status and biology:** Uncommon resident (Zimbabwe) or br. summer migrant in open grassland, especially near damp areas; sometimes in flocks of hundreds. Brood hosts include Tawny-flanked Prinia and a range of cisticolas. (Koekoekvink)

DUSKY INDIGOBIRD

br. ♂
bright
red
legs

♀

non-br. ♂

br. ♂

pinkish-
white legs

PURPLE INDIGOBIRD

♀

VILLAGE INDIGOBIRD

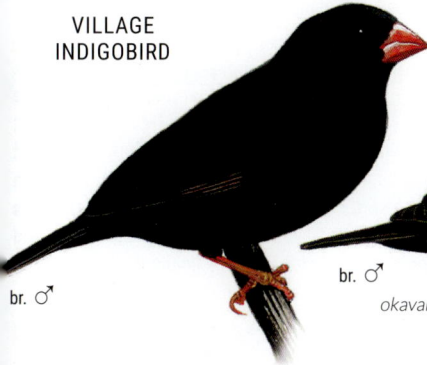

bill red in east,
white in west

br. ♂

br. ♂

okavangensis

transitional ♂

♀

ZAMBEZI INDIGOBIRD

ark
ot brown)
g panel

br. ♂

♀

short,
robust
bill

CUCKOO-FINCH

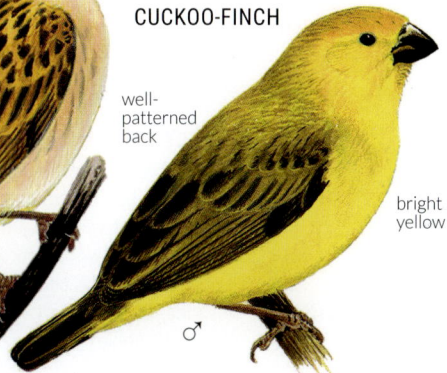

well-
patterned
back

bright
yellow

♀

♂

MANNIKINS, WAXBILLS AND FINCHES

A diverse group of small, colourful, granivorous finches. Feed mainly on grass seeds, coming into the open to feed and retiring to adjacent cover when disturbed. Most species are in flocks when not br., but some forest species occur singly or in pairs year-round. Sexes alike in some species, differ in others; juvs typically much duller. Breed in untidy, ball-shaped nests; some species use old weaver or sparrow nests.

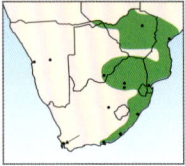

BRONZE MANNIKIN   *Lonchura cucullata*

B	B	B	B	B			b	b	b	b	B
J	F	M	A	M	J	J	A	S	O	N	D

9 cm; 7–12 g A small, brownish mannikin; usually the most abundant species. Lacks the chestnut back and barred wings of Red-backed Mannikin. Smaller than Magpie Mannikin, with a much shorter bill, finely barred (not heavily blotched) flanks and barred (not blackish) rump. Juv. is uniformly dull brown, with barred undertail coverts; lacks contrast between a darker face and pale throat and plain undertail coverts of other juv. mannikins. **Voice:** Soft, buzzy *'chizza, chizza'*; also a dry *'krrr krrr'*. **Status and biology:** Abundant resident in grassy areas in woodland, forest edges and gardens. Breeds singly or in loose colonies. (Gewone Fret)

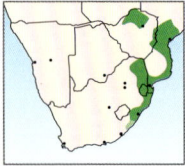

RED-BACKED MANNIKIN   *Lonchura nigriceps*

b	B	B	B	B				b	b	b	
J	F	M	A	M	J	J	A	S	O	N	D

10 cm; 8–11 g A striking mannikin with a black head and breast, chestnut back and wing coverts. White spots on the outer webs of its flight feathers form a finely barred black-and-white wing panel. Bill pale blue-grey, lacking darker upper mandible of other mannikins. Juv. much duller, but differs from other juv. mannikins by its reddish-brown back. **Voice:** Sharp, high-pitched *'whit weet weet'*; soft *'seeet-seeet'* when flushed. **Status and biology:** Locally common resident of moist woodland and forest edge. (Rooirugfret)

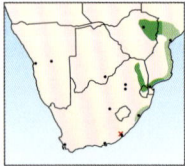

MAGPIE MANNIKIN   *Lonchura fringilloides*

b	b	B	B	B	b	b			b	B	B
J	F	M	A	M	J	J	A	S	O	N	D

13 cm; 16–19 g The largest mannikin, with a much longer bill than Bronze Mannikin. Has black pectoral patches and broad blotches (not fine barring) on its chestnut-washed flanks; rump blackish (not barred). Juv. is grey-brown above, larger than other juv. mannikins, with plain undertail coverts. **Voice:** Loud *'peeoo-peeoo'*, deeper than other mannikins, given in flight. **Status and biology:** Uncommon resident and local nomad in moist, tall-grass savanna and bamboo at forest edge; locally in gardens, where it sometimes visits bird feeders with other mannikins. (Dikbekfret)

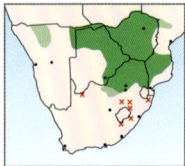

CUT-THROAT FINCH   *Amadina fasciata*

b	B	B	B	B		b					b
J	F	M	A	M	J	J	A	S	O	N	D

12 cm; 17–18 g A heavily barred, brown finch with a whitish bill. Male easily identified by its pinkish-red throat band. Female is smaller than female Red-headed Finch, with a barred crown and nape, mottled back, buffy wash on breast and more widely spaced, irregular barring on its flanks. Juv. like female. Imm. male has a narrow, pink throat band. Occasionally hybridises with Red-headed Finch; hybrids have underpart plumage of one and head plumage of the other. **Voice:** Melodic, buzzy song; also *'eee-eee-eee'* flight call. **Status and biology:** Locally common resident and nomad in dry, broadleafed woodland. Often in flocks, frequently with other seedeaters. (Bandkeelvink)

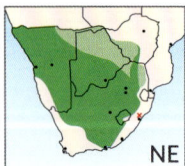

RED-HEADED FINCH   *Amadina erythrocephala*

b	B	B	B	B	b	b	b	b	b	b	
J	F	M	A	M	J	J	A	S	O	N	D

14 cm; 18–27 g Male has a distinctive red head; distinguished from male queleas (p. 448) by its barred or scalloped underparts and whitish bill. Female is larger than female Cut-throat Finch, with a plain, grey-brown (not barred and mottled) crown and mantle, and more uniformly barred underparts, lacking a warm buffy wash to the breast. Juv. is duller than female. Occasionally hybridises with Cut-throat Finch; hybrids have underpart plumage of one and head plumage of the other. **Voice:** Dry, buzzy *'chuk-chuk'* song; *'zree, zree'* flight call. **Status and biology:** Near-endemic. Common resident and nomad in dry grassland, acacia and broadleafed woodland. Often in large flocks, frequently with other seedeaters; occasionally irrupts outside its normal range. (Rooikopvink)

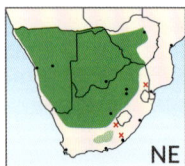

SCALY-FEATHERED WEAVER (FINCH)   *Sporopipes squamifrons*

B	B	B	B	b	b	b	b	B	B	b	
J	F	M	A	M	J	J	A	S	O	N	D

10 cm; 9–13 g A small, finch-like bird related to sparrows and weavers. Ads are easily identified by their petite pink bill, broad, black malar stripes (which give distinct, moustached appearance), finely speckled forecrown and white-edged wing feathers. Tail black, with white edges. Juv. lacks black malar stripes and freckled forecrown. **Voice:** Song is a rather shrill, sparrow-like, monotonous *'creep-creep'* or *'creep-crop'*; also a soft *'chizz, chizz, chizz'* flight call. **Status and biology:** Near-endemic. Common resident and nomad in dry savanna, bushy, desert watercourses and fields. Often in small groups, frequently with other seedeaters. (Baardmannetjie)

BRONZE MANNIKIN

two-toned bill

dark brown back

ad.

juv.

RED-BACKED MANNIKIN

reddish-brown back

juv.

pale bill

ad.

MAGPIE MANNIKIN

juv.

heavy, dark bill

boldly barred flanks

ad.

CUT-THROAT FINCH

barred head

♀

scaling below

♂

RED-HEADED FINCH

plain head

heavy bill

♀

♂

large white spots

SCALY-FEATHERED WEAVER

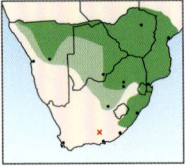

BLUE WAXBILL *Uraeginthus angolensis*

B B B b b b b b b b b
J F M A M J J A S O N D

12 cm; 8–13 g Easily identified by its powder-blue face, rump and underparts. Female is usually slightly paler blue, often with buffy flanks and vent. Juv. is paler than female, with blue restricted to its face and throat. **Voice:** Piercing *'sree-seee-seee'*, sometimes with melodic warbles; also dry *'krrt'* notes. **Status and biology:** Common resident and local nomad in dry woodland, savanna and gardens. Breeds singly or in loose colonies. (Gewone Blousysie)

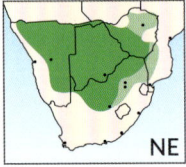

VIOLET-EARED WAXBILL *Uraeginthus granatinus*

B B B B B b b b b B
J F M A M J J A S O N D

13 cm; 8–14 g A large waxbill with violet cheeks, a warm brown body, electric-blue rump and frons, and a long, black, graduated tail; bill red. Male deeper chestnut, with a black throat; female paler, with buffy underparts. Juv. has a black bill and pale tan face. **Voice:** Dry, buzzy *'tziit'* contact call; song is a whistled *'tu-weoowee'*. **Status and biology:** Near-endemic. Common resident and local nomad in acacia woodland, savanna and dry thickets. Pairs remain together year-round. Brood host of Shaft-tailed Whydah. (Koningblousysie)

COMMON WAXBILL *Estrilda astrild*

B B b b b b b b B B B B
J F M A M J J A S O N D

11 cm; 7–10 g A grey-brown waxbill with a red eye stripe and bill, and a pinkish-red belly. At close range, shows finely barred upperparts and flanks. Juv. is duller, with less barred plumage and a black bill. **Voice:** Song is a harsh, nasal *'di-di-CHEERRR'*, descending slightly on last note. Typical flight call is a distinctive, high-pitched *'pink, pink'*. **Status and biology:** Common resident in long grass, rank vegetation, reeds and scrub, often near water. Often in flocks outside br. season. Main brood host of Pin-tailed Whydah. (Rooibeksysie)

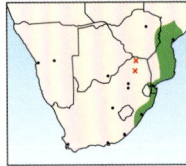

GREY WAXBILL *Estrilda perreini*

b b b b b b
J F M A M J J A S O N D

11 cm; 7–9 g A slate-grey waxbill with a black eye stripe and deep crimson rump. Differs from Cinderella Waxbill in having a black-and-grey (not reddish) bill and grey (not red) thighs; ranges do not overlap. Longer tailed than Yellow-bellied and Swee Waxbills (p. 460), with a grey (not olive) back. Juv. duller; lacks black eye stripe. **Voice:** Soft, whistled *'pseeu, pseeu'* contact call; song is a mournful *'fweeeeeee'*. **Status and biology:** Locally common resident along forest margins, and dense coastal and riverine thickets. (Gryssysie)

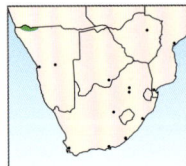

CINDERELLA WAXBILL *Estrilda thomensis*

B B B b b
J F M A M J J A S O N D

11 cm; 7–9 g Shorter tailed and slightly paler than Grey Waxbill, with a red-based bill and red (not grey) thighs; ranges do not overlap in region. Male has pinkish wash to breast and a blackish vent; female duller, with grey vent. Juv. duller grey brown; lacks red on rump and flanks; bill pink and brown. **Voice:** Thin but penetrating, reedy *'sweee-sweee-SWEEEEEEE-woooo'*; short, repeated *'trrt-tsoo'* or *'trrrt-trrrt'* contact call (lower pitched than co-occurring Blue Waxbill). **Status and biology:** Uncommon resident and local nomad in semi-arid savanna and scrub. Most easily seen along small watercourses feeding the Kunene River west of Ruacana, Namibia, where it comes to drink at small pools in the heat of the day. (Angolasysie)

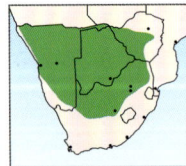

BLACK-FACED WAXBILL *Estrilda erythronotos*

B B B b b b b
J F M A M J J A S O N D

12 cm; 7–11 g A large, grey-and-crimson waxbill with barred wings, a prominent black face and fairly long, black tail. Bill black, with a blue-grey base. Female duller, with grey vent. Juv. is mostly grey; barring more diffuse; bill black. **Voice:** High-pitched *'chuloweee'* contact call; song is a short warble. **Status and biology:** Common resident and nomad in arid savanna and riverine thickets. Often in small flocks; drinks regularly. (Swartwangsysie)

ad.

juv.

**BLUE
WAXBILL**

**VIOLET-EARED
WAXBILL**

♀

♂

blue
rump

**COMMON
WAXBILL**

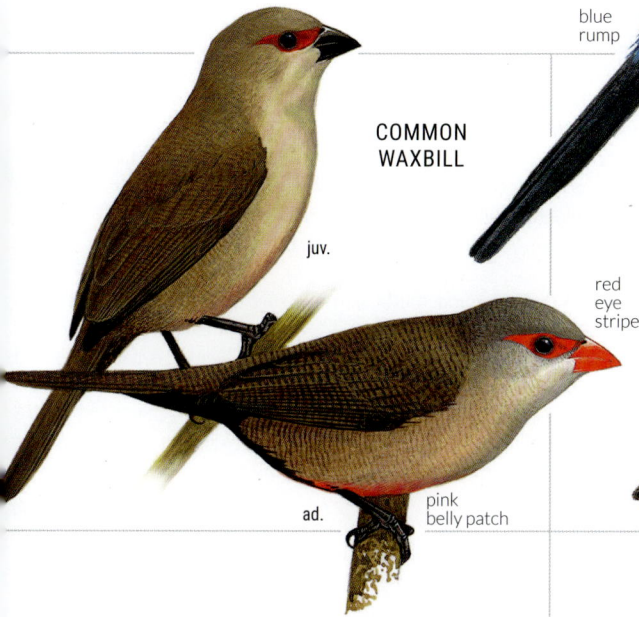

juv.

red
eye
stripe

ad.

pink
belly patch

**GREY
WAXBILL**

grey
flanks

**CINDERELLA
WAXBILL**

red flanks

dull red
rump

barred
wings

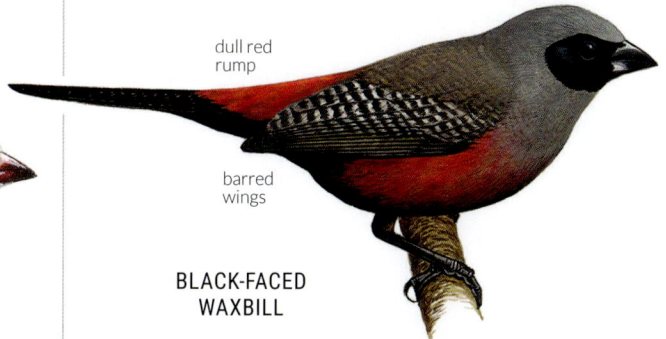

**BLACK-FACED
WAXBILL**

459

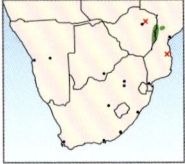

YELLOW-BELLIED WAXBILL *Coccopygia quartinia*

B	b	b	b								B
J	F	M	A	M	J	J	A	S	O	N	D

9 cm; 5–9 g A tiny, olive-backed waxbill with a bright red rump, black tail, greyish head and breast, and yellow-washed belly. Sexes alike. Male lacks black face of Swee Waxbill or Angolan Waxbill; range does not overlap with other species. Ads have a black upper mandible and red lower mandible. Juv. is duller, with a black bill; told from other juv. waxbills by its olive-tinged back. **Voice:** 'Swee-swee' contact call, usually given in flight; song is a soft, high-pitched warble. **Status and biology:** Locally common resident of forest edge and plantations with rank vegetation. (Tropiese Swie)

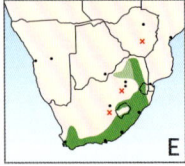

SWEE WAXBILL *Coccopygia melanotis*

B	b	b	b							b	B	B
J	F	M	A	M	J	J	A	S	O	N	D	

9 cm; 5–9 g Often treated as a race of Yellow-bellied Waxbill, but male has a distinctive, black face. Female resembles Yellow-bellied Waxbill, but typically has paler underparts; ranges do not overlap. Juv. is duller, with an all-black bill. **Voice:** Soft, high-pitched 'swie swie sweeeeeuu' song; also 'swee-swee' contact call. **Status and biology:** Endemic. Locally common resident of forest edge and wooded areas with rank vegetation; also gardens. (Suidelike Swie)

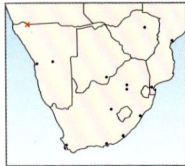

ANGOLAN WAXBILL *Coccopygia bocagei*

								X			
J	F	M	A	M	J	J	A	S	O	N	D

9–10 cm; 5–9 g Vagrant. Unmistakable small waxbill with a black face (males only). Cannot be mistaken for Swee Waxbill as the ranges do not overlap. Both sexes have diagnostic fine barring on the mantle. **Voice:** Not recorded in the region, however, likely to be similar to Swee Waxbill. **Status and biology:** Vagrant to Namibian border with Angola, where several birds have been seen and photographed. (Angolaswie)

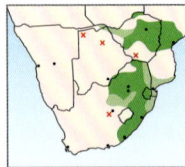

ORANGE-BREASTED WAXBILL *Amandava subflava*

b	B	B	B	B	b						
J	F	M	A	M	J	J	A	S	O	N	D

9 cm; 6–10 g A tiny, short-tailed waxbill with a prominent orange-red rump and supercilium, separating it from Quailfinch. Lacks grey head and olive back of Yellow-bellied and Swee Waxbills. At close range, the orange-yellow underparts with olive-barred flanks are diagnostic. Female is duller, with paler yellow underparts. Juv. is duller than female, with dark bill and plain flanks; browner than juv. Yellow-bellied Waxbill. **Voice:** Soft, clinking 'zink zink zink' flight call; rapid 'trip-trip' on take-off. **Status and biology:** Fairly common resident in grassland and weedy areas, especially near water; range and abundance appears to have decreased in recent years in S Africa. Regular brood host of Pin-tailed Whydah. (Rooiassie)

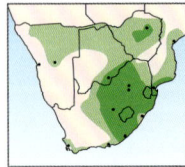

QUAILFINCH *Ortygospiza atricollis*

B	B	B	b	b	b					b	b
J	F	M	A	M	J	J	A	S	O	N	D

10 cm; 9–14 g A small, compact finch with dark grey upperparts, barred breast and flanks, and an orange-buff central belly and vent; bill entirely red (br. male) or upper mandible brown (female and non-br. male). At close range, shows a narrow, white 'spectacle' around its eye and a white chin. Female is paler, with less distinct barring; lacks black on face and throat. Juv. is paler than female, with only faint barring on thighs; bill dark brown. **Voice:** Distinctive, tinny 'chink-chink' in flight. **Status and biology:** Common resident and nomad in short grass and bare areas in grassland and fields, often near water. Rarely seen unless in flight, when it gives distinctive contact call; normally disappears from view on landing. (Gewone Kwartelvinkie)

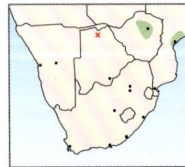

LOCUST FINCH *Paludipasser locustella*

B	B	b	b	b							
J	F	M	A	M	J	J	A	S	O	N	D

9 cm; 8–12 g A compact finch with reddish wings; smaller and darker than Quailfinch. Male appears black and red, with a bright red face, throat and breast. At close range, shows fine, white bars on thighs and white spots on back. Female is whitish beneath, with barred flanks; told from female Quailfinch by its rufous wings and plain face (lacking white 'spectacles'). Juv. is browner than female; wing feathers edged brown (not rufous); bill black. **Voice:** Fast 'tinka-tinka-tinka', higher pitched than Quailfinch. **Status and biology:** Scarce resident and local nomad in moist grassland and vleis; often in small groups. (Rooivlerkkwartelvinkie)

YELLOW-BELLIED WAXBILL

grey throat (both sexes)

yellow belly

SWEE WAXBILL

pale belly

♀

ANGOLAN WAXBILL

dull yellow belly

♂

♂

red rump

barred flanks

♂

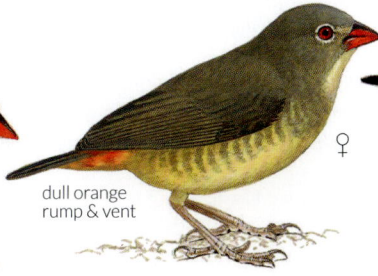

dull orange rump & vent

♀

ORANGE-BREASTED WAXBILL

juv.

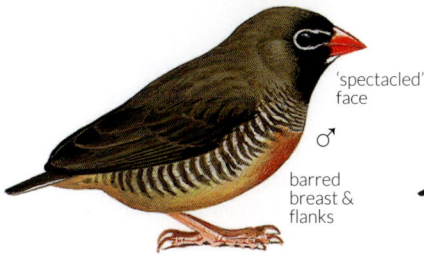

'spectacled' face

♂

barred breast & flanks

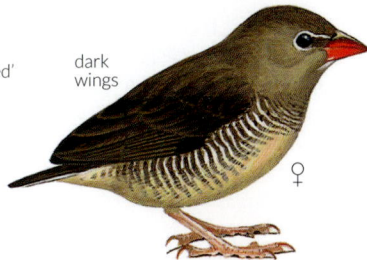

dark wings

♀

QUAILFINCH

dark rump & wings

♂

♂

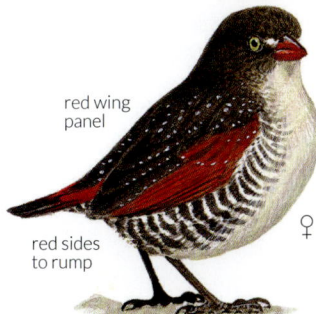

red wing panel

red sides to rump

♀

LOCUST FINCH

red wings

♂

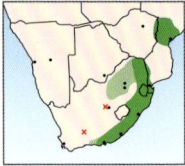

AFRICAN FIREFINCH *Lagonosticta rubricata*

B B B B b b   b b
J F M A M J J A S O N D

11 cm; 8–13 g A dark, richly coloured firefinch with blue-grey bill and legs, and blackish vent and tail. Male is redder (less pink) than male Jameson's Firefinch, with brown back and wings lacking any pink wash. S African *L. r. rubricata* has a distinctive grey (not pink) crown and nape, but *L. r. haematocephala* (Zimbabwe and central Mozambique) is richly coloured, with a red wash to its crown and reddish cheeks (at a glance could be confused with Lesser Seedcracker, p. 464). Female is pinkish-brown below, darker than female Jameson's Firefinch. Juv. is brown above, buffy-yellow below; red coloration confined to rump. **Voice:** Dry *'trrt-trrt-trrt-trrt'* and high-pitched alarm trills; whistled song includes clear *'wink-wink-wink'* and fast trills. **Status and biology:** Common resident of thickets in woodland, savanna and riverine scrub. Brood host of Dusky Indigobird. (Kaapse Vuurvinkie)

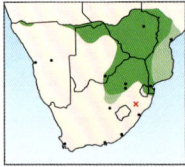

JAMESON'S FIREFINCH *Lagonosticta rhodopareia*

B B B B b b b b b b b B
J F M A M J J A S O N D

11 cm; 8–13 g Paler pink than African Firefinch, with the crown, nape, back and wing coverts suffused pink. Dark bill, legs, tail and vent separate it from Red-billed Firefinch. Male is pink below; female pinkish-buff, with red lores. Juv. is duller than female; lacks red lores. **Voice:** Long trills, interspersed with sharp *'vit-vit-vit'* and *'sweet sweet sweet'* whistles. **Status and biology:** Common resident and local nomad in thickets and grassy tangles in savanna and dry woodland. Brood host of Purple Indigobird. (Jamesonvuurvinkie)

RED-BILLED FIREFINCH *Lagonosticta senegala*

B B B B b b b b b b b B
J F M A M J J A S O N D

10 cm; 7–10 g Ad. identified by mostly red bill, pinkish legs, narrow, yellow eye-ring and brown (not blackish) vent and tail; rump red (not grey-brown as Brown Firefinch). Male has head and underparts pink. Female is pale brown, with only lores, rump and uppertail pink. Juv. has dark bill; lacks yellow eye-ring and white breast spots. Regularly mistaken for Brown Firefinch because dull, reddish rump can be difficult to see, but juvs usually accompanied by ads. **Voice:** Fairly melodic, slurred *'sweet er-urrrrrr'* song; also sharp, fast *'vut-vut chit-it-errrr'* and dry, rattling *'trrrt'* call. **Status and biology:** Common resident and local nomad in semi-arid woodland, especially near water; also in gardens. Brood host of Village Indigobird. (Rooibekvuurvinkie)

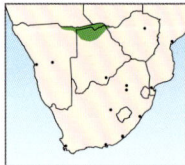

BROWN FIREFINCH *Lagonosticta nitidula*

B B B b   b b b
J F M A M J J A S O N D

10 cm; 9–12 g A rather drab firefinch with a diagnostic grey-brown (not red) rump; vent pale grey; bill and legs pinkish. Male has pinkish-red face, throat and upper breast, with small, white spots on sides of breast. Female is duller, with only a faint pink wash to the face and throat. Juv. is all-brown, with a dark bill. Could be confused with juv. Red-billed Firefinch, but latter has dull, reddish rump (not always easy to see). **Voice:** High-pitched, chittering song; also *'tsiep, tsiep'* flight call. **Status and biology:** Locally common resident of thick scrub and reeds close to water. Visits bird baths at riverside lodges. Brood host for Village Indigobird. (Bruinvuurvinkie)

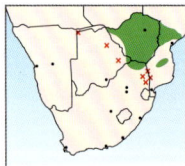

ORANGE-WINGED PYTILIA *Pytilia afra*

b b b
J F M A M J J A S O N D

13 cm; 15–18 g Darker and shorter tailed than Green-winged Pytilia. Golden-yellow edges to its wing feathers form a golden panel on the folded wing; belly and flanks are mottled olive (not barred grey). Male has less crisply defined red face patch extending onto the ear coverts, and green back colour extends onto the nape. Juv. is browner above, with only a faint yellowish wing panel; bill brown with a pink base. **Voice:** Piping *'seee, seee'* whistle; also sharp *'tsik'* call. **Status and biology:** Fairly common resident and local nomad in thick, tangled scrub and understorey of miombo woodland. Brood host of Broad-tailed Paradise Whydah. (Oranjevlerkmelba)

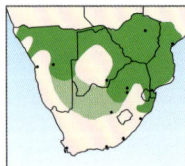

GREEN-WINGED PYTILIA *Pytilia melba*

B B B B b   b b B
J F M A M J J A S O N D

13 cm; 9–19 g A fairly large finch with a grey head, green back and red rump; bill red, with dark top to upper mandible. Told from Orange-winged Pytilia by its uniform olive-green wings (lacks golden feather edges) and more boldly barred (not mottled) underparts. Male has scarlet frons and throat; female head all-grey. Juv. is browner above and plain buff below; bill pinkish. Rare yellow morph has yellow (not red) face and rump. In central Mozambique, male *P. m. grotei* has breast washed pink (not olive-green). **Voice:** Stuttering song with trilling whistles, often rising and falling in pitch; also short *'wick-ick-ick'* call. **Status and biology:** Common resident in acacia savanna and dry woodland. Brood host of Long-tailed Paradise Whydah. (Gewone Melba)

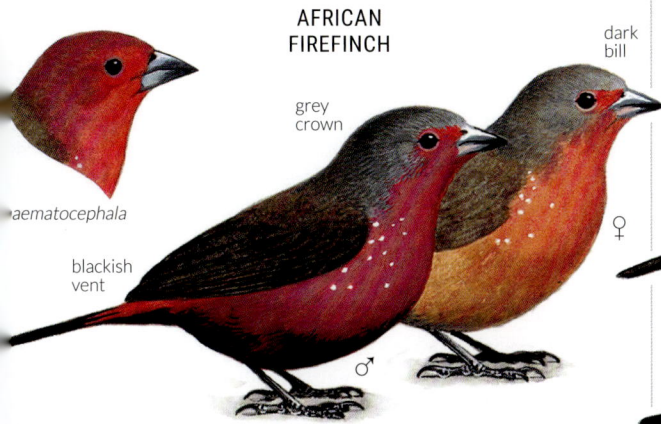

AFRICAN FIREFINCH

dark bill

grey crown

aematocephala

blackish vent

♀

♂

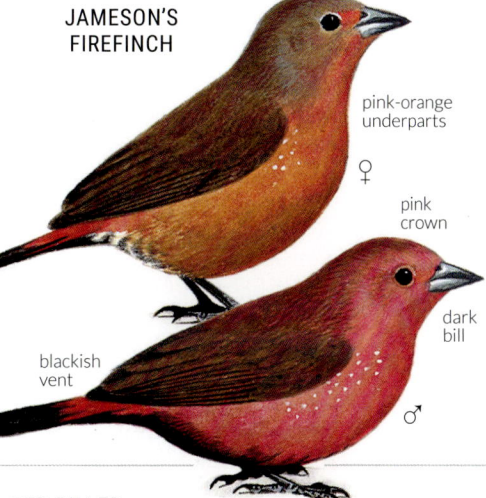

JAMESON'S FIREFINCH

pink-orange underparts

♀

pink crown

dark bill

blackish vent

♂

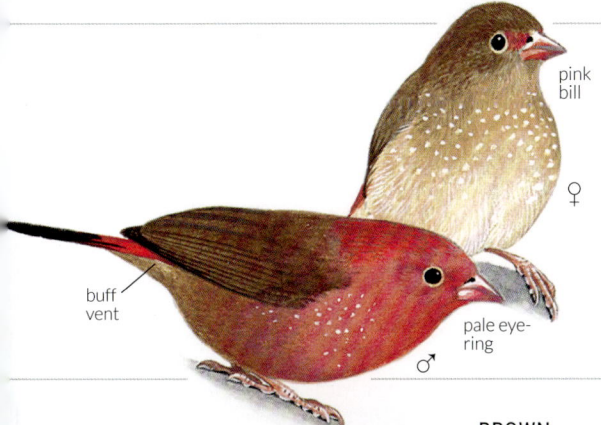

pink bill

♀

RED-BILLED FIREFINCH

buff vent

pale eye-ring

♂

juv.

BROWN FIREFINCH

pink bill

♀

grey rump

pale vent

♂

clear crisp spots

ORANGE-WINGED PYTILIA

red mask around eye

♂

orange wing panel

♀

green-barred flanks

GREEN-WINGED PYTILIA

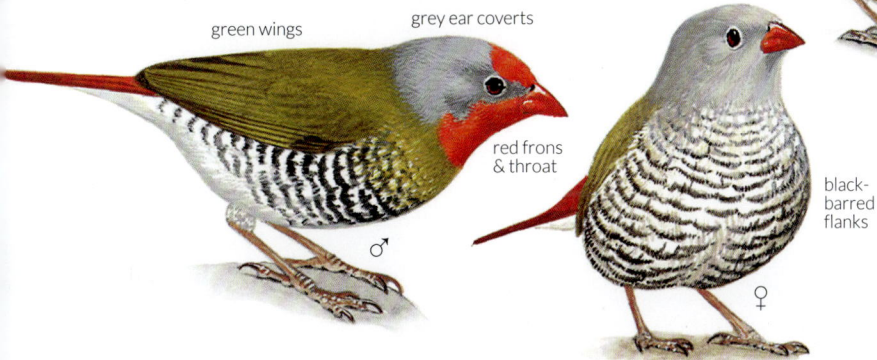

green wings

grey ear coverts

red frons & throat

♂

black-barred flanks

♀

463

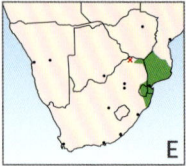

PINK-THROATED TWINSPOT   *Hypargos margaritatus*

b | J F M A M J J A S O N D

12 cm; 12–14 g Male is paler than male Red-throated Twinspot, with a pinkish face, throat and breast, and a brown (not grey) crown. Female has a pale grey-brown throat, breast and belly; lacks orange or pink wash on throat and breast of female Red-throated Twinspot; belly much paler. Juv. is brown above, pale grey-buff below, lacking spots. Paler than juv. Red-throated Twinspot. **Voice:** Soft, high-pitched, insect-like trill. **Status and biology:** Endemic. Locally common, but inconspicuous resident in thornveld thickets and coastal scrub. Often drinks at puddles early in the morning. Suspected to be the brood host of an undescribed indigobird species in ne KZN and s Mozambique. (Rooskeelkolpensie)

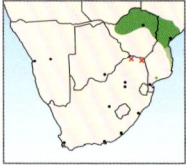

RED-THROATED TWINSPOT   *Hypargos niveoguttatus*

b b b b b b | J F M A M J J A S O N D

13 cm; 13–17 g Male is darker than male Pink-throated Twinspot, with a deep red face, throat and breast and a grey (not brown) crown. Female has pinkish-orange wash across breast and sooty-grey central belly. Juv. is dull brown, lacking flank spots. Darker than juv. Pink-throated Twinspot, with a blackish (not grey-buff) belly. **Voice:** Insect-like trill, longer and deeper than Pink-throated Twinspot. Song includes mournful, slurred whistles and *'chink'* notes. **Status and biology:** Locally common resident of dense thickets, rank growth and forest edge; prefers moister habitats than Pink-throated Twinspot. Brood host of Zambezi Indigobird. (Rooikeelkolpensie)

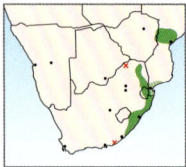

GREEN TWINSPOT   *Mandingoa nitidula*

b b b b | | b | b b b | J F M A M J J A S O N D

10 cm; 8–11 g A tiny twinspot with green (not brown) upperparts. Boldly spotted flanks and green (not rufous) back separate it from Red-faced Crimsonwing. Male has a red face and black belly, boldly spotted white. Female has a buff face and olive-grey belly, spotted white. Juv. is duller than female, with unspotted, pale olive-grey underparts. **Voice:** Soft, rolling, insect-like *'zrrreet'* and whistled notes; contact call is a very high-pitched *'zeet'*. **Status and biology:** Locally common resident and altitudinal migrant in forest, usually near clearings or areas with dense undergrowth; also woodland and coastal scrub in winter. Secretive and easily overlooked. (Groenkolpensie)

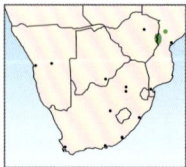

RED-FACED CRIMSONWING   *Cryptospiza reichenovii*

b b b | | | | b b b b | J F M A M J J A S O N D

12 cm; 12–17 g A forest finch with a dull red back, rump and wings. Face of male is red, female yellow-buff. Larger than Green Twinspot, with plain (not spotted) olive flanks and belly. Smaller than Lesser Seedcracker, with a black (not red) tail and dull red (not olive-brown) back and wings. Juv. has browner upperparts; face dull olive. **Voice:** Soft, very high-pitched trills or *'zeet'* calls. **Status and biology:** Locally common resident of forest understorey, but easily overlooked. Flight very fast and direct. (Rooiwangwoudsysie)

LESSER SEEDCRACKER   *Pyrenestes minor*

b b b b b | | | | | b | J F M A M J J A S O N D

14 cm; 18 g A large, chunky finch that appears all-dark in poor light. Larger than Red-faced Crimsonwing, with red head, olive-brown (not reddish) back and wings, and red (not black) tail. Male has more extensive red on its head that extends onto its breast. In e Zimbabwe, could be confused with richly coloured African Firefinch (p. 462), but is larger, with an all-red (not blackish) tail; lacks white spotting on sides of breast. Juv. is dull brown, with reddish wash on rump and tail. **Voice:** A soft, chittering trill; also deep *'chat-chat chat'* and *'tseet'* calls. **Status and biology:** Uncommon resident in tangled growth in moist miombo woodland, thickets and forest edge; also in tea plantations. Bill size varies, with larger-billed birds taking larger seeds. (Oostelike Saadbrekertjie)

brown
crown

**PINK-THROATED
TWINSPOT**

♀

grey
crown

**RED-THROATED
TWINSPOT**

russet
breast

♂

♀

**GREEN
TWINSPOT**

green
upperparts

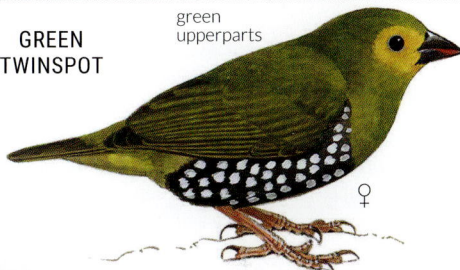

♂

♀

**RED-FACED
CRIMSONWING**

reddish back
& wings

plain green
underparts

♀

♂

dark brown
upperparts

♀

red
tail

**LESSER
SEEDCRACKER**

♂

heavy
bill

SEEDEATERS, CANARIES AND SISKINS
A diverse group of granivorous birds (although diet often quite varied, including fruits and insects). The distinction between seedeaters and canaries is somewhat arbitrary: seedeaters tend to have large bills and most are rather drab, whereas canaries have smaller bills and typically have some yellow plumage (which should make the Protea Canary a seedeater). Occur singly or in flocks. Sexes similar in most species, but differ in some, especially in species that flock when not br. Juvs duller and more streaked.

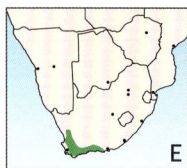

PROTEA CANARY (SEEDEATER) *Crithagra leucoptera*
15 cm; 18–25 g A drab brown canary with a diagnostic, blackish face and chin that contrast with its pale bill and whitish throat. Narrow, white edgings to secondary coverts result in 2 diagnostic white wingbars, but these are only visible at close range. Lacks the white supercilium of Streaky-headed Seedeater. In flight, lacks the greenish-yellow rump of White-throated Canary. **Voice:** Contact call is *'tree-dili-eeee'*; song intersperses the contact call with harsh, repetitive elements. **Status and biology:** NEAR-THREATENED. Endemic. Uncommon to locally common resident and nomad in thick, tangled scrub and dense fynbos, not especially near proteas. Most abundant in drier areas. (Witvlerkkanarie)

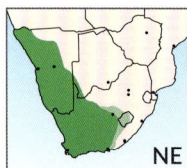

WHITE-THROATED CANARY *Crithagra albogularis*
16 cm; 18–38 g A large, pale grey-brown canary with a heavy bill, white throat, small, white supercilium and diagnostic, greenish-yellow rump. Lacks black face of Protea Canary. Paler than Streaky-headed Seedeater, with a paler face, less prominent supercilium and shorter, heavier bill. Larger than female Yellow Canary (p. 468), with a much larger bill and an unstreaked breast. Juv. is lightly streaked above and on breast. **Voice:** Song is a rich, jumbled mix of melodious notes; contact call is a querulous *'tsuu-eeeee'*. **Status and biology:** Near-endemic. Common resident and local nomad in coastal thicket, Karoo scrub and semi-desert. (Witkeelkanarie)

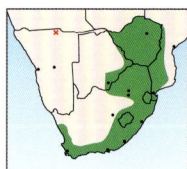

STREAKY-HEADED SEEDEATER *Crithagra gularis*
15 cm; 12–25 g A fairly large, grey-brown canary with a broad, creamy or buff supercilium and relatively long, slender bill. Grey or dark grey (not black) face, plain (or very lightly streaked) breast and warmer brown upperparts separate it from Black-eared Seedeater; streaked crown and absence of wingbars distinguish it from Yellow-throated Bush Sparrow (p. 440). Juv. is more heavily streaked above and below. **Voice:** Short, rather deep, melodious song; soft *'trrreet'* contact call. **Status and biology:** Common resident in woodland, thickets and dense scrub, often in hilly areas. Frequently associated with aloes. (Streepkopkanarie)

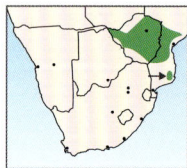

BLACK-EARED SEEDEATER *Crithagra mennelli*
14 cm; 13–18 g Slightly paler and colder grey above than Streaky-headed Seedeater, with a white (not buff-tinged) supercilium and more heavily streaked breast; bill heavier. Male has distinct black cheeks and a more crisply streaked black-and-white crown. Female and juv. have dark grey cheeks. **Voice:** Long, rambling song *'teeu-twee-teeu, twiddy-twee-twee'*; soft *'see-see-see'* contact call. **Status and biology:** Uncommon to locally common resident in miombo and mopane woodland. Often joins bird parties. (Swartoorkanarie)

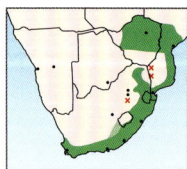

BRIMSTONE CANARY *Crithagra sulphurata*
14 cm; 15–23 g A large, greenish-yellow canary with a yellow supercilium and stout bill. Larger than Yellow Canary (p. 468), with a much heavier bill; in area of overlap, *C. s. sulphurata* (southern S Africa) has an olive-washed breast and flanks (not bright yellow of male Yellow Canary), contrasting with its yellow throat. Brimstone Canary shows a dark green frons (yellow in male Yellow Canary); an important identification feature between the two species. Other races are smaller, with yellower faces and underparts, more closely resembling Yellow Canary. Larger than Yellow-fronted Canary (p. 468), with a much heavier bill; lacks grey-washed crown, bright yellow rump and white tail tips. Female is duller. Juv. is duller and greyer than female, with lightly streaked flanks. **Voice:** Rich *'zwee zwee duid duid duid tweer weerr'* song; deeper and slower than other yellow canaries. **Status and biology:** Common resident in woodland, mesic thickets and gardens. Usually feeds in the canopy. (Dikbekkanarie)

PROTEA CANARY

indistinct supercilium

dark chin, pale throat

white wingbars

WHITE-THROATED CANARY

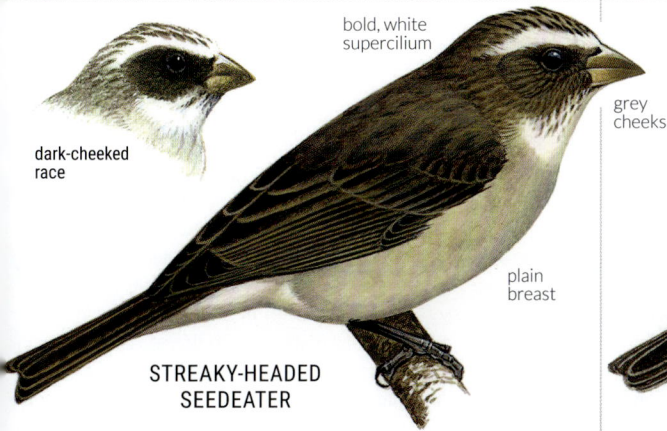

pale supercilium

heavy bill

pale chin & throat

greenish-yellow rump

STREAKY-HEADED SEEDEATER

bold, white supercilium

grey cheeks

dark-cheeked race

plain breast

BLACK-EARED SEEDEATER

black cheeks

streaked breast

BRIMSTONE CANARY

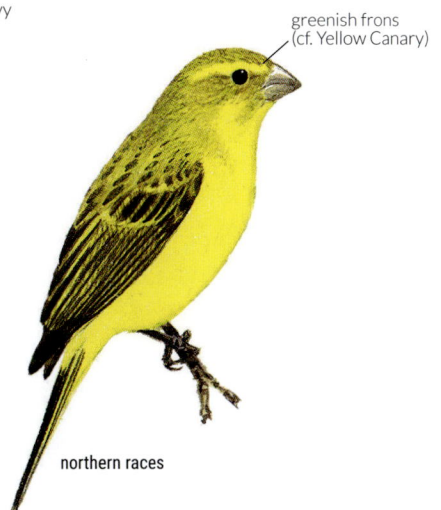

heavy bill

greenish frons (cf. Yellow Canary)

sulphurata

northern races

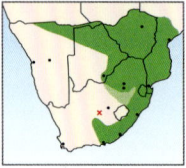

YELLOW-FRONTED CANARY *Crithagra mozambica*

B B B | | | | | | b b b B
J F M A M J J A S O N D

11 cm; 9–16 g A small canary with a blackish malar stripe and bright yellow frons and supercilium that contrast with the grey-green crown. Smaller than Brimstone (p. 466) and Yellow Canaries, with a more strongly marked face, bright yellow rump and white tail tips. In flight, the conspicuous yellow rump and white tail tips recall Black-throated and Lemon-breasted Canaries, but rump is much yellower. Female is duller, with paler yellow underparts. Juv. has buffy-yellow, lightly streaked underparts. **Voice:** Sweet, rather monotonous *'zeee-zereee, zeee-zereee chereeo'*, with occasional trills and flourishes. Occasionally mimics other birds. **Status and biology:** Common to abundant resident in open woodland, savanna and gardens. Often in flocks with other seedeaters. (Geeloogkanarie)

BLACK-THROATED CANARY *Crithagra atrogularis*

B B b b b b b b b B B
J F M A M J J A S O N D

11 cm; 9–16 g A small, pale grey-brown canary, streaked dark brown above, with a bright lemon-yellow rump and white tail tips. Underparts pale grey-brown, variably streaked darker brown. Blackish throat is often hard to see; most obvious when br. Lacks white face patches and dark malar stripe of Lemon-breasted Canary. Female is duller, sometimes lacking black on throat. Juv. has a buffy-grey throat. **Voice:** A prolonged series of wheezy whistles and chirrups. **Status and biology:** Common resident and local nomad in acacia savanna, dry woodland and fields. Usually in flocks. (Bergkanarie)

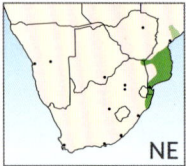

LEMON-BREASTED CANARY *Crithagra citrinipectus*

B b | | | | | | | | | B
J F M A M J J A S O N D

11 cm; 10–13 g Appears intermediate between Yellow-fronted and Black-throated Canaries. Male has a pale lemon throat and upper breast, peachy-buff flanks and a white belly. Has bold head markings of Yellow-fronted Canary, but with white (not yellow) wingbars and face patches; upperparts grey-brown (not greenish). Female duller, lacking lemon throat and breast; told from Black-throated Canary by its dark malar stripe and peachy-buff-washed underparts. **Voice:** A rather short, warbled song, higher pitched and less melodious than Black-throated Canary, with shorter phrases. **Status and biology:** Near-endemic. Locally common resident in palm savanna and adjacent acacia woodland and grassland. Often occurs in flocks with Yellow-fronted Canary. (Geelborskanarie)

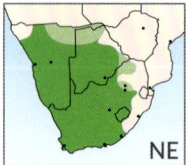

YELLOW CANARY *Crithagra flaviventris*

B B B b b b b B B B B
J F M A M J J A S O N D

13 cm; 13–20 g A strongly dimorphic canary. Male is bright yellow below and olive-green above, finely streaked darker. Southern and eastern races are darker green above, with strongly contrasting face patterns; in the north, *C. f. damarensis* has a yellower head. Smaller than Brimstone Canary (p. 466), with a smaller bill; has brighter yellow underparts than the subspecies of Brimstone Canary found in the area of overlap and is usually found in drier or more open habitats. Male Yellow Canary shows a yellow frons (dark green in Brimstone Canary); an important identification feature between the two species. Larger than Yellow-fronted Canary, lacking a grey-washed crown, bright yellow rump and white tail tips. Female is much duller olive-grey above, and pale grey or lemon below, streaked brown. Juv. is more heavily streaked than female. **Voice:** A warbling medley, faster and more varied than Yellow-fronted Canary. **Status and biology:** Near-endemic. Common resident and local nomad in Karoo, coastal scrub and semi-desert. Often in large flocks with sparrows and other seedeaters. (Geelkanarie)

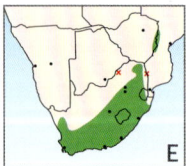

CAPE CANARY *Serinus canicollis*

b b b b b | | B B B B B
J F M A M J J A S O N D

12 cm; 12–22 g A slender canary with a smooth, blue-grey hind crown, nape and mantle. Male has mustard-yellow frons, face and throat. Female is duller, lightly streaked above and on the flanks, with a grey wash on the face and breast. Juv. is browner above than female and greenish-yellow below with heavy brown streaking. Lacks yellow supercilium of Forest Canary or juv. Yellow Canary. **Voice:** Male gives protracted, warbling song from a prominent perch or in 'butterfly' display flight; flight call is distinctive *'peeet'* or *'pee-weee'*. **Status and biology:** Endemic. Common resident in fynbos, grassland, coastal dunes and gardens; often in small flocks. (Kaapse Kanarie)

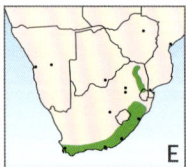

FOREST CANARY *Crithagra scotops*

b b b | | | | | | b B B
J F M A M J J A S O N D

13 cm; 12–20 g A dark canary, streaked above and below, with a black chin and greyish cheeks that contrast with the yellow supercilium and pale bill. Female and juv. have little or no black bib and are more heavily streaked below. **Voice:** High-pitched, warbling song, often with querulous notes; also a very high-pitched *'tseeek'* contact call. **Status and biology:** Endemic. Locally common resident in forest, forest edge and clearings. (Gestreepte Kanarie)

YELLOW-FRONTED CANARY

broad yellow supercilium

prominent malar stripe

yellow belly

white tip

BLACK-THROATED CANARY

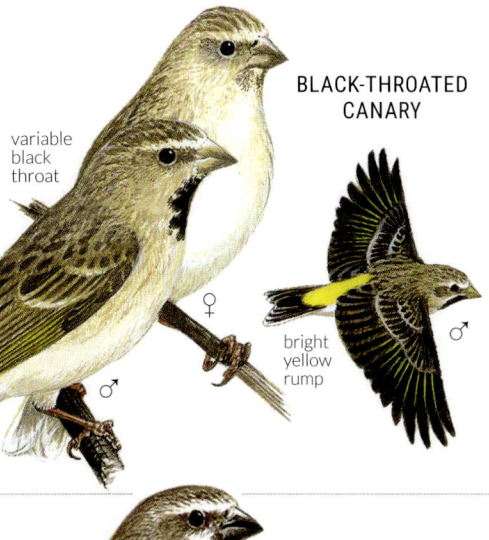

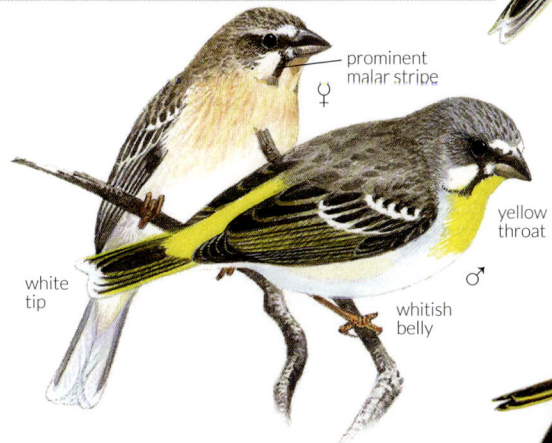

variable black throat

bright yellow rump

♀

♂

white tail tip

prominent malar stripe

♀

white tip

yellow throat

♂

whitish belly

LEMON-BREASTED CANARY

YELLOW CANARY

yellow frons

small bill

♀

♂

damarensis

♂

grey nape

♀

CAPE CANARY

FOREST CANARY

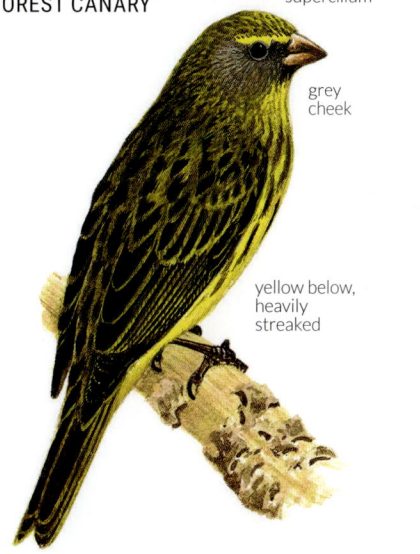

prominent yellow supercilium

grey cheek

yellow below, heavily streaked

juv.

dull brown below, heavily streaked

♂

469

BLACK-HEADED CANARY *Serinus alario*

b b b b b B B B B B B b
J F M A M J J A S O N D

12 cm; 11–13 g A small, slender canary. Male superficially recalls male Cape Sparrow (p. 440), but is much smaller, with plain, black head, no white wingbar and much brighter rump and tail. *S. a. leucolaemus* has distinct white face markings. Female much duller, with a streaky grey-brown head and back, but still has the distinctive chestnut wing coverts, rump and tail. Juv. resembles female, but is more heavily streaked. **Voice:** Soft *'sweea'* or *'tweet'* contact call; song is a rambling warble incorporating call notes. **Status and biology:** Endemic. Locally common nomad in Karoo scrub, usually in better vegetated areas and in fields. Drinks regularly. *S. a. leucolaemus* is split as a full species by some authorities as 'Damara Canary'; not known to hybridise with nominate subspecies despite broadly overlapping ranges and occurrence in mixed flocks. (Swartkopkanarie)

CAPE SISKIN *Crithagra totta*

b b B B b
J F M A M J J A S O N D

12 cm; 10–16 g A brown-backed canary with rather dull, yellow underparts and a blackish tail with white tips. Differs from Drakensberg Siskin by having white tips to its primaries and tail feathers (not white outer-tail feathers); ranges do not overlap. Female is duller, with streaked throat and less extensive white tips to primaries and tail. Juv. is heavily streaked on head and breast; lacks white in wing and tail. **Voice:** Diagnostic, querulous *'voyp-veeyr'* contact call, often given in flight; wispy, canary-like song often incorporates its contact call. **Status and biology:** Endemic. Locally common resident in mountain fynbos, forest margins, and sometimes along the coast; also exotic pine and eucalyptus plantations and well-vegetated gullies in the w Karoo. Favours rocky areas and recently burnt fynbos. (Kaapse Pietjiekanarie)

DRAKENSBERG SISKIN *Crithagra symonsi*

b b B
J F M A M J J A S O N D

13 cm; 12–18 g Similar to Cape Siskin, but has white outer-tail feathers (not white tail tips) and lacks white tips to primaries; ranges do not overlap. Female is duller, lacking yellow in plumage. Juv. is drab brown, heavily streaked on head, mantle and breast. **Voice:** Similar to Cape Siskin, but more strident; contact call is sharper *'voyp-wvip'*. **Status and biology:** Endemic. Locally common resident and altitudinal migrant in mountain scrub and grassland, mostly above 2 000 m. (Bergpietjiekanarie)

BUNTINGS Sparrow-like granivorous birds with rather small bills and boldly marked head patterns in most species. Occur singly, in pairs or in small to large flocks when not br. Sexes alike or similar. Juvs duller.

LARK-LIKE BUNTING *Emberiza impetuani*

b B B B b b b b B B b b
J F M A M J J A S O N D

14 cm; 13–20 g A rather nondescript, superficially lark-like bunting, but with a short, stout bill, long tail, short legs and hopping (not walking) gait. Best told from vagrant Ortolan Bunting by its broad, buffy supercilium, lack of a prominent pale eye-ring, its pale cinnamon-washed breast, and wing feathers broadly edged rufous. Sexes alike. Juv. is paler, with a mottled breast. **Voice:** Song is a short, rapid series of buzzy notes, accelerating and ending in a dry trill; also a soft *'tuc-tuc'* call. **Status and biology:** Near-endemic. Common to abundant nomad in semi-desert plains, Karoo scrub and arid savanna; occasional irruptions into coastal lowlands of w W Cape and the grassland and savanna regions. Gathers in huge numbers to breed following good rains, sometimes in loose associations with sparrow-larks. Regularly visits waterholes to drink. (Vaalstreepkoppie)

ORTOLAN BUNTING *Emberiza hortulana*

X X
J F M A M J J A S O N D

15–16 cm 17–28 g Vagrant. A rather plain bunting with a mostly grey head, striking pale eye-ring, pinkish bill and pale yellowish submoustachial stripe at all ages. Male has a yellow throat, plain olive-grey head and breast, and chestnut belly; intensity of yellow-olive coloration varies individually. Female is duller, with a streaked breast, and finely streaked crown, nape and flanks. Imm. is even duller with more streaking on the breast and flanks than female; bill greyish-pink; submoustachial stripes reduced in imm. female. **Voice:** Jolting metallic click or two-syllable note. **Status and biology:** Palearctic-br. migrant that winters in the Sahel. 4 records: one from NamibRand Nature Reserve, Namibia (Nov 2013), one in Kgalagadi Transfrontier Park, N Cape (Dec 2014), one in Etosha National Park, Namibia (Nov 2018), and one in Mudumo National Park, Namibia (Nov 2019). (Ortolanstreepkoppie)

BLACK-HEADED CANARY

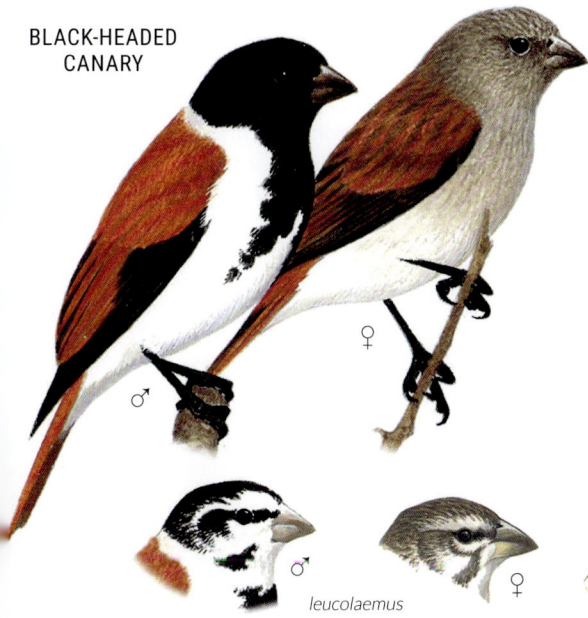

♂

♀

leucolaemus

♂

♀

CAPE SISKIN

♀

♂

white primary & tail tips

DRAKENSBERG SISKIN

♂

♀

white outer tail

LARK-LIKE BUNTING

buffy supercilium

rufous-edged wing feathers

ORTOLAN BUNTING

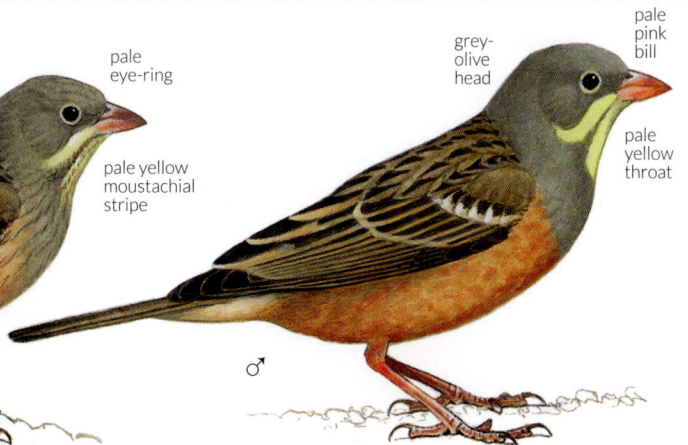

grey head

pale eye-ring

pale yellow moustachial stripe

♀

grey-olive head

pale pink bill

pale yellow throat

♂

CINNAMON-BREASTED BUNTING *Emberiza tahapisi*

14 cm; 12–20 g A dark, richly coloured bunting. Male has chestnut upperparts and warm cinnamon-brown underparts, black chin and two-tone yellow and black bill. Female is paler, with grey wash to nape and mantle. Both sexes differ from Cape Bunting by their black or greyish (not whitish) throat, cinnamon (not pale grey) underparts and presence of a pale median crown stripe. Juv. resembles female, but with a mottled breast. **Voice:** Short, dry rattling song *'cher-ippity-prrreee'*; also a querulous *'where-wheer'* call. **Status and biology:** Common resident and partial migrant on rocky slopes in grassland and open woodland. (Klipstreepkoppie)

GOLDEN-BREASTED BUNTING *Emberiza flaviventris*

15 cm; 15–22 g Differs from Cabanis's Bunting in having a white stripe below the eye, entirely yellow (not yellow-and-white) throat, chestnut (not greyish) mantle and a warmer, orange-yellow breast. Female is duller, with a sooty-brown head. Juv. resembles female, but is duller, with a slightly streaked breast. **Voice:** Song is a loud, whistled *'weechee, weechee, weechee'* or *'sweet-cher, sweet-cher, sweet-cher'*; also a nasal, buzzy *'zzhrrrr'*. **Status and biology:** Common resident in woodland and moist savanna. (Rooirugstreepkoppie)

CAPE BUNTING *Emberiza capensis*

16 cm; 17–27 g Paler than Cinnamon-breasted Bunting, with a whitish throat and pale grey breast; crown lacks a pale median stripe. Chestnut wing coverts contrast with greyish nape and mantle. Female has face stripes washed buff. Juv. is duller, with mottled breast. Intensity of grey wash on underparts varies geographically, and is slightly buffy in some races. **Voice:** Song is a cheerful *'chrip chip chup chip tur-twee'*, accelerating and fading away at the end. Typical call is a nasal, ascending *'wer we-wer'*. **Status and biology:** Near-endemic. Common resident in south, becoming local and uncommon in the north. Occurs on rocky hill slopes, usually with bushes or shrubs. (Rooivlerkstreepkoppie)

CABANIS'S BUNTING *Emberiza cabanisi*

16 cm; 15–25 g Differs from Golden-breasted Bunting by its solid black cheeks, lacking a white stripe below the eye, white sides to its throat, greyish (not chestnut) mantle and plain yellow breast. Female is duller, with a mottled crown and sooty-brown cheeks; upperparts browner. Juv. resembles a drab female. **Voice:** Song is a high-pitched *'swee-swee-swee'* or *'swi-swee-er, swe-er, swee-er'*; also a clear *'tsseeoo'* contact note. **Status and biology:** Uncommon to locally common resident in miombo woodland. (Geelstreepkoppie)

CHAFFINCHES A small Eurasian genus of finches in the same family as the siskins, canaries and seedeaters. 1 species introduced to the sw Cape. Sexes differ; juvs duller.

COMMON CHAFFINCH *Fringilla coelebs*

14–16 cm; 16–21 g Introduced from Europe. A sparrow-sized finch with 2 distinctive white wingbars that are prominent in flight. Male is striking, with a pinkish face and breast, blue-grey head and nape, and a vinaceous back. Female is duller, olive-brown above and pale grey-brown below. **Voice:** Short, hurried song, descending in pitch and ending in a short trill, is delivered monotonously from high in a tree; also a sharp *'pink, pink, pink'* call. **Status and biology:** Fairly common resident on Cape Peninsula in pine plantations and well-wooded gardens; increasingly moving into suburbia as alien plantations are felled. Easily overlooked when not calling. (Gryskoppie)

CINNAMON-BREASTED BUNTING

♂

black throat

chestnut below

♀

grey throat

CAPE BUNTING

chestnut mantle

white stripe below eye

broad white outer tail

♂

pale throat

pale grey below

GOLDEN-BREASTED BUNTING

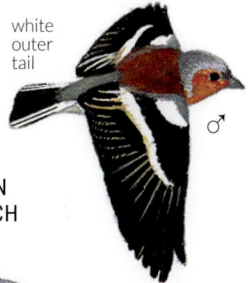

♀

COMMON CHAFFINCH

white outer tail

♂

CABANIS'S BUNTING

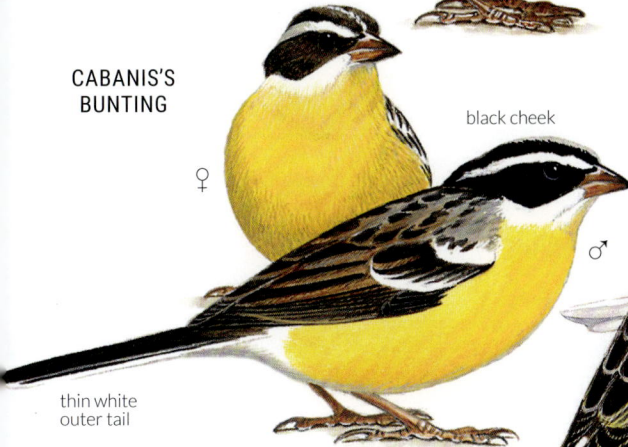

♀

black cheek

♂

♂

♀

thin white outer tail

white wingbars

473

INDEX TO SCIENTIFIC NAMES

NDEX TO AFRIKAANS NAMES

INDEX TO COMMON NAMES

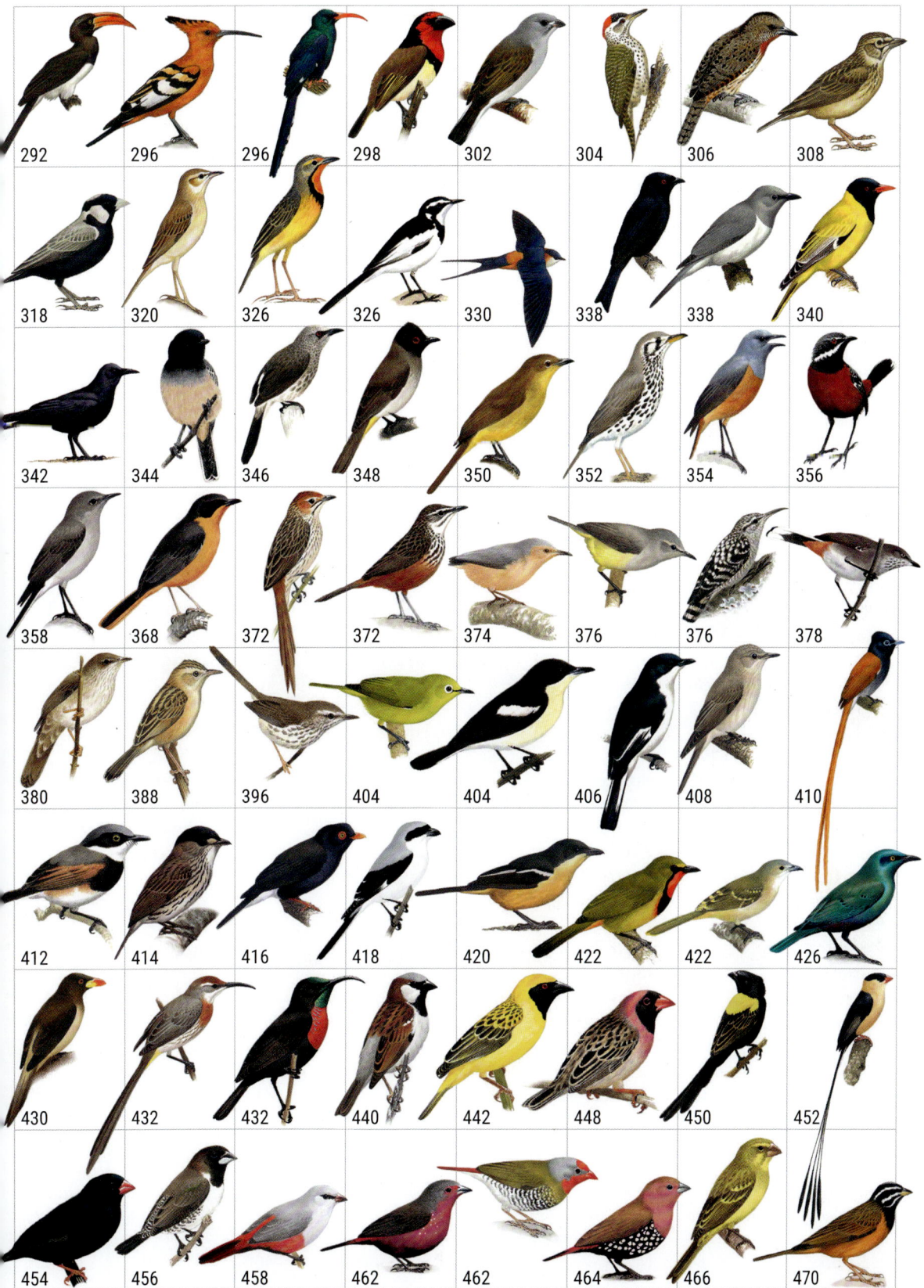

292	296	296	298	302	304	306	308
318	320	326	326	330	338	338	340
342	344	346	348	350	352	354	356
358	368	372	372	374	376	376	378
380	388	396	404	404	406	408	410
412	414	416	418	420	422	422	426
430	432	432	440	442	448	450	452
454	456	458	462	462	464	466	470

ALPHABETICAL QUICK REFERENCE TO BIRD GROUPS